Stochastic Mechanics
Random Media
Signal Processing
and Image Synthesis
Mathematical Economics and Finance
Stochastic Optimization
Stochastic Control
Stochastic Models in Life Sciences

Stochastic Modelling and Applied Probability

(Formerly:
Applications of Mathematics)

67

For further volumes:
http://www.springer.com/series/602

Jean Jacod · Philip Protter

Discretization of Processes

Springer

Jean Jacod
Institut de Mathématiques
Université Paris VI – Pierre et Marie Curie
Place Jussieu 4
Paris Cedex, Paris 75252
France
jean.jacod@upmc.fr

Philip Protter
Department of Statistics
Columbia University
Amsterdam Av. 1255
New York, NY 10027
USA
pep2117@columbia.edu

ISSN 0172-4568 Stochastic Modelling and Applied Probability
ISBN 978-3-642-24126-0 e-ISBN 978-3-642-24127-7
DOI 10.1007/978-3-642-24127-7
Springer Heidelberg Dordrecht London New York

Library of Congress Control Number: 2011941186

Mathematics Subject Classification (2010): 60F05, 60G44, 60H10, 60H35, 60J75, 60G51, 60G57, 60H05, 60J65, 60J25

Cover design: deblik

Printed on acid-free paper

Springer is part of Springer Science+Business Media (www.springer.com)

Preface

Two decades ago the authors of this book undertook the study of the errors one makes when numerically approximating the solutions of stochastic differential equations driven by Lévy processes. In particular we were interested in the normalized asymptotic errors of approximations via an Euler scheme, and it turned out we needed sophisticated laws of large numbers and central limit theorems that did not yet exist. While developing such tools, it became apparent that they would be useful in a wide range of applications.

One usually explains the difference between probability and statistics as being that probability theory lays the basis for a family of models, and statistics uses data to infer which member or members of that family best fit the data. Often this reduces to parameter estimation, and estimators are shown to be consistent via a Law of Large Numbers (LLN), and the accuracy of an estimator is determined using a Central Limit Theorem (CLT), when possible. The case of stochastic processes, and even stochastic dynamical systems, is of course more difficult, since often one is no longer estimating just a parameter, but rather one is estimating a stochastic process, or—worse—trying to tell which family of models actually does fit the data. Examples include using data to determine whether or not a model governing a dynamical system has continuous paths or has jumps, or trying to determine the dimension of the driving Brownian forces in a system of stochastic differential equations. This subject, broadly speaking, is a very old subject, especially as concerns asymptotic studies when the time parameter tends to infinity. The novelty presented here in this book is a systematic study of the case where the time interval is fixed and compact (also known as the *finite horizon* case). Even in the finite horizon case however, efforts predate the authors' study of numerical methods for stochastic differential equations, and go back 5 years earlier to attempts to find the volatility coefficient of an Itô process, via a fine analysis of its quadratic variation, by the first author joint with Valentine Genon-Catalot. This in turn builds on the earlier work of G. Dohnal, which itself builds on earlier work; it is indeed an old yet still interesting subject.

There are different variations of LLNs and CLTs one might use to study such questions, and over the last two decades substantial progress has been made in finding such results, and also in applying them via data to delve further into the unknown, and to reveal structures governing complicated stochastic systems. The most common examples used in recent times are those of financial models, but these ideas can be used in models of biological, chemical, and electrical applications as well. In this book we establish, in a systematic way, many of the recent results. The ensuing theorems are often complicated both to state, and especially to prove, and the technical level of the book is (inevitably, it seems) quite demanding. This is a theory book, and we do not treat applications, although we do reference papers that use these kinds of results for applications, and we do indicate at the end of most chapters how this theory can be used for applications.

An introduction explaining our approach, and an outline of how we have organized the book, can be found in the Introductory Chapter 1. In addition, in Chap. 1 we present several sketches of frameworks for potential applications of our theory, and indeed, these frameworks have inspired much of the development of the theory we present in this book.

If one were to trace back how we came to be interested in this theory, the history would have to center on the work and personality of Denis Talay and his “équipe” at INRIA in Sophia-Antipolis, as well as that of Jean Mémin at the University of Rennes. Both of these researchers influenced our taste in problems in enduring ways. We would also like to thank our many collaborators in this area over the years, with a special mention to Tom Kurtz, whose work with the second author started this whole enterprise in earnest, and also to Yacine Aït-Sahalia, who has provided a wealth of motivations through applications to economics. We also wish to thank O.E. Barndorff-Nielsen, S. Delattre, J. Douglas, Jr., V. Genon-Catalot, S.E. Graversen, T. Hayashi, Yingying Li, Jin Ma, S. Méléard, P. Mykland, M. Podolskij, J. San Martin, N. Shephard, V. Todorov, S. Torres, M. Vetter, and N. Yoshida, as well as A. Diop, for his careful reading of an earlier version of the manuscript.

The authors wish to thank Hadda and Diane for their forbearance and support during the several years involved in the writing of this book.

The second author wishes to thank the Fulbright Foundation for its support for a one semester visit to Paris, and the National Science Foundation, whose continual grant support has made this trans-Atlantic collaboration possible.

Paris, France
New York, USA

Jean Jacod
Philip Protter

Contents

Basic Notation

$\mathbb{R}$ = the set of real numbers, $\mathbb{R}_+ = [0, \infty)$
$\mathbb{Q}$ = the set of rational numbers, $\mathbb{Q}_+ = \mathbb{Q} \cap \mathbb{R}_+$
$\mathbb{N}$ = the set $\{0, 1, \dots\}$ of natural integers, $\mathbb{N}^* = \{1, 2, \dots\}$
$\mathbb{Z}$ = the set of relative integers
$\mathbb{R}^d$ = the Euclidean d-dimensional space;
the components of $x \in \mathbb{R}^d$ are $(x^1, \dots, x^d)$
$\mathcal{R}^d$ = the Borel σ-field of $\mathbb{R}^d$, $\mathcal{R} = \mathcal{R}^1$
$\mathcal{M}^+_{d\times d}$ = the set of nonnegative symmetric $d \times d$ matrices
$\mathcal{M}^{++}_{d\times d}$ = the set of nonnegative symmetric invertible $d \times d$ matrices
$|x|$ = the absolute value of $x \in \mathbb{R}$, $\|x\|$ = Euclidean norm of $x \in \mathbb{R}^d$
$d(x, B)$ = the distance between a point x and a subset B, in a metric space
y^* = the transpose of the vector or matrix y
$[x]$ = the integer part of $x \in \mathbb{R}$ (biggest $n \in \mathbb{Z}$ such that $n \leq x$)
$a \vee b$ = $\sup(a, b)$, $a \wedge b = \inf(a, b)$, if $a, b \in \mathbb{R}$
x^+ = $x \vee 0$, $x^- = (-x) \vee 0$, if $x \in \mathbb{R}$
1_A = the indicator function of the set A
A^c = the complement of the set A
ε_a = the Dirac measure sitting at point a
δ^{ij} = the Kronecker symbol, equal to 1 if $i = j$ and to 0 otherwise

Notation for convergences:

$\xrightarrow{\text{a.s.}}$ a.s. (almost sure) convergence for random variables
$\xrightarrow{\mathbb{P}}$ convergence in probability for random variables
$\xrightarrow{\mathcal{L}}$ convergence in law for random variables
$\xrightarrow{\mathcal{L}\text{-s}}$ stable convergence in law for random variables
$\stackrel{\text{a.s.}}{\Longrightarrow}$ a.s. convergence for processes, for Skorokhod's topology
$\stackrel{\mathbb{P}}{\Longrightarrow}$ convergence in probability for processes, for Skorokhod's topology
$\stackrel{\text{u.c.p.}}{\Longrightarrow}$ convergence in probability for processes, for the local uniform topology

$\stackrel{\mathcal{L}}{\Longrightarrow}$ convergence in law for processes, for Skorokhod's topology
$\stackrel{\mathcal{L}\text{-}s}{\Longrightarrow}$ stable convergence in law for processes, for Skorokhod's topology

Miscellaneous:
$f(x) = o(g(x))$ as $x \to x_0$ if $f(x)/g(x) \to 0$ as $x \to x_0$
$f(x) = O(g(x))$ as $x \to x_0$ if $\limsup_{x \to x_0} |f(x)/g(x)| < \infty$
càdlàg = "right-continuous with left limits"
càglàd = "left-continuous with right limits"
$u_n \asymp v_n$ means that both sequences u_n/v_n and v_n/u_n are bounded

Part I
Introduction and Preliminary Material

This introductory part contains two chapters. The first one is a detailed introduction: the whole book is devoted to many variations around two basic theorems, under various conditions and with various degrees of generality, and we explain how it can be used by a reader interested in a specific result or a special topic. The first chapter also contains (without proofs) a simplified version of the two basic theorems in three very special cases, when the underlying process is a Brownian motion, or a Brownian motion plus a drift, or when it is a Brownian motion plus a drift and a Poisson process: this part could be skipped, but its aim is to give a flavor of the subsequent material, without complicated assumptions or notation or technical details.

The second chapter mainly is a record of known facts about semimartingales and limit theorems. By "known" we mean that they can be found in a number of books. Some of these facts are elementary, others are more sophisticated, but it would take too much space and be outside the scope of this book to present the proofs. A few properties in this chapter are new, at least in a book form, and their proofs are given in the Appendix.

Chapter 1
Introduction

Discretization of stochastic processes indexed by the interval $[0, T]$ or by the half-line $[0, \infty)$ occurs very often. Historically it has been first used to deduce results on continuous-time processes from similar and often simpler results for discrete-time processes: for example Markov processes may be considered as limits of Markov chains, which are much simpler to analyze; or, stable processes as limits of random walks. This also applies to the theory of stochastic integration: the first constructions of stochastic integrals, by N. Wiener and K. Itô, were based on a Riemann-type approximation, which is a kind of discretization in time. More recently but still quite old, and a kind of archetype of what is done in this book, is the approximation of the quadratic variation process of a semimartingale by the approximate quadratic variation process: this result, due to P.A. Meyer [76] in its utmost generality, turns out to be one of the most useful results for applications.

Discretization of processes has become an increasingly popular tool in practical applications, for mainly (but not only) two reasons: one is the overwhelming extension of Monte-Carlo methods, which serve to compute numerically the expectations of a wide range of random variables which are often very complicated functions of a stochastic process: this is made available by the increasing power of computers. The second reason is related to statistics: although any stochastic process can only be observed at finitely many times, with modern techniques the frequency of observations increases steadily: in finance for example one observes and records prices every second, or even more frequently; in biology one measures electrical or chemical activity at an even higher frequency.

Let us be more specific, by describing a simple but fundamental example of some of the problems at hand. Suppose that we have a one-dimensional diffusion process X of the form

$$dX_t = a(X_t)\,dt + \sigma(X_t)\,dW_t, \qquad X_0 = x_0. \tag{1.0.1}$$

Here the initial value $x_0 \in \mathbb{R}$ is given, and W denotes a Brownian motion defined on some probability space, about which we do not care in this introduction. The drift and diffusion coefficients a and σ are nice enough, so the above equation has a unique solution.

J. Jacod, P. Protter, *Discretization of Processes*,
Stochastic Modelling and Applied Probability 67,
DOI 10.1007/978-3-642-24127-7_1, © Springer-Verlag Berlin Heidelberg 2012

Problem 1) We know a and σ, and we are interested in the law of the variable X_1. This law is usually not explicitly known, so to compute it, that is to compute the expected value $\mathbb{E}(f(X_1))$ for various test functions f, one may use a Monte-Carlo technique (other techniques based on PDEs are also available, especially in the one-dimensional case, but do not work so well in high dimensions). To implement this we simulate on a computer a number N of independent variables $X(j)_1$ having the law of X_1, and an approximation of $\mathbb{E}(f(X_1))$ is

$$Z_N = \frac{1}{N}\sum_{j=1}^{N} f\big(X(j)_1\big). \tag{1.0.2}$$

Indeed, by the law of large numbers the sequence Z_N converges almost surely to $\mathbb{E}(f(X_1))$ as $N \to \infty$, and moreover the central limit theorem tells us that, when f is for example bounded, the error made in replacing $\mathbb{E}(f(X_1))$ by Z_N is of order $1/\sqrt{N}$.

This presumes that one knows how to simulate X_1, which is about as scarce as the cases when $\mathbb{E}(f(X_1))$ can be explicitly computed. (More accurately some recent techniques due to A. Beskos, O. Papaspiliopoulos and G.O. Roberts, see [16] and [17] for example, allow to simulate X_1 exactly, but they require that σ does not vanish and, more important, that the dimension is 1; moreover, in contrast to what follows, they cannot be extended to equations driven by processes other than a Brownian motion.) Hence we have to rely on approximations, and the simplest way for this is to use an Euler scheme. That is, for any integer $n \geq 1$ we recursively define the approximation $X^n_{i/n}$ for $i = 1, \dots, n$, by setting

$$X^n_0 = x_0, \qquad X^n_{i/n} = X^n_{(i-1)/n} + \frac{1}{n}\, a\big(X^n_{(i-1)/n}\big) + \sigma\big(X^n_{(i-1)/n}\big)(W_{i/n} - W_{(i-1)/n}),$$

the increments of the Brownian motion being easily simulated. Other, more sophisticated, schemes can be used, but they all rely upon the same basic ideas.

Then in (1.0.2) we substitute the $X(j)_1$'s with N independent copies of the simulated variables X^n_1, giving rise to an average Z^n_N which now converges to $\mathbb{E}(f(X^n_1))$ for each given n. Therefore we need to assert how close $\mathbb{E}(f(X^n_1))$ and $\mathbb{E}(f(X_1))$ are, and this more or less amounts to estimating the difference $(X_1 - X^n_1)^2$. Some calculations show that this boils down to evaluating the difference

$$\sum_{i=1}^{n} g_n\big(\omega, (i-1)/n\big)\left((W_{i/n} - W_{(i-1)/n})^2 - \frac{1}{n}\right)$$

for suitable functions $g_n(\omega, t)$, where $\omega \mapsto g_n(\omega, t)$ is measurable with respect to the σ-field $\mathcal{F}^W_t$ of the past of W before time t. That is, we have to determine the behavior of "functionals" of the increments of W of the form above: do they converge when $n \to \infty$? And if so, what is the rate of convergence?

Problem 2) The setting is the same, that is we know a and σ, but we want to find the law of $Y = \int_0^1 h(X_s)\,ds$ for some known function h. Again, one can use a Monte-

Carlo technique, coupled with a preliminary Euler method: we set

$$Y^n = \frac{1}{n}\sum_{i=1}^{n} h(X^n_{i/n}),$$

where X^n is the Euler approximation introduced above. We can then simulate N independent versions $Y^n(1), \dots, Y^n(N)$ of the variable Y^n above, and

$$\frac{1}{N}\sum_{j=1}^{N} h\big(Y^n(j)\big)$$

is our approximation of $\mathbb{E}(h(Y))$. If X^n is a good approximation of X, then certainly Y^n is a good approximation of $\frac{1}{n}\sum_{i=1}^{n} h(X_{i/n})$, provided h satisfies some suitable smoothness assumptions. However we have an additional problem here, namely to evaluate the difference

$$\frac{1}{n}\sum_{i=1}^{n} h(X_{i/n}) - \int_0^1 h(X_s)\,ds.$$

The convergence to 0 of this difference is ensured by Riemann approximation, but the rate at which it takes place is not clear, in view of the fact that the paths of X are not smooth, albeit continuous. This is another discretization problem.

Problem 3) Suppose now that the functions a and σ are known, but depend on an additional parameter, say θ, so we have $a = a(x,\theta)$ and $\sigma = \sigma(x,\theta)$. We observe the process $X = X^\theta$, which now depends on θ, over $[0,1]$, and we want to infer θ. However, in any realistic situation we cannot really observe the whole path $t \mapsto X_t(\omega)$ for $t \in [0,1]$, and we simply have "discrete" observations, say at times $0, \frac{1}{n}, \dots, \frac{n}{n}$, so we have $n+1$ observations.

We are here in the classical setting of a parametric statistical problem. For any given n there is no way exactly to infer θ, unless a and σ have a very special form. But we may hope for good asymptotic estimators as $n \to \infty$. All estimation methods, and there are many, are based on the behavior of functionals of the form

$$\sum_{i=1}^{n} f_n\big(\theta, \omega, (i-1)/n, X_{i/n} - X_{(i-1)/n}\big) \tag{1.0.3}$$

for suitable functions $f_n(\theta,\omega,t,x)$, where again $\omega \mapsto f_n(\theta,\omega,t,x)$ is $\mathcal{F}^W_t$ measurable. The consistency of the estimators is deduced from the convergence of functionals as above, and rates of convergence are deduced from associated central limit theorems for those functionals.

Problem 4) Here the functions a and σ are unknown, and they may additionally depend on (ω,t), as for example $\sigma = \sigma(\omega,t,x)$. We observe X at the same discrete times $0, \frac{1}{n}, \dots, \frac{n}{n}$ as above. We want to infer some knowledge about the coefficients a and σ. As is well known, we usually can say nothing about a in this setting, but the convergence of the approximate quadratic variation mentioned before says that:

$$\sum_{i=1}^{[nt]} (X_{i/n} - X_{(i-1)/n})^2 \to \int_0^t \sigma(X_s)^2 ds$$

(convergence in probability, for each t; here, $[nt]$ denotes the integer part of the real nt). This allows us in principle to determine asymptotically the function $t \mapsto \sigma(\omega, t, X_t(\omega))$ on $[0, 1]$, and under suitable assumptions we even have rates of convergence. Here again, everything hinges upon functionals as in the left side above. Note that here we have a statistical problem similar to Problem 3, except that we do not want to infer a parameter θ but a quantity which is fundamentally random: this occurs for example in finance, for the estimation of the so-called stochastic volatility.

Problem 5) A more basic problem is perhaps the following one, which deals directly with discretized processes. Namely, let us call an n-discretized process of X the process defined by $X_t^{(n)} = X_{[nt]/n}$. Then of course $X^{(n)} \to X$ pointwise in ω, locally uniformly in time when X is continuous and for the Skorokhod topology when X is right-continuous and with left limits. But, what is the rate of convergence?

The common feature of all the problems described above, as different as they may appear, is the need to consider the asymptotic behavior of functionals like (1.0.3). And, when the process X is discontinuous, many other problems about the jumps can also be solved by using functionals of the same type.

1.1 Content and Organization of the Book

In the whole book we consider a basic underlying d-dimensional process X, always a *semimartingale*. This process is sampled at discrete times, most of the time regularly spaced: that is, we have a mesh $\Delta_n > 0$ and we consider the increments

$$\Delta_i^n X = X_{i\Delta_n} - X_{(i-1)\Delta_n}$$

and two types of *functionals*, where f is a function on $\mathbb{R}^d$:

$$\begin{aligned} V^n(f,X)_t &= \sum_{i=1}^{[t/\Delta_n]} f(\Delta_i^n X) && \text{“non-normalized functional”} \\ V'^n(f,X)_t &= \Delta_n \sum_{i=1}^{[t/\Delta_n]} f\big(\Delta_i^n X/\sqrt{\Delta_n}\big) && \text{“normalized functional”}. \end{aligned} \tag{1.1.1}$$

The aim of this book is to provide a comprehensive treatment of the mathematical results about functionals of this form, when the mesh Δ_n goes to 0. We will not restrict ourselves to the simple case of (1.1.1), and will also consider more general (but similar) types of functionals:

- f may depend on k successive increments of X for $k \geq 2$.
- $f = f_n$ may depend on n, and also on k_n successive increments, with $k_n \to \infty$.
- $f = f(\omega, t, x)$ may be a function on $\Omega \times \mathbb{R}_+ \times \mathbb{R}^d$, so that $f(\Delta_i^n X)$ is replaced by $f(\omega, (i-1)\Delta_n, \Delta_i^n X)$ in the first formula (1.1.1), for example.
- The sampling times are not necessarily equally spaced.

Basically, there are two different levels of results:

Level 1: We have (under appropriate assumptions, of course, and sometimes after normalization) convergence of the functionals to a limiting process, say for example $V^n(f, X) \to V(f, X)$. This convergence typically takes place in probability, either for a fixed time t, or "functionally" for the local uniform (in time) topology, or for the Skorokhod topology. We call this type of convergence a *Law of Large Numbers*, or LLN.

Level 2: There is a "second order" type of results, which we qualify as *Central Limit Theorems*, or CLT. Namely, for a proper normalizing factor $u_n \to \infty$ the sequence $u_n(V^n(f, X) - V(f, X))$ for example converges to a limiting process. In this case, the convergence (for a given time t, or functionally as above) is typically in law, or more accurately "stably in law" (the definition of stable convergence in law is recalled in detail in Chap. 2).

In connection with the previous examples, it should be emphasized that, even though the mathematical results given below have some interest from a purely theoretical viewpoint, the main motivation is *practical*. This motivation is stressed by the fact that the last section of most chapters contains a brief account of possible applications. These applications have indeed been the reason for which all this theory has been developed.

As it is written, one can hardly consider this book as "applied". Nevertheless, we hope that the reader will get some feeling about the applications, through the last sections mentioned above. In particular, the problem of estimating the *volatility* is recurrent through the whole book, and appears in Chaps. 3, 5, 8, 9, 11, 13, 14 and 16.

Two last general comments are in order:

1. A special feature of this book is that it concentrates on the case where the underlying process X has a non-trivial continuous martingale part X^c, which is $X_t^c = \int_0^t \sigma(X_s)\, dW_s$ in the case of (1.0.1). All results are of course still true in the degenerate situation where the continuous martingale part vanishes identically, but most of them become "trivial", in the sense that the limiting processes are also vanishing. That is, in this degenerate situation one should employ other normalization, and use different techniques for the proofs.

2. We are concerned with the behavior of functionals like (1.1.1) as $\Delta_n \to 0$, but *not* as the time t goes to infinity. That is, we only consider the "finite horizon" case. When $t \to \infty$ the results for these functionals requires some ergodicity assumptions on the process X: the results, as well as the techniques needed for the proofs, are then fundamentally different.

Synopsis of the Book: Chapter 2 is devoted to recalling the basic necessary results about semimartingales and the various notions of convergence used later (Skorokhod topology, stable convergence in law, and a few useful convergence criteria). The rest of the book is divided into four main parts:

Part II: This part is about the "simple" functionals, as introduced in (1.1.1):

- Chapter 3 is devoted to the Laws of Large Numbers (first level).
- Chapter 4 contains the technical results needed for Central Limit Theorems. To avoid fastidious repetitions, these technical results are general enough to provide for the proofs of the CLTs for more general functionals than those of (1.1.1).
- Chapter 5 is about Central Limit Theorems (second level). For $V'^n(f,X)$ it requires few assumptions on the function f but quite a lot about the jumps of X, if any; for $V^n(f,X)$ it requires little of X, but (in, say, the one-dimensional case) it basically needs either $f(x) \sim x^2$ or $f(x)/|x|^3 \to 0$ as $x \to 0$.
- Chapter 6 gives another kind of Central Limit Theorems (in the extended sense used in this book) for $V^n(f,X)$, when $f(x) = x$: this is a case left out in the previous Chap. 5, but it is also important because $V^n(f,X)_t$ is then $X_t^{(\Delta_n)} - X_0$, where $X^{(\Delta_n)}$ is the "discretized process" $X_t^{(\Delta_n)} = X_{\Delta_n[t/\Delta_n]}$.

Part III: This part concerns various extensions of the Law of Large Numbers:

- In Chap. 7 the test function f is random, that is, it depends on (ω, t, x).
- In Chap. 8 the test function $f = f_n$ may depend on n and on k (fixed) or k_n (going to infinity) successive increments.
- In Chap. 9 the test function f is truncated at a level u_n, with u_n going to 0 as Δ_n does; that is, instead of $f(\Delta_i^n X)$ we consider $f(\Delta_i^n X) 1_{\{|\Delta_i^n X| \le u_n\}}$ or $f(\Delta_i^n X) 1_{\{|\Delta_i^n X| > u_n\}}$, for example. The function f can also depend on several successive increments.

Part IV: In this part we study the Central Limit Theorems associated with the extended LLNs of the previous part:

- Chapter 10 gives the CLTs associated with Chap. 7 (random test functions).
- Chapter 11 gives the CLTs associated with Chap. 8 when the test function depends on k successive increments.
- Chapter 12 gives the CLTs associated with Chap. 8 when the test function depends on k_n successive increments, with $k_n \to \infty$.
- Chapter 13 gives the CLTs associated with Chap. 9 (truncated test functions).

Part V: The last part is devoted to three problems which do not fall within the scope of the previous chapters, but are of interest for applications:

- In Chap. 14 we consider the situation where the discretization scheme is not regular. This is of fundamental importance for applications, but only very partial results are provided here, and only when the process X is continuous.
- In Chap. 15 we study some degenerate situations where the rate of convergence is not the standard $1/\sqrt{\Delta_n}$ one.
- In Chap. 16 we consider a situation motivated again by practical applications: we replace the process X by a "noisy" version, that is by $Z_t = X_t + \varepsilon_t$ where ε_t is a noise, not necessarily white but subject to some specifications. Then we examine how the functionals (based on the observations $Z_{i\Delta_n}$ instead of $X_{i\Delta_n}$) should be modified, in order to obtain limits which are basically the same as in the non-noisy case, and in particular do not depend on the noise.

1.2 When X is a Brownian Motion

Before proceeding to the main stream of the book, we give in some detail and with heuristic explanations, but without formal proofs, the simplest form of the results: we suppose that the one-dimensional process X is either a Brownian motion, or a Brownian motion with a drift, or a Brownian motion plus a drift plus a compound Poisson process.

Although elementary, these examples essentially show most qualitative features found later on, although of course the simple structure accounts for much simpler statements. So the remainder of this chapter may be skipped without harm, and its aim is to exhibit the class of results given in this book, and their variety, in an especially simple situation.

We start with the Brownian case, that is

$$X = \sigma W, \quad \text{where } W \text{ is a Brownian motion and } \sigma > 0; \text{ we set } c = \sigma^2. \tag{1.2.1}$$

We will also use, for any process Y, its "discretized" version at stage n:

$$Y_t^{(\Delta_n)} = Y_{\Delta_n[t/\Delta_n]}.$$

1.2.1 The Normalized Functionals $V'^n(f, X)$

Recalling (1.1.1), the functionals $V'^n(f, X)$ are easier than $V^n(f, X)$ to analyze. Indeed, the summands $f(\Delta_i^n X/\sqrt{\Delta_n})$ are not only i.i.d. as i varies, but they also have the same law as n varies. We let ρ_c be the centered Gaussian law $\mathcal{N}(0, c)$ and write $\rho_c(f) = \int f(x)\rho_c(dx)$ when the integral exists. Then, as soon as f is Borel and integrable, resp. square integrable, with respect to ρ_c, then $f(\Delta_i^n X/\sqrt{\Delta_n})$ has expectation $\rho_c(f)$ and variance $\rho_c(f^2) - \rho_c(f)^2$.

The ordinary Law of Large Numbers (LLN) and Central Limit Theorem (CLT) readily give us the following two convergence results:

$$\begin{aligned} &V'^n(f,X)_t \stackrel{\mathbb{P}}{\longrightarrow} t\rho_c(f) \\ &\frac{1}{\sqrt{\Delta_n}}\big(V'^n(f,X)_t - t\rho_c(f)\big) \stackrel{\mathcal{L}}{\longrightarrow} \mathcal{N}\big(0, t\big(\rho_c\big(f^2\big) - \rho_c(f)^2\big)\big), \end{aligned} \tag{1.2.2}$$

where $\stackrel{\mathbb{P}}{\longrightarrow}$ and $\stackrel{\mathcal{L}}{\longrightarrow}$ stand for the convergence in probability and the convergence in law, respectively. This example shows why we have put the normalizing factor $1/\sqrt{\Delta_n}$ inside the function f.

The first subtle point we encounter, even in this basic case, is that, contrary to the usual LLN, we get convergence in probability but *not* almost surely in the first part of (1.2.2). The reason is as follows: let ζ_i be a sequence of i.i.d. variables with the same law as $f(X_1)$. The LLN implies that $Z_n = \frac{t}{[t/\Delta_n]}\sum_{i=1}^{[t/\Delta_n]}\zeta_i$ converges a.s. to $t\rho_c(f)$. Since $V'^n(f,X)_t$ has the same law as Z_n we deduce the convergence in probability in (1.2.2) because, for a deterministic limit, convergence in probability and convergence in law are equivalent. However the variables $V'^n(f,X)_t$ are connected one with the others in a way we do not really control when n varies, so we cannot conclude that $V'^n(f,X)_t \to t\rho_c(f)$ a.s.

(1.2.2) gives us the convergence for any time t, but we also have a "functional" convergence:

1) First, recall that a sequence g_n of nonnegative increasing functions on $\mathbb{R}_+$ converging pointwise to a *continuous* function g also converges locally uniformly; then, from the first part of (1.2.2) applied separately for the positive and negative parts f^+ and f^- of f and using a "subsequence principle" for the convergence in probability, we obtain

$$V'^n(f,X)_t \stackrel{\text{u.c.p.}}{\Longrightarrow} t\rho_c(f) \tag{1.2.3}$$

where $Z_t^n \stackrel{\text{u.c.p.}}{\Longrightarrow} Z_t$ means "convergence in probability, locally uniformly in time": that is, $\sup_{s\le t}|Z_s^n - Z_s| \stackrel{\mathbb{P}}{\longrightarrow} 0$ for all t finite.

2) Next, if instead of the one-dimensional CLT we use the "functional CLT", or Donsker's Theorem, we obtain

$$\left(\frac{1}{\sqrt{\Delta_n}}\big(V'^n(f,X)_t - t\rho_c(f)\big)\right)_{t\ge0} \stackrel{\mathcal{L}}{\Longrightarrow} \sqrt{\rho_c\big(f^2\big) - \rho_c(f)^2}\, B \tag{1.2.4}$$

where B is another standard Brownian motion, and $\stackrel{\mathcal{L}}{\Longrightarrow}$ stands for the convergence in law of processes (for the Skorokhod topology, see later for details on this topology, even though in this special case we could also use the "local uniform topology", since the limit is continuous).

In (1.2.4) we see a new Brownian motion B appear. What is its connection with the basic underlying Brownian motion W? To study that, one can try to prove the "joint convergence" of the processes on the left side of (1.2.4) together with W (or equivalently X) itself.

This is an easy task: consider the 2-dimensional process Z^n whose first component is W and the second component is the left side of (1.2.4). The discretized version of Z^n is $(Z^n)_t^{(\Delta_n)} = \sqrt{\Delta_n}\sum_{i=1}^{[t/\Delta_n]}\zeta_i^n$, where the ζ_i^n are 2-dimensional i.i.d. variables as i varies, with the same distribution as $(W_1, f(\sigma W_1) - \rho_c(f))$. Then the 2-dimensional version of Donsker's Theorem gives us that the pair of processes with components $W^{(\Delta_n)}$ and $\frac{1}{\sqrt{\Delta_n}}(\Delta_n V'^n(f,X)_t - t\rho_c(f))$ converges in law to a 2-dimensional Brownian motion with variance-covariance matrix at time 1 given by

$$\begin{pmatrix} 1 & \rho_c(g) \\ \rho_c(g) & \rho_c(f^2) - \rho_c(f)^2 \end{pmatrix}, \quad \text{where } g(x) = xf(x)/\sigma.$$

We write this as

$$\begin{gathered} \left(W_t^{(\Delta_n)}, \frac{1}{\sqrt{\Delta_n}}\left(V'^n(f,X)_t - t\rho_c(f)\right)\right)_{t\geq 0} \stackrel{\mathcal{L}}{\Longrightarrow} (W, aW + a'W'), \\ \text{where } a = \rho_c(g),\ a' = \left(\rho_c(f^2) - \rho_c(f)^2 - \rho_c(g)^2\right)^{1/2}, \end{gathered} \tag{1.2.5}$$

where W' is a standard Brownian motion independent of W.

In (1.2.5) we could have used another symbol in place of W since what really matters is the joint law of the pair (W, W'). However for the first component, not only do we have convergence in law but pathwise convergence $W^{(\Delta_n)} \to W$. This explains why we use the notation W here, and in fact this results in a stronger form of convergence for the second component as well. This mode of convergence, called *stable convergence in law*, will be explained in detail in the next chapter.

Remark 1.2.1 We can even make $f = f_n$ depend on n, in such a way that f_n converges to some limit f fast enough. This is straightforward, and useful in some applications.

Remark 1.2.2 (1.2.5) is stated in a unified way, but there are really two—quite different—types of results here, according to the parity properties of f:

a) If f is an even function then $\rho_c(f) \neq 0$ in general, and $a = 0$. The limit in the CLT is $(W, a'W')$, with two independent components.

b) If f is an odd function then $\rho_c(f) = 0$ and $a \neq 0$ in general. The limit in the CLT has two dependent components. A special case is $f(x) = x$: then $a = \sigma$ and $a' = 0$, so the limit is $(W, X) = (W, \sigma W)$. This was to be anticipated, since in this case $V'^n(f,X) = \sqrt{\Delta_n}\, X^{(\Delta_n)}$, and the convergence in (1.2.5) takes place not only in law, but even in probability.

In general, the structure of the limit is thus much simpler in case (a), and most applications use this convergence for test functions f which are even.

1.2.2 The Non-normalized Functionals $V^n(f, X)$

We now turn to the processes $V^n(f, X)$. Their behavior results from the behavior of the processes $V'^n(f, X)$, but already in this simple case they show some distinctive features that will be encountered in more general situations. Basically, all increments $\Delta_i^n X$ become small as n increases, so the behavior of f near 0 is of the utmost importance, and in fact it conditions the normalization we have to use for the convergence.

To begin with, we consider power functions:

$$f_r(x) = |x|^r, \qquad \overline{f}_r(x) = |x|^r \operatorname{sign}(x),$$

where $r > 0$ and where $\operatorname{sign}(x)$ takes the value $+1$, 0 or -1, according to whether $x > 0$, $x = 0$ or $x < 0$. Note that

$$V^n(f_r, X) = \Delta_n^{r/2-1}\, V'^n(f_r, X)$$

and the same for $\overline{f}_r$. Moreover, if m_p denotes the p absolute moment of $\mathcal{N}(0,1)$, that is $m_p = \rho_1(f_p)$, and if $h_r(x) = x f_r(x)/\sigma$ and $\overline{h}_r(x) = x f_r'(x)/\sigma$ (recall $\sigma > 0$), we have

$$\begin{array}{lll} \rho_c(f_r) = m_r\sigma^r, & \rho_c\big(f_r^2\big) = m_{2r}\sigma^{2r}, & \rho_c(h_r) = 0, \\ \rho_c(\overline{f}_r) = 0, & \rho_c\big(\overline{f}_r^2\big) = m_{2r}\sigma^{2r}, & \rho_c(\overline{h}_r) = m_{r+1}\sigma^r. \end{array}$$

Hence we can rewrite (1.2.3) and (1.2.5) as follows, where W' denotes a standard Brownian motion independent of W (we single out the two cases f_r and $\overline{f}_r$, which correspond to cases (a) and (b) in Remark 1.2.2):

$$\left.\begin{array}{l} \Delta_n^{1-r/2} V^n(f_r, X)_t \stackrel{\text{u.c.p.}}{\Longrightarrow} t m_r \sigma^r, \\ \left(W_t^{(\Delta_n)}, \dfrac{1}{\sqrt{\Delta_n}}\left(\Delta_n^{1-r/2} V^n(f_r, X)_t - t m_r \sigma^r\right)\right)_{t\geq 0} \stackrel{\mathcal{L}}{\Longrightarrow} \left(W, \sigma^r\sqrt{m_{2r} - m_r^2}\, W'\right), \end{array}\right. \tag{1.2.6}$$

$$\left.\begin{array}{l} \Delta_n^{1-r/2} V^n(\overline{f}_r, X) \stackrel{\text{u.c.p.}}{\Longrightarrow} 0, \\ \left(W^{(\Delta_n)}, \Delta_n^{1/2-r/2} V^n(\overline{f}_r, X)\right) \stackrel{\mathcal{L}}{\Longrightarrow} \left(W, \sigma^r\left(m_{r+1} W + \sqrt{m_{2r} - m_{r+1}^2}\, W'\right)\right). \end{array}\right. \tag{1.2.7}$$

Note that the second statement implies the first one in these two properties.

Next, we consider functions f which vanish on a neighborhood of 0, say over some interval $[-\varepsilon, \varepsilon]$. Since X is continuous, we have $\sup_{i\leq[t/\Delta_n]} |\Delta_i^n X| \to 0$ pointwise for all t, and thus for each t there is a (random) integer A_t such that

$$n \geq A_t \ \Rightarrow\ V^n(f, X)_s = 0 \quad \forall s \leq t. \tag{1.2.8}$$

Finally, we consider "general" functions f, say Borel and with polynomial growth. If we combine (1.2.6) or (1.2.7) with (1.2.8), we see that the behavior of f

far from 0 does not matter at all, whereas the behavior near 0 is crucial for $V^n(f,X)$ to converge (with or without normalization). So it is no wonder that we get the following result:

$$\begin{aligned} &f(x)\sim f_r(x) \text{ as } x\to 0 \quad\Rightarrow\quad \Delta_n^{1-r/2}V^n(f,X)_t \overset{\text{u.c.p.}}{\Longrightarrow} tm_r\sigma^r,\\ &f(x)\sim \overline{f}_r(x) \text{ as } x\to 0 \quad\Rightarrow\quad \Delta_n^{1-r/2}V^n(f,X)_t \overset{\text{u.c.p.}}{\Longrightarrow} 0. \end{aligned} \tag{1.2.9}$$

These results are trivial consequences of the previous ones when f coincides with f_r or $\overline{f}_r$ on a neighborhood of 0, whereas if they are only equivalent one needs an (easy) additional argument. As for the CLT, we need f to coincide with f_r or $\overline{f}_r$ on a neighborhood of 0 ("close enough" would be sufficient, but how "close" is difficult to express, and "equivalent" is not enough). So we have, for any $\varepsilon>0$ (recall that f is of polynomial growth):

$$\begin{aligned} &f(x)=f_r(x) \text{ if } |x|\le\varepsilon \quad\Rightarrow\\ &\Big(W_t^{(\Delta_n)},\frac{1}{\sqrt{\Delta_n}}\big(\Delta_n^{1-r/2}V^n(f,X)_t-tm_r\sigma^r\big)\Big)_{t\ge0} \overset{\mathcal{L}}{\Longrightarrow}\big(W,\sigma^r\sqrt{m_{2r}-m_r^2}\,W'\big), \end{aligned} \tag{1.2.10}$$

$$\begin{aligned} &f(x)=\overline{f}_r(x) \text{ if } |x|\le\varepsilon \quad\Rightarrow\\ &\big(W^{(\Delta_n)},\Delta_n^{1/2-r/2}V^n(f,X)\big) \overset{\mathcal{L}}{\Longrightarrow}\Big(W,\sigma^r\big(m_{r+1}W+\sqrt{m_{2r}-m_{r+1}^2}\,W'\big)\Big) \end{aligned} \tag{1.2.11}$$

where again W' is a standard Brownian motion independent of W.

These results do not exhaust all possibilities for the convergence of $V^n(f,X)$. For example one can prove the following:

$$f(x)=|x|^r\log|x| \quad\Rightarrow\quad \frac{\Delta_n^{1-r/2}}{\log(1/\Delta_n)}V^n(f,X) \overset{\text{u.c.p.}}{\Longrightarrow} -\frac{1}{2}\,tm_r\sigma^r,$$

and a CLT is also available in this situation. Or, we could consider functions f which behave like x^r as $x\downarrow\downarrow 0$ and like $(-x)^{r'}$ as $x\uparrow\uparrow 0$. However, we essentially restrict our attention to functions behaving like f_r or $\overline{f}_r$ near the origin: for simplicity, and because more general functions do not really occur in applications, and also because the extension to processes X more general than the Brownian motion is not easy, or not available at all, for other functions.

Example 1.2.3 Convergence of the approximate quadratic variation. The functional

$$V^n(f_2,X)_t = \sum_{i=1}^{[t/\Delta_n]}\big(\Delta_i^nX\big)^2$$

is called the "approximate quadratic variation", and "realized quadratic variation" or "realized volatility" in the econometrics literature. It is of course well known, and a consequence of (1.2.6), that it converges in probability, locally uniformly in time, to the "true" quadratic variation which here is $\sigma^2 t$. Then (1.2.6) also gives the rate of convergence, namely that $\frac{1}{\sqrt{\Delta_n}}(V^n(f_2,X)_t - t\sigma^2)$ converges in law to $2\sigma^4 W'$; and we even have the joint convergence with X itself, and in the limit W' and X (or W) are independent.

1.3 When X is a Brownian Motion Plus Drift

Here we replace (1.2.1) by

$$X_t = bt + \sigma W_t, \qquad \text{where } \sigma \geq 0 \text{ and } b \neq 0.$$

1.3.1 The Normalized Functionals $V'^n(f, X)$

We first assume that $\sigma > 0$. The normalized increments $\Delta_i^n X/\sqrt{\Delta_n}$ are still i.i.d. when i varies, but now their laws depend on n. However, $\Delta_i^n X/\sqrt{\Delta_n} = Y_i^n + b\sqrt{\Delta_n}$ with Y_i^n being $\mathcal{N}(0,\sigma^2)$ distributed. Then, clearly enough, $f(\Delta_i^n X/\sqrt{\Delta_n})$ and $f(Y_i^n)$ are almost the same, at least when f is continuous, and it is no wonder that (1.2.3) remains valid (with the same limit) here, that is

$$V'^n(f,X)_t \stackrel{\text{u.c.p.}}{\Longrightarrow} t\rho_c(f).$$

Moreover, it turns out that the continuity of f is not even necessary for this, being Borel with some growth condition is again enough.

For the CLT, things are more complicated. When $X = \sigma W$ the CLT (1.2.4) boils down to the ordinary (functional) CLT, or Donsker's theorem, for the i.i.d. centered variables $\zeta_i^n = f(\Delta_i^n X/\sqrt{\Delta_n}) - \rho_c(f)$, but now while these variables are still i.i.d. when i varies, they are no longer centered, and their laws depend on n.

In fact ζ_i^n is distributed as $f(\sigma U + b\sqrt{\Delta_n}) - \rho_c(f)$, where U denotes an $\mathcal{N}(0,1)$ variable. Now, assume that f is C^1, with a derivative f' having at most polynomial growth. Then $f(\sigma U + b\sqrt{\Delta_n}) - f(\sigma U)$ is approximately equal to $f'(\sigma U)b\sqrt{\Delta_n}$. It follows that the variables ζ_i^n satisfy

$$\mathbb{E}(\zeta_i^n) = \sqrt{\Delta_n}\big(b\rho_c(f') + \mathrm{o}(1)\big)$$
$$\mathbb{E}\big((\zeta_i^n)^2\big) = \rho_c(f^2) - \rho_c(f)^2 + \mathrm{o}(1)$$
$$\mathbb{E}\big((\zeta_i^n)^4\big) = \mathrm{O}(1).$$

A CLT for triangular arrays of i.i.d. variables (see the next chapter) gives us

$$\left(\frac{1}{\sqrt{\Delta_n}}\big(V'^n(f,X)_t - t\rho_c(f)\big)\right)_{t\geq 0} \stackrel{\mathcal{L}}{\Longrightarrow} \Big(b\,\rho_c(f')\,t + \sqrt{\rho_c(f^2) - \rho_c(f)^2}\,B_t\Big)_{t\geq 0}. \tag{1.3.1}$$

Comparing with (1.2.4), we see an additional bias coming in here. Exactly as in (1.2.5), we also have a joint convergence (and stable convergence in law as well). With the notation a, a' and W' of (1.2.5), the expression is

$$\left(W_t^{(\Delta_n)}, \frac{1}{\sqrt{\Delta_n}}\left(bV'^n(f,X)_t - t\rho_c(f)\right)\right)_{t\geq 0} \stackrel{\mathcal{L}}{\Longrightarrow} \left(W_t, b\,\rho_c(f')\,t + aW_t + a'W'_t\right)_{t\geq 0}. \tag{1.3.2}$$

Remark 1.3.1 We have the same dichotomy as in Remark 1.2.2. When f is an even function, the limit in (1.3.2) is simply $(W, a'W')$, with $a' = \sqrt{\rho_c(f^2) - \rho_c(f)^2}$, and in particular there is no bias (observe that f' is then odd, so $\rho_c(f') = 0$). When f is an odd function, we do have $\rho_c(f') \neq 0$ in general, and the bias does appear. A special case again is when $f(x) = x$, so $a = \sigma$ and $a' = 0$ and $\rho_c(f') = 1$, so the limit is (W, X) again, as it should be from the property $V'^n(f,X) = \sqrt{\Delta_n}\, X^{(\Delta_n)}$.

Suppose now $\sigma = 0$, that is $X_t = bt$. Then of course there is no more randomness, and all results ought to be elementary, but they are different from the previous ones. For example if f is differentiable at 0, we have

$$\frac{1}{\sqrt{\Delta_n}}\left(V'^n(f,X)_t - tf(0)\right) \to b\,f'(0)\,t,$$

locally uniformly in t. This can be considered as a special case of (1.3.1), with ρ_0 being the Dirac mass at 0. Note that the normalization $1/\sqrt{\Delta_n}$ inside the test function f is not really adapted to this situation, a more natural normalization would be $1/\Delta_n$.

1.3.2 The Non-normalized Functionals $V^n(f, X)$

For the functionals $V^n(f,X)$ we deduce the results from the previous subsection, exactly as for Brownian motion, at least when $\sigma > 0$. We have (1.2.8) when f vanishes on a neighborhood of 0, because this property holds for any continuous process X. Then we have (1.2.9), and also (1.2.10) when $r \geq 1$ (use Remark 1.3.1, the condition $r \geq 1$ ensures that f_r is C^1, except at 0 when $r = 1$). Only (1.2.11) needs to be modified, as follows, and again with $r \geq 1$:

$$\begin{aligned} f(x) = \overline{f}_r(x) \text{ if } |x| \leq \varepsilon \quad &\Rightarrow \quad \left(W^{(\Delta_n)}, \Delta_n^{1/2-r/2} V^n(f,X)\right) \\ &\stackrel{\mathcal{L}}{\Longrightarrow} \left(W_t, r\,m_{r-1}bt + \sigma^r\left(m_{r+1}W_t + \sqrt{m_{2r} - m_{r+1}^2}\;W'_t\right)\right)_{t\geq 0}. \end{aligned} \tag{1.3.3}$$

The case of the approximate quadratic variation is exactly as in Example 1.2.3.

Finally when $\sigma = 0$ we have $V^n(f,X)_t = f(b\Delta_n)\,\Delta_n[t/\Delta_n]$, and thus trivially

$$f \text{ differentiable at } 0 \;\Rightarrow\; \frac{1}{\Delta_n}\left(V^n(f,X)_t - f(0)\,t\right) \to b\,f'(0)\,t.$$

1.4 When X is a Brownian Motion Plus Drift Plus a Compound Poisson Process

In this section the structure of the process X is

$$X = Y + Z, \qquad Y_t = bt + \sigma W_t, \qquad Z_t = \sum_{n\geq 1} \Psi_n 1_{\{T_n \leq t\}}, \tag{1.4.1}$$

where $b \in \mathbb{R}$, $\sigma \geq 0$ and W is a Brownian motion, and Z is a compound Poisson process: that is, the times $T_1 < T_2 < \cdots$ are the arrival times of a Poisson process on $\mathbb{R}_+$, say with parameter $\lambda > 0$, and independent of W, and the Ψ_n's are i.i.d. variables with law F, say, and independent of everything else. For convenience, we put $T_0 = 0$ and $N_t = \sum_{n\geq 1} 1_{\{T_n \leq t\}}$ (which is the Poisson process mentioned above). To avoid trivial complications, we assume $\lambda > 0$ and $F(\{0\}) = 0$.

Before proceeding, we state an important remark:

Remark 1.4.1 The Poisson process N, hence X as well, has a.s. infinitely many jumps on the whole of $\mathbb{R}_+$. However, in practice we are usually interested in the behavior of our functionals on a given fixed finite interval $[0, T]$. Then the subset Ω_T of Ω on which N and X have no jump on this interval has a positive probability. On Ω_T we have $X_t = Y_t$ for all $t \leq T$, hence for example $V^n(f, X)_t = V^n(f, Y)_t$ for $t \leq T$ as well. Then, *in restriction to the set* Ω_T, $(V^n(f, X)_t)_{t\in[0,T]}$ behaves as $(V^n(f, Y)_t)_{t\in[0,T]}$, as described in the previous section: there is no problem for (1.2.9) since the convergence in probability is well defined in restriction to the subset Ω_T. For the convergence in law in (1.2.10) and (1.2.11) saying that it holds "in restriction to Ω_T" makes *a priori* no sense; however, as mentioned before, we do have also the stronger stable convergence in law, for which it makes sense to speak of the convergence in restriction to Ω_T: this is our first example of the importance of stable convergence, from a purely theoretical viewpoint.

The functionals $V'^n(f, X)$ are particularly ill-suited when X has jumps, because the normalized increment $\Delta_i^n X/\sqrt{\Delta_n}$ "explodes" as $n \to \infty$ if we take $i = i_n$ such that the interval $((i-1)\Delta_n, i\Delta_n]$ contains a jump. More precisely, $\Delta_i^n X/\sqrt{\Delta_n}$ is equivalent to $\Psi/\sqrt{\Delta_n}$ if Ψ is the size of the jump occurring in this interval. So general results for these functionals ask for very specific properties of f near infinity. Therefore, below we restrict our attention to $V^n(f, X)$.

1.4.1 The Law of Large Numbers

The key point now is that (1.2.8) fails. In the situation at hand, for any t there are at most finitely many q's with $T_q \leq t$, or equivalently $N_t < \infty$. The difference $V^n(f, X)_t - V^n(f, Y)_t$ is constant in t on each interval $[i\Delta_n, j\Delta_n)$ such that $(i\Delta_n, (j-1)\Delta_n]$ contains no jump. Moreover, let us denote by Ω_t^n the subset of

Ω on which $T_q - T_{q-1} \geq \Delta_n$ for all q such that $T_q \leq t$, and by $i(n,q)$ the unique (random) integer i such that $(i-1)\Delta_n < T_q \leq i\Delta_n$. Note that Ω_t^n tends to Ω as $n \to \infty$, for all t. Then if we set

$$\zeta_q^n = f\big(\Psi_q + \Delta_{i(n,q)}^n Y\big) - f\big(\Delta_{i(n,q)}^n Y\big), \qquad \overline{V}^n(f)_t = \sum_{q=1}^{N_t^{(\Delta_n)}} \zeta_q^n,$$

where Ψ_q is as in (1.4.1), we have

$$V^n(f,X)_s = V^n(f,Y)_s + \overline{V}^n(f)_s, \quad \forall s \leq t, \quad \text{on the set } \Omega_t^n. \tag{1.4.2}$$

Observe that $\Delta_{i(n,q)}^n Y \to 0$ for all q, because Y is continuous. Then as soon as f is continuous and vanishes at 0, we have $\zeta_q^n \to f(\Psi_q)$, hence $\zeta_q'^n \to f(\Psi_q)$ as well. Since $N_t^{(\Delta_n)} \leq N_t < \infty$ and since $\mathbb{P}(\Delta X_t \neq 0) = 0$ for any given t (because the Poisson process N has no fixed time of discontinuity), we deduce

$$\overline{V}^n(f)_t \xrightarrow{\text{a.s.}} \sum_{q=1}^{N_t} f(\Psi_q) = \sum_{s\leq t} f(\Delta X_s),$$

where $\Delta X_s = X_s - X_{s-}$ denotes the size of the jump of X at time s. This convergence is not local uniform in time. However, it holds for the Skorokhod topology (see Chap. 2 for details), and we write

$$\overline{V}^n(f)_t \overset{\text{a.s.}}{\Longrightarrow} \sum_{s\leq t} f(\Delta X_s). \tag{1.4.3}$$

When f vanishes on a neighborhood of 0 and is continuous, and if we combine the above with (1.2.8) for Y, with (1.4.2) and with $\Omega_t^n \to \Omega$, we see that (1.2.8) ought to be replaced by

$$V^n(f,X)_t \overset{\mathbb{P}}{\Longrightarrow} \sum_{s\leq t} f(\Delta X_s) \tag{1.4.4}$$

(convergence in probability for the Skorokhod topology).

The general case is also a combination of (1.4.2) and (1.4.4) with (1.2.9) applied to the process Y: it all depends on the behavior of the normalizing factor $\Delta_n^{1-r/2}$ in front of $V^n(f,Y)$, which ensures the convergence. If $r > 2$ the normalizing factor blows up, so $V^n(f,Y)$ goes to 0; when $r < 2$ then $V^n(f,Y)$ blows up (at least in the first case of (1.2.9)) and when $r = 2$ the functionals $V^n(f,Y)$ go to a limit, without normalization. Therefore we end up with the following LLNs (we always suppose f continuous, and is of polynomial growth in the last statement below; this means that $|f(x)| \leq K(1+|x|^p)$ for some constants K and p):

$$f(x) = o\big(|x|^2\big) \text{ as } x \to 0 \;\Rightarrow\; V^n(f,X)_t \overset{\mathbb{P}}{\Longrightarrow} \sum_{s\leq t} f(\Delta X_s)$$

$$
\begin{aligned}
&f(x) \sim x^2 \text{ as } x \to 0 \;\Rightarrow\; V^n(f,X)_t \stackrel{\mathbb{P}}{\Longrightarrow} ct + \sum_{s\le t} f(\Delta X_s) \\
&f(x) \sim |x|^r \text{ as } x \to 0 \Rightarrow \begin{cases} V^n(f,X)_t \stackrel{\mathbb{P}}{\longrightarrow} +\infty & \text{if } r \in (0,2) \text{ and } t>0 \\ \Delta_n^{1-r/2} V^n(f,X)_t \stackrel{\text{u.c.p.}}{\Longrightarrow} t m_r \sigma^r. \end{cases}
\end{aligned}
\tag{1.4.5}
$$

Once more, this does not cover all possible test functions f.

1.4.2 The Central Limit Theorem

We have different CLTs associated with the different LLNs in (1.4.5). The results rely again upon the decomposition (1.4.2). In view of (1.4.2), and since we already have the CLT for $V^n(f,Y)$, we basically need to establish a CLT for $\overline{V}^n(f)$, for which the LLN takes the form (1.4.3). Due to some peculiarity of the Skorokhod topology, (1.4.3) does *not* imply that the difference $\overline{V}^n(f)_t - \sum_{s\le t} f(\Delta X_s)$ goes to 0 for this topology. However we do have Skorokhod convergence to 0 of the discretized processes, that is

$$
\widehat{V}^n(f)_t := \overline{V}^n(f)_t - \sum_{s\le \Delta_n[t/\Delta_n]} f(\Delta X_s) \stackrel{\text{a.s.}}{\Longrightarrow} 0,
$$

and we are looking for a CLT for these processes $\widehat{V}^n(f)$.

The key steps of the argument are as follows:

Step 1) We rewrite $\widehat{V}^n(f)_t$ as $\widehat{V}^n(f)_t = \sum_{q=1}^{N_t^{(\Delta_n)}} \eta_q^n$, where

$$
\eta_q^n = f\big(\Psi_q + \Delta_{i(n,q)}^n Y\big) - f(\Psi_q) - f\big(\Delta_{i(n,q)}^n Y\big).
$$

Assuming that f is C^1 with $f(0)=0$, and recalling $\Delta_{i(n,q)}^n Y \to 0$, a Taylor expansion gives

$$
\eta_q^n = \big(f'(\Psi_q) - f'(0)\big)\, \Delta_{i(n,q)}^n Y \big(1 + \mathrm{o}\big(\Delta_{i(n,q)}^n Y\big)\big).
$$

Since $\Delta_{i(n,q)}^n Y = b\Delta_n + \sigma\sqrt{\Delta_n}\; \Delta_{i(n,q)}^n W$, we deduce (this has to be justified, of course):

$$
\eta_q^n = \big(f'(\Psi_q) - f'(0)\big)\, \sigma\, \Delta_{i(n,q)}^n W + \mathrm{o}(\sqrt{\Delta_n}\,). \tag{1.4.6}
$$

Step 2) The jump times T_q and sizes Ψ_q, hence the random integers $i(n,q)$, are independent of W. Moreover one can check that the sequence $(\Delta_{i(n,q)}^n W)_{q\ge 1}$ is asymptotically independent of the process X as $n\to\infty$, whereas in restriction to the set Ω_t^n the variables $\Delta_{i(n,q)}^n W$ for $q = 1,\dots,N_t$ are independent and $\mathcal{N}(0,\Delta_n)$.

Therefore, if $(\Phi_q)_{q\geq 1}$ denotes a sequence of independent $\mathcal{N}(0,1)$ variables, independent of the process X, we deduce the following joint convergence in law, as $n\to\infty$:

$$\left(X,\left(\frac{1}{\sqrt{\Delta_n}}\,\eta_q^n\right)_{q\geq 1}\right)\stackrel{\mathcal{L}}{\longrightarrow}\left(X,\left(\left(f'(\Psi_q)-f'(0)\right)\sigma\,\Phi_q\right)_{q\geq 1}\right).$$

Step 3) The previous step and (1.4.6) give

$$\left(X,\frac{1}{\sqrt{\Delta_n}}\,\widehat{V}^n(f)\right)\stackrel{\mathcal{L}}{\Longrightarrow}\left(X,\widehat{V}(f)\right),\quad\text{where }\widehat{V}(f)_t=\sum_{q=1}^{N_t}\left(f'(\Psi_q)-f'(0)\right)\sigma\,\Phi_q \tag{1.4.7}$$

(we also have the stable convergence in law). This is the desired CLT for $\widehat{V}^n$.

Step 4) It remains to combine (1.4.7) with the result of the previous section, in the light of the decomposition (1.4.2). In order to stay simple, although keeping the variety of possible results, we only consider the absolute power functions $f_r(x)=|x|^r$. The results strongly depend on r, as did the LLNs (1.4.5) already, but here we have more cases.

For getting a clear picture of what happens, it is useful to rewrite (1.4.7) in a somewhat loose form (in particular the "equality" below is in law only), as follows, at least when $r>1$ so f_r is C^1 and $f_r'(0)=0$:

$$\begin{aligned}\overline{V}^n(f_r)_t &= A_t^n+B_t^n+\mathrm{o}(\sqrt{\Delta_n})\quad\text{"in law", where}\\ A_t^n &= \sum_{q=1}^{N_t^{(\Delta_n)}} f_r(\Psi_q),\qquad B_t^n=\sqrt{\Delta_n}\sum_{q=1}^{N_t} f_r'(\Psi_q)\,\sigma\,\Phi_q.\end{aligned}\tag{1.4.8}$$

Analogously, we can rewrite (1.2.10) for Y as follows:

$$\begin{aligned}V^n(f_r,Y)_t &= A_t'^n+B_t'^n+\mathrm{o}\big(\Delta_n^{r/2-1/2}\big)\quad\text{"in law", where}\\ A_t'^n &= \Delta_n^{r/2-1}m_r\sigma^r\,t,\qquad B_t'^n=\Delta_n^{r/2-1/2}\sigma^r\sqrt{m_{2r}-m_r^2}\,W_t'.\end{aligned}$$

Note that $A_t^n\gg B_t^n$ (meaning $B_t^n/A_t^n\stackrel{\mathbb{P}}{\longrightarrow}0$ as $n\to\infty$), and $A_t'^n\gg B_t'^n$. Then we can single out seven (!) different cases. For simplicity we do not write the joint convergence with the process X itself, but this joint convergence nevertheless always holds.

1) If $r>3$: We have $B_t^n\gg A_t'^n$, hence

$$\frac{1}{\sqrt{\Delta_n}}\left(V^n(f_r,X)_t^n-\sum_{q=1}^{N_t^{(\Delta_n)}} f_r(\Psi_q)\right)\stackrel{\mathcal{L}}{\Longrightarrow}\sum_{q=1}^{N_t} f_r'(\Psi_q)\,\sigma\,\Phi_q.$$

2) If $r=3$**:** Both terms B_t^n and $A_t'^n$ are of the same order of magnitude, hence

$$\frac{1}{\sqrt{\Delta_n}}\left(V^n(f_3,X)_t^n-\sum_{q=1}^{N_t^{(\Delta_n)}} f_3(\Psi_q)\right) \stackrel{\mathcal{L}}{\Longrightarrow} m_3\sigma^3 t+\sum_{q=1}^{N_t} f_3'(\Psi_q)\,\sigma\,\Phi_q.$$

3) If $2<r<3$**:** We have $A_t^n \gg A_t'^n \gg B_t^n$. Then we do *not* have a proper CLT here, but the following two properties:

$$\frac{1}{\Delta_n^{r/2-1}}\left(V^n(f_r,X)_t^n-\sum_{q=1}^{N_t^{(\Delta_n)}} f_r(\Psi_q)\right) \stackrel{\text{u.c.p.}}{\Longrightarrow} m_r\sigma^r\, t,$$

$$\frac{1}{\sqrt{\Delta_n}}\left(V^n(f_r,X)_t^n-\sum_{q=1}^{N_t^{(\Delta_n)}} f_r(\Psi_q)-\Delta_n^{r/2-1}m_r\sigma^r\, t\right) \stackrel{\mathcal{L}}{\Longrightarrow} \sum_{q=1}^{N_t} f_r'(\Psi_q)\,\sigma\,\Phi_q.$$

4) If $r=2$**:** Both terms A_t^n and $A_t'^n$, resp. B_t^n and $B_t'^n$, are of the same order of magnitude, and one can show that

$$\frac{1}{\sqrt{\Delta_n}}\left(V^n(f_2,X)_t^n-\sigma^2 t+\sum_{q=1}^{N_t^{(\Delta_n)}} (\Psi_q)^2\right) \stackrel{\mathcal{L}}{\Longrightarrow} \sqrt{2}\sigma^2\, W_t'+2\sum_{q=1}^{N_t} \Psi_q\,\sigma\,\Phi_q$$

(recall $m_2=1$ and $m_4=3$ and $f_2'(x)=2x$). Here W' is a Brownian motion independent of X, and also of the sequence (Φ_q). Note that, if we replace t by $\Delta_n[t/\Delta_n]$ in the left side above, which does not affect the convergence, this left side is the difference between the approximate quadratic variation and the discretized true quadratic variation.

5) If $1<r<2$**:** We have $A_t'^n \gg A_t^n \gg B_t'^n \gg B_t^n$. Then as in Case 3 we have two results:

$$\frac{1}{\Delta_n^{1-r/2}}\left(\Delta_n^{1-r/2}V^n(f_r,X)_t^n-m_r\sigma^r\, t\right) \stackrel{\mathbb{P}}{\Longrightarrow} \sum_{q=1}^{N_t} f_r(\Psi_q),$$

$$\frac{1}{\Delta_n^{r/2-1/2}}\left(V^n(f_r,X)_t^n-\Delta_n^{r/2-1}m_r\sigma^r\, t-\sum_{q=1}^{N_t^{(\Delta_n)}} f_r(\Psi_q)\right) \stackrel{\mathcal{L}}{\Longrightarrow} \sigma^r\sqrt{m_{2r}-m_r^2}\; W_t'.$$

6) If $r=1$**:** The function f_1 is not differentiable at 0, but one can show that $\overline{V}^n(f_1)$ has a decomposition (1.4.8) with the same A_t^n and with a B_t^n satisfying $A_t^n \gg B_t^n$. Now, A_t^n and $B_t'^n$ have the same order of magnitude, so we get

$$\frac{1}{\sqrt{\Delta_n}}\left(\sqrt{\Delta_n}\; V^n(f_1,X)_t^n-m_1\sigma\, t\right) \stackrel{\mathcal{L}}{\Longrightarrow} \sum_{q=1}^{N_t} |\Psi_q|+|\sigma|\sqrt{1-m_1^2}\; W_t'.$$

7) If $0 < r < 1$**:** Again the function f_r is not differentiable at 0, but obviously $\overline{V}^n(f_r)_t$ stays bounded in probability. Then we have:

$$\frac{1}{\sqrt{\Delta_n}}\left(\Delta_n^{1-r/2}\, V^n(f_r, X)_t^n - m_r\sigma^r\, t\right) \stackrel{\mathcal{L}}{\Longrightarrow} \sigma^r\sqrt{m_{2r} - m_r^2}\; W_t'.$$

The jumps have disappeared from the picture in this case, which is as in (1.2.10).

From this brief description, we are able to conclude a moral that pervades the theory: including processes with jumps complicate matters more than one might naively suspect.

Chapter 2
Some Prerequisites

This second preliminary chapter is very different from the first one. Its aim is to establish notation to be used throughout the book, and to recall some properties of semimartingales and a few limit theorems which are basic to our study. Most of these results are available in book form already, and the proofs are omitted: we refer to the books of Jacod and Shiryaev [57] for most results, and of Protter [83] or Ikeda and Watanabe [50] for some specific results on semimartingales and stochastic calculus. A few results are new in book form, and those are mostly proved in the Appendix.

2.1 Semimartingales

The basic process whose "discretization" is studied in this book is a d-dimensional semimartingale, say X. This means a process indexed by nonnegative times t, with d components $X^1, \dots, X^d$, and such that each component $X^i = (X^i_t)_{t\geq 0}$ is a semimartingale.

We need to be a bit more specific: we start with a probability space $(\Omega, \mathcal{F}, \mathbb{P})$ endowed with a *filtration* $(\mathcal{F}_t)_{t\geq 0}$, that is an increasing (meaning $\mathcal{F}_s \subset \mathcal{F}_t$ if $s \leq t$) and right-continuous (meaning $\mathcal{F}_t = \cap_{s>t}\mathcal{F}_s$) family of sub-$\sigma$-fields $\mathcal{F}_t$ of $\mathcal{F}$. We say that $(\Omega, \mathcal{F}, (\mathcal{F}_t)_{t\geq 0}, \mathbb{P})$ is a *filtered probability space*. We do not make the usual assumption that the filtration is "complete", since this property does not play any role in the sequel.

A real-valued process Y on this filtered probability space is called a *semimartingale* if

(i) it is adapted (to the underlying filtration, *i.e.*, each Y_t is $\mathcal{F}_t$ measurable);
(ii) it has càdlàg (the acronym for "right-continuous with left limits", in French) paths;
(iii) there is a sequence (T_n) of stopping times increasing to ∞ such that for any n the stopped process $Y(n) = Y_{t\wedge T_n}$ is the sum of a martingale plus a process whose paths have bounded variation over each finite interval (such a process is called "process of finite variation").

J. Jacod, P. Protter, *Discretization of Processes*,
Stochastic Modelling and Applied Probability 67,
DOI 10.1007/978-3-642-24127-7_2, © Springer-Verlag Berlin Heidelberg 2012

Property (iii) is the crucial one, and it may be expressed equivalently by saying that Y is the sum of a local martingale plus a process of finite variation. A d-dimensional process is a semimartingale if its components are real-valued semimartingales.

Among all processes, semimartingales play a very special role. For example they are the most general processes with respect to which a (stochastic) integration theory, having the usual "nice" properties like a Lebesgue convergence theorem, can be constructed. This fact may even be used as the definition of the semimartingales, see e.g. Protter [83]. In mathematical finance they also play a special role, since one of the most basic results (the so-called "fundamental asset pricing theorem") says that if no arbitrage is allowed, then the price process should at least be a semimartingale.

We are not going to describe the properties of semimartingales at large, as this constitutes a whole theory by itself, and one may for example consult Dellacherie & Meyer [25] for a comprehensive study. Rather, we will focus our attention on the properties which are most useful for our purposes.

Before proceeding, we recall two useful notions:

- a *localizing sequence of stopping times* is a sequence of stopping times which increases to $+\infty$,
- an $\mathbb{R}^d$-valued process H is *locally bounded* if $\sup_{\omega\in\Omega,0<t\leq T_n(\omega)} \|H_t(\omega)\| < \infty$ for some localizing sequence (T_n) of stopping times ($\|.\|$ denotes the Euclidean norm on $\mathbb{R}^d$).

Note that in the second definition above we take $\sup_{0<t\leq T_n} \|H_t\|$ instead of the most customary $\sup_{0\leq t\leq T_n} \|H_t\|$: when H is right-continuous this makes no difference on the set $\{T_n > 0\}$, and when H is the integrand of a (stochastic or ordinary) integral the value H_0 plays no role at all. The reason for this slightly weaker definition is the following: saying that $\sup_{0\leq t\leq T_n} \|H_t\|$ is bounded automatically implies that the initial variable H_0 is bounded, and in most cases we do not want such a restriction.

2.1.1 First Decompositions and the Basic Properties of a Semimartingale

Let X be a real semimartingale on a filtered space $(\Omega, \mathcal{F}, (\mathcal{F}_t)_{t\geq 0}, \mathbb{P})$.

1) We have a first decomposition of X, which is

$$X_t = X_0 + A_t + X^c_t + M_t \tag{2.1.1}$$

where $A_0 = X^c_0 = M_0 = 0$ and A is an adapted process whose paths are of finite variation, and X^c is a *continuous* local martingale, and M is a local martingale which is orthogonal to all continuous local martingales, meaning that the product MN is a local martingale for any continuous local martingale N. One says that M is "purely discontinuous", although this does not refer to sample path behavior: for example $Y_t - t$ where Y is a standard Poisson process is a purely discontinuous martingale in this sense.

The decomposition (2.1.1) is by no way unique. However, any other decomposition $X_t = X_0 + A'_t + X'^c_t + M'_t$ of the same type is such that $X'^c = X^c$ outside a null set. We usually identify two processes whose paths are a.s. the same, so we say that X^c is *the* continuous local martingale part of X.

2) Next, we define "stochastic integrals" with respect to X. This is first done for integrands H which are "simple", that is of the form $H_t = \sum_{m\geq 1} U_m 1_{(T_m, T_{m+1}]}(t)$ (where 1_A denotes the indicator function of any set A), for a sequence (T_m) of times increasing to $+\infty$ and random variables U_m. The integral process is then defined as

$$\int_0^t H_s\, dX_s = \sum_{m\geq 1} U_m\, (X_{t\wedge T_{m+1}} - X_{t\wedge T_m}). \tag{2.1.2}$$

This is the "naive" integral, taken ω-wise, of a piecewise constant function $t \mapsto H_t(\omega)$ with respect to the "measure" having the distribution function $t \mapsto X_t(\omega)$.

Of course, there is no such measure in general. Nevertheless, the above elementary integral can be extended to the set of all *predictable and bounded* (*or, locally bounded,*) processes. For this, we first recall that the *predictable* σ-field $\mathcal{P}$ is the σ-field on $\Omega \times \mathbb{R}_+$ which is generated by all processes Y which are adapted and continuous (or only left-continuous, the σ-field is the same). A predictable process is a process which, considered as a map from $\Omega \times \mathbb{R}_+$ into $\mathbb{R}$, is $\mathcal{P}$ measurable. Note that a simple process H as above is predictable as soon as the T_m's are stopping times and the U_m's are $\mathcal{F}_{T_m}$ measurable.

The extendability of (2.1.2) means that, for any predictable and locally bounded process H, one can define (in a unique way, up to null sets) a process called the "stochastic integral process" and denoted as

$$Z_t = \int_0^t H_s\, dX_s,$$

in such a way that it coincides with (2.1.2) when H is simple and predictable, and that we further have a "dominated convergence theorem" which is stated as Proposition 2.2.7 in Sect. 2.2 (in which all limit theorems are gathered). The above notation implicitly means the integral is taken over the interval $(0, t]$, with t included in, and 0 excluded from, the domain of integration.

When t varies, this defines a process $Z = (Z_t)$ which itself is a semimartingale, as we see below: indeed, if we consider any decomposition like (2.1.1), and since all three processes A, X^c and M are semimartingales, we can integrate H with respect to each of these, and we have

$$\int_0^t H_s\, dX_s = \int_0^t H_s\, dA_s + \int_0^t H_s\, dX^c_s + \int_0^t H_s\, dM_s. \tag{2.1.3}$$

This provides a decomposition of the type (2.1.1) for Z, and in particular $Z^c = \int_0^t H_s\, dX^c_s$.

One may in fact define stochastic integrals for a class of integrands H larger than the set of predictable locally bounded processes, and a precise description of those H which are "integrable" with respect to a given semimartingale X is available, although not immediately obvious. We do not need this here, except when X is a Brownian motion: in this case the set of integrable processes is the set of all *progressively measurable* processes H (meaning that for any t the map $(\omega, s) \mapsto H_s(\omega)$ on $\Omega \times (0, t]$ is $\mathcal{F}_t \otimes \mathcal{B}([0, t])$ measurable), such that

$$\int_0^t H_s^2 \, ds < \infty \quad \text{a.s. for all } t \in \mathbb{R}_+.$$

For example any adapted càdlàg process H is integrable with respect to the Brownian motion, and its integral coincides with the integral of the left-continuous process H_-.

3) Now we look at the "jumps" of a càdlàg process Y, say $\mathbb{R}^d$-valued. We set

$$\begin{aligned} Y_{t-} &= \lim_{s \uparrow\uparrow t} Y_s \quad \text{(with the convention } Y_{0-} = Y_0\text{)}, \\ \Delta Y_t &= Y_t - Y_{t-}, \qquad D(Y) = \big\{(\omega, t) : \Delta Y_t(\omega) \neq 0\big\}. \end{aligned}$$

The jump process (ΔY_t) is $\mathbb{R}^d$-valued, and for each ω the set $D(Y)(\omega) = \{t : (\omega, t) \in D(Y)\}$ of all times at which Y jumps is at most countable, although typically it may be a dense subset of $\mathbb{R}_+$. However, even in this case, the set $\{t : \|\Delta Y_t\| > \varepsilon\}$ is locally finite for any $\varepsilon > 0$, because of the càdlàg property.

If X is a semimartingale and $Z_t = \int_0^t H_s \, dX_s$ is the stochastic integral of a predictable process H, then we can find a version of the integral process Z satisfying identically:

$$\Delta Z_t = H_t \, \Delta X_t. \tag{2.1.4}$$

4) At this point we can introduce the *quadratic variation* of X. First if Y is a continuous local martingale with $Y_0 = 0$, there is a unique increasing adapted and continuous process, null at 0, and denoted by $\langle Y, Y \rangle$, such that $Y^2 - \langle Y, Y \rangle$ is a local martingale (this is the Doob-Meyer decomposition of the local submartingale Y^2). Next, for X a one-dimensional semimartingale, we set

$$[X, X]_t = \big\langle X^c, X^c \big\rangle_t + \sum_{s \le t} (\Delta X_s)^2. \tag{2.1.5}$$

The sum above makes sense, since it is a sum of positive numbers on the countable set $D(X)(\omega) \cap [0, t]$. What is not immediately obvious is that it is a.s. finite, but this fact is one of the main properties of semimartingales. Hence the process $[X, X]$ is increasing and càdlàg, and also adapted (another not immediately obvious property), and it is called the *quadratic variation process* of X, or sometimes the "square bracket". Note that $[X, X] = \langle X, X \rangle$ when X is a continuous local martingale. Also note that, for any semimartingale, $[X^c, X^c] = \langle X^c, X^c \rangle$ is the "continuous part" of the increasing process $[X, X]$ (not to be confused with its "continuous martingale part", which is identically 0).

When Y and Y' are two continuous local martingales, null at 0, we analogously have a unique process $\langle Y, Y'\rangle$ which is continuous, adapted, null at 0 and of locally finite variation, and such that $YY' - \langle Y, Y'\rangle$ is a local martingale. We have $\langle Y, Y'\rangle = \langle Y', Y\rangle$. When X and X' are two real-valued semimartingales we then set

$$[X, X']_t = \langle X^c, X'^c\rangle_t + \sum_{s\le t} \Delta X_s \, \Delta X'_s.$$

Here again the sum above is a.s. absolutely convergent, by the finiteness in (2.1.5) for X and X' and the Cauchy-Schwarz inequality. The process $[X, X'] = [X', X]$ is adapted and locally of finite variation, but not necessarily increasing any more, and is called the *quadratic covariation process* of X and X'. We also have for any real a and any other semimartingale X'':

$$[X + aX', X''] = [X, X''] + a[X', X'']. \tag{2.1.6}$$

Another useful property, which immediately follows from this, is the *polarization identity*:

$$[X, X'] = \frac{1}{4}\big([X + X', X + X'] - [X - X', X - X']\big). \tag{2.1.7}$$

To put an end to this topic, let us mention a fundamental property of the quadratic variation. Take any sequence of subdivisions of $\mathbb{R}_+$ with meshes going to 0: we can even consider *random* subdivisions, that is for each n we have a sequence $(T(n,i) : i \ge 0)$ of stopping times, which strictly increases to $+\infty$, and with $T(n,0) = 0$, and such that $\sup(T(n,i+1) \wedge t - T(n,i) \wedge t : i \ge 0)$ goes to 0 in probability for all $t > 0$ as $n \to \infty$. Then we have the following convergence in probability, for all $t \ge 0$, and as $n \to \infty$:

$$\sum_{i\ge 1} (X_{t\wedge T(n,i)} - X_{t\wedge T(n,i-1)})\big(X'_{t\wedge T(n,i)} - X'_{t\wedge T(n,i-1)}\big) \xrightarrow{\mathbb{P}} [X, X']_t. \tag{2.1.8}$$

This is a very simple consequence of the forthcoming Itô's formula and the dominated convergence theorem for stochastic integrals, and in view of its importance for this book we will prove it later.

5) Let now A be an increasing adapted càdlàg process, null at 0, and which is *locally integrable*: the latter means that $\mathbb{E}(A_{T_n}) < \infty$ for all n, for a localizing sequence (T_n) of stopping times. Then A is a local submartingale and by the Doob-Meyer decomposition again there is an almost surely unique, increasing càdlàg *predictable* process A' with $A'_0 = 0$, such that $A - A'$ is a local martingale. The same holds when A is adapted, càdlàg and of locally integrable variation (meaning: it is the difference of two increasing locally integrable processes), except that A' is no longer increasing but of finite (and even locally integrable) variation.

In these two cases, A' is called the *compensator*, or "predictable compensator", of A. When A is of locally finite variation adapted and continuous with $A_0 = 0$, then it is necessarily of locally integrable variation, and its compensator is $A' = A$.

6) The above notion applies in particular to the quadratic variation process of a semimartingale X. Suppose that $[X,X]$ is locally integrable. In this case, we denote by $\langle X,X\rangle$, and call "angle bracket" or "predictable quadratic variation process", the compensator of $[X,X]$. This notation does not conflict with the notation $\langle X,X\rangle$ previously defined as the quadratic variation when X is a continuous local martingale: indeed, in this case the quadratic variation is continuous increasing adapted, hence predictable and locally integrable, hence its own compensator.

More generally if X and X' are two semimartingales with both $[X,X]$ and $[X',X']$ locally integrable, then $[X,X']$ is of locally integrable variation, and $\langle X,X'\rangle$ denotes its compensator.

Note that the local integrability of $[X,X]$ may fail, in which case the predictable quadratic variation is *not defined.*

7) Now we consider a d-dimensional semimartingale $X=(X^i)_{i\le d}$. First, we still have (many) decompositions like (2.1.1), which should be read component by component: that is, we have $A=(A^i)_{i\le d}$ and $X^c=(X^{i,c})_{i\le d}$ and $M=(M^i)_{i\le d}$. Next, we can integrate locally bounded predictable processes H which are d-dimensional, say $H=(H^i)_{i\le d}$, and the stochastic integral process is (with $H^\star$ denoting the transpose):

$$Z_t = \int_0^t H_s^\star\, dX_s := \sum_{i=1}^d \int_0^t H_s^i\, dX_s^i,$$

where on the right side we have a collection of "one-dimensional" integrals defined as before. We still have a formula as in (2.1.3), which gives a decomposition of the type (2.1.1) for Z, and (2.1.4) holds as well. And, if H is "simple", we again have (2.1.2) with the summands $U_m^\star(X_{t\wedge T_{m+1}}-X_{t\wedge T_m})$.

Turning to the quadratic variation, we now have a collection $[X,X]=([X^i,X^j]: 1\le i,j\le d)$ of adapted càdlàg processes of locally finite variation. By (2.1.6) applied twice it is easy to check that $[X,X]$ takes its values in the set $\mathcal{M}_{d\times d}^+$ of all nonnegative symmetric $d\times d$ matrices, and it is non-decreasing for the strong order of this set (the last qualifier means that $[X,X]_{t+s}-[X,X]_t$ belongs to $\mathcal{M}_{d\times d}^+$ for all $s,t\ge 0$).

If further all increasing processes $[X^j,X^j]$ are locally integrable, we have the "angle bracket" $\langle X,X\rangle=(\langle X^j,X^k\rangle : 1\le j,k\le d)$, which again takes its values in the set $\mathcal{M}_{d\times d}^+$, and is non-decreasing for the strong order of this set.

8) We end this subsection with a statement of Itô's formula. If X is a d-dimensional semimartingale and if f is a C^2 function on $\mathbb{R}^d$ (C^2 = twice continuously differentiable), then the process $f(X)$ is also a semimartingale. Moreover, with $\partial_i f$ and $\partial_{ij}^2 f$ denoting the first and second partial derivatives of f, we have

$$f(X_t)=f(X_0)+\sum_{i=1}^d \int_0^t \partial_i f(X_{s-})\,dX_s^i + \frac{1}{2}\sum_{i,j=1}^d \int_0^t \partial_{ij}^2 f(X_{s-})\,d\langle X^{ic},X^{jc}\rangle_s$$

$$+\sum_{s\le t}\left(f(X_{s-}+\Delta X_s)-f(X_{s-})-\sum_{i=1}^d \partial_i f(X_{s-})\Delta X_s^i\right). \tag{2.1.9}$$

The integrals above are meaningful because $\partial_i f(X_-)$ and $\partial^2_{ij} f(X_-)$ are predictable and locally bounded, and it turns out that the last sum is absolutely convergent for all t, even though, separately, the sums $\sum_{s\le t}(f(X_{s-}+\Delta X_s)-f(X_{s-}))$ and $\sum_{s\le t}\partial_i f(X_{s-})\Delta X_s^i$ may diverge.

Important Warning: We have often seen the qualifier "up to a null set" appear in the text above. And indeed, the brackets $[X,X]$, and $\langle X,X\rangle$ when it exists, and the stochastic integrals, and the predictable compensators, are all defined uniquely, up to a $\mathbb{P}$ null set. Therefore it is convenient—and without harm—to identify two processes X and X' which have the same paths outside a $\mathbb{P}$ null set: this identification will be made, usually without further mention, in the whole book.

2.1.2 Second Decomposition and Characteristics of a Semimartingale

Here again the filtered probability space $(\Omega,\mathcal{F},(\mathcal{F}_t)_{t\ge0},\mathbb{P})$ and the semimartingale X are fixed.

1) First, we associate with X the following d-dimensional process

$$X'_t = X_t - X_0 - J_t, \quad \text{where } J_t = \sum_{s\le t}\Delta X_s\, 1_{\{\|\Delta X_s\|>1\}}.$$

The sum defining J_t is for all ω and t a finite sum, and the process J is adapted and càdlàg and obviously of finite variation. Hence it is a semimartingale, and so is X'. Moreover $\|\Delta X'\|\le 1$ by construction. Then X' is, or rather each of its d components are, "special" semimartingales. This implies that among all decompositions (2.1.1) for X', there is one and only one (recall the above warning: the uniqueness is up to null sets) such that the process A is predictable, in addition to being of finite variation. We write this decomposition as

$$X'_t = X_0 + B_t + X^c_t + M_t,$$

where $B_0=M_0=0$ and B is predictable and (component by component) of locally finite variation, and M is (component by component again) a purely discontinuous local martingale, and X^c is the same as in (2.1.1) (recall that X^c does not depend on the decomposition). This yields

$$X_t = X_0 + B_t + X^c_t + M_t + \sum_{s\le t}\Delta X_s\, 1_{\{\|\Delta X_s\|>1\}}. \tag{2.1.10}$$

It is important to mention that this decomposition is *unique* (up to null sets). Note that B is not necessarily continuous, but its jump process satisfies $\|\Delta B\| \leq 1$, and thus $\|\Delta M\| \leq 2$.

2) Next, we associate with X a random measure μ (or, μ^X if we want to emphasize the dependency on X), called the *jump measure* of X, by the formula

$$\mu(\omega; dt, dx) = \sum_{(\omega,s)\in D(X)} \varepsilon_{(s,\Delta X_s(\omega))}(dt, dx).$$

Here ε_a denotes the Dirac measure charging $a \in \mathbb{R}_+ \times \mathbb{R}^d$, so for each ω, $\mu(\omega; .)$ is an integer-valued measure on $\mathbb{R}_+ \times \mathbb{R}^d$, which charges neither the set $\{0\} \times \mathbb{R}^d$, nor the set $\mathbb{R}_+ \times \{0\}$, and such that $\mu(\{t\} \times \mathbb{R}^d)$ equals 1 if $t \in D(X)$ and 0 otherwise. For any Borel subset A of $\mathbb{R}^d$ we have

$$1_A \star \mu_t := \mu\big((0,t] \times A\big) = \sum_{s\leq t} 1_A(\Delta X_s). \tag{2.1.11}$$

The process $1_A \star \mu$ is non-decreasing and adapted, although it may take the value $+\infty$ at some time $t > 0$, or even at all times $t > 0$. However when A lies at a positive distance from 0 the process $1_A \star \mu$ is càdlàg, $\mathbb{N}$-valued and with jumps of size 1, hence locally integrable; then it admits a predictable compensator which we denote by $1_A \star \nu$, and which is a predictable increasing process, null at 0, and also locally integrable.

Moreover $A \mapsto 1_A \star \mu_t$ is σ-additive, and it follows that $A \mapsto 1_A \star \nu_t$ is almost surely σ-additive. So it is no wonder that there exists a genuine positive random measure $\nu(\omega; dt, dx)$ on $\mathbb{R}_+ \times \mathbb{R}^d$ such that

$$1_A \star \nu(\omega)_t = \nu(\omega; (0,t] \times A) \tag{2.1.12}$$

for all A as above (the compensator being defined up to a null set, one should rather say: there is a measure ν such that the formula (2.1.12) defines a version $1_A \star \nu$ of the compensator of $1_A \star \mu$). Not surprisingly, the measure ν is called the (*predictable*) *compensator* of μ.

We extend the notation (2.1.11) or (2.1.12) to more general integrands. If $\delta = \delta(\omega, t, x)$ is a real function on $\Omega \times \mathbb{R}_+ \times \mathbb{R}^d$, we write

$$\begin{aligned} \delta \star \mu_t &= \int_{[0,t]\times\mathbb{R}^d} \delta(\omega, s, x)\mu(\omega; ds, dx), \\ \delta \star \nu_t &= \int_{[0,t]\times\mathbb{R}^d} \delta(\omega, s, x)\nu(\omega; ds, dx), \end{aligned} \tag{2.1.13}$$

as soon as these integrals make sense. In particular if the first one makes sense, we have

$$\delta \star \mu_t = \sum_{s\leq t} \delta(s, \Delta X_s). \tag{2.1.14}$$

3) At this point, we can define the *characteristics*, also known as "predictable characteristics" or "local characteristics", of the semimartingale X. These characteristics consist of the triplet (B, C, ν), where

$$\begin{cases} \bullet\ B = (B^i)_{1 \le i \le d} & \text{is the predictable process of locally finite variation} \\ & \text{with } B_0 = 0, \text{ occurring in (2.1.10)}, \\ \bullet\ C = (C^{ij})_{1 \le i,j \le d}, & \text{where } C^{ij} = \langle X^{i,c}, X^{j,c} \rangle, \\ \bullet\ \nu & \text{is the compensator of the jump measure } \mu, \text{ as defined above.} \end{cases}$$

It is useful to express the quadratic variation process $[X, X]$, and also the angle bracket $\langle X, X \rangle$ when it exists, in terms of the previous quantities. First we have always

$$\left[X^i, X^j\right] = C^{ij} + \left(x^i x^j\right) \star \mu.$$

Here the last process is $\delta \star \mu$ (notation (2.1.14)) for the function $\delta(\omega, t, x) = x^i x^j$, where $(x^i)_{1 \le i \le d}$ denote the components of $x \in \mathbb{R}^d$).

Second, it can be shown that the angle bracket exists if and only if $(x^i)^2 \star \nu_t < \infty$ a.s. for all t and i, or equivalently if $(x^i)^2 \star \mu$ is locally integrable for all i. In this case we have

$$\left\langle X^i, X^j \right\rangle_t = C_t^{ij} + \left(x^i x^j\right) \star \nu_t - \sum_{s \le t} \Delta B_s^i \Delta B_s^j. \tag{2.1.15}$$

4) The integrals in (2.1.13) are Lebesgue integrals with respect to two positive measures, for any fixed ω. Now, the signed measure $\mu - \nu$ is a "martingale measure," in the sense that for any Borel subset A of $\mathbb{R}^d$ at a positive distance of 0 the process $1_A \star (\mu - \nu) = 1_A \star \mu - 1_A \star \nu$ is a local martingale. So we also have a notion of stochastic integral with respect to $\mu - \nu$, which proves quite useful.

We will say that a function δ on $\Omega \times \mathbb{R}_+ \times \mathbb{R}^d$ is *predictable* if it is $\widetilde{\mathcal{P}}$ measurable, where $\widetilde{\mathcal{P}} = \mathcal{P} \otimes \mathcal{R}^d$, where $\mathcal{P}$ is the predictable σ-field on $\Omega \times \mathbb{R}_+$ and $\mathcal{R}^d$ the Borel σ-field of $\mathbb{R}^d$. Clearly $\delta(\omega, t, x) = 1_A(x)$ is predictable in this sense when $A \in \mathcal{R}^d$.

Let us take a predictable function δ as above. If

$$\left(\delta^2 \wedge |\delta|\right) \star \nu_t < \infty \quad \forall t > 0, \tag{2.1.16}$$

we can define a process denoted by

$$\int_0^t \int_{\mathbb{R}^d} \delta(s, x)(\mu - \nu)(ds, dx), \quad \text{or} \quad \delta \star (\mu - \nu)_t,$$

and called the *stochastic integral* of δ with respect to $\mu - \nu$: this is the unique (up to null sets) purely discontinuous local martingale whose jumps are given by

$$\Delta\left(\delta \star (\mu - \nu)\right)_t = \int \delta(t, x)(\mu - \nu)\left(\{t\}, dx\right) = \delta(t, \Delta X_t) - \int \delta(t, x)\nu\left(\{t\}, dx\right), \tag{2.1.17}$$

and moreover it coincides with the difference $\delta \star \mu - \delta \star \nu$ when both processes $\delta \star \mu$ and $\delta \star \nu$ are well-defined and finite-valued. (Here again one could define the stochastic integral $\delta * (\mu - \nu)$ for predictable integrands δ satisfying a condition slightly weaker than (2.1.16), but more complicated to state; the condition (2.1.16) will however be enough for our purposes.)

5) With this notion of stochastic integral, we arrive at the final decomposition of a semimartingale. Namely, we have

$$X = X_0 + B + X^c + (x 1_{\{\|x\| \le 1\}}) \star (\mu - \nu) + (x 1_{\{\|x\| > 1\}}) \star \mu. \tag{2.1.18}$$

This is called the *Lévy-Itô decomposition* of the semimartingale, by analogy with the formula bearing the same name for Lévy processes, see the next subsection. It is in fact another way of writing the decomposition (2.1.10), with the same B and X^c, and the last two terms in each of the two decompositions are the same. So M in (2.1.10) is equal to the stochastic integral $(x 1_{\{\|x\| \le 1\}}) \star (\mu - \nu)$, which should be read component by component and is $\delta * (\mu - \nu)$ with the function $\delta(\omega, t, x) = x 1_{\{\|x\| \le 1\}}$. Note that this function δ is predictable, and it satisfies (2.1.16) because the third characteristic ν of X always satisfies

$$\left(\|x\|^2 \wedge 1\right) \star \nu_t \ < \infty \quad \forall t > 0. \tag{2.1.19}$$

The latter property comes from the fact that $(\|x\|^2 \wedge 1) \star \mu_t = \sum_{s \le t} \|\Delta X_s\|^2 \wedge 1$ is finite-valued and with bounded jumps, hence locally integrable.

6) Finally we give a version of Itô's formula based on the characteristics. This version is a straightforward consequence of the classical formula (2.1.9) and of the previous properties of random measures and the decomposition (2.1.18). If f is a C^2 function,

$$\begin{aligned} f(X_t) = {} & f(X_0) + \sum_{i=1}^d \int_0^t \partial_i f(X_{s-})\, dB_s^i + \frac{1}{2} \sum_{i,j=1}^d \int_0^t \partial_{ij}^2 f(X_{s-})\, dC_s^{ij} \\ & + \left(\left(f(X_- + x) - f(X_-) - \sum_{i=1}^d \partial_i f(X_-) x^i \right) 1_{\{\|x\| \le 1\}} \right) \star \nu_t \\ & + \sum_{i=1}^d \int_0^t \partial_i f(X_{s-})\, dX_s^{ic} + \left(\left(f(X_- + x) - f(X_-) \right) 1_{\{\|x\| \le 1\}} \right) \star (\mu - \nu)_t \\ & + \left(\left(f(X_- + x) - f(X_-) \right) 1_{\{\|x\| > 1\}} \right) \star \mu_t. \end{aligned} \tag{2.1.20}$$

This formula looks complicated, but it turns out to be quite useful. We use a short hand notation here, for example $(f(X_- + x) - f(X_-)) 1_{\{\|x\| > 1\}}$ stands for the predictable function $\delta(\omega, t, x) = (f(X_{t-}(\omega) + x) - f(X_{t-}(\omega))) 1_{\{\|x\| > 1\}}$. The right side gives a decomposition of the semimartingale $f(X)$ which is somewhat similar to (2.1.18): apart from the initial value $f(X_0)$, the sum of the first three terms is a predictable process of locally finite variation, the fourth term is the continuous martingale part $f(X)^c$, the fifth one is a purely discontinuous local martingale with locally bounded jumps, and the last one is a finite sum of "big" jumps of $f(X)$.

2.1.3 A Fundamental Example: Lévy Processes

An adapted d-dimensional process X on $(\Omega, \mathcal{F}, (\mathcal{F}_t)_{t\geq 0}, \mathbb{P})$ is said to have $(\mathcal{F}_t)$-*independent increments*, if for all $s,t \geq 0$, the increment $X_{t+s} - X_s$ is independent of the σ-field $\mathcal{F}_t$. We have the following general result: if X is a d-dimensional semimartingale, it is a process with $(\mathcal{F}_t)$-independent increments if and only if its characteristics (B, C, ν) have a deterministic version, that is B is a d-dimensional càdlàg function with locally finite variation, C is a continuous function with values in the set $\mathcal{M}^+_{d\times d}$ and increasing in this set, and ν is a (non-random) positive measure on $\mathbb{R}_+ \times \mathbb{R}^d$.

It turns out that there exist processes with independent increments, even càdlàg ones, which are *not* semimartingales. For example a deterministic process $X_t = f(t)$ is always a process with independent increments, whereas it is a semimartingale if and only if the function f is càdlàg *and* of locally finite variation.

This, however, cannot happen if we assume in addition that the process has stationary increments. We say that X is an $(\mathcal{F}_t)$-*Lévy process* if it is càdlàg adapted with $X_0 = 0$ and if the increments $X_{t+s} - X_t$ are independent of $\mathcal{F}_t$ and with a law depending on s of course, but not on t. When $(\mathcal{F}_t)$ is the filtration generated by X, we simply say *Lévy process*. We then have the following fundamental result:

$$\left\{\begin{array}{l} \bullet \text{ Any } (\mathcal{F}_t)\text{-Lévy process } X \text{ is a semimartingale on } (\Omega, \mathcal{F}, (\mathcal{F}_t)_{t\geq 0}, \mathbb{P}), \\ \bullet \text{ A } d\text{-dimensional semimartingale } X \text{ is an } (\mathcal{F}_t)\text{-Lévy process if and} \\ \quad \text{only if } X_0 = 0 \text{ and its characteristics have the form} \\ \qquad B_t(\omega) = bt, \qquad C_t(\omega) = ct, \qquad \nu(\omega; dt, dx) = dt \otimes F(dx). \end{array}\right. \tag{2.1.21}$$

Here we have:

$$\begin{array}{l} b = (b^i)_{i\leq d} \in \mathbb{R}^d, \qquad c = (c^{ij})_{i,j\leq d} \in \mathcal{M}^+_{d\times d}, \qquad F \text{ is a positive} \\ \text{measure on } \mathbb{R}^d \text{ with } F(\{0\}) = 0 \text{ and } \int (\|x\|^2 \wedge 1)\, F(dx) < \infty. \end{array} \tag{2.1.22}$$

The term (b, c, F) is called the *characteristic triplet* of X; b is the "drift", c is the covariance of the "Gaussian part", and F is the "Lévy measure". Conversely, with any triplet (b, c, F) satisfying (2.1.22) one associates a Lévy process X, and the triplet (b, c, F) completely characterizes the law of the process X (hence the name "characteristics") via the independence and stationarity of the increments, and the *Lévy-Khintchine* formula which gives the characteristic function of the variable X_t. With vector notation, the Lévy-Khintchine formula reads as follows, for all $u \in \mathbb{R}^d$:

$$\mathbb{E}\big(e^{iu^\star X_t}\big) = \exp t\left(u^\star b - \frac{1}{2}\, u^\star c\, u + \int \big(e^{iu^\star x} - 1 - iu^\star x 1_{\{\|x\|\leq 1\}}\big) F(dx)\right).$$

Now, assuming that X is an $(\mathcal{F}_t)$-Lévy process with the above triplet, we can look at the specific form taken by the decomposition (2.1.18). We have $X_0 = 0$ and $B_t = bt$, as said before. The continuous local martingale X^c has quadratic variation $\langle X^c, X^c\rangle_t = ct$, and by one of Lévy's theorems this implies that X^c is in fact a Gaussian martingale; more specifically, if σ denotes a square-root of c, that is a $d \times d$ matrix such that $\sigma\sigma^\star = c$, we can write $X^c = \sigma W$ where W is a standard d-dimensional Brownian motion (to be more accurate one should say that, if k is the rank of c, we can choose σ such that its $d-k$ last columns vanish, so that only the first k components of W really matter; in particular when $c = 0$ then $X^c = 0$ and no Brownian motion at all comes into the picture).

As for the jump measure $\mu = \mu^X$, it turns out to be a *Poisson random measure* on $\mathbb{R}_+ \times \mathbb{R}^d$, with (deterministic) intensity measure ν. This means that for any finite family (A_i) of Borel subsets of $\mathbb{R}_+ \times \mathbb{R}^d$ which are pairwise disjoint, the variables $\mu(A_i)$ are independent, with $\mathbb{E}(\mu(A_i)) = \nu(A_i)$. Moreover $\mu(A_i) = \infty$ a.s. if $\nu(A_i) = \infty$, and otherwise $\mu(A_i)$ is a Poisson random variable with parameter $\nu(A_i)$.

The Lévy-Itô decomposition (2.1.18) of X takes the form

$$X_t = bt + \sigma W_t + (x 1_{\{\|x\|\leq 1\}}) \star (\mu - \nu) + (x 1_{\{\|x\|>1\}}) \star \mu. \tag{2.1.23}$$

The four terms in the right side are independent, and each one is again an $(\mathcal{F}_t)$-Lévy process. The last term is also a compound Poisson process.

Lévy processes have a lot of other nice properties. Some will be mentioned later in this book, and the reader can consult the books of Bertoin [15] or Sato [87] for much more complete accounts.

Semimartingales do not necessarily behave like Lévy processes, however a special class of semimartingales does: this class is introduced and studied in the next subsection. It is the most often encountered class of semimartingales in applications: for example the solutions of stochastic differential equations often belong to this class.

We end this subsection with some facts about Poisson random measures which are not necessarily the jump measure of a Lévy process. Let $(E, \mathcal{E})$ be a Polish space (= a metric, complete and separable space) endowed with its Borel σ-field. In this book we will call an $(\mathcal{F}_t)$-*Poisson random measure* on $\mathbb{R}_+ \times E$ a random measure $\mathfrak{p} = \mathfrak{p}(\omega; dt, dx)$ on $\mathbb{R}_+ \times E$, which is a sum of Dirac masses, no two such masses lying on the same "vertical" line $\{t\} \times E$, and such that for some σ-finite measure λ on $(E, \mathcal{E})$ and all $A \in \mathcal{E}$ with $\lambda(A) < \infty$ we have

$$\begin{aligned} &\bullet\ 1_A \star \mathfrak{p}_t = \mathfrak{p}\big([0,t] \times A\big) \text{ is an } (\mathcal{F}_t)\text{-Lévy process} \\ &\bullet\ \mathbb{E}(1_A \star \mathfrak{p}_t) = t\lambda(A). \end{aligned} \tag{2.1.24}$$

Note that when $\lambda(A) = \infty$ then $1_A \star \mathfrak{p}_t = \infty$ a.s. for all $t > 0$, and otherwise $1_A \star \mathfrak{p}$ is an ordinary Poisson process with parameter $\lambda(A)$.

The measure λ is called the *Lévy measure* of $\mathfrak{p}$, by analogy with the case of Lévy processes: indeed, the jump measure μ of an $(\mathcal{F}_t)$-Lévy process is of this type, with $E = \mathbb{R}^d$ and the same measure $\lambda = F$ in (2.1.21) and (2.1.24).

With $\mathfrak{p}$ as above, we set $\mathfrak{q}(A) = \mathbb{E}(\mathfrak{p}(A))$ for any $A \in \mathcal{R}_+ \otimes \mathcal{E}$ (this is the intensity measure). Then $\mathfrak{q}$ is also the "compensator" of $\mathfrak{p}$ in the sense that the process $1_A \star \mathfrak{q}_t = \mathfrak{q}([0,t] \times A)$ is the (predictable) compensator of the process $1_A \star \mathfrak{p}$ for all $A \in \mathcal{E}$ such that $\lambda(A) < \infty$.

At this point, one may introduce stochastic integrals $\delta \star (\mathfrak{p} - \mathfrak{q})$ as in (2.1.17) for all predictable functions δ on $\Omega \times \mathbb{R}_+ \times \mathcal{E}$ (with $\widetilde{\mathcal{P}} = \mathcal{P} \otimes \mathcal{E}$ here) which satisfy (2.1.16) for $\mathfrak{q}$ instead of ν. Everything in this respect works as in the previous subsection, with $\mu = \mu^X$ substituted with $\mathfrak{p}$.

2.1.4 Itô Semimartingales

1) In this subsection we will be slightly more formal and go into more details than before, since what follows is not as standard as what precedes. We start with a definition:

Definition 2.1.1 A d-dimensional semimartingale X is an *Itô semimartingale* if its characteristics (B, C, ν) are absolutely continuous with respect to Lebesgue measure, that is

$$B_t = \int_0^t b_s\,ds, \qquad C_t = \int_0^t c_s\,ds, \qquad \nu(dt,dx) = dt\;F_t(dx), \tag{2.1.25}$$

where $b = (b_t)$ is an $\mathbb{R}^d$-valued process, $c = (c_t)$ is an $\mathcal{M}^+_{d\times d}$-valued process, and $F_t = F_t(\omega, dx)$ is for each (ω, t) a measure on $\mathbb{R}^d$.

These b_t, c_t and F_t necessarily have some additional measurability properties, so that (2.1.25) makes sense. It is always possible to choose versions of them such that b_t and c_t are predictable processes, as well as $F_t(A)$ for all $A \in \mathcal{R}^d$. Further, since (2.1.19) holds, we can also choose a version of F which satisfies identically

$$\int \big(\|x\|^2 \wedge 1\big)\;F_t(\omega, dx) \;<\; \infty.$$

However, the predictability of b, c and F is not necessary, the minimal assumption being that they are progressively measurable (for F that means that the process $F_t(A)$ is progressively measurable for all $A \in \mathcal{R}^d$): this property, which will always be assumed in the sequel when we speak of Itô semimartingales, is enough to ensure the predictability of (B, C, ν), as given by (2.1.25).

Obviously, an $(\mathcal{F}_t)$-Lévy process with characteristic triplet (b, c, F) is an Itô semimartingale, with $b_t(\omega) = b$ and $c_t(\omega) = c$ and $F_t(\omega, .) = F$.

2) Our next aim is to give a representation of all d-dimensional Itô semimartingales in terms of a d-dimensional $(\mathcal{F}_t)$-Brownian motion W (that is an $(\mathcal{F}_t)$-Lévy process which is a Brownian motion) and of an $(\mathcal{F}_t)$-Poisson random measure $\mathfrak{p}$.

For this we have to be careful, because the space $(\Omega, \mathcal{F}, (\mathcal{F}_t)_{t\geq 0}, \mathbb{P})$ may be too small to support a Brownian motion or a Poisson random measure. In the extreme case, $X_t = t$ is a semimartingale on the space $(\Omega, \mathcal{F}, (\mathcal{F}_t)_{t\geq 0}, \mathbb{P})$ when Ω contains a single point ω, with the probability $\mathbb{P}(\{\omega\}) = 1$, and evidently there is no Brownian motion on this space. This example is perhaps too trivial, but we may also have the following situation: suppose that $X_t = W_{(t-1)^+}$ where W is a Brownian motion; then X is a semimartingale, relative to the filtration $(\mathcal{F}_t)$ which it generates. Obviously $\mathcal{F}_t$ is the trivial σ-algebra when $t < 1$, so again there is no $(\mathcal{F}_t)$-Brownian motion on this space.

Hence to solve our problem we need to enlarge the space $(\Omega, \mathcal{F}, (\mathcal{F}_t)_{t\geq 0}, \mathbb{P})$. This question will arise quite often in this book, so here we give some details about the procedure.

The space $(\Omega, \mathcal{F}, (\mathcal{F}_t)_{t\geq 0}, \mathbb{P})$ is fixed and given. We consider another measurable space $(\Omega', \mathcal{F}')$ and a transition probability $\mathbb{Q}(\omega, d\omega')$ from $(\Omega, \mathcal{F})$ into $(\Omega', \mathcal{F}')$. Then we define the products

$$\widetilde{\Omega} = \Omega \times \Omega', \qquad \widetilde{\mathcal{F}} = \mathcal{F} \otimes \mathcal{F}', \qquad \widetilde{\mathbb{P}}\big(d\omega, d\omega'\big) = \mathbb{P}(d\omega)\, \mathbb{Q}\big(\omega, d\omega'\big). \tag{2.1.26}$$

The probability space $(\widetilde{\Omega}, \widetilde{\mathcal{F}}, \widetilde{\mathbb{P}})$ is called an *extension* of $(\Omega, \mathcal{F}, \mathbb{P})$. Any variable or process which is defined on either Ω or Ω' can, as usual, be considered as defined on $\widetilde{\Omega}$: for example $X_t(\omega, \omega') = X_t(\omega)$ if X_t is defined on Ω. In the same way, a set $A \subset \Omega$ is identified with the set $A \times \Omega' \subset \widetilde{\Omega}$, and we can thus identify $\mathcal{F}_t$ with $\mathcal{F}_t \otimes \{\emptyset, \Omega'\}$, so $(\widetilde{\Omega}, \widetilde{\mathcal{F}}, (\mathcal{F}_t)_{t\geq 0}, \widetilde{\mathbb{P}})$ is a filtered space.

The filtration $(\mathcal{F}_t)$ on the extended space is not enough, because it does not incorporate any information about the second factor Ω'. To bridge this gap we consider another filtration $(\widetilde{\mathcal{F}}_t)_{t\geq 0}$ on $(\widetilde{\Omega}, \widetilde{\mathcal{F}})$, with the inclusion property

$$\mathcal{F}_t \subset \widetilde{\mathcal{F}}_t \quad \forall t \geq 0.$$

The filtered space $(\widetilde{\Omega}, \widetilde{\mathcal{F}}, (\widetilde{\mathcal{F}}_t)_{t\geq 0}, \widetilde{\mathbb{P}})$ is called a *filtered extension* of the filtered space $(\Omega, \mathcal{F}, (\mathcal{F}_t)_{t\geq 0}, \mathbb{P})$.

In many, but not all, cases the filtration $(\widetilde{\mathcal{F}}_t)$ has the product form

$$\widetilde{\mathcal{F}}_t = \cap_{s>t} \mathcal{F}_s \otimes \mathcal{F}'_s \tag{2.1.27}$$

where $(\mathcal{F}'_t)$ is a filtration on $(\Omega', \mathcal{F}')$. In many, but not all, cases again the transition probability $\mathbb{Q}$ has the simple form $\mathbb{Q}(\omega, d\omega') = \mathbb{P}'(d\omega')$ for some probability on $(\Omega', \mathcal{F}')$. In the latter case we say that the extension is a *product extension*, and if further (2.1.27) holds we say that we have a *product filtered extension*: this is simply the product of two filtered spaces.

A filtered extension is called *very good* if it satisfies

$$\omega \mapsto \int 1_A\big(\omega, \omega'\big)\, \mathbb{Q}\big(\omega, d\omega'\big) \quad \text{is } \mathcal{F}_t \text{ measurable for all } A \in \widetilde{\mathcal{F}}_t, \text{ all } t \geq 0. \tag{2.1.28}$$

Under (2.1.27), this is equivalent to saying that $\omega \mapsto \mathbb{Q}(\omega, A')$ is $\mathcal{F}_t$ measurable for all $A' \in \mathcal{F}'_t$, all $t \geq 0$. A very good filtered extension is very good because it has the following nice properties:

$$\left.\begin{array}{l}\bullet\ \text{any martingale, local martingale, submartingale, supermartingale}\\ \quad\text{on } (\Omega,\mathcal{F},(\mathcal{F}_t)_{t\ge0},\mathbb{P}) \text{ is also a martingale, local martingale,}\\ \quad\text{submartingale, supermartingale on } (\widetilde{\Omega},\widetilde{\mathcal{F}},(\widetilde{\mathcal{F}}_t)_{t\ge0},\widetilde{\mathbb{P}}),\\ \bullet\ \text{a semimartingale on } (\Omega,\mathcal{F},(\mathcal{F}_t)_{t\ge0},\mathbb{P}) \text{ is a semimartingale on}\\ \quad(\widetilde{\Omega},\widetilde{\mathcal{F}},(\widetilde{\mathcal{F}}_t)_{t\ge0},\widetilde{\mathbb{P}}), \text{ with the same characteristics}\end{array}\right\} \tag{2.1.29}$$

(in fact, (2.1.28) is equivalent to the fact that any bounded martingale on $(\Omega,\mathcal{F},(\mathcal{F}_t)_{t\ge0},\mathbb{P})$ is a martingale on $(\widetilde{\Omega},\widetilde{\mathcal{F}},(\widetilde{\mathcal{F}}_t)_{t\ge0},\widetilde{\mathbb{P}})$). For example a Brownian motion on $(\Omega,\mathcal{F},(\mathcal{F}_t)_{t\ge0},\mathbb{P})$ is also a Brownian motion on $(\widetilde{\Omega},\widetilde{\mathcal{F}},(\widetilde{\mathcal{F}}_t)_{t\ge0},\widetilde{\mathbb{P}})$ if the extension is very good, and the same holds for Poisson measures.

Note that many extensions are *not* very good: let for example $\mathbb{Q}(\omega,.)$ be the Dirac mass $\varepsilon_{U(\omega)}$, on the space $(\Omega',\mathcal{F}')=(\mathbb{R},\mathcal{R})$ endowed with the filtration $\mathcal{F}'_t=\mathcal{F}'$ for all t, and where U is an $\mathbb{R}$-valued variable on $(\Omega,\mathcal{F})$ which is not measurable with respect to the $\mathbb{P}$ completion of $\mathcal{F}_1$, say. Then $\mathbb{Q}(\omega,A')=1_{A'}(U(\omega))$ is not $\mathcal{F}_1$ measurable in general, whereas $A'\in\mathcal{F}'_1$, and the extension (with the product filtration (2.1.27)) is not very good.

3) Now we are ready to give our representation theorem. The difficult part comes from the jumps of our semimartingale, and it is fundamentally a representation theorem for integer-valued random measures in terms of a Poisson random measure, a result essentially due to Grigelionis [39]. The form given below is Theorem (14.68) of [52]. In this theorem, d' is an arbitrary integer with $d'\ge d$, and E is an arbitrary Polish space with a σ-finite and infinite measure λ having no atom, and $\mathfrak{q}(dt,dx)=dt\otimes\lambda(dx)$.

Theorem 2.1.2 *Let X be a d-dimensional Itô semimartingale on the space $(\Omega,\mathcal{F},(\mathcal{F}_t)_{t\ge0},\mathbb{P})$, with characteristics (B,C,ν) given by* (2.1.25). *There is a very good filtered extension, say $(\widetilde{\Omega},\widetilde{\mathcal{F}},(\widetilde{\mathcal{F}}_t)_{t\ge0},\widetilde{\mathbb{P}})$, on which are defined a d'-dimensional Brownian motion W and a Poisson random measure $\mathfrak{p}$ on $\mathbb{R}_+\times E$ with Lévy measure λ, such that*

$$X_t=X_0+\int_0^t b_s\,ds+\int_0^t \sigma_s\,dW_s+(\delta 1_{\{\|\delta\|\le1\}})\star(\mathfrak{p}-\mathfrak{q})_t+(\delta 1_{\{\|\delta\|>1\}})\star\mathfrak{p}_t, \tag{2.1.30}$$

and where σ_t is an $\mathbb{R}^d\otimes\mathbb{R}^{d'}$-valued predictable (or simply progressively measurable) process on $(\Omega,\mathcal{F},(\mathcal{F}_t)_{t\ge0},\mathbb{P})$, and δ is a predictable $\mathbb{R}^d$-valued function on $\Omega\times\mathbb{R}_+\times E$.

Moreover, outside a null set, we have $\sigma_t\sigma_t^\star=c_t$, and $F_t(\omega,.)$ is the image of the measure λ restricted to the set $\{x:\delta(\omega,t,x)\neq0\}$ by the map $x\mapsto\delta(\omega,t,x)$.

Conversely, any process of the form (2.1.30) (*with possibly b, σ and δ defined on the extension instead of on $(\Omega,\mathcal{F},(\mathcal{F}_t)_{t\ge0},\mathbb{P})$*) *is an Itô semimartingale on $(\widetilde{\Omega},\widetilde{\mathcal{F}},(\widetilde{\mathcal{F}}_t)_{t\ge0},\widetilde{\mathbb{P}})$, and also on $(\Omega,\mathcal{F},(\mathcal{F}_t)_{t\ge0},\mathbb{P})$ if it is further adapted to $(\mathcal{F}_t)$.*

Once more, (2.1.30) should be read component by component, and the ith component of $\int_0^t \sigma_s \, dW_s$ is the integral of $\sigma_t^{i\cdot} = (\sigma_t^{ij} : 1 \leq j \leq d')$ with respect to the d'-dimensional process W.

There is a lot of freedom for choosing the extension, of course, but also the space E, and the function δ, and even the dimension d' and the process σ (the requirement being that $\sigma_t \sigma_t^\star = c_t$): we can always take an arbitrary $d' \geq d$, or more generally not smaller than the maximal rank of the matrices $c_t(\omega)$. A natural choice for E consists in taking $E = \mathbb{R}^d$, but this is not compulsory and we may take in all cases $E = \mathbb{R}$ with λ being the Lebesgue measure. For example if we have several Itô semimartingales, and even countably many of them, we can use the same measure $\mathfrak{p}$ for representing all of them at once. Any decomposition as in (2.1.30) will be called a *Grigelionis decomposition* of X.

Remark 2.1.3 Even when the measure λ has atoms, or is finite, (2.1.30) gives an Itô semimartingale. Moreover, in the same spirit as for the choice of the dimension d' above, when $A = \sup_{(\omega,t)} F_{\omega,t}(\mathbb{R}^d)$ is finite, we can find a Grigelionis representation for any choice of the measure λ without atom and total mass $\lambda(E) \geq A$.

Note that the fact that an extension of the space is needed is a rather common fact in stochastic calculus. For example the celebrated Dubins theorem according to which any continuous local martingale M null at 0 is a time-changed Brownian motion also necessitates an extension of the space to be true, unless $\langle M, M \rangle_\infty = \infty$ a.s. Here we have a similar phenomenon: when for example X is continuous and the rank of c_t is everywhere d, the extension is not needed, but it is otherwise.

Example 2.1.4 Lévy processes: Let X be an $(\mathcal{F}_t)$-Lévy process with triplet (b, c, F), and take $E = \mathbb{R}^d$ and $\lambda = F$ (even though this measure may have atoms and/or may be finite, or even null). Then (2.1.30) holds with $\delta(\omega, t, x) = x$ and $\mathfrak{p} = \mu$, and it is then nothing else than the Lévy-Itô decomposition (2.1.23). More generally, for an Itô semimartingale the decompositions (2.1.18) and (2.1.30) agree, term by term.

As a matter of fact, the representation (2.1.30) may serve as a definition for an Itô semimartingale, if we do not mind about extending the space. This is in line with the processes that are solutions of stochastic differential equations driven by a Brownian motion and a Poisson measure: the "strong" solutions have the representation (2.1.30), whereas the "weak" solutions are Itô semimartingales in the sense of Definition 2.1.1.

In any case, and since in the questions studied below it is innocuous to enlarge the underlying probability space, throughout the remainder of this book all Itô semimartingales will be of the form (2.1.30), and we assume that both W and the Poisson measure $\mathfrak{p}$ are defined on $(\Omega, \mathcal{F}, (\mathcal{F}_t)_{t \geq 0}, \mathbb{P})$. By analogy with the stochastic differential equation case, the terms b, σ and δ will be called the *coefficients* of X, respectively the drift, the diffusion, and the jump coefficients.

Finally, as soon the process

$$\int_0^t ds \int \|x\|^2 F_s(dx) = \int_0^t ds \int \|\delta(s,z)\|^2 \lambda(dz)$$

is finite-valued for all t, the angle bracket of X exists, and it is given by

$$\langle X^i, X^j\rangle_t = C_t^{ij} + \int_0^t ds \int x^i x^j F_s(dx) = C_t^{ij} + \int_0^t ds \int \bigl(\delta^i(s,z)\delta^j(s,z)\bigr)\lambda(dz) \tag{2.1.31}$$

(compare with (2.1.15)).

2.1.5 Some Estimates for Itô Semimartingales

This subsection is devoted to various estimates for Itô semimartingales. Most of them are rather standard, but scarcely appear in book form, and a few of them are new. So although only the results are presented in this chapter, the proofs are fully given in the Appendix.

Before starting, we recall the *Burkholder-Davis-Gundy inequalities*. They play a key role here, and can be found for example in Protter [83]: for each real $p \geq 1$ there are two constants $0 < c_p < C_p < \infty$ such that, for any local martingale M starting at $M_0 = 0$ and any two stopping times $S \leq T$, we have

$$\begin{aligned} &c_p\, \mathbb{E}\bigl(([M,M]_T - [M,M]_S)^{p/2} \mid \mathcal{F}_S\bigr) \\ &\quad\leq \mathbb{E}\Bigl(\sup_{t\in\mathbb{R}_+:\, S<t\leq T} |M_t - M_S|^p \mid \mathcal{F}_S\Bigr) \\ &\quad\leq C_p\, \mathbb{E}\bigl(([M,M]_T - [M,M]_S)^{p/2} \mid \mathcal{F}_S\bigr) \end{aligned} \tag{2.1.32}$$

(most often, these inequalities are stated in expectation only, and with $S = 0$; the meaning of $[M,M]_T$ on the set $\{T = \infty\}$ is $[M,M]_T = \lim_{t\to\infty}[M,M]_t$, an increasing limit which may be infinite; when $p > 1$ these inequalities are simply Burkholder-Gundy inequalities.)

The results below are stated, and will be used, in the d-dimensional setting. But, as seen from the proofs, they are fundamentally one-dimensional estimates. They are moment estimates, typically of the form $E(|Z_t|^p) \leq z_t$, where Z_t is the variable of interest (a stochastic integral, or some specific semimartingale) and z_t is an appropriate bound. However, a semimartingale, even of the form (2.1.30), has no moments in general; so it may very well happen in the forthcoming inequalities that both members are infinite; however, if the right member z_t is finite, then so is the left member $\mathbb{E}(|Z_t|^p)$.

Below, constants appear everywhere. They are usually written as K, *and change from line to line*, or even within a line. If they depend on a parameter of interest, say p, they are written as K_p (for example in estimates for the p moment, they usually depend on p and it may be useful to keep track of this). On the other hand, constants occurring in estimates for a given X may also depend on its characteristics, or at least on some bounds for the characteristics, and we are usually not interested in keeping track of this and so they are just written as K, unless explicitly stated.

The Itô semimartingale X has the Grigelionis decomposition (2.1.30), and below we give estimates about, successively, the four terms (besides the initial value X_0) occurring in this decomposition. We start with estimates which require no specific assumptions, and we consider a finite-valued stopping time T and some $s > 0$. The constants often depend on p, but neither on T nor on s.

1) The drift term. The first estimate is simple (it is Hölder's inequality) and given for completeness. Note that it is "ω-wise", and valid for $p \geq 1$:

$$\sup_{0 \leq u \leq s} \left\| \int_T^{T+u} b_r\, dr \right\|^p \leq \left(\int_T^{T+s} \|b_u\|\, du \right)^p \leq s^p \left(\frac{1}{s} \int_T^{T+s} \|b_u\|^p\, du \right). \tag{2.1.33}$$

The way the last term above is written may look strange: it has the advantage of singling out the term s^p which is typically the order of magnitude of the whole expression, times a term which is typically of "order 1" (here, the average of $\|b_u\|^p$ over $[T, T+s]$).

Note that this term of "typical order 1" may in some cases be infinite: the inequality becomes trivial but totally useless. The same comment applies to all forthcoming inequalities.

2) Continuous martingales. Here we consider the continuous martingale part $\int_0^t \sigma_s\, dW_s$, where W is a d'-dimensional Brownian motion and σ_t is $\mathbb{R}^d \otimes \mathbb{R}^{d'}$-valued. Applying the Burkholder-Davis-Gundy inequality for each component, we deduce that for all $p \geq 1$ we have

$$\mathbb{E}\left(\sup_{0 \leq u \leq s} \left\| \int_T^{T+u} \sigma_r\, dW_r \right\|^p \mid \mathcal{F}_T \right) \leq K_p\, s^{p/2}\, \mathbb{E}\left(\left(\frac{1}{s} \int_T^{T+s} \|\sigma_u\|^2 du \right)^{p/2} \mid \mathcal{F}_T \right). \tag{2.1.34}$$

Note that the constant K_p here depends on C_p in (2.1.32), and also (implicitly) on the dimensions d and d'.

3) Purely discontinuous martingales. The next estimates are less classical. We state them as lemmas, to be proved in the Appendix. We consider the integral of a d-dimensional predictable function δ on $\Omega \times \mathbb{R}_+ \times E$, without any reference to the semimartingale X.

To keep in line with the above way of singling out a "main term" which is a power of s and a factor which is "typically of order 1", we introduce a series of notation associated with a given function δ. Below, $p \in [0, \infty)$ and $a \in (0, \infty]$:

$$
\begin{aligned}
\widehat{\delta}(p,a)_{t,s} &= \frac{1}{s}\int_t^{t+s} du \int_{\{\|\delta(u,z)\|\leq a\}} \|\delta(u,z)\|^p\,\lambda(dz), \qquad \widehat{\delta}(p)=\widehat{\delta}(p,\infty)\\
\widehat{\delta}'(p)_{t,s} &= \widehat{\delta}(p,1)+\frac{1}{s}\int_t^{t+s} du \int_{\{\|\delta(u,z)\|>1\}} \|\delta(u,z)\|\,\lambda(dz) \qquad (2.1.35)\\
\widehat{\delta}''(p)_{t,s} &= \widehat{\delta}(p,1)+\frac{1}{s}\int_t^{t+s} \lambda\big(\big\{z: \|\delta(u,z)\|>1\big\}\big)\,du.
\end{aligned}
$$

Lemma 2.1.5 *Suppose that $\int_0^t ds \int \|\delta(s,z)\|^2\lambda(dz)<\infty$ for all t. Then the process $Y=\delta\star(\mathfrak{p}-\mathfrak{q})$ is a locally square integrable martingale, and for all finite stopping times T and $s>0$ and $p\in[1,2]$ we have*

$$
\mathbb{E}\Big(\sup_{0\leq u\leq s}\|Y_{T+u}-Y_T\|^p \mid \mathcal{F}_T\Big) \leq K_p\, s\, \mathbb{E}\big(\widehat{\delta}(p)_{T,s}\mid \mathcal{F}_T\big) \qquad (2.1.36)
$$

and also for $p\geq 2$:

$$
\mathbb{E}\Big(\sup_{0\leq u\leq s}\|Y_{T+u}-Y_T\|^p \mid \mathcal{F}_T\Big) \leq K_p\big(s\mathbb{E}\big(\widehat{\delta}(p)_{T,s}\mid \mathcal{F}_T\big)+s^{p/2}\mathbb{E}\big(\widehat{\delta}(2)_{T,s}^{p/2}\mid \mathcal{F}_T\big)\big). \qquad (2.1.37)
$$

These two inequalities agree when $p=2$. The variables $\widehat{\delta}(p)_{T,s}$ may be infinite, and are neither increasing nor decreasing as p increases, in general. However, as soon as δ is bounded, our assumption implies that $\widehat{\delta}(p)_{T,s}$ is finite for all $p\geq 2$, whereas it may be infinite for $p<2$, and typically is so, for all p small. Hence it is often the case that the right side of (2.1.37) (where $p\geq 2$) is finite, whereas the right side of (2.1.36) (where $p\leq 2$) is infinite, contrary to what one would perhaps think at first glance.

There is a fundamental difference between the estimates given so far, as $s\to 0$. In (2.1.33) the right side is basically of order s^p, in (2.1.34) it is $s^{p/2}$, and in (2.1.36) and (2.1.37) it is s, irrespective of the value of $p\geq 1$. This phenomenon already occurs when $Y_t=N_t-t$, where N is a standard Poisson process: in this case, for each integer $p\geq 1$ we have $\mathbb{E}(|Y_s|^p)\sim\alpha_p s$ as $s\to 0$, for some constant $\alpha_p>0$.

The inequality (2.1.36), when the right side is finite for some $p<2$, implies that

$$
(Y_{T+s}-Y_T)/\sqrt{s} \xrightarrow{\mathbb{P}} 0 \qquad (2.1.38)
$$

as $s\to 0$. As said before the right side of (2.1.36) is often infinite, but nevertheless (2.1.38) holds when $Y=\delta*(\mathfrak{p}-\mathfrak{q})$ under quite general circumstances. We provide a useful lemma to this effect: (2.1.38) follows from (2.1.39) below by taking $q=1/2$ and $r=2$.

Lemma 2.1.6 *Let $r\in[1,2]$. There exists a constant $K>0$ depending on r,d, such that for all $q\in[0,1/r]$ and $s\in[0,1]$, all finite stopping times T, and all* *d-dimensional processes* $Y = \delta \star (\mathfrak{p} - \mathfrak{q})$, *we have*

$$\mathbb{E}\bigg(\sup_{u\le s}\Big(\frac{\|Y_{T+u}-Y_T\|}{s^q}\bigwedge 1\Big)^r \mid \mathcal{F}_T\bigg) \\ \le Ks^{1-qr}\,\mathbb{E}\big(\widehat{\delta}(r,s^{\frac{q}{2}})_{T,s} + s^{\frac{q(r-1)}{2}}\widehat{\delta}'(r)_{T,s}\mid \mathcal{F}_T\big), \tag{2.1.39}$$

where $\widehat{\delta}(r,a)$ *and* $\widehat{\delta}'(r)$ *are associated with* δ *by* (2.1.35).

4) Purely discontinuous processes of finite variation. Here we consider the case of $\delta \star \mathfrak{p}$, where again δ is predictable. The results (and the proofs as well) parallel those for $\delta \star (\mathfrak{p} - \mathfrak{q})$, with different powers.

Lemma 2.1.7 *a) If* $\int_0^t \lambda(\{z : \delta(r,z) \neq 0\})\,dr < \infty$ *for all* t, *the process* $Y = \delta \star \mathfrak{p}$ *has almost surely finitely many jumps on any finite interval.*

b) Suppose that $\int_0^t ds \int \|\delta(s,z)\|\lambda(dz) < \infty$ *for all* t. *Then the process* $Y = \delta \star \mathfrak{p}$ *is of locally integrable variation, and for all finite stopping times* T *and* $s > 0$ *and* $p \in (0,1]$ *we have*

$$\mathbb{E}\Big(\sup_{0\le u\le s}\|Y_{T+u}-Y_T\|^p \mid \mathcal{F}_T\Big) \le K_p\, s\, \mathbb{E}\big(\widehat{\delta}(p)_{T,s}\mid \mathcal{F}_T\big), \tag{2.1.40}$$

and also for $p \ge 1$

$$\mathbb{E}\Big(\sup_{0\le u\le s}\|Y_{T+u}-Y_T\|^p \mid \mathcal{F}_T\Big) \le K_p\big(s\mathbb{E}\big(\widehat{\delta}(p)_{T,s}\mid \mathcal{F}_T\big) + s^p\mathbb{E}\big(\widehat{\delta}(1)^p_{T,s}\mid \mathcal{F}_T\big)\big). \tag{2.1.41}$$

Next, we have a result similar to Lemmas 2.1.6:

Lemma 2.1.8 *Let* $r \in (0,1]$. *There exists a constant* $K > 0$ *depending on* r, d, *such that for all* $q \in [0, 1/r]$ *and* $s \in [0,1]$, *all finite stopping times* T, *and all* d*-dimensional processes* $Y = \delta \star \mathfrak{p}$, *we have*

$$\mathbb{E}\bigg(\sup_{u\le s}\Big(\frac{\|Y_{T+u}-Y_T\|}{s^q}\bigwedge 1\Big)^r \mid \mathcal{F}_T\bigg) \\ \le Ks^{1-qr}\,\mathbb{E}\big(\widehat{\delta}(r,s^{\frac{q}{2}})_{T,s} + s^{\frac{rq}{2}}\widehat{\delta}''(r)_{T,s}\mid \mathcal{F}_T\big), \tag{2.1.42}$$

where $\widehat{\delta}(r,a)$ *and* $\widehat{\delta}''(r)$ *are associated with* δ *by* (2.1.35).

5) Itô semimartingales. The previous estimates will be used under various hypotheses and with various choices of the powers under consideration. However, the most useful results are about Itô semimartingales X of the form (2.1.30). Then if T is a finite stopping time, and $s > 0$, and $p \ge 2$, we have

$$
\begin{aligned}
&\mathbb{E}\Big(\sup_{u\le s}\|X_{T+u}-X_T\|^p \mid \mathcal{F}_T\Big)\\
&\quad\le K\mathbb{E}\Bigg(\bigg(\int_T^{T+s}\|b_u\|\,du\bigg)^p\\
&\qquad+\bigg(\int_T^{T+s}\|\sigma_u\|^2\,du\bigg)^{p/2}+\int_T^{T+s}du\int\big\|\delta(r,z)\big\|^p\lambda(dz)\\
&\qquad+\bigg(\int_T^{T+s}du\int_{\{z:\|\delta(u,z)\|\le1\}}\big\|\delta(u,z)\big\|^2\lambda(dz)\bigg)^{p/2}\\
&\qquad+\bigg(\int_T^{T+s}du\int_{\{z:\|\delta(u,z)\|>1\}}\big\|\delta(u,z)\big\|\lambda(dz)\bigg)^p \mid \mathcal{F}_T\Bigg).
\end{aligned}
\tag{2.1.43}
$$

In the "bounded" case, we get for $p\ge2$ again

$$
\begin{aligned}
&\big\|b_t(\omega)\big\|\le\beta,\quad \big\|\sigma_t(\omega)\big\|\le\alpha,\quad \big\|\delta(\omega,t,z)\big\|\le\Gamma(z)\quad\Longrightarrow\\
&\quad\mathbb{E}\Big(\sup_{u\le s}\|X_{T+u}-X_T\|^p\mid\mathcal{F}_T\Big)\;\le\;K\bigg(s^p\beta^p+s^{p/2}\alpha^p+s\int\Gamma(z)^p\lambda(dz)\\
&\qquad+s^{p/2}\bigg(\int\Gamma(z)^2 1_{\{\Gamma(z)\le1\}}\,\lambda(dz)\bigg)^{p/2}+s^p\bigg(\int\Gamma(z)1_{\{\Gamma(z)>1\}}\,\lambda(dz)\bigg)^p\bigg).
\end{aligned}
\tag{2.1.44}
$$

These are applications of the "non-normalized" estimates (2.1.33), (2.1.34), (2.1.37) and (2.1.41). In the "bounded" case we also have a simple statement for the normalized estimates (2.1.39) and (2.1.42). These normalized estimates are useful typically for a power $p\le2$. In view of their usefulness, we state these estimates in a corollary, proved in the Appendix.

Corollary 2.1.9 *Let the d-dimensional predictable function δ be such that $\|\delta(\omega,t,z)\|\le\Gamma(z)$ for some measurable function Γ on E, and let $p>0$, $r\in(0,2]$ and $q\in[0,1/r)$.*

a) If $r\in(1,2]$ and $\int(\Gamma(z)^r\wedge\Gamma(z))\,\lambda(dz)<\infty$, the process $Y=\delta*(\mathfrak{p}-\mathfrak{q})$ satisfies

$$
\mathbb{E}\bigg(\sup_{u\le s}\bigg(\frac{\|Y_{T+u}-Y_T\|}{s^q}\bigwedge 1\bigg)^p\mid\mathcal{F}_T\bigg)\le\begin{cases}Ks^{p(1-qr)/r}\,\phi(s) & \text{if } p\le r\\ Ks^{1-qr}\,\phi(s) & \text{if } p\ge r\end{cases}
\tag{2.1.45}
$$

for all $s\in(0,1]$ and all finite stopping times T, where K and ϕ depend on r,p,q, Γ and λ, and $\phi(s)\to0$ as $s\to0$ when $q>0$, and $\sup\phi<\infty$ when $q=0$.

b) If $r \in (0,1]$ and $\int(\Gamma(z)^r \vee \Gamma(z))\,\lambda(dz) < \infty$, the process $Y = \delta * (\mathfrak{p} - \mathfrak{q})$ satisfies

$$\mathbb{E}\Big(\sup_{u\le s}\Big(\frac{\|Y_{T+u} - Y_T\|}{s^q}\bigwedge 1\Big)^p \mid \mathcal{F}_T\Big) \le K s^{1-qr}\,\phi(s) \quad \text{if } p>1,\ q<\frac{p-1}{p-r} \tag{2.1.46}$$

for all $s \in (0,1]$ and all finite stopping times T, with K and ϕ as in (a).

c) If $r \in (0,1]$ and $\int(\Gamma(z)^r \wedge 1)\,\lambda(dz) < \infty$, the process $Y = \delta * \mathfrak{p}$ satisfies

$$\mathbb{E}\Big(\sup_{u\le s}\Big(\frac{\|Y_{T+u} - Y_T\|}{s^q}\bigwedge 1\Big)^p \mid \mathcal{F}_T\Big) \le \begin{cases} K s^{p(1-qr)/r}\,\phi(s) & \text{if } p\le r\\ K s^{1-qr}\,\phi(s) & \text{if } p\ge r\end{cases} \tag{2.1.47}$$

for all $s \in (0,1]$ and all finite stopping times T, with K and ϕ as in (a).

2.1.6 Estimates for Bigger Filtrations

The estimates of the previous subsection hold when T is a stopping time for a filtration $(\mathcal{F}_t)$ with respect to which W and $\mathfrak{p}$ are a Brownian motion and a Poisson random measure. In practice the filtration $(\mathcal{F}_t)$ is usually given *a priori*, but it turns out that those estimates are also valid for (suitable) bigger filtrations. This will be sometimes a useful tool.

More specifically, we start with $(\mathcal{F}_t)$, W and $\mathfrak{p}$ as above. Consider a (non-random) measurable subset A of E. We denote by $\mathcal{H}^A$ the σ-field generated by the restriction of the measure $\mathfrak{p}$ to $\mathbb{R}_+ \times A$, that is the σ-field generated by all the random variables $\mathfrak{p}(B)$, where B ranges through all measurable subsets of $\mathbb{R}_+ \times A$. We also denote by $\mathcal{H}^W$ the σ-field generated by the process W, that is $\mathcal{H}^W = \sigma(W_t : t \ge 0)$. Then we set

$$\begin{aligned} (\mathcal{G}_t^A) &= \text{the smallest filtration containing } (\mathcal{F}_t) \text{ and with } \mathcal{H}^A \subset \mathcal{G}_0^A\\ (\mathcal{G}_t^{A,W}) &= \text{the smallest filtration containing } (\mathcal{F}_t) \text{ and with } \mathcal{H}^A \cup \mathcal{H}^W \subset \mathcal{G}_0^{A,W}. \end{aligned} \tag{2.1.48}$$

Proposition 2.1.10 *In the above setting, we have:*

a) The process W is a Brownian motion relative to the filtration $(\mathcal{G}_t^A)$, and (2.1.34) *holds if σ is $(\mathcal{F}_t)$-optional and T is a stopping time relative to the filtration $(\mathcal{G}_t^A)$ and the conditional expectations are taken relative to $\mathcal{G}_T^A$.*

b) The restriction $\mathfrak{p}'$ of $\mathfrak{p}$ to the set $\mathbb{R}_+ \times A^c$ is a Poisson random measure with respect to the filtration $(\mathcal{G}_t^{A,W})$, and its Lévy measure λ' is the restriction of λ to A^c. Moreover if δ is $(\mathcal{F}_t)$-predictable and satisfies $\delta(\omega,t,z) = 0$ for all (ω,t,z) with $z \in A$, Lemmas 2.1.5, 2.1.6, 2.1.7 *and* 2.1.8 *hold if T is a stopping time relative to the filtration $(\mathcal{G}_t^{A,W})$ and the conditional expectations are taken relative to $\mathcal{G}_T^{A,W}$.*

2.1.7 The Lenglart Domination Property

We end this section with a property that is quite useful for limit theorems. It is about the comparison between two càdlàg adapted processes Y and A, where A is increasing with $A_0 = 0$. We say that Y is *Lenglart-dominated by* A if $\mathbb{E}(|Y_T|) \leq E(A_T)$ for all finite stopping times T, both expectations being possibly infinite.

The following result has been proved by Lenglart in [70], and a proof can be found in Lemma I.3.30 of [57]. Suppose that Y is Lenglart-dominated by A. Then for any (possibly infinite) stopping time T and all $\varepsilon, \eta > 0$ we have

$$\mathbb{P}\Big(\sup_{s\leq T} |Y_s| \geq \varepsilon\Big) \leq \begin{cases} \frac{\eta}{\varepsilon} + \mathbb{P}(A_T \geq \eta) & \text{if } A \text{ is predictable} \\ \frac{1}{\varepsilon}(\eta + \mathbb{E}(\sup_{s\leq T} \Delta A_s)) + \mathbb{P}(A_T \geq \eta) & \text{otherwise.} \end{cases} \tag{2.1.49}$$

This result is typically applied when Y is also an increasing process, and either A is the predictable compensator of Y (the first inequality is used), or Y is the predictable compensator of A (the second inequality is used).

2.2 Limit Theorems

The aims of this section are twofold: first we define stable convergence in law. Second, we recall a few limit theorems for partial sums of triangular arrays of random variables.

Before getting started, and in view of a comparison with stable convergence in law, we recall the notion of convergence in law. Let (Z_n) be a sequence of E-valued random variables, where E is some topological space endowed with its Borel σ-field $\mathcal{E}$, and each Z_n is defined on some probability space $(\Omega_n, \mathcal{F}_n, \mathbb{P}_n)$; these spaces may differ as n varies. We say that Z_n converges in law if there is a probability measure μ on $(E, \mathcal{E})$ such that

$$\mathbb{E}\big(f(Z_n)\big) \rightarrow \int f(x)\mu(dx) \tag{2.2.1}$$

for all functions f on E that are bounded and continuous. It is customary to "realize" the limit as a random variable Z with law μ, on some space $(\Omega, \mathcal{F}, \mathbb{P})$ (for example on $(\Omega, \mathcal{F}, \mathbb{P}) = (E, \mathcal{E}, \mu)$ with the canonical variable $Z(x) = x$), and then (2.2.1) reads as

$$\mathbb{E}\big(f(Z_n)\big) \rightarrow \mathbb{E}\big(f(Z)\big) \tag{2.2.2}$$

for all f as before, and we usually write $Z_n \xrightarrow{\mathcal{L}} Z$.

The above definition only requires E to be a topological space, but as soon as one wants to prove results one needs additional properties, at least that E is a metric space, and very often that E is a Polish space. In the latter case, it is a known fact that the convergence (2.2.2) for all functions f which are bounded and Lipschitz is enough to imply the convergence in law, see e.g. Parthasarathy [78].

2.2.1 Stable Convergence in Law

The notion of stable convergence in law has been introduced by Rényi [84], for the very same "statistical" reason as we need it here and which we shortly explain just below. We refer to [4] for a very simple exposition and to [57] for more details, and also to the book [43] of Hall and Heyde for some different insights on the subject. However the same notion or very similar ones appear in different guises in control theory, in the theory of fuzzy random variables and randomized times, and also for solving stochastic differential equations in the weak sense.

In an asymptotic statistical context, stable convergence in law appears in the following situation: we wish to estimate some parameter with a sequence of statistics, say Z_n, or use such a sequence to derive some testing procedure. Quite often the variables Z_n converge in law to a limit Z which has, say, a mixed centered normal distribution: that is, $Z = \Sigma U$ where U is an $\mathcal{N}(0,1)$ variable and Σ is a positive variable independent of U. This poses no problem other than computational when the law of Σ is known. However, in many instances the law of Σ is unknown, but we can find a sequence of statistics Σ_n such that the pair (Z_n, Σ_n) converges in law to (Z, Σ). So, although the law of the pair (Z, Σ) is unknown, the variable Z_n/Σ_n converges in law to $\mathcal{N}(0,1)$ and we can base estimation or testing procedures on these new statistics Z_n/Σ_n. This is where stable convergence in law comes into play.

The formal definition is a bit involved. It applies to a sequence of random variables Z_n, all defined on the *same* probability space $(\Omega, \mathcal{F}, \mathbb{P})$, and taking their values in the same state space $(E, \mathcal{E})$, assumed to be a Polish space. We say that Z_n *stably converges in law* if there is a probability measure η on the product $(\Omega \times E, \mathcal{F} \otimes \mathcal{E})$, such that

$$\mathbb{E}\big(Y f(Z_n)\big) \to \int Y(\omega) f(x) \eta(d\omega, dx) \tag{2.2.3}$$

for all bounded continuous functions f on E and all bounded random variables Y on $(\Omega, \mathcal{F})$. Taking $f = 1$ and $Y = 1_A$ above yields in particular that $\eta(A \times E) = \mathbb{P}(A)$ for all $A \in \mathcal{F}$.

This is an "abstract" definition, similar to (2.2.1). Now, exactly as we prefer to write the convergence in law as in (2.2.2), it is convenient to "realize" the limit Z for the stable convergence in law as well. Since, in contrast with mere convergence in law, all Z_n here are necessarily on the same space $(\Omega, \mathcal{F}, \mathbb{P})$, it is natural to realize Z on an arbitrary extension $(\widetilde{\Omega}, \widetilde{\mathcal{F}}, \widetilde{\mathbb{P}})$ of $(\Omega, \mathcal{F}, \mathbb{P})$, as defined by (2.1.26). Letting Z be an E-valued random variable defined on this extension, (2.2.3) is equivalent to saying (with $\widetilde{\mathbb{E}}$ denoting the expectation w.r.t. $\widetilde{\mathbb{P}}$)

$$\mathbb{E}\big(Y f(Z_n)\big) \to \widetilde{\mathbb{E}}\big(Y f(Z)\big) \tag{2.2.4}$$

for all f and Y as above, as soon as $\widetilde{\mathbb{P}}(A \cap \{Z \in B\}) = \eta(A \times B)$ for all $A \in \mathcal{F}$ and $B \in \mathcal{E}$. We then say that Z_n converges stably to Z, and this convergence is denoted by $Z_n \xrightarrow{\mathcal{L}\text{-s}} Z$. Note that, exactly as for (2.2.3), the stable convergence in law holds as soon as we have (2.2.4) for all Y as above and all functions f which are bounded and Lipschitz.

We can always do this in the following simple way: take $\widetilde{\Omega} = \Omega \times E$ and $\widetilde{\mathcal{F}} = \mathcal{F} \otimes \mathcal{E}$ and endow $(\widetilde{\Omega}, \widetilde{\mathcal{F}})$ with the probability η, and put $Z(\omega, x) = x$. However, exactly as in the case of the convergence in law where usually (2.2.2) is stated with an "arbitrary" Z with law μ, here we prefer to write (2.2.4) with an arbitrary Z, defined on an arbitrary extension of the original space.

Clearly, when η is given, the property $\widetilde{\mathbb{P}}(A \cap \{Z \in B\}) = \eta(A \times B)$ for all $A \in \mathcal{F}$ and $B \in \mathcal{E}$ simply amounts to specifying the law of Z, conditionally on the σ-field $\mathcal{F}$, that is under the measures $\mathbb{Q}(\omega, .)$ of (2.1.26). Therefore, saying $Z_n \stackrel{\mathcal{L}\text{-s}}{\longrightarrow} Z$ amounts to saying that we have stable convergence in law towards a variable Z, defined on any extension $(\widetilde{\Omega}, \widetilde{\mathcal{F}}, \widetilde{\mathbb{P}})$ of $(\Omega, \mathcal{F}, \mathbb{P})$, and with a specified conditional law, knowing $\mathcal{F}$.

Stable convergence in law obviously implies convergence in law. But it implies much more, and in particular the following crucial result: if Y_n and Y are variables defined on $(\Omega, \mathcal{F}, \mathbb{P})$ and with values in the same Polish space F, then

$$Z_n \stackrel{\mathcal{L}\text{-s}}{\longrightarrow} Z, \quad Y_n \stackrel{\mathbb{P}}{\longrightarrow} Y \quad \Rightarrow \quad (Y_n, Z_n) \stackrel{\mathcal{L}\text{-s}}{\longrightarrow} (Y, Z). \tag{2.2.5}$$

Let us mention the following useful extensions of (2.2.2) and (2.2.4):

$$\left.\begin{array}{ll} \bullet\ Z_n \stackrel{\mathcal{L}}{\longrightarrow} Z \ \Rightarrow & \mathbb{E}(f(Z_n)) \to \mathbb{E}(f(Z)) \text{ for all } f \text{ bounded, Borel,} \\ & \text{and } \mu\text{-a.s. continuous,} \\ \bullet\ Z_n \stackrel{\mathcal{L}\text{-s}}{\longrightarrow} Z \ \Rightarrow & \mathbb{E}(F(., Z_n)) \to \widetilde{\mathbb{E}}(F(., Z)) \text{ for all } F \text{ bounded,} \\ & \mathcal{F} \otimes \mathcal{E}\text{-measurable on } \Omega \times E \text{ and } x \mapsto F(\omega, x) \text{ is} \\ & \text{continuous at each point } (\omega, x) \in A \text{ for some set} \\ & A \in \mathcal{F} \otimes \mathcal{E} \text{ with } \widetilde{\mathbb{P}}(\{(\omega, \omega') : (\omega, Z(\omega, \omega') \in A\}) = 1. \end{array}\right\} \tag{2.2.6}$$

Among all criteria for stable convergence in law, the following one is quite useful. The σ-field generated by all Z_n is separable, that is generated by a countable algebra, say $\mathcal{G}$. Then if, for any finite family $(A_p : 1 \le p \le q)$ in $\mathcal{G}$, the sequence $(Z_n, (1_{A_p})_{1 \le p \le q})$ of $E \times \mathbb{R}^q$-valued variables converges in law as $n \to \infty$, then necessarily Z_n converges stably in law. Also, if Z is defined *on the same space as* all Z_n's, we have

$$Z_n \stackrel{\mathbb{P}}{\longrightarrow} Z \iff Z_n \stackrel{\mathcal{L}\text{-s}}{\longrightarrow} Z \tag{2.2.7}$$

(the implication from left to right is obvious, the converse is much more involved but will not be used in this book).

Finally, let us mention a slightly different setting in which stable convergence can occur. Let Z_n be a sequence of E-valued variables, each one being defined on some extension $(\overline{\Omega}_n, \overline{\mathcal{F}}_n, \overline{\mathbb{P}}_n)$ of *the same space* $(\Omega, \mathcal{F}, \mathbb{P})$. These extensions may be different when n varies. Then we say that Z_n converges stably in law to Z, also defined on still another extension $(\widetilde{\Omega}, \widetilde{\mathcal{F}}, \widetilde{\mathbb{P}})$ of $(\Omega, \mathcal{F}, \mathbb{P})$, if for all Y and f as in

(2.2.4) we have

$$\overline{\mathbb{E}}_n\big(Yf(Z_n)\big) \to \widetilde{\mathbb{E}}\big(Yf(Z)\big). \tag{2.2.8}$$

We still write $Z_n \xrightarrow{\mathcal{L}\text{-s}} Z$ for this notion, which is the same as the previous one when $(\overline{\Omega}_n, \overline{\mathcal{F}}_n, \overline{\mathbb{P}}_n) = (\Omega, \mathcal{F}, \mathbb{P})$ for all n. However, one has to be careful when applying the properties of stable convergence in law in this extended setting. For example (2.2.5) still holds, but only with the following interpretation of $Y_n \xrightarrow{\mathbb{P}} Y$: each variable Y_n is defined on $(\overline{\Omega}_n, \overline{\mathcal{F}}_n, \overline{\mathbb{P}}_n)$ and takes its values in some Polish space F with a distance d; the variable Y is also F-valued, and defined on $(\Omega, \mathcal{F}, \mathbb{P})$; and $Y_n \xrightarrow{\mathbb{P}} Y$ means that for all $\varepsilon > 0$ we have $\overline{\mathbb{P}}_n(d(Y_n, Y) > \varepsilon) \to 0$ as $n \to \infty$. In a similar way, (2.2.7) holds with the same interpretation of the convergence in probability.

A special case of this setting is as follows: we have a probability space $(\Omega, \mathcal{G}, \mathbb{P})$ and a sub-σ-field $\mathcal{F}$ of $\mathcal{G}$, and $(\overline{\Omega}_n, \overline{\mathcal{F}}_n, \overline{\mathbb{P}}_n) = (\Omega, \mathcal{G}, \mathbb{P})$ for all n. Then, in this situation, $Z_n \xrightarrow{\mathcal{L}\text{-s}} Z$ amounts to the "ordinary" stable convergence in law when $\mathcal{F} = \mathcal{G}$, and to $Z_n \xrightarrow{\mathcal{L}} Z$ ("ordinary" convergence in law) when $\mathcal{F} = \{\Omega, \emptyset\}$ is the trivial σ-field, and to something "in between" when $\mathcal{F}$ is a non-trivial proper sub-σ-field of $\mathcal{G}$.

2.2.2 Convergence for Processes

In this book we study sequences of random variables, but also of processes, aiming to prove the convergence in probability, or in law, or stably in law.

If we consider a sequence $Z^n = (Z^n_t)_{t\geq 0}$ of $\mathbb{R}^d$-valued stochastic processes, we first have the so-called "finite-dimensional convergence": this means the convergence of $(Z^n_{t_1}, \ldots, Z^n_{t_k})$ for any choice of the integer k and of the times $t_1, \ldots, t_k$. When the convergence is in probability, the convergence for any single fixed t obviously implies finite-dimensional convergence, but this is no longer true when the convergence is in law, or stably in law.

There is another, more restricted, kind of convergence for processes, called "functional" convergence. That means that we consider each process Z^n as taking its values in some functional space (= a space of functions from $\mathbb{R}_+$ into $\mathbb{R}^d$), and we endow this functional space with a topology which makes it a Polish space.

The simplest functional space having this structure is the space $\mathbb{C}^d = \mathbb{C}(\mathbb{R}_+, \mathbb{R}^d)$ of all continuous functions from $\mathbb{R}_+$ into $\mathbb{R}^d$, endowed with the local uniform topology corresponding for example to the distance $d(x,y) = \sum_{n\geq 1} 2^{-n}(1 \wedge \sup_{s\leq n} \|x(s) - y(s)\|)$. The Borel σ-field for this topology is $\sigma(x(s) : s \geq 0)$.

However, although many of our limiting processes will be continuous, it is (almost) never the case of the pre-limiting processes which typically are partial sums of the form $\sum_{i=1}^{[nt]} \zeta^n_i$ for suitable random variables ζ^n_i and $[nt]$ denotes the integer part of nt. Such a process has discontinuous, although càdlàg, paths.

This is why we need to consider, in an essential way, the space $\mathbb{D}^d = \mathbb{D}(\mathbb{R}_+, \mathbb{R}^d)$ of all càdlàg functions from $\mathbb{R}_+$ into $\mathbb{R}^d$. This space is called the *Skorokhod space*, and it may be endowed with the *Skorokhod topology*, introduced by Skorokhod in [89] under the name "J1-topology". Under this topology $\mathbb{D}^d$ is a Polish space, and again the Borel σ-field is $\sigma(x(s) : s \geq 0)$. We are not going to define this topology here, and the reader is referred to the books of Billingsley [18] or Ethier and Kurtz [31] or Jacod and Shiryaev [57]. The convergence of a sequence (x_n) towards x, for this topology, will be denoted by $x_n \stackrel{\text{Sk}}{\longrightarrow} x$.

Let us just point out a small number of important properties which will be used in the sequel. The main drawback of the Skorokhod topology is that it is *not* compatible with the natural linear structure of the space: we may have $x_n \stackrel{\text{Sk}}{\longrightarrow} x$ and $y_n \stackrel{\text{Sk}}{\longrightarrow} y$ *without* $x_n + y_n$ converging to $x + y$. We do have, however (with $x(t-)$ denoting the left limit of x at time t):

$$x_n \stackrel{\text{Sk}}{\longrightarrow} x, \quad y_n \stackrel{\text{Sk}}{\longrightarrow} y, \quad t \mapsto y(t) \text{ is continuous} \Rightarrow x_n + y_n \stackrel{\text{Sk}}{\longrightarrow} x + y, \tag{2.2.9}$$

$$x_n \stackrel{\text{Sk}}{\longrightarrow} x, \quad x(t) = x(t-), \quad \Rightarrow \quad x_n(t) \to x(t), \tag{2.2.10}$$

$$x_n \stackrel{\text{Sk}}{\longrightarrow} x, \quad t \mapsto x(t) \text{ is continuous} \Rightarrow \sup_{s \in [0,T]} \left\| x_n(s) - x(s) \right\| \to 0 \ \forall T. \tag{2.2.11}$$

For the last property that we want to recall, we need to introduce a sequence of subdivisions: for each n, we have a sequence $(t(n,i) : i \geq 0)$ which increases strictly to $+\infty$ and has $t(n,0) = 0$. With $x \in \mathbb{D}^d$ we associate the "discretized" function $x^{(n)}$ as follows:

$$x^{(n)}(t) = x\big(t(n,i)\big) \quad \text{if } t(n,i) \leq t < t(n,i+1). \tag{2.2.12}$$

This defines a new function $x^{(n)} \in \mathbb{D}^d$. Then for any sequence y_n of functions, we have (this is Proposition VI.6.37 of Jacod and Shiryaev [57] applied with non-random functions):

$$\lim_{n \to \infty} \sup_{i \geq 1} \big(t \wedge t(n,i) - t \wedge t(n,i-1)\big) = 0 \ \forall t > 0, \ y_n \stackrel{\text{Sk}}{\longrightarrow} y \ \Rightarrow \ y_n^{(n)} \stackrel{\text{Sk}}{\longrightarrow} y. \tag{2.2.13}$$

Now we consider a sequence of càdlàg $\mathbb{R}^d$-valued processes (X^n) and another càdlàg $\mathbb{R}^d$-valued process X. They can be considered as random variables taking their values in the space $\mathbb{D}^d$, and then we have the notion of convergence of X^n towards X in law, or stably in law, or in probability (in the first case X is defined on an arbitrary probability space, in the second case it is defined on an extension, and in the third case it is defined on the same space as are all the X^n's). When the underlying topology on $\mathbb{D}^d$ under which the convergence takes place is the Skorokhod topology, we write these three convergences, respectively, as follows

$$X^n \stackrel{\mathcal{L}}{\Longrightarrow} X, \qquad X^n \stackrel{\mathcal{L}\text{-s}}{\Longrightarrow} X, \qquad X^n \stackrel{\mathbb{P}}{\Longrightarrow} X. \tag{2.2.14}$$

For convenience of notation, we sometimes write the above as $X_t^n \stackrel{\mathcal{L}}{\Longrightarrow} X_t$ or $X_t^n \stackrel{\mathcal{L}\text{-s}}{\Longrightarrow} X_t$ or $X_t^n \stackrel{\mathbb{P}}{\Longrightarrow} X_t$: these *should not be confused with* $X_t^n \stackrel{\mathcal{L}}{\longrightarrow} X_t$ or $X_t^n \stackrel{\mathcal{L}\text{-s}}{\longrightarrow} X_t$ or $X_t^n \stackrel{\mathbb{P}}{\longrightarrow} X_t$, which mean the convergence, in law or stably in law or in probability, of the variables X_t^n towards X_t for some fixed t.

Since we are establishing notation, and with X^n and X being càdlàg processes, we continue with the following conventions:

$$\left.\begin{array}{lll}
\bullet\ U_n \stackrel{\text{a.s.}}{\longrightarrow} U & \text{for random variables means almost sure convergence} \\
\bullet\ X^n \stackrel{\text{a.s.}}{\Longrightarrow} X & (\text{or } X_t^n \stackrel{\text{a.s.}}{\Longrightarrow} X_t) \text{ means almost sure} \\
 & \quad \text{convergence for the Skorokhod topology} \\
\bullet\ X^n \stackrel{\mathcal{L}_f}{\longrightarrow} X & \text{means finite-dimensional convergence in law} \\
\bullet\ X^n \stackrel{\mathcal{L}_f\text{-s}}{\longrightarrow} X & \text{means finite-dimensional stable convergence in law} \\
\bullet\ X^n \stackrel{\text{u.c.p.}}{\Longrightarrow} X & (\text{or } X_t^n \stackrel{\text{u.c.p.}}{\Longrightarrow} X_t) \text{ means } \sup_{s\le t} \|X_s^n - X_s\| \stackrel{\mathbb{P}}{\longrightarrow} 0 \text{ for all } t.
\end{array}\right\} \tag{2.2.15}$$

2.2.3 *Criteria for Convergence of Processes*

In this subsection we gather a few criteria which help to prove that a sequence of processes converges in probability, or in law, or stably in law. These criteria will often be used below.

To begin with, there is a simple result for real-valued processes with increasing paths:

$$\begin{array}{l}
\text{if } X^n \text{ and } X \text{ have increasing paths and } X \text{ is continuous, then} \\
X_t^n \stackrel{\mathbb{P}}{\longrightarrow} X_t \quad \forall t \in D, \text{ with } D \text{ a dense subset of } \mathbb{R}_+ \quad \Rightarrow \quad X^n \stackrel{\text{u.c.p.}}{\Longrightarrow} X.
\end{array} \tag{2.2.16}$$

Second, there is a well known trick, called the *subsequences principle*. It concerns a sequence (Z_n) of E-valued random variables with E a Polish space, and thus also applies to processes viewed as variables taking values in the functional spaces $\mathbb{C}^d$ or $\mathbb{D}^d$. This "principle" goes as follows:

$$\begin{array}{l}
\text{we have } Z_n \stackrel{\mathbb{P}}{\longrightarrow} Z, \text{ resp. } Z_n \stackrel{\mathcal{L}}{\longrightarrow} Z, \text{ resp. } Z_n \stackrel{\mathcal{L}\text{-s}}{\longrightarrow} Z \text{ if and only if,} \\
\text{from any subsequence } n_k \to \infty \text{ we can extract a sub-subsequence } n_{k_l} \\
\text{such that } Z_{n_{k_l}} \stackrel{\mathbb{P}}{\longrightarrow} Z, \text{ resp. } Z_{n_{k_l}} \stackrel{\mathcal{L}}{\longrightarrow} Z, \text{ resp. } Z_{n_{k_l}} \stackrel{\mathcal{L}\text{-s}}{\longrightarrow} Z.
\end{array} \tag{2.2.17}$$

For the convergence in probability or in law, this simply comes from the fact that those are convergence for a metric. For the stable convergence in law, it comes from the fact that for any given Y and f as in (2.2.4) the "subsequences principle" holds for the convergence of the real numbers $\mathbb{E}(Y f(Z_n))$.

In connection with this, we state two well known and useful results. The second one is called the *Skorokhod representation theorem*. The setting is as before.

$$Z_n \xrightarrow{\mathbb{P}} Z \;\Rightarrow\; \text{there is a sequence } n_k \to \infty \text{ with } Z_{n_k} \xrightarrow{\text{a.s.}} Z \tag{2.2.18}$$

$$Z_n \xrightarrow{\mathcal{L}} Z \;\Rightarrow\; \left.\begin{array}{l}\text{there exist variables } Z'_n \text{ and } Z' \text{ defined on the}\\ \text{same probability space, having the same laws}\\ \text{as } Z_n \text{ and } Z \text{ respectively, and with } Z'_n \xrightarrow{\text{a.s.}} Z'.\end{array}\right\} \tag{2.2.19}$$

In many instances we have a sequence (X^n) of d-dimensional processes, whose convergence towards X we seek, and for which a natural decomposition arises for any integer $m \geq 1$

$$X^n = X(m)^n + X'(m)^n \tag{2.2.20}$$

and for which the convergence of $X(m)^n$ (as $n \to \infty$) to a limit $X(m)$ can be easily proved. Then, showing that $X'(m)^n$ goes to 0 as $m \to \infty$, "uniformly in n", and that $X(m)$ converges to X, allows one to prove the desired result. Depending on which kind of convergence we are looking for, the precise statements are as follows.

Proposition 2.2.1 *Let X^n and X be defined on the same probability space. For $X^n \overset{\mathbb{P}}{\Longrightarrow} X$ it is enough that there are decompositions* (2.2.20) *and also $X = X(m) + X'(m)$, with the following properties*:

$$\forall m \geq 1, \quad X(m)^n \overset{\mathbb{P}}{\Longrightarrow} X(m), \quad \text{as } n \to \infty \tag{2.2.21}$$

$$X(m) \overset{\text{u.c.p.}}{\Longrightarrow} X, \quad \text{as } m \to \infty, \tag{2.2.22}$$

$$\forall \eta, t > 0, \quad \lim_{m\to\infty} \limsup_{n\to\infty} \mathbb{P}\Big(\sup_{s\leq t} \|X'(m)^n_s\| > \eta\Big) = 0. \tag{2.2.23}$$

Proposition 2.2.2 *For $X^n \overset{\mathcal{L}}{\Longrightarrow} X$ it is enough that there are decompositions* (2.2.20) *satisfying* (2.2.23) *and*

$$\forall m \geq 1, \quad X(m)^n \overset{\mathcal{L}}{\Longrightarrow} X(m), \quad \text{as } n \to \infty \tag{2.2.24}$$

for some limiting processes $X(m)$, which in turn satisfy

$$X(m) \overset{\mathcal{L}}{\Longrightarrow} X, \quad \text{as } m \to \infty.$$

Remark 2.2.3 In the previous proposition, the processes X^n may be defined on different spaces $(\Omega_n, \mathcal{F}_n, \mathbb{P}_n)$, the decompositions (2.2.20) taking place on those spaces, of course. If this is the case, in (2.2.23) one should have $\mathbb{P}_n$ instead of $\mathbb{P}$. To avoid even more cumbersome notation, we still write $\mathbb{P}$ instead of $\mathbb{P}_n$. The same comment applies also to the next results.

The similar statement about stable convergence needs more care. The most common situation is when all processes X^n are defined on the same probability space $(\Omega, \mathcal{F}, \mathbb{P})$ and have the decompositions (2.2.20), and the limit X is defined on an extension $(\widetilde{\Omega}, \widetilde{\mathcal{F}}, \widetilde{\mathbb{P}})$. However, (2.2.21) or (2.2.24) are replaced by the stable convergence in law. This means that for each m we have a (possibly different) extension $(\widetilde{\Omega}_m, \widetilde{\mathcal{F}}_m, \widetilde{\mathbb{P}}_m)$, on which the limit process $X(m)$ is defined.

With our "extended" notion of stable convergence, as given in (2.2.8), each X^n is defined on some extension $(\overline{\Omega}_n, \overline{\mathcal{F}}_n, \overline{\mathbb{P}}_n)$ of the same space $(\Omega, \mathcal{F}, \mathbb{P})$, and X^n has the decomposition (2.2.20) on $(\overline{\Omega}_n, \overline{\mathcal{F}}_n, \overline{\mathbb{P}}_n)$. And again, for each m we have an extension $(\widetilde{\Omega}_m, \widetilde{\mathcal{F}}_m, \widetilde{\mathbb{P}}_m)$ of $(\Omega, \mathcal{F}, \mathbb{P})$, on which the limit process $X(m)$ is defined.

Proposition 2.2.4 *In the above two settings, for $X^n \stackrel{\mathcal{L}\text{-s}}{\Longrightarrow} X$ it is enough that there are decompositions* (2.2.20) *satisfying* (2.2.23), *and*

$$\forall m \geq 1, \quad X(m)^n \stackrel{\mathcal{L}\text{-s}}{\Longrightarrow} X(m), \quad \text{as } n \to \infty$$

for some limiting processes $X(m)$, which in turn satisfy

$$X(m) \stackrel{\mathcal{L}\text{-s}}{\Longrightarrow} X, \quad \text{as } m \to \infty.$$

These three results are proved in the Appendix, as well as the forthcoming propositions which are all well known, but used constantly.

Proposition 2.2.5 *Let (M^n) be a sequence of local martingales on the space $(\Omega, \mathcal{F}, (\mathcal{F}_t)_{t\geq 0}, \mathbb{P})$, with $M_0^n = 0$. Then $M^n \stackrel{\text{u.c.p.}}{\Longrightarrow} 0$ as soon as one of the following two conditions holds*:

(*i*) *each M^n admits an angle bracket and $\langle M^n, M^n\rangle_t \stackrel{\mathbb{P}}{\longrightarrow} 0$ for all $t > 0$*,

(*ii*) *we have $|\Delta M_s^n| \leq K$ for a constant K, and $[M^n, M^n]_t \stackrel{\mathbb{P}}{\longrightarrow} 0$ for all $t > 0$.*

In the next result, μ is the jump measure of a càdlàg d-dimensional process (in which case $E = \mathbb{R}^d$ below), or a Poisson random measure on $\mathbb{R}_+ \times E$ for E a Polish space, and in both cases ν is the compensator of μ.

Proposition 2.2.6 *Let (δ_n) be a sequence of predictable functions on $\Omega \times \mathbb{R}_+ \times E$, each δ_n satisfying* (2.1.16). *Then*

$$\big((\delta_n)^2 \wedge |\delta_n|\big) \star \nu_t \stackrel{\mathbb{P}}{\longrightarrow} 0 \quad \forall t > 0 \quad \Rightarrow \quad \delta_n \star (\mu - \nu) \stackrel{\text{u.c.p.}}{\Longrightarrow} 0. \tag{2.2.25}$$

The "dominated convergence theorem for stochastic integrals" mentioned earlier is:

Proposition 2.2.7 *Let X be a semimartingale and (H^n) a sequence of predictable processes satisfying $|H^n| \leq H'$ for some predictable and locally bounded process*

H'. Then if outside a null set we have $H_t^n \to H_t$ for all t, where H is another predictable process, we have

$$\int_0^t H_s^n \, dX_s \stackrel{\text{u.c.p.}}{\Longrightarrow} \int_0^t H_s \, dX_s.$$

Finally, we also have the convergence of "Riemann sums" for stochastic integrals, under some regularity of the integrand process. Below, X and H may be r and $d \times r$-dimensional, respectively, so that the integral process is d-dimensional.

Proposition 2.2.8 *Let X be a semimartingale and H be a càglàd adapted process. For each n let $(T(n,i) : i \geq 0)$ be a sequence of stopping times, which strictly increases to $+\infty$, with $T(n,0) = 0$, and such that* $\sup(T(n,i+1) \wedge t - T(n,i) \wedge t :$ *$i \geq 0)$ goes to 0 in probability for all t as $n \to \infty$. Then*

$$\sum_{i \geq 1,\ T(n,i) \leq t} H_{T(n,i-1)} \left(X_{T(n,i)} - X_{T(n,i-1)} \right) \stackrel{\mathbb{P}}{\Longrightarrow} \int_0^t H_s \, dX_s \tag{2.2.26}$$

(convergence for the Skorokhod topology). If further X is continuous the same holds also when H is adapted càdlàg, and we even have the local uniform convergence in probability.

2.2.4 Triangular Arrays: Asymptotic Negligibility

In this subsection we give a few limit theorems for sums of triangular arrays: the results are stated in a somewhat abstract setting, but the connection with discretized processes is explained in Remark 2.2.3 below, and they will be of constant use in this book.

A d-dimensional *triangular array* is a double sequence $(\zeta_i^n : n, i \geq 1)$ of d-dimensional variables $\zeta_i^n = (\zeta_i^{n,j})_{1 \leq j \leq d}$. Associated with the nth row $(\zeta_i^n)_{i \geq 1}$, we have a *stopping rule*, that is a process N_n satisfying:

$$t \mapsto N_n(t) \quad \text{is } \mathbb{N}\text{-valued, càdlàg increasing, with jumps equal to 1} \\ \text{and } N_n(0) = 0 \text{ and } \lim_{t \to \infty} N_n(t) = \infty. \tag{2.2.27}$$

We are interested in the behavior of the partial sums

$$S_t^n = \sum_{i=1}^{N_n(t)} \zeta_i^n, \tag{2.2.28}$$

which are càdlàg processes, with finite variation.

To accommodate the applications we have in mind, we have to be careful about the structure of those triangular arrays and associated stopping rules. Each row

$(\zeta_i^n)_{i\geq 1}$ and its associated stopping rule $N_n(t)$ are defined on some probability space $(\Omega_n, \mathcal{G}^n, \mathbb{P}_n)$, which is endowed with a discrete-time filtration $(\mathcal{G}_i^n)_{i\in\mathbb{N}}$, and the basic assumptions are as follows:

$$\begin{aligned} &\bullet\ n\geq 1,\ i\geq 1\ \Rightarrow\ \zeta_i^n \text{ is } \mathcal{G}_i^n \text{ measurable} \\ &\bullet\ n\geq 1,\ t\geq 0\ \Rightarrow\ N_n(t) \text{ is a } \big(\mathcal{G}_i^n\big)\text{-stopping time.} \end{aligned} \tag{2.2.29}$$

We also consider the continuous-time filtration $\overline{\mathcal{F}}_t^n = \mathcal{G}_{N_n(t)}^n$, and we set

$$T(n,i) = \inf\big(t : N_n(t)\geq i\big). \tag{2.2.30}$$

The following (easy) properties are proved in §II.3b of [57]: we have $T(n,0)=0$, and for $i\geq 1$ the variable $T(n,i)$ is a predictable $(\overline{\mathcal{F}}_t^n)$-stopping time such that $\overline{\mathcal{F}}_{T(n,i)}^n = \mathcal{G}_i^n$ and $\overline{\mathcal{F}}_{T(n,i)-}^n = \mathcal{G}_{i-1}^n$. Then we can rewrite S^n as

$$S_t^n = \sum_{i\geq 1} \zeta_i^n\, 1_{\{T(n,i)\leq t\}}$$

and obviously S^n is $(\overline{\mathcal{F}}_t^n)$-adapted. If further each ζ_i^n is integrable this process S^n admits a predictable compensator (relative to $(\overline{\mathcal{F}}_t^n)$) which, since each $T(n,i)$ is predictable and $\overline{\mathcal{F}}_{T(n,i)-}^n = \mathcal{G}_{i-1}^n$, takes the form

$$S_t'^n = \sum_{i\geq 1} \mathbb{E}\big(\zeta_i^n \mid \mathcal{G}_{i-1}^n\big)\, 1_{\{T(n,i)\leq t\}} \tag{2.2.31}$$

(we use the notation $\mathbb{E}$ for the expectation with respect to $\mathbb{P}_n$ when no confusion may arise, otherwise we write $\mathbb{E}_n$). The form (2.2.31) is indeed the key ingredient for proving the limit theorems for triangular arrays, and for this the condition (2.2.29) is crucial.

Remark 2.2.9 Most of the time in this book, triangular arrays occur as follows. On an underlying (continuous-time) filtered space $(\Omega, \mathcal{F}, (\mathcal{F}_t)_{t\geq 0}, \mathbb{P})$, and for each n, there is a strictly increasing sequence $(T(n,i) : i\geq 0)$ of finite $(\mathcal{F}_t)$-stopping times with limit $+\infty$ and $T(n,0)=0$. The stopping rule is

$$N_n(t) = \sup\big(i : T(n,i)\leq t\big) = \sum_{i\geq 1} 1_{\{T(n,i)\leq t\}}, \tag{2.2.32}$$

which satisfies (2.2.27): note that (2.2.32) and (2.2.30) are indeed equivalent. Finally, we have a double sequence (ζ_i^n) such that each ζ_i^n is $\mathcal{F}_{T(n,i)}$ measurable: for example, ζ_i^n may be a function of the increment $Y_{T(n,i)} - Y_{T(n,i-1)}$, for some underlying adapted càdlàg process Y.

To fit this into the previous setting, we take $(\Omega_n, \mathcal{G}^n, \mathbb{P}_n) = (\Omega, \mathcal{F}, \mathbb{P})$. There is a problem, though, for defining the discrete-time filtration $(\mathcal{G}_i^n)$ in such a way that (2.2.29) holds. Two situations may arise:

1 All $T(n,i)$ are deterministic, or more generally they are "strongly predictable" stopping times, that is stopping times such that $T(n,i+1)$ is $\mathcal{F}_{T(n,i)}$ measurable. Then (2.2.28) holds with $\mathcal{G}_i^n = \mathcal{F}_{T(n,i)}$, and the σ-field $\overline{\mathcal{F}}_t^n = \mathcal{G}_{N_n(t)}^n$ is the σ-field satisfying

$$\overline{\mathcal{F}}_t^n \cap \{T(n,i) \le t < T(n,i+1)\} = \mathcal{F}_{T(n,i)} \cap \{T(n,i) \le t < T(n,i+1)\}$$

for all $i \ge 0$. In particular $\overline{\mathcal{F}}_t^n \subset \mathcal{F}_t$ and $\overline{\mathcal{F}}_{T(n,i)}^n = \mathcal{F}_{T(n,i)}$ for all i.

2 The $T(n,i)$'s are arbitrary stopping times. Then (2.2.28) does not usually hold with $\mathcal{G}_i^n = \mathcal{F}_{T(n,i)}$. It holds with $\mathcal{G}_i^n = \mathcal{F}_{T(n,i)} \vee \sigma(T(n,i+1))$, and

$$\begin{aligned}&\overline{\mathcal{F}}_t^n \cap \{T(n,i) \le t < T(n,i+1)\}\\ &\quad = \mathcal{F}_{T(n,i)} \vee \sigma\big(T(n,i+1)\big) \cap \{T(n,i) \le t < T(n,i+1)\},\end{aligned}$$

but the inclusion $\overline{\mathcal{F}}_t^n \subset \mathcal{F}_t$ is no longer valid.

This is one of the reasons why discretization along stopping times is significantly more difficult to study than discretization along deterministic times.

In the rest of this subsection we give criteria for a triangular array to be *asymptotically negligible*, or *AN* in short, in the sense that

$$\sum_{i=1}^{N_n(t)} \zeta_i^n \stackrel{\text{u.c.p.}}{\Longrightarrow} 0, \quad \forall t > 0. \tag{2.2.33}$$

In other words, $\sup_{s \le t} |\sum_{i=1}^{N_n(s)} \zeta_i^n| \stackrel{\mathbb{P}}{\longrightarrow} 0$, or equivalently $\sup_{s \le t} |\sum_{i=1}^{N_n(s)} \zeta_i^n| \stackrel{\mathcal{L}}{\longrightarrow} 0$. This makes sense even when each row $(\zeta_i^n : i \ge 1)$ and the associated stopping rule N_n are defined on some probability spaces $(\Omega_n, \mathcal{G}^n, \mathbb{P}_n)$ depending on n. The AN property is about the array (ζ_i^n) together with the stopping rules $N_n(t)$, although we just write that the array is AN: usually $N_n(t)$ is clear from the context.

The AN property is a property for each component, so the following criteria are all stated for one-dimensional arrays. The first property below is a trivial consequence of (2.2.16), whereas the last implication is obvious:

$$\sum_{i=1}^{N_n(t)} |\zeta_i^n| \stackrel{\mathbb{P}}{\longrightarrow} 0 \quad \forall t > 0 \quad \Leftrightarrow \quad \big(|\zeta_i^n|\big) \text{ is AN} \quad \Rightarrow \quad \big(\zeta_i^n\big) \text{ is AN.}$$

The following lemmas are proved in the Appendix, and we always assume (2.2.29).

Lemma 2.2.10 *The array* (ζ_i^n) *is AN as soon as the array* $(\mathbb{E}(|\zeta_i^n| \mid \mathcal{G}_{i-1}^n))$ *is AN.*

Lemma 2.2.11 *Let* (ζ_i^n) *be a triangular array such that each* ζ_i^n *is square-integrable. Then the array* $(\zeta_i^n - \mathbb{E}(\zeta_i^n \mid \mathcal{G}_{i-1}^n))$ *is AN under each of the following three conditions:*

(a) *The array* $(\mathbb{E}(|\zeta_i^n|^2 \mid \mathcal{G}_{i-1}^n))$ *is AN.*

(b) *The sequence of variables* $(\sum_{i=1}^{N_n(t)} \mathbb{E}(|\zeta_i^n|^2 1_{\{|\zeta_i^n|>1\}} \mid \mathcal{G}_{i-1}^n))_{n\geq 1}$ *is bounded in probability for each* $t > 0$, *and the array* $(|\zeta_i^n|^2)$ *is AN.*

(c) *We have* $|\zeta_i^n| \leq K$ *for a constant* K, *and the array* $(|\zeta_i^n|^2)$ *is AN.*

In particular if (ζ_i^n) *is a "martingale difference" array, that is* $\mathbb{E}(\zeta_i^n \mid \mathcal{G}_{i-1}^n) = 0$ *for all* $i, n \geq 1$, *then either one of the above conditions implies that it is AN.*

As a simple consequence of this lemma, we get:

Lemma 2.2.12 *Let* A *be an* $\mathbb{R}^d$*-valued (deterministic) function and* (ζ_i^n) *a* d*-dimensional array. If*

$$\sum_{i=1}^{N_n(t)} \mathbb{E}(\zeta_i^n \mid \mathcal{G}_{i-1}^n) \stackrel{\text{u.c.p.}}{\Longrightarrow} A_t \tag{2.2.34}$$

and if the array (ζ_i^n) *satisfies any one of* (a), (b), (c) *of the previous lemma, we have* $\sum_{i=1}^{N_n(t)} \zeta_i^n \stackrel{\text{u.c.p.}}{\Longrightarrow} A_t$. *The same holds when* A *is a process, provided all spaces* $(\Omega_n, \mathcal{G}^n, \mathbb{P}_n)$ *are the same.*

Another (trivial, but extremely convenient) consequence is that the array (ζ_i^n) is AN as soon as at least one of the following two properties is satisfied:

$$\begin{aligned} &\mathbb{E}\left(\sum_{i=1}^{N_n(t)} |\zeta_i^n|\right) \to 0 \quad \forall t \\ &\mathbb{E}\left(\sum_{i=1}^{N_n(t)} |\zeta_i^n|^2\right) \to 0 \quad \forall t \quad \text{and} \quad \mathbb{E}(\zeta_i^n \mid \mathcal{G}_{i-1}^n) = 0 \quad \forall i, n. \end{aligned} \tag{2.2.35}$$

2.2.5 Convergence in Law of Triangular Arrays

Here we study the convergence in law. We do not give the most general results available, and only state a special case of the results of §VIII.3c of [57]. Here again we assume (2.2.29).

Theorem 2.2.13 *Assume* (2.2.34) *for some (deterministic) continuous* $\mathbb{R}^d$*-valued function of locally finite variation* A, *and also the following two conditions, for all* $j, k = 1, \ldots d$ *for the first one and for some* $p > 2$ *for the second one*:

$$\sum_{i=1}^{N_n(t)} \left(\mathbb{E}(\zeta_i^{n,j} \zeta_i^{n,k} \mid \mathcal{G}_{i-1}^n) - \mathbb{E}(\zeta_i^{n,j} \mid \mathcal{G}_{i-1}^n)\mathbb{E}(\zeta_i^{n,k} \mid \mathcal{G}_{i-1}^n)\right) \stackrel{\mathbb{P}}{\longrightarrow} C_t^{jk} \quad \forall t > 0, \tag{2.2.36}$$

$$\sum_{i=1}^{N_n(t)} \mathbb{E}(\|\zeta_i^n\|^p \mid \mathcal{G}_{i-1}^n) \xrightarrow{\mathbb{P}} 0 \quad \forall t > 0, \tag{2.2.37}$$

where $C = (C^{jk})$ is a (deterministic) continuous $\mathcal{M}_{d\times d}^+$-valued function (it is then necessarily increasing for the strong order in $\mathcal{M}_{d\times d}^+$), *then we have*

$$\sum_{i=1}^{N_n(t)} \zeta_i^n \stackrel{\mathcal{L}}{\Longrightarrow} A + Y, \tag{2.2.38}$$

where Y is a continuous centered Gaussian $\mathbb{R}^d$-valued process with independent increments having $\mathbb{E}(Y_t^j Y_t^k) = C_t^{jk}$.

The above conditions, of course, completely characterize the law of the process Y. Equivalently we could say that Y is a Gaussian martingale (relative to the filtration it generates), starting from 0, and with quadratic variation process C.

Note that if (2.2.37) holds for some t, it also holds for all $t' \leq t$, but this is not true of (2.2.36), which should hold for each t (in fact, it would be enough that it holds for all t in a dense subset of $\mathbb{R}_+$): if (2.2.36) and (2.2.37) and $\sum_{i=1}^{N_n(t)} \mathbb{E}(\zeta_i^n \mid \mathcal{G}_{i-1}^n) \xrightarrow{\mathbb{P}} A_t$ hold for a single time t, we cannot conclude (2.2.38), and even the convergence $\sum_{i=1}^{N_n(t)} \zeta_i^n \xrightarrow{\mathcal{L}} A_t + Y_t$ for this particular t fails in general. There is an exception, however, when the variables ζ_i^n are independent by rows and N_t^n is non-random. Indeed, as seen for example in Theorem VII-2-36 of [57], we have:

Theorem 2.2.14 *Assume that for each n the variables $(\zeta_i^n : i \geq 1)$ are independent, and let l_n be integers, possibly infinite. Assume also that, for all $j, k = 1, \ldots d$ and for some $p > 2$,*

$$\sum_{i=1}^{l_n} \mathbb{E}(\zeta_i^{n,j}) \xrightarrow{\mathbb{P}} A^j,$$

$$\sum_{i=1}^{l_n} \big(\mathbb{E}(\zeta_i^{n,j} \zeta_i^{n,k}) - \mathbb{E}(\zeta_i^{n,j})\mathbb{E}(\zeta_i^{n,k})\big) \xrightarrow{\mathbb{P}} C^{jk},$$

$$\sum_{i=1}^{l_n} \mathbb{E}(\|\zeta_i^n\|^p) \xrightarrow{\mathbb{P}} 0,$$

where C^{jk} and A^j are (deterministic) numbers. Then the variables $\sum_{i=1}^{l_n} \zeta_i^n$ converge in law to a Gaussian vector with mean $A = (A^j)$ and covariance matrix $C = (C^{jk})$.

Finally we turn to stable convergence in law. The reader will have observed that the conditions (2.2.34) and (2.2.36) are very restrictive, because the limits are non-random. Such a situation rarely occurs, whereas quite often these conditions are satisfied with A and C random. Then we need an extra condition, under which it turns out that the convergence holds not only in law, but even stably in law.

As before, for each n we have the sequence $(\zeta_i^n)_{i\geq 1}$, the stopping rules $N_n(t)$ and the associated stopping times $T(n,i)$, all defined on some space $(\Omega_n, \mathcal{G}^n, \mathbb{P}_n)$ with the discrete-time and the continuous-time filtrations $(\mathcal{G}_i^n)$ and $(\overline{\mathcal{F}}_t^n)$. But here we need some more structure on these objects. Namely, we assume that we have a filtered space $(\Omega, \mathcal{F}, (\mathcal{F}_t)_{t\geq 0}, \mathbb{P})$ such that, for each n,

$$\big(\Omega_n, \mathcal{G}^n, (\overline{\mathcal{F}}_t^n), \mathbb{P}_n\big) \text{ is a very good filtered extension of } \big(\Omega, \mathcal{F}, (\mathcal{F}_t), \mathbb{P}\big). \tag{2.2.39}$$

Quite often, but not always, we will have $(\Omega_n, \mathcal{G}^n, (\overline{\mathcal{F}}_t^n), \mathbb{P}_n) = (\Omega, \mathcal{F}, (\mathcal{F}_t), \mathbb{P})$.

We also single out, among all martingales on $(\Omega, \mathcal{F}, (\mathcal{F}_t)_{t\geq 0}, \mathbb{P})$, a q-dimensional Brownian motion W, and a subset $\mathcal{N}$ of bounded martingales, all orthogonal (in the martingale sense) to W and such that the set $\{N_\infty : N \in \mathcal{N}\}$ is total for $\mathbb{L}^1$ convergence in the set of the terminal variables of all bounded martingales orthogonal to W (when $q = 0$, we have $W = 0$).

Theorem 2.2.15 *We suppose that* (2.2.29) *and* (2.2.39) *hold. Assume that* (2.2.34) *holds for some $\mathbb{R}^d$-valued process A, and* (2.2.36) *for some continuous adapted process $C = (C^{jk})$ with values in $\mathcal{M}_{d\times d}^+$, both A and C being defined on $(\Omega, \mathcal{F}, (\mathcal{F}_t)_{t\geq 0}, \mathbb{P})$. Assume also* (2.2.37) *for some $p > 2$, and*

$$\sum_{i=1}^{N_n(t)} \mathbb{E}\big(\zeta_i^n (M_{T(n,i)} - M_{T(n,i-1)}) \mid \mathcal{G}_{i-1}^n\big) \xrightarrow{\mathbb{P}} 0 \quad \forall t > 0 \tag{2.2.40}$$

whenever M is one of the components of W or is in a set $\mathcal{N}$ as described before. Then we have

$$\sum_{i=1}^{N_n(t)} \zeta_i^n \stackrel{\mathcal{L}\text{-}s}{\Longrightarrow} A + Y,$$

where Y is a continuous process defined on a very good filtered extension $(\widetilde{\Omega}, \widetilde{\mathcal{F}}, (\widetilde{\mathcal{F}}_t)_{t\geq 0}, \widetilde{\mathbb{P}})$ of $(\Omega, \mathcal{F}, (\mathcal{F}_t)_{t\geq 0}, \mathbb{P})$ and which, conditionally on the σ-field $\mathcal{F}$, is a centered Gaussian $\mathbb{R}^d$-valued process with independent increments satisfying $\widetilde{\mathbb{E}}(Y_t^j Y_t^k \mid \mathcal{F}) = C_t^{jk}$.

Proof When $(\Omega_n, \mathcal{G}^n, (\overline{\mathcal{F}}_t^n), \mathbb{P}_n) = (\Omega, \mathcal{F}, (\mathcal{F}_t), \mathbb{P})$, this is a particular case of Theorem IX.7.28 of [57], except that this theorem is stated when $T(n,i) = i/n$, but the extension to the present situation is totally straightforward.

When $(\Omega_n, \mathcal{G}^n, (\overline{\mathcal{F}}_t^n), \mathbb{P}_n) \neq (\Omega, \mathcal{F}, (\mathcal{F}_t), \mathbb{P})$, the result does not formally follow from the above-quoted theorem. However, this theorem is based on Theorem IX.7.3 of [57]: the proof of this latter theorem can be reproduced word for word: indeed, the only two differences are in the last display of p. 587, where the first equality holds here because of (2.2.39), and in Step 5 of the proof where X_t^n should be excluded from the definition of $\mathcal{H}$. □

In view of the comments made before (2.2.5), the conditions stated above completely specify the conditional law of Y, knowing $\mathcal{F}$, so the stable convergence in law is well defined. Processes like Y above will be studied more thoroughly in Chap. 4, but we mention right away some of their nice properties: Y is a continuous local martingale on $(\widetilde{\Omega}, \widetilde{\mathcal{F}}, (\widetilde{\mathcal{F}}_t)_{t\geq 0}, \widetilde{\mathbb{P}})$, whose quadratic variation process is C. Moreover it is orthogonal to any martingale on the space $(\Omega, \mathcal{F}, (\mathcal{F}_t)_{t\geq 0}, \mathbb{P})$, but of course it is no longer a (unconditionally) Gaussian process in general.

Bibliographical Notes

An historical sketch of the development of the theory of semimartingales and stochastic integration with respect to semimartingales and random measures is outside the scope of these short bibliographical notes, although one cannot avoid to mention at least the two pioneering names of K. Itô and P. Lévy, and also P.A. Meyer who, with many collaborators, established the final state of the theory. An interested reader should consult Dellacherie and Meyer [25], or Jarrow and Protter [65] for an historical account.

The characteristics of a semimartingale are implicit in the work of K. Itô, and have formally been introduced by Grigelionis [40] in the case of what is called an Itô semimartingale in this book, and by Jacod and Mémin [51] in general. What we call the "Grigelionis form" of an Itô semimartingale is basically due to Grigelionis [39] of course, see also El Karoui and Lepeltier [30], and the form given here is in Jacod [52]. Section 2.1.5 about estimates for semimartingales contains some new material, but related estimates or special cases are quite common in the literature, in which they have been proved when need arose, and repeatedly, by many authors. The Burkholder-Davis-Gundy inequalities were proved by Burkholder and Gundy in [20] for the case $p > 1$, and in [21] when $p = 1$. The Lenglart inequalities were proved in [70].

Likewise, all results of Sect. 2.2 are taken from [57], which contains a relatively detailed account on the (long) history of the subject. Let us just emphasize the facts that stable convergence in law has been introduced by Rényi [84] and largely developed by Aldous and Eagleson [4] and Hall and Heyde [43]. The Skorokhod convergence was introduced by Skorokhod in [89] and subsequently developed and expounded in book form by Skorokhod himself [90], Billingsley [18] or Ethier and Kurtz [31].

Part II
The Basic Results

We now reach the main core of this book, that is the behavior of processes, or "functionals", which have the form (1.1.1) for a test function f and a process X, under assumptions on f and X, and also on the discretization scheme, which will be chosen to be as weak as possible.

The first chapter of this part is devoted to the convergence in probability of our functionals, that is, the "Laws of Large Numbers".

In the second chapter we introduce some tools for the "Central Limit Theorems", or distributional results, associated with the convergences obtained in the first chapter, and the central limit theorems themselves are presented in the third chapter.

The fourth chapter of this part concerns the case where the test function is $f(x) = x$, a case not covered (at least for the CLT) by the previous results. This is important because the associated functional is then simply the "discretized" process, and the type of CLT obtained in this situation is quite different from the previous ones.

Chapter 3
Laws of Large Numbers: The Basic Results

In this chapter we prove the "Law of Large Numbers", LLN in short, for the two types of functionals introduced in (1.1.1). By this, we mean their convergence in probability. One should perhaps call these results "weak" laws of large numbers, but in our setting there is never a result like the "strong" law of large numbers, featuring almost sure convergence. Two important points should be mentioned: unlike in the usual LLN setting, the limit is usually not deterministic, but random; and, whenever possible, we consider functional convergence (as processes).

The first type of LLNs concerns raw sums, without normalization, and the results essentially do not depend on the discretization schemes, as soon as the discretization mesh goes to 0. The second type is about normalized sums of functions of normalized increments, and it requires the underlying process to be an Itô semimartingale and also the discretization scheme to be regular (irregular schemes in this context will be studied in Chap. 14, and are much more difficult to analyze).

We start with two preliminary sections: the first one is about "general" discretization schemes, to set up notation and a few simple properties. The second one studies semimartingales which have p-summable jumps, meaning that $\sum_{s\leq t}\|\Delta X_s\|^p$ is almost surely finite for all t: this is always true for $p\geq 2$, but it may fail when $0\leq p<2$.

3.1 Discretization Schemes

1) A *discretization grid* is a strictly increasing sequence of times, starting at 0 and with limit $+\infty$, and which in practice represents the times at which an underlying process is sampled. In most cases these times are non-random, and quite often regularly spaced. In some instances it is natural to assume that they are random, perhaps independent of the underlying process, or perhaps not.

For a given discretization grid, very little can be said. Things become interesting when we consider a *discretization scheme*, that is a sequence of discretization grids indexed by n, and such that the meshes of the grids go to 0 as $n\to\infty$. This notion has already appeared at some places in the previous chapter, for example in (2.1.8)

J. Jacod, P. Protter, *Discretization of Processes*,
Stochastic Modelling and Applied Probability 67,
DOI 10.1007/978-3-642-24127-7_3,

and (2.2.26), and we formalize it as follows. Below, when we speak of a "random" discretization scheme, we assume that the filtered space $(\Omega, \mathcal{F}, (\mathcal{F}_t)_{t\geq 0}, \mathbb{P})$ is given.

Definition 3.1.1 a) A *random discretization scheme* is a sequence $\mathcal{T} = (\mathcal{T}_n)_{n\geq 1}$ defined as follows: each $\mathcal{T}_n$ consists of a strictly increasing sequence $(T(n,i) : i \geq 0)$ of *finite* stopping times, with $T(n,0) = 0$ and $T(n,i) \to \infty$ as $i \to \infty$, and also

$$\forall t > 0, \qquad \sup_{i\geq 1} \big(T(n,i)\wedge t - T(n,i-1)\wedge t\big) \stackrel{\mathbb{P}}{\longrightarrow} 0. \tag{3.1.1}$$

b) A *discretization scheme* is as above, with all $T(n,i)$ deterministic, and to emphasize this fact we usually write $T(n,i) = t(n,i)$ with lower case letters.

c) The scheme is called *regular* if $t(n,i) = i\Delta_n$ for a sequence Δ_n of positive numbers going to 0 as $n \to \infty$.

The condition (3.1.1) expresses the fact that the mesh goes to 0. With any (random) discretization scheme we associate the quantities (where $t \geq 0$ and $i \geq 1$):

$$\begin{array}{ll} \Delta(n,i) = T(n,i) - T(n,i-1), & N_n(t) = \sum_{i\geq 1} 1_{\{T(n,i)\leq t\}}, \\ T_n(t) = T\big(n, N_n(t)\big), & I(n,i) = \big(T(n,i-1), T(n,i)\big]. \end{array} \tag{3.1.2}$$

In the regular case $\Delta(n,i) = \Delta_n$ and $N_n(t) = [t/\Delta_n]$. Random schemes $T(n,i)$ and the associated notation $N_n(t)$ have been already encountered in Sect. 2.2.4 of Chap. 2.

Regular schemes are the most common in practice, but it is also important to consider non-regular ones to deal with "missing data" and, more important, with cases where a process is observed at irregularly spaced times, as is often the case in finance.

Note that in many applications the time horizon T is fixed and observations occur before or at time T. In this context, for each n we have finitely many $T(n,i)$ only, all smaller than T. However one can always add fictitious extra observation times after T so that we are in the framework described here.

2) We consider a d-dimensional process $X = (X_t)_{t\geq 0}$ defined on the space $(\Omega, \mathcal{F}, (\mathcal{F}_t)_{t\geq 0}, \mathbb{P})$, with components X^i for $i = 1, \ldots, d$. Suppose that a random discretization scheme $\mathcal{T} = (\mathcal{T}_n)$ is given, with $\mathcal{T}_n = (T(n,i) : i \geq 0)$. We will use the following notation:

$$\Delta_i^n X = X_{T(n,i)} - X_{T(n,i-1)}.$$

Note that this notation is relative to the discretization scheme $\mathcal{T}$, although this does not show explicitly. We will study various sums of functions of the above increments, with or without normalization. The most basic one, called *non-normalized functional*, is as follows. We have an arbitrary function f on $\mathbb{R}^d$ (it may be real-valued, or $\mathbb{R}^q$-valued, in which case the following should be read component by component):

$$V^n(f,X)_t = V(\mathcal{T}_n; f,X)_t = \sum_{i=1}^{N_n(t)} f(\Delta_i^n X). \tag{3.1.3}$$

The second notation emphasizes the dependency on the discretization grid $\mathcal{T}_n$.

By construction the process $V^n(f,X)$ is piecewise constant, and in some applications it may be useful to consider an "interpolated" version, which we define as follows:

$$V^n_{int}(f,X)_t = V^n(f,X)_t + f(X_t - X_{T_n(t)}) = \sum_{i\geq 1} f(X_{t\wedge T(n,i)} - X_{t\wedge T(n,i-1)}). \tag{3.1.4}$$

We have $V^n_{int}(f,X)_{T(n,i)} = V^n(f,X)_{T(n,i)}$, but unlike $V^n(f,X)$ the process $V^n_{int}(f,X)$ varies on each interval $I(n,i)$ along the path of X transformed by f.

3) We have already encountered the functionals $V^n(f,X)$ in two special cases. The first case is when $f(x)=x$, the identity map on $\mathbb{R}^d$ for which we have

$$V^n(f,X)_t = X_{T_n(t)} - X_0, \qquad V^n_{int}(f,X)_t = X_t - X_0.$$

Then, up to the initial value X_0, the process $V^n(f,X)$ is the *discretized version* of the process X. The discretization $y^{(n)}(t) = y(t(n,i))$ of a given càdlàg function y along a (non-random) discretization scheme $(t(n,i): i\geq 0)$ always converges to y in the Skorokhod sense as the mesh of the scheme goes to 0 in the sense of (3.1.1): this is a special case of (2.2.13). We thus deduce by the subsequence principle that

$$V^n(f,X) \stackrel{\mathbb{P}}{\Longrightarrow} X - X_0, \quad \text{for } f(x)=x \tag{3.1.5}$$

(recall that this is convergence in probability, for the Skorokhod topology). When the convergence (3.1.1) holds for all ω instead of in probability, for example when the $T(n,i)=t(n,i)$ are non-random, we get of course $V^n(f,X)(\omega) \stackrel{\text{Sk}}{\longrightarrow} X(\omega) - X_0(\omega)$ for each ω.

Another particular case is when $f(x) = x^j x^k$ (the product of two components of $x\in\mathbb{R}^d$). Then, provided X is a semimartingale, we can rewrite (2.1.8) as follows:

$$V^n_{int}(f,X)_t \stackrel{\mathbb{P}}{\longrightarrow} \left[X^j, X^k\right]_t.$$

This is wrong in general for $V^n(f,X)_t$, unless t is not a fixed time of discontinuity of X, that is $\mathbb{P}(\Delta X_t = 0) = 1$. This fact is a reason to consider the interpolated functionals, although in practical applications we only know, or can use, the variables $X_{T(n,i)}$, the other X_t's being unavailable. More generally, if $f(x) = x^j x^k$, later we will see that $V^n_{int}(f,X) \stackrel{\text{u.c.p.}}{\Longrightarrow} [X^j, X^k]$ (convergence in probability, locally uniformly in time) and $V^n(f,X) \stackrel{\mathbb{P}}{\Longrightarrow} [X^j,X^k]$, the latter implying $V^n(f,X)_t \stackrel{\mathbb{P}}{\longrightarrow} [X^j,X^k]_t$ only when $\mathbb{P}(\Delta X_t = 0) = 1$.

4) Another case is of special interest. Suppose that $d=1$. Then with $f(x)=|x|$ we have

$$V^n(f,X)_t = \sum_{i=1}^{N_n(t)} |\Delta_i^n X|$$
$$V_{int}^n(f,X)_t = \sum_{i\geq 1} |X_{t\wedge T(n,i)} - X_{t\wedge T(n,i-1)}|,$$

which are two versions of the "approximate total variation" at time t and stage n (or, for the grid $\mathcal{T}_n$) of X. In general these variables explode as $n\to\infty$. However, if X is of finite variation (meaning, with paths of finite variation over finite intervals), and if $\mathrm{Var}(X)_t$ denotes the total variation of the path $s\mapsto X_s$ over $(0,t]$, then both $V^n(f,X)_t$ and $V_{int}^n(f,X)_t$ are smaller than $\mathrm{Var}(X)_t$ (note that $\mathrm{Var}(X)=\mathrm{Var}(X-X_0)$ here). We have more, namely the following result, whose proof is given in the Appendix (although it is, except for the Skorokhod convergence, an old result, see e.g. Grosswald [42]):

Proposition 3.1.2 *Suppose that the one-dimensional process X is of finite variation, and let $f(x)=|x|$. Then for any random discretization scheme we have*:

$$\begin{aligned} &V^n(f,X) \overset{\mathbb{P}}{\Longrightarrow} \mathrm{Var}(X) \\ &V^n(f,X)_t - \mathrm{Var}(X)_{T_n(t)} \overset{\text{u.c.p.}}{\Longrightarrow} 0 \\ &V_{int}^n(f,X) \overset{\text{u.c.p.}}{\Longrightarrow} \mathrm{Var}(X). \end{aligned} \tag{3.1.6}$$

This result is stated for convergence in probability, because (3.1.1) holds in probability. But this is really a "pathwise" result, that is the convergence holds for any ω for which the left side of (3.1.1) goes to 0. The property (3.1.6) emphasizes the differences between $V^n(f,X)$ and $V_{int}^n(f,X)$. We do *not* have $V^n(f,X)\overset{\text{u.c.p.}}{\Longrightarrow}\mathrm{Var}(X)$ in general, unless X is continuous: indeed if X jumps at time S then $\mathrm{Var}(X)$ has the jump $|\Delta X_S|$ at time S, and $V^n(f,X)$ has a jump of approximate size $|\Delta X_S|$ at time $T_n(S)=\inf(T(n,i): i\geq 1, T(n,i)\geq S)$, and $T_n(S)$ converges to S but in general is not equal to S.

3.2 Semimartingales with p-Summable Jumps

In this section X is a d-dimensional semimartingale on $(\Omega,\mathcal{F},(\mathcal{F}_t)_{t\geq 0},\mathbb{P})$, and we denote by μ its jump measure and by (B,C,ν) its characteristics. The processes

$$\Sigma(p,X)_t \;=\; \sum_{s\leq t}\|\Delta X_s\|^p \;=\; \|x\|^p \star \mu_t \tag{3.2.1}$$

will play a central role. Here, $p\geq 0$, and we use the convention $0^0=0$: hence $\Sigma(0,X)_t=\sum_{s\leq t} 1_{\{\Delta X_s\neq 0\}}$ is simply the number of jumps of X before time t.

Each process $\Sigma(X,p)$ is well-defined, with values in $[0,\infty]$. It is adapted, null at 0, has left limits everywhere and is right-continuous on $[0,T)\cup(T,\infty)$, where $T=\inf(t:\Sigma(p,X)_t=\infty)$. Then we associate with X the following subset of $[0,\infty)$:

$$\mathcal{I}(X) = \big\{p\geq 0:\ \text{the process } \Sigma(p,X) \text{ is a.s. finite-valued}\big\}. \tag{3.2.2}$$

We will also say that X has *p-summable jumps* if $p\in\mathcal{I}(X)$.

Lemma 3.2.1 a) *We have $2\in\mathcal{I}(X)$ always, and $\mathcal{I}(X)$ is an interval of the form (p_0,∞) or $[p_0,\infty)$, for some $p_0\in[0,2]$.*

b) *A real $p<2$ belongs to $\mathcal{I}(X)$ if and only if the process $(\|x\|^p\wedge 1)\star\nu$ is a.s. finite-valued.*

In particular when X is a Lévy process, with Lévy measure F, then $V(p,X)_t<\infty$ a.s. for all t if and only if $\int(\|x\|^p\wedge 1)F(dx)<\infty$ (once more, this is always true when $p\geq 2$), and otherwise it can be shown (see the Appendix) that $V(p,X)_t=\infty$ a.s. for all $t>0$: this is an old result of Blumenthal and Getoor in [19], and the infimum p_0 of the set $\mathcal{I}(X)$ in the Lévy case is called the *Blumenthal-Getoor* index of the process. When X is a stable process, the Blumenthal-Getoor index is the stability index.

Coming back to semimartingales, when $1\in\mathcal{I}(X)$ the jumps are summable (on finite time intervals, of course) and the process $(\|x\|\wedge 1)\star\nu$ is a.s. finite-valued. Then we may rewrite the Lévy-Itô decomposition (2.1.18) as

$$X = X_0 + B' + X^c + x\star\mu, \quad \text{where } B' = B - (x1_{\{\|x\|\leq 1\}})\star\nu. \tag{3.2.3}$$

Proof a) The property $2\in\mathcal{I}(X)$ follows from (2.1.5). If $p\geq q$ we have $\|x\|^p\leq\|x\|^q+\|x\|^p 1_{\{\|x\|>1\}}$, hence

$$\Sigma(p,X)_t \leq \Sigma(q,X)_t + \sum_{s\leq t}\|\Delta X_s\|^p 1_{\{\|\Delta X_s\|>1\}}.$$

The last sum above is a finite sum, so $q\in\mathcal{I}(X)$ implies $p\in\mathcal{I}(X)$. This proves (a).

b) Let $p<2$. Exactly as above, we see that $\Sigma(p,X)_t<\infty$ if and only if $A_t:=(\|x\|^p 1_{\{\|x\|\leq 1\}})\star\mu_t<\infty$. Set also $A'=(\|x\|^p 1_{\{\|x\|\leq 1\}})\star\nu$. The processes A and A' are increasing with bounded jumps, so they are finite-valued if and only if they are locally integrable. Since for any stopping time T we have $\mathbb{E}(A_T)=\mathbb{E}(A'_T)$, the claim follows. □

We end this section with an extension of Itô's formula to a wider class of test functions f, when we have $p\in\mathcal{I}(X)$ for some $p<2$ and $X^c=0$.

The usual Itô's formula (2.1.9) requires f to be C^2, and indeed the second derivatives $\partial^2_{ij}f$ appear in the right side. The C^2 property can be somewhat relaxed, and replaced by the fact that f is convex (or the difference of two convex functions), and there is still an Itô's formula, but it involves the local times of X and is then called Itô-Tanaka Formula or Itô-Meyer Formula.

However, when $X^c=0$, or equivalently when the second characteristic C is identically 0, then the second derivatives do not show up any more, and one may expect a similar formula with less smoothness on f, provided of course the last term involving the jumps converges. This is what we will do now: when $X^c=0$ we extend Itô's formula for functions f that are C^p for $p<2$, as soon as $p\in\mathcal{I}(X)$: when $p>0$ is not an integer, saying that a function is C^p means that it is $[p]$ times differentiable, and its $[p]$th partial derivatives are Hölder with index $p-[p]$ on every compact set. In particular when $0<p<1$, C^p is the same as "locally Hölder with index p".

Theorem 3.2.2 *Assume that $X^c=0$ and that $p\in\mathcal{I}(X)$. Then*

a) *If $1\le p<2$ and if f is a C^p function (in the sense above), the process $f(X)$ is a semimartingale satisfying*

$$f(X_t)=f(X_0)+\sum_{i=1}^d\int_0^t\partial_i f(X_{s-})\,dX^i_s$$
$$+\sum_{s\le t}\left(f(X_{s-}+\Delta X_s)-f(X_{s-})-\sum_{i=1}^d\partial_i f(X_{s-})\Delta X^i_s\right). \tag{3.2.4}$$

b) *If $0\le p\le 1$ and if further the process B' of* (3.2.3) *vanishes, then for any function f which is locally Hölder with index p when $p>0$, and which is measurable when $p=0$, the process $f(X)$ is a semimartingale satisfying*

$$f(X_t)\;=\;f(X_0)+\sum_{s\le t}\big(f(X_{s-}+\Delta X_s)-f(X_{s-})\big). \tag{3.2.5}$$

Before proceeding to the proof, we give some comments. If f is C^p, the sth summand in either one of these two formulas is of order of magnitude $\|\Delta X_s\|^p$, or more precisely there is a locally bounded process (H_t) such that the sth summand is smaller than $H_s\|\Delta X_s\|^p$ as soon as $\|\Delta X_s\|\le 1$ (see the proof below). So if $p\in\mathcal{I}(X)$, the hypothesis that f is C^p is exactly what we need for the series in (3.2.4) and in (3.2.5) to be absolutely convergent.

This extended Itô's formula is *not* the same as the Itô-Tanaka formula for convex functions. Note that (3.2.5) when $p=0$ is trivial, and added here for completeness (under the assumptions of (b) we simply have $X_t=X_0+\sum_{s\le t}\Delta X_s$, and when $0\in\mathcal{I}(S)$ this sum is a finite sum).

In (b) with $p=1$, the function f needs to be locally Lipschitz, a weaker assumption than in (a) with $p=1$, but the process X is more restricted. Both formulas (3.2.4) and (3.2.5) are the same when $p=1$ and the assumptions of (a) and (b) are all satisfied.

Proof (a) We assume that $1\le p<2$ and that f is C^p. We associate with f the function $g_f(x,y)=f(x+y)-f(x)-\sum_{i=1}^d\partial_i f(x)y^i$. For each integer $q\ge 1$ there is a constant Γ_q such that $|\partial_i f(x)|\le\Gamma_q$ and (by the Hölder property of index $p-1$ for the derivatives) $|\partial_i f(x+y)-\partial_i fi(x)|\le\Gamma_q\|y\|^{p-1}$ whenever $\|x\|,\|y\|\le q$.

There is a continuous increasing function h on $\mathbb{R}_+$ satisfying $h(q-1) \geq \Gamma_q d$ for all $q \geq 1$, and we deduce that for all $x, y \in \mathbb{R}^d$:

$$\left|\partial_i f(x)\right| \leq h\left(\|x\|\right), \qquad \|y\| \leq 1 \Rightarrow \left|g_f(x,y)\right| \leq h\left(\|x\|\right)\|y\|^p. \tag{3.2.6}$$

Now, the summand in the last sum of (3.2.4), at any time s such that $\|\Delta X_s\| \leq 1$, is smaller than $h(\|X_{s-}\|)\|\Delta X_s\|^p$. Since $p \in \mathcal{I}(X)$, this implies that the last term in (3.2.4) is an absolutely convergent sum.

Denote by f_n the convolution of f with a C^∞ nonnegative function ϕ_n on $\mathbb{R}^d$, with support in $\{x : \|x\| \leq 1/n\}$ and $\int \phi_n(x)\,dx = 1$. We deduce from (3.2.6) that

$$\begin{gathered} f_n \to f, \qquad \partial_i f_n \to \partial_i f, \qquad g_{f_n} \to g_f \\ \left|\partial_i f_n(x)\right| \leq h\left(\|x\|+1\right), \qquad \|y\| \leq 1 \Rightarrow \left|g_{f_n}(x,y)\right| \leq h\left(\|x\|+1\right)\|y\|^p. \end{gathered} \tag{3.2.7}$$

Each f_n is C^∞, so the usual Itô's formula and $X^c = 0$ yield

$$f_n(X_t) - f_n(X_0) = \sum_{i=1}^d \int_0^t \partial_i f_n(X_{s-})\,dX_s^i + \int_0^t \int_{\mathbb{R}^d} g_{f_n}(X_{s-}, y)\,\mu(ds, dy). \tag{3.2.8}$$

By (3.2.7) the left side of (3.2.8) converges to $f(X_t) - f(X_0)$, and the dominated convergence theorem for stochastic integrals yields $\int_0^t \partial_i f_n(X_{s-})\,dX_s^i \xrightarrow{\mathbb{P}} \int_0^t \partial_i f(X_{s-})\,dX_s^i$. Furthermore, since $\Sigma(p,X)_t < \infty$ for all t, the convergence $g_{f_n} \to g_f$ and the last estimate in (3.2.7) yield, together with the (ordinary) dominated convergence theorem, that $\int_0^t \int_{\mathbb{R}^d} g_{f_n}(X_{s-}, y)\,\mu(ds, dy) \to \int_0^t \int_{\mathbb{R}^d} g_f \times (X_{s-}, y)\,\mu(ds, dy)$. Then we deduce (3.2.4) from (3.2.8), and the semimartingale property of $f(X)$ follows from (3.2.4).

(b) Since the case $p = 0$ is trivial, we assume that $0 < p \leq 1$ and that f is locally Hölder with index p, and also that $p \in \mathcal{I}(X)$ and $X_t = X_0 + \sum_{s \leq t} \Delta X_s$. As in (a), there is a continuous increasing function h on $\mathbb{R}_+$ such that $|f(x+y) - f(x)| \leq h(\|x\|)|y|^{p-1}$ if $\|y\| \leq 1$, so the right side of (3.2.5) is an absolutely convergent sum. Both sides of (3.2.5) are processes which are sums of their jumps, they have the same jumps, and the same initial value, so they are equal. Finally, the semimartingale property is again obvious. □

3.3 Law of Large Numbers Without Normalization

3.3.1 The Results

Here again X is a d-dimensional semimartingale on $(\Omega, \mathcal{F}, (\mathcal{F}_t)_{t\geq 0}, \mathbb{P})$, with jump measure μ and characteristics (B, C, ν). We are also given a random discretization scheme $\mathcal{T} = (\mathcal{T}_n)$ with $\mathcal{T}_n = (T(n,i) : i \geq 0)$. Apart from being subject to (3.1.1), this scheme is totally arbitrary.

Our aim is to prove that $V^n(f,X)$ converges in probability, for the Skorokhod topology. When $f=(f^i)_{i\geq q}$ is q-dimensional the Skorokhod-convergence of each $V^n(f^i,X)$ does *not* imply the Skorokhod-convergence of $V^n(f,X)$, so below we state the multidimensional result.

We obviously need some conditions on f. Typically, most increments $\Delta_i^n X$ are (very) small, so the behavior of f near the origin plays a crucial role. On the other hand, at least when X has jumps, some $\Delta_i^n X$ are big, hence the behavior of f outside the origin cannot be totally ignored: this explains the continuity assumption made below.

The results are rather diverse, and far from covering all possible behaviors of f near 0, especially in the multidimensional case $d\geq 2$, but they at least completely describe what happens for the functions $f(x)=|x|^p$ for all $p>0$ when $d=1$. For a better understanding of (C) below, remember that when $1\in\mathcal{I}(X)$ and $X^c=0$, we have

$$X_t = X_0 + B'_t + \sum_{s\leq t} \Delta X_s \tag{3.3.1}$$

where B' is the "genuine" drift, which is a continuous process of locally finite variation.

Theorem 3.3.1 *Let X be a d-dimensional semimartingale and $\mathcal{T}=(\mathcal{T}_n)$ be any random discretization scheme. Let also f be a continuous function from $\mathbb{R}^d$ into $\mathbb{R}^q$.*
A) *Under either one of the following four conditions on f and X:*

(A-a) $f(x)=\mathrm{o}(\|x\|^2)$ *as* $x\to 0$,
(A-b) $X^c=0$ *and there is a* $p\in\mathcal{I}(X)\cap(1,2]$ *such that* $f(x)=\mathrm{O}(\|x\|^p)$ *as* $x\to 0$,
(A-c) $X^c=0$ *and* $1\in\mathcal{I}(X)$ *and* $f(x)=\mathrm{o}(\|x\|)$ *as* $x\to 0$,
(A-d) $X^c=0$ *and there is a* $p\in\mathcal{I}(X)\cap[0,1]$ *such that* $f(x)=\mathrm{O}(\|x\|^p)$ *as* $x\to 0$, *and further* $B'=0$ *in* (3.3.1),

we have the following Skorokhod convergence in probability:

$$V^n(f,X) \stackrel{\mathbb{P}}{\Longrightarrow} V(f,X) := f\star\mu. \tag{3.3.2}$$

B) *If* $f(x)=\sum_{i,j=1}^d \alpha_{ij}x^ix^j+o(\|x\|^2)$ *as* $x\to 0$, *for some* $\alpha_{ij}\in\mathbb{R}^q$, *then*

$$V^n(f,X) \stackrel{\mathbb{P}}{\Longrightarrow} V(f,X) := \sum_{i,j=1}^d \alpha_{ij}C^{ij} + f\star\mu. \tag{3.3.3}$$

C) *Assume* $1\in\mathcal{I}(X)$ *and* $X^c=0$. *If* $f(x)=\sum_{i=1}^d \alpha_i|x^i|+\mathrm{o}(\|x\|)$ *as* $x\to 0$, *for some* $\alpha_i\in\mathbb{R}^q$, *then*

$$V^n(f,X) \stackrel{\mathbb{P}}{\Longrightarrow} V(f,X) := \sum_{i=1}^d \alpha_i \operatorname{Var}\bigl(B'^i\bigr) + f\star\mu. \tag{3.3.4}$$

D) *In all these cases, we also have*

$$V^n_{int}(f,X) \overset{\text{u.c.p.}}{\Longrightarrow} V(f,X). \tag{3.3.5}$$

E) *Suppose that* $f(x) = \|x\|^p$ *on a neighborhood of the origin. Then there exists a time* $t > 0$ *such that the sequence of variables* $V^n(f,X)_t$ *is not bounded in probability, as soon as one of the following (non-exclusive) conditions is satisfied*:

(E-a) $p < 2$ *and* X^c *is not identically* 0,
(E-b) $p \notin \mathcal{I}(X)$,
(E-c) $p < 1 \in \mathcal{I}(X)$ *and* B' *is not identically* 0.

When further X is a Lévy process, we even have $V^n(f,X)_t \xrightarrow{\mathbb{P}} +\infty$ *for all* $t > 0$ *under either one of* (E-a), (E-b) *or* (E-c) *above.*

The limit $V(f,X)$ in this theorem is defined unambiguously: when f satisfies two different assumptions at once, the process $V(f,X)$ is the same in the two convergence statements.

In (B) the condition on f is equivalent to saying that it is twice differentiable at 0, with $f(0) = 0$ and $\partial_i f(0) = 0$. In (C-a) the condition on f is equivalent to saying that it is once differentiable at 0, with $f(0) = 0$. Note that (E) shows—for a very special type of test functions f—that the hypotheses on the process X in (A,B,C) are necessary as well as sufficient to have convergence. For a general f, however, no necessary conditions are available.

Remark 3.3.2 The key result is (3.3.5), which by virtue of (2.2.13) and the subsequences principle immediately implies (3.3.2)–(3.3.4). It also implies the following: for a given finite (random) time T, we have $V^n(f,X)_T \xrightarrow{\mathbb{P}} V(f,X)_T$ in two non-exclusive cases:

(i) we have $\mathbb{P}(\Delta X_T \neq 0) = 0$;
(ii) for each n large enough the time T belongs to the discretization scheme, in the sense that $\cup_{i\geq 1}\{T(n,i) = T\} = \Omega$ almost surely.

Otherwise, $V^n(f,X)_T \xrightarrow{\mathbb{P}} V(f,X)_T$ may fail, even when $T = t$ is not random: for example if $X_t = 1_{[s,\infty)}(t)$ is non-random and $T(n,i) = i/n$ and $f(x) = |x|^p$, then $V^n(f,X)_s = 0$ does not converge to $V(f,X)_s = 1$ if s is irrational.

Remark 3.3.3 When $d = 1$ and $f(x) = |x|^p$ for $p > 1$, the convergence $V^n(f,X)_t \xrightarrow{\mathbb{P}} V(f,X)_t$ when t belongs to the (non-random) discretization scheme goes back to Lépingle [71], and the case $p = 2$ is simply the convergence (2.1.8). Indeed, Lépingle proved the almost sure convergence in this setting, when $p > 2$ and also when $p > 1$ and $X^c = 0$, provided of course (3.1.1) holds almost surely. If the latter hypothesis holds, we have in fact the almost sure convergence in all cases of (A) and (C) (but *not* in (B)).

In the setting of the previous remark, Lépingle also shows that (3.3.2) holds not only when $X^c = 0$ and $p \in \mathcal{I}(X)$ but also, without these "global" assumptions, in restriction to the set on which $X^c_s = 0$ for all $s \le t$ and $\Sigma(p, X)_t < \infty$. This property is of importance for applications, so we state a result in this direction. We improve on (A-b,d) only, but (A-c) and (C) could be improved accordingly:

Corollary 3.3.4 *In the setting of Theorem* 3.3.1, *let T be a finite stopping time with $\mathbb{P}(\Delta X_T \neq 0) = 0$ and f be a continuous function on $\mathbb{R}^d$ with $f(x) = \mathrm{O}(\|x\|^p)$ as $x \to 0$, for some $p \in [0, 2)$. Then the stopped processes $V^n(f, X)_{t\wedge T}$ converge in probability, for the Skorokhod topology, to the stopped process $V(f, X)_{t\wedge T}$, in restriction to the set*

$$\Omega_T = \begin{cases} \{C_T = 0\} \cap \{\Sigma(p, X)_T < \infty\} & \text{if } p > 1 \\ \{C_T = 0\} \cap \{\Sigma(p, X)_T < \infty\} \cap \{B'_s = 0 \ \forall s \le T\} & \text{if } p \le 1. \end{cases}$$

There is also an improvement of the theorem in another direction, namely when one (partially) drops out the continuity of the test function. This is quite useful, because in some applications one needs to use a test function f of the form $f = g1_A$, where A is a Borel subset of $\mathbb{R}^d$ and g satisfies the conditions of the theorem.

Theorem 3.3.5 *Statements* (A, B, C, D) *of Theorem* 3.3.1 *hold if we replace the continuity of the test function f by the fact that it is Borel and that either one of the following two equivalent conditions holds, with D_f denoting the set of all $x \in \mathbb{R}^d$ at which f is not continuous*:

(i) $\mathbb{P}(\exists t > 0 : \Delta X_t \in D_f) = 0$;
(ii) $1_{D_f} * \nu_\infty = 0$ *almost surely.*

Accordingly, the statement of Corollary 3.3.4 *holds when f is not continuous, in restriction to the set $\Omega_T \cap \{1_{D_f} * \nu_T = 0\}$.*

Remark 3.3.6 When $q = d$ and $f(x) = x$ we have (3.1.5). As a matter of fact, if we set $g(x) = x$ and take f as in the above theorem, we could prove that the pairs $(V^n(f, X), V^n(g, X))$ converge in probability to $(V(f, X), X - X_0)$ for the Skorokhod topology on $\mathbb{D}^{q+d}$, and the proof would be essentially the same. As a consequence, if A is any $q \times d$ matrix, we deduce that $V^n(f + Ag, X) \stackrel{\mathbb{P}}{\Longrightarrow} V(f) + A(X - X_0)$, and this allows one to get limits for functions satisfying weaker hypotheses. For example we may deduce from (C) that if $1 \in \mathcal{I}(X)$ and $X^c = 0$ and $f(x) = \sum_{i=1}^d \alpha_i x^i + \mathrm{o}(\|x\|)$ as $x \to 0$ for some $\alpha_i \in \mathbb{R}^q$, then

$$V^n(f, X) \stackrel{\mathbb{P}}{\Longrightarrow} V(f, X) := \sum_{i=1}^d \alpha_i B'^i + f \star \mu.$$

The same comments apply for the interpolated functionals $V^n_{int}(f, X)$. These extensions, which do not seem to be particularly useful in practice, are left to the reader.

3.3.2 The Proofs

Proving Theorems 3.3.1 and 3.3.5 requires first to show the convergence in a variety of special cases, which will be studied in several lemmas. Before proceeding, we observe that it is no restriction to suppose that our discretization scheme satisfies

$$\forall t > 0, \quad \sup_{i \geq 1} \big(T(n,i) \wedge t - T(n,i-1) \wedge t\big) \to 0 \tag{3.3.6}$$

for all $\omega \in \Omega$, instead of a mere convergence in probability as in (3.1.1). Indeed, the left side of (3.1.1) is non-decreasing in t, and using (2.2.18), we see that from any subsequence of integers one may extract a further subsequence along which (3.1.1) holds, simultaneously for all t, pointwise in ω outside a null set. Then, upon using the subsequences principle (2.2.17) and upon discarding a null set, it is enough to prove the results when (3.3.6) holds identically.

Omitting the mention of X, we write

$$W^n(f) = V^n_{int}(f,X) - V(f,X). \tag{3.3.7}$$

The key step of the proof consists in showing that, under the appropriate conditions on f, we have

$$W^n(f) \overset{\text{u.c.p.}}{\Longrightarrow} 0. \tag{3.3.8}$$

This is (D), which implies (A), (B), (C), as said before. The proof of (E) will be given at the end. As for the proof of (3.3.8), it is carried out under two different sets of assumptions on the test function f, and then the two cases are pasted together:

Case 1: The function f vanishes on a neighborhood of 0 and satisfies

$$\text{outside a } \mathbb{P} \text{ null set, } f \text{ is continuous at each point } \Delta X_t(\omega). \tag{3.3.9}$$

Case 2: The function f is of the class C^p, as described before Theorem 3.2.2, for some $p > 0$; furthermore $f(x) = 0$ if $\|x\| > 1$, and f has an appropriate behavior at 0, depending on the properties of X.

Lemma 3.3.7 *In Case* 1 *above, $W^n(f)_t \to 0$ locally uniformly in t, and for each ω.*

Proof Let $S_1, S_2, \ldots$ be the successive jump times of X corresponding to jumps with norm bigger than $\varepsilon/2$, where $\varepsilon > 0$ is such that $f(x) = 0$ when $\|x\| \leq \varepsilon$. We have $S_q \to \infty$ as $q \to \infty$. Set $X' = X - (x 1_{\{\|x\| > \varepsilon/2\}}) \star \mu$. By construction $\|\Delta X'\| \leq \varepsilon/2$, hence for any $T > 0$ we have

$$\limsup_{\theta \to 0} \sup_{0 \leq t < s \leq T,\ s-t \leq \theta} \big\|X'_t(\omega) - X'_s(\omega)\big\| \leq \frac{\varepsilon}{2}.$$

for all ω. In view of (3.3.6) it follows that for all $T>0$ and $\omega\in\Omega$ there is an integer $M_T(\omega)$ with the following properties: if $n\geq M_T$, for all $i\leq N_n(T)+1$, we have

- either the interval $I(n,i)$ contains no S_q (recall (3.1.2)), and
$$s\in I(n,i)\ \Rightarrow\ \|X_s-X_{T(n,i-1)}\|\leq\varepsilon$$
- or $I(n,i)$ contains exactly one S_q, and we write $i=i(n,q)$, and
$$s\in I(n,i)\ \Rightarrow\ \|X_s-X_{T(n,i-1)}-\Delta X_{S_q}\,1_{\{S_q\leq s\}}\|\leq\varepsilon. \tag{3.3.10}$$

Therefore, with Q_t denoting the number of q such that $S_q\leq t$, and since $f(x)=0$ when $\|x\|\leq\varepsilon$, when $n\geq M_T$ and $t\leq T$ we have

$$W^n(f)_t=\sum_{q=1}^{Q_T}\big(f(X_{t\wedge T(n,i(n,q))}-X_{T(n,i(n,q)-1)})-f(\Delta X_{S_q})\big).$$

Now, on the set $\{S_q\leq t\}$ we have $X_{t\wedge T(n,i(n,q))}-X_{T(n,i(n,q)-1)}\to\Delta X_{S_q}$, because X is càdlàg. Since f satisfies (3.3.9) and Q_T is finite, we deduce $\sup_{t\leq T}\|W^n(f)\|\to 0$. □

For Case (2) above, we single out the three possible situations: we may have no restriction on X (so $p=2$ and X^c is arbitrary), or we may have $p\in(1,2]$ and $p\in\mathcal{I}(X)$ and $X^c=0$, or we may have $p\in[0,1]$ and $p\in\mathcal{I}(X)$ and $X^c=0$ and $B'=0$.

Lemma 3.3.8 *Let f be a C^2 function from $\mathbb{R}^d$ into $\mathbb{R}^q$ with $f(0)=0$ and $\partial_i f(0)=0$ for all i, and also $f(x)=0$ if $\|x\|\geq 1$. Then (3.3.8) holds.*

Proof Our assumptions imply that f satisfies (B) of Theorem 3.3.1 with $\alpha_{ij}=\frac{1}{2}\,\partial^2_{ij}f(0)$. Therefore in the definition (3.3.7) of $W^n(f)$ we have $V(f,X)=\frac{1}{2}\sum_{i,j=1}^d\partial^2_{ij}f(0)C^{ij}+f*\mu$, where $f*\mu$ is well defined because $\|f(x)\|\leq K(\|x\|^2\wedge 1)$.

For each n we introduce the following adapted left-continuous process Y^n:

$$Y^n_0=0,\qquad Y^n_t=X_{t-}-X_{T(n,i-1)}\quad\text{if } T(n,i-1)<t\leq T(n,i).$$

We associate with f three functions: $\mathbb{R}^d\times\mathbb{R}^d\to\mathbb{R}^q$ by setting

$$\begin{aligned}
k(y,x)&=f(y+x)-f(y)-f(x)\\
g(y,x)&=f(y+x)-f(y)-f(x)-\Big(\sum_{i=1}^d\partial_i f(y)x^i\Big)1_{\{\|x\|\leq 1\}}\\
h(y,x)&=f(y+x)-f(y)-f(x)-\sum_{i=1}^d\partial_i f(y)x^i.
\end{aligned}$$

The following estimates are easy:

$$\|k(y,x)\| \le K(\|x\| \wedge 1), \qquad \|g(y,x)\| \le K(\|x\|^2 \wedge 1). \tag{3.3.11}$$

Recall that $V^n_{int}(f,X)_t$ is the sum of all $f(X_{t\wedge T(n,i)} - X_{T(n,i-1)})$ for $i \le N_n(t)+1$. Since f is C^2 we can evaluate each of these summands by applying Itô's formula for the process $X_{T(n,i-1)+s} - X_{T(n,i-1)}$. By subtracting $V(f,X)_{t\wedge T(n,i)} - V(f,X)_{t\wedge T(n,i-1)}$ to the ith summand, we obtain

$$\begin{aligned} W^n(f)_t &= \sum_{j=1}^d \int_0^t \partial_j f(Y^n_s)\, dX^j_s \\ &\quad + \frac{1}{2}\sum_{j,k=1}^d \int_0^t \big(\partial^2_{jk} f(Y^n_s) - \partial^2_{jk} f(0)\big)\, dC^{jk}_s + h(Y^n,x) \star \mu_t. \end{aligned} \tag{3.3.12}$$

By (3.3.11) the process $g(Y^n,x)\star\mu$ has locally integrable variation, with compensator $g(Y^n,x)\star\nu$. Then if we use the Lévy-Itô decomposition (2.1.18) for X, we deduce that $W^n(f) = A^n + M^n$, where

$$\begin{aligned} A^n_t &= \sum_{j=1}^d \int_0^t \partial_j f(Y^n_s)\, dB^j_s \\ &\quad + \frac{1}{2}\sum_{j,k=1}^d \int_0^t \big(\partial^2_{jk} f(Y^n_s) - \partial^2_{jk} f(0)\big)\, dC^{jk}_s + g(Y^n,x)\star\nu_t, \\ M^n_t &= \sum_{j=1}^d \int_0^t \partial_j f(Y^n_s)\, dX^{jc}_s + k(Y^n,x)\star(\mu-\nu)_t. \end{aligned}$$

Now we observe that, outside a null set, we have $Y^n_s \to 0$ for all s (we use (3.3.6) once more here). In view of the assumptions on f, we deduce that outside a null set again we have $\partial_j f(Y^n_s) \to 0$ and $\partial^2_{jk} f(Y^n_s) - \partial^2_{jk} f(0) \to 0$ and $g(Y^n_s,x) \to 0$ and $k(Y^n_s,x) \to 0$ for all s and x, whereas (3.3.11) holds and $\partial_j f$ and $\partial^2_{jk} f$ are bounded. Then the Lebesgue dominated convergence theorem (for ordinary and stochastic integrals) and (2.2.25) yield $W^n(f) \stackrel{\text{u.c.p.}}{\Longrightarrow} 0$. □

Lemma 3.3.9 *Assume that* $p \in [1,2] \cap \mathcal{I}(X)$ *and* $X^c = 0$. *Let* f *be a* C^p *function from* $\mathbb{R}^d$ *into* $\mathbb{R}^q$ *with* $f(0) = 0$ *and* $\partial_i f(0) = 0$ *for all* i, *and also* $f(x) = 0$ *when* $\|x\| \ge 1$. *Then* (3.3.8) *holds.*

Proof We use the notation of the previous proof. Instead of (3.3.11) we have

$$\|k(y,x)\| \le K(\|x\| \wedge 1), \qquad \|g(y,x)\| \le K(\|x\|^p \wedge 1). \tag{3.3.13}$$

The process $f\star\mu$ is well defined because $\|f(x)\| \le K(\|x\|^p \wedge 1)$ in this case.

The extension (3.2.4) of Itô's formula yields, instead of (3.3.12),

$$W^n(f)_t = \sum_{j=1}^{d} \int_0^t \partial_j f(Y_s^n)\, dX_s^j + h(Y^n, x) \star \mu_t.$$

By (3.3.13) the process $g(Y^n, x) \star \mu$ is of locally integrable variation, with compensator $g(Y^n, x) \star \nu$. Then as in the previous proof we deduce that $W^n(f) = A^n + M^n$, where

$$A_t^n = \sum_{j=1}^{d} \int_0^t \partial_j f(Y_s^n)\, dB_s^j + g(Y^n, x) \star \nu_t, \quad M_t^n = k(Y^n, x) \star (\mu - \nu)_t. \tag{3.3.14}$$

We end the proof exactly as in the previous lemma. Namely, $Y_s^n \to 0$ for all s, hence $\partial_j f(Y_s^n) \to 0$. Moreover $g(Y_s^n, x) \to 0$ and $k(Y_s^n, x) \to 0$, hence in view of (3.3.13) and of $(\|x\|^p \wedge 1) \star \nu_t < \infty$ (see Lemma 3.2.1) we have $\|g(Y^n, x)\| \star \nu_t \to 0$ and $k(Y^n, x)^2 \star \nu_t \to$ a.s. Then (3.3.14) and (2.2.25) again yield $W^n(f) \overset{\text{u.c.p.}}{\Longrightarrow} 0$. □

Lemma 3.3.10 *Assume that* $1 \in \mathcal{I}(X)$ *and* $X^c = 0$. *Let* $f(x) = \sum_{j=1}^d \alpha_j |x^j|$ *for some* $\alpha_j \in \mathbb{R}^q$. *Then* (3.3.8) *holds.*

Proof The assumptions imply that each component X^j is of finite variation, see (3.3.1), so if $h(y) = |y|$, Proposition 3.1.2 yields $W^n(h, X^j) := V^n(h, X^j) - \operatorname{Var}(X^j) \overset{\text{u.c.p.}}{\Longrightarrow} 0$. Since $f(x) = \sum_{j=1}^d \alpha_j h(x_j)$, we have $W^n(f) = \sum_{j=1}^d \alpha_j \times W^n(h, X^j)$, and the result follows. □

Lemma 3.3.11 *Assume that* $p \in [0, 1] \cap \mathcal{I}(X)$ *and* $X^c = 0$ *and* $B' = 0$. *Let* f *be a function from* $\mathbb{R}^d$ *into* $\mathbb{R}^q$ *with* $f(0) = 0$ *and* $f(x) = 0$ *if* $\|x\| \geq 1$, *and which is Hölder with index* p. *Then* (3.3.8) *holds.*

Proof Again we use the notation of the previous proofs. The function g is no longer defined, but k is, and f and k satisfy

$$\|f(x)\| \leq K(\|x\|^p \wedge 1), \qquad \|k(x, y)\| \leq K(\|x\|^p \wedge 1). \tag{3.3.15}$$

Then $U = f \star \mu$ is well defined. The extension (3.2.5) of Itô's formula shows that $W^n(f) = k(Y^n, x) \star \mu$. Here again, outside a null set we have $k(Y_s^n, x) \to 0$ for all s and x, so (3.3.15) yields $k(Y^n, x) \star \mu \overset{\text{u.c.p.}}{\Longrightarrow} 0$, and we have the result. □

Proof of Theorem 3.3.5 1) The equivalence of two properties (i) and (ii) in the statement of the theorem is easy to prove: indeed (i), which is the same as (3.3.9), amounts to saying that $\mathbb{E}(1_{D_f} * \mu_\infty) = 0$. Now, D_f is a Borel subset of $\mathbb{R}^d$, hence $1_{D_j} * \nu$ is the predictable compensator of $1_{D_f} * \mu$, and thus $\mathbb{E}(1_{D_f} * \mu_\infty) = \mathbb{E}(1_{D_f} * \nu_\infty)$.

2) As said before, we can assume (3.3.6) and we will prove (3.3.8), which is (D) and implies (A), (B) and (C).

Let us introduce another notation, to be used throughout this proof and also throughout the whole book:

$$\begin{aligned}&\psi \text{ is a } C^\infty \text{ function: } \mathbb{R}_+ \to [0,1], \text{ with } 1_{[1,\infty)}(x) \le \psi(x) \le 1_{[1/2,\infty)},\\&\psi_\varepsilon(x) = \psi\big(\|x\|/\varepsilon\big), \qquad \psi'_\varepsilon = 1 - \psi_\varepsilon.\end{aligned} \tag{3.3.16}$$

We set $g_p(x) = \|x\|^p$, and for each $\varepsilon \in (0,1]$ we introduce the two functions

$$f_\varepsilon = f - f'_\varepsilon, \qquad f'_\varepsilon(x) = \begin{cases} f(x)\psi'_\varepsilon(x) & \text{for (A)} \\ \sum_{i,j=1}^d \alpha_{ij} x^i x^j \psi'_\varepsilon(x) & \text{for (B)} \\ \sum_{i=1}^d \alpha_i |x^i| & \text{for (C).} \end{cases}$$

Since $W^n(f) = W^n(f_\varepsilon) + W^n(f'_\varepsilon)$, in order to obtain (3.3.8) it is enough (as a particular case of Proposition 2.2.1) to prove the following two properties:

$$\begin{aligned}&\varepsilon > 0 \quad \Rightarrow \quad W^n(f_\varepsilon) \stackrel{\text{u.c.p.}}{\Longrightarrow} 0 \qquad (\text{as } n \to \infty)\\&t, \eta > 0 \quad \Rightarrow \quad \lim_{\varepsilon \to 0} \limsup_n \mathbb{P}\Big(\sup_{s \le t} \big\|W^n(f'_\varepsilon)_s\big\| > \eta\Big) = 0.\end{aligned} \tag{3.3.17}$$

3) Here we prove (3.3.17) in case (A). The function f_ε satisfies (3.3.9) and vanishes on a neighborhood of 0, so the first part of (3.3.17) follows from Lemma 3.3.7. In case (A-a), resp. (A-c), we have $\|f'_\varepsilon\| \le \theta(\varepsilon)h$, where $\theta(\varepsilon) \to 0$ as $\varepsilon \to 0$, and $h = g_2\psi'_1$, resp. $h = g_1\psi'_1$. Coming back to the definition of $W^n(f'_\varepsilon)$ we see that this implies

$$\big\|W^n(f'_\varepsilon)\big\| \le \theta(\varepsilon)V^n_{int}(h, X) + \theta(\varepsilon)h \star \mu.$$

The function h and the process X satisfy the assumptions of Lemma 3.3.8 in Case (A-a), whereas in Case (A-c) we have the estimate $V^n_{int}(h, X) \le \mathrm{Var}(X)$, hence in both cases the sequence of variables $\sup_{s \le t} V^n_{int}(h, X)_s$ is bounded in probability, whereas $h \star \mu_t < \infty$ because $2 \in \mathcal{I}(X)$, resp. $1 \in \mathcal{I}(X)$. Then the second part of (3.3.17) becomes obvious.

In Case (A-b) we have $1 < p \le 2$ and $X^c = 0$ and the function $h'_\varepsilon = g_p\psi'_\varepsilon$ is of class C^p with support in $\{x : \|x\| \le \varepsilon\}$, and $h'_\varepsilon(0) = \partial_i h'_\varepsilon(0) = 0$. Then $W^n(h'_\varepsilon)_t = V^n(h_\varepsilon)_t - h'_\varepsilon \star \mu_{T_n(t)}$. We also have $\|f'_\varepsilon\| \le K h'_\varepsilon$, and thus

$$\big\|W^n(f'_\varepsilon)\big\| \le K V^n_{int}\big(h'_\varepsilon, X\big) + K h'_\varepsilon \star \mu \le K\big|W^n\big(h'_\varepsilon\big)\big| + 2K h'_\varepsilon \star \mu.$$

Lemma 3.3.9 yields $W^n(h'_\varepsilon) \stackrel{\text{u.c.p.}}{\Longrightarrow} 0$, whereas $h'_\varepsilon \star \mu_t \to 0$ as $\varepsilon \to 0$ by the dominated convergence theorem, because $p \in \mathcal{I}(X)$. Then the previous estimate gives us the second part of (3.3.17) in this case.

Case (A-d) is treated in the same way, except that $0 \le p \le 1$, so now $V(h'_\varepsilon) = h'_\varepsilon \star \mu$, and $X^c = 0$ and $B' = 0$. We then apply Lemma 3.3.11 instead of Lemma 3.3.9, to deduce the second part of (3.3.17).

4) Next we prove (3.3.8) in case (B). In view of the definition of f_ε, we see that $f_\varepsilon(x) = \mathrm{o}(\|x\|^2)$. Then we deduce the first part of (3.3.17) from case (A-a).

Moreover, Lemma 3.3.8 yields, in view of the definition of G_ε, that $W^n(f'_\varepsilon) \stackrel{\text{u.c.p.}}{\Longrightarrow} 0$ for each $\varepsilon > 0$, so the second part of (3.3.17) *a fortiori* holds.

5) Finally we prove (3.3.8) in case (C). We have $f_\varepsilon(x) = \mathrm{o}(\|x\|)$, hence the first part of (3.3.17) follows from case (A-c). Moreover, f'_ε does not depend on ε and Lemma 3.3.10 gives the second part of (3.3.17). □

Proof of Theorem 3.3.1 Since Theorem 3.3.5 has already been proved, it remains to show (E). Therefore we suppose that $f = g_p$ on a neighborhood of the origin, say on $\{x : \|x\| \le 2\eta\}$ for some $\eta \in (0,1]$. By (A-a), and since $f - g_p\psi'_\eta$ satisfies (3.3.9) and vanishes on a neighborhood of 0, the sequence of variables $(V^n(f - g_p\psi'_\eta, X)_t)$ is bounded in probability for any $t \ge 0$. Hence it suffices to prove the result when $f = g_p\psi'_\eta$. We single out three cases which, although phrased differently, cover the three cases in the theorem.

Case (1): we assume here that $p < 2$ and that X^c does not vanish identically, so there is a $t > 0$ such that $\mathbb{P}(\Delta X_t \ne 0) = 0$ and $\mathbb{P}(D_t > 0)$, where $D_t = \sum_{i=1}^d C_t^{ii}$. Set $h_\varepsilon = g_2\psi'_\varepsilon$. Observe that $f \ge \varepsilon^{p-2}h_\varepsilon$ if $\varepsilon \in (0,\eta]$, hence by (B) we have

$$V^n(f,X)_t \;\ge\; \varepsilon^{p-2}V^n(h_\varepsilon,X)_t \;\stackrel{\mathbb{P}}{\longrightarrow}\; \varepsilon^{p-2}(D_t + h_\varepsilon \star \mu_t) \;\ge\; \varepsilon^{p-2}D_t.$$

This implies that, for all $A > 0$, $\liminf_n \mathbb{P}(V^n(f,X)_t > A) \ge \mathbb{P}(D_t > A\varepsilon^{2-p})$, which goes to $\mathbb{P}(D_t > 0) > 0$ when $\varepsilon \to 0$. Hence the sequence $(V^n(f,X)_t)$ is not bounded in probability.

When X is a Lévy process, then $X^c \ne 0$ implies that $\mathbb{P}(D_t > 0) = 1$ for all $t > 0$, hence $\liminf_n \mathbb{P}(V^n(f,X)_t > A) = 1$ for all A, which means that $V^n(f,X)_t \stackrel{\mathbb{P}}{\longrightarrow} +\infty$.

Case (2): we assume here $X^c = 0$ and $p \notin \mathcal{I}(X)$ and either $1 \le p < 2$, or $0 \le p < 1$ and $1 \in \mathcal{I}$ and $B' = 0$. There is $t > 0$ such that $\mathbb{P}(\Delta X_t \ne 0) = 0$ and $\mathbb{P}(\Sigma(p,X)_{t-s} = \infty)$ for some $s \in (0,t)$. We have $f\psi_\varepsilon \le f$, hence by (A-a) we get

$$V^n(f,X)_t \;\ge\; V^n(f\psi_\varepsilon,X)_t \;\stackrel{\mathbb{P}}{\longrightarrow}\; (f\psi_\varepsilon)\star\mu_t,$$

and $(f\psi_\varepsilon)\star\mu_t$ increases to $+\infty$ as ε decreases to 0 by the monotone convergence theorem if $\Sigma(p,X)_{t-s} = \infty$. For all $A > 0$ we thus have $\liminf_n \mathbb{P}(V^n(f,X)_t > A) \ge \mathbb{P}(\Sigma(p,X)_{t-s} = \infty) > 0$. Therefore the sequence $(V^n(f,X)_t)$ is not bounded in probability.

When further X is a Lévy process, in addition to the other assumptions, we have $\mathbb{P}(\Sigma(p,X)_{t-s} = \infty) = 1$ for all $t > s$ and we conclude as above that $V^n(f,X)_t \stackrel{\mathbb{P}}{\longrightarrow} +\infty$.

Case (3): we assume here $X^c = 0$ and $1 \in \mathcal{I}(X)$ and $B' \ne 0$, and $0 \le p < 1$. There is a $t > 0$ such that $\mathbb{P}(\Delta X_t \ne 0) = 0$ and $\mathbb{P}(B'_t \ne 0)$. Set $h'_\varepsilon(x) = \frac{1}{d}\sum_{j=1}^d |x^j|\psi'_\varepsilon(x)$. Observe that $f \ge \varepsilon^{p-1}h'_\varepsilon$ if $\varepsilon \in (0,\eta]$, hence by (C-b) we have with $D =$

$\frac{1}{d}\sum_{j=1}^{d}\mathrm{Var}(B'^j)$:

$$V^n(f,X)_t \geq \varepsilon^{p-1}V^n\big(h'_\varepsilon,X\big)_t \xrightarrow{\mathbb{P}} \varepsilon^{p-1}(D_t+h_\varepsilon\star\mu_t) \geq \varepsilon^{p-2}D_t.$$

Since $D_t>0$ if $B'_t\neq 0$, we conclude that the sequence $(V^n(f_p,X)_t)$ is not bounded in probability, and also that it converges in probability to $+\infty$ when $\mathbb{P}(B'_t\neq 0)=1$ (for example in the Lévy case), exactly as for case (1). □

Proof of Corollary 3.3.4 Let

$$T_q = \inf(t:C_t\neq 0)\wedge\inf\big(t:\Sigma(p,X)_t\geq q\big),$$

and denote by $X(q)_t=X_{t\wedge T_q}$ the process X stopped at time T_q. First, $X(q)$ is a semimartingale with $p\in\mathcal{I}(X(q))$ and $X(q)^c_t=0$ for all t. Therefore under our standing assumptions on f, Theorem 3.3.1 yields $V^n(f,X(q))\overset{\mathbb{P}}{\Longrightarrow}V(f,X(q))$ as $n\to\infty$ for each fixed q. Moreover $\mathbb{P}(\Delta X(q)_T\neq 0)=0$ by hypothesis, so the continuity of the stopping mapping for the Skorokhod topology when the stopping time is a time of continuity for the limit yields that $V^n(f,X(q))_{t\wedge T}\overset{\mathbb{P}}{\Longrightarrow}V(f,X(q))_{t\wedge T}$ as $n\to\infty$.

It remains to observe that $V^n(f,X)_s=V^n(f,X(q))_s$ and $V(f,X)_s=V(f,X(q))_s$ for all $s\leq T_q$, whereas $\Omega_T\cap\{T_q\leq T\}\downarrow\emptyset$ as $q\to\infty$, and the result readily follows. □

3.4 Law of Large Numbers with Normalization

3.4.1 Preliminary Comments

We turn now to a second—very different—type of LLN. In Theorem 3.3.1, the behavior of f near 0 is the determining factor, and even the only one when X is continuous, for obtaining a limit. In what follows, we consider another type of functionals, whose behavior depends on the entire test function f. More specifically, instead of the functionals $V^n(f,X)$ of (3.1.3), we consider functionals of the "normalized" increments:

$$\sum_{i=1}^{N_n(t)} f\big(\Delta_i^n X/u(n,i)\big) \tag{3.4.1}$$

perhaps with an "outside" normalization as well. Here, f is an arbitrary function on $\mathbb{R}^d$, and $u(n,i)>0$ is a "suitable" normalizing factor chosen in such a way that "most" of the variables $\Delta_i^n X/u(n,i)$ are neither going to 0 nor exploding. This cannot be achieved in general, unless $u(n,i)$ strongly depends on the structure of the process X itself over the interval $I(n,i)$.

For example, let X be a one-dimensional continuous Gaussian martingale with angle bracket $C=\langle X,X\rangle$ (a deterministic continuous increasing function). With a (non-random) scheme $\mathcal{T}=(t(n,i):i\geq 0)_{n\geq 1}$, the variables $\Delta_i^n X/u(n,i)$ are $\mathcal{N}(0,1)$ if we take $u(n,i)=\sqrt{\Delta_i^n C}$, whereas $|\Delta_i^n X|/u(n,i)$ goes to 0, resp. ∞

if $u(n,i)/\sqrt{\Delta_i^n C}$ goes to ∞, resp. 0. When the scheme is random, we do not know how to choose *a priori* a normalization ensuring that $|\Delta_i^n X|/u(n,i)$ and its inverse are tight. Even worse: with the function $f \equiv 1$ (so $u(n,i)$ no longer enters the picture), (3.4.1) is equal to $N_n(t)$, and in general there is no normalizing factor v_n such that $v_n N_n(t)$ converges *for all* t to a finite and non-vanishing limit, even for a non-random scheme.

These considerations lead us to consider *regular discretization schemes* below, that is $t(n,i) = i\Delta_n$ for a sequence $\Delta_n \to 0$ (see however Chap. 14 for some special cases of irregular schemes). In this case $N_n(t) = [t/\Delta_n]$. With the view of obtaining results holding at least for Brownian motion (see Chap. 1), we introduce the following functionals, with both an "inside" and an "outside" normalization:

$$V'^n(f,X)_t = V'(\Delta_n; f, X)_t = \Delta_n \sum_{i=1}^{[t/\Delta_n]} f\big(\Delta_i^n X/\sqrt{\Delta_n}\big) \tag{3.4.2}$$

(we use the first notation most of the time, and $V'(\Delta_n; f, X)$ when we want to emphasize the dependency on Δ_n.) The normalizing factor $\sqrt{\Delta_n}$ is designed for the Brownian term, but it works for all Itô semimartingales having a non-vanishing continuous martingale part.

In the particular case $f(x) = \|x\|^p$, or more generally when f is *positively homogeneous of degree* $p > 0$, meaning that $f(\lambda x) = \lambda^p f(x)$ for all $x \in \mathbb{R}^d$ and $\lambda \geq 0$, the functionals $V'(\Delta_n; f, X)$ and $V(\mathcal{T}_n; f, X)$, which will be written as $V(\Delta_n; f, X)$ when $\mathcal{T}_n$ is a regular grid with stepsize Δ_n, are essentially the same object: namely, we have

$$\begin{aligned} &f \text{ positively homogeneous of degree } p \\ &\quad\Rightarrow\quad V'(\Delta_n; f, X) = \Delta_n^{1-p/2}\, V(\Delta_n; f, X). \end{aligned} \tag{3.4.3}$$

Exactly as for $V^n(f,X)$, there is an *interpolated* version of $V'^n(f,X)$:

$$V'^n_{int}(f,X)_t = \Delta_n \sum_{i\geq 1} f\left(\frac{X_{t\wedge(i\Delta_n)} - X_{t\wedge((i-1)\Delta_n)}}{\sqrt{\Delta_n}}\right).$$

However this has little interest, and although the forthcoming results holds for $V'^n_{int}(f,X)$ under exactly the same assumptions as for $V'^n(f,X)$, we will not pursue this case here.

3.4.2 The Results

Before stating the results, we need some additional notation, recall that $\mathcal{M}^+_{d\times d}$ is the set of all $d \times d$ symmetric nonnegative matrices:

$$\left.\begin{array}{l} \text{if} \in \mathcal{M}^+_{d\times d}, \text{ then } \rho_a \text{ denotes the centered Gaussian law} \\ \text{with covariance matrix } a, \text{ and } \rho_a(f) = \int f(x)\rho_a(dx). \end{array}\right\} \tag{3.4.4}$$

In contrast with Theorem 3.3.1, we state the next result for a one-dimensional test function. This automatically implies the same result when f is q-dimensional, because when all components of a sequence of multidimensional processes converge in the u.c.p. sense, the same is true of the multidimensional processes themselves.

Below, X is an Itô semimartingale. Its characteristics have the form (2.1.25), that is

$$B_t = \int_0^t b_s\,ds, \qquad C_t = \int_0^t c_s\,ds, \qquad \nu(dt,dx) = dt\,F_t(dx). \tag{3.4.5}$$

In particular the process c_t plays a crucial role below. The notation $\|c_t\|$ denotes the Euclidean norm of the $d \times d$ matrix c_t, considered as a vector in $\mathbb{R}^{d^2}$.

Theorem 3.4.1 *Assume that X is a d-dimensional Itô semimartingale and that the discretization scheme is regular with stepsize Δ_n. Let f be a continuous function, which satisfies one of the following three conditions:*

(a) $f(x) = \mathrm{o}(\|x\|^2)$ as $\|x\| \to \infty$,

(b) $f(x) = \mathrm{O}(\|x\|^2)$ as $\|x\| \to \infty$, and X is continuous,

(c) $f(x) = \mathrm{O}(\|x\|^p)$ as $\|x\| \to \infty$ for some $p > 2$, and X is continuous and satisfies

$$\int_0^t \|b_s\|^{2p/(2+p)}\,ds < \infty, \qquad \int_0^t \|c_s\|^{p/2}\,ds < \infty. \tag{3.4.6}$$

Then

$$V'^n(f,X) \stackrel{\text{u.c.p.}}{\Longrightarrow} V'(f,X)_t = \int_0^t \rho_{c_s}(f)\,ds. \tag{3.4.7}$$

Conversely, if for some $p \geq 2$ the processes $V'^n(g_p,X)$ converge in probability locally uniformly in time to a continuous process, where $g_p(x) = \|x\|^p$, then X is continuous.

Condition (3.4.6) for $p = 2$ holds for *any* Itô semimartingale (cf. (2.1.25)), so (b) is exactly (c) with $p = 2$. Observe also that $2p/(2+p) < 2$ always.

The last claim is not a complete converse for the cases (b) and (c). However, if $V'(g_p,X)$ is finite-valued for some $p \geq 2$, then obviously the second part of (3.4.6) holds. Moreover, the next example shows that for any $p > 2$ and $\varepsilon > 0$ there is an Itô semimartingale such that $V'^n(g_p,X) \stackrel{\text{u.c.p.}}{\Longrightarrow} V'(g_p,X)$, and the first part of (3.4.6) holds for p but *not* for $p + \varepsilon$. So the exponent $\frac{2p}{2+p}$ is in some sense "optimal", although we do not know whether the first part of (3.4.6) is actually necessary for having $V'^n(g_p,X) \stackrel{\text{u.c.p.}}{\Longrightarrow} V'(g_p,X)$.

Example 3.4.2 This example is a very simple "Itô semimartingale", namely an absolutely continuous increasing function (no randomness here). So $d = 1$, and $X = B$ is the function

$$B_t = t^\alpha, \quad \text{hence } b_t = \alpha t^{\alpha-1},$$

for some $\alpha \in (0, \frac{1}{2})$. Here $V'(f, X) = 0$ for any function f. We take $p > 2$. We have

$$V'^n(g_p, B)_t = \Delta_n^{1-p/2} \sum_{i=1}^{[t/\Delta_n]} (\Delta_i^n B)^p = \Delta_n^{1-p/2+p\alpha} \left(1 + \sum_{i=1}^{[t/\Delta_n]} u_i\right), \tag{3.4.8}$$

where $u_i = i^{p\alpha}((1+1/i)^\alpha - 1)^p$. One easily checks that the partial sums of the series $\sum u_i$ behave as follows, as $n \to \infty$, and for a suitable constant $A = A(p, \alpha) > 0$:

$$\sum_{i=1}^n u_i \begin{cases} \to A & \text{if } p(1-\alpha) > 1 \\ \sim A \log n & \text{if } p(1-\alpha) = 1 \\ \sim A n^{1-(1-\alpha)p} & \text{if } p(1-\alpha) < 1. \end{cases}$$

Substituting this in (3.4.8) gives $V'^n(g_p, B)_t \to 0$ if and only if $p < \frac{2}{1-2\alpha}$.

On the other hand, we have $\int_0^t |b_s|^q ds < \infty$ for all t if and only if $q < \frac{1}{1-\alpha}$. Then, if $p < \frac{2}{1-2\alpha}$, (3.4.6) holds for p, but not for $p + \varepsilon$, where $\varepsilon = \frac{2-p(1-2\alpha)}{1-2\alpha}$ is as small as one wishes when α is close to $\frac{p-2}{4}$.

The result for $V'^n(f, X)$ can be transformed into a result for $V^n(f, X)$, even when f is not homogeneous, in some cases. These cases cover some of the situations in Theorem 3.3.1 where the convergence does not take place.

Corollary 3.4.3 *Assume that the discretization scheme is regular with stepsize Δ_n. Let f be a Borel function which satisfies $f(x) \sim h(x)$ as $x \to 0$, where h is a positively homogeneous continuous function of degree $p \in (0, 2)$ on $\mathbb{R}^d$. Then*

$$\Delta_n^{1-p/2} V^n(f, X)_t \overset{\text{u.c.p.}}{\Longrightarrow} \int_0^t \rho_{c_s}(h)\, ds, \tag{3.4.9}$$

and the same holds for the interpolated functionals $V^n_{int}(f, X)$.

We see once more that for $V^n(f, X)$, only the behavior of f near 0 matters, since the limit depends on h only.

Finally, the continuity assumption in Theorem 3.4.1 can be relaxed, in the same spirit as Theorem 3.3.5 extends Theorem 3.3.1:

Corollary 3.4.4 *All statements of Theorem 3.4.1 remain valid if we replace the continuity of the test function f by the fact that it is Borel and locally bounded and, with D_f denoting the set of all $x \in \mathbb{R}^d$ where f is not continuous,*

$$\mathbb{E}\left(\int_0^\infty \rho_{c_s}(D_f)\, ds\right) = 0. \tag{3.4.10}$$

When the matrix $c_s(\omega)$ is $\mathbb{P}(d\omega) \otimes ds$ almost everywhere invertible, (3.4.10) amounts to saying that the Lebesgue measure (on $\mathbb{R}^d$) of the set D_f is null, because in this case the measures ρ_{c_s} have positive densities. Otherwise, this property is more complicated to interpret, except when $d=1$: in this case, (3.4.10) means that f is Lebesgue-almost everywhere continuous, and further it is continuous at 0 when $\mathbb{P}(\int_0^\infty 1_{\{c_s=0\}}\,ds>0)>0$.

3.4.3 The Proofs

When $X=\sigma W$ with σ a constant matrix, the variable $V'^n(f,X)_t$ is Δ_n times the sum of $[t/\Delta_n]$ i.i.d. random variables with a law not depending on n and expectation $\rho_{\sigma\sigma^*}(f)$: so the result amounts to the LLN for i.i.d. variables. In the general case, the proof takes the following steps:

1) It is enough to show the result when X satisfies some strengthened assumptions, mainly boundedness of its characteristics, in a suitable sense; this is called the "localization step".

2) The jumps do not matter (the "elimination of jumps step"), under appropriate assumptions on f: in the present case, the assumption (a).

3) If X is continuous with the additional boundedness assumptions of Step 1, then it is "locally" sufficiently close to being a Brownian motion, so that the aforementioned trivial LLN in the "pure Brownian" case holds in a "local" sense.

This is the scheme of the present proof, and of many forthcoming proofs as well, when the processes of interest are of the type $V'^n(f,X)$ or extensions of these, and including the proofs of the central limit theorems: usually Step 1 is easy and it is the same or almost the same for most theorems; Step 2 is often more complicated, and the difficulty of Step 3 greatly varies with the problem at hand.

Before proceeding, and with $\varepsilon\in(0,1]$ and $p\geq 2$, we associate with any Itô semimartingale X with characteristics given by (3.4.5) the following processes:

$$\gamma(\varepsilon)_t^X=\gamma(\varepsilon)_t=\int_{\{\|x\|\leq\varepsilon\}}\|x\|^2\,F_t(dx),\qquad \gamma_t'^X=\gamma_t'=\int\big(\|x\|^2\wedge 1\big)\,F_t(dx),\tag{3.4.11}$$

$$A(p)_t^X=A(p)_t=\int_0^t\big(\|b_s\|^{2p/(2+p)}+\|c_s\|^{p/2}+\gamma_s'\big)\,ds.\tag{3.4.12}$$

Then $A(2)$ is always finite-valued, and in the continuous case $A(p)$ is finite-valued if and only (3.4.6) holds. Below, we set $p=2$ in Cases (a) and (b) of the theorem. Recall also that we can write a Grigelionis decomposition for X:

$$X_t=X_0+\int_0^t b_s\,ds+\int_0^t\sigma_s\,dW_s+(\delta 1_{\{\|\delta\|\leq 1\}})\star(\mathfrak{p}-\mathfrak{q})_t+(\delta 1_{\{\|\delta\|>1\}})\star\mathfrak{p}_t\tag{3.4.13}$$

(see Sect. 2.1.4 for details about the Brownian motion W and the Poisson measure $\mathfrak{p}$ with compensator $\mathfrak{q}(dt,dz)=dt\otimes\lambda(dz)$, which may require an extension of the underlying space, still denoted by $(\Omega,\mathcal{F},(\mathcal{F}_t)_{t\geq 0},\mathbb{P})$). Note that we can take $d'=d$ and choose a square-root σ_t of c_t which is symmetric nonnegative to obtain a process satisfying $\|\sigma_t\|^2\leq K\|c_t\|$ for a constant K.

Lemma 3.4.5 (Localization) *Let f be a function on $\mathbb{R}^d$ and $p \geq 2$ be such that the convergence $V'^n(f,X) \overset{\text{u.c.p.}}{\Longrightarrow} V'(f,X)$ holds for all Itô semimartingales X which satisfies*

$$\sup_{\omega} A(p)^X_\infty(\omega) < \infty, \qquad \sup_{(\omega,t)} \|\Delta X_t(\omega)\| \begin{cases} < \infty & \text{in case } (1) \\ = 0 & \text{in case } (2). \end{cases} \tag{3.4.14}$$

Then $V'^n(f,X) \overset{\text{u.c.p.}}{\Longrightarrow} V'(f,X)$ for all Itô semimartingales X satisfying $A(p)^X_t < \infty$ for all finite t, and which further are continuous in case (2).

Proof Let X be given by (3.4.13) and satisfy $A(p)^X_t < \infty$ for all finite t, and be further continuous in case (2). Then $T_m = \inf(t : A(p)^X_t + \|\Delta X_t\| > m)$ for $m \in \mathbb{N}^*$ defines a sequence of stopping times increasing to $+\infty$. Consider the semimartingale

$$X(m)_t = X_0 + \int_0^{t\wedge T_m} b_s\,ds + \int_0^{t\wedge T_m} \sigma_s\,dW_s$$
$$+ (\delta 1_{\{\|\delta\|\leq 1\}}) \star (\mathfrak{p} - \mathfrak{q})_{t\wedge T_m} + (\delta 1_{\{1<\|\delta\|\leq m\}}) \star \mathfrak{p}_{t\wedge T_m}.$$

Then $X_t = X(m)_t$ for all $t < T_m$, hence

$$\mathbb{P}\Big(\sup_{s\leq t} \big|V'^n(f,X)_s - V'(f,X)_s\big| > \eta\Big)$$
$$\leq \mathbb{P}(T_m \leq t) + \mathbb{P}\Big(\sup_{s\leq t} \big|V'^n\big(f,X(m)\big)_s - V'\big(f,X(m)\big)_s\big| > \eta\Big).$$

Moreover $X(m)$ satisfies (3.4.14), since $A(p)^{X(m)}_\infty \leq A(p)^X_{T_m} \leq m$ and $\|\Delta X(m)\| \leq m$, and $X(m)$ is continuous when X is continuous. Hence $\mathbb{P}(\sup_{s\leq t} |V'^n(f,X(m))_s - V'(f,X(m))_s| > \eta) \to 0$ as $n \to \infty$ for all m by hypothesis. Since $\mathbb{P}(T_m \leq t) \to 0$ as $m \to \infty$, we deduce

$$\mathbb{P}\Big(\sup_{s\leq t} \big|V'^n(f,X)_s - V'(f,X)_s\big| > \eta\Big) \to 0,$$

and the result is proved. □

From now on, we suppose that X satisfies (3.4.14) for the relevant p (recall $p = 2$ in cases (a,b)). We can then take a version of δ satisfying $\|\delta\| \leq K$ for some constant K, and we can rewrite (3.4.13) as

$$X = X' + X'', \quad \text{where } X'_t = X_0 + \int_0^t b''_s\,ds + \int_0^t \sigma_s\,dW_s, \quad X'' = \delta \star (\mathfrak{p} - \mathfrak{q}), \tag{3.4.15}$$

where $b''_t = b_t + \int_{\{\|\delta(t,z)\|>1\}} \delta(t,z)\,\lambda(dz)$. Note that the processes of (3.4.11) are

$$\gamma(\varepsilon)_t = \int_{\{\|\delta(t,z)\|\leq\varepsilon\}} \big\|\delta(t,z)\big\|^2 \lambda(dz), \qquad \gamma'_t = \int \big(\big\|\delta(t,z)\big\|^2 \wedge 1\big)\,\lambda(dz).$$

Lemma 3.4.6 (Elimination of Jumps) *Assume* (3.4.14) *for* $p=2$ *and let* f *be continuous with* $f(x)=o(\|x\|^2)$ *as* $\|x\|\to\infty$. *The property* $V'^n(f,X')\stackrel{\text{u.c.p.}}{\Longrightarrow}V'(f,X')$ *implies* $V'^n(f,X)\stackrel{\text{u.c.p.}}{\Longrightarrow}V'(f,X)$.

Proof By hypothesis we have for $0<\varepsilon<1<A$:

$$\begin{aligned}
&\|x\|\le 2A \quad\Rightarrow\quad \bigl|f(x)\bigr|\le\Phi(A)\\
&\|x\|\le 2A,\ \|y\|\le\varepsilon \quad\Rightarrow\quad \bigl|f(x+y)-f(x)\bigr|\le\Phi'_A(\varepsilon)\\
&\|x\|>A \quad\Rightarrow\quad \bigl|f(x)\bigr|\le\Phi''(A)\|x\|^2
\end{aligned}$$

where $\Phi(A)<\infty$, and $\Phi'_A(\varepsilon)\to 0$ as $\varepsilon\to 0$, and $\Phi''(A)\to 0$ as $A\to\infty$. These properties easily yield that for all $x,y\in\mathbb{R}^d$,

$$\bigl|f(x+y)-f(x)\bigr| \le \Phi'_A(\varepsilon)+\frac{2\Phi(A)(\|y\|^2\wedge 1)}{\varepsilon^2}+2\Phi''(A)\bigl(\|x\|^2+\|y\|^2\bigr). \tag{3.4.16}$$

Now we provide some estimates. First, from (2.1.33) and (2.1.34), plus the facts that $\|b''_t\|\le\|b_t\|+K\gamma'_t$ and that $A(2)_\infty\le K$, we get

$$\mathbb{E}\bigl(\bigl\|\Delta_i^n X'\bigr\|^2\bigr) \le \mathbb{E}\bigl(\Delta_i^n A(2)+\bigl(\Delta_i^n A(2)\bigr)^2\bigr) \le KE\bigl(\Delta_i^n A(2)\bigr). \tag{3.4.17}$$

Next, with the notation (2.1.35) and since $\|\delta\|\le K$, we have

$$\widehat{\delta'}(2)_{t,s}\le\widehat{\delta}(2)_{t,s}\le\frac{K}{s}\int_t^{t+s}\gamma'_u\,du,\qquad \widehat{\delta}(2,\varepsilon)_{t,s}=\frac{1}{s}\int_t^{t+s}\gamma(\varepsilon)_u\,du.$$

Then (2.1.36) with $p=2$ and (2.1.39) with $r=2$ and $q=1/2$ yield

$$\begin{aligned}
&\mathbb{E}\bigl(\bigl\|\Delta_i^n X''\bigr\|^2\bigr) \le K\mathbb{E}\bigl(\Delta_i^n A(2)\bigr)\\
&\mathbb{E}\biggl(\biggl(\frac{\|\Delta_i^n X''\|}{\sqrt{\Delta_n}}\bigwedge 1\biggr)^2\biggr)\le\frac{K}{\Delta_n}\,\mathbb{E}\biggl(\Delta_n^{1/4}\ \Delta_i^n A(2)+\int_{(i-1)\delta_n}^{i\Delta_n}\gamma\bigl(\Delta_n^{1/4}\bigr)_s\,ds\biggr).
\end{aligned}\tag{3.4.18}$$

Therefore if we take $x=\Delta_i^n X'/\sqrt{\Delta_n}$ and $y=\Delta_i^n X''/\sqrt{\Delta_n}$ in (3.4.16), we deduce from the estimates (3.4.17) and (3.4.18) that

$$\begin{aligned}
&\mathbb{E}\bigl(\bigl|f(\Delta_i^n X/\sqrt{\Delta_n})-f\bigl(\Delta_i^n X'/\sqrt{\Delta_n}\bigr)\bigr|\bigr)\\
&\quad\le\Phi'_A(\varepsilon)+K\biggl(\frac{\Phi''(A)}{\Delta_n}+\frac{\Phi(A)}{\varepsilon^2\,\Delta_n^{3/4}}\biggr)\mathbb{E}\bigl(\Delta_i^n A(2)\bigr)\\
&\qquad+\frac{K\Phi(A)}{\varepsilon^2\,\Delta_n}\,\mathbb{E}\biggl(\int_{(i-1)\Delta_n}^{i\Delta_n}\gamma\bigl(\Delta_n^{1/4}\bigr)_s\,ds\biggr).
\end{aligned}$$

Applying $A(2)_\infty \le K$ once more, plus the property $V(f,X) = V(f,X')$, we deduce

$$a_n(t) := \mathbb{E}\Big(\sup_{s\le t}\big|V'^n(f,X)_s - V'^n(f,X)_s|\Big)$$
$$\le K\left(\Phi'_A(\varepsilon)t + \Phi''(A) + \frac{\Phi(A)\Delta_n^{1/4}}{\varepsilon^2} + \frac{\Phi(A)}{\varepsilon^2}\mathbb{E}\left(\int_0^t \gamma\big(\Delta_n^{1/4}\big)_s\,ds\right)\right).$$

The variables $\gamma(\Delta_n^{1/4})_s$ are smaller than γ'_s and go to 0 as $n\to\infty$, hence by the dominated convergence theorem $\limsup_n a_n(t) \le a'(t,\varepsilon,A) = K(\Phi'_A(\varepsilon)t + \Phi''(A))$. This is true for all $\varepsilon\in(0,1)$ and $A>1$, and since $\lim_{A\to\infty}\limsup_{\varepsilon\to 0} a'(t,\varepsilon,A) = 0$, the result follows. □

So far, we have achieved Steps 1 and 2. For Step 3 we begin with two lemmas.

Lemma 3.4.7 *If X is continuous and $A(2)_\infty$ is bounded, if σ is bounded and continuous, and if f is bounded and uniformly continuous, then $V'^n(f,X) \stackrel{\text{u.c.p.}}{\Longrightarrow} V'(f,X)$.*

Proof 1) Consider the processes

$$U_t^n = \Delta_n \sum_{i=1}^{[t/\Delta_n]} f\big(\beta_i^n\big), \quad \text{where } \beta_i^n = \sigma_{(i-1)\Delta_n}\Delta_i^n W/\sqrt{\Delta_n}$$
$$U_t'^n = \Delta_n \sum_{i=1}^{[t/\Delta_n]} \rho_{c_{(i-1)\Delta_n}}(f).$$

On the one hand $\rho_{c_t}(f) = \widetilde{\mathbb{E}}(f(\sigma_t U))$, where the expectation is taken for the d'-dimensional variable U, which is $\mathcal{N}(0,I_{d'})$. Hence the function $t\mapsto \rho_{c_t}(f)$ is bounded continuous, and by Riemann integration we have

$$U_t'^n \stackrel{\text{u.c.p.}}{\Longrightarrow} \int_0^t \rho_{c_s}(f)\,ds.$$

On the other hand $U_t^n - U_t'^n = \sum_{i=1}^{[t/\Delta_n]}(\zeta_i^n - \mathbb{E}(\zeta_i^n \mid \mathcal{F}_{(i-1)\Delta_n}))$, where $\zeta_i^n = \Delta_n f(\beta_i^n)$ is $\mathcal{F}_{i\Delta_n}$ measurable (because β_i^n is independent of $\mathcal{F}_{(i-1)\Delta_n}$, with law $\rho_{c_{(i-1)\Delta_n}}$), and $|\zeta_i^n| \le K\Delta_n$, so the array $(|\zeta_i^n|^2)$ is asymptotically negligible. Then (2.2.29) holds with $\mathcal{G}_i^n = \mathcal{F}_{i\Delta_n}$ and $N_n(t) = [t/\Delta_n]$, and by case (c) of Lemma 2.2.11 we have $U^n - U'^n \stackrel{\text{u.c.p.}}{\Longrightarrow} 0$. Hence we deduce

$$U_t^n \stackrel{\text{u.c.p.}}{\Longrightarrow} \int_0^t \rho_{c_s}(f)\,ds. \tag{3.4.19}$$

2) In view of (3.4.19), it remains to prove that

$$\mathbb{E}\left(\Delta_n \sum_{i=1}^{[t/\Delta_n]} \big|\chi_i^n\big|\right) \to 0, \quad \text{where } \chi_i^n = f\big(\Delta_i^n X/\sqrt{\Delta_n}\big) - f\big(\beta_i^n\big). \tag{3.4.20}$$

To this end, we note that our assumption on f yields a constant K and a positive function θ satisfying $\theta(\varepsilon) \to 0$ as $\varepsilon \to 0$, such that

$$\left|f(x+y) - f(x)\right| \le \theta(\varepsilon) + K\|y\|^2/\varepsilon^2 \tag{3.4.21}$$

for all $\varepsilon > 0$ and $x, y \in \mathbb{R}^d$. This, applied with $x = \beta_i^n$ and $y = \Delta_i^n X/\sqrt{\Delta_n} - \beta_i^n$, yields

$$\mathbb{E}\big(\left|\chi_i^n\right|\big) \le \theta(\varepsilon) + \frac{K}{\varepsilon^2}\,\mathbb{E}\big(\big\|\Delta_i^n X^c/\sqrt{\Delta_n} - \beta_i^n\big\|^2\big).$$

Now, if $\eta_t^n = \sup_{i \le [t/\Delta_n]} \Delta_i^n A(2)$, we deduce from (2.1.34) that, for $i \le [t/\Delta_n]$,

$$\mathbb{E}\big(\big\|\Delta_i^n X/\sqrt{\Delta_n} - \beta_i^n\big\|^2\big) \le \frac{K}{\Delta_n}\,\mathbb{E}\left(\eta_t^n\,\Delta_i^n A(2) + \int_{(i-1)\Delta_n}^{i\Delta_n} \|\sigma_s - \sigma_{\Delta_n[s/\Delta_n]}\|^2\,ds\right). \tag{3.4.22}$$

Therefore, since $A(2)_\infty \le K$,

$$\mathbb{E}\left(\Delta_n \sum_{i=1}^{[t/\Delta_n]} \left|\chi_i^n\right|\right) \le t\theta(\varepsilon) + \frac{K}{\varepsilon^2}\,\mathbb{E}\left(\eta_t^n + \int_0^t \|\sigma_s - \sigma_{\Delta_n[s/\Delta_n]}\|^2\,ds\right). \tag{3.4.23}$$

In the right side above, the second term goes to 0 by the dominated convergence theorem, because σ is continuous and bounded, and η_t^n goes to 0 and is smaller than K. Then if $\varepsilon \in (0,1]$ we have $\limsup_n \mathbb{E}(\Delta_n \sum_{i=1}^{[t/\Delta_n]} |\chi_i^n|) \le t\theta(\varepsilon)$ and (3.4.20) follows from $\theta(\varepsilon) \to 0$ as $\varepsilon \to 0$. □

Lemma 3.4.8 *If X is continuous and $A(2)_\infty$ is bounded, and if f is bounded and uniformly continuous, then $V'^n(f,X) \stackrel{\text{u.c.p.}}{\Longrightarrow} V'(f,X)$.*

Proof By a classical density argument, and in view of $A(2)_\infty \le K$, one can find a sequence $\sigma(m)$ of adapted bounded and continuous processes satisfying

$$\mathbb{E}\left(\int_0^\infty \big\|\sigma(m)_s - \sigma_s\big\|^2\,ds\right) \to 0 \tag{3.4.24}$$

as $m \to \infty$. With each m we associate the semimartingale

$$X(m)_t = X_0 + \int_0^t b_s\,ds + \int_0^t \sigma(m)_s\,dW_s.$$

The previous lemma yields that $V'^n(f,X(m)) \stackrel{\text{u.c.p.}}{\Longrightarrow} V'(f,X(m))$ as $n \to \infty$, for each m. Therefore to obtain $V'^n(f,X) \stackrel{\text{u.c.p.}}{\Longrightarrow} V'(f,X)$, and by Proposition 2.2.1, it suffices to prove the following two properties, where $c(m) = \sigma(m)\sigma(m)^\star$:

$$\sup_{s \le t}\left|\int_0^s \rho_{c(m)_r}(f)\,dr - \int_0^s \rho_{c_r}(f)\,dr\right| \stackrel{\mathbb{P}}{\longrightarrow} 0, \tag{3.4.25}$$

$$\lim_{m\to\infty} \limsup_{n\to\infty} \mathbb{E}\Big(\sup_{s\le t} \big|V'^n(f,X)_s - V'^n\big(f,X(m)\big)_s\big|\Big) \to 0, \quad \forall t>0. \quad (3.4.26)$$

By the subsequence principle, in order to prove (3.4.25) for a given t, it is no restriction to assume as a consequence of (3.4.24) that $\sigma(m)_s \to \sigma_s$ almost everywhere for the measure $\mathbb{P}(d\omega)\otimes ds$, on $\Omega\times[0,t]$. Since $a\mapsto \rho_a(f)$ is continuous we obtain (3.4.25).

Next, with $\chi(m)_i^n = f(\Delta_i^n X/\sqrt{\Delta_n}) - f(\Delta_i^n X(m)/\sqrt{\Delta_n})$, (3.4.26) will follow from

$$\lim_{m\to\infty} \limsup_{n\to\infty} \mathbb{E}\Bigg(\Delta_n \sum_{i=1}^{[t/\Delta_n]} \big|\chi(m)_i^n\big|\Bigg) = 0, \quad \forall t>0.$$

To see this, we use $X_t - X(m)_t = \int_0^t (\sigma_s - \sigma(m)_s)\,dW_s$ and apply (2.1.34) and (3.4.21) with $y=0$ and $x = \Delta_i^n X/\sqrt{\Delta_n}$ to get

$$\mathbb{E}\Bigg(\Delta_n \sum_{i=1}^{[t/\Delta_n]} \big|\chi(m)_i^n\big|\Bigg) \le t\theta(\varepsilon) + \frac{K}{\varepsilon^2}\,\mathbb{E}\bigg(\int_0^t \big\|\sigma_s - \sigma(m)_s\big\|^2 ds\bigg).$$

Then the result readily follows from (3.4.24) and the property $\theta(\varepsilon)\to 0$ as $\varepsilon\to 0$. □

Proof of Theorem 3.4.1 1) By the localization lemma we may assume that X satisfies (3.4.14) for the relevant p. Then (3.4.7) under (b) is a particular case of (3.4.7) under (c), whereas under (a) it follows from Lemma 3.4.6 if we know it under (b). So it is enough to consider the case (c).

For $A>1$ we use the notation ψ_A and ψ'_A of (3.3.16), that is

$$\begin{aligned} &\psi \text{ is a } C^\infty \text{ function: } \mathbb{R}_+ \to [0,1], \text{ with } 1_{[1,\infty)}(x) \le \psi(x) \le 1_{[1/2,\infty)}, \\ &\psi_A(x) = \psi\big(\|x\|/A\big), \qquad \psi'_A = 1-\psi_A. \end{aligned}$$

Recalling $g_p(x) = \|x\|^p$, we have for a constant independent of $A>1$:

$$|f\psi_A| \le K g_p. \quad (3.4.27)$$

The function $f\psi'_A$ is bounded and uniformly continuous, so $V'^n(f\psi'_A, X)_t \overset{\text{u.c.p.}}{\Longrightarrow} \int_0^t \rho_{c_s}(f\psi'_A)\,ds$ by Lemma 3.4.8. Thus by Proposition 2.2.1 it is enough to prove the following two properties:

$$\int_0^t \rho_{c_s}(f\psi_A)\,ds \overset{\text{u.c.p.}}{\Longrightarrow} 0 \quad \text{as } A\to\infty, \quad (3.4.28)$$

$$\forall \eta>0,\ t>0, \qquad \lim_{A\to\infty} \limsup_{n\to\infty} \mathbb{P}\Big(\sup_{s\le t} \big|V'^n(f\psi_A,X)_s\big| > \eta\Big) = 0. \quad (3.4.29)$$

The first property (3.4.28) is an obvious consequence of (3.4.27) and $f\psi_A\to 0$ as $A\to\infty$ and $|\rho_c(g_p)| \le K\|c\|^{p/2}$. As for (3.4.29), since $X = X_0 + B + X^c$ and

$|f\psi_A| \le Kg_p\psi_A$ and $(g_p\psi_A)(x+y) \le K_p(g_p\psi_{2A})(x) + K_pg_p(y)$, it suffices to prove the two properties

$$\forall \eta > 0,\ t > 0, \qquad \lim_{n\to\infty} \mathbb{P}\big(V'^n(g_p, B)_t > \eta\big) = 0 \tag{3.4.30}$$

$$\forall \eta > 0,\ t > 0, \qquad \lim_{A\to\infty} \limsup_{n\to\infty} \mathbb{P}\big(V'^n\big(g_p\psi_A, X^c\big)_t > \eta\big) = 0. \tag{3.4.31}$$

2) We start with (3.4.30). We set $\eta_t^n = \sup_{i\le[t/\Delta_n]}(\Delta_i^n A(p))$, which goes to 0 as $n\to\infty$. Using Hölder's inequality for the second inequality below, we have

$$\begin{aligned} V'^n(g_p, B)_t &\le \Delta_n^{1-p/2} \sum_{i\ge1} \bigg\| \int_{t\wedge((i-1)\Delta_n)}^{t\wedge(i\Delta_n)} b_s\, ds \bigg\|^p \\ &\le \sum_{i\ge1} \bigg(\int_{t\wedge((i-1)\Delta_n)}^{t\wedge(i\Delta_n)} \|b_s\|^{2p/(2+p)}\, ds \bigg)^{1+p/2} \le A(p)_t \big(\eta_t^n\big)^{p/2}. \end{aligned}$$

This goes to 0, and (3.4.30) is proved.

3) Now we prove (3.4.31). We set $\sigma(m) = \sigma 1_{\{\|\sigma\|\le m\}}$ and $Y(m)_t = \int_0^t \sigma(m)_s\, dW_s$ and $Y'(m) = X^c - Y(m)$. Using again $(g_p\psi_A)(x+y) \le \frac{1}{2\alpha}((g_p\psi_{2A})(x) + g_p(y))$ for some constant $\alpha > 0$ (depending on p), we obtain

$$\begin{aligned} \mathbb{P}\big(V'^n\big(g_p\psi_A, X^c\big)_t > \eta\big) &\le \mathbb{P}\big(V'^n\big(g_p\psi_{2A}, Y(m)\big)_t > \alpha\eta\big) \\ &\quad + \mathbb{P}\big(V'^n\big(g_p, Y'(m)\big)_t > \alpha\eta\big). \end{aligned} \tag{3.4.32}$$

On the one hand (2.1.34) and the property $(g_p\psi_{2A})(x) \le 4\|x\|^{p+1}/A$ yield

$$\begin{aligned} &E\big(V'^n\big(g_p\psi_{A/2}, Y(m)\big)_t\big) \\ &\quad\le \frac{4\Delta_n^{1/2-p/2}}{A} \sum_{i\ge1} \mathbb{E}\bigg(\bigg\| \int_{t\wedge((i-1)\Delta_n)}^{t\wedge(i\Delta_n)} \sigma(m)_s\, dW_s \bigg\|^{p+1} \bigg) \\ &\quad\le \frac{K\Delta_n^{1/2-p/2}}{A} \sum_{i\ge1} \mathbb{E}\bigg(\bigg(\int_{t\wedge((i-1)\Delta_n)}^{t\wedge(i\Delta_n)} \|\sigma(m)_s\|^2\, ds \bigg)^{p/2+1/2} \bigg) \le \frac{Ktm^{p+1}}{A}, \end{aligned}$$

and thus

$$\lim_{A\to\infty} \limsup_{n\to\infty} \mathbb{P}\big(V'^n\big(g_p\psi_{A/2}, Y(m)\big)_t > \alpha\eta\big) = 0. \tag{3.4.33}$$

On the other hand, (2.1.34) and Hölder's inequality give us the following string of inequalities:

$$\begin{aligned} &E\big(V'^n\big(g_p, Y'(m)\big)_t\big) \\ &\quad\le K\Delta_n^{1-p/2} \sum_{i\ge1} \mathbb{E}\bigg(\bigg\| \int_{t\wedge((i-1)\Delta_n)}^{t\wedge(i\Delta_n)} \big(\sigma_s - \sigma(m)_s\big)\, dW_s \bigg\|^p \bigg) \end{aligned}$$

$$\le K\Delta_n^{1-p/2}\sum_{i\ge1}\mathbb{E}\left(\left(\int_{t\wedge((i-1)\Delta_n)}^{t\wedge(i\Delta_n)}\left\|\sigma_s-\sigma(m)_s\right\|^2 ds\right)^{p/2}\right)$$

$$\le K\sum_{i\ge1}\mathbb{E}\left(\int_{t\wedge((i-1)\Delta_n)}^{t\wedge(i\Delta_n)}\left\|\sigma_s-\sigma(m)_s\right\|^p ds\right)$$

$$\le K\mathbb{E}\left(\int_0^t\left\|\sigma_s-\sigma(m)_s\right\|^p ds\right).$$

Since $\sigma_s-\sigma(m)_s=\sigma_s 1_{\{\|\sigma_s\|>m\}}$, the last expression above goes to 0 as $m\to\infty$ because $A(2)_\infty\le K$. Hence

$$\lim_{m\to\infty}\sup_n\mathbb{P}\left(V'^n\left(g_p,Y'(m)\right)_t>\alpha\eta\right)=0.$$

Combining this with (3.4.32) and (3.4.33) readily gives (3.4.31).

4) Finally, we prove the last claim of the theorem. Let $p\ge2$ and suppose that $V'^n(g_p,X)\overset{\text{u.c.p.}}{\Longrightarrow}Y$, where Y is a continuous process. When $p=2$, we know that (3.4.6) holds, and $V'^n(g_2,X)=V^n(g_2,X)$ (recall (3.4.2)) converges in probability to the process $V(g_2,X)$ of (3.3.3). Therefore $V(g_2,X)=Y$ is continuous, which implies that the process X itself is continuous.

Now assume $p>2$. Suppose that X has a jump at some (random) time T. If $i(n,T)$ is the unique (random) integer such that T belongs to the interval $I(n,i(n,T))$, when $T<\infty$, we have $\Delta^n_{i(n,T)}X\to\Delta X_T$. Therefore on the set $\{T<\infty\}$ the process $V'^n(g_p,X)$ has a jump at time $T_{i(n,T)}$ which is equivalent, as $n\to\infty$, to $\Delta_n^{1-p/2}\|\Delta X_T\|^p$, whereas $T_{i(n,T)}\overset{\mathbb{P}}{\longrightarrow}T$. Since $p>2$ this implies that $V'^n(g_p,X)_t\overset{\mathbb{P}}{\longrightarrow}+\infty$ on the set $\{T<\infty\}$ and we obtain a contradiction. Therefore X should be continuous. □

Proof of Corollary 3.4.3 Since h is continuous and positively homogeneous of degree $p\in(0,2)$, the same is true of $|h|$, and we have $|h|\le Kg_p$. Thus (3.4.3) and (a) of the previous theorem give us

$$\begin{aligned}\Delta_n^{1-p/2}V^n(h,X)_t&=V'^n(h,X)_t\overset{\text{u.c.p.}}{\Longrightarrow}\int_0^t\rho_{c_s}(h)\,ds\\ \Delta_n^{1-p/2}V^n\left(|h|,X\right)_t&=V'^n\left(|h|,X\right)_t\overset{\text{u.c.p.}}{\Longrightarrow}\int_0^t\rho_{c_s}\left(|h|\right)ds.\end{aligned}\tag{3.4.34}$$

For each $\varepsilon>0$, the function $(|f|+|h|)\psi_\varepsilon$ vanishes on a neighborhood of 0 (recall the notation (3.3.16) for ψ_ε), so (3.3.2) yields that $V^n((|f|+|h|)\psi_\varepsilon,X)$ converges in probability in the Skorokhod sense. Since $p<2$ it follows that

$$\Delta_n^{1-p/2}V^n\left(\left(|f|+|h|\right)\psi_\varepsilon,X\right)\overset{\text{u.c.p.}}{\Longrightarrow}0.\tag{3.4.35}$$

Now, the assumption $f \sim h$ near 0 implies the existence of a nonnegative function k on $\mathbb{R}_+$ such that $k(\varepsilon) \to 0$ as $\varepsilon \to 0$ and $|f-h| \le k(\varepsilon)|h| + (|f|+|h|)\psi_\varepsilon$. Therefore

$$\begin{aligned}&\left|\Delta_n^{1-p/2} V^n(f,X) - \Delta_n^{1-p/2} V^n(h,X)\right|\\&\quad\le k(\varepsilon)\Delta_n^{1-p/2} V^n\big(|h|,X\big) + \Delta_n^{1-p/2} V^n\big((|f|+|h|)\psi_\varepsilon, X\big).\end{aligned}$$

Then (3.4.9) follows from (3.4.34) and (3.4.35). The same argument holds for the processes $V^n_{int}(f,X)$ as well. □

Proof of Corollary 3.4.4 Let f be a Borel locally bounded function on $\mathbb{R}^d$, satisfying (3.4.10), and also the conditions (a), (b) or (c) of the theorem, according to the case. There is a positive function f' which is continuous, and $|f| \le f'$ identically, and which satisfies (a), (b) or (c) as well.

Up to using the decomposition $f = f^+ - f^-$ and (2.2.16), plus the fact that $V'(f,X)$ is continuous in time, it is in fact enough to prove that for any fixed t,

$$V'^n(f,X)_t \overset{\mathbb{P}}{\longrightarrow} \int_0^t \rho_{c_s}(f)\,ds. \tag{3.4.36}$$

Below we fix $t \ge 0$. The formula $m_n(A) = V'^n(f'1_A, X)_t$ for all $A \in \mathcal{R}^d$ defines a (random) positive finite measure $m_n = m_n(\omega, dx)$ on $\mathbb{R}^d$, and accordingly $m(A) = \int_0^t \rho_{c_s}(f'1_A)\,ds$ defines a random measure m. For any continuous bounded function g the product $f'g$ satisfies (a), (b) or (c). Then Theorem 3.4.1 yields $m_n(g) \overset{\mathbb{P}}{\longrightarrow} m(g)$ for any such g. Moreover, (3.4.10) yields that for all ω outside a $\mathbb{P}$ null set N the bounded function f/f' is $m(\omega, dx)$ almost surely continuous.

Now, we know that there exists a *countable* family $\mathcal{G}$ of continuous bounded functions on $\mathbb{R}^d$ which is *convergence determining*, that is if $\eta_n(g) \to \eta(g)$ for all $g \in \mathcal{G}$, where η_n and η are (positive) finite measures on $\mathbb{R}^d$, then $\eta_n \to \eta$ weakly. Because $\mathcal{G}$ is countable, from any sequence $n_k \to \infty$ one can extract a subsequence $n_{k_m} \to \infty$ such that, for all ω outside a $\mathbb{P}$ null set N' containing the set N described above, we have $m_{n_{k_m}}(\omega, g) \to m(\omega, g)$ for all $g \in \mathcal{G}$. It follows that, still when $\omega \notin N$, we have $m_{n_{k_m}}(\omega, .) \to m(\omega, .)$ weakly. Therefore, since the function f/f' is bounded and $m(\omega, .)$ almost everywhere continuous when $\omega \notin N'$, we deduce $mn_{k_m}(f/f') \to m(f/f')$ outside N', that is almost surely.

Observe that the left and right sides of (3.4.36) are respectively equal to $m_n(f/f')$ and $m(f/f')$. Hence, so far, we have proved that from any subsequence n_k we can extract a further subsequence n_{k_m} along which the convergence (3.4.36) is almost sure. By the subsequences principle (2.2.17), we deduce (3.4.36). □

3.5 Applications

In this last section we introduce two of the fundamental problems which motivate this entire book. We explain how the results of this chapter contribute to their solution, and also why they are insufficient for a complete solution: the same examples, together with a few others, will be pursued in a systematic way at the end of most forthcoming chapters.

3.5.1 Estimation of the Volatility

Our first example is of particular interest in mathematical finance. In this field, the price or the log-price of an asset is typically a semimartingale, and indeed an Itô semimartingale, under virtually all models that have been used to date. That is, we have

$$X_t = X_0 + \int_0^t b_s\,ds + \int_0^s \sigma_s\,dW_s + (\delta 1_{\{\|\delta\|\leq 1\}}) \star (\mathfrak{p} - \mathfrak{q})_t + (\delta 1_{\{\|\delta\|>1\}}) \star \mathfrak{p}_t, \quad (3.5.1)$$

when we write its Grigelionis form. In this equation X represents the logarithm of a price, the quantity σ_t is called the *volatility* (or *spot volatility*) of the price, and it turns out to be of primary importance for many purposes in finance.

Actually, it should be clear from (3.5.1) that if we replace σ_t by $-\sigma_t$ for all t in some (random, progressively measurable) set, the model is not changed, so really it is $|\sigma_t|$ that is important. In most models one imposes the condition $\sigma_t \geq 0$ which, as we have just seen, is not a restriction. More generally, in the multivariate case where X is a vector of (log)-prices, the important quantity is the diffusion matrix $c_t = \sigma_t \sigma_t^*$.

Quite often the volatility is a random process, as above, but even when it is non-random it varies quite significantly with time, typically with seasonal variations within each day (for instance it is usually quite smaller at lunch time than at the beginning or at the end of the day). So perhaps an even more useful quantity to evaluate is the "average volatility" for some given period, typically a day. In the multidimensional case, one wants to evaluate the average of c_t.

Evidently the average of σ_s^2 over, say, the interval $[0,t]$, is not the same as the squared average of $|\sigma_s|$, and after all the qth-root of the average of $|\sigma_s|^q$ is also, for any $q>0$, a kind of average of $|\sigma_s|$ over $[0,t]$. However the power $q=2$ has, here as in many other places, very special and nice properties, and it is the only power for which there is a straightforward multivariate extension because in this case $c=\sigma\sigma^*$ is uniquely determined. This explains why most of the interest has been focused on the evaluation of the averaged squared-volatility, or equivalently the so-called (somewhat misleadingly) *integrated volatility*:

$$\int_0^t \sigma_s^2\,ds, \quad \text{or} \quad \int_0^t c_s\,ds \quad \text{in the multivariate case.}$$

So, as said before, we will study this problem throughout the book. At this stage, we can say only one thing, namely

$$X \text{ is continuous} \;\Rightarrow\; \sum_{i=1}^{[t/\Delta_n]} \Delta_i^n X^j\, \Delta_i^n X^k \stackrel{\text{u.c.p.}}{\Longrightarrow} \int_0^t c_s^{jk}\,ds, \quad (3.5.2)$$

which comes from Theorem 3.3.1, and even from the long known result (2.1.8). This is far from fully solving the problem, for two reasons: one is that it requires X to be continuous, otherwise the result is wrong, and to overcome this difficulty we will need to "truncate" the increments, see Chap. 9. The second reason is that, even when (3.5.2) holds, it is useful in practice only if we know whether the approximation is accurate. Mathematically speaking this means that we need a "rate of convergence" in (3.5.2): this is the object of the Central Limit Theorem which we start developing in the next chapter.

Finally, other integrated powers $\int_0^t |\sigma_s|^p\,ds$ for $p>0$ may also sometimes be of interest, for instance this quantity with $p=4$ arises naturally in the CLT for (3.5.2). Then the normalized LLN proved above gives us the following result, say in the one-dimensional case. We introduce the notation

$$D(X,p,\Delta_n)_t = \sum_{i=1}^{[t/\Delta_n]} |\Delta_i^n X|^p \tag{3.5.3}$$

which is $V^n(f,X)_t$, or equivalently $\Delta_n^{p/2} V'^n(f,X)$ for the function $f(x)=|x|^p$. We will see later the reason for this new notation, which emphasizes the time step Δ_n. Then when X is continuous and $p\geq 2$ and (3.4.6) holds, and also for all Itô semimartingales when $p<2$, we have

$$\Delta_n^{1-p/2} D(X,p,\Delta_n)_t \stackrel{\text{u.c.p.}}{\Longrightarrow} m_p \int_0^t |\sigma_s|^p\,ds, \tag{3.5.4}$$

where $m_p=\mathbb{E}(|U|^p)$ is the p absolute moment of an $\mathcal{N}(0,1)$ random variable U.

3.5.2 Detection of Jumps

The second example is about the problem of deciding whether a discretely observed process is continuous or not. More specifically, the one-dimensional process X is observed at all times $i\Delta_n$ (with $i\in\mathbb{N}$) within a fixed time interval $[0,t]$, and we want to decide whether the partially observed path is continuous or not.

This seems an impossible question to answer, since any discrete set of observations is compatible (in many different ways, obviously) with a continuous path, and also with a discontinuous path. However, when X is an Itô semimartingale, there is a procedure which allows one to solve this question in a *consistent* way, as Δ_n becomes small.

By "procedure" we mean the following: at stage n, the set of all possible (relevant) observations is $\mathbb{R}^{N_n(t)+1}$. Recall that $N_n(t)+1$ is the number of integers i such that $i\Delta_n$ lies in the interval $[0,t]$. Then a *procedure* is the assessment, for each n, of a subset A_n of the set $\mathbb{R}^{N_n(t)+1}$, such that if the observations fall in A_n we decide that the observed path is continuous, and if they fall outside A_n we decide that the observed path is discontinuous. The procedure is called *consistent*, and one should indeed say "weakly consistent", if the probability of taking the right decision goes to 1 as $n \to \infty$.

Mathematically speaking we divide the sample space Ω into a 2-sets partition:

$$\Omega_t^{(c)} = \{\omega : s \mapsto X_s(\omega) \text{ is continuous on } [0,t]\}$$
$$\Omega_t^{(d)} = \{\omega : s \mapsto X_s(\omega) \text{ is not continuous on } [0,t]\}.$$

It may be that $\Omega_t^{(c)} = \Omega$, when the specific model for X does not allow jumps, that is $\delta \equiv 0$ in (3.5.1). It may be also that $\Omega_t^{(c)} = \emptyset$, for example when the "purely discontinuous" part of X (the last two terms in (3.5.1)) is a Lévy process with infinite Lévy measure: in these two cases we have a classical statistical testing problem to solve. But it may also be that neither $\Omega_t^{(c)}$ nor $\Omega_t^{(d)}$ are empty, when for example the purely discontinuous part of X is a Poisson or a compound Poisson process. In this case the statistical problem is not quite classical: we do not have to test whether some parameter of the model takes a specific value or lies in a specific domain. Instead we have to test whether the (random) outcome ω lies in some specific subset of the sample space Ω.

In the face of such a problem, the statistician has to come up with a procedure A_n as described above: if C_n is the subset of Ω on which $(X_{i\Delta_n})_{0\le i\le N_n(t)} \in A_n$, the statistician decides for "continuous" if $\omega \in C_n$ and for "discontinuous" otherwise. One may see C_n as the critical (rejection) region for testing the null hypothesis that X is discontinuous. The procedure A_n, or C_n, is consistent if

$$\begin{aligned} \mathbb{P}(\Omega_t^{(c)}) > 0 \;&\Rightarrow\; \mathbb{P}(C_n \mid \Omega_t^{(c)}) \to 1 \\ \mathbb{P}(\Omega_t^{(d)}) > 0 \;&\Rightarrow\; \mathbb{P}((C_n)^c \mid \Omega_t^{(d)}) \to 1 \end{aligned} \tag{3.5.5}$$

(here $(C_n)^c$ is the complement of C_n in Ω, and $\mathbb{P}(. \mid A)$ is the ordinary conditional probability).

The two LLNs proved in this chapter provide a simple solution to this problem. To see that, we take some real $p > 2$ and some integer $k \ge 2$ and, recalling (3.5.3), we set

$$S_n = \frac{D(X,p,k\Delta_n)_t}{D(X,p,\Delta_n)_t}. \tag{3.5.6}$$

Theorem 3.5.1 *Let $p > 2$ and let $k \ge 2$ be an integer, and $t > 0$ and S_n given by* (3.5.6).

a) *If X is a semimartingale and if $\mathbb{P}(\Delta X_t \ne 0) = 0$, then*

$$S_n \xrightarrow{\mathbb{P}} 1 \quad \text{in restriction to the set } \Omega_t^{(d)}. \tag{3.5.7}$$

b) *If X is an Itô semimartingale and the process $\int_{\{|\delta(s,z)|\le 1\}} \delta(s,z)^2\,\lambda(dz)$ is locally bounded, and b_s and c_s satisfy* (3.4.6), *and if further $\int_0^t c_s\,ds > 0$ a.s., then*

$$S_n \xrightarrow{\mathbb{P}} k^{p/2-1} \quad \text{in restriction to the set } \Omega_t^{(c)}. \tag{3.5.8}$$

Before providing the proof, we show how this result solves our problem, at least partially. Define the critical region C_n by

$$C_n = \{S_n < x\}, \quad \text{where } x \in \left(1, k^{p/2-1}\right). \tag{3.5.9}$$

(3.5.8) gives us $\mathbb{P}(C_n \cap \Omega_t^{(c)}) \to 0$ and $\mathbb{P}((C_n)^c \cap \Omega_t^{(d)}) \to 0$, from which (3.5.5) follows. That is, we have thus constructed a *consistent procedure* for our problem.

This is a purely asymptotic result, in which the choice of the cut-off x is arbitrary in $(1, k^{p/2-1})$. But it should be clear that if, for example, we choose x very close to 1, the second convergence in (3.5.5) may be very slow, and in practice for finite samples $\mathbb{P}((C_n)^c \mid \Omega_t^{(d)})$ may be close to 0 instead of 1. Therefore we need to improve on the procedure by choosing an x appropriate to the size Δ_n, or to the number $N_n(t)+1$ of observations. Typically the choice $x = x_n$ will depend on n, and is based (as usual in asymptotic statistics) on a rate of convergence in (3.5.8). This is why we need some kind of CLT, in both cases (3.5.7) and (3.5.8).

Another comment: if we relax the assumption $\int_0^t c_s\,ds > 0$, the convergence in (3.5.8) holds on the set $\Omega_t^{(c)} \cap \Omega_t^W$, where

$$\Omega_t^W = \left\{\omega : \int_0^t c_s\,ds > 0\right\}. \tag{3.5.10}$$

And, of course, in virtually all models the set $\Omega_t^{(c)} \cap (\Omega_t^W)^c$ is empty (on this set, the path of X is a "pure drift" over the whole interval $[0,t]$).

Proof of Theorem 3.5.1 a) (3.3.2) applied with $f(x) = |x|^p$, plus the property $\mathbb{P}(\Delta X_t \neq 0) = 0$, gives us that both $D(p,\Delta_n)_t$ and $D(p,k\Delta_n)_t$ converge in probability to the same limit $\sum_{s\le t} |\Delta X_s|^p$, which is positive on the set $\Omega_t^{(d)}$. This yields the convergence (3.5.7).

b) Suppose that X is *continuous*. By (3.5.6), the variables $\Delta_n^{1-p/2} D(p,\Delta_n)_t$ and $\Delta_n^{1-p/2} D(p,k\Delta_n)_t$ converge in probability to $D_t = m_p \int_0^t |\sigma_s|\,ds$ and to $k^{p/2-1} D_t$ respectively. Since $D_t > 0$ a.s. by hypothesis, we deduce $S_n \xrightarrow{\mathbb{P}} k^{p/2-1}$.

For proving the same convergence in restriction to $\Omega_t^{(c)}$ when X is not continuous, we need some preparation, somewhat similar to the proof of Corollary 3.3.4. We set

$$u_s = \lambda\left(\{z : \delta(s,z) \neq 0\}\right), \qquad v_s = \int_{\{|\delta(s,z)|\le 1\}} \delta(s,z)^2\,\lambda(dz),$$

$$T_q = \inf\left(s : \int_0^s u_r\,dr \ge q\right), \qquad w_s = \begin{cases} \int_{\{|\delta(s,z)|\le 1\}} \delta(s,z)\,\lambda(dz) & \text{if } u_s < \infty \\ +\infty & \text{otherwise.} \end{cases}$$

We have

$$\mathbb{E}(1_{\{|\delta|>0\}} * \mathfrak{p}_{T_q}) = \mathbb{E}(1_{\{|\delta|>0\}} * \mathfrak{q}_{T_q}) = \mathbb{E}\left(\int_0^{T_q} u_s\,ds\right) \le q, \tag{3.5.11}$$

hence $1_{\{|\delta|>0\}} * p_{T_q} < \infty$ a.s. and X has almost surely finitely many jumps on $[0, T_q]$. The Cauchy-Schwarz inequality yields $|w_s| \leq \sqrt{u_s v_s}$, hence $\int_0^{t\wedge T_q} w_s^2\, ds < \infty$ because by hypothesis the process v_s is locally bounded. Therefore if we stop both members of (3.5.1) at time T_q, we get

$$X_{t\wedge T_q} = X(q)_{t\wedge T_q} + \sum_{s\leq t\wedge T_q} \Delta X_s,$$

$$\text{where } X(q)_t = X_0 + \int_0^{t\wedge T_q} (b_s - w_s)\, ds + \int_0^t \sigma_s\, dW_s.$$

The process $X(q)$ is a continuous Itô semimartingale which satisfies (3.4.2) (for this we apply (3.4.2) for b and c, and also $\int_0^{t\wedge T_q} w_s^2\, ds < \infty$ and $\frac{2p}{2+p} < 2$), and with the same process c as X. Therefore what precedes yields, as $n \to \infty$:

$$S_n(q) = \frac{D(X(q), p, k\Delta_n)_t}{D(X(q), p, \Delta_n)_t} \xrightarrow{\mathbb{P}} k^{p/2-1}. \tag{3.5.12}$$

On the set $\Omega_t^{(c)} \cap \{T_q \geq t\}$ we see that $X_s = X(q)_s$ for all $s \leq t$, hence also $S_n = S_n(q)$. Thus, in view of (3.5.12), it remains to prove that $\Omega_t^{(c)} \subset \cup_{q\geq 1}\{T_q \geq t\}$ almost surely. To see this, we set $S = \inf(s : \Delta X_s \neq 0)$ and we observe that $\Omega_t^{(c)} \subset \{S \geq t\}$ (note that we may have $S = t$ on $\Omega_t^{(c)}$, if there is a sequence of jump times decreasing strictly to t). The same argument as for (3.5.11) gives $\mathbb{E}(\int_0^S u_s\, ds) = \mathbb{E}(1 * \mu_S) \leq 1$ (here μ is the jump measure of X). Therefore $\int_0^S u_s\, ds < \infty$ a.s., implying that $\{S \geq t\} \subset \cup_{q\geq 1}\{T_q \geq t\}$ almost surely. This completes the proof. □

Bibliographical Notes

The generalized Itô's formula of Theorem 3.2.2 is taken from Jacod, Jakubowski and Mémin [56]. The basic Law of Large Numbers (Theorem 3.3.1) has a relatively long history: it has been proved by Greenwood and Fristedt [38] for Lévy processes, for some classes of test functions f, and with an almost sure convergence in some cases. The same result for semimartingales, when the test function is a power and absolute power function, is due to Lépingle [71], including again the almost sure convergence in some cases. This result builds on the convergence of the approximate quadratic variation, which is a result as old as martingales and stochastic calculus in special cases, and due to Meyer [76] for arbitrary semimartingales. The second basic Law of Large Numbers (Theorem 3.4.1) can be found in many places, under various special hypotheses, especially in the statistical literature for diffusion processes; the general version given here is taken from Jacod [60].

Chapter 4
Central Limit Theorems: Technical Tools

Usually, Laws of Large Numbers are adorned with an associated Central Limit Theorem, which describes the "rate of convergence" at which they take place, and also the limiting variable or process that are obtained after normalization. The two LLNs of Chap. 3 are no exceptions: we have a rate, that is a sequence v_n of positive numbers going to infinity, such that the processes $v_n(V^n(f,X) - V(f,X))$ in the setting of Theorem 3.3.1, or $v_n(V'^n(f,X) - V'(f,X))$ in the setting of Theorem 3.4.1, converge to a limiting process which is not degenerate, which means that it is finite-valued but not identically 0.

For the first case, the LLN was obtained without any restriction on the random discretization scheme, and the semimartingale X was arbitrary, whereas in the second case we needed regular schemes and X to be an Itô semimartingale. For the CLT, a regular scheme and that X is an Itô semimartingale is needed in both cases. Then, not surprisingly, the rate of convergence will be $v_n = 1/\sqrt{\Delta_n}$ always.

Central Limit Theorems have lengthy proofs. This is why we start with a preliminary chapter which sets up the basic notions and tools that we will need. The CLTs themselves will be stated and proved in the next chapter. This means that a reader can skip the present chapter, except perhaps the first section which describes the limiting processes, and can come back to it when a specific technical result is needed.

4.1 Processes with $\mathcal{F}$-Conditionally Independent Increments

We have seen in Theorem 2.2.15 a situation where a triangular array of random variables is defined on some space $(\Omega, \mathcal{F}, (\mathcal{F}_t)_{t\geq 0}, \mathbb{P})$, and the associated rows of partial sums converge stably in law to a limit $A + Y$, where A is defined on Ω and Y is defined on a very good filtered extension of $(\Omega, \mathcal{F}, (\mathcal{F}_t)_{t\geq 0}, \mathbb{P})$ and is, conditionally on $\mathcal{F}$, a continuous Gaussian martingale with a quadratic variation C being $\mathcal{F}$ measurable.

In this section, we give a "concrete" construction of Y, at least when the process C is absolutely continuous with respect to Lebesgue measure. We also extend this construction to some cases where Y is discontinuous.

J. Jacod, P. Protter, *Discretization of Processes*,
Stochastic Modelling and Applied Probability 67,
DOI 10.1007/978-3-642-24127-7_4,

4.1.1 The Continuous Case

We consider the following problem: let $\widetilde{C}$ be a process on $(\Omega, \mathcal{F}, (\mathcal{F}_t)_{t\geq 0}, \mathbb{P})$, of the form $\widetilde{C}_t = \int_0^t \widetilde{c}_s\, ds$, where $\widetilde{c}$ takes its values in the set $\mathcal{M}^+_{q\times q}$ of all $q \times q$ symmetric nonnegative matrices and is progressively measurable (or predictable). We want to construct a q-dimensional process Y, defined on a very good filtered extension $(\widetilde{\Omega}, \widetilde{\mathcal{F}}, (\widetilde{\mathcal{F}}_t)_{t\geq 0}, \widetilde{\mathbb{P}})$ of the original space, which conditionally on $\mathcal{F}$ is a *centered Gaussian process with independent increments* and (conditional) covariance

$$\widetilde{\mathbb{E}}\big(Y_t^i Y_t^j \mid \mathcal{F}\big) = \widetilde{C}_t^{ij} = \int_0^t \widetilde{c}_s^{ij}\, ds. \tag{4.1.1}$$

Recall that this is equivalent to saying that, conditionally on $\mathcal{F}$, the process Y is a continuous martingale with "deterministic" quadratic variation-covariation process $\widetilde{C}$, and since $\widetilde{C}$ is continuous and Y is conditionally Gaussian, it is also necessarily a.s. continuous. This problem is studied in Sect. II.7 of [57] but, in view of its importance in the present book, we repeat the construction in detail here.

For solving this problem, we consider an auxiliary filtered probability space $(\Omega', \mathcal{F}', (\mathcal{F}'_t)_{t\geq 0}, \mathbb{P}')$, and we consider the product filtered extension $(\widetilde{\Omega}, \widetilde{\mathcal{F}}, (\widetilde{\mathcal{F}}_t)_{t\geq 0}, \widetilde{\mathbb{P}})$, as defined by (2.1.26) with $\mathbb{Q}(\omega, d\omega') = \mathbb{P}'(d\omega')$ and (2.1.27). This is a "very good extension", and we start with a general result, of independent interest.

Proposition 4.1.1 *In the above setting, assume further that $\mathcal{F}'_{t-} = \vee_{s<t}\mathcal{F}'_s$ is a separable σ-field for each $t > 0$. Let $Z = Z_t(\omega, \omega')$ be a martingale on the extended space, which is orthogonal to all bounded martingales on $(\Omega, \mathcal{F}, (\mathcal{F}_t)_{t\geq 0}, \mathbb{P})$. Then for $\mathbb{P}$ almost all ω the process $(\omega', t) \mapsto Z_t(\omega, \omega')$ is a martingale on $(\Omega', \mathcal{F}', (\mathcal{F}'_t)_{t\geq 0}, \mathbb{P}')$.*

Proof Any M in the set $\mathcal{M}_b$ of all bounded martingales on $(\Omega, \mathcal{F}, (\mathcal{F}_t)_{t\geq 0}, \mathbb{P})$ is also a martingale on $(\widetilde{\Omega}, \widetilde{\mathcal{F}}, (\widetilde{\mathcal{F}}_t)_{t\geq 0}, \widetilde{\mathbb{P}})$ because the extension is very good. Our assumption on Z means that the product MZ is again a martingale on the extended space for all $M \in \mathcal{M}_b$.

We want to prove that for all ω outside a $\mathbb{P}$ null set, we have

$$\int Z_t\big(\omega, \omega'\big) 1_{A'}\big(\omega'\big)\, \mathbb{P}'\big(d\omega'\big) = \int Z_s\big(\omega, \omega'\big) 1_{A'}\big(\omega'\big)\, \mathbb{P}'\big(d\omega'\big) \tag{4.1.2}$$

for all $0 \leq s \leq t$ and $A' \in \mathcal{F}'_s$. Since Z is right-continuous, it is enough to prove this when s, t are rational and $A \in \mathcal{F}'_{s-}$. Since further $\mathcal{F}'_{s-}$ is separable, it is enough to prove it when A' ranges through a countable algebra generating $\mathcal{F}'_{s-}$. At this stage, we can permute "for $\mathbb{P}$ almost all ω" and "for all s, t, A' "; that is we need to prove that, for any given s, t, A' as above we have (4.1.2) for $\mathbb{P}$ almost all ω. Since both sides of (4.1.2) are $\mathcal{F}_t$ measurable, when considered as functions of ω, we are then left to show that for all $s \leq t$ and $A' \in \mathcal{F}'_s$ and $A \in \mathcal{F}_t$, we have

$$\widetilde{\mathbb{E}}(Z_t\, 1_{A\times A'}) = \widetilde{\mathbb{E}}(Z_s\, 1_{A\times A'}). \tag{4.1.3}$$

Consider the bounded martingale $M_r = \mathbb{P}(A \mid \mathcal{F}_r)$ on $(\Omega, \mathcal{F}, (\mathcal{F}_t)_{t\geq 0}, \mathbb{P})$. We have

$$\widetilde{\mathbb{E}}(Z_t\, 1_{A\times A'}) = \widetilde{\mathbb{E}}(Z_t\, M_t\, 1_{A'}) = \widetilde{\mathbb{E}}(Z_s\, M_s\, 1_{A'}) = \widetilde{\mathbb{E}}(Z_s\, M_t\, 1_{A'}) = \widetilde{\mathbb{E}}(Z_s\, 1_{A\times A'})$$

where the second and third equalities come from $A' \in \widetilde{\mathcal{F}}_s$ and, respectively, the facts that ZM and M are martingales on $(\widetilde{\Omega}, \widetilde{\mathcal{F}}, (\widetilde{\mathcal{F}}_t)_{t\geq 0}, \widetilde{\mathbb{P}})$. Hence (4.1.3) holds. □

Next, we assume that $(\Omega', \mathcal{F}', (\mathcal{F}'_t)_{t\geq 0}, \mathbb{P}')$ supports a q-dimensional $(\mathcal{F}'_t)$-Brownian motion W': we use a "prime" here because it is defined on Ω', and should not be confused with the Brownian motion W on $(\Omega, \mathcal{F}, (\mathcal{F}_t)_{t\geq 0}, \mathbb{P})$ which enters the Grigelionis decomposition of the basic semimartingale X in which we are interested. Finally, the process $\widetilde{c}$ admits a progressively measurable square-root $\widetilde{\sigma}$, that is a $q \times q$ matrix-valued process having $\widetilde{c} = \widetilde{\sigma}\widetilde{\sigma}^\star$, and with the additional property that $\|\widetilde{\sigma}\|^2 \leq K\|\widetilde{c}\|$ for some constant K (one may take for example a symmetric square-root).

Proposition 4.1.2 *In the above setting, the process*

$$Y_t = \int_0^t \widetilde{\sigma}_s\, dW'_s \quad \textit{with components}\quad Y^i_t = \sum_{j=1}^q \int_0^t \widetilde{\sigma}^{ij}_s\, dW'^j_s, \tag{4.1.4}$$

is well defined on the extension, and defines a process which, conditionally on $\mathcal{F}$, is a centered continuous Gaussian process with independent increments satisfying (4.1.1).

Proof The progressive measurability of $\widetilde{\sigma}$ and $\|\widetilde{\sigma}\|^2 \leq K\|\widetilde{c}\|$ ensures that the stochastic integral (4.1.4) is well defined, and defines a continuous q-dimensional local martingale Y on the extended space, with angle bracket $\widetilde{C}$. This holds regardless of the filtration $(\mathcal{F}'_t)$ on the second factor Ω', as soon as it makes the process W' an $(\mathcal{F}'_t)$-Brownian motion, so it is no restriction here to assume that $(\mathcal{F}'_t)$ is indeed the filtration generated by W', hence in particular each $\mathcal{F}'_{t-}$ is a separable σ-field.

We set $T_n = \inf(t : \int_0^t \|\widetilde{c}_s\|\, ds \geq n)$, which is a sequence of $(\mathcal{F}_t)$-stopping times increasing to infinity. Each stopped process $Z(n,i)_t = Y^i_{t\wedge T_n}$ is a martingale (and not only a local martingale), as well as each process $Z'(n,i,j)_t = Z(n,i)_t Z(n,j)_t - \widetilde{C}^{ij}_{t\wedge T_n}$. Moreover since W' is obviously orthogonal to each element of $\mathcal{M}_b$ because we have taken a product extension, and since $Z(n,i)$ and $Z'(n,i,j)$ are stochastic integrals with respect to W' (use Itô's formula to check this for $Z'(n,i,j)$), those processes $Z(n,i)$ and $Z'(n,i,j)$ are orthogonal to all elements of $\mathcal{M}_b$.

At this stage, the previous proposition yields that for $\mathbb{P}$ almost all ω, the processes $Z(n,i)(\omega,.)$ and $Z'(n,i,j)(\omega,.)$ are continuous martingales on $(\Omega', \mathcal{F}', (\mathcal{F}'_t)_{t\geq 0}, \mathbb{P}')$. Using the fact that the stopping times T_n are functions of ω only, we deduce that $Y^i(\omega,.)$ and $Y^i(\omega,.)Y^j(\omega,.) - \widetilde{C}^{ij}(\omega)$ are continuous martingales on $(\Omega', \mathcal{F}', (\mathcal{F}'_t)_{t\geq 0}, \mathbb{P}')$. Since $\widetilde{C}(\omega)$ does not depend on ω', it follows that (for $\mathbb{P}$

almost all ω again) the process $Y(\omega,.)$ is a centered Gaussian process with independent increments on $(\Omega',\mathcal{F}',(\mathcal{F}'_t)_{t\ge 0},\mathbb{P}')$, satisfying $\int (Y^i_t Y^j_t)(\omega,\omega')\mathbb{P}'(d\omega') = \widetilde{C}^{ij}_t(\omega)$, and the result is proved. □

4.1.2 The Discontinuous Case

So far, we have constructed continuous processes which, $\mathcal{F}$-conditionally, have independent increments. Here we will again construct processes with $\mathcal{F}$-conditionally independent increments, but they will be "purely discontinuous", with the unusual feature that they will jump only at "fixed times of discontinuity". Even though this does not cover the most general case of discontinuous processes which conditionally on $\mathcal{F}$ have independent increments, it will be enough for our purposes.

The situation is as follows: we have a probability measure η on $\mathbb{R}^r$ for some integer $r\ge 1$, and an optional process V on the space $(\Omega,\mathcal{F},(\mathcal{F}_t)_{t\ge 0},\mathbb{P})$, taking its values in the set of all $q\times r$ matrices, and such that the sets $D(V,\omega)=\{t: V_t(\omega)\neq 0\}$ are at most countable for all $\omega\in\Omega$ and do not contain 0. Our aim is to construct a q-dimensional process Y on an extension of the initial probability space, such that conditionally on $\mathcal{F}$ (that is, loosely speaking, for each $\omega\in\Omega$) the process $Y(\omega,.)$ can be written as

$$Y(\omega,.)_t = \sum_{s\le t} V_s(\omega)U_s \tag{4.1.5}$$

(with matrix notation), where the U_s's are i.i.d. with law η. In a more mathematical way, this amounts to constructing a process Y which conditionally on $\mathcal{F}$ has independent increments and a characteristic functions for the increments (where $u\in\mathbb{R}^q$ and u^* denotes the transpose, so u^*v is the scalar product when $u,v\in\mathbb{R}^q$) given by

$$\widetilde{\mathbb{E}}\big(e^{iu^*(Y_{t+s}-Y_t)}\mid\mathcal{F}\big) = \prod_{v\in(t,t+s]\cap D(V,\omega)} \int e^{iu^* V_v(\omega)x}\,\eta(dx). \tag{4.1.6}$$

Typically, this will be used with $V_t=f(\Delta X_t)$, for a matrix-valued function on $\mathbb{R}^d$ vanishing at 0 and for X a d-dimensional semimartingale.

(4.1.5) is not really meaningful, but (4.1.6) makes sense, under appropriate conditions ensuring that the possibly infinite product on the right side converges. Below we give two different sets of conditions for this, and we reformulate (4.1.5) so that it becomes meaningful.

The construction is based on the existence of *weakly exhausting sequences for the set* $D(V,.)$: by this we mean a sequence $(T_n)_{n\ge 1}$ of stopping times on $(\Omega,\mathcal{F},(\mathcal{F}_t)_{t\ge 0},\mathbb{P})$, such that for all ω outside a null set, we have $D(V,\omega)\subset\{T_n(\omega): n\ge 1, T_n(\omega)<\infty\}$ and also $T_n(\omega)\neq T_m(\omega)$ whenever $n\neq m$ and $T_n(\omega)<\infty$. When further $D(V,\omega)=\{T_n(\omega): n\ge 1, T_n(\omega)<\infty\}$ we have an *exhausting sequence*. The existence of exhausting, and *a fortiori* of weakly exhausting, sequences is a well known fact of the "general theory of processes", see e.g. Dellacherie [24], and as a rule there are many different exhausting sequences for a given set $D(V)$.

In view of this, we consider an auxiliary probability space $(\Omega', \mathcal{F}', \mathbb{P}')$, endowed with a sequence $(U_n)_{n\geq 1}$ of i.i.d. r-dimensional variables with law η. We take the product $(\widetilde{\Omega}, \widetilde{\mathcal{F}}, \widetilde{\mathbb{P}})$ of $(\Omega, \mathcal{F}, \mathbb{P})$ and $(\Omega', \mathcal{F}', \mathbb{P}')$, and we interpret (4.1.5) as

$$Y_t = \sum_{n:\, T_n \leq t} V_{T_n} U_n \tag{4.1.7}$$

(we still need conditions for this "series" to converge in a suitable sense). The natural filtration to consider on the extended space is the following one:

$(\widetilde{\mathcal{F}}_t)$ is the smallest filtration containing $(\mathcal{F}_t)$
and such that U_n is $\widetilde{\mathcal{F}}_{T_n}$ measurable for all n.

We thus get a filtered extension $(\widetilde{\Omega}, \widetilde{\mathcal{F}}, (\widetilde{\mathcal{F}}_t)_{t\geq 0}, \widetilde{\mathbb{P}})$, and it is immediate to verify that it is very good.

Proposition 4.1.3 *Assume that the variables U_n have a finite first moment $\mu_1 \in \mathbb{R}^r$. As soon as*

$$\sum_{s\leq t} \|V_s\| \leq \infty \quad a.s.\ \forall t > 0, \tag{4.1.8}$$

for any weakly exhausting sequence (T_n) for $D(V,.)$ the series $\sum_{n:\, T_n\leq t} V_{T_n} U_n$ in (4.1.7) *is a.s. absolutely convergent for all $t > 0$, and this formula defines a càdlàg adapted process with finite variation on the extended space, which conditionally on $\mathcal{F}$ has independent increments, and satisfies*

$$\widetilde{\mathbb{E}}(Y_t \mid \mathcal{F}) = \mu_1 \sum_{s\leq t} V_s \tag{4.1.9}$$

and also (4.1.6). *In particular the $\mathcal{F}$-conditional law of Y does not depend on the weakly exhausting sequence (T_n).*

Proof The formula $Y'_t = \sum_{n:\, T_n\leq t} \|V_{T_n}\|\, \|U_n\|$ defines a $[0,\infty]$-valued increasing process on the extension, and obviously

$$\mathbb{E}\big(Y'_t \mid \mathcal{F}\big) = \mu'_1 \sum_{n:\, T_n\leq t} \|V_{T_n}\|,$$

where μ'_1 is the expected value of $\|U_n\|$. Then (4.1.8) implies that the above is a.s. finite, hence $Y'_t < \infty$ a.s. This implies the first claim, and (4.1.9) is obvious, as is the independent increments property of Y conditionally on $\mathcal{F}$. By construction the left side of (4.1.6) equals

$$\prod_{n:\, t<T_n(\omega)\leq t+s} \int e^{iu^\star V_{T_n}(\omega)x}\, \eta(dx),$$

hence (4.1.6) holds. Finally the last claim follows. □

Proposition 4.1.4 *Assume that the variables U_n are centered and have a finite variance-covariance matrix $\mu_2 = (\mu_2^{ij})_{1\le i,j\le r}$. As soon as*

$$\sum_{s\le t} \|V_s\|^2 \ < \ \infty \quad a.s.\ \forall t>0, \tag{4.1.10}$$

for any weakly exhausting sequence (T_n) for $D(V,.)$ the series $\sum_{n:\,T_n\le t} V_{T_n}U_n$ in (4.1.7) *converges in probability for all $t>0$, and this formula defines an adapted process on the extended space, which is a.s. càdlàg and is $\mathcal{F}$-conditionally centered with independent increments and finite second moments (hence a square-integrable martingale) and satisfies*

$$\widetilde{\mathbb{E}}\big(Y_t^i Y_t^j \mid \mathcal{F}\big) \ = \ \sum_{s\le t}\ \sum_{k,l=1}^{r} V_s^{ik}\mu_2^{kl}V_s^{jl}, \tag{4.1.11}$$

and also (4.1.6). *In particular the $\mathcal{F}$-conditional law of Y does not depend on the weakly exhausting sequence (T_n). Moreover:*

a) *If the random vectors U_n are Gaussian, the process Y is $\mathcal{F}$-conditionally Gaussian.*
b) *If the process $\sum_{s\le t}\|V_s\|^2$ is locally integrable, then Y is a locally square-integrable martingale on the extended space.*

Proof 1) We denote by N the $\mathbb{P}$ null set outside of which $\sum_{s\le t}\|V_s\|^2<\infty$ for all t. For all $m\ge 1$ and $t\ge 0$ we set $I_m(t)=\{n:\ 1\le n\le m,\ T_n\le t\}$ and

$$Z(m)_t \ = \ \sum_{n\in I_m(t)} V_{T_n}U_n,$$

which obviously is well defined and $\widetilde{\mathcal{F}}_t$ measurable. We write $Z^\omega(m)_t(\omega')$ for $Z(m)_t(\omega,\omega')$, and we put $c_s^{ij}=\sum_{k,l=1}^r V_s^{ik}\mu_2^{kl}V_s^{jl}$. The following properties are elementary:

$$\left.\begin{array}{l}\text{under } \mathbb{P}' \text{ the process } Z^\omega(m) \text{ is centered with independent}\\ \qquad\qquad \text{increments and is a square-integrable martingale,}\\ \mathbb{E}'\big(Z^\omega(m)_t^i\, Z^\omega(m)_t^j\big)=\sum_{n\in I_m(t)(\omega)} c_{T_n}^{ij}(\omega),\\ \mathbb{E}'\big(e^{iu^\star(Z^\omega(m)_{t+s}-Z(m)_t)}\big)=\prod_{n\in I_m(t+s)(\omega)\setminus I_m(t)(\omega)}\int e^{iu^\star V_{T_n}(\omega)x}\,\eta(dx).\end{array}\right\} \tag{4.1.12}$$

2) By classical results on sums of independent centered random variables, for any fixed ω such that $\sum_{s\le t}\|V_s(\omega)\|^2<\infty$, the sequence $Z^\omega(m)_t$ converges in $\mathbb{L}^2(\mathbb{P}')$, as $m\to\infty$. This being true for all $\omega\notin N$, for any $\varepsilon>0$ we have

$$\widetilde{\mathbb{P}}\big(\|Z(m)_t-Z(m')_t\|>\varepsilon\big)=\int \mathbb{P}(d\omega)\,\mathbb{P}'\big(\|Z^\omega(m)_t-Z^\omega(m')_t\|>\varepsilon\big)\ \to\ 0$$

as $m, m' \to \infty$. Hence the sequence $Z(m)_t$ is a Cauchy sequence for convergence in probability on $(\widetilde{\Omega}, \widetilde{\mathcal{F}}, \widetilde{\mathbb{P}})$ and thus it converges in probability to a limit Z_t which is $\widetilde{\mathcal{F}}_t$ measurable. Moreover, for each $t \geq 0$ and $\omega \notin N$,

$$\begin{aligned} & Z^{\omega}(m)_t \to Z_t(\omega, .) \quad \text{in } \mathbb{L}^2(\Omega', \mathcal{F}', \mathbb{P}'), \\ & \int Z_t^i(\omega, \omega') Z_t^j(\omega, \omega') \mathbb{P}'(d\omega') = \sum_{s \leq t} c_s^{ij}(\omega) \end{aligned} \tag{4.1.13}$$

(use (4.1.12) for the last equality). Then we define the $(\widetilde{\mathcal{F}}_t)$-adapted process Y by

$$Y_t = \limsup_{(s \downarrow\downarrow t, s \in \mathbb{Q})} Z_s. \tag{4.1.14}$$

3) (4.1.14) looks like the construction of a càdlàg version of a martingale, which should not come as a surprise. Indeed, by the first part of (4.1.13), the process $(Z_t(\omega, .))_{t \geq 0}$ is a (not necessarily càdlàg) square-integrable martingale on $(\Omega', \mathcal{F}', \mathbb{P}')$, relative to the filtration it generates (because under $\mathbb{P}'$ the variables U_n are centered and independent), and also a process with independent increments. By the regularization scheme for martingales we get that the process $(Y_t(\omega, .))_{t \geq 0}$ is a modification of $(Z_t(\omega, .))_{t \geq 0}$ under the measure $\mathbb{P}'$, with $\mathbb{P}'$ a.s. càdlàg paths. Then we readily deduce the following properties:

- The process Y is $\widetilde{\mathbb{P}}$ a.s. càdlàg.
- We have (4.1.13) with Y in place of Z.

It follows that conditionally on $\mathcal{F}$ the process Y is centered with independent increments, and a square-integrable martingale, and satisfies (4.1.6) and (4.1.11): all these are consequences of (4.1.12) and (4.1.13) applied with Y.

4) It remains to prove (a) and (b). When η is a Gaussian measure, the processes $Z^{\omega}(m)$ are Gaussian under $\mathbb{P}'$, so (a) again follows from (4.1.13) applied with Y.

Finally, for (b) we assume the existence of a localizing sequence (S_p) of $(\mathcal{F}_t)$-stopping times, such that $a_p := \mathbb{E}(\sum_{s \leq S_p} \|V_s\|^2) < \infty$ for all p. Now, (4.1.13) implies that $Z(m)_{t \wedge S_p}$ converges in $\widetilde{\mathbb{P}}$-probability to $Y_{t \wedge S_p}$, and $\widetilde{\mathbb{E}}(\|Z(m)_{t \wedge S_p}\|^2) \leq a_p$, so in fact the convergence holds in $\mathbb{L}^1(\widetilde{\mathbb{P}})$ as well. Then it is enough to prove that each stopped process $Z(m)_{t \wedge S_p}$ is a martingale. Since $Z(m) = \sum_{n=1}^m N(n)$, where $N(n) = V_{T_n} U_n 1_{\{T_n \leq t\}}$, it is even enough to prove that each $N(n)_{t \wedge S_p}$ is a martingale.

For this we take $0 \leq t < s$ and we observe that $\widetilde{\mathcal{F}}_t$ is contained in the σ-field generated by the sets of the form $A \cap B \cap C$, where $A \in \mathcal{F}_t$ and $B \in \sigma(U_p : p \neq n)$ and $C \in \sigma(U_n)$ is such that $C \cap \{T_n > t\}$ is either empty or equal to $\{T_n > t\}$. Hence is it enough to prove that $z := \mathbb{E}(1_{A \cap B \cap C}(N(n)_{s \wedge S_p} - N(n)_{t \wedge S_p}))$ vanishes. This is obvious when $C \cap \{T_n > t\} = \emptyset$, and otherwise $C \cap \{T_n > t\} = \{T_n > t\}$ and we have

$$z = \mathbb{E}(1_{A \cap \{t \wedge S_p < T_n \leq s \wedge S_p\}} V_{T_n}) \int U_n(\omega') 1_B(\omega') \mathbb{P}'(d\omega') = 0$$

because U_n is centered and independent of B under $\mathbb{P}'$. This ends the proof. $\square$

4.1.3 The Mixed Case

In Propositions 4.1.2, 4.1.3 and 4.1.4, the $\mathcal{F}$-conditional law of Y is completely characterized by the independent increments property, plus the conditional characteristic function of the increments, given by (4.1.6) in the latter cases, and in the first case by

$$\widetilde{\mathbb{E}}\big(e^{iu^\star(Y_{t+s}-Y_t)} \mid \mathcal{F}\big) = e^{-\frac{1}{2}u^\star(\widetilde{C}_{t+s}-\widetilde{C}_t)u}.$$

In several applications one needs to "mix" the two kinds of processes: first, we want a q-dimensional process Y on a very good filtered extension $(\widetilde{\Omega},\widetilde{\mathcal{F}},(\widetilde{\mathcal{F}}_t)_{t\ge0},\widetilde{\mathbb{P}})$ of the space, with

$$\left.\begin{array}{l}\bullet\ Y \text{ has independent increments, conditionally on } \mathcal{F}\\ \bullet\ Y_0=0 \text{ and the paths of } Y \text{ are a.s. càdlàg}\\ \bullet\ \text{the } \mathcal{F}\text{-conditional characteristic function of the increment } Y_{t+s}-Y_t \text{ is}\\ \widetilde{\mathbb{E}}\big(e^{iu^\star(Y_{t+s}-Y_t)} \mid \mathcal{F}\big)=e^{-\frac{1}{2}u^\star(\widetilde{C}_{t+s}-\widetilde{C}_t)u}\prod_{v\in(t,t+s]\cap D(V,\omega)}\int e^{iu^\star V_v(\omega)x}\eta(dx).\end{array}\right\}\tag{4.1.15}$$

Here, $\widetilde{C}$, V and η are as in the previous two subsections.

Solving this problem simply amounts to pasting together the previous constructions. We consider an auxiliary probability space $(\Omega',\mathcal{F}',\mathbb{P}')$ endowed with a q-dimensional Brownian motion W', and with a sequence (U_n) of i.i.d. variables with law η, and independent of the process W'. Then we define $(\widetilde{\Omega},\widetilde{\mathcal{F}},(\widetilde{\mathcal{F}}_t)_{t\ge0},\widetilde{\mathbb{P}})$ as follows:

$$\left.\begin{array}{l}\widetilde{\Omega}=\Omega\times\Omega',\qquad \widetilde{\mathcal{F}}=\mathcal{F}\otimes\mathcal{F}',\qquad \widetilde{\mathbb{P}}=\mathbb{P}\otimes\mathbb{P}'\\ (\widetilde{\mathcal{F}}_t) \text{ is the smallest filtration containing } (\mathcal{F}_t), \text{ to which}\\ W' \text{ is adapted, and such that } U_n \text{ is } \widetilde{\mathcal{F}}_{T_n} \text{ measurable for all } n.\end{array}\right\}\tag{4.1.16}$$

Again, $(\widetilde{\Omega},\widetilde{\mathcal{F}},(\widetilde{\mathcal{F}}_t)_{t\ge0},\widetilde{\mathbb{P}})$ is a very good filtered extension of $(\Omega,\mathcal{F},(\mathcal{F}_t)_{t\ge0},\mathbb{P})$, and W' is an $(\widetilde{\mathcal{F}}_t)$-Brownian motion. A mix of the proofs of Propositions 4.1.2 and 4.1.4 gives the following (the—easy—details are left to the reader):

Proposition 4.1.5 *In the previous setting, let (T_n) be any weakly exhausting sequence (T_n) for $D(V,.)$, and assume that $\widetilde{C}$ has the form* (4.1.1) *with $\widetilde{c}_t=\widetilde{\sigma}_t\widetilde{\sigma}_t^*$, that V satisfies* (4.1.10), *and that η is centered with variance-covariance matrix μ_2. In the formula*

$$Y_t = \int_0^t \widetilde{\sigma}_s\,dW'_s + \sum_{n=1}^\infty V_{T_n}U_n\,1_{\{T_n\le t\}},$$

the series converges in probability for all t, and this formula defines an adapted process on the extended space, which is a.s. càdlàg and is $\mathcal{F}$-conditionally centered with independent increments and finite second moments (hence a square-integrable *martingale) and satisfies*

$$\widetilde{\mathbb{E}}(Y_t^i Y_t^j \mid \mathcal{F}) = \widetilde{C}_t^{ij} + \sum_{s\le t} \sum_{k,l=1}^{r} V_s^{ik} \mu_2^{kl} V_s^{jl},$$

and also (4.1.15). *In particular the $\mathcal{F}$-conditional law of Y does not depend on the weakly exhausting sequence (T_n). Moreover:*

a) *If the random vectors U_n are Gaussian, the process Y is $\mathcal{F}$-conditionally Gaussian.*
b) *If the process $\sum_{s\le t} \|V_s\|^2$ is locally integrable, then Y is a locally square-integrable martingale on the extended space.*

4.2 Stable Convergence Result in the Continuous Case

In this section, we have a d'-dimensional Brownian motion W on the filtered probability space $(\Omega, \mathcal{F}, (\mathcal{F}_t)_{t\ge 0}, \mathbb{P})$. The main problem which we want to solve is the following: let σ be a càdlàg adapted $d \times d'$-dimensional process, and $c = \sigma\sigma^*$. For all $i, n \ge 1$, and in the setting of a regular discretization scheme with time step Δ_n, we define the following d-dimensional variables:

$$\beta_i^n = \frac{1}{\sqrt{\Delta_n}} \sigma_{(i-1)\Delta_n} \Delta_i^n W. \tag{4.2.1}$$

For any Borel function g from $\mathbb{R}^d$ into $\mathbb{R}^q$ with at most polynomial growth, and recalling that, if $a \in \mathcal{M}_{d\times d}^+$, then ρ_a denotes the law $\mathcal{N}(0, a)$ on $\mathbb{R}^d$, we introduce the processes

$$\overline{U}^n(g)_t = \sqrt{\Delta_n} \sum_{i=1}^{[t/\Delta_n]} \left(g\left(\beta_i^n\right) - \rho_{c_{(i-1)\Delta_n}}(g)\right). \tag{4.2.2}$$

Our aim to describe the limiting behavior of the processes $\overline{U}^n(g)$.

It turns out that the techniques for solving this problem work in a much more general context, and sometimes we need to consider more complex processes than $\overline{U}^n(g)$ above. This is why we extend the setting to a situation which may seem *a priori* disturbingly general, but will prove useful later. The ingredients are as follows:

1. a sequence $u_n > 0$ of numbers which goes to 0;
2. an adapted càdlàg $q \times q'$ matrix-valued process θ and an adapted càdlàg $\mathbb{R}^w$-valued process Y on $(\Omega, \mathcal{F}, (\mathcal{F}_t)_{t\ge 0}, \mathbb{P})$;
3. a function Φ from Ω^W into $\mathbb{R}^{q'}$ (where $(\Omega^W, \mathcal{F}^W, (\mathcal{F}_t^W), \mathbb{P}^W)$ denotes the canonical d'-dimensional Wiener space and $\mathbb{E}^W$ is the associated expectation).

We suppose that Φ satisfies

$$\Phi \text{ is } \mathcal{F}_1^W \text{ measurable, and } \mathbb{E}^W\left(\|\Phi\|^p\right) < \infty \text{ for all } p > 0. \tag{4.2.3}$$

In (4.2.1) and (4.2.2) we then replace the increments $\Delta_i^n W$ by the processes

$$w(n,i)_s = \frac{1}{\sqrt{u_n}} (W_{(i-1+s)u_n} - W_{(i-1)u_n}). \tag{4.2.4}$$

These are Brownian motions, and will be considered as Ω^W-valued random variables.

We need some notation. Let α be a $q \times q'$ matrix and $F = F(x,y)$ and $G = G(x,y)$ be two real Borel functions on $\mathbb{R}^w \times \mathbb{R}^q$ with at most polynomial growth. The function Φ is as in (4.2.3), the canonical process on Ω^W is denoted by W, and its transpose (a row vector) is W^*. Then we set

$$\left.\begin{array}{ll} \gamma_\alpha^\Phi(x,F) = \mathbb{E}^W\big(F(x,\alpha\Phi)\big), & \widehat{\gamma}_\alpha^\Phi(x,F) = \mathbb{E}^W\big(F(x,\alpha\Phi)W_1^*\big), \\ \overline{\gamma}_\alpha^\Phi(x,F,G) = \mathbb{E}^W\big(\big(F(x,\alpha\Phi) - \widehat{\gamma}_\alpha^\Phi(x,F)W_1\big)\big(G(x,\alpha\Phi) - \widehat{\gamma}_\alpha^\Phi(x,G)W_1\big)\big) & \\ \qquad - \gamma_\alpha^\Phi(x,F)\,\gamma_\alpha^\Phi(x,G). & \end{array}\right\} \tag{4.2.5}$$

Note that $F(x,\alpha\Phi) - \widehat{\gamma}_\alpha^\Phi(x,F)W_1$ is the orthogonal projection of the variable $F(x,\alpha\Phi)$, in the space $\mathbb{L}^2(\Omega^W,\mathbb{P}^W)$, on the subspace orthogonal to all variables W_1^j, and its mean value is $\gamma_\alpha^\Phi(x,F)$, so $\overline{\gamma}_\alpha^\Phi(F,G)$ is a covariance. In particular, if $(G^j)_{1\le j\le q}$ is a family of Borel functions, we have:

$$\text{the matrix } \big(\overline{\gamma}_\alpha^\Phi\big(x,G^j,G^k\big)\big)_{1\le j,k\le r} \quad \text{is symmetric nonnegative.}$$

Observe that $\alpha \mapsto \gamma_\alpha^\Phi(x,F)$, $\alpha \mapsto \widehat{\gamma}_\alpha^\Phi(x,F)$ and $\alpha \mapsto \overline{\gamma}_\alpha^\Phi(x,F,G)$ are measurable, and even continuous when F and G are continuous in y, whereas $x \mapsto \gamma_\alpha^\Phi(x,F)$, $x \mapsto \widehat{\gamma}_\alpha^\Phi(x,F)$ and $x \mapsto \overline{\gamma}_\alpha^\Phi(x,F,G)$ are continuous in x when F and G are continuous in x.

Our functionals of interest are

$$\overline{U}^n(G)_t = \sqrt{u_n} \sum_{i=1}^{[t/u_n]} \big(G\big(Y_{(i-1)u_n}, \theta_{(i-1)u_n}\Phi\big(w(n,i)\big)\big) - \gamma_{\theta_{(i-1)u_n}}^\Phi(Y_{(i-1)u_n},G)\big). \tag{4.2.6}$$

The similitude of notation with (4.2.2) is not by chance: if $q = d$ and $q' = d'$ and $\Phi(y) = y(1)$ and $\theta_t = \sigma_t$ and $G(x,y) = g(y)$ and $u_n = \Delta_n$, then $\overline{U}^n(g) = \overline{U}^n(G)$.

We recall that a (possibly multi-dimensional) function f on $\mathbb{R}^q$ for some integer q is of *polynomial growth* if $\|f(x)\| \le K(1+\|x\|^p)$ for some positive constants K and p.

Theorem 4.2.1 *Let Φ satisfy* (4.2.3) *and $G = (G^j)_{1\le j\le r}$ be continuous with polynomial growth, and suppose that θ and Y are adapted and càdlàg. Then the processes $\overline{U}^n(G)$ of* (4.2.6) *converge stably in law to an r-dimensional continuous process $\overline{U}(G) = (\overline{U}(G)^j)_{1\le j\le r}$ which can be written as*

$$\overline{U}(G)_t^j = \sum_{k=1}^{d'} \int_0^t \widehat{\gamma}_{\theta_s}^\Phi\big(Y_s,G^j\big)^k\, dW_s^k + \overline{U}'(G)_t^j, \tag{4.2.7}$$

where $\overline{U}'(G)$ is a continuous process on a very good extension $(\widetilde{\Omega}, \widetilde{\mathcal{F}}, (\widetilde{\mathcal{F}}_t), \widetilde{\mathbb{P}})$ of $(\Omega, \mathcal{F}, (\mathcal{F}_t), \mathbb{P})$, and which conditionally on $\mathcal{F}$ is a centered Gaussian process with independent increments satisfying

$$\widetilde{\mathbb{E}}\big(\overline{U}'(G)_t^j\, \overline{U}'(G)^k \mid \mathcal{F}\big) = \int_0^t \overline{\gamma}_{\theta_s}^{\Phi}\big(Y_s, G^j, G^k\big)\, ds. \tag{4.2.8}$$

The process $\overline{U}'(G)$ above can be "realized" as explained in Proposition 4.1.2.

Proof By a localization argument similar (and indeed quite simpler) to what is done in Lemma 3.4.5, we may assume that θ is bounded. In view of (4.2.5), and with the notation

$$\eta_i^{n,j} = G^j\big(Y_{(i-1)u_n}, \theta_{(i-1)u_n}\Phi\big(w(n,i)\big)\big) - \sum_{k=1}^{d'} \widehat{\gamma}_{\theta_{(i-1)u_n}}^{\Phi}\big(Y_{(i-1)u_n}, G^j\big)^k\, w(n,i)_1^k,$$

(recall $w(n,i)_1 = \frac{1}{\sqrt{u_n}}(W_{iu_n} - W_{(i-1)u_n})$), we can write $\overline{U}(G)_t^n = \sum_{i=1}^{[t/u_n]}(\zeta_i^n + \zeta_i'^n)$, where

$$\zeta_i^{n,j} = \sum_{k=1}^{d} \widehat{\gamma}_{\theta_{(i-1)u_n}}^{\Phi}\big(Y_{(i-1)u_n}, G^j\big)^k \big(W_{iu_n}^k - W_{(i-1)u_n}^k\big),$$

$$\zeta_i'^{n,j} = \sqrt{u_n}\,\big(\eta_i^{n,j} - \gamma_{\theta_{(i-1)u_n}}^{\Phi}\big(Y_{(i-1)u_n}, G^j\big)\big),$$

Observe that $\sum_{i=1}^{[t/\Delta_n]} \zeta_i^{n,j}$ is a Riemann sum, for the sum over k of the stochastic integrals occurring in the right side of (4.2.7), whereas the "integrand" processes $\widehat{\gamma}_{\theta_t}^{\Phi}(Y_t, G^j)^k$ are càdlàg. Then as a consequence of the Riemann approximation for stochastic integrals (Proposition 2.2.8), we deduce that

$$\sum_{i=1}^{[t/u_n]} \zeta_i^{n,j} \stackrel{\text{u.c.p.}}{\Longrightarrow} \sum_{k=1}^{d'} \int_0^t \widehat{\gamma}_{\theta_s}^{\Phi}\big(Y_s, G^j\big)^k\, dW_s^k.$$

So, in view of the fact that $(Z_n, Y_n) \stackrel{\mathcal{L}\text{-s}}{\longrightarrow} (Z, Y)$ when $Z_n \stackrel{\mathcal{L}\text{-s}}{\longrightarrow} Z$ and $Y_n \stackrel{\mathbb{P}}{\longrightarrow} Y$, see (2.2.5), it remains to prove that

$$\sum_{i=1}^{[t/u_n]} \zeta_i'^n \stackrel{\mathcal{L}\text{-s}}{\Longrightarrow} \overline{U}'(G)_t. \tag{4.2.9}$$

We will deduce (4.2.9) from Theorem 2.2.15 applied to $\zeta_i'^n$, with $N_n(t) = [t/u_n]$ and $T(n,i) = iu_n$ and $(\Omega_n, \mathcal{G}^n, (\overline{\mathcal{F}}_t^n), \mathbb{P}_n) = (\Omega, \mathcal{F}, (\mathcal{F}_t), \mathbb{P})$ and $\mathcal{G}_i^n = \mathcal{F}_{iu_n}$, so we trivially have (2.2.29) and (2.2.39). The heart of the proof is that, conditionally on $\mathcal{F}_{(i-1)u_n}$, the process $w(n,i)$ is a d'-dimensional Brownian motion. In view of the definition of $\eta_i^{n,j}$ and of (4.2.3) and (4.2.5), plus the boundedness of θ and the polynomial growth of all g^j's, we obtain $\mathbb{E}(\zeta_i'^n \mid \mathcal{F}_{(i-1)u_n}) = 0$ and $\mathbb{E}(\|\zeta_i'^n\|^4 \mid \mathcal{F}_{(i-1)u_n}) \le \Gamma_t u_n^2$ if $iu_n \le t$ for some locally bounded process Γ, and also

$$\mathbb{E}\big(\zeta_i'^{n,j}\zeta_i'^{n,k} \mid \mathcal{F}_{(i-1)u_n}\big) = u_n\,\overline{\gamma}^{\Phi}_{\theta_{(i-1)u_n}}\big(Y_{(i-1)u_n}, G^j, G^k\big).$$

Then (2.2.34) holds with $A=0$, and (2.2.37) holds. Since $t \mapsto \overline{\gamma}^{\Phi}_{\theta_t}(Y_t, G^j, G^k)$ is càdlàg, we also deduce (2.2.36) with C_t^{jk} being the right side of (4.2.8). It thus remains to prove (2.2.40), which will of course follow from

$$\mathbb{E}\big(\zeta_i'^{n,j}(M_{iu_n} - M_{(i-1)u_n}) \mid \mathcal{F}_{(i-1)u_n}\big) = 0 \tag{4.2.10}$$

when $M = W^m$ for $m = 1, \dots, d'$, and when M is a bounded martingale orthogonal to W.

The left side of (4.2.10) is $\sqrt{u_n}\,\mathbb{E}(\eta_i^{n,j}(M_{iu_n} - M_{(i-1)u_n}) \mid \mathcal{F}_{(i-1)u_n})$, because M is a martingale. When $M = W^m$, this equals $\mathbb{E}^W(G^j(x, \alpha\Phi)W_1^m) - \widehat{\gamma}^{\Phi}_{\alpha}(x, G^j)^m$ evaluated at $\alpha = \theta_{(i-1)u_n}$ and $x = Y_{(i-1)u_n}$, and thus (4.2.10) holds.

Assume now that M is a bounded martingale, orthogonal to W. The variable $\zeta_i'^n$ is integrable and depends only on $\theta_{(i-1)u_n}$, $Y_{(i-1)u_n}$ and $w(n,i)$, hence by the martingale representation theorem for the Brownian motion (see e.g. [57]) it can be written as $\zeta_i'^n = Y + Y'$, where Y is $\mathcal{F}_{(i-1)u_n}$ measurable and Y' is the value at iu_n of a martingale which vanishes at time $(i-1)u_n$ and is a stochastic integral with respect to W. Since M and W are orthogonal, we deduce that $\mathbb{E}(Y'(M_{iu_n} - M_{(i-1)u_n}) \mid \mathcal{F}_{(i-1)u_n}) = 0$, whereas $\mathbb{E}(Y(M_{iu_n} - M_{(i-1)u_n}) \mid \mathcal{F}_{(i-1)u_n}) = 0$ is obvious because of the martingale property of M. This completes the proof of (4.2.10). □

4.3 A Stable Convergence Result in the Discontinuous Case

In this section we suppose that the space $(\Omega, \mathcal{F}, (\mathcal{F}_t)_{t\ge0}, \mathbb{P})$ is endowed with the d'-dimensional Brownian motion W as before, and also with a Poisson random measure $\mathfrak{p}$ on $\mathbb{R}_+ \times E$, where $(E, \mathcal{E})$ is a Polish space. Its intensity measure, or predictable compensator, is $\mathfrak{q}(dt, dz) = dt \otimes \lambda(dz)$ for some σ-finite measure λ on E, see (2.1.24).

The set $D(\omega) = \{t : \mathfrak{p}(\omega; \{t\} \times E) = 1\}$ is countable, and we choose an "exhausting sequence" (S_p) of stopping times for this set, in the sense of Sect. 4.1.2, in a special way. Namely, let $A_0 = \emptyset$ and $(A_m)_{m\ge1}$ be a sequence of Borel subsets of E, increasing to E, and such that $\lambda(A_m) < \infty$ for all m. The sequence (S_p) is then constructed as follows:

$$\left.\begin{array}{l} (S_p)_{p\ge1} \text{ is a reordering of the double sequence} \\ (S(m,j) : m, j \ge 1), \text{ where } S(m,1), S(m,2), \dots \text{ are the} \\ \text{successive jump times of the Poisson process } 1_{\{A_m\setminus A_{m-1}\}} * \mathfrak{p}. \end{array}\right\} \tag{4.3.1}$$

Next, we pick two sequences $u_n > 0$ and $v_n > 0$ of numbers going to 0. We associate with u_n the processes $w(n,i)$ by (4.2.4), and we set for $n, p \ge 1$:

$$\left.\begin{array}{l}\overline{w}(n,p)_s=\frac{1}{\sqrt{v_n}}\left(W_{(i-1+s)v_n}-W_{(i-1)v_n}\right)\\ \kappa(n,p)=\frac{S_p}{v_n}-i\end{array}\right\}\text{ on the set }\left\{iv_n<S_p\le(i+1)v_n\right\}. \tag{4.3.2}$$

With this notation, each $\overline{w}(n,p)$ is a Brownian motion, as is each $w(n,i)$, and $\kappa(n,p)$ is a $(0,1]$-valued variable.

Our aim is toward a *joint* limit theorem for the variables or processes of (4.3.2), together with the processes $\overline{U}^n(G)$ associated by (4.2.6) with an r-dimensional continuous function $G=(G^j)$ having polynomial growth on $\mathbb{R}^w\times\mathbb{R}^q$, and with a q'-dimensional function Φ satisfying (4.2.3), and a càdlàg adapted $\mathbb{R}^q\otimes\mathbb{R}^{q'}$-valued process θ and a càdlàg adapted $\mathbb{R}^w$-valued process Y.

For describing the limit, we consider an auxiliary space $(\Omega',\mathcal{F}',\mathbb{P}')$ on which are defined an r-dimensional Brownian motion W', a sequence $(\kappa_p)_{p\ge1}$ of variables uniformly distributed over $(0,1]$, and a sequence $(W''(p)=(W''(p)_t)_{t\ge0})_{p\ge1}$ of d'-dimensional Brownian motions, all these being mutually independent. The very good filtered extended space $(\widetilde{\Omega},\widetilde{\mathcal{F}},(\widetilde{\mathcal{F}}_t)_{t\ge0},\widetilde{\mathbb{P}})$ is defined by (4.1.16), where T_n and U_n are substituted with S_n and $(\kappa_n,W''(n))$. The process $\overline{U}(G)$, which by Theorem 4.2.1 is the limit of $\overline{U}^n(G)$, can be realized on the extension by setting (with matrix notation)

$$\overline{U}(G)_t=\int_0^t\widehat{\gamma}_{\theta_s}^{\Phi}(Y_s,G)\,dW_s+\overline{U}'(G)_t,\quad \overline{U}'(G)_t=\int_0^t H_s\,dW'_s, \tag{4.3.3}$$

where H is an adapted càdlàg $r\times r$ matrix-valued process whose square $H_tH_t^\star$ equals the matrix with entries $\overline{\gamma}_{\theta_t}^{\Phi}(Y_t,G^j,G^k)$. Note that $\overline{U}(G)$ is independent of the sequence $(\kappa_p,W''(p))$.

Theorem 4.3.1 *With the previous notation and assumptions, we have*

$$\left(\overline{U}^n(G),\left(\overline{w}(n,p),\kappa(n,p)\right)_{p\ge1}\right)\stackrel{\mathcal{L}\text{-s}}{\longrightarrow}\left(\overline{U}(G),\left(W''(p),\kappa_p\right)_{p\ge1}\right),$$

for the product topology on the space $\mathbb{D}^r\times(\mathbb{D}^{d'}\times\mathbb{R})^{\mathbb{N}^\star}$.

Proof *Step 1)* As mentioned after (2.2.4), it suffices for the stable convergence in law to prove this convergence for test functions f which are bounded and Lipschitz. We can even further restrict the set of test functions to be any convergence-determining set of functions (for the weak convergence). By virtue of the properties of the product topology and of the Skorokhod topology, it is then enough to prove the following convergence:

$$\mathbb{E}\left(Z\,F\left(\overline{U}^n(G)\right)\prod_{p=1}^{l}F_p\left(\overline{w}(n,p)\right)f_p\left(\kappa(n,p)\right)\right)$$

$$\to\widetilde{\mathbb{E}}\left(Z\,F\left(\overline{U}(G)\right)\prod_{p=1}^{l}F_p\left(W''(n)\right)f_p(\kappa_p)\right),$$

where Z is any bounded $\mathcal{F}$ measurable variable, and $l \in \mathbb{N}^*$, and the f_p's are bounded C^∞ functions on $\mathbb{R}$, and F and the F_p's are bounded Lipschitz functions on $\mathbb{D}^q$ and $\mathbb{D}^{d'}$ respectively, and further $F_p(x) = F_p(y)$ for any two $x, y \in \mathbb{D}^{d'}$ such that $x(t) = y(t)$ for all $t \leq T$, for some $T \geq 1$, and the same for F on $\mathbb{D}^q$.

We will reduce this problem to simpler ones, along several steps, and the first reduction is elementary. We use the simplifying notation $V_n = \prod_{p=1}^l F_p(\overline{w}(n,p)) \times f_p(\kappa(n,p))$. Then, since under $\widetilde{\mathbb{P}}$ the variables or processes W', $W''(p)$, κ_p are all independent, and independent of $\mathcal{F}$, what precedes amounts to

$$\mathbb{E}\big(Z\, F\big(\overline{U}^n(G)\big)\, V_n\big) \to \widetilde{\mathbb{E}}\big(Z\, F\big(\overline{U}(G)\big)\big) \prod_{p=1}^{l} \mathbb{E}'\big(F_p(W''(p))\big)\, \mathbb{E}'\big(f_p(\kappa_p)\big). \tag{4.3.4}$$

Step 2) The next reduction consists in showing that it is enough to prove (4.3.4) when Z is measurable with respect to a suitable sub-σ-field $\mathcal{H}$ of $\mathcal{F}$ which is *separable*, that is, generated by a countable algebra. We take for $\mathcal{H}$ the σ-field generated by the processes W, θ, Y and the measure $\mathfrak{p}$. Observing that $\overline{U}'^n(G)_t$ and V_n are $\mathcal{H}$ measurable, and $\overline{U}'(G)_t$ is $\mathcal{H} \otimes \mathcal{F}'$ measurable, we can substitute Z with $\mathbb{E}(Z \mid \mathcal{H})$ in both members of (4.3.4), so it is indeed enough to prove (4.3.4) when Z is $\mathcal{H}$ measurable.

In fact, we can even replace the filtration $(\mathcal{F}_t)$ by the smaller filtration $(\mathcal{H} \cap \mathcal{F}_t)$, without changing any of the properties of $(W, \mathfrak{p})$, nor those of the extended space. So, below, we can and will assume that $(\mathcal{F}_t)$ is a filtration of the separable σ-field $\mathcal{F}$.

Step 3) Let us introduce some notation. The integer l and the constant T are fixed. We denote by $\overline{S}$ the l-dimensional vector $\overline{S} = (S_1, \ldots, S_l)$. For any $j \geq 1$ we introduce the random set $B_j = \cup_{p=1}^l [(S_p - 1/j)^+, S_p + 1/j]$, and the (random) family $\mathcal{I}(n,j)$ of all integers $i \geq 1$ such that $((i-1)u_n, iu_n] \cap B_j \neq \emptyset$. We also denote by $(\mathcal{G}_t)$ the smallest filtration containing $(\mathcal{F}_t)$ and such that $\overline{S}$ is $\mathcal{G}_0$ measurable.

The processes $1_{\{A_m \setminus A_{m-1}\}} * \mathfrak{p}$ in (4.3.1) are independent Poisson processes, also independent of W. Therefore on the one hand the l-dimensional vector $\overline{S}$ admits a density h on $\mathbb{R}^l$, which is C^∞ in the interior of its support. On the other hand the process W is a $(\mathcal{G}_t)$-Brownian motion (because it is independent of the variable $\overline{S}$) and $\{s \in B_j\} \in \mathcal{G}_0$ for all s, hence we can define the following two processes:

$$W(j)_t = \int_0^t 1_{B_j}(s)\, dW_s, \qquad \overline{W}(j)_t = \int_0^t 1_{B_j^c}(s)\, dW_s \; = \; W_t - W(j)_t.$$

With the notation (4.3.3), we also define on the extended space:

$$\overline{U}_t^{(j)} = \int_0^t \widehat{\gamma}_{\theta_s}^{\Phi}(Y_s, G)\, 1_{B_j^c}\, dW_s + \int_0^t H_s\, 1_{B_j^c}(s)\, dW'_s$$

(compare with (4.3.3)). We end the set of notation by putting

$$\zeta_i^n = \sqrt{u_n}\big(G\big(Y_{(i-1)u_n}, \theta_{(i-1))u_n}\Phi\big(w(n,i)\big)\big) - \gamma_{\theta_{(i-1))u_n}}^{\Phi}(Y_{(i-1)u_n}, G)\big)$$

$$\zeta_i'^{n,(j)} = \zeta_i^n \, 1_{\{i \in \mathcal{I}(n,j)\}}$$

$$\overline{U}_t^{n,(j)} = \sum_{i=1}^{[t/u_n]} \zeta_i^n \, 1_{\{i \notin \mathcal{I}(n,j)\}}, \qquad \overline{U}_t'^{n,(j)} = \sum_{i=1}^{[t/u_n]} \zeta_i'^{n,(j)} = \overline{U}^n(G)_t - \overline{U}_t^{n,(j)}.$$

Step 4) Since B_j decreases to the finite set $\{S_1, \dots, S_l\}$, we have

$$\overline{U}^{(j)} \overset{\text{u.c.p.}}{\Longrightarrow} \overline{U}(G) \quad \text{as } j \to \infty. \tag{4.3.5}$$

On the other hand, we have $\{i \in \mathcal{I}(n,j)\} \in \mathcal{G}_0$, hence $\mathbb{E}(\zeta_i'^{n,(j)} \mid \mathcal{G}_{(i-1)u_n}) = 0$ and the process $\overline{U}'^{n,(j)}$ is a partial sum of martingale increments. Furthermore, since G is of polynomial growth, $\mathbb{E}(\|\zeta_i'^{n,(j)}\|^2 \mid \mathcal{G}_{(i-1)u_n}) \le H_t u_n \, 1_{\{i \in \mathcal{I}(n,j)\}}$ when $iu_n \le t$ for a locally bounded process H: there is a localizing sequence of stopping times R_m such that $H_t \le m$ if $t \le m$. Then by Doob's inequality and the fact that the cardinal of $\mathcal{I}(n,j)$ is at most $l(1 + [2/ju_n])$, and that the set $\mathcal{I}(n,j)$ is $\mathcal{G}_0$-measurable, we get

$$\mathbb{E}\Big(\sup_{s \le R_m} \|\overline{U}_s'^{n,(j)}\|^2\Big) \le K\mathbb{E}\Bigg(\sum_{i=1}^{[R_m/u_n]} \|\zeta_i'^{n,(j)}\|^2\Bigg) \le \frac{Klm}{j}.$$

Hence, since $y \mapsto F(y)$ is Lipschitz and bounded and depends on the restriction of y to $[0,T]$ only, whereas $\overline{U}^n(G) = \overline{U}^{n,(j)} + \overline{U}'^{n,(j)}$, we deduce that

$$\mathbb{E}\big(\big|F(\overline{U}^{n,(j)}) - F(\overline{U}^n(G))\big|\big) \le K\mathbb{P}(R_m < T) + K\sqrt{lm/j}$$

for all $m \ge 1$. Now $\mathbb{P}(R_m < T) \to 0$ as $m \to \infty$, so

$$\lim_{j \to \infty} \sup_n \mathbb{E}\big(\big|F(\overline{U}^{n,(j)}) - F(\overline{U}^n(G))\big|\big) = 0.$$

Thus, by using also (4.3.5) we see that, instead of (4.3.4), it is enough to prove that for each j, we have

$$\mathbb{E}\big(Z\, F(\overline{U}^{n,(j)})\, V_n\big) \to \widetilde{\mathbb{E}}\big(Z\, F(\overline{U}^{(j)})\big) \prod_{p=1}^{l} \mathbb{E}'\big(F_p(W''(p))\big) \mathbb{E}'\big(f_p(\kappa_p)\big). \tag{4.3.6}$$

Step 5) In the sequel we fix j. Introduce the σ-fields $\mathcal{H}^{W(j)}$ generated by the variables $W(j)_s$ for $s \ge 0$, and the filtration $(\mathcal{G}(j)_t)$ which is the smallest filtration containing $(\mathcal{G}_t)$ and such that $\mathcal{H}^{W(j)} \subset \mathcal{G}(j)_0$. Since $\mathcal{F}$ is separable, there is a regular version $\mathbb{Q}_\omega(.)$ of the probability $\mathbb{P}$ on $(\Omega, \mathcal{F})$, conditional on $\mathcal{G}(j)_0$, and we set $\widetilde{\mathbb{Q}}_\omega = \mathbb{Q}_\omega \otimes \mathbb{P}'$.

Under $\mathbb{Q}_\omega$, the processes $w(n,i)$ restricted to the time interval $[0,1]$, and for $i \notin \mathcal{I}(n,j)(\omega)$ (recall that $\mathcal{I}(n,j)$ is $\mathcal{G}_0$ measurable), are constructed via (4.2.4) on the basis of a $(\mathcal{G}(j)_t)$-Brownian motion. Hence $\overline{U}^{n,(j)}$ is exactly like $\overline{U}^n(G)$, except that we discard the summands for which $i \in I(n,j)$, because the set $I(n,j)$ is a.s.

deterministic under $\mathbb{Q}_\omega$. Then we can reproduce the proof of Theorem 4.2.1, with $\mathbb{P}$ and $\widetilde{\mathbb{P}}$ and $(\mathcal{F}_t)$ substituted with $\mathbb{Q}_\omega$ and $\widetilde{\mathbb{Q}}_\omega$ and $(\mathcal{G}(j)_t)$, and with the process in (4.2.7) being substituted with $\overline{U}^{(j)}$ (since B_j^c is a finite union of intervals). We deduce that $\overline{U}^{n,(j)}$ converges stably in law under the measure $\mathbb{Q}_\omega$ towards $\overline{U}^{(j)}$, hence

$$\mathbb{E}_{\mathbb{Q}_\omega}\big(Z\, F\big(\overline{U}^{n,(j)}\big)\big) \to \mathbb{E}_{\widetilde{\mathbb{Q}}_\omega}\big(Z\, F\big(\overline{U}^{(j)}\big)\big). \tag{4.3.7}$$

Now, as soon as $v_n T < 1/2j$, that is for all n large enough, we have $(i-1)v_n > S_p - 1/j$ and $(i-1+T)v_n < S_p + 1/j$ if $i v_n < S_p \le (i+1)v_n$. Thus the variable V_n is $\mathcal{G}(j)_0$ measurable and

$$\begin{aligned} &\mathbb{E}\big(Z\, F\big(\overline{U}^{n,(j)}\big)\, V_n\big) - \mathbb{E}\big(V_n\, \mathbb{E}_{\widetilde{\mathbb{Q}}_.}\big(Z\, F\big(\overline{U}^{(j)}\big)\big)\big) \\ &\quad = \mathbb{E}\big(V_n\, \big(\mathbb{E}_{\mathbb{Q}_.}\big(Z\, F\big(\overline{U}^{n,(j)}\big)\big) - \mathbb{E}_{\widetilde{\mathbb{Q}}_.}\big(Z\, F\big(\overline{U}^{(j)}\big)\big)\big)\big) \to 0, \end{aligned}$$

where the last convergence comes from (4.3.7) (all variables and functions are uniformly bounded, here). Moreover, $Z' = \mathbb{E}_{\widetilde{\mathbb{Q}}_.}(Z\, F(\overline{U}^{(j)}))$ is $\mathcal{G}(j)_0$ measurable and $\widetilde{\mathbb{E}}(Z\, F(\overline{U}^{(j)})) = \mathbb{E}(Z')$. Thus (4.3.6) amounts to having, for any bounded $\mathcal{G}(j)_0$ measurable variable Z',

$$\mathbb{E}\big(Z'\, V_n\big) \to \mathbb{E}\big(Z'\big) \prod_{p=1}^{l} \mathbb{E}'\big(F_p\big(W''(p)\big)\big)\, \mathbb{E}'\big(f_p(\kappa_p)\big). \tag{4.3.8}$$

Step 6) In this step we show that it is enough to prove (4.3.8) when Z' is $\mathcal{G}_0$ measurable. Since $\mathcal{G}(j)_0 = \mathcal{G}_0 \vee \mathcal{H}^{W(j)}$, the set $\mathcal{A}$ of all products $Z' = Z'' L(W(j))$, where Z'' is bounded $\mathcal{G}_0$ measurable and L is a bounded Lipschitz function on $\mathbb{D}^{d'}$, is total in the set $\mathbb{L}^1(\Omega, \mathcal{G}(j)_0, \mathbb{P})$. Hence it suffices to prove (4.3.8) when $Z' = Z'' L(W(j)) \in \mathcal{A}$.

Let F_n be the union of all intervals $(i v_n, (i+T)v_n]$ which contain at least one S_p for $p \le l$, and set $W^n(j)_t = \int_0^t 1_{F_n^c}(s)\, dW(j)_s$. Since $v_n \to 0$ we have $W^n(j) \overset{\text{u.c.p.}}{\Longrightarrow} W(j)$ as $n \to \infty$, hence $L(W^n(j)) \overset{\mathbb{P}}{\longrightarrow} L(W(j))$, and thus since Z'', L and V_n are bounded,

$$\begin{aligned} &\mathbb{E}\big(Z''\, L\big(W^n(j)\big)\, V_n\big) - \mathbb{E}\big(Z''\, L\big(W(j)\big)\, V_n\big) \to 0 \\ &\mathbb{E}\big(L\big(W^n(j)\big)\big) \to \mathbb{E}\big(L\big(W(j)\big)\big). \end{aligned} \tag{4.3.9}$$

Therefore we can substitute the left side of (4.3.8) with $\mathbb{E}(Z''\, L(W^n(j))\, V_n)$, which equals $\mathbb{E}(Z''\, V_n)\, \mathbb{E}(L(W^n(j)))$ because (V_n, Z'') and $W^n(j)$ are independent. The second part of (4.3.9) shows that indeed we can even substitute the left side of (4.3.8) with the product $\mathbb{E}(Z''\, V_n)\, \mathbb{E}(L(W(j)))$.

Since $\mathbb{E}(Z') = \mathbb{E}(Z'')\mathbb{E}(L(W(j)))$ because Z'' and $W(j)$ are independent, it now suffices to prove (4.3.8) with Z'' instead of Z', or equivalently to prove (4.3.8) when Z' is $\mathcal{G}_0$ measurable.

Step 7) Let Ω_l^n be the $\mathcal{G}_0$ measurable set on which $v_n T$ is smaller than all differences $|S_p - S_q|$ for $0 \leq p < q \leq l$ (with $S_0 = 0$). Conditionally on $\mathcal{G}_0$, and in restriction to Ω_l^n, the stopped processes $\overline{w}(n,p)_{t\wedge T}$ are independent Brownian motion stopped at time T, hence

$$\mathbb{E}(Z' 1_{\Omega_l^n} V_n) = \mathbb{E}\left(Z' 1_{\Omega_l^n} \prod_{p=1}^l f_p\big(\kappa(n,p)\big)\right) \prod_{p=1}^l \mathbb{E}'\big(F_p\big(W''(p)\big)\big).$$

Since $\mathbb{P}(\Omega_l^n) \to 1$, we are thus left to prove

$$\mathbb{E}\left(Z' \prod_{p=1}^l f_p\big(\kappa(n,p)\big)\right) \to \mathbb{E}(Z') \prod_{p=1}^l \mathbb{E}'\big(f_p(\kappa_p)\big)$$

when Z' is $\mathcal{G}_0$ measurable. We can even go further: recalling that $\mathcal{G}_0$ is generated by $\mathcal{F}_0$ and the random vector $\overline{S}$, which is independent of $\mathcal{F}_0$, and since all $\kappa(n,p)$ are functions of $\overline{S}$, it is even enough to prove that for any bounded measurable function f on $\mathbb{R}^l$,

$$\mathbb{E}\left(f(\overline{S}) \prod_{p=1}^l f_p\big(\kappa(n,p)\big)\right) \to \mathbb{E}\big(f(\overline{S})\big) \prod_{p=1}^l \mathbb{E}'\big(f_p(\kappa_p)\big). \tag{4.3.10}$$

Yet another density argument (as in the beginning of Step 6) yields that it even enough to check (4.3.10) with f Lipschitz with compact support and satisfying $f(s_1,\dots,s_l) = 0$ on the set $\cup_{1\leq p<q\leq l}(\{|s_p - s_q| \leq \varepsilon\} \cup \{s_p \leq \varepsilon\})$.

Step 8) Finally we prove (4.3.10), which when $l = 1$ is very close to an old result of Kosulajeff [67] and Tuckey [92]. We use the notation $I'(n,i) = (iv_n, (i+1)v_n]$ and recall that h is the density of $\overline{S}$. For any family $\mathcal{I} = (i_1,\dots,i_l)$, as soon as $v_n < \varepsilon$ we have

$$\begin{aligned}
\gamma_n(\mathcal{I}) &= \mathbb{E}\left(f(\overline{S}) \prod_{p=1}^l f_p\big(\kappa(n,p)\big) 1_{I'(n,i_p)}(S_p)\right) \\
&= \mathbb{E}\left(f(\overline{S}) \prod_{p=1}^l f_p\left(\frac{S_p}{u_n} - i_p\right) 1_{I'(n,i_p)}(S_p)\right) \\
&= \int_{\prod_{p=1}^l I'(n,i_p)} (fh)(s_1,\dots,s_r) \prod_{p=1}^l f_p\big((s_p - i_p v_n)/v_n\big)\, ds_1 \dots ds_r \\
&= v_n^l\, (fh)(i_1 v_n,\dots,i_l v_n) \prod_{p=1}^l \int_0^1 f_p(s)\, ds + \gamma_n'(\mathcal{I})
\end{aligned}$$

by a change of variable, and where $\gamma_n'(\mathcal{I})$ vanishes if at least one of the i_p is bigger than A/v_n for some A, and $|\gamma_n'(\mathcal{I})| \leq K v_n^{l+1}$ always (due to the Lipschitz property of the product fh). It remains to observe that the left side of (4.3.10) is $\sum_{\mathcal{I}} \gamma_n(\mathcal{I})$, where the sum is extended over all possible families $\mathcal{I}$ of indices. By what precedes, this sum converges to the Lebesgue integral of fh on $\mathbb{R}^l$, times $\prod_{p=1}^{l} \mathbb{E}'(f_p(\kappa_p))$, and this product is the right side of (4.3.10). □

4.4 An Application to Itô Semimartingales

In the rest of this chapter, we have an underlying process X which is a d-dimensional Itô semimartingale on $(\Omega, \mathcal{F}, (\mathcal{F}_t)_{t\geq 0}, \mathbb{P})$. Its characteristics have the form (2.1.25) and the jump measure of X is called μ, but we mainly use a *Grigelionis representation* (2.1.30) for it, possibly defined on an extended space which is still denoted as $(\Omega, \mathcal{F}, (\mathcal{F}_t)_{t\geq 0}, \mathbb{P})$:

$$X_t = X_0 + \int_0^t b_s\, ds + \int_0^t \sigma_s\, dW_s + (\delta 1_{\{\|\delta\|\leq 1\}}) \star (\mathfrak{p} - \mathfrak{q})_t + (\delta 1_{\{\|\delta\|>1\}}) \star \mathfrak{p}_t, \quad (4.4.1)$$

where W is a Brownian motion and $\mathfrak{p}$ is a Poisson measure with compensator $\mathfrak{q}(dt, dz) = dt \otimes \lambda(dz)$. As we know, it is always possible to take the dimension of W to be d. However in many applications the semimartingale of interest is directly given in the form (4.4.1), with W and $\mathfrak{p}$ given, and there is no reason to assume that the dimensions of X and W are the same. We thus denote by d' the (*a priori* arbitrary) dimension of W. Then of course σ is $d \times d'$-dimensional, whereas $c = \sigma\sigma^\star$ is necessarily $d \times d$-dimensional. Note that W and $\mathfrak{p}$ are exactly as in the previous section.

In this section, we develop two technical tools, in constant use in the sequel. One concerns a *localization procedure* which has already been used in a simple situation (Lemma 3.4.5), but below we extend it to a wider setting. The other one concerns a consequence of Theorem 4.3.1 which will be one of our main tools later.

4.4.1 The Localization Procedure

The localization procedure is a commonly used technique in stochastic calculus. For example, let M be a continuous local martingale starting at $M_0 = 0$ and suppose that one wants to prove that the difference $N = M^2 - [M, M]$ is a local martingale (a well known result, of course). Typically, the proof goes as follows. We start by proving the property for all bounded M, that is when $\sup_t |M_t| \leq K$ for some constant K. Then, assuming this, the result for an arbitrary continuous local martingale is proved via the localization procedure: letting $T_q = \inf(t : |M_t| > p)$, the stopped process $M(q)_t = M_{t\wedge T_q}$ is a bounded martingale, so $N(q) = M(q)^2 - [M(q), M(q)]$ is a local martingale (and even a martingale, indeed). Next, observing that $[M(q), M(q)]_t = [M, M]_{t\wedge T_q}$, we see that $N(q)_t = N_{t\wedge T_q}$. Thus, since each

$N(q)$ is a martingale and since the sequence T_q of stopping times increases to ∞, we deduce from the definition of local martingales that N is indeed a local martingale.

The same procedure can be put to use in much more complex situations. In this book, we use it extensively in the following setting. We have a space $(\Omega, \mathcal{F}, (\mathcal{F}_t)_{t\geq 0}, \mathbb{P})$ on which we consider the class $\mathcal{S}$ of all semimartingales X satisfying some specific assumption, say (ABC). With any X in $\mathcal{S}$ we associate a sequence of q-dimensional càdlàg processes $U^n(X)$ on $(\Omega, \mathcal{F}, (\mathcal{F}_t)_{t\geq 0}, \mathbb{P})$, and another q-dimensional càdlàg process $U(X)$ which may be defined on $(\Omega, \mathcal{F}, (\mathcal{F}_t)_{t\geq 0}, \mathbb{P})$, or on an extension $(\widetilde{\Omega}, \widetilde{\mathcal{F}}, (\widetilde{\mathcal{F}}_t)_{t\geq 0}, \widetilde{\mathbb{P}})$ of it; this extension may depend on X itself.

The processes $U^n(X)$ and $U(X)$ are subject to the following conditions, where X and X' are any two semimartingales in the class $\mathcal{S}$, and S is any $(\mathcal{F}_t)$-stopping time:

$$
\begin{aligned}
&X_t = X'_t \text{ a.s. } \forall t < S \quad \Rightarrow \\
&\quad \bullet\ t < S \Rightarrow U^n(X)_t = U^n(X')_t \text{ a.s.} \\
&\quad \bullet\ \text{the } \mathcal{F}\text{-conditional laws of } (U(X)_t)_{t<S} \text{ and } (U(X')_t)_{t<S} \text{ are a.s. equal.}
\end{aligned}
\tag{4.4.2}
$$

When $U(X)$ is defined on $(\Omega, \mathcal{F}, (\mathcal{F}_t)_{t\geq 0}, \mathbb{P})$ itself, the second condition above amounts to $U(X)_t = U(X')_t$ a.s. for all t, on $\{t < S\}$: this is the case below, when we speak about convergence in probability.

The properties of interest for us are either one of the following properties:

$$
\begin{aligned}
&\bullet \text{ The processes } U^n(X) \text{ converge in probability to } U(X) \\
&\bullet \text{ The variables } U^n(X)_t \text{ converge in probability to } U(X)_t \\
&\bullet \text{ The processes } U^n(X) \text{ converge stably in law to } U(X) \\
&\bullet \text{ The variables } U^n(X)_t \text{ converge stably in law to } U(X)_t.
\end{aligned}
\tag{4.4.3}
$$

Definition 4.4.1 In the previous setting, and if (SABC) is an assumption stronger than (ABC), we say that the **localization procedure** "from (SABC) to (ABC)" applies when we have the following: if one of the properties (4.4.3) holds for all semimartingales satisfying (SABC), it also holds for semimartingales satisfying (ABC).

The terminology "localization procedure" may not be clear from the previous definition; however it is substantiated by the facts that, on the one hand the properties (4.4.2) mean that the functionals $U^n(X)$ are "local" in some sense, and on the other hand the assumption (ABC) to which we apply the procedure is typically a "localized version" of the assumption (SABC), in the same sense as the class of local martingales, for example, is the localized version of the class of martingales.

Now we turn to some concrete forms of the assumption (ABC). In this subsection we only consider three assumptions, among many others in this book. These three assumptions exhibit all the difficulties involved by the localization procedure, and the extension to the other assumptions which we will encountered is obvious.

The first assumption is the most basic one:

Assumption 4.4.2 (or **(H)**) X is an Itô semimartingale of the form (4.4.1), and we have:

(i) The process b is locally bounded.
(ii) The process σ is càdlàg.
(iii) There is a localizing sequence (τ_n) of stopping times and, for each n, a *deterministic* nonnegative function Γ_n on E satisfying $\int \Gamma_n(z)^2 \lambda(dz) < \infty$ and such that $\|\delta(\omega,t,z)\| \wedge 1 \leq \Gamma_n(z)$ for all (ω,t,z) with $t \leq \tau_n(\omega)$.

The second assumption is basically (H) plus the fact that the process σ occurring in (5.0.1) is itself an Itô semimartingale, hence the pair (X,σ) as well. We could then write various Grigelionis representations of this pair, globally driven by a Brownian motion and a Poisson random measure.

However, it is more convenient to express our needed regularity assumptions in terms of the following decomposition of σ, which is a kind of "projection" of the process σ on the Brownian motion W which drives (together with the Poisson measure $\mathfrak{p}$) the process X in (5.0.1). Namely, as soon as it is an Itô semimartingale, σ can be written as

$$\left.\begin{array}{l} \sigma_t = \sigma_0 + \int_0^t \widetilde{b}_s\,ds + \int_0^t \widetilde{\sigma}_s\,dW_s + M_t + \sum_{s\leq t} \Delta\sigma_s\, 1_{\{\|\Delta\sigma_s\|>1\}}, \\ \bullet\ M \text{ is a local martingale with } \|\Delta M_t\| \leq 1, \text{ orthogonal to } W, \\ \quad \text{and its predictable quadratic covariation process} \\ \quad \text{has the form } \langle M,M\rangle_t = \int_0^t a_s\,ds \\ \bullet\ \text{the compensator of } \sum_{s\leq t} 1_{\{\|\Delta\sigma_s\|>1\}} \text{ is } \int_0^t \widetilde{a}_s\,ds. \end{array}\right\} \tag{4.4.4}$$

This is in matrix form: $\widetilde{b}$ and M are $d \times d'$-dimensional, and $\widetilde{\sigma}$ is $d \times d' \times d'$-dimensional (for example the (ij)th component of the stochastic integral with respect to W is $\sum_{k=1}^{d'} \int_0^t \widetilde{\sigma}_s^{ijk}\,dW_s^k$), and a is d'^4-dimensional and $\widetilde{a}$ is one-dimensional nonnegative.

Assumption 4.4.3 (or (**K**)) We have (H) and the process σ is also an Itô semimartingale. Furthermore, with the notation of (4.4.4), we have:

(i) the processes $\widetilde{b}$, a and $\widetilde{a}$ are locally bounded and progressively measurable;
(ii) the processes $\widetilde{\sigma}$ and b are càdlàg or càglàd (and adapted, of course).

Note that the distinction as to whether b and $\widetilde{\sigma}$ are càdlàg or are càglàd is irrelevant here, because these processes are determined up to a Lebesgue-null set in time anyway.

(K) implies that the process $c = \sigma\sigma^\star$ is also an Itô semimartingale. On the other hand, assuming that c is an Itô semimartingale is not quite enough to obtain a "square-root" σ which is also an Itô semimartingale, unless both processes c_t and c_{t-} are everywhere invertible. This invertibility property is sometimes important, so we state:

Assumption 4.4.4 (or (**K'**)) We have (K) and both processes c_t and c_{t-} take their values in the set $\mathcal{M}_{d\times d}^{++}$ of all symmetric positive definite $d \times d$ matrices.

Remark 4.4.5 (H) will be in force for most of the CLTs proved in this book, and it is rather mild, in the setting of Itô semimartingales.

For example

$$X_t = X_0 + \int_0^t H_s \, dZ_s,$$

where Z is a multidimensional Lévy process and H is a predictable and locally bounded process (we use here matrix notation), defines an Itô semimartingale X which automatically satisfies (H). Note that if Z has no Gaussian part then $\sigma \equiv 0$. Otherwise, the Gaussian part of Z has the form αW for some multidimensional Brownian motion and a (non-random) matrix α, and a version of σ is $\sigma_t = H_t \alpha$, which is càglàd and not càdlàg. However, $\sigma'_t = H_{t+}\alpha$ is a càdlàg version of σ, in the sense that $\sigma'_t = \sigma_t$ for all t outside a countable set and thus $\int_0^t \sigma'_s \sigma'^*_s \, ds = \int_0^t \sigma_s \sigma^*_s \, ds$.

Assumption (K) is stronger than (H), but nevertheless very often satisfied. It is the case, for example, when X is the solution (weak or strong, when it exists) of a stochastic differential equation of the form

$$X_t = X_0 + \int_0^t f(s, X_{s-}) \, dZ_s,$$

with Z again a Lévy process, and f a $C^{1,2}$ function on $\mathbb{R}_+ \times \mathbb{R}^d$.

As the reader will have noticed already, estimates of various kinds play an important role in our topic, and will do so even more in the sequel. Now, as seen in Sect. 2.1.5 for example, "good" estimates are only available under suitable boundedness assumptions on (b, σ, δ), when X has the form (4.4.1). These boundedness hypotheses are not satisfied under (H), (K) or (K'), so we strengthen those assumptions as follows.

Assumption 4.4.6 (or **(SH)**) We have (H), and there are a constant A and a nonnegative function Γ on E, such that

$$\begin{aligned} &\|b_t(\omega)\| \le A, \qquad \|\sigma_t(\omega)\| \le A, \qquad \|X_t(\omega)\| \le A \\ &\|\delta(\omega, t, z)\| \le \Gamma(z), \quad \Gamma(z) \le A, \quad \int \Gamma(z)^2 \lambda(dz) \le A. \end{aligned} \tag{4.4.5}$$

Assumption 4.4.7 (or **(SK)**) We have (K), and there are a constant A and a nonnegative function Γ on E, such that (4.4.5) holds, and also

$$\|\widetilde{b}_t(\omega)\| \le A, \quad \|\widetilde{\sigma}_t(\omega)\| \le A, \quad \|a_t(\omega)\| \le A, \quad \|a_t(\omega)\| \le A. \tag{4.4.6}$$

Assumption 4.4.8 (or **(SK')**) We have (SK), and the process c_t has an inverse which is (uniformly) bounded.

The localization lemma is then the following one:

Lemma 4.4.9 (Localization) *The localization procedure applies, from* (SH) *to* (H), *and from* (SK) *to* (K), *and from* (SK') *to* (K').

Proof 1) We take (ABC) to be equal to (H) or (K) or (K'), and the corresponding strengthened assumption (SABC) is (SH) or (SK) or (SK'), accordingly. We suppose that one of the properties in (4.4.3) holds under (SABC) and we want to show that it also holds under (ABC). So we take $U^n(X)$ and $U(X)$ which satisfy (4.4.2).

Since by (2.2.7) stable convergence in law and convergence in probability are the same when the limit is defined on the original space, the last two properties in (4.4.3) are the same as the first two ones when $U(X)$ is defined on $(\Omega, \mathcal{F}, \mathbb{P})$. Also, observing that for any fixed t the processes $U'^n(X)_s = U^n(X)_t \, 1_{\{s \geq t\}}$ and $U'(X)_s = U(X)_t \, 1_{\{s \geq t\}}$ satisfy (4.4.2), and $U'^n(X) \stackrel{\mathcal{L}\text{-s}}{\Longrightarrow} U'(X)$ (functional stable convergence in law) if and only if $U'^n(X)_t \stackrel{\mathcal{L}\text{-s}}{\longrightarrow} U'(X)_t$ (convergence of variables), it is enough below to consider the first property in (4.4.3).

2) We let X be any semimartingale satisfying (ABC). Suppose for a moment the existence of a localizing sequence S_p of stopping times, such that

$$\begin{array}{l} \bullet\ t < S_p \quad \Rightarrow \quad X(p)_t \;=\; X_t \ \text{a.s.} \\ \bullet\ \text{each } X(p) \text{ satisfies (SABC).} \end{array} \tag{4.4.7}$$

We want to prove that $U^n(X) \stackrel{\mathcal{L}\text{-s}}{\Longrightarrow} U(X)$, which means that for any bounded measurable variable Z on $(\Omega, \mathcal{F}, \mathbb{P})$ and any bounded continuous function F on $\mathbb{D}^q = \mathbb{D}(\mathbb{R}_+, \mathbb{R}^q)$ for the Skorokhod topology, we have

$$\mathbb{E}\big(Z\, F\big(U^n(X)\big)\big) \;\to\; \widetilde{\mathbb{E}}\big(Z\, F\big(U(X)\big)\big). \tag{4.4.8}$$

We reformulate (4.4.8) as follows: let Q_X be a regular version of the $\mathcal{F}$-conditional distribution of the process $U(X)$. This is a transition probability $Q_X(\omega, dx)$ from $(\Omega, \mathcal{F})$ into $(\mathbb{D}^q, \mathcal{D}^q)$, where $\mathcal{D}^q$ is the Borel σ-field of $\mathbb{D}^q$. Then if $Q_X(F)$ denotes the expectation of F with respect to $Q_X(\omega, .)$, we have $\widetilde{\mathbb{E}}(Z\, F(U(X))) = \mathbb{E}(Z\, Q_X(F))$. In other words, (4.4.8) is exactly

$$\mathbb{E}\big(Z\, F\big(U^n(X)\big)\big) \;\to\; \mathbb{E}\big(Z\, Q_X(F)\big). \tag{4.4.9}$$

Since the Skorokhod topology is "local" in time, it is enough to prove (4.4.9) when F satisfies $F(x) = F(y)$ whenever $x(s) = y(s)$ for all $s \leq t$, with t arbitrarily large but finite, and we suppose further on that this is the case. Since $S_p \to \infty$ and since Z and F are bounded, it is then enough to prove that for any $p \geq 1$ we have

$$\mathbb{E}\big(Z\, F\big(U^n(X)\big)\, 1_{\{t<S_p\}}\big) \;\to\; \mathbb{E}\big(Z\, Q_X(F)\, 1_{\{t<S_p\}}\big). \tag{4.4.10}$$

Now, taking into account the special structure of F, the first part of (4.4.2) yields that on the set $\{t < S_p\}$ we have $F(U^n(X)) = F(U^n(X(p))$ a.s., whereas the second part yields that on the same set we have $Q_X(F) = Q_{X(p)}(F)$. Hence (4.4.10) amounts to

$$\mathbb{E}\big(Z\, F\big(U^n\big(X(p)\big)\big)\, 1_{\{t<S_p\}}\big) \;\to\; \mathbb{E}\big(Z\, Q_{X(p)}(F)\, 1_{\{t<S_p\}}\big). \tag{4.4.11}$$

It remains to apply the hypothesis, which says that for each $p \geq 1$ we have $U^n(X(p)) \stackrel{\mathcal{L}\text{-s}}{\Longrightarrow} U(X(p))$ as $n \to \infty$. This gives (4.4.11), hence at this stage it remains to prove the existence of the processes $X(p)$ satisfying (4.4.7).

3) We start with the case (ABC) = (H), so here X satisfies (H). The process b is locally bounded and without loss of generality we may take $b_0 = 0$, so there is a localizing sequence T_p such that $\|b_t\| \leq p$ if $0 \leq t \leq T_p$. Next, the stopping times $R_p = \inf(t : \|X_t\| + \|\sigma_t\| \geq p)$ increase to $+\infty$ as well. Then, letting τ_n be the stopping times occurring in (H)-(iii), we set $S_p = T_p \wedge R_p \wedge \tau_p$ and we choose a $d \times d'$ non-random matrix with $\|\alpha\| < 1$. Then we set

$$\begin{gathered} b_t^{(p)} = b_{t\wedge S_p}, \qquad \sigma_t^{(p)} = \sigma_{t\wedge S_p}\, 1_{\{\|\sigma_{t\wedge S_p}\|\leq p\}} \\ \delta^{(p)}(t,z) \;=\; \delta(t\wedge S_p, z)\; 1_{\{\|\delta(t\wedge S_p,z)\|\leq 2p\}}. \end{gathered} \tag{4.4.12}$$

This defines three terms $(b^{(p)}, \sigma^{(p)}, \delta^{(p)})$ analogous to (b, σ, δ). By construction $\|b^{(p)}\| \leq p$ and $\|\sigma^{(p)}\| \leq p$ and $\sigma^{(p)}$ is càdlàg, and also $\|\delta^{(p)}(\omega,t,z)\| \leq \Gamma^{(p)}(z)$, where $\Gamma^{(p)} = p(\Gamma_p \wedge 1)$. Hence the process $X(p)$ given by

$$X(p)_t = \begin{cases} 0 & \text{if } S_p = 0 \\ X_0 + \int_0^t b_s^{(p)}\, ds + \int_0^s \sigma_s^{(p)}\, dW_s & \\ \quad + (\delta^{(p)} 1_{\{\|\delta^{(p)}\|\leq 1\}}) \star (\mathfrak{p} - \mathfrak{q})_t & \\ \quad + (\delta^{(p)} 1_{\{\|\delta^{(p)}\|> 1\}}) \star \mathfrak{p}_t & \text{if } S_p > 0 \end{cases} \tag{4.4.13}$$

satisfies (SH) (note that $\|X(p)_t\| \leq 3p$), and it remains to prove the first part of (4.4.7).

We denote by $Y(p,i)_t$ for $i = 1,2,3,4$ the four last terms in the right side of (4.4.13), and in the same way the $Y(i)_t$'s denote the four last terms in the right side of (4.4.1). Then (4.4.7) will follow from the properties

$$t < S_p \quad \Rightarrow \quad Y(p,i)_t \;=\; Y(i)_t \quad \text{a.s.} \tag{4.4.14}$$

for $i = 1,2,3,4$. That (4.4.14) holds for $i = 1$ is obvious, and for $i = 2$ it follows from the fact that $\sigma_t^{(p)} = \sigma_t$ when $t < S_p$. For $i = 3,4$ we first recall the following representation for the Poisson random measure $\mathfrak{p}$: this is a random measure on $\mathbb{R}_+ \times E$, and with Δ being an additional point outside E, we can find an $E \cup \{\Delta\}$-valued optional process θ_t such that

$$\mathfrak{p}(dt,dz) \;=\; \sum_{s:\theta_s(\omega)\in E} \varepsilon_{(s,\theta_s(\omega))}(dt,dz)$$

where ε_a denotes the Dirac mass sitting at a. We also extend δ and $\delta^{(p)}$ by setting $\delta(\omega,t,\Delta) = \delta^{(p)}(\omega,t,\Delta) = 0$. Then, outside a $\mathbb{P}$ null set N, we have $\Delta X_s = \delta(s,\theta_s)$ and $\Delta X(p)_s = \delta^{(p)}(s,\theta_s)$ for all s. If $s < S_p$ we also have $s < R_p$ and thus $\Delta X_s \leq 2p$, therefore outside N, and in view of (4.4.12), we must have $\delta(s,\theta_s) = \delta^{(p)}(s,\theta_s)$ for all $s < S_p$. Then (4.4.14) for $i = 4$ readily follows. Furthermore outside N, we also have $\delta(s,\theta_s) = \delta^{(p)}(s,\theta_s)$ for all $s \leq S_p$ as soon as either

$\|\delta(s,\theta_s)\|$ or $\|\delta^{(p)}(s,\theta_s)\|$ belongs to $(0,1]$. Therefore the two purely discontinuous local martingales $Y(3)$ and $Y(p,3)$ have the same jumps on the interval $(0,S_p]$, hence they must be a.s. equal on this interval (because the difference $Y(3)-Y(p,3)$, stopped at time S_p, is a purely discontinuous local martingale which is also a.s. continuous, hence it vanishes). This proves (4.4.14) for $i=3$.

4) Now we consider the case (ABC) = (K), so here X satisfies (K). We can always use the left continuous version of the process $\widetilde{\sigma}$, which is locally bounded, as well as $\widetilde{b}$, a and $\widetilde{a}$, which can also be all taken equal to 0 at time $t=0$. Hence we first have a localizing sequence T_p such that

$$0<t\le T_p \Rightarrow \|b_t\|\le p, \quad t\le T_p \Rightarrow \|\widetilde{b}_t\|+\|\widetilde{\sigma}_t\|+\|a_t\|+\widetilde{a}_t\le p$$

(we cannot assume here $b_0=0$, in the case it is càdlàg). The stopping times R_p are as in the previous step, as is $S_p=T_p\wedge R_p\wedge\tau_p$ and we choose a $d\times d'$ non-random matrix with $\|\alpha\|\le 1$. Then we define $\delta^{(p)}$ as in (4.4.12) and set $b_t^{(p)}=b_{t\wedge S_p}\,1_{\{S_p>0\}}$ and

$$\sigma_t^{(p)}=\begin{cases}\alpha & \text{if } S_p=0\\ \sigma_0^{(p)}+\int_0^{t\wedge S_p}\widetilde{b}_s\,ds+\int_0^{t\wedge S_p}\widetilde{\sigma}_s\,dW_s+M_{t\wedge S_p} & \\ \quad+\sum_{s\le t,\,s<S_p}\Delta\sigma_s\,1_{\{\|\Delta\sigma_s\|>1\}} & \text{if } S_p>0.\end{cases} \tag{4.4.15}$$

By construction $\|\sigma^{(p)}\|\le p+1$ (recall that $\|\Delta M\|\le 1$). Moreover $\sigma^{(p)}$ is again of the form (4.4.4), with the associated processes $\|\widetilde{b}^{(p)}\|$, $\|\widetilde{\sigma}^{(p)}\|$, $\|a^{(p)}\|$ and $\widetilde{a}^{(p)}$ bounded by p (this is simple, except for $\widetilde{a}^{(p)}$: for this case we consider the two processes $Y_t=\sum_{s\le t\wedge S_p}1_{\{\|\Delta\sigma\|>1\}}$ and $Y'_t=\sum_{s\le t,\,s<S_p}1_{\{\|\Delta\sigma^{(p)}\|>1\}}$, with respective compensators $\widetilde{Y}$ and $\widetilde{Y}'$; observing that $Y_t-Y'_t=1_{\{\|\Delta\sigma_{S_p}\|>1\}}\,1_{\{t\ge S_p\}}$ is nondecreasing, the same is true of $\widetilde{Y}-\widetilde{Y}'$; therefore $\widetilde{Y}'_t=\int\widetilde{a}_s^{(p)}\,ds$ for some $\widetilde{a}^{(p)}$ satisfying $0\le\widetilde{a}_t^{(p)}\le\widetilde{a}_{t\wedge S_p}\le p$).

Then, the process $X(p)$ defined by (4.4.4) satisfies (SK) and, since $\sigma_t^{(p)}=\sigma_t$ when $t<S_p$, we obtain the first part of (4.4.7) exactly as in Step 3.

5) Finally we let (ABC) = (K'), so X satisfies (K'). The processes c_t and c_{t-} are everywhere invertible, so $\gamma_t=\sup_{s\le t}\|c_s^{-1}\|$ is finite-valued (if $\gamma_t=\infty$ for some t, there is a sequence s_n converging to a limit $s\le t$, either increasingly or decreasingly, and with $\|c_{s_n}^{-1}\|\to\infty$; this implies $\|c_s^{-1}\|=\infty$ in the first case, and $\|c_{s-}^{-1}\|=\infty$ in the second case, thus bringing a contradiction).

We repeat the construction of Step 4, with the following changes. We keep the same T_p and R_p and set $T'_p=\inf(t:\gamma_t\ge p)$, which increases to ∞, and we now take $S_p=T_p\wedge R_p\wedge T'_p\wedge\tau_p$. We choose α such that $a=\alpha\alpha^*$ is invertible, and set $\rho=\|a^{-1}\|$. We also choose another non-random $d\times d'$ matrix α'_p such that for any matrix α'' with $\|\alpha''\|\le p+1$, then $A=(\alpha+\alpha'_p)(\alpha+\alpha'_p)^*$ is invertible with a (Euclidean) norm of the inverse satisfying $\|A^{-1}\|\le 1$: an easy computation shows that it is possible for some α' satisfying $\|\alpha'\|\le\rho'=(p+1)(d'+1)$. Then at this stage, we replace the second part of (4.4.15) by

$$\sigma_t^{(p)} = \sigma_0^{(p)} + \int_0^{t\wedge S_p} \widetilde{b}_s\,ds + \int_0^{t\wedge S_p} \widetilde{\sigma}_s\,dW_s$$
$$+ M_{t\wedge S_p} + \sum_{s\le t,\, s<S_p} \Delta\sigma_s\, 1_{\{\|\Delta\sigma_s\|>1\}} + \alpha'\, 1_{\{t\ge S_p\}}.$$

Hence $\sigma_t^{(p)} = \sigma_t$ for all $t < S_p$, and $\|\sigma_t^{(p)}\| \le p+1+\rho'$ and also, due to our choices of α and α', such that $c_t^{(p)} = \sigma_t^{(p)}\sigma_t^{(p)*}$ is invertible, with $\|c_t^{(p)-1}\| \le p+1+\rho$. So the process $X(p)$ defined by (4.4.13) satisfies (SK'), and the proof of the first part of (4.4.7) is as in the two previous steps. □

4.4.2 A Stable Convergence for Itô Semimartingales

We will now show how Theorem 4.3.1 can be applied to the semimartingale X satisfying the strengthened Assumption (SH), that is Assumption 4.4.6. In particular the function Γ on E satisfies (4.4.5). Then we may define the sequence (S_p) of stopping times by (4.3.1), upon taking $A_m = \{z : \Gamma(z) > 1/m\}$. This sequence weakly exhausts the jump of X, in the sense of Sect. 4.1.2. We consider a regular discretization scheme with time step Δ_n, and the d-dimensional variables

$$\left.\begin{array}{l} S_-(n,p) = (i-1)\Delta_n, \quad S_+(n,p) \;=\; i\Delta_n \\ R_-(n,p) = \frac{1}{\sqrt{\Delta_n}}\left(X_{S_p-} - X_{(i-1)\Delta_n}\right) \\ R_+(n,p) = \frac{1}{\sqrt{\Delta_n}}\left(X_{i\Delta_n} - X_{S_p}\right) \end{array}\right\} \text{ if } (i-1)\Delta_n < S_p \le i\Delta_n. \qquad (4.4.16)$$

We want to describe the limiting behavior of these variables, together with processes $\overline{U}^n(G)$ as in (4.2.6). Those processes are associated with a sequence $u_n > 0$ going to 0, and a function Φ satisfying (4.2.3) and an adapted càdlàg $\mathbb{R}^q \otimes \mathbb{R}^{q'}$-valued process θ, and an adapted càdlàg $\mathbb{R}^w$-valued process Y, and an r-dimensional continuous function $G = (G^j)$ on $\mathbb{R}^w \times \mathbb{R}^q$ with polynomial growth.

For describing the limit, and as in Sect. 4.3, we consider an auxiliary space $(\Omega', \mathcal{F}', \mathbb{P}')$ endowed with an r-dimensional Brownian motion W', a sequence $(\kappa_p)_{p\ge1}$ of variables uniformly distributed over (0, 1], and two sequences $(\Psi_{p-})_{p\ge1}$ and $(\Psi_{p+})_{p\ge1}$ of d'-dimensional centered Gaussian variables with covariance the identity matrix, all mutually independent. The very good filtered extension is defined by (4.1.16), where T_n and U_n are substituted with S_n and $(\kappa_n, \Psi_{n-}, \Psi_{n+})$. The limiting process $\overline{U}(G)$ of the sequence $\overline{U}^n(G)$ is defined on the extension by (4.3.3). Finally the limits of the variables $R_\pm(n,p)$ will also be defined on the extension, as the following d-dimensional variables:

$$R_{p-} \;=\; \sqrt{\kappa_p}\,\sigma_{T_p-}\Psi_{p-}, \qquad R_{p+} \;=\; \sqrt{1-\kappa_p}\,\sigma_{T_p}\Psi_{p+}. \qquad (4.4.17)$$

Proposition 4.4.10 *If X satisfies* (SH), *and with the previous notation and assumptions, we have*

$$\left(\overline{U}^n(G), \left(R_-(n,p), R_+(n,p)\right)_{p\ge 1}\right) \stackrel{\mathcal{L}\text{-s}}{\longrightarrow} \left(\overline{U}(G), (R_{p-}, R_{p+})_{p\ge 1}\right) \quad (4.4.18)$$

for the product topology on the space $\mathbb{D}^r \times (\mathbb{R}^2)^{\mathbb{N}^*}$.

Proof 1) We have the extended space $(\widetilde{\Omega}, \widetilde{\mathcal{F}}, (\widetilde{\mathcal{F}}_t)_{t\ge 0}, \widetilde{\mathbb{P}})$ as defined above, and another extended space $(\widetilde{\Omega}', \widetilde{\mathcal{F}}', (\widetilde{\mathcal{F}}'_t)_{t\ge 0}, \widetilde{\mathbb{P}}')$ which supports the limit described in Theorem 4.3.1: this limit is denoted as $(\overline{U}(G)', (W''(p)', \kappa'_p)_{p\ge 1})$. On the second extension, we set

$$\Psi'_{p-} = \frac{1}{\sqrt{\kappa'_p}}\left(W''(p)'_{1+\kappa'_p} - W''(p)'_1\right), \qquad \Psi'_{p+} = \frac{1}{\sqrt{1-\kappa'_p}}\left(W''(p)'_2 - W''(p)'_{1+\kappa'_p}\right)$$

$$R'_{p-} = \sqrt{\kappa'_p}\,\sigma_{T_p-}\,\Psi'_{p-}, \qquad R'_{p+} = \sqrt{1-\kappa'_p}\,\sigma_{T_p}\,\Psi'_{p+}.$$

Since $W''(p)'$ is a Brownian motion independent of κ'_p, one easily checks that the three variables $\Psi'_{p-}, \Psi'_{p+}, \kappa'_p$ are independent, the first two being $\mathcal{N}(0, I_{d'})$. Hence the $\mathcal{F}$-conditional law of the family $(\overline{U}(G)', (\kappa'_p, \Psi'_{p-}, \Psi'_{p+}))$ on the second extension is the same as the $\mathcal{F}$-conditional law of the family $(\overline{U}(G), (\kappa_p, \Psi_{p-}, \Psi_{p+}))$ on the first extension. Thus, the $\mathcal{F}$-conditional laws of the two families $(\overline{U}(g)', (R'_{p-}, R'_{p+}))$ and $(\overline{U}(G), (R_{p-}, R_{p+}))$ are also the same, and it is enough to prove (4.4.18) with $(\overline{U}(G)', (R'_{p-}, R'_{p+}))$ in the right side.

2) Now, we put

$$\alpha_-(n,p) = \frac{1}{\sqrt{\Delta_n}}(W_{S_p} - W_{S_-(n,p)}), \qquad \alpha_+(n,p) = \frac{1}{\sqrt{\Delta_n}}(W_{S_+(n,p)} - W_{S_p}).$$

With the notation (4.3.2), and with $v_n = \Delta_n$, we have

$$\alpha_-(n,p) = \overline{w}(n,p)_{\kappa(n,p)+1} - \overline{w}(n,p)_1, \quad \alpha_-(n,p) = \overline{w}(n,p)_2 - \overline{w}(n,p)_{\kappa(n,p)+1}.$$

Since the map $(t,w) \mapsto w(t)$ from $\mathbb{R}_+ \times \mathbb{D}^{d'}$ is continuous at any (t,w) such that $w(t) = w(t-)$, and recalling the definition of $\Psi'_{p\pm}$, we deduce from Theorem 4.3.1 that

$$\left(\overline{U}^n(G), \left(\alpha_-(n,p), \alpha_-(n,p)\right)_{p\ge 1}\right) \stackrel{\mathcal{L}\text{-s}}{\longrightarrow} \left(\overline{U}(G)', \left(\sqrt{\kappa'_p}\,\Psi'_{p-}, \sqrt{1-\kappa'_p}\,\Psi'_{p+}\right)_{p\ge 1}\right).$$

By virtue of the definition of $R_\pm(n,p)$ and $R'_{p\pm}$, and from the fact that σ is càdlàg, it remains to prove that for any $p \ge 1$ we have

$$R_-(n,p) - \sigma_{S_-(n,p)}\alpha_-(n,p) \stackrel{\mathbb{P}}{\longrightarrow} 0, \quad R_+(n,p) - \sigma_{S_p}\alpha_+(n,p) \stackrel{\mathbb{P}}{\longrightarrow} 0. \quad (4.4.19)$$

3) For proving (4.4.19) we need some notation. First, recalling $A_m = \{z : \Gamma(z) > 1/m\}$, we set

$$\left.\begin{array}{l} b(m)_t = b_t - \int_{A_m \cap \{z: \|\delta(t,z)\| \le 1\}} \delta(t,z)\,\lambda(dz), \\ X(m)_t \;=\; X_0 + \int_0^t b(m)_s\,ds + \int_0^t \sigma_s\,dW_s + (\delta\,1_{A_m^c}) \star (\mathfrak{p} - \mathfrak{q})_t, \\ X'(m) \;=\; X - X(m) \;=\; (\delta\,1_{A_m}) \star \mathfrak{p}. \end{array}\right\} \tag{4.4.20}$$

Next, with $I(n,i) = ((i-1)\Delta_n, i\Delta_n]$, we write

$$\begin{array}{l} \Omega_n(T,m) = \text{the set of all } \omega \text{ such that each interval } [0,T] \cap I(n,i) \\ \qquad \text{contains at most one jump of } X'(m)(\omega), \text{ and that} \\ \qquad \|X(m)(\omega)_{t+s} - X(m)(\omega)_t\| \le 2/m \text{ for all } t \in [0,T],\ s \in [0,\Delta_n]. \end{array} \tag{4.4.21}$$

By construction, $X(m)$ is a càdlàg process with jumps smaller than $1/m$, whereas $X'(m)$ has finitely many jumps on $[0,T]$. Therefore, for all $T > 0$ and $m \ge 1$ we have

$$\mathbb{P}\big(\Omega_n(T,m)\big) \to 1 \quad \text{as } n \to \infty. \tag{4.4.22}$$

By (4.3.1) we have $S_p = S(m,j)$ for some $m, j \ge 1$. Then on the set $\Omega_n(T,m) \cap \{S_p < T\}$ we have $R_-(n,p) = (X(m)_{S_p} - X(m)_{S_-(n,p)})/\sqrt{\Delta_n}$ and $R_+(n,p) = (X(m)_{S_+(n,p)} - X(m)_{S_p})/\sqrt{\Delta_n}$. Hence in view of (4.4.22), the property (4.4.19) follows from

$$\begin{array}{l} w_p^n := \left| \dfrac{1}{\sqrt{\Delta_n}} \big(X(m)_{S_p} - X(m)_{S_-(n,p)}\big) - \sigma_{S_-(n,p)} \alpha_-(n,p) \right| \\ \qquad + \left| \dfrac{1}{\sqrt{\Delta_n}} \big(X(m)_{S_+(n,p)} - X(m)_{S_p}\big) - \sigma_{S_p} \alpha_+(n,p) \right| \stackrel{\mathbb{P}}{\longrightarrow} 0. \end{array} \tag{4.4.23}$$

Let $(\mathcal{G}_t)$ be the filtration $(\mathcal{G}_t^{A_m})$ associated with the set A_m by (2.1.48). Note that $S_+(n,p)$ and $S_-(n,p)$ and S_p are stopping times with respect to $(\mathcal{G}_t)$, so by Proposition 2.1.10 we can apply (2.1.33) and (2.1.34) and (2.1.39) (with $r = 2$ and $q = 1/2$), twice to the three integral processes in the definition of $X(m)$ with the stopping times $S_-(n,p)$ and S_p, to deduce from the definition of w_p^n that

$$\begin{array}{l} \mathbb{E}\big((w_p^n)^2 \wedge 1\big) \le K\sqrt{\Delta_n} + K \displaystyle\int_{\{\Gamma(z) \le \Delta_n^{1/4}\}} \Gamma(z)^2 \lambda(dz) \\ \qquad + K\mathbb{E}\Big(\displaystyle\sup_{0<u\le\Delta_n} \big(\|\sigma_{S_p-u} - \sigma_{S_p-}\|^2 + \|\sigma_{S_p+u} - \sigma_{S_p}\|^2\big)\Big). \end{array}$$

All three terms on the right above go to 0 as $n \to \infty$ (use the dominated convergence theorem for the last one, in which σ is càdlàg and bounded). Then we have (4.4.23). □

Chapter 5
Central Limit Theorems: The Basic Results

After having established some key tools in the previous chapter, we are now ready for the Central Limit Theorems associated with the Laws of Large Numbers of Chap. 3. We begin with the CLT for the non-normalized functionals and continue with the CLT for the normalized ones. The first CLT is significantly more complicated to state than the second one, but the proof is somewhat simpler. We end the chapter with a CLT for the "approximate" quadratic variation, that is for the basic convergence exhibited in (2.1.8).

For all these results we need X to be a d-dimensional *Itô semimartingale*, defined on a filtered probability space $(\Omega, \mathcal{F}, (\mathcal{F}_t)_{t\geq 0}, \mathbb{P})$. Its characteristics have the form (2.1.25) and the jump measure of X is called μ. We will also use a *Grigelionis representation* (2.1.30) for it, possibly defined on an extended space, still denoted as $(\Omega, \mathcal{F}, (\mathcal{F}_t)_{t\geq 0}, \mathbb{P})$. That is,

$$X_t = X_0 + \int_0^t b_s\,ds + \int_0^t \sigma_s\,dW_s + (\delta 1_{\{\|\delta\|\leq 1\}}) \star (\mathfrak{p} - \mathfrak{q})_t + (\delta 1_{\{\|\delta\|>1\}}) \star \mathfrak{p}_t, \quad (5.0.1)$$

where W is a d'-dimensional Brownian motion and $\mathfrak{p}$ is a Poisson measure on $\mathbb{R}_+ \times E$. Here, $(E, \mathcal{E})$ is some Polish space, and the intensity measure, or compensator, of $\mathfrak{p}$ is of the form $\mathfrak{q}(dt, dz) = dt \otimes \lambda(dz)$, where λ is a σ-finite measure on E. Then σ is $d \times d'$-dimensional, whereas $c = \sigma\sigma^\star$ is $d \times d$-dimensional.

We also need, in a fundamental way, that the discretization scheme is a *regular scheme*, whose time step we denote by Δ_n.

5.1 The Central Limit Theorem for Functionals Without Normalization

We consider here the non-normalized functionals

$$V^n(f, X)_t = \sum_{i=1}^{[t/\Delta_n]} f\big(\Delta_i^n X\big).$$

J. Jacod, P. Protter, *Discretization of Processes*,
Stochastic Modelling and Applied Probability 67,
DOI 10.1007/978-3-642-24127-7_5,

The CLT will be available when $f(x) = \mathrm{o}(\|x\|^3)$ as $x \to 0$, plus some mild smoothness conditions: this corresponds to a special case of (A-a) of Theorem 3.3.1, when $V^n(f, X) \stackrel{\mathbb{P}}{\Longrightarrow} f * \mu$. In Sect. 5.4 below we will also obtain a CLT corresponding to case (B). For the other cases of this theorem no CLT is available in general: this was already shown in the introductory chapter 1 for the example of a Brownian motion plus a compound Poisson process and $f(x) = |x|^p$: there is no CLT in this case when $p < 2$ or $2 < p < 3$, and a CLT with a bias term when $p = 3$.

5.1.1 The Central Limit Theorem, Without Normalization

We need Assumption (H), introduced as Assumption 4.4.2 in the previous chapter, and which we briefly recall below for the reader's convenience:

Assumption (H) X is an Itô semimartingale of the form (5.0.1), where the process b is locally bounded, the process σ is càdlàg, and $\|\delta(\omega, t, z)\| \wedge 1 \le \Gamma_n(z)$ for all (ω, t, z) with $t \le \tau_n(\omega)$, where (τ_n) is a localizing sequence (τ_n) of stopping times and each Γ_n is a nonnegative function on E satisfying $\int \Gamma_n(z)^2 \lambda(dz) < \infty$.

We also describe the ingredients coming into the limiting process, which is in fact as described in (4.1.7). We have an auxiliary space $(\Omega', \mathcal{F}', \mathbb{P}')$ endowed with a triple sequence $(\Psi_{n-}, \Psi_{n+}, \kappa_n)_{n\ge 1}$ of variables, all independent, and with the following laws:

$$\Psi_{n\pm} \text{ are } d'\text{-dimensional}, \mathcal{N}(0, I_{d'}), \quad \kappa_n \text{ is uniform on } [0,1]. \tag{5.1.1}$$

We also consider an arbitrary weakly exhausting sequence $(T_n)_{n\ge 1}$ of stopping times for the jumps of X, which means that $T_n \neq T_m$ is $n \neq m$ and $T_n < \infty$, and that for any (ω, s) with $\Delta X_s(\omega) \neq 0$ there is some n with $T_n = s$. The *very good filtered extension* $(\widetilde{\Omega}, \widetilde{\mathcal{F}}, (\widetilde{\mathcal{F}}_t)_{t\ge 0}, \widetilde{\mathbb{P}})$ of $(\Omega, \mathcal{F}, (\mathcal{F}_t)_{t\ge 0}, \mathbb{P})$ is defined by (4.1.16), that is:

$$\left.\begin{array}{l} \widetilde{\Omega} = \Omega \times \Omega', \qquad \widetilde{\mathcal{F}} = \mathcal{F} \otimes \mathcal{F}', \qquad \widetilde{\mathbb{P}} = \mathbb{P} \otimes \mathbb{P}' \\ (\widetilde{\mathcal{F}}_t) \text{ is the smallest filtration containing } (\mathcal{F}_t) \text{ and such that} \\ \qquad\qquad (\Psi_{n-}, \Psi_{n+}, \kappa_n) \text{ is } \widetilde{\mathcal{F}}_{T_n} \text{ measurable for all } n \end{array}\right\} \tag{5.1.2}$$

(an additional Brownian motion W' is not necessary here). We also introduce the following d-dimensional random variables, see (4.4.17), with σ being the process occurring in (5.0.1):

$$R_n = R_{n-} + R_{n+}, \quad \text{where } R_{n-} = \sqrt{\kappa_n}\, \sigma_{T_n -} \Psi_{n-}, \quad R_{n+} = \sqrt{1 - \kappa_n}\, \sigma_{T_n} \Psi_{n+}. \tag{5.1.3}$$

The following proposition describes some properties of the limiting process, and some intuition about why it takes this form will be given after the statement of the CLT.

Proposition 5.1.1 *Assume* (H). *Let f be a C^1 function from $\mathbb{R}^d$ into $\mathbb{R}^q$, with $\partial_i f(x) = \mathrm{O}(\|x\|)$ as $x \to 0$, where $\partial_i f$ for $i = 1, \dots, d$ denotes the q-dimensional partial derivatives of f. The formula*

$$\overline{V}(f, X)_t = \sum_{n=1}^{\infty} \left(\sum_{i=1}^{d} \partial_i f(\Delta X_{T_n}) R_n^i \right) 1_{\{T_n \le t\}} \tag{5.1.4}$$

defines a q-dimensional process $\overline{V}(f, X)$ on $(\widetilde{\Omega}, \widetilde{\mathcal{F}}, (\widetilde{\mathcal{F}}_t)_{t \ge 0}, \widetilde{\mathbb{P}})$ which is a.s. càdlàg, adapted, and conditionally on $\mathcal{F}$ has centered and independent increments and satisfies

$$\widetilde{\mathbb{E}}\big(\overline{V}(f, X)_t^i \, \overline{V}(f, X)_t^j \mid \mathcal{F}\big) = \frac{1}{2} \sum_{s \le t} \sum_{k,l=1}^{d} \big(\partial_k f^i \, \partial_l f^j\big)(\Delta X_s)\big(c_{s-}^{kl} + c_s^{kl}\big), \tag{5.1.5}$$

and its $\mathcal{F}$-conditional law does not depend on the choice of the weakly exhausting sequence T_n. If further the processes X and σ have no common jumps, the process $\overline{V}(f, X)$ is $\mathcal{F}$-conditionally Gaussian.

Proof As in Proposition 4.1.4, writing that (5.1.4) "defines" a process means that for all t the series converge in probability.

For the first claim, the only thing we have to do is to show that (5.1.5) is nothing else than (4.1.7) for a suitable process V_t and suitable random variables U_n. We take $r = 2d'$ and $U_n^i = \sqrt{\kappa_n} \, \Phi_{n-}^i$ if $1 \le i \le d'$ and $U_n^i = \sqrt{1 - \kappa_n} \, \Phi_{n+}^{d'+i}$ if $d' + 1 \le i \le r$. The U_n's are i.i.d. centered with covariance $\mu_2^{ij} = \frac{1}{2}$ when $i = j$ and $\mu_2^{ij} = 0$ otherwise. Next, we take

$$V_s^{ij} = \begin{cases} \sum_{k=1}^{d} \partial_k f^i(\Delta X_s) \sigma_{s-}^{kj} & \text{if } 1 \le j \le d' \\ \sum_{k=1}^{d} \partial_k f^i(\Delta X_s) \sigma_s^{k,j-d'} & \text{if } d' + 1 \le j \le r. \end{cases}$$

The hypothesis on f implies $\|V_s\| \le K \|\Delta X_s\| (\|\sigma_{s-}\| + \|\sigma_s\|)$ as soon as $\|\Delta X_s\| \le 1$. Since $\sup_{s \le t} \|\sigma_s\| < \infty$ for all t by (H), and $\sum_{s \le t} \|\Delta X_s\|^2 < \infty$, we clearly have that $\sum_{s \le t} \|V_s\|^2 < \infty$ for all t, which is the condition (4.1.10) of Proposition 4.1.4. Since (5.1.4) is of the form $\sum_{n: T_n \le t} V_{T_n} U_{T_n}$, the result is proved, upon observing that because $c = \sigma \sigma^\star$, (5.1.5) amounts to (4.1.11).

We now prove the final claim, assuming that X and σ have no common jump. This means that $\sigma_{T_n} = \sigma_{T_n -}$ on the set $\{\Delta X_{T_n} \ne 0\}$, for all n. In this case $R_n = \sigma_{T_n} U_n'$ on the same set, where $U_n' = \sqrt{\kappa_n} \, \Psi_{n-} + \sqrt{1 - \kappa_n} \, \Psi_{n+}$ has obviously the law $\mathcal{N}(0, I_{d'})$. Moreover if $V_s' = (V_s^{ij})_{1 \le i \le q, 1 \le j \le d'}$ we also have $V_{T_n} U_n = V_{T_n}' U_n'$, and of course $\sum_{s \le t} \|V_s'\|^2 < \infty$. Then the claim follows from (a) of Proposition 4.1.4. □

We are now ready to state the main result:

Theorem 5.1.2 (CLT without Normalization) *Assume* (H), *and let f be a C^2 function from $\mathbb{R}^d$ into $\mathbb{R}^q$, with $f(0) = 0$ and $\partial_i f(0) = 0$ and $\partial_{ij}^2 f(x) = \mathrm{o}(\|x\|)$ for all*

$i, j = 1, \dots, d$ as $x \to 0$. Then the processes

$$\overline{V}^n(f, X)_t = \frac{1}{\sqrt{\Delta_n}} \left(V^n(f, X)_t - f \star \mu_{\Delta_n[t/\Delta_n]} \right) \tag{5.1.6}$$

converge stably in law to $\overline{V}(f, X)$, as defined by (5.1.4), *and for each fixed t we also have*

$$\frac{1}{\sqrt{\Delta_n}} \left(V^n(f, X)_t - f \star \mu_t \right) \xrightarrow{\mathcal{L}\text{-s}} \overline{V}(f, X)_t. \tag{5.1.7}$$

(stable convergence in law of q-dimensional variables). Moreover, if $X_t^{(n)} = X_{\Delta_n[t/\Delta_n]}$ is the discretized version of the process X we also have the stable convergence in law $(X^{(n)}, \overline{V}^n(f, X)) \stackrel{\mathcal{L}\text{-s}}{\Longrightarrow} (X, \overline{V}(f, X))$ for the Skorokhod topology.

When $f(x) = \|x\|^p$, our assumption on f is satisfied if and only if $p > 3$. The really new result here is $\overline{V}^n(f, X) \stackrel{\mathcal{L}\text{-s}}{\Longrightarrow} \overline{V}(f, X)$, whereas we already know that $X^{(n)} \xrightarrow{\text{Sk}} X$ pathwise. The joint convergence of $(X^{(n)}, \overline{V}^n(f, X))$ is usually not a trivial consequence of the convergence of $X^{(n)}$ and $\overline{V}^n(f, X)$ separately, but here we get it almost for free, and it turns out to be necessary for an important application, the convergence of Euler schemes for stochastic differential equations.

Remark 5.1.3 (Important comment) It would be nice to have the stable convergence in law of the processes $Y^n = \frac{1}{\sqrt{\Delta_n}} (V^n(f, X) - f \star \mu)$ to $\overline{V}(f, X)$, but this is unfortunately not true. Suppose for example that $X = W + N$, where W is a Brownian motion and N a standard Poisson process (so $d = 1$), and let S the first jump time of N and $S_n^- = \Delta_n[S/\Delta_n]$ and $S_n^+ = S_n^- + \Delta_n$. We have $S_n^- < S < S_n^+$ almost surely, so if $f(x) = x^4$, say, Y^n has a jump of size $-1/\sqrt{\Delta_n}$ at time S. This rules out the convergence in law of Y^n, and we even have $\sup_{t \le S} |Y_t'^n| \to \infty$. By the same token, the process $Y'^n = \frac{1}{\sqrt{\Delta_n}} (V_{int}^n(f, X) - f \star \mu)$ using the interpolated functional satisfies $Y_{S_n^+}'^n - Y_S'^n = ((1 + W_{S_n^+} - W_{S_n^-})^4 - (1 + W_S - W_{S_n^-})^4)/\sqrt{\Delta_n}$ (for n large enough, so the second jump time of N is bigger than S_n^+), which does not go to 0, whereas $S_n^+ - S \to 0$: again, this rules out the convergence in law of Y'^n.

On the other hand, if t is fixed, the difference $f \star \mu_t - f \star \mu_{\Delta_n[t/\Delta_n]}$ involves only the jumps between $\Delta_n[t/\Delta_n]$ and t, and those are "very small" with a large probability, in such a way that this difference divided by $\sqrt{\Delta_n}$ becomes asymptotically negligible; so we have the convergence in law of Y_t^n, as random variables.

Remark 5.1.4 (The intuition behind the limiting process) The process $\overline{V}(f, X)$ comes as a sum over all jump times of X, but the idea underlying the form of each summand is quite simple. To see it, we suppose that $q = d = d' = 1$ and that there is a single jump for X, say at time $T_1 = T$, and we set $X_t' = X_t - \Delta X_T \, 1_{\{T \le t\}}$. For each n we have a (random) integer i_n such that the interval $I(n, i_n) = ((i_n - 1)\Delta_n, i_n\Delta_n]$ contains T. We see that $\overline{V}^n(f, X)_t$ is the sum of the variables $f(\Delta_i^n X')/\sqrt{\Delta_n}$ for all $i \leq [t/\Delta_n]$, plus the variable

$$Z_n = \frac{1}{\sqrt{\Delta_n}}\big(f(\Delta X_T + \Delta_{i_n}^n X') - f(\Delta_{i_n}^n X') - f(\Delta X_T)\big)$$

if $t \geq i_n\Delta_n$. Since X' is a continuous Itô semimartingale, each $\Delta_i^n X'$ is "of order" $\sqrt{\Delta_n}$ and thus $f(\Delta_i^n X')/\sqrt{\Delta_n}$ is of order $\mathrm{o}(\Delta_n)$ because $f(x) = \mathrm{o}(|x|^3)$ as $x \to 0$. The sum of the $[t/\Delta_n]$ summands of this type is thus negligible (this needs a proof, of course, and this is why we make a rather strong assumption on the behavior of f near 0). Thus $\overline{V}^n(f, X)_t$ is close to Z_n, which by a Taylor's expansion is approximately $\partial f(\Delta X_T)\Delta_{i_n}^n X'/\sqrt{\Delta_n}$. It remains to observe that $\Delta_{i_n}^n X'/\sqrt{\Delta_n}$ is approximately $(\sigma_{T-}(W_T - X_{(i_n-1)\Delta_n}) + \sigma_T(W_{i_n\Delta_n} - W_T))/\sqrt{\Delta_n}$. Moreover $\kappa(n) = (T - (i_n - 1)\Delta_n)/\Delta_n$ and $\Psi(n)_- = (W_T - X_{(i_n-1)\Delta_n})/\sqrt{\Delta_n\,\kappa(n)}$ and $\Psi(n)_+ = (W_{i_n\Delta_n} - W_T)/\sqrt{\Delta_n(1-\kappa(n))}$ are asymptotically independent, converging in law to κ_1 and Ψ_{1-} and Ψ_{1+} (even stably in law, as proved in Theorem 4.3.1). In other words, when $t > T$ the variable $\overline{V}^n(f, X)_t$ is approximately $\partial f(\Delta X_T)R_1$, which in our simple case is $\overline{V}(f, X)_t$.

5.1.2 Proof of the Central Limit Theorem, Without Normalization

The proof of this theorem requires several steps, which basically straighten out the intuitive argument explained in Remark 5.1.4. We start with a useful observation: the processes $V^n(f, X)$ and $\overline{V}(f, X)$ satisfy the condition (4.4.2), so we deduce from the localization Lemma 4.4.9 that it is enough to prove the result when X satisfies (SH), that is Assumption 4.4.6, recalled below:

Assumption (SH) We have (H), and there are a constant A and a nonnegative function Γ on E, such that

$$\begin{array}{lll} \|b_t(\omega)\| \leq A, & \|\sigma_t(\omega)\| \leq A, & \|X_t(\omega)\| \leq A \\ \|\delta(\omega,t,z)\| \leq \Gamma(z) \leq A, & \int \Gamma(z)^2\lambda(dz) \leq A. & \end{array} \tag{5.1.8}$$

Step 1) We set $\overline{Y}^n(f, X) = (X^{(n)}, \overline{V}^n(f, X))$, and we use the notation of the previous Chapter: first, (4.3.1) with $A_m = \{z : \Gamma(z) > 1/m\}$, that is, $(S(m, j) : j \geq 1)$ are the successive jump times of the Poisson process $1_{A_m \setminus A_{m-1}} * \mu$ and $(S_p)_{p\geq 1}$ is a reordering of the double sequence $(S(m, j))$, and we let $\mathcal{P}_m$ denote the set of all indices p such that $S_p = S(m', j)$ for some $j \geq 1$ and some $m' \leq m$; second, (4.4.16), (4.4.20) and (4.4.21), which are

$$\left.\begin{array}{l} S_-(n,p) = (i-1)\Delta_n, \quad S_+(n,p) = i\Delta_n \\ R_-(n,p) = \frac{1}{\sqrt{\Delta_n}}(X_{S_p-} - X_{(i-1)\Delta_n}) \\ R_+(n,p) = \frac{1}{\sqrt{\Delta_n}}(X_{i\Delta_n} - X_{S_p}) \end{array}\right\} \text{ if } (i-1)\Delta_n < S_p \leq i\Delta_n. \tag{5.1.9}$$

$$\left.\begin{array}{l} b(m)_t = b_t - \int_{A_m \cap \{z:\|\delta(t,z)\|\leq 1\}} \delta(t,z)\,\lambda(dz) \\ X(m)_t \;=\; X_0 + \int_0^t b(m)_s\,ds + \int_0^t \sigma_s\,dW_s + (\delta\,1_{A_m^c}) \star (\mathfrak{p}-\mathfrak{q})_t \\ X'(m) \;=\; X - X(m) \;=\; (\delta\,1_{A_m}) \star \mathfrak{p} \end{array}\right\} \tag{5.1.10}$$

$$\begin{array}{l} \Omega_n(T,m) = \text{the set of all } \omega \text{ such that each interval } [0,T]\cap I(n,i) \\ \qquad \text{contains at most one jump of } X'(m)(\omega), \text{ and that} \\ \qquad \left\|X(m)(\omega)_{t+s} - X(m)(\omega)_t\right\| \leq 2/m \text{ for all } t\in[0,T],\ s\in[0,\Delta_n]. \end{array} \tag{5.1.11}$$

We also set

$$\begin{aligned} & R(n,p) = R_-(n,p) + R_+(n,p) \;=\; \frac{1}{\sqrt{\Delta_n}}\,(X_{S_+(n,p)} - X_{S_-(n,p)} - \Delta X_{S_p}) \\ & \zeta_p^n = \frac{1}{\sqrt{\Delta_n}}\big(f(\Delta X_{S_p} + \sqrt{\Delta_n}\,R(n,p)) - f(\Delta X_{S_p}) - f\big(\sqrt{\Delta_n}\,R(n,p)\big)\big) \\ & \overline{\zeta}_p^n = \big(\Delta X_{S_p}, \zeta_p^n\big) \\ & Y^n(m)_t = \sum_{p\in\mathcal{P}_m: S_p\leq \Delta_n[t/\Delta_n]} \zeta_p^n, \qquad \overline{Y}^n(m)_t = \sum_{p\in\mathcal{P}_m: S_p\leq \Delta_n[t/\Delta_n]} \overline{\zeta}_p^n. \end{aligned} \tag{5.1.12}$$

The d first components of $\overline{Y}^n(m)_t$ are exactly $X'(m)_t^{(n)}$, whereas

$$\overline{V}^n(f,X)_t = \overline{V}^n\big(f, X(m)\big)_t + Y^n(m)_t \quad \forall t\leq T, \text{ on the set } \Omega_n(T,m). \tag{5.1.13}$$

Step 2) With p fixed, the sequence $R(n,p)$ is bounded in probability (use Proposition 4.4.10). Since f is C^2 and $f(x) = \mathrm{o}(\|x\|^3)$ as $x\to 0$, the definition of ζ_p^n and a Taylor expansion of f around ΔX_{S_p} yield that $\zeta_p^n - \sum_{i=1}^d \partial_i f(\Delta X_{S_p})\,R(n,p)^i \stackrel{\mathbb{P}}{\longrightarrow} 0$ as $n\to\infty$. Therefore, another application of Proposition 4.4.10 yields

$$\big(\overline{\zeta}_p^n\big)_{p\geq 1} \stackrel{\mathcal{L}\text{-}s}{\longrightarrow} (\overline{\zeta}_p)_{p\geq 1}, \quad \text{where } \overline{\zeta}_p = \left(\Delta X_{S_p}, \sum_{i=1}^d \partial_i f(\Delta X_{S_p}) R_p^i\right), \tag{5.1.14}$$

and where the variables $R_p = (R_p^i)_{1\leq i\leq d}$ are as in (5.1.3). Since the set $\{S_p : p\in \mathcal{P}_m\}\cap[0,t]$ is finite for all t, it follows that $\overline{Y}^n(m) \stackrel{\mathcal{L}\text{-}s}{\Longrightarrow} \overline{Y}(m)$, where $\overline{Y}(m)_t = \sum_{p\in\mathcal{P}_m: S_p\leq t} \overline{\zeta}_p$. Note that the first d components of $\overline{Y}(m)$ equal $X'(m)$, whereas the last q ones are the process $\overline{V}(f, X'(m))$ which is defined in (5.1.4), with $X'(m)$ instead of X. In other words, we have proved

$$\big(X'(m)^{(n)}, Y^n(m)\big) \stackrel{\mathcal{L}\text{-}s}{\Longrightarrow} \big(X'(m), \overline{V}\big(f, X'(m)\big)\big).$$

On the other hand, we know that $X(m)^{(n)} \stackrel{\text{Sk}}{\longrightarrow} X(m)$ (pathwise), whereas $X(m)$ has no jump at the same time as $(X'(m), \overline{V}(f, X'(m)))$. Therefore a well known property of the Skorokhod topology allows us to deduce from the above that

$$\big(X^{(n)}, Y^n(m)\big) \stackrel{\mathcal{L}\text{-}s}{\Longrightarrow} \big(X, \overline{V}\big(f, X'(m)\big)\big). \tag{5.1.15}$$

Step 3) Now we vary m. With our definition of $\overline{V}(f, X'(m))$, these processes are all defined on the same extension. Since $\overline{V}(f, X) - \overline{V}(f, X'(m)) = \overline{V}(f, X(m))$ is a square-integrable martingale, conditionally on $\mathcal{F}$, we deduce from (5.1.5) and Doob's inequality that

$$\begin{aligned}
&\widetilde{\mathbb{E}}\Big(\sup_{s\le t} \big\|\overline{V}(f, X)_s - \overline{V}\big(f, X'(m)\big)_s\big\|^2\Big) \\
&\quad = \mathbb{E}\Big(\widetilde{\mathbb{E}}\Big(\sup_{s\le t} \big\|\overline{V}\big(f, X(m)\big)_s\big\|^2 \mid \mathcal{F}\Big)\Big) \\
&\quad \le K\mathbb{E}\left(\sum_{s\le t}\sum_{i=1}^{d} \|\partial_i f\|^2\big(\Delta X(m)_s\big)\big(\|c_{s-}\| + \|c_s\|\big)\right).
\end{aligned}$$

Since c_t is bounded and the $\partial_i f$'s are bounded on $\{x : \|x\| \le 1\}$, the variable in the last expectation above is smaller than $K\sum_{s\le t} \|\Delta X_s\|^2 1_{\{\|\Delta X_s\|\le 1/m\}}$. Moreover

$$\begin{aligned}
\mathbb{E}\left(\sum_{s\le t} \big\|\Delta X(m)_s\big\|^2\right) &= E\big(\big(\|x\|^2 1_{\{\|x\|\le 1/m\}}\big) \star \nu_t\big) \\
&\le t \int_{\{z:\Gamma(z)\le 1/m\}} \Gamma(z)^2 \lambda(dz),
\end{aligned}$$

which goes to 0 as $m \to \infty$. Therefore

$$\overline{V}\big(f, X'(m)\big) \stackrel{\text{u.c.p.}}{\Longrightarrow} \overline{V}(f, X). \tag{5.1.16}$$

Now, by Proposition 2.2.4, and in view of (5.1.13), (5.1.15) and (5.1.16), and since $\lim_n \mathbb{P}(\Omega_n(t, m)) = 1$ by (4.4.22), for proving $(X^{(n)}, \overline{V}^n(f, X)) \stackrel{\mathcal{L}\text{-s}}{\Longrightarrow} (X, \overline{V}(f, X))$ it remains to show that, for all $\eta > 0$, we have

$$\lim_{m\to\infty} \limsup_{n\to\infty} \mathbb{P}\Big(\Omega_n(t, m) \cap \Big\{\sup_{s\le t} \big\|\overline{V}^n\big(f, X(m)\big)_s\big\| > \eta\Big\}\Big) = 0. \tag{5.1.17}$$

Step 4) Here we prove (5.1.17), and it is enough to do this when f is one-dimensional. We set

$$k(x, y) = f(x + y) - f(x) - f(y), \qquad g(x, y) = k(x, y) - \sum_{i=1}^{d} \partial_i f(x) y^i. \tag{5.1.18}$$

Recall (4.4.20) and that f is C^2. Applying Itô's formula to the process $X(m)_t - X(m)_{(i-1)\Delta_n}$ and the function f, we get for $t > (i-1)\Delta_n$:

$$\begin{aligned}
&f\big(X(m)_t - X(m)_{(i-1)\Delta_n}\big) - \sum_{(i-1)\Delta_n < s\le t} f\big(\Delta X(m)_s\big) \\
&\quad = A(n, m, i)_t + M(n, m, i)_t,
\end{aligned} \tag{5.1.19}$$

where $M(n,m,i)$ is a square-integrable martingale with predictable bracket $A'(n,m,i)$, and

$$A(n,m,i)_t = \int_{(i-1)\Delta_n}^t a(n,m,i)_u\,du, \qquad A'(n,m,i)_t = \int_{(i-1)\Delta_n}^t a'(n,m,i)_u\,du, \tag{5.1.20}$$

and

$$\begin{cases} a(n,m,i)_t &= \sum_{j=1}^d \partial_j f\big(X(m)_t - X(m)_{(i-1)\Delta_n}\big) b(m)_t^j \\ &\quad + \frac{1}{2}\sum_{j,l=1}^d \partial_{jl}^2 f\big(X(m)_t - X(m)_{(i-1)\Delta_n}\big) c_t^{jl} \\ &\quad + \int_{A_m} g\big(X(m)_t - X(m)_{(i-1)\Delta_n}, \delta(t,z)\big)\,\lambda(dz) \\ a'(n,m,i)_t &= \sum_{j,l=1}^d (\partial_j f\,\partial_l f)\big(X(m)_t - X(m)_{(i-1)\Delta_n}\big) c_t^{jl} \\ &\quad + \int_{A_m} k\big(X(m)_t - X(m)_{(i-1)\Delta_n}, \delta(t,z)\big)^2\,\lambda(dz). \end{cases}$$

Now we set $T(n,m,i) = \inf(s > (i-1)\Delta_n : \|X(m)_s - X(m)_{(i-1)\Delta_n}\| > 2/m)$. On the set $\Omega_n(t,m)$ we have by construction $T(n,m,i) > i\Delta_n$ for all $i \le [t/\Delta_n]$. Therefore in view of (5.1.19), and on this set, the variable $|\overline{V}^n(f, X(m))_t|$ is smaller than:

$$\frac{1}{\sqrt{\Delta_n}} \sum_{i=1}^{[t/\Delta_n]} \big|A(n,m,i)_{(i\Delta_n)\wedge T(n,m,i)}\big| + \frac{1}{\sqrt{\Delta_n}} \left| \sum_{i=1}^{[t/\Delta_n]} M(n,m,i)_{(i\Delta_n)\wedge T(n,m,i)} \right|.$$

Henceforth in order to get (5.1.17), it is enough to prove the following:

$$\begin{aligned} &\lim_{m\to\infty} \limsup_n \frac{1}{\sqrt{\Delta_n}}\,\mathbb{E}\left(\sum_{i=1}^{[t/\Delta_n]} \big|A(n,m,i)_{(i\Delta_n)\wedge T(n,m,i)}\big| \right) = 0, \\ &\lim_{m\to\infty} \limsup_n \frac{1}{\Delta_n}\,\mathbb{E}\left(\sum_{i=1}^{[t/\Delta_n]} A'(n,m,i)_{(i\Delta_n)\wedge T(n,m,i)} \right) = 0. \end{aligned} \tag{5.1.21}$$

Recall that $f(0) = 0$ and $\partial_i f(0) = 0$ and $\partial_{ij}^2 f(x) = o(\|x\|)$ as $x \to 0$, so

$$\|x\| \le \frac{3}{m} \Rightarrow \big|f(x)\big| \le \alpha_m \|x\|^3, \quad \big|\partial_i f(x)\big| \le \alpha_m \|x\|^2, \quad \big|\partial_{ij}^2 f(x)\big| \le \alpha_m \|x\|$$

for some α_m going to 0 as $m \to \infty$. By singling out the two cases $\|y\| \le \|x\|$ and $\|x\| < \|y\|$ and by Taylor's formula, one deduces

$$\|x\| \le \frac{3}{m}, \quad \|y\| \le \frac{1}{m} \Rightarrow \big|k(x,y)\big| \le K\alpha_m \|x\|\,\|y\|, \quad \big|g(x,y)\big| \le K\alpha_m \|x\|\,\|y\|^2. \tag{5.1.22}$$

We have $\|X(m)_{s\wedge T(n,m,i)} - X(m)_{(i-1)\Delta_n}\| \le 3/m$ for $s \ge i\Delta_n$, because the jumps of $X(m)$ are smaller than $1/m$. Moreover the definition of $b(m)$ in (4.4.20) allows us to deduce, by the Markov inequality and $\|\delta(t,z)\| \le \Gamma(z)$, that $\|b(m)_t\| \le Km$. Then in view of (SH) and (5.1.22) we obtain for $(i-1)\Delta_n \le t \le T(n,m,i)$ (then $\|X(m)_t - X(m)_{(i-1)\Delta_n}\| \le 3/m$):

$$\begin{aligned} \big|a(n,m,i)_t\big| &\le K\alpha_m\big(\big\|X(m)_t - X(m)_{(i-1)\Delta_n}\big\| + m\big\|X(m)_t - X(m)_{(i-1)\Delta_n}\big\|^2\big) \\ a'(n,m,i)_t &\le K\alpha_m^2\big\|X(m)_t - X(m)_{(i-1)\Delta_n}\big\|^2. \end{aligned} \tag{5.1.23}$$

Now, applying (2.1.44) to $X(m)$ when $p=2$, plus the Cauchy-Schwarz inequality when $p=1$, and using again $\|b(m)\| \le Km$, we deduce that for $p=1,2$ and $s \in [0,1]$,

$$\mathbb{E}\big(\big\|X(m)_{t+s} - X(m)_t\big\|^p\big) \le K\big(s^{(p/2)\wedge 1} + m^p s^p\big). \tag{5.1.24}$$

This gives that the two "lim sup" in (5.1.21) are smaller than $Kt\alpha_m$ and $Kt\alpha_m^2$ respectively. Hence (5.1.21) holds, and the last claim of the theorem (which obviously implies the first one) is proved.

Step 5) It remains to prove the second claim, that is (5.1.7). Since t is not a fixed time of discontinuity of the process $\overline{V}(f,X)$, we deduce from the first claim that $\overline{V}^n(f,X)_t \xrightarrow{\mathcal{L}\text{-s}} \overline{V}(f,X)_t$ for any given t. Therefore, (5.1.7) will follow if we can prove that

$$U_n := \frac{1}{\sqrt{\Delta_n}}\,|f\star\mu_{\Delta_n[t/\Delta_n]} - f\star\mu_t| \xrightarrow{\mathbb{P}} 0. \tag{5.1.25}$$

Let Ω_n be the set on which there is no jump of X bigger than 1 between the times $t-\Delta_n$ and t. We have $\Omega_n \uparrow \Omega$ as $n\to\infty$, so it is enough to prove that $\mathbb{E}(U_n 1_{\Omega_n}) \to 0$. The properties of f yield that $|f(x)| \le K\|x\|^2$, as soon as $\|x\| \le 1$, hence $U_n \le \frac{K}{\sqrt{\Delta_n}}\sum_{t-\Delta_n<s\le t}\|\Delta X_s\|^2$ on the set Ω_n. Therefore

$$\begin{aligned} \mathbb{E}(U_n 1_{\Omega_n}) &\le \frac{K}{\sqrt{\Delta_n}}\,\mathbb{E}\Big(\sum_{t-\Delta_n<s\le t}\|\Delta X_s\|^2\Big) \\ &= \frac{K}{\sqrt{\Delta_n}}\,\mathbb{E}\Big(\int_{t-\Delta_n}^t ds\int\big\|\delta(s,z)\big\|^2\lambda(dz)\Big), \end{aligned}$$

which is smaller than $KA\sqrt{\Delta_n}$ by (SH), and the result follows.

5.2 The Central Limit Theorem for Normalized Functionals: Centering with Conditional Expectations

Now we turn to the processes $V'^n(f,X)_t = \Delta_n\sum_{i=1}^{[t/\Delta_n]} f(\Delta_i^n X/\sqrt{\Delta_n})$, for which the LLN has been given in Theorem 3.4.1 under appropriate assumptions. The associated CLT should be a limit theorem for the processes

$$\overline{V}'^n(f,X)_t = \frac{1}{\sqrt{\Delta_n}}\left(V'^n(f,X)_t - \int_0^t \rho_{c_s}(f)\,ds\right). \tag{5.2.1}$$

Theorem 4.2.1 gives a CLT for the processes $\overline{U}^n(f)$, defined by (4.2.2) and which look similar to the $\overline{V}'^n(f,X)$'s. However, these processes differ in two main ways: the first – obvious – one is that $\overline{U}^n(f)$ involves the summands $f(\beta_i^n) = f(\sigma_{(i-1)\Delta_n}\Delta_i^n W/\sqrt{\Delta_n})$ instead of $f(\Delta_i^n X/\sqrt{\Delta_n})$. The second difference is more subtle: the centering in $\overline{U}^n(f)$ is not the integral $\int_0^t \rho_{c_s}(f)\,ds$, but the sum of the terms $\Delta_n\,\rho_{c_{(i-1)\Delta_n}}(f)$, a Riemann sum for the previous integral, and at the same time it makes each summand a martingale difference, a property which was key for obtaining the CLT.

Henceforth there are two kinds of CLTs for $V'^n(f,X)$. A first kind concerns the case where the centering is the conditional expectation of $f(\Delta_i^n X/\sqrt{\Delta_n})$, and this type of CLT requires minimal assumptions on b and σ. The other kind is the CLT for $\overline{V}'^n(f,X)$ itself, and it requires stronger assumptions to ensure that $\mathbb{E}(f(\Delta_i^n X/\sqrt{n}) \mid \mathcal{F}_{(i-1)\Delta_n})$ is close enough to $\rho_{c_{(i-1)\Delta_n}}(f)$, and also that the Riemann sums hinted at above converge to the integral at a rate faster than $1/\sqrt{\Delta_n}$. Although the second kind of CLT is more natural in some sense, and much more useful for applications, it needs more assumptions and is more difficult to prove: so we begin with the first kind of CLT, to which this section is devoted.

5.2.1 Statement of the Results

Below, X is again a d-dimensional Itô semimartingale, which we write in its Grigelionis form (5.0.1). We also use its Lévy measure F_t and the process $c=\sigma\sigma^\star$.

In order to "center" each increment $f(\Delta_i^n X/\sqrt{\Delta_n})$ by its conditional expectation, we need it to be integrable. This is the case in some situations, when f is bounded of course, or when it has polynomial growth and the coefficients in (5.0.1) are bounded, but in general this is not the case. Hence we need to truncate those increments before taking their conditional expectations. For this, and since the test function $f=(f^j)_{1\le j\le q}$ may be multi-dimensional, we define a component by component truncation based on the functions $\psi'_A(x)=\psi'(x/A)$ of (3.3.16), in which ψ' is a C^∞ function on $\mathbb{R}_+$ with $1_{[0,1/2]}(x)\le\psi'(x)\le 1_{[0,1]}(x)$:

$$f_n(x)^j = f(x)^j\,\psi'_{1/\sqrt{\Delta_n}}\big(f(x)^j\big). \tag{5.2.2}$$

Then f_n is bounded continuous, whatever the continuous function f, and $f_n(x)^j = f(x)^j$ when $|f(x)^j|\le 1/\sqrt{\Delta_n}$. We consider the following two processes:

$$Y^n(f,X)_t = \sqrt{\Delta_n}\sum_{i=1}^{[t/\Delta_n]}\big(f(\Delta_i^n X/\sqrt{\Delta_n}) - \mathbb{E}(f(\Delta_i^n X/\sqrt{\Delta_n}) \mid \mathcal{F}_{(i-1)\Delta_n})\big)$$

$$Y'^n(f,X)_t = \sqrt{\Delta_n}\sum_{i=1}^{[t/\Delta_n]}\big(f(\Delta_i^n X/\sqrt{\Delta_n}) - \mathbb{E}(f_n(\Delta_i^n X/\sqrt{\Delta_n}) \mid \mathcal{F}_{(i-1)\Delta_n})\big). \tag{5.2.3}$$

The first process is well defined when the $f(\Delta_i^n X/\sqrt{\Delta_n})$'s are integrable, the second one is always well defined.

Next, in order to describe the limit we need some notation. If α is a $d \times d'$ matrix and $a = \alpha\alpha^*$ and f, g are Borel function on $\mathbb{R}^d$ with polynomial growth and U is an $\mathcal{N}(0, I_{d'})$ distributed variable, we set

$$\begin{aligned} \widehat{\gamma}_\alpha(f) &= \mathbb{E}\big(f(\alpha U)U^\star\big) \qquad \text{(a } d'\text{-dimensional row vector)} \\ \overline{\gamma}_\alpha(f,g) &= \mathbb{E}\big(\big(f(\alpha U) - \widehat{\gamma}_\alpha(f)U\big)\big(g(\alpha U) - \widehat{\gamma}_\alpha(g)U\big)\big) - \rho_a(f)\,\rho_a(g). \end{aligned} \tag{5.2.4}$$

When f and g are globally even, i.e. $f(x) = f(-x)$ and $g(x) = g(-x)$ for all x, the following is obvious:

$$\widehat{\gamma}_\alpha(f) = \widehat{\gamma}_\alpha(g) = 0, \qquad \overline{\gamma}_\alpha(f,g) = \rho_a(fg) - \rho_a(f)\rho_a(g). \tag{5.2.5}$$

For a q-dimensional function $(f^j)_{1\le j\le q}$, the matrix $(\overline{\gamma}_\alpha(f^j, f^k))_{1\le j,k\le q}$ is symmetric nonnegative, and $\alpha \mapsto \widehat{\gamma}_\alpha(f)$ and $\alpha \mapsto \overline{\gamma}_\alpha(f,g)$ are measurable, and even continuous when f and g are continuous.

Remark 5.2.1 These are notation used in (4.2.5) when $G(x,y) = g(y)$ and the process Y does not enter the picture and Φ in (4.2.3) is $\Phi(y) = y(1)$, so the integers q and q' in (4.2.5) are here $q = d$ and $q' = d'$. Then we have $\gamma_\alpha^\Phi(f) = \rho_a(f)$ and $\widehat{\gamma}_\alpha^\Phi(f) = \widehat{\gamma}_\alpha(f)$ and $\overline{\gamma}_\alpha^\Phi(f,g) = \overline{\gamma}_\alpha(f,g)$.

We can now state the result. It is remarkable that the assumptions in the following are exactly the same as in the LLN (Theorem 3.4.1), except for the rate at which f can grow at infinity, whose exponent is exactly divided by 2.

Theorem 5.2.2 *Let X be a d-dimensional Itô semimartingale. Let f be a continuous q-dimensional function which satisfies one of the following three conditions:*

(a) $\|f(x)\| = \mathrm{o}(\|x\|)$ *as* $\|x\| \to \infty$,

(b) $\|f(x)\| = \mathrm{O}(\|x\|)$ *as* $\|x\| \to \infty$, *and X is continuous,*

(c) $\|f(x)\| = \mathrm{O}(\|x\|^p)$ *as* $\|x\| \to \infty$ *for some* $p > 1$, *and X is continuous and satisfies*

$$\int_0^t \|b_s\|^{2p/(p+1)}\,ds < \infty, \qquad \int_0^t \|c_s\|^p\,ds < \infty. \tag{5.2.6}$$

Then:

(i) *The sequence of q-dimensional processes* $Y'^n(f,X)$ *converges stably in law to a q-dimensional process of the form* $\overline{Y}(f,X) = \overline{U}(f,X) + \overline{U}'(f,X)$, *where*

$$\overline{U}(f,X)^j_t = \sum_{k=1}^r \int_0^t \widehat{\gamma}_{\sigma_s}(f^j)^k \, dW^k_s \tag{5.2.7}$$

and $\overline{U}'(f,X)$, *defined on a very good filtered extension* $(\widetilde{\Omega}, \widetilde{\mathcal{F}}, (\widetilde{\mathcal{F}}_t)_{t\geq 0}, \widetilde{\mathbb{P}})$ *of the space* $(\Omega, \mathcal{F}, (\mathcal{F}_t)_{t\geq 0}, \mathbb{P})$, *is continuous and conditionally on* $\mathcal{F}$ *is a centered Gaussian process with independent increments satisfying*

$$\widetilde{\mathbb{E}}\big(\overline{U}'(f,X)^j_t \, \overline{U}'(f,X)^k_t \mid \mathcal{F}\big) = \int_0^t \overline{\gamma}_{\sigma_s}(f^j, f^k) \, ds. \tag{5.2.8}$$

When further the function f *is globally even, we have* $\overline{U}(f,X) = 0$, *and* $\overline{\gamma}_{\sigma_s}(f^j, f^k) = \rho_{c_s}(f^j f^k) - \rho_{c_s}(f^j)\rho_{c_s}(f^k)$.

(ii) *The sequence* $Y^n(f,X)$ *also converges stably in law to* $\overline{Y}(f,X)$ *when the function* f *is bounded, or in case* (a) *when*

$$t < \infty \quad \Rightarrow \quad \mathbb{E}\bigg(\int_0^t \Big(\|b_s\| + \|c_s\| + \int \big(\|x\|^2 \wedge \|x\|\big) F_t(dx)\Big) ds\bigg) < \infty, \tag{5.2.9}$$

or in cases (b) *and* (c) *when b and c are bounded.*

(iii) *Conversely, and with the notation* $g_p(x) = \|x\|^p$, *if either for some* $p > 1$ *the sequence of processes* $Y'^n(g_p, X)$ *is tight, or the sequence* $Y'^n(g_1, X)$ *converges in law to a continuous process and the three processes b and* σ *and* $\int (\|x\|^2 \wedge \|x\|) F_t(dx)$ *are bounded, then necessarily the process X is continuous.*

The conditions on f and X in (ii) ensure the integrability of all the variables $f(\Delta^n_i X/\sqrt{\Delta_n})$, hence the conditional expectations in the definition of $Y^n(f,X)$ do exist. Again, (iii) is not a genuine converse to (i), but it indicates that when X is discontinuous one cannot substantially weaken the growth assumption on the test function f if one wants the convergence of $Y'^n(f,X)$ towards $\overline{Y}(f,X)$.

Remark 5.2.3 The second property in (5.2.6) is exactly the condition needed for the process $\overline{Y}(f,X)$ to be well defined. Indeed, by (5.2.4) we have $\|\widehat{\gamma}_\alpha(f^j)\| \leq K(1+\|\alpha\|^p)$ and $|\overline{\gamma}_\alpha(f^j, f^k)| \leq K(1+\|\alpha\|^{2p})$ if f is as in the theorem. Therefore this second property implies that the stochastic integral (5.2.7) makes sense and the integral (5.2.8) is finite.

Conversely, considering the one-dimensional case $d = d' = q = 1$ for example and the (even) function $f(x) = |x|^p$, we see that the right side of (5.2.8) is exactly $(m_{2p} - m_p^2) \int_0^t c_s^p \, ds$, so the second property (5.2.6) is necessary for the theorem to hold for this f.

As to the first property (5.2.6), which is the same as the first part of (3.4.6), with $2p$ instead of p, and exactly as for Theorem 3.4.1, we do not know whether it is necessary.

Remark 5.2.4 The truncation in (5.2.2) could be made differently, for example one could truncate globally as $f_n(x) = f(x)\psi'_{1/\sqrt{\Delta_n}}(\|f(x)\|)$. The results and the proofs would be similar.

5.2.2 The Proof

Below, p is taken to be $p = 1$ in cases (a) and (b), and it is the number occurring in the statement of case (c) otherwise. Recall that X is continuous, except in case (a). As in (3.4.12), for any $r \geq 2$ we use the notation:

$$A(r)_t^X = A(r)_t = \int_0^t \Big(\|b_s\|^{2r/(r+2))} + \|c_s\|^{r/2} + \int \big(\|x\|^2 \wedge 1\big) F_s(dx) \Big) ds.$$

Here, $A(2)$ is always finite-valued, and $A(2p)$ is finite if and only if (5.2.6) holds.

The proof of (i) and (ii) basically follows the same route as for Theorem 3.4.1. However the three steps of that proof are not so well identified here: for example, the localization procedure is not as straightforward as in Lemma 3.4.5, mainly because the localization does not "commute" with the conditional expectations. Nevertheless we heavily use the following additional conditions, similar to (3.4.14):

$$\sup_\omega \big(A(2p)\big)_\infty < \infty, \qquad \sup_{(\omega,t)} \|\Delta X_t(\omega)\| < \infty. \tag{5.2.10}$$

So we postpone the localization procedure, and start proving the results under (5.2.10), and when f is *bounded*: in this case $f_n = f$ for n large enough, so $Y'^n(f,X) = Y^n(f,X)$.

Before getting started, we recall some notation and properties of the previous two chapters. First, under (5.2.10) we have the decomposition $X = X' + X''$, where

$$X'_t = X_0 + \int_0^t b''_s\, ds + \int_0^t \sigma_s\, dW_s, \qquad X'' = \delta \star (\mathfrak{p} - \mathfrak{q}),$$

and $b''_t = b_t + \int_{\{\|\delta(t,z)\|>1\}} \delta(t,z)\,\lambda(dz)$. With $\eta^n_t = \sup_{i \leq [t/\Delta_n} \Delta^n_i A(2)$, the estimates (3.4.17), (3.4.18) and (3.4.22) read as

$$\left.\begin{array}{l}
\mathbb{E}\big(\|\Delta^n_i X'\|^2\big) \leq \mathbb{E}\big(\Delta^n_i A(2) + (\Delta^n_i A(2))^2\big) \\
\mathbb{E}\big(\|\Delta^n_i X'/\sqrt{\Delta_n} - \beta^n_i\|^2\big) \\
\quad \leq \frac{K}{\Delta_n}\,\mathbb{E}\big(\eta^n_t\, \Delta^n_i A(2) + \int_{(i-1)\Delta_n}^{i\Delta_n} \|\sigma_s - \sigma_{\Delta_n[s/\Delta_n]}\|^2 ds\big) \\
\mathbb{E}\Big(\big(\frac{\|\Delta^n_i X''\|}{\sqrt{\Delta_n}} \bigwedge 1\big)^2\Big) \\
\quad \leq \frac{K}{\Delta_n}\,\mathbb{E}\big(\Delta_n^{1/4}\, \Delta^n_i A(2) + \int_{(i-1)\delta_n}^{i\Delta_n} ds \int_{\{\|x\| \leq \Delta_n^{1/4}\}} \|x\|^2 F_s(dx)\big).
\end{array}\right\} \tag{5.2.11}$$

Next, we set

$$\beta_i^n = \frac{1}{\sqrt{\Delta_n}}\,\sigma_{(i-1)\Delta_n}\Delta_i^n W, \qquad \overline{U}^n(f)_t = \sqrt{\Delta_n}\sum_{i=1}^{[t/\Delta_n]}\big(f(\beta_i^n) - \rho_{c_{(i-1)\Delta_n}}(f)\big). \tag{5.2.12}$$

Lemma 5.2.5 *If* (5.2.10) *holds for* $p=1$ *and* σ *is bounded continuous and* f *is bounded and uniformly continuous, we have* $Y^n(f,X) \stackrel{\mathcal{L}\text{-}s}{\Longrightarrow} \overline{Y}(f,X)$.

Proof With G as in Remark 5.2.1, the processes $\overline{U}^n(f)$ et $\overline{Y}(f,X)$ are exactly $\overline{U}^n(G)$ and $\overline{U}(G)$, as defined by (4.2.6) and (4.2.5), provided we take $\theta=\sigma$ and $u_n=\Delta_n$. Thus in view of Theorem 4.2.1, it suffices to prove that $Y^n(f,X) - \overline{U}^n(f) \stackrel{\text{u.c.p.}}{\Longrightarrow} 0$. We have $Y^n(f,X)_t - \overline{U}^n(f)_t = \sum_{i=1}^{[t/\Delta_n]}(\zeta_i^n - \mathbb{E}(\zeta_i^n \mid \mathcal{F}_{(i-1)\Delta_n}))$, where

$$\zeta_i^n = \sqrt{\Delta_n}\big(f\big(\Delta_i^n X/\sqrt{\Delta_n}\big) - f\big(\beta_i^n\big)\big).$$

Hence by virtue of Lemma 2.2.11, case (a), it is enough to show that

$$a_t^n = \mathbb{E}\left(\sum_{i=1}^{[t/\Delta_n]}\|\zeta_i^n\|^2\right) \to 0. \tag{5.2.13}$$

Our assumptions on f yield a constant K and a positive function θ satisfying $\theta(\varepsilon)\to 0$ as $\varepsilon\to 0$, such that for all $\varepsilon>0$ and $x,y,z\in\mathbb{R}^d$ we have:

$$\big\|f(x+y+z)-f(x)\big\|^2 \le \theta(\varepsilon) + K 1_{\{\|y+z\|>\varepsilon\}} \le \theta(\varepsilon) + \frac{K(\|y\|^2\wedge 1)}{\varepsilon^2} + \frac{K\|z\|^2}{\varepsilon^2}. \tag{5.2.14}$$

We apply this with $x=\beta_i^n$ and $z=\Delta_i^n X'/\sqrt{\Delta_n} - \beta_i^n$ and $y=\Delta_i^n X''/\sqrt{\Delta_n}$ and use the estimates (5.2.11). We deduce that, similar with (3.4.23), and since $A(2)_\infty \le K$:

$$a_t^n \le t\theta(\varepsilon) + \frac{K}{\varepsilon^2}\,\mathbb{E}\Bigg(\eta_t^n + \Delta_n^{1/4} + \int_0^t\Bigg(\|\sigma_s - \sigma_{\Delta_n[s/\Delta_n]}\|^2 + \int_{\{\|x\|\le\Delta_n^{1/4}\}}\|x\|^2 F_s(dx)\Bigg)\,ds\Bigg).$$

The expectation above goes to 0 by, because all three sequences of variables η_t^n, $\int_0^t\|\sigma_s - \sigma_{\Delta_n[s/\Delta_n]}\|^2\,ds$ and $\int_0^t ds\int_{\{\|x\|\le\Delta_n^{1/4}\}}\|x\|^2 F_s(dx)$ go to 0 as $n\to\infty$ and are uniformly bounded (use again $A(2)_\infty\le K$ and also $\|\sigma_t\|\le K$, and the dominated convergence theorem). Since $\theta(\varepsilon)\to$ as $\varepsilon\to 0$, (5.2.13) follows. □

Lemma 5.2.6 *If* (5.2.10) *holds for* $p=1$ *and* f *is bounded uniformly continuous, we have* $Y^n(f,X) \stackrel{\mathcal{L}\text{-}s}{\Longrightarrow} \overline{Y}(f,X)$.

Proof 1) As in Lemma 3.4.8, there is a sequence $\sigma(m)$ of adapted bounded continuous processes such that $a_m = \mathbb{E}(\int_0^\infty \|\sigma(m)_s - \sigma_s\|^2 ds) \to 0$. We associate the semimartingales

$$X(m)_t = X_0 + \int_0^t b_s'' ds + \int_0^t \sigma(m)_s dW_s + \delta \star (\mathfrak{p} - \mathfrak{q})_t.$$

The previous lemma yields $Y^n(f, X(m)) \stackrel{\mathcal{L}\text{-s}}{\Longrightarrow} \overline{Y}(f, X(m))$ as $n \to \infty$, for each m. Hence, by Proposition 2.2.4, it suffices to prove the following two properties, for all $t > 0$, $\eta > 0$ for the first one:

$$\lim_{m\to\infty} \limsup_{n\to\infty} \mathbb{P}\Big(\sup_{s\le t} \big\|Y^n(f,X)_s - Y^n\big(f,X(m)\big)_s\big\| > \eta\Big) = 0, \tag{5.2.15}$$

$$\overline{Y}\big(f, X(m)\big) \stackrel{\mathcal{L}\text{-s}}{\Longrightarrow} \overline{Y}(f, X) \quad \text{as } m \to \infty. \tag{5.2.16}$$

(Recall (2.2.8) for the definition of the stable convergence for processes which are defined on extensions of the original space.)

2) We set $\zeta(m)_i^n = \sqrt{\Delta_n}(f(\Delta_i^n X/\sqrt{\Delta_n}) - f(\Delta_i^n X(m)/\sqrt{\Delta_n}))$, so

$$Y^n(f,X)_t - Y^n\big(f,X(m)\big)_t = \sum_{i=1}^{[t/\Delta_n]} \big(\zeta(m)_i^n - \mathbb{E}\big(\zeta(m)_i^n \mid \mathcal{F}_{(i-1)\Delta_n}\big)\big). \tag{5.2.17}$$

Therefore by Doob's inequality,

$$a(m)_t^n = \mathbb{E}\Big(\sup_{s\le t} \big\|Y^n(f,X)_s - Y^n\big(f,X(m)\big)_s\big\|^2\Big) \le K \sum_{i=1}^{[t/\Delta_n]} \mathbb{E}\big(\big\|\zeta(m)_i^n\big\|^2\big).$$

Exactly as in the previous lemma, upon using (5.2.14) with $x = \Delta_i^n X/\sqrt{\Delta_n}$ and $z = \Delta_i^n(X(m) - X)/\sqrt{\Delta_n}$ and $y = 0$, we obtain

$$a(m)_t^n \le Kt\theta(\varepsilon) + \frac{K}{\varepsilon^2}\mathbb{E}\bigg(\int_0^t \big\|\sigma_s - \sigma(m)_s\big\|^2 ds\bigg) \le Kt\theta(\varepsilon) + \frac{Ka_m}{\varepsilon^2}.$$

Then $\sup_n a(m)_t^n \to 0$ as $m \to \infty$ because $a_m \to 0$ and $\theta(\varepsilon) \to 0$ as $\varepsilon \to 0$, and (5.2.15) follows.

3) It remains to prove (5.2.16). First, $\alpha \mapsto \widehat{\gamma}_\alpha(f)$ is continuous and bounded by the same bound as f, so $\mathbb{E}(\int_0^t \|\widehat{\gamma}_{\sigma(m)_s}(f) - \widehat{\gamma}_{\sigma_s}(f)\|^2 ds) \to 0$ by $a_m \to 0$ and the dominated convergence theorem. Hence $\overline{U}(f, X(m)) \stackrel{\text{u.c.p.}}{\Longrightarrow} \overline{U}(f, X)$, and by (2.2.5) it is enough to show that

$$\overline{U}'\big(f, X(m)\big) \stackrel{\mathcal{L}\text{-s}}{\Longrightarrow} \overline{U}'(f, X). \tag{5.2.18}$$

Recall that the processes $\overline{U}'(f, X(m))$ and $\overline{U}'(f, X)$ are possibly defined on different extensions of our original space, but conditionally on $\mathcal{F}$ they are centered Gaussian with independent increments and covariances given by (5.2.8), with $\sigma(m)_s$ in place of σ_s for the former. For (5.2.18) it is clearly enough to show that the $\mathcal{F}$-conditional laws of the processes $\overline{U}'(f, X(m))$ weakly converge, in $\mathbb{P}$-probability, to the $\mathcal{F}$-conditional law of the process $\overline{U}'(f, X)$, and for this it is even enough to prove the convergence in probability of the covariances, conditional on $\mathcal{F}$, by a well known property of continuous Gaussian processes with independent increments.

Now, $\alpha \mapsto \overline{\gamma}_\alpha(f^j, f^k)$ is continuous and bounded. Therefore, the same argument as above shows that the right side of (5.2.8) for $\sigma(m)_s$ converges in probability, and even in $\mathbb{L}^1(\mathbb{P})$, to the same with σ_s, and (5.2.18) follows. □

The next lemma is what replaces the localization Lemma 3.4.5.

Lemma 5.2.7 *We have $Y^n(f, X) \stackrel{\mathcal{L}\text{-}s}{\Longrightarrow} \overline{Y}(f, X)$ if f is bounded and uniformly continuous.*

Proof We no longer assume (5.2.10). However, the stopping times $T_m = \inf(t : A(2)_t \geq m \text{ or } \|\Delta X_s\| > m)$ increase to infinity, and the processes $X(m)$ defined by

$$X(m)_t = X_0 + \int_0^{t\wedge T_m} b_s\,ds + \int_0^{t\wedge T_m} \sigma_s\,dW_s$$
$$+ (\delta 1_{\{\|\delta\|\leq 1\}}) \star (\mathfrak{p} - \mathfrak{q})_{t\wedge T_m} + (\delta 1_{\{1<\|\delta\|\leq m\}}) \star \mathfrak{p}_{t\wedge T_m}$$

(warning: these are not the same as in the proof of the previous lemma) satisfy (5.2.10). Thus $Y^n(f, X(m)) \stackrel{\mathcal{L}\text{-}s}{\Longrightarrow} \overline{Y}(f, X(m))$ for each fixed m, and exactly as in the previous proof it then suffices to prove (5.2.15) and (5.2.16).

For (5.2.16) the argument is easy: in view of the special structure of $X(m)$, one may realize the limit $\overline{Y}(f, X(m))$ on the same extension as $\overline{Y}(f, X)$, simply by stopping, that is $\overline{Y}(f, X(m))_t = \overline{Y}(f, X)_{t\wedge T_m}$, and then (5.2.16) is obvious because $T_m \to \infty$.

For (5.2.15) things are a bit more complicated. We use the same notation $\zeta(m)_i^n$ as in the previous proof, and observe that $\|\zeta(m)_i^n\| \leq K\sqrt{\Delta_n}$ because f is bounded, and also $\zeta(m)_i^n = 0$ if $T_m > i\Delta_n$. Therefore on the set $\{T_m > t\}$ we have for all $s \leq t$:

$$\left\|Y^n\big(f, X(m)\big)_s - Y^n(f, X)_s\right\| = \left\|\sum_{i=1}^{[s/\Delta_n]} \mathbb{E}\big(\zeta(m)_i^n\, 1_{\{T_m \leq i\Delta_n\}} \mid \mathcal{F}_{(i-1)\Delta_n}\big)\right\|$$
$$\leq K\sqrt{\Delta_n} \sum_{i=1}^{[t/\Delta_n]} \mathbb{P}(T_m \leq i\Delta_n) \mid \mathcal{F}_{(i-1)\Delta_n})$$

Therefore, we deduce that for all $\eta > 0$,

$$\begin{aligned}
&\mathbb{P}\Big(\sup_{s\le t} \|Y^n(f,X)_s - Y^n(f,X(m))_s\| > \eta\Big) \\
&\quad \le \mathbb{P}(T_m \le t) + \frac{K\sqrt{\Delta_n}}{\eta} \sum_{i=1}^{[t/\Delta_n]} \mathbb{E}\big(\mathbb{P}(T_m \le i\Delta_n) \mid \mathcal{F}_{(i-1)\Delta_n}\big)\, 1_{\{T_m > t\}}\big) \\
&\quad \le \mathbb{P}(T_m \le t) + \frac{K\sqrt{\Delta_n}}{\eta} \sum_{i=1}^{[t/\Delta_n]} \mathbb{E}\big(\mathbb{P}((i-1)\Delta_n < T_m \le i\Delta_n) \mid \mathcal{F}_{(i-1)\Delta_n}\big) \\
&\quad \le \mathbb{P}(T_m \le t) + \frac{K\sqrt{\Delta_n}}{\eta}.
\end{aligned}$$

Since $T_m \to \infty$ as $m \to \infty$, (5.2.15) follows. □

Lemma 5.2.8 *Under the assumptions of Theorem* 5.2.2, *and recalling the functions* f_n *of* (5.2.2), *we have* $Y^n(f_n,X) \stackrel{\mathcal{L}\text{-}s}{\Longrightarrow} \overline{Y}(f,X)$.

Proof For each m the function f_m is bounded and uniformly continuous, so the previous lemma yields $Y^n(f_m,X) \stackrel{\mathcal{L}\text{-}s}{\Longrightarrow} \overline{Y}(f_m,X)$ as $n \to \infty$. With $Y(m)^n_t = \sup_{s\le t} \|Y^n(f_m,X)_s - Y^n(f_n,X)_s\|$, it is thus enough to show, as in Lemma 5.2.6, the following two properties:

$$\eta, t > 0 \ \Rightarrow\ \lim_{m\to\infty} \limsup_{n\to\infty} \mathbb{P}\big(Y(m)^n_t > \eta\big) = 0, \tag{5.2.19}$$

$$\overline{Y}(f_m,X) \stackrel{\mathcal{L}\text{-}s}{\Longrightarrow} \overline{Y}(f,X) \quad \text{as } m \to \infty. \tag{5.2.20}$$

1) Let us first prove (5.2.19). Observe that $Y^n(f_m,X)_t - Y^n(f_n,X)_t$ is equal to the right side of (5.2.17), if we take $\zeta(m)^n_i = \sqrt{\Delta_n}\,(f_m - f_n)(\Delta^n_i X/\sqrt{\Delta_n})$. Hence Doob's inequality yields that, for any stopping time T, and with the notation $D(m)^n_t = \sum_{i=1}^{[t/\Delta_n]} \|\zeta(m)^n_i\|^2$,

$$\mathbb{E}\big(\big(Y(m)^n_T\big)^2\big) \le K\,\mathbb{E}\big(D(m)^n_T\big).$$

In other words, the process $(Z(m)^n)^2$ is Lenglart-dominated (see Sect. 2.1.7) by the increasing adapted process $D(m)^n$, and the second part of (2.1.49) yields for all $\eta, \varepsilon, t > 0$:

$$\mathbb{P}\big(Y(m)^n_t > \eta\big) \le \frac{1}{\eta^2}\Big(\varepsilon + \mathbb{E}\Big(\sup_{s\le t} \Delta D(m)^n_s\Big)\Big) + \mathbb{P}\big(D(m)^n_t \ge \varepsilon\big). \tag{5.2.21}$$

Now we deduce from (5.2.2) that $\|f_m - f_n\|^2 \le h_m$ for all $n > m$, where h_m is some continuous function on $\mathbb{R}^d$, satisfying $h_m(x) \le K\|x\|^{2p} 1_{\{\|x\|>a_m\}}$ for some sequence a_m going to ∞. Then clearly, with the notation (3.4.2), we have

$$D(m)^n \leq V'^n(h_m, X).$$

The assumptions of Theorem 3.4.1 are satisfied by X and h_m, with $2p$ in place of p, so $V'^n(h_m, X) \stackrel{\text{u.c.p.}}{\Longrightarrow} H(m)_t = \int_0^t \rho_{c_s}(h_m)\,ds$. Moreover $h_m \to 0$ pointwise to 0 as $m \to \infty$ and $h_m(x) \leq K\|x\|^{2p}$, so by the dominated convergence theorem (use also (5.2.6) here) we have $H(m) \stackrel{\text{u.c.p.}}{\Longrightarrow} 0$ as $m \to \infty$. Then for any $\theta, \varepsilon > 0$ there are integers m_0 and $n_0(m) \geq m$ such that

$$m \geq m_0, \;\; n \geq n_0(m) \quad \Rightarrow \quad \mathbb{P}\big(D(m)_t^n \geq \varepsilon\big) \leq \theta.$$

Taking (5.2.2) into consideration, we see that $\|f_n(x)\| \leq K/\sqrt{\Delta_n}$ for all $x \in \mathbb{R}^d$, hence $\|\zeta(m)_i^n\| \leq K$, and the jumps of $D(m)^n$ are bounded by a constant K. Hence $\mathbb{E}(\sup_{s\leq t} \Delta D(m)_s^n) \leq \varepsilon + K\mathbb{P}(D(m)_t^n \geq \varepsilon)$. Plugging these in (5.2.21) yields

$$m \geq m_0, \;\; n \geq n_0(m) \quad \Rightarrow \quad \mathbb{P}\big(Y(m)_t^n > \eta\big) \leq \frac{2\varepsilon + K\theta}{\eta^2} + \theta,$$

and since ε and θ are arbitrary, (5.2.19) follows.

2) Next we prove (5.2.20). We have $f_m \to f$ pointwise, and $\|f_m(x)\| \leq K(1 + \|x\|^p)$. Then $\widehat{\gamma}_{\sigma_s}(f_m) \to \widehat{\gamma}_{\sigma_s}(f)$ and $\overline{\gamma}_{\sigma_s}(f_m^j, f_m^k) \to \overline{\gamma}_{\sigma_s}(f^j, f^k)$ pointwise, together with the estimates $\|\widehat{\gamma}_{\sigma_s}(f_m)\| \leq K(1+\|c_s\|^{p/2})$ and $\|\overline{\gamma}_{\sigma_s}(f_m^j, f^k)\| \leq K(1+\|c_s\|^p)$. Hence the dominated convergence theorem and the second part of (5.2.6) yield

$$\int_0^t \big\|\widehat{\gamma}_{\sigma_s}(f_m) - \widehat{\gamma}_{\sigma_s}(f)\big\|^2 ds \to 0, \qquad \int_0^t \overline{\gamma}_{\sigma_s}\big(f_m^j, f_m^k\big)\,ds \to \int_0^t \overline{\gamma}_{\sigma_s}\big(f^j, f^k\big)\,ds.$$

The first property above implies that $\overline{U}''(f_m, X) \stackrel{\text{u.c.p.}}{\Longrightarrow} \overline{U}''(f, X)$, and at this point we conclude (5.2.20) exactly as in Step 3 of the proof of Lemma 5.2.6. □

Proof of Theorem 5.2.2 (i) The last claims concerning even functions follow from (5.2.5). In order to prove $Y'^n(f, X) \stackrel{\mathcal{L}\text{-s}}{\Longrightarrow} \overline{Y}(f, X)$, by Lemma 5.2.8 it is enough to show that $Y'^n(f, X) - Y^n(f_n, X) \stackrel{\text{u.c.p.}}{\Longrightarrow} 0$. To check this, we set

$$\zeta_i^n = \sqrt{\Delta_n}\big(f\big(\Delta_i^n X/\sqrt{\Delta_n}\big) - f_n\big(\Delta_i^n X/\sqrt{\Delta_n}\big)\big),$$

and we observe that $Y'^n(f, X)_t - Y^n(f_n, X)_t = \sum_{i=1}^{[t/\Delta_n]} \zeta_i^n$. Recalling the definition (5.2.2) of f_n, we see that by virtue of the growth assumption on f the difference $f(x) - f_n(x)$ vanishes when $\|x\| < 1/K\Delta_n^{1/2p}$ for some constant K and, if it is not 0, its modulus is smaller than $\|f(x)\|$, which itself is bigger than $1/2\sqrt{\Delta_n}$. Therefore for $A > 0$ we have, for n large enough:

$$\big\|f(x) - f_n(x)\big\| \leq K\sqrt{\Delta_n}\,\big\|f(x)\big\|^2 \psi_A(x).$$

Then

$$\sup_{s\le t}\big\|Y'^n(f,X)_s - Y^n(f_n,X)_s\big\| \;\le\; K\,V'^n\big(\|f\|^2\psi_A, X\big). \tag{5.2.22}$$

Once more, the assumptions of Theorem 3.4.1 are met by X and the function $\|f\|^2\psi'_A$, with again $2p$ instead of p. Then $V'^n(\|f\|^2\psi_A, X)_t \stackrel{\text{u.c.p.}}{\Longrightarrow} H(A)_t = \int_0^t \rho_{c_s}(\|f\|\psi_A)\,ds$, whereas $H(A) \stackrel{\text{u.c.p.}}{\Longrightarrow} 0$ as $A\to\infty$ because $\|f\|^2\psi_A$ converges pointwise to $\|f\|$ and is smaller than $K\|x\|^{2p}$. Combining this with (5.2.22) and letting first $n\to\infty$ and then $A\to\infty$, we conclude $Y'^n(f,X) - Y^n(f_n,X) \stackrel{\text{u.c.p.}}{\Longrightarrow} 0$.

(ii) As said before, only the unbounded case for f needs to be proved, and proving $Y^n(f,X) - Y'^n(f,X) \stackrel{\text{u.c.p.}}{\Longrightarrow} 0$ is enough. We have

$$Y'^n(f,X)_t - Y^n(f,X)_t \;=\; \sum_{i=1}^{[t/\Delta_n]} \mathbb{E}\big(\zeta_i^n \mid \mathcal{F}_{(i-1)\Delta_n}\big). \tag{5.2.23}$$

As in (i) above, the assumption on f yields, for some sequence $\varepsilon_n\to 0$ and constant K:

$$\big\|f(x) - f_n(x)\big\| \;\le\; \begin{cases} \varepsilon_n\|x\|1_{\{\|x\|>1/K\sqrt{\Delta_n}\}} & \text{in case (a)}\\ K\|x\|^p 1_{\{\|x\|>1/K\Delta_n^{1/2p}\}} & \text{in cases (b), (c)}\end{cases}$$

Then we use the inequality $\|x+y\|^p 1_{\{\|x+y\|>a\}} \le K(\|x\|^p 1_{\{\|x\|>a/2\}} + \|y\|^p 1_{\{\|y\|>a/2\}})$, and we decompose $X - X_0$ into the sum of the four terms B, X^c, $Z = (\delta 1_{\{\|\delta\|\le 1\}})\star(\mathfrak{p}-\mathfrak{q})$ and $Z' = (\delta 1_{\{\|\delta\|>1\}})\star\mathfrak{p}$, to get

$$\|\zeta_i^n\| \;\le\; \begin{cases} \varepsilon_n K\big(\|\Delta_i^n B\| + \|\Delta_i^n X^c\|^2 + \|\Delta_i^n Z\|^2 + \|\Delta_i^n Z'\|\big) & \text{in case (a)}\\ K\Delta_n^{1+1/2p}\,\|\Delta_i^n X/\sqrt{\Delta_n}\|^{p+1} & \text{in cases (b), (c).}\end{cases}$$

In case (a) we deduce from (2.1.33) and (2.1.40) with $p=1$ on the one hand, from (2.1.34) and (2.1.36) with $p=2$ on the other hand, that

$$\sum_{i=1}^{[t/\Delta_n]} \mathbb{E}\big(\|\zeta_i^n\|\big) \le \varepsilon_n K\mathbb{E}\bigg(\int_0^t \Big(\|b_s\| + \|c_s\| + \int \big(\|\delta(s,z)\|^2 \wedge \|\delta(s,z)\|\big)\,\lambda(dz)\Big)\,ds\bigg)$$

and the expectation on the right is exactly the one in (5.2.9). In view of (5.2.23), the condition (5.2.9) obviously implies that $Y^n(f,X) - Y'^n(f,X) \stackrel{\text{u.c.p.}}{\Longrightarrow} 0$.

In cases (b) and (c), (2.1.44) gives $\mathbb{E}(\|\zeta_i^n\|) \le K\Delta_n^{1+1/2p}$, and the result follows as above.

(iii) By construction $(g_p)_n \le K/\sqrt{\Delta_n}$, and when further $p=1$ and the processes b and c and $\int(\|x\|^2\wedge\|x\|)F_t(dx)$ are bounded we can apply (2.1.33), (2.1.34), (2.1.36) and (2.1.40) to get $\mathbb{E}((g_1)_n(\Delta_i^n X/\sqrt{\Delta_n}) \mid \mathcal{F}_{(i-1)\Delta_n}) \le K$.

Suppose now that X has a jump at T, of absolute size $Z = \|\Delta X_T\| > 0$ on the set $\{T < \infty\}$. The jump of $Y'^n(f,X)$ at time $\Delta_n[T/\Delta_n]$ is equal to

$$Z_n = \sqrt{\Delta_n}\big(g_p\big(\Delta_i^n X/\sqrt{\Delta_n}\big) - \mathbb{E}\big(g_p\big(\Delta_i^n X/\sqrt{\Delta_n}\big) \mid \mathcal{F}_{(i-1)\Delta_n}\big)\big), \qquad (5.2.24)$$

taken at $i = [T/\Delta_n]$. We then deduce from the previous estimates that Z_n is approximately equal to $Z^p/\Delta_n^{p/2}$ (in all cases the conditional expectation in (5.2.24) is always asymptotically negligible in front of the first term). It follows that when $p > 1$ the sequence of processes $Y'^n(g_p, X)$ cannot be tight, and that when $p = 1$ and when it is tight, any of its limiting processes has a jump. This finishes the proof. □

5.3 The Central Limit Theorem for the Processes $\overline{V}'^n(f,X)$

Now we turn to the processes $\overline{V}'^n(f,X)$ of (5.2.1). We have

$$\overline{V}'^n(f,X) = Y^n(f,X) + A^n(f,X), \quad \text{where}$$
$$A^n(f,X)_t = \frac{1}{\sqrt{\Delta_n}}\left(\Delta_n \sum_{i=1}^{[t/\Delta_n]} \mathbb{E}\big(f\big(\Delta_i^n X/\sqrt{\Delta_n}\big) \mid \mathcal{F}_{(i-1)\Delta_n}\big) - \int_0^t \rho_{c_s}(f)\,ds\right) \qquad (5.3.1)$$

as soon as the process $Y^n(f,X)$ is well defined, and the previous section provides a CLT for $Y^n(f,X)$. Hence at this point, finding the asymptotic behavior of $\overline{V}'^n(f,X)$ essentially amounts to determining the asymptotic behavior of $A^n(f,X)$.

To this end, and because those processes have an exploding factor $1/\sqrt{\Delta_n}$ in front, we need some assumptions on the coefficients b and especially σ, significantly stronger than the weak ones under which Theorem 5.2.2 holds. We also need stronger assumptions on the test function f, leading to a trade-off between the assumptions on (b,σ) and those on f. In order to get some insight as to why rather strong assumptions on σ are necessary, we give an elementary example.

Example 5.3.1 We suppose that X is continuous, one-dimensional, with $b_t = 0$ and σ_t *deterministic*, and bounded as well as $1/\sigma_t$. We take $f(x) = |x|^p$. In this case $\Delta_i^n X/\sqrt{\Delta_n}$ is independent of $\mathcal{F}_{(i-1)\Delta_n}$ and centered normal with variance $\frac{1}{\Delta_n}\int_{(i-1)\Delta_n}^{i\Delta_n} c_s\,ds$. The process $A^n(f,X)$ is non-random and takes the simple form $A^n(f,X)_t = m_p\sqrt{\Delta_n}\sum_{i=1}^{[t/\Delta_n]} a_i^n + a'_n(t)$, where m_p is the p absolute moment of the law $\mathcal{N}(0,1)$, and

$$a_i^n = \left(\frac{1}{\Delta_n}\int_{(i-1)\Delta_n}^{i\Delta_n} c_s\,ds\right)^{p/2} - \frac{1}{\Delta_n}\int_{(i-1)\Delta_n}^{i\Delta_n} c_s^{p/2}\,ds,$$

and $a'_n(t)$ is a "boundary term" corresponding to the integral of $\rho_{c_s}(f)$ between $\Delta_n[t/\Delta_n]$ and t, and which is $O(\sqrt{\Delta_n})$. If c_t (or equivalently $\sigma_t = \sqrt{c_t}$) is, say,

Hölder with index $\alpha \in [0, 1]$, the typical magnitude of each a_i^n is Δ_n^α, so we conclude $A^n(f, X)_t \to 0$ if $\alpha > 1/2$, whereas $A^n(f, X)$ does not go to 0 if $\alpha = 1/2$ and even explodes if $\alpha < 1/2$.

A special case, though, is $p = 2$, because we then have $a_i^n = 0$. This explains why, when considering $f(x) = x^2$, which amounts to looking at the convergence of the "approximate quadratic variation" towards the quadratic variation, the assumptions on σ which we need are much less stringent, as we will see in Sect. 5.4.

This example is somewhat artificial, since when σ_t is deterministic it is often natural to assume that it is also differentiable, or at least Lipschitz. When X is a continuous Markov process σ is typically of the form $\sigma_t = g(X_t)$: in this case, even if g is a C^∞ function, the process σ inherits the path properties of X itself, and thus it is typically Hölder in time with any index $\alpha < 1/2$, but not Hölder with index $1/2$, not to speak of the case where X, hence σ, have jumps. It is also customary to consider situations where σ is a itself the solution of another stochastic differential equation, as in the case of "stochastic volatility".

5.3.1 Assumptions and Results

The process X has the form (5.0.1), and another—fundamental—structural assumption is that the process σ occurring in (5.0.1) is itself an Itô semimartingale, hence the pair (X, σ) as well. We will use the assumption (K), already introduced as Assumption 4.4.3, and which we recall:

Assumption (K) We have (H) and

$$\sigma_t = \sigma_0 + \int_0^t \widetilde{b}_s \, ds + \int_0^t \widetilde{\sigma}_s \, dW_s + M_t + \sum_{s \le t} \Delta\sigma_s \, 1_{\{\|\Delta\sigma_s\| > 1\}}, \tag{5.3.2}$$

where

- M is a local martingale with $\|\Delta M_t\| \le 1$, orthogonal to W, and an angle bracket of the form $\langle M, M\rangle_t = \int_0^t a_s \, ds$
- the compensator of $\sum_{s \le t} 1_{\{\|\Delta\sigma_s\| > 1\}}$ has the form $\int_0^t \widetilde{a}_s \, ds$.

Moreover, the processes $\widetilde{b}$, a and $\widetilde{a}$ are locally bounded, and the processes $\widetilde{\sigma}$ and b are càdlàg or càglàd.

The next assumption involves a number $r \in [0, 1]$.

Assumption 5.3.2 (or (**K-***r*)) We have (K), except that b is not required to be càdlàg or càglàd, and

(i) There is a localizing sequence (τ_n) of stopping times and, for each n, a *deterministic* nonnegative function Γ_n on E satisfying $\int \Gamma_n(z)^r \lambda(dz) < \infty$ (with the convention $0^0 = 0$) and such that $\|\delta(\omega,t,z)\| \wedge 1 \leq \Gamma_n(z)$ for all (ω,t,z) with $t \leq \tau_n(\omega)$;

(ii) The process

$$b'_t = b_t - \int_{\{\|\delta(t,z)\|\leq 1\}} \delta(t,z)\,\lambda(dz) \tag{5.3.3}$$

(which is well defined under (i) with $r \leq 1$) is càdlàg or càglàd.

When X is continuous then (K-r) is the same as (K) for all r. Here again, whether b' is càdlàg or càglàd does not matter. However, b' is typically càglàd rather than càdlàg, because of the predictability requirement on δ. The condition $r \leq 1$, necessary for the process b' to be well defined, implies that the process X has r-summable jumps, that is $\sum_{s\leq t} \|\Delta X_s\|^r < \infty$ a.s. for all t, see Lemma 3.2.1. Observe also that (K-r) for some $r \in [0,1]$ implies (K-r') for all $r' \in [r,1]$.

Remark 5.3.3 When the process X is the solution of the stochastic differential equation

$$X_t = X_0 + \int_0^t f(s, X_{s-})\,dZ_s,$$

where Z is a multidimensional Lévy process and f is a $C^{1,2}$ function on $\mathbb{R}_+ \times \mathbb{R}^d$, we have seen that (K) holds in Remark 4.4.5. If further Z has r-summable jumps with $r \leq 1$, then Assumption (K-r) is also satisfied. In particular, if X itself is a Lévy process with r-summable jumps, it satisfies (K-r).

In some circumstances, we also need the following assumptions, the first one having been already introduced as Assumption 4.4.4:

Assumption (K') We have (K) and both processes c_t and c_{t-} take their values in the set $\mathcal{M}^{++}_{d\times d}$ of all symmetric positive definite $d \times d$ matrices.

Assumption 5.3.4 (or (**K'-r**)) We have (K-r) and both processes c_t and c_{t-} take their values in $\mathcal{M}^{++}_{d\times d}$.

We are now ready to state the main results of this section. They are given in two theorems, which only differ by their assumptions: the first one has quite simple assumptions, and it describes the most useful case. The second one is also sometimes useful, but unfortunately is rather cumbersome to state.

Each theorem contains two parts: the most useful and simple one concerns test functions f that are globally even, the other one is about "general" functions f. Exactly as in Theorem 5.2.2 these statements differ by the description of the limiting process, which is simple in the globally even case and complicated otherwise. In order to describe the limit in the non-even case, we need the notation $\widehat{\gamma}_\alpha(f)$ and $\overline{\gamma}_\alpha(f,g)$ of (5.2.4) and to introduce a further one. For any Borel function g with polynomial growth and any $d \times d'$ matrix α we define a $d' \times d'$ matrix with entries

$$\widehat{\gamma}'_\alpha(g)^{jk} = \mathbb{E}\left(g(\alpha W_1) \int_0^1 W_s^j \, dW_s^k \right). \tag{5.3.4}$$

Below f is q-dimensional and ∇f denotes the gradient, that is the $q \times d$ matrix-valued function $(\partial_i f^j)$. The reader will observe that the process b'_t of (5.3.3) explicitly occurs in the limit below, so the summability $\sum_{s \le t} \|\Delta X_s\| < \infty$ seems to be a necessary condition for the validity of the following claim.

Theorem 5.3.5 (CLT with Normalization – I) *Let X be a d-dimensional Itô semimartingale and f be a C^1 function from $\mathbb{R}^d$ into $\mathbb{R}^q$, with polynomial growth as well as ∇f. Assume either one of the following three properties:*

(α) (K) *holds and X is continuous,*
(β) *(K-1) holds and f and ∇f are bounded,*
(γ) *For some $r, r' \in [0, 1)$, we have (K-r) and $\|f(x)\| \le K(1 + \|x\|^{r'})$.*

(i) *When the function f is globally even, the sequence of q-dimensional processes $\overline{V}'^n(f, X)$ in* (5.2.1) *converges stably in law to a continuous process $\overline{V}'(f, X) = \overline{U}'(f, X)$ which is defined on a very good filtered extension $(\widetilde{\Omega}, \widetilde{\mathcal{F}}, (\widetilde{\mathcal{F}}_t)_{t \ge 0}, \widetilde{\mathbb{P}})$ of $(\Omega, \mathcal{F}, (\mathcal{F}_t)_{t \ge 0}, \mathbb{P})$, and conditionally on $\mathcal{F}$ is a centered Gaussian process with independent increments satisfying*

$$\widetilde{\mathbb{E}}\big(\overline{U}'(f, X)_t^j \, \overline{U}'(f, X)_t^k \mid \mathcal{F}\big) = \int_0^t \big(\rho_{c_s}(f^j f^k) - \rho_{c_s}(f^j)\rho_{c_s}(f^k)\big) \, ds. \tag{5.3.5}$$

(ii) *Otherwise, the sequence $\overline{V}'^n(f, X)$ converges stably in law to a process of the form*

$$\overline{V}'(f, X) = \overline{U}'(f, X) + \overline{A}(f, X) + \overline{A}'(f, X) + \overline{U}(f, X), \tag{5.3.6}$$

where $\overline{U}'(f, X)$ is as above, except that (5.3.5) *is replaced by*

$$\widetilde{\mathbb{E}}\big(\overline{U}'(f, X)_t^j \, \overline{U}'(f, X)_t^k \mid \mathcal{F}\big) = \int_0^t \overline{\gamma}_{\sigma_s}(f^j, f^k) \, ds \tag{5.3.7}$$

and where, with $\widetilde{\sigma}_t$ and b'_t as given in (5.3.2) *and* (5.3.3),

$$\left.\begin{aligned} \overline{A}(f, X)_t^i &= \textstyle\sum_{j=1}^d \int_0^t b_s'^j \, \rho_{c_s}(\partial_j f^i) \, ds \\ \overline{A}'(f, X)_t^i &= \textstyle\sum_{j=1}^d \sum_{m,k=1}^{d'} \int_0^t \widetilde{\sigma}_s^{jkm} \, \widehat{\gamma}'_{\sigma_s}(\partial_j f^i)^{mk} \, ds \\ \overline{U}(f, X)_t^i &= \textstyle\sum_{k=1}^{d'} \int_0^t \widehat{\gamma}_{\sigma_s}(f^i)^k \, dW_s^k. \end{aligned}\right\} \tag{5.3.8}$$

When f is even, the three processes in (5.3.8) vanish since $\rho_{c_t}(\partial_k f^i) = \widehat{\gamma}'_{\sigma_t}(\partial_k f^i) = \widehat{\gamma}_{\sigma_t}(f^i) = 0$, and also $\overline{\gamma}_{\sigma_s}(f^j, f^k) = \rho_{c_s}(f^j f^k) - \rho_{c_s}(f^j)\rho_{c_s}(f^k)$, so (i) is a special case of (ii). The two processes $\overline{U}(f, X)$ and $\overline{U}'(f, X)$ are those appearing in Theorem 5.2.2. The process $\overline{A}(f, X) + \overline{A}'(f, X)$ is a drift term adapted to the filtration $(\mathcal{F}_t)$, and $\overline{U}(f,X)$ is an $(\mathcal{F}_t)$-martingale which, conditionally on $\mathcal{F}$, may also be interpreted as a "drift term": namely, the limit $\overline{V}'(f,X)$ is, conditionally on $\mathcal{F}$, a Gaussian process with independent increments and mean value $\overline{A}(f,X)_t + \overline{A}'(f,X)_t + \overline{U}(f,X)_t$ and variance given by (5.3.7).

The main drawback of this result is this: it holds for the power function $f(x) = |x^j|^w$ only when $w > 1$ and X is continuous. Relaxing these restrictions is the primary aim of the next theorem, whose statement is unfortunately somewhat cumbersome. For this statement, we recall that an *affine hyperplane* of $\mathbb{R}^d$ is a set of the form $\{x \in \mathbb{R}^d : x^*y = a\}$ for some $a \in \mathbb{R}$ and some unitary vector y in $\mathbb{R}^d$.

Theorem 5.3.6 (CLT with Normalization – II) *Let X be a d-dimensional Itô semimartingale. Let f be a function from $\mathbb{R}^d$ into $\mathbb{R}^q$, with polynomial growth, and C^1 outside a subset B of $\mathbb{R}^d$ which is a finite union of affine hyperplanes. With $d(x,B)$ denoting the distance between $x \in \mathbb{R}^d$ and B, assume also that for some $w \in (0,1]$ and $p \geq 0$ we have*

$$x \in B^c \;\Rightarrow\; \|\nabla f(x)\| \leq K\big(1+\|x\|^p\big)\left(1+\frac{1}{d(x,B)^{1-w}}\right), \tag{5.3.9}$$

$$\begin{aligned} & x \in B^c,\ \|y\| \leq 1 \bigwedge \frac{d(x,B)}{2} \\ & \quad \Rightarrow \|\nabla f(x+y) - \nabla f(x)\| \leq K\|y\|\big(1+\|x\|^p+\|y\|^p\big)\left(1+\frac{1}{d(x,B)^{2-w}}\right). \end{aligned} \tag{5.3.10}$$

Finally, assume either (K') *and X to be continuous, or that for some $0 < r \leq r' < 1$ and some $p \geq 0$ we have (K'-r) and*

$$\|f(x+y) - f(x)\| \;\leq\; K\big(1+\|x\|^p\big)\big(\|y\|^r + \|y\|^{r'}\big). \tag{5.3.11}$$

Then the sequence of q-dimensional processes $\overline{V}'^n(f,X)$ converges stably in law to the q-dimensional process $\overline{V}'(f,X)$, as described in Theorem 5.3.5, *in both cases* (i) *and* (ii).

Although f is not everywhere differentiable, the variables $\rho_{c_s}(\partial_k f^i)$ and $\widehat{\gamma}'_{\sigma_s}(\partial_k f^i)$ are well defined, hence the limiting processes as well: this is because ∇f exists outside the set B, which has vanishing Lebesgue measure, hence $\rho_{c_s}(B) = 0$.

Remark 5.3.7 The condition (5.3.11) should be compared with the condition $\|f(x)\| \leq K(1+\|x\|^{r'})$, called (A) in this remark, and which is used instead of (5.3.11) in Theorem 5.3.5 when X is discontinuous. Recall that in both conditions r' should be smaller than 1.

In one direction, (5.3.11) implies (A) with the same r' (but of course a different constant K). In the other direction, if f is C^1 on $\mathbb{R}^d$ and $\|\nabla f(x)\| \leq K(1+\|x\|^q)$ for some $q \geq 0$, then (A) implies (5.3.11) with $r = r'$ and $p = q \vee r'$. The latter implication, however, is no longer true when f is C^1 outside a non-empty set B, but not on $\mathbb{R}^d$ itself.

Example 5.3.8 The conditions on f, in connection with the assumptions on X, are especially designed to accommodate the functions

$$f(x) = \prod_{j=1}^{d} |x^j|^{w_j}, \quad w_j \geq 0, \quad v = w_1 + \cdots + w_d > 0,$$

with the convention $0^0 = 1$, and we set u to be the minimum of all w_i which do not vanish. This function is C^1 if $u > 1$, in which case Theorem 5.3.5 does *not* apply when X jumps.

Suppose that $u \leq 1$, so f is C^1 outside the set B which is the union of the hyperplanes $\{x \in \mathbb{R}^d : x^i = 0\}$ for all i such that $0 < w_i \leq 1$. Then (5.3.9) and (5.3.10) are satisfied with $w = u$ and $p = v - u$: so the result holds for these functions, if X is continuous and satisfies (K').

When X jumps, we further need (5.3.11) for some $0 < r \leq r' < 1$: this is the case if and only if $v < 1$ and then (5.3.11) holds with $p = r' = v$ and $r = u$. Then the theorem applies under (K'-r).

Remark 5.3.9 As seen in the previous example, when X has jumps the assumptions on f are quite restrictive, and subsequently we will consider other normalized functionals ("truncated", or depending on several successive increments), with the aim of weakening these assumptions, and with some success.

However, there is also a very strong assumption on the jumps of X, somewhat hidden in (K-1): the jumps are *summable*. This assumption can *never be relaxed* in CLTs for normalized functionals, even for the extensions hinted at above: see for example the recent work of Vetter [93] for functionals depending on several increments, or Mancini [75] for truncated functionals.

We now proceed to the proof of these two theorems, through several steps.

5.3.2 Localization and Elimination of Jumps

As for the proof of Theorem 5.0.1, we begin by stating some strengthened assumptions, partially introduced in the previous chapter already:

Assumption (SK) We have (K), and there is a constant A such that

$$\|b_t(\omega)\| + \|\widetilde{b}_t(\omega)\| + \|\sigma_t(\omega)\| + \|\widetilde{\sigma}_t(\omega)\| + \|X_t(\omega)\| + \widetilde{a}_t(\omega) + \|a_t(\omega)\| \leq A. \tag{5.3.12}$$

Assumption 5.3.10 (or **(SK-r)**) We have (K-r), and there are a constant A and a function Γ on E, such that (5.3.12) holds, and also

$$\|\delta(\omega, t, z)\| \leq \Gamma(z), \quad \Gamma(z) \leq A, \quad \int \Gamma(z)^r \lambda(dz) \leq A.$$

Assumption (SK') We have (SK) and c_t^{-1} is bounded.

Assumption 5.3.11 (or **(SK'-r)**) We have (SK-r) and c_t^{-1} is bounded.

Then by the localization Lemma 4.4.9 and a trivial extension of it when (ABC) = (K-r) or (ABC) = (K'-r), we have the following:

Lemma 5.3.12 (Localization) *It is enough to prove Theorems* 5.3.5 *and* 5.3.6 *when* (K), (K-1), (K') *and* (K'-r) *are replaced by* (SK), (SK-1), (SK') *and* (SK'-r), *respectively.*

From now on, we assume at least (SK-1). Recalling (5.3.3), we have

$$X = X_0 + X' + X'', \quad \text{where } X'_t = \int_0^t b'_s\,ds + \int_0^t \sigma_s\,dW_s, \quad X'' = \delta * \mathfrak{p} \tag{5.3.13}$$

and b'_t is bounded. The following lemma shows that it is enough to prove the results for the process X', because in the case (β) of Theorem 5.3.5 we have $\|f(x+y) - f(x)\| \leq K(\|y\| \wedge 1)$, whereas in case (γ) we have (5.3.11), as mentioned in Remark 5.3.7.

Lemma 5.3.13 *Assume either* (SK-1) *and* $\|f(x+y) - f(x)\| \leq K(\|y\| \wedge 1)$, *or* (SK-$r$) *and* (5.3.11) *with* $0 < r \leq r' < 1$. *Then* $\frac{1}{\sqrt{\Delta_n}}(V'^n(f,X) - V'^n(f,X')) \stackrel{\text{u.c.p.}}{\Longrightarrow} 0$.

Proof Observe that $\frac{1}{\sqrt{\Delta_n}}(V'^n(f,X)_t - V'^n(f,X')_t) = \sum_{i=1}^{[t/\Delta_n]} \zeta_i^n$, where

$$\zeta_i^n = \sqrt{\Delta_n}\,\mathbb{E}\big(f\big(\Delta_i^n X/\sqrt{\Delta_n}\big) - f\big(\Delta_i^n X'/\sqrt{\Delta_n}\big) \mid \mathcal{F}_{(i-1)\Delta_n}\big).$$

Therefore it is enough to prove that for all $t > 0$ we have $\sum_{i=1}^{[t/\Delta_n]} \mathbb{E}(\|\zeta_i^n\|) \to 0$.

In the first case, (SK-1) allows us to apply (2.1.47) with $p = r = 1$ and $q = 1/2$ to obtain

$$\mathbb{E}\big(\|\zeta_i^n\|\big) \leq K\sqrt{\Delta_n}\,\mathbb{E}\big(\big(\|\Delta_i^n X''\|/\sqrt{\Delta_n}\big) \wedge 1\big) \leq K\Delta_n\,\theta_n$$

for some sequence θ_n of numbers going to 0, and this in turn implies $\sum_{i=1}^{[t/\Delta_n]} \mathbb{E}(\|\zeta_i^n\|) \to 0$.

In the second case, with the notation $\eta_i^n = \|\Delta_i^n X'\|/\sqrt{\Delta_n}$, (5.3.11) yields

$$\mathbb{E}\big(\|\zeta_i^n\|\big) \leq K\Delta_n^{\frac{1-r}{2}}\,\mathbb{E}\big(\big(1 + (\eta_i^n)^p\big)\|\Delta_i^n X''\|^r\big) + K\Delta_n^{\frac{1-r'}{2}}\,\mathbb{E}\big(\big(1 + (\eta_i^n)^p\big)\|\Delta_i^n X''\|^{r'}\big)$$

(SK-r) together with (2.1.33), (2.1.34) and Lemma 2.1.7 imply $\mathbb{E}((\eta_i^n)^q) \leq K_q$ for all $q > 0$ and $\mathbb{E}(\|\Delta_i^n X''\|^q) \leq K_q\Delta_n$ for all $q \in [r, 1]$. Then, since $r \leq r' < 1$, Hölder's inequality with the exponent $\frac{4}{3+r}$ for $\|\Delta_i^n X''\|^r$ and $\frac{4}{3+r'}$ for $\|\Delta_i^n X''\|^{r'}$ respectively yields $\mathbb{E}(\|\zeta_i^n\|) \leq K(\Delta_n^{1+\frac{1-r}{4}} + \Delta_n^{1+\frac{1-r'}{4}})$, and again $\sum_{i=1}^{[t/\Delta_n]} \mathbb{E}(\|\zeta_i^n\|) \to 0$ follows. □

We end this subsection by stating some consequences of (SK-1). Since σ_t is bounded, the process $N_t = \sum_{s\le t} \Delta\sigma_s 1_{\{\|\Delta\sigma_s\|>1\}}$ admits a compensator of the form $N'_t = \int_0^t \widetilde{a}'_s\,ds$ with $\widetilde{a}'_t$ bounded, and $M' = M + N - N'$ is again a local martingale orthogonal to W and with predictable bracket $\int_0^t a'_s\,ds$ with a'_s bounded. So (5.3.2) takes the form

$$\sigma_t = \sigma_0 + \int_0^t \widetilde{b}'_s\,ds + \int_0^t \widetilde{\sigma}_s\,dW_s + M'_t, \tag{5.3.14}$$

where $\widetilde{b}'_t = \widetilde{b}_t + \widetilde{a}'_t$ is bounded. We also have a Grigelionis representation of the following form for the Itô martingale M':

$$M'_t = \int_0^t \sigma'_s\,dW'_s + \delta' \star (\mathfrak{p}' - \mathfrak{q}')_t, \tag{5.3.15}$$

relative to some Poisson random measure $\mathfrak{p}'$ on $\mathbb{R}_+ \times E'$ (which could indeed be taken equal to $\mathfrak{p}$, although this is not necessary) and its deterministic compensator $\mathfrak{q}'(dt,dz) = dt \otimes \lambda'(dz)$, and with respect to a $d \times d'$-dimensional Brownian motion W', which can be taken orthogonal to W because M' is so. Moreover since a'_t is bounded, it is clearly possible to choose σ'_t and δ' above in such a way that

$$\|\sigma'_t(\omega)\| \le K, \qquad \|\delta'(\omega,t,z)\| \le K, \qquad \int_{E'} \|\delta'(\omega,t,z)\|^2 \lambda'(dz) \le K. \tag{5.3.16}$$

Observe that putting (5.3.14) and (5.3.15) together, we get the following Grigelionis representation for σ_t:

$$\sigma_t = \sigma_0 + \int_0^t \widetilde{b}'_s\,ds + \int_0^t \widetilde{\sigma}_s\,dW_s + \int_0^t \sigma'_s\,dW'_s + \delta' \star (\mathfrak{p}' - \mathfrak{q}')_t. \tag{5.3.17}$$

5.3.3 Proof of the Central Limit Theorem for $V'^n(f,X)$

A – Outline of the proof As already mentioned, (i) is a special case of (ii) in Theorem 5.3.5. As seen in Remark 5.3.7, the assumptions on f in Lemma 5.3.13 are satisfied under the hypothesis (β) or (γ) of Theorem 5.3.5. Hence this lemma and the localization Lemma 5.3.12 yield that we only need to prove the CLT for the processes $V'^n(f,X')$, and under the strengthened assumptions.

In other words, we may and will assume in the sequel that $X = X'$ is *continuous*. We prove both theorems together, in this continuous case and under the strengthened assumptions. For this we will set $B = \emptyset$ and $d(x,\emptyset) = \infty$ for all x when f is C^1 on $\mathbb{R}^d$, so the next notation always makes sense:

$$\phi_B(x) = 1 + \frac{1}{d(x,B)}. \tag{5.3.18}$$

Next, if $A>1$ we set

$$\begin{aligned}\mathcal{M}_A &= \text{the set of all } d\times d' \text{ matrices } \alpha \text{ with } \|\alpha\|\le A\\ \mathcal{M}'_A &= \big\{\alpha\in\mathcal{M}_A:\ \alpha\alpha^* \text{ is invertible, with } \big\|(\alpha\alpha^*)^{-1}\big\|\le A\big\}.\end{aligned}\tag{5.3.19}$$

Then we have to study the following two cases, where $A>1$ is some constant, and which correspond to Theorems 5.3.5 and 5.3.6, respectively:

$$\begin{array}{l}\text{(a) } B=\emptyset,\ \text{(SK)},\ f \text{ is } C^1,\ \nabla f \text{ has polynomial growth},\ \sigma_t\in\mathcal{M}_A\\ \text{(b) } B\neq\emptyset,\ \text{(SK')},\ f \text{ has polynomial growth with (5.3.9)}\\ \qquad \text{and 5.3.10)},\ \sigma_t\in\mathcal{M}'_A.\end{array}\tag{5.3.20}$$

We can now rewrite our assumptions on f. With $w=1$ in case (a) and $w\in(0,1]$ as in (5.3.9) and (5.3.10) in case (b), the function f is C^1 outside of B and satisfies

$$x\notin B\quad\Rightarrow\quad \big\|\nabla f(x)\big\|\ \le\ K\big(1+\|x\|^p\big)\phi_B(x)^{1-w},\tag{5.3.21}$$

which reduces to $K(1+\|x\|^p)$ in case (a). Moreover, recalling that f is C^1 in case (a), we have

$$\big\|\nabla f^j(x+y)-\nabla f^j(x)\big\|\le\begin{cases}\phi'_C(\varepsilon)+\frac{K}{C}\,(\|x\|^{p+1}+\|y\|^{p+1})+\frac{KC^p\|y\|}{\varepsilon}\\ \qquad\text{in case (a)}\\ K(1+\|x\|^p+\|y\|^p)\phi_B(x)^{2-w}\|y\|\\ \qquad\text{in case (b) and if } x\notin B,\ \|y\|\le\frac{d(x,B)}{2}\end{cases}\tag{5.3.22}$$

for some p and for all $C\ge1$ and $\varepsilon\in(0,1]$, and where $\phi'_C(\varepsilon)\to0$ as $\varepsilon\to0$ for all C.

Recalling the notation (5.3.1), we can now outline the scheme of the proof:

1. In the decomposition (5.3.1) we use Theorem 5.2.1-(ii), whose assumption (c) is satisfied here, because (5.2.6) holds for all p, to get $Y^n(f,X)\stackrel{\mathcal{L}\text{-}s}{\Longrightarrow}\overline{Y}(f,X)=\overline{U}(f,X)+\overline{U}'(f,X)$.
2. Taking advantage of the property $\rho_{c_{(i-1)\Delta_n}}(f)=\mathbb{E}(f(\beta_i^n)\mid\mathcal{F}_{(i-1)\Delta_n})$ (recall $\beta_i^n=\sigma_{(i-1)\Delta_n}\Delta_i^nW/\sqrt{\Delta_n}$), we write $A^n(f,X)=A^n(1)+A^n(2)$, where

$$\begin{aligned}A^n(1)_t &= \sqrt{\Delta_n}\sum_{i=1}^{[t/\Delta_n]}\mathbb{E}\big(f\big(\Delta_i^nX/\sqrt{\Delta_n}\big)-f(\beta_i^n)\mid\mathcal{F}_{(i-1)\Delta_n}\big)\\ A^n(2)_t &= \frac{1}{\sqrt{\Delta_n}}\left(\sum_{i=1}^{[t/\Delta_n]}\Delta_n\,\rho_{c_{(i-1)\Delta_n}}(f)-\int_0^t\rho_{c_s}(f)\,ds\right).\end{aligned}\tag{5.3.23}$$

3. At this stage, it remains to prove the following two properties:

$$A^n(2)\stackrel{\text{u.c.p.}}{\Longrightarrow}0\tag{5.3.24}$$

$$A^n(1)\stackrel{\text{u.c.p.}}{\Longrightarrow}\overline{A}(f,X)+\overline{A}'(f,X).\tag{5.3.25}$$

B – Proof of (5.3.24) Among the two properties (5.3.24) and (5.3.25) which remain to be proved, the hard one is the latter, so we begin with (5.3.24), which is simple enough and goes through two steps:

Step 1) Upon arguing component by component, we can assume that f is one-dimensional. Set

$$\psi(\alpha) = \rho_{\alpha\alpha^\star}(f) = \mathbb{E}\big(f(\alpha U)\big), \tag{5.3.26}$$

where U is an $\mathcal{N}(0, I_{d'})$ random vector.

In case (a) we deduce from Lebesgue's theorem that the function ψ is C_b^1 on the set $\mathcal{M}_A$ of (5.3.19). In case (b) this is no longer necessarily true, however if $\alpha \in \mathcal{M}'_A$, we have

$$\psi(\alpha) = \frac{1}{(2\pi)^{d/2}\det\big(\alpha\alpha^\star\big)^{1/2}} \int f(x)\, \exp\bigg(-\frac{1}{2}\, x^\star(\alpha\alpha^\star)^{-1}x\bigg)\, dx.$$

We deduce that ψ is C_b^∞ on the set $\mathcal{M}'_A$ (actually, we only need f to be Borel with polynomial growth here). Thus in both cases (a) and (b) we have (here $\nabla\psi$ is $\mathbb{R}^{d\times d'}$-valued):

$$\begin{aligned} &\big|\psi(\sigma_t)\big| + \big\|\nabla\psi(\sigma_t)\big\| \le K \\ &\big|\psi(\sigma_t) - \psi(\sigma_s)\big| \le K\|\sigma_t - \sigma_s\| \\ &\big|\psi(\sigma_t) - \psi(\sigma_s) - \nabla\psi(\sigma_s)(\sigma_t - \sigma_s)\big| \le \Psi\big(\|\sigma_t - \sigma_s\|\big)\, \|\sigma_t - \sigma_s\| \end{aligned} \tag{5.3.27}$$

for some constant K and some increasing function Ψ on $\mathbb{R}_+$, continuous and null at 0.

A Taylor expansion gives $A^n(2)_t = -\overline{\eta}^n_t - \sum_{i=1}^{[t/\Delta_n]}(\eta^n_i + \eta'^n_i)$, where

$$\begin{aligned} \overline{\eta}^n_t &= \frac{1}{\sqrt{\Delta_n}} \int_{[t/\Delta_n]\Delta_n}^t \psi(\sigma_s)\, ds \\ \eta^n_i &= \frac{1}{\sqrt{\Delta_n}}\, \nabla\psi(\sigma_{(i-1)\Delta_n}) \int_{I(n,i)} (\sigma_u - \sigma_{(i-1)\Delta_n})\, du \\ \eta'^n_i &= \frac{1}{\sqrt{\Delta_n}} \int_{I(n,i)} \big(\psi(\sigma_u) - \psi(\sigma_{(i-1)\Delta_n}) - \nabla\psi(\sigma_{(i-1)\Delta_n})(\sigma_u - \sigma_{(i-1)\Delta_n})\big)\, du. \end{aligned}$$

Step 2) (5.3.16) implies $\int \|\delta'(t,z)\|^l \lambda'(dz) \le K_l$ for any $l \ge 2$. Then in view of (5.3.17) we deduce from (2.1.33), (2.1.34) and (2.1.37) that for $u \in I(n,i)$ and $l \ge 2$:

$$\mathbb{E}\big(\|\sigma_u - \sigma_{(i-1)\Delta_n}\|^l \mid \mathcal{F}_{(i-1)\Delta_n}\big) \le K_l\, \Delta_n. \tag{5.3.28}$$

Moreover

$$\mathbb{E}(\sigma_u - \sigma_{(i-1)\Delta_n} \mid \mathcal{F}_{(i-1)\Delta_n}) = \mathbb{E}\bigg(\int_{(i-1)\Delta_n}^u \widetilde{b}'_v\, dv \mid \mathcal{F}_{(i-1)\Delta_n}\bigg),$$

whose norm is smaller than $K\Delta_n$. Then, taking (5.3.27) into consideration, we get

$$\big|\mathbb{E}(\eta_i^n \mid \mathcal{F}_{(i-1)\Delta_n})\big| \le K\Delta_n^{3/2}, \qquad \mathbb{E}\big(|\eta_i^n|^2 \mid \mathcal{F}_{(i-1)\Delta_n}\big) \le K\Delta_n^2$$

(use (5.3.28) and the Cauchy-Schwarz inequality for the second estimate). From this and Doob's inequality for the discrete time martingale $\sum_{i\le j}(\eta_i^n - \mathbb{E}(\eta_i^n \mid \mathcal{F}_{(i-1)\Delta_n}))$, we deduce

$$\mathbb{E}\Bigg(\sup_{s\le t}\Bigg|\sum_{i=1}^{[s/\Delta_n]} \eta_i^n\Bigg|\Bigg) \le K(t+\sqrt{t})\sqrt{\Delta_n}.$$

Next, (5.3.27) yields for any $\varepsilon \in (0,1)$:

$$\big|\eta_i'^n\big| \le \frac{\Psi(\varepsilon)}{\sqrt{\Delta_n}} \int_{I(n,i)} \|\sigma_u - \sigma_{(i-1)\Delta_n}\|\,du + \frac{1}{\varepsilon\sqrt{\Delta_n}} \int_{I(n,i)} \|\sigma_u - \sigma_{(i-1)\Delta_n}\|^2\,du.$$

Then (5.3.28) and the Cauchy-Schwarz inequality yield $\mathbb{E}(|\eta_i'^n|) \le K\Psi(\varepsilon)\Delta_n + \frac{K\Delta_n^{3/2}}{\varepsilon}$. Since moreover $|\overline{\eta}_i^n| \le K\sqrt{\Delta_n}$, we deduce from all that precedes that

$$\mathbb{E}\Big(\sup_{s\le t}\big|A^n(2)_s\big|\Big) \le K(t+\sqrt{t})\bigg(\Psi(\varepsilon) + \frac{\sqrt{\Delta_n}}{\varepsilon}\bigg). \tag{5.3.29}$$

So $\limsup_n \mathbb{E}(\sup_{s\le t}|A^n(2)_s|) \le K(t+\sqrt{t})\Psi(\varepsilon)$, and $\lim_{\varepsilon\to 0}\Psi(\varepsilon) = 0$ gives (5.3.24).

C – Proof of (5.3.25) For this last claim, we have many more steps to go. It suffices to prove the claim for each component, hence we suppose that $q=1$, and in case (a) f is C^1, and in case (b) f is C^1 outside the non-empty set B.

Step 1) (Preliminaries and notation.) Set

$$\theta_i^n = \frac{\Delta_i^n X}{\sqrt{\Delta_n}} - \beta_i^n = \frac{1}{\sqrt{\Delta_n}} \int_{I(n,i)} b_s\,ds + \frac{1}{\sqrt{\Delta_n}} \int_{I(n,i)} (\sigma_s - \sigma_{(i-1)\Delta_n})\,dW_s. \tag{5.3.30}$$

We will constantly use the following estimates on β_i^n and θ_i^n, which follow from (SK) (the one about β_i^n is obvious from the definition of β_i^n and the scaling property of the Brownian motion; for θ_i^n we use (5.3.28) and we apply (2.1.33) and (2.1.34) plus Hölder's inequality when $l<2$, recall X is continuous):

$$l>0 \;\Rightarrow\; \mathbb{E}\big(\|\beta_i^n\|^l\big) \le K_l, \qquad \mathbb{E}\big(\|\theta_i^n\|^l\big) \le K_l\Delta_n^{(l/2)\wedge 1}. \tag{5.3.31}$$

Next, we can and will always assume that $\Delta_n \le 1$ for all n. In view of (5.3.14), we have the decomposition $\theta_i^n = \frac{1}{\sqrt{\Delta_n}}\sum_{j=1}^4 \zeta(j)_i^n$, where (with matrix notation)

$$\zeta(1)_i^n = \Delta_n\, b_{(i-1)\Delta_n}$$

$$\zeta(2)_i^n = \int_{I(n,i)} \big(\widetilde{\sigma}_{(i-1)\Delta_n}(W_s - W_{(i-1)\Delta_n})\big)\,dW_s$$

$$\zeta(3)_i^n = \int_{I(n,i)} \left(M'_s - M'_{(i-1)\Delta_n}\right) dW_s$$

$$\zeta(4)_i^n = \int_{I(n,i)} (b_s - b_{(i-1)\Delta_n})\, ds + \int_{I(n,i)} \left(\int_{(i-1)\Delta_n}^s \widetilde{b}'_u du\right) dW_s$$
$$+ \int_{I(n,i)} \left(\int_{(i-1)\Delta_n}^s (\widetilde{\sigma}_u - \widetilde{\sigma}_{(i-1)\Delta_n})\, dW_u\right) dW_s.$$

We also set

$$A_i^n = \left\{\|\theta_i^n\| > d\big(\beta_i^n, B\big)/2\right\}.$$

We can express the difference $f(\beta_i^n + \theta_i^n) - f(\beta_i^n)$, using a Taylor expansion if we are on the set $(A_i^n)^c$, to get

$$f(\beta_i^n + \theta_i^n) - f(\beta_i^n) = \nabla f(\beta_i^n)\, \theta_i^n + \left(f(\beta_i^n + \theta_i^n) - f(\beta_i^n)\right) 1_{A_i^n}$$
$$- \nabla f(\beta_i^n)\, \theta_i^n\, 1_{A_i^n} + \left(\nabla f(\beta_i^n + u_i^n \theta_i^n) - \nabla f(\beta_i^n)\right) \theta_i^n 1_{(A_i^n)^c} \tag{5.3.32}$$

where u_i^n is some (random) number between 0 and 1: note that $\nabla f(\beta_i^n + u_i^n \theta_i^n)$ is well defined on $(A_i^n)^c$ because then $\beta_i^n + u_i^n \theta_i^n$ belongs to B^c. As to $\nabla f(\beta_i^n)$, it is well defined in case (a), and a.s. well defined in case (b) because in this case B has Lebesgue measure 0 and β_i^n has a density by (SK'). Observe also that $A_i^n = \emptyset$ in case (a).

Then, taking (5.3.23) and (5.3.32) into consideration, we have

$$A^n(1) = \sum\nolimits_{j=1}^7 D^n(j), \tag{5.3.33}$$

where

$$D^n(j)_t = \sum_{i=1}^{[t/\Delta_n]} \delta(j)_i^n, \quad \delta(j)_i^n = \mathbb{E}\big(\delta'(j)_i^n \mid \mathcal{F}_{(i-1)\Delta_n}\big), \quad \text{and}$$

$$\delta'(j)_i^n = \sum_{k=1}^d \partial_k f(\beta_i^n)\, \zeta(j)_i^{n,k} \quad \text{for } j = 1, 2, 3, 4$$

$$\delta'(5)_i^n = \sqrt{\Delta_n} \sum_{k=1}^d \partial_k f(\beta_i^n)\, \theta_i^{n,k}\, 1_{A_i^n} \tag{5.3.34}$$

$$\delta'(6)_i^n = \sqrt{\Delta_n} \sum_{k=1}^d \left(\partial_k f(\beta_i^n + u_i^n \theta_i^n) - \partial_k f(\beta_i^n)\right) \theta_i^{n,k}\, 1_{(A_i^n)^c}$$

$$\delta'(7)_i^n = \sqrt{\Delta_n}\, \left(f(\beta_i^n + \theta_i^n) - f(\beta_i^n)\right) 1_{A_i^n}.$$

Finally, we need an additional notation:

$$\alpha_i^n = \Delta_n^{3/2} + \mathbb{E}\bigg(\int_{I(n,i)} \big(\|b_s - b_{(i-1)\Delta_n}\|^2 + \|\widetilde{\sigma}_s - \widetilde{\sigma}_{(i-1)\Delta_n}\|^2\big)\, ds\bigg). \tag{5.3.35}$$

Step 2) (Estimates on $\zeta(j)_i^n$.) A repeated use of (2.1.34) gives for all $l \geq 2$:

$$\mathbb{E}\big(\|\zeta(1)_i^n\|^l\big) + \mathbb{E}\big(\|\zeta(2)_i^n\|^l\big) \leq K_l \Delta_n^l, \qquad \mathbb{E}\big(\|\zeta(4)_i^n\|^l\big) \leq K_l\, \Delta_n^{l-1} \alpha_i^n. \tag{5.3.36}$$

As seen before, $\mathbb{E}(\|M'_{t+s} - M'_t\|^l) \leq K_l\, s$ for any $l \geq 1$, and it follows from the Burkholder-Davis-Gundy inequality (2.1.32) that

$$\mathbb{E}\big(\|\zeta(3)_i^n\|^l\big) \;\leq\; K\Delta_n^{l/2+(1\wedge(l/2))}. \tag{5.3.37}$$

Next, recalling the function $\phi_B(x) = 1 + 1/d(x, B)$ of (5.3.18), we introduce the following variables:

$$\gamma_i^n \;=\; \begin{cases} 1 & \text{if } w = 1 \\ \phi_B(\beta_i^n) & \text{if } w < 1. \end{cases} \tag{5.3.38}$$

Lemma 5.3.14 *For any $t \in [0, 1)$ we have*

$$\mathbb{E}\big((\gamma_i^n)^t \mid \mathcal{F}_{(i-1)\Delta_n}\big) \;\leq\; K_t. \tag{5.3.39}$$

Proof In case (a) we have $\gamma_i^n = 1$, hence the result. In case (b), $B = \cup_{l=1}^L B_l$ where $B_l = \{x \in \mathbb{R}^d : x^* y_l = z_l\}$ for some $z_l \in \mathbb{R}$ and $y_l \in \mathbb{R}_1^d := \{x \in \mathbb{R}^d : \|x\| = 1\}$. Since $d(x, B) = \min_{1\leq l\leq L} d(x, B_l)$ we have $\gamma_i^n \leq 1 + \sum_{l=1}^L 1/d(\beta_i^n, B_l)$. Moreover $d(x, B_l) = |x^* y_l - z_l|$, and conditionally on $\mathcal{F}_{(i-1)\Delta_n}$ the variable β_i^n has the law $\rho_{c_{(i-1)\Delta_n}}$. Thus it is enough to show the following property (too strong for what we need here, but it is for free), where $\mathcal{M}_{d\times d}^{(A)} = \{a \in \mathcal{M}_{d\times d}^{++} : \|a^{-1}\| \leq A\}$:

$$t \in (0, 1),\ A > 0 \Rightarrow \qquad \sup_{z\in\mathbb{R},\, y\in\mathbb{R}_1^d,\, a\in\mathcal{M}_{d\times d}^{(A)}} \int \frac{1}{|x^* y - z|^t}\, \rho_a(dx) < \infty.$$

With U standard normal, the integral above is $E(|\sqrt{y^* a y}\, U - z|^{-t})$, always smaller than $(y^* a y)^{-t/2}\, \mathbb{E}(|U|^{-t})$, and the result readily follows. □

Lemma 5.3.15 *Let Φ_i^n be arbitrary variables, satisfying for all n, i:*

$$s > 0 \;\Rightarrow\; \mathbb{E}\big(|\Phi_i^n|^s\big) \;\leq\; K_s. \tag{5.3.40}$$

Then for all $s > 0$ and $m \in [0, 1)$, we have:

$$j = 1, 2, 3,\ l < 2 \;\Rightarrow\; \mathbb{E}\big(|\Phi_i^n|^s\ \|\zeta(j)_i^n\|^l\ (\gamma_i^n)^m\big) \;\leq\; K_{s,l,m}\, \Delta_n^l. \tag{5.3.41}$$

$$u \in \left(0, (1-m) \wedge \frac{l}{2}\right) \quad \Rightarrow$$
$$\mathbb{E}\big(|\Phi_i^n|^s \, \|\zeta(4)_i^n\|^l \, (\gamma_i^n)^m\big) \le K_{s,l,m,u} \, \Delta_n^{l-u} \, (\alpha_i^n)^u \le K_{s,l,m,u} \Delta_n^l. \quad (5.3.42)$$

Proof 1) The last inequality in (5.3.42) follows from $\alpha_i^n \le K\Delta_n$. Suppose that we know that

$$\begin{aligned} & p<2,\ r\in[0,1),\ j=1,2,3 \quad \Rightarrow \quad \mathbb{E}\big(\|\zeta(j)_i^n\|^p \, (\gamma_i^n)^r\big) \le K_{p,r}\,\Delta_n^p \\ & p\ge 1,\ r\in[0,1),\ v\in\left(0,(1-r)\wedge\frac{p}{2}\right) \Rightarrow \\ & \quad \mathbb{E}\big(\|\zeta(4)_i^n\|^p \, (\gamma_i^n)^r\big) \le K_{p,r,v}\,\Delta_n^{p-v} \, (\alpha_i^n)^v. \end{aligned} \quad (5.3.43)$$

By Hölder's inequality and (5.3.40), we have for all j and all $\gamma>1$:

$$\mathbb{E}\big(|\Phi_i^n|^s \, \|\zeta(j)_i^n\|^l \, (\gamma_i^n)^m\big) \le K_{s,\gamma}\big(\mathbb{E}\big(\|\zeta(j)_i^n\|^{l\gamma} \, (\gamma_i^n)^{m\gamma}\big)\big)^{1/\gamma}.$$

This, together with the first part of (5.3.43) applied with $p=l\gamma$ and $r=m\gamma$, yields (5.3.41) if we choose $\gamma\in(1,(1/m)\wedge(2/l))$. Together with the second part of (5.3.43) with again $p=l\gamma$ and $r=m\gamma$ and $v=u\gamma$, this also gives (5.3.43) upon choosing $\gamma\in(1,1/(u+m))$. Therefore, we are left to prove (5.3.43).

2) Hölder's inequality and a combination of (5.3.36) and (5.3.39) give (5.3.43) for $j=1,2,4$. However, this does not work for $j=3$, because (5.3.37) only holds for the exponent 2, and we need a more sophisticated argument.

We denote by $(\mathcal{F}'_t)$ the filtration (depending on (n,i), although this is not shown in the notation) such that $\mathcal{F}'_t=\mathcal{F}_t$ if $t<(i-1)\Delta_n$ and $\mathcal{F}'_t=\mathcal{F}_t\bigvee\sigma(W_s:s\ge 0)$ otherwise. Recall the Grigelionis representation (5.3.15) for M'. The three terms W, W' and $\mathfrak{p}'$ are independent, and by the integration by parts formula we see that

$$\begin{aligned} \zeta(3)_i^n = {} & \Delta_i^n M' \, \Delta_i^n W - \int_{I(n,i)} \sigma'_s(W_s - W_{(i-1)\Delta_n})\, dW'_s \\ & - \int_{I(n,i)}\int_{E'} \delta'(s,z)(W_s - W_{(i-1)\Delta_n})(\mathfrak{p}'-\mathfrak{q}')(ds,dz). \end{aligned} \quad (5.3.44)$$

The last two terms above, as well as the two terms on the right of (5.3.15), are stochastic integrals with respect to the original filtration $(\mathcal{F}_t)$, and also with respect to the *augmented filtration* $(\mathcal{F}'_t)$ because, due to the independence of W' and $\mathfrak{p}'$ from W, those are an $(\mathcal{F}'_t)$-Brownian motion and an $(\mathcal{F}'_t)$-Poisson measure, respectively. It follows that all three terms on the right of (5.3.44) are $(\mathcal{F}'_t)$-martingale increments. Therefore, if $\widetilde{W}_i^n=\sup_{s\in I(n,i)}\|W_s-W_{(i-1)\Delta_n}\|$, we deduce from (2.1.34) and (2.1.37) plus Hölder's inequality, and from (5.3.16), that

$$l\le 2 \quad \Rightarrow \quad \mathbb{E}\big(\|\zeta(3)_i^n\|^l \mid \mathcal{F}'_{(i-1)\Delta_n}\big) \le K_l\,\Delta_n^{l/2}\,\big(\widetilde{W}_i^n\big)^l. \quad (5.3.45)$$

Since γ_i^n is $\mathcal{F}'_{(i-1)\Delta_n}$ measurable, we deduce

$$\mathbb{E}\big(\|\zeta(3)_i^n\|^l\,(\gamma_i^n)^m\big)\le K_l\,\Delta_n^{l/2}\,\mathbb{E}\big((\gamma_i^n)^m\,(\widetilde{W}_i^n)^l\big).$$

Finally we have $\mathbb{E}((\widetilde{W}_i^n)^p)\le K_p\Delta_n^{p/2}$ for all $p>0$, so another application of Hölder's inequality and (5.3.39) gives (5.3.43) for $j=3$. □

Step 3) The aim of this step is to prove the following lemma.

Lemma 5.3.16 *We have* $D^n(j)\stackrel{\text{u.c.p.}}{\Longrightarrow}0$ *for* $j=4,5,6,7$.

Proof 1) We begin with a consequence of Hölder's inequality and (5.3.35): if $v\in(0,1]$,

$$\begin{aligned}\Delta_n^{1-v}\sum_{i=1}^{[t/\Delta_n]}\big(\alpha_i^n\big)^v &\le t^{1-v}\Bigg(\sum_{i=1}^{[t/\Delta_n]}\alpha_i^n\Bigg)^v\\ &\le Kt\Delta_n^v+Kt^{1-v}\Bigg(\mathbb{E}\bigg(\int_0^t\big(\|b_s-b_{\Delta_n[s/\Delta_n]}\|^2+\|\widetilde{\sigma}_s-\widetilde{\sigma}_{\Delta_n[s/\Delta_n]}\|^2\big)\,ds\bigg)\Bigg)^v.\end{aligned}$$

Then, since b' and $\widetilde{\sigma}$ are bounded and càdlàg or càglàd, we deduce from Lebesgue's Theorem that

$$0<v\le 1\quad\Rightarrow\quad\Delta_n^{1-v}\sum_{i=1}^{[t/\Delta_n]}\big(\alpha_i^n\big)^v\ \to\ 0. \tag{5.3.46}$$

In view of (5.3.34), and since $\mathbb{E}(|\delta(j)_i^n|)\le\mathbb{E}(|\delta'(j)_i^n|)$, it is enough to prove that for any $t>0$ and $j=4,5,6,7$ we have

$$\sum_{i=1}^{[t/\Delta_n]}\mathbb{E}\big(\big|\delta'(j)_i^n\big|\big)\ \to\ 0. \tag{5.3.47}$$

2) Consider $j=4$. We deduce from (5.3.21) that

$$\big|\delta'(4)_i^n\big|\ \le\ K\Phi_i^n\,\big\|\zeta(4)\big\|\,\big(\gamma_i^n\big)^{1-w},\quad\text{where }\ \Phi_i^n=1+\big\|\beta_i^n\big\|^p, \tag{5.3.48}$$

and where γ_i^n is given by (5.3.38). By (5.3.31), Φ_i^n satisfies (5.3.40). Then (5.3.42) with $l=1$ and $m=1-w$ yields

$$\sum_{i=1}^{[t/\Delta_n]}\mathbb{E}\big(\big|\delta'(4)_i^n\big|\big)\ \le\ K\Delta_n^{1-u}\sum_{i=1}^{[t/\Delta_n]}\big(\alpha_i^n\big)^u,$$

where $u\in(0,1)$. Then (5.3.47) for $j=4$ follows from (5.3.46).

3) Next we prove the result for $j=5,6,7$ in case (a). When $j=5,7$ there is nothing to prove, since then $A_i^n=\emptyset$ and thus $\delta'(j)_i^n=0$. For $j=6$ we use the first

part of (5.3.22) and recall that the random variable u_i^n in the definition of $\delta'(6)_i^n$ is $[0,1]$-valued, to deduce

$$|\delta'(6)_i^n| \leq \sqrt{\Delta_n}\Big(\phi'_C(\varepsilon)\|\theta_i^n\| + \frac{K}{C}\big(\|\beta_i^n\|^{p+1} + \|\theta_i^n\|^{p+1}\big)\|\theta_i^n\| + \frac{KC^p}{\varepsilon}\,\|\theta_i^n\|^2\Big).$$

Then we deduce from (5.3.31) and an application of Hölder's inequality that

$$\mathbb{E}\big(|\delta'(6)_i^n|\big) \leq K\Delta_n\phi'_C(\varepsilon) + \frac{K}{C}\,\Delta_n + \frac{KC^p}{\varepsilon}\,\Delta_n^{3/2}. \tag{5.3.49}$$

Thus the $\limsup$ of the left side of (5.3.47) for $j=6$ is smaller than $Kt\phi'_C(\varepsilon) + Kt/C$, which can be made arbitrarily small by choosing first C large and then ε small. Then we have the result.

4) Now we turn to $j=5,6,7$ in case (b). By (5.3.21) and (5.3.22), and recalling that $u_i^n \in [0,1]$, we have

$$|\delta'(j)_i^n| \leq \begin{cases} K\sqrt{\Delta_n}\,\big(1+\|\beta_i^n\|^p + \|\theta_i^n\|^p\big)\,\|\theta_i^n\|\,\phi_B(\beta_i^n)^{1-w}\,1_{A_i^n} & \text{if } j=5 \\ K\sqrt{\Delta_n}\,\big(1+\|\beta_i^n\|^p + \|\theta_i^n\|^p\big)\,\|\theta_i^n\|^2\,\phi_B(\beta_i^n)^{2-w}\,1_{(A_i^n)^c} & \text{if } j=6 \\ K\sqrt{\Delta_n}\,\big(1+\|\beta_i^n\|^p + \|\theta_i^n\|^p\big)\,\|\theta_i^n\|1_{A_i^n} & \text{if } j=7. \end{cases}$$

On the set A_i^n we have $\|\theta_i^n\|\phi_B(\beta_i^n) \geq 1/2$, whereas on the complement $(A_i^n)^c$ we have $\|\theta_i^n\|\phi_B(\beta_i^n) \leq \|\theta_i^n\| + 1/2$. Since $\frac{w}{2} < 1 - \frac{w}{2}$ and $\phi_B \geq 1$, we see that in all three cases,

$$|\delta'(j)_i^n| \leq K\sqrt{\Delta_n}\,\big(1+\|\beta_i^n\|^{p+1} + \|\theta_i^n\|^{p+1}\big)\,\|\theta_i^n\|^{1+w/2}\,\phi_B(\beta_i^n)^{1-w/2}. \tag{5.3.50}$$

Note that $1+\|\beta_i^n\|^{p+1} + \|\theta_i^n\|^{p+1}$ satisfies (5.3.40) by (5.3.31), and $\sqrt{\Delta_n}\,\theta_i^n = \sum_{j=1}^4 \zeta(j)_i^n$ and $\phi_B(\beta_i^n) = \gamma_i^n$ here. Then (5.3.41) and (5.3.42) applied with $m = 1-w/2$ and $l = 1+w/2$ yield $\mathbb{E}(|\delta'(j)_i^n|) \leq K\Delta_n^{1+w/4}$. Thus (5.3.47) holds, and the proof is complete. □

Step 4) In view of Lemma 5.3.16 and of (5.3.33), the property (5.3.25) will follow from the next three lemmas. Note that, because of Lemma 5.3.13, this will end the proof of both Theorems 5.3.5 and 5.3.6.

Lemma 5.3.17 *We have* $D^n(3) \stackrel{\text{u.c.p.}}{\Longrightarrow} 0$.

Proof As seen in the proof of Lemma 5.3.15, see (5.3.44), $\zeta(3)_i^n$ is a martingale increment over the interval $((i-1)\Delta_n, i\Delta_n]$, relative to the augmented filtration $(\mathcal{F}'_t)$ (which depends on (n,i)), whereas β_i^n is $\mathcal{F}'_{(i-1)\Delta_n}$ measurable. Henceforth by successive conditioning we have $\mathbb{E}(\nabla f(\beta_i^n)\,\zeta(j)_i^n \mid \mathcal{F}_{(i-1)\Delta_n}) = 0$, and thus $D^n(3)$ is identically 0. □

Lemma 5.3.18 *We have* $D^n(1) \stackrel{\text{u.c.p.}}{\Longrightarrow} \overline{A}(f, X)$.

Proof Since

$$\mathbb{E}\big(\partial_j f(\beta_i^n)^l\, \zeta(1)_i^{n,j} \mid \mathcal{F}_{(i-1)\Delta_n}\big) = \Delta_n\, b^j_{(i-1)\Delta_n}\, \rho_{c_{(i-1)\Delta_n}}\big(\partial_j f^l\big), \tag{5.3.51}$$

and all $s \mapsto \rho_{c_s}(\partial_j f^l)$ are càdlàg the result follows from Riemann integration. □

Lemma 5.3.19 *We have* $D^n(2) \overset{\text{u.c.p.}}{\Longrightarrow} \overline{A}'(f, X)$.

Proof By the scaling property of the Brownian motion and (5.3.4), we have

$$\mathbb{E}\big(\partial_j f(\beta_i^n)^l\, \zeta(2)_i^{n,j} \mid \mathcal{F}_{(i-1)\Delta_n}\big) = \Delta_n \sum_{k,m=1}^{r} \widetilde{\sigma}^{jkm}_{(i-1)\Delta_n}\, \widehat{\gamma}'_{\sigma_{(i-1)\Delta_n}}\big(\partial_j f^l\big)^{mk}. \tag{5.3.52}$$

Moreover all $s \mapsto \widehat{\gamma}'_{\sigma_s}(\partial_j f^l)$ are càdlàg, so again Riemann integration yields the result. □

5.4 The Central Limit Theorem for Quadratic Variation

So far, and for a power function $f(x) = |x|^p$ in the one-dimensional case, say, we have two Central Limit Theorems for the processes $V^n(f, X)$:

- Theorem 5.1.2, which holds for $p > 3$ and requires the normalizing factor $\Delta_n^{-1/2}$ and for which the limit $Z(f, X)$ essentially depends on the jumps of X, and in particular vanishes if there is no jump.
- Theorem 5.3.5, which applies for such an f when X is continuous only, and which is also a CLT for the process $V^n(f, X) = \Delta_n^{p/2-1} V'^n(f, X)$, with the normalizing factor $\Delta_n^{1/2-p/2}$ in front of $V^n(f, X)$.

These two normalizing factors coincide when $p = 2$, although in this case neither one of the two theorems applies when X jumps. Nevertheless, we have a CLT for this case, even in the discontinuous case. Although there is a result available for all test functions which are "close enough to quadratic" near the origin, we give the result for the quadratic variation only, that is for the test function f on $\mathbb{R}^d$ with components $f^{ij}(x) = x^i x^j$.

We start again with a d-dimensional semimartingale. Recall that $[X, X]$ is the $\mathcal{M}^+_{d\times d}$-valued process whose components are the quadratic covariation processes $[X^j, X^k]$. The "approximate" quadratic variation associated with the stepsize Δ_n is

$$[X, X]^n = \big([X^j, X^k]^n\big)_{1\le j,k\le d}, \quad \text{where } \big[X^j, X^k\big]^n_t = \sum_{i=1}^{[t/\Delta_n]} \Delta_i^n X^j\, \Delta_i^n X^k.$$

We know that $[X, X]^n \overset{\mathbb{P}}{\Longrightarrow} [X, X]$, see e.g. (B) of Theorem 3.3.1, and with no assumption whatsoever on the semimartingale X. The associated CLT needs the following assumption, which is significantly weaker than (H) when X is continuous.

Assumption 5.4.1 (or **(Q)**) X is an Itô semimartingale of the form (4.4.1), with:

(i) If X is continuous, then $\int_0^t (\|b_s\|^2 + \|c_s\|^2)\,ds < \infty$ for all $t > 0$.
(ii) If X has jumps, then it satisfies (H), that is Assumption 4.4.2.

The CLT is about the convergence of the normalized and discretized processes

$$\overline{Z}_t^n = \frac{1}{\sqrt{\Delta_n}}\big([X,X]_t^n - [X,X]_{\Delta_n[t/\Delta_n]}\big). \tag{5.4.1}$$

As said before, one has $[X,X]^n = V^n(f,X) = V'^n(f,X)$, for the d^2-dimensional globally even test function f with components $f^{ij}(x) = x^i x^j$. It is thus not a surprise that the limiting process of $\overline{Z}^n$ is

$$\overline{Z} = \overline{V}(f,X) + \overline{V}'(f,X) \tag{5.4.2}$$

for this particular test function, where $\overline{V}(f,X)$ given by (5.1.4) accounts for the jump part, and $\overline{V}'(f,X)$ (as defined in Theorem 5.2.2) accounts for the continuous part.

However, we need these two limiting processes to be defined together on the same extension. For this we do as in Proposition 4.1.5. We choose a progressively measurable "square-root" $\widehat{\sigma}_s$ of the $\mathcal{M}^+_{d^2\times d^2}$-valued process $\widehat{c}_s$ with entries $\widehat{c}_s^{ij,kl} = c_s^{ik} c_s^{jl}$, so the matrix with entries $\frac{1}{\sqrt{2}}(\widehat{\sigma}_s^{ij,kl} + \widehat{\sigma}_s^{ji,kl})$ is a square-root of the matrix with entries $\widehat{c}_s^{ij,kl} + \widehat{c}_s^{il,jk}$ (due to the symmetry of the matrix c_s). We also choose a weakly exhausting sequence (T_n) for the jumps of X, see before (5.1.2), with $T_n = \infty$ for all n when X is continuous.

Next, let $(\Omega', \mathcal{F}', \mathbb{P}')$ be an auxiliary space supporting a triple sequence $(\Psi_{n-}, \Psi_{n+}, \kappa_n)$ of variables, all independent and satisfying (5.1.1), and a d^2-dimensional Brownian motion W' independent of the above sequence. The very good filtered extension $(\widetilde{\Omega}, \widetilde{\mathcal{F}}, (\widetilde{\mathcal{F}}_t)_{t\ge 0}, \widetilde{\mathbb{P}})$ is defined by (5.1.2), except that we additionally require W' to be $(\widetilde{\mathcal{F}}_t)$-adapted. Then when X has jumps, so σ_t is càdlàg, we define R_n by (5.1.3), and in all cases the limiting process will have the components:

$$\overline{Z}_t^{ij} = \frac{1}{\sqrt{2}} \sum_{k,l=1}^d \int_0^t \big(\widehat{\sigma}_s^{ij,kl} + \widehat{\sigma}_s^{ji,kl}\big)\,dW_s'^{kl} + \sum_{p=1}^\infty \big(\Delta X_{T_p}^j R_p^i + \Delta X_{T_p}^i R_p^j\big) 1_{\{T_p \le t\}}. \tag{5.4.3}$$

When X is continuous the last sum above is absent, and otherwise it makes sense by Proposition 5.1.1, and it is indeed $\overline{V}(f^{ij},X)$, with f^{ij} as above. The stochastic integral in (5.4.3) is the $(ij)^{\text{th}}$ component of $\overline{V}'(f,X)$, as characterized in Theorem 5.2.2, because $\overline{\gamma}_{\sigma_s}(f^{ij}, f^{kl}) = \widehat{c}_s^{ij,kl} + \widehat{c}_s^{ji,kl}$ by an elementary computation on Gaussian vectors. Therefore we have (5.4.2), where conditionally on $\mathcal{F}$ the two processes $\overline{V}(f,X)$ and $\overline{V}'(f,X)$ are independent, centered, with independent increments, and moreover

$$\begin{aligned}\mathbb{E}\big(\overline{Z}_t^{ij}\,\overline{Z}_t^{kl}\mid\mathcal{F}\big) &= \frac{1}{2}\sum_{s\le t}\big(\Delta X_s^i\Delta X_s^k\big(c_{s-}^{jl}+c_s^{jl}\big)+\Delta X_s^i\Delta X_s^l\big(c_{s-}^{jk}+c_s^{jk}\big)\\ &\quad+\Delta X_s^j\Delta X_s^k\big(c_{s-}^{il}+c_s^{il}\big)+\Delta X_s^j\Delta X_s^l\big(c_{s-}^{ik}+c_s^{ik}\big)\big)\\ &\quad+\int_0^t\big(c_s^{ik}c_s^{jl}+c_s^{il}c_s^{jk}\big)\,ds \end{aligned} \tag{5.4.4}$$

(use (5.1.5) and (5.2.8); again, when X is continuous, σ_{s-} does not necessarily exist under (Q), but only the last integral above appears in the formula). The CLT is as follows.

Theorem 5.4.2 *Let X be an Itô semimartingale satisfying Assumption (Q). Then the $d\times d$-dimensional processes $\overline{Z}^n$ defined in* (5.4.1) *converge stably in law to a process $\overline{Z}=(\overline{Z}^{ij})_{1\le i,j\le d}$, defined on a very good filtered extension $(\widetilde{\Omega},\widetilde{\mathcal{F}},(\widetilde{\mathcal{F}}_t)_{t\ge0},\widetilde{\mathbb{P}})$ of $(\Omega,\mathcal{F},(\mathcal{F}_t)_{t\ge0},\mathbb{P})$ and which, conditionally on $\mathcal{F}$, is centered with independent increments and finite second moments given by* (5.4.4). *This process $\overline{Z}$ can be realized as* (5.4.3), *and it is $\mathcal{F}$-conditionally Gaussian if further the processes X and σ have no common jumps.*

Moreover, the same is true of the processes $\frac{1}{\sqrt{\Delta_n}}([X,X]^n-[X,X])$ when X is continuous, and otherwise for each t we have the following stable convergence of variables

$$\frac{1}{\sqrt{\Delta_n}}\big([X,X]_t^n-[X,X]_t\big)\ \stackrel{\mathcal{L}\text{-}s}{\longrightarrow}\ \overline{Z}_t. \tag{5.4.5}$$

Remark 5.4.3 Exactly as in Remark 5.1.3, and for the same reason, when X jumps the processes $\frac{1}{\sqrt{\Delta_n}}([X,X]^n-[X,X])$ do *not* converge in law for the Skorokhod topology.

Remark 5.4.4 This result is surprising by its assumptions, when compared with Theorem 5.3.5 in the (simple) case X is continuous: the results are formally the same, but the assumptions are deeply different, since (K) is *much* stronger than (Q). This is of course due to the very special properties of the quadratic test function $f(x)^{ij}=x^ix^j$.

Remark 5.4.5 Assumption (Q) is obviously not the weakest possible assumption, at least when X is discontinuous. What is really needed is that (i) of (Q) holds and that σ is càdlàg *at each jump time of* X. However, unless X has locally finitely many jumps, this latter property is "almost" the same as σ being càdlàg everywhere.

The ijth and jith components of the process (5.4.1) are the same, as are $\overline{Z}^{ij}$ and $\overline{Z}^{ji}$ in (5.4.3). By Itô's formula,

$$\overline{Z}^{n,ij} = Z_t^{n,ij}+Z_t^{n,ji}, \tag{5.4.6}$$

where

$$Z_t^{n,jk} = \frac{1}{\sqrt{\Delta_n}} \sum_{i=1}^{[t/\Delta_n]} \int_{I(n,i)} \left(X_{s-}^j - X_{(i-1)\Delta_n}^j\right) dX_s^k. \tag{5.4.7}$$

Therefore it is natural to give also a CLT for the process $Z^n = (Z^{n,ij})_{1\le i,j\le d}$, and in fact the previous theorem will be a simple consequence of the following one (as for Theorem 5.1.2 we give a joint convergence with the discretized process $X^{(n)}$). The limiting process can be defined on the same very good filtered extension described before (5.4.3), and with the notation $R_{n\pm}$ of (5.1.3), as

$$Z_t^{ij} = \frac{1}{\sqrt{2}} \sum_{k,l=1}^{d} \int_0^t \widehat{\sigma}_s^{ij,kl}\, dW_s^{\prime kl} + \sum_{p=1}^{\infty} \left(\Delta X_{T_p}^j R_{p-}^i + \Delta X_{T_p}^i R_{p+}^j\right) 1_{\{T_n \le t\}}. \tag{5.4.8}$$

The process Z has the same properties as $\overline{Z}$, except that (5.4.4) is replaced by

$$\mathbb{E}\left(Z_t^{ij} Z_t^{kl} \mid \mathcal{F}\right) = \frac{1}{2} \sum_{s\le t} \left(\Delta X_s^i\, \Delta X_s^k\, c_s^{jl} + \Delta X_s^j\, \Delta X_s^l\, c_{s-}^{ik}\right) + \frac{1}{2} \int_0^t c_s^{ik}\, c_s^{jl}\, ds. \tag{5.4.9}$$

Theorem 5.4.6 *Let X be an Itô semimartingale satisfying Assumption (Q). Then the $(d+d^2)$-dimensional processes $(X^{(n)}, Z^n)$ with Z^n given by (5.4.7) converge stably in law to (X, Z), where the process $Z = (Z^{ij})_{1\le i,j\le d}$ is defined on a very good filtered extension $(\widetilde{\Omega}, \widetilde{\mathcal{F}}, (\widetilde{\mathcal{F}}_t)_{t\ge0}, \widetilde{\mathbb{P}})$ of $(\Omega, \mathcal{F}, (\mathcal{F}_t)_{t\ge0}, \mathbb{P})$ and, conditionally on $\mathcal{F}$, is centered with independent increments and finite second moments given by (5.4.9). This process Z can be realized as (5.4.8), and it is $\mathcal{F}$-conditionally Gaussian if further the processes X and σ have no common jumps.*

We will first prove Theorem 5.4.6 and then deduce Theorem 5.4.2. The proof proceeds by several lemmas, which basically establish the result for bigger and bigger classes of processes X.

First we establish some notation: The processes Z^n and Z depend on X, and we make this explicit by writing $Z^n(X)$ and $\overline{Z}(X)$. For any two real semimartingales Y, Y' we set

$$\zeta\left(Y, Y'\right)_i^n = \frac{1}{\sqrt{\Delta_n}} \int_{I(n,i)} (Y_{s-} - Y_{(i-1)\Delta_n})\, dY_s'.$$

By a localization argument similar to Lemma 4.4.9, and which we omit, we can and will replace (Q) by the following strengthened assumption.

Assumption 5.4.7 (or (**SQ**)) We have (Q) and

(i) if X is continuous, then $\int_0^\infty (\|b_s\|^2 + \|c_s\|^2)\, ds \le A$ for some constant A;
(ii) if X is discontinuous, then (SH) (that is, Assumption 4.4.6) holds.

As usual the $\mathcal{F}$-conditional distribution of the limiting process $Z(X)$ does not depend on the choice of the exhausting sequence (T_p). So below we choose the sequence (S_p) described before (5.1.9), and $R_{p\pm}$ is associated by (5.1.3) with $T_p = S_p$. In connection with this, and on the extended space described above, we define the d^2-dimensional variables η_p with components:

$$\eta_p^{jk} = R_{p-}^j \Delta X_{S_p}^k + R_{p+}^k \Delta X_{S_p}^j. \tag{5.4.10}$$

We start with an auxiliary result in the continuous case. For $l, r = 1, \dots, d'$ and $j, k = 1, \dots, d$ we set

$$\begin{aligned} \xi_i^{n,lr} &= \frac{1}{\sqrt{\Delta_n}} \int_{I(n,i)} \big(W_{s-}^l - W_{(i-1)\Delta_n}^l\big)\, dW_s^r \\ \zeta_i^{\prime n,jk} &= \sum_{l,r=1}^{d'} \sigma_{(i-1)\Delta_n}^{jl} \sigma_{(i-1)\Delta_n}^{kr} \xi_i^{n,lr} \\ \rho_i^{n,jk} &= \zeta\big(X^j, X^k\big)_i^n - \zeta_i^{\prime n,jk}. \end{aligned} \tag{5.4.11}$$

Lemma 5.4.8 *Assume that X is continuous, with σ càdlàg bounded and b bounded and piecewise constant, in the sense that*

$$b_s = \sum_{q \geq 0} b_{t_q} 1_{[t_q, t_{q+1})}(t) \tag{5.4.12}$$

for a sequence t_q of (deterministic) times increasing strictly to $+\infty$. Then for all $j, k = 1, \dots, d$ the array $(\rho_i^{n,jk})$ is asymptotically negligible (or, AN: see (2.2.33)).

Proof We have $\rho_i^{n,jk} = \sum_{m=1}^4 \zeta(m)_i^{n,jk}$, where

$$\begin{aligned} \zeta(1)_i^{n,jk} &= \zeta\big(B^j, B^k\big)_i^n, & \zeta(2)_i^{n,jk} &= \zeta\big(X^{c,j}, B^k\big)_i^n \\ \zeta(3)_i^{n,jk} &= \zeta\big(B^j, X^{c,k}\big)_i^n, & \zeta(4)_i^{n,jk} &= \zeta\big(X^{c,j}, X^{c,k}\big)_i^n - \zeta_i^{\prime n,jk} \end{aligned}$$

(recall $B_t = \int_0^t b_s\, ds$ and $X_t^c = \int_0^t \sigma_s\, dW_s$.) Thus it is enough to prove the AN property of the array $\zeta(m)_i^{n,jk}$ for each $m = 1, 2, 3, 4$.

We have $|\zeta(1)_i^{n,jk}| \leq K \Delta_n^{3/2}$, so the AN property for $m = 1$ readily follows. Next, $\mathbb{E}((\zeta(m)_i^{n,jk})^2) \leq K \Delta_n^2$ for $m = 2, 3$, and also

$$E\big(\zeta(m)_i^{n,jk} \mid \mathcal{F}_{(i-1)\Delta_n}\big) = 0 \tag{5.4.13}$$

for all i when $m = 3$: then (2.2.35) yields the AN property for $m = 3$. When $m = 2$ the equality (5.4.13) holds for all i such that the interval $I(n,i)$ contains no times t_p, because on these intervals $b_t = b_{(i-1)\Delta_n}$. That is, (5.4.13) holds for all $i \leq [t/\Delta_n]$ except at most N_t of them, where $N_t = \sup(p : t_p \leq t)$, and we then deduce from

$\mathbb{E}((\zeta(2)_i^{n,jk})^2) \le K\Delta_n^2$ that

$$\mathbb{E}\left(\left(\sum_{i=1}^{[t/\Delta_n]} \zeta(2)_i^{n,jk}\right)^2\right) \le K\Delta_n t(1+N_t),$$

thus obtaining the AN property for $m=2$. Finally, we have

$$\zeta(4)_i^{n,jk} = \frac{1}{\sqrt{\Delta_n}} \int_{I(n,i)} \left(\int_{(i-1)\Delta_n}^t \sum_{l=1}^{d'} (\sigma_s^{jl} - \sigma_{(i-1)\Delta_n}^{jl})\, dW_s^l\right) dX_t^{c,k}$$
$$+ \frac{1}{\sqrt{\Delta_n}} \int_{I(n,i)} \sum_{l,r=1}^{d'} \sigma_{(i-1)\Delta_n}^{jl} \left(W_t^l - W_{(i-1)\Delta_n}^l\right) \left(\sigma_s^{kr} - \sigma_{(i-1)\Delta_n}^{kr}\right) dW_s^r.$$

By (2.1.34) applied repeatedly, plus the Cauchy-Schwarz inequality, we deduce

$$\mathbb{E}\left((\zeta(4)_i^{n,jk})^2\right) \le K\sqrt{\Delta_n}\left(\mathbb{E}\left(\int_{I(n,i)} \|\sigma_t - \sigma_{(i-1)\Delta_n}\|^4\, dt\right)\right)^{1/2},$$

which in turn implies

$$\sum_{i=1}^{[t/\Delta_n]} \mathbb{E}\left((\zeta(4)_i^{n,jk})^2\right) \le K\sqrt{t}\left(\mathbb{E}\left(\int_0^t \|\sigma_s - \sigma_{\Delta_n[s/\Delta_n]}\|^4\, ds\right)\right)^{1.2}.$$

Since σ is right-continuous and bounded, the above goes to 0 by the dominated convergence theorem. Since (5.4.13) holds for $m=4$, the AN property for $m=4$ follows from (2.2.35) again. □

Lemma 5.4.9 *Assume that X is continuous, with σ càdlàg bounded and $\int_0^\infty \|b_s\|^2\, ds \le A$ for a constant A. Then for all $j,k=1,\dots,d$ the array $(\rho_i^{n,jk})$ is asymptotically negligible.*

Proof Exactly as in Lemma 3.4.8, there is a sequence $b(p)$ of adapted processes, all bounded and of the form (5.4.12), and such that

$$\mathbb{E}\left(\int_0^\infty \|b(p)_s\|^2\right) \le 2A, \qquad \alpha(p) =: \mathbb{E}\left(\int_0^\infty \|b(p)_s - b_s\|^2\, ds\right) \to 0 \text{ as } p\to\infty. \tag{5.4.14}$$

Recall $X = X_0 + B + X^c$ with $B_t = \int_0^t b_s\, ds$ and $X_t^c = \int_0^t \sigma_s\, dW_s$, and set

$$B(p)_t = \int_0^t b(p)_s\, ds, \qquad X(p) = X_0 + B(p) + X^c, \qquad \overline{B}(p) = B - B(p).$$

The previous lemma applies to each process $X(p)$, and the variables $\zeta_i^{\prime n,jk}$ are the same for X and $X(p)$. Therefore it is enough to show that, for any T,

$$\lim_{p\to\infty} \sup_n \mathbb{E}\left(\sum_{i=1}^{[t/\Delta_n]} \left|\zeta(X^j, X^k)_i^n - \zeta(X(p)^j, X(p)^k)_i^n\right|\right) \to 0. \tag{5.4.15}$$

We have

$$\begin{aligned}
\zeta(X^j, X^k)_i^n - \zeta(X(p)^j, X(p)^k)_i^n &= \frac{1}{\sqrt{\Delta_n}} \sum_{r=1}^{3} \chi(p,r)_i^n, \quad \text{where} \\
\chi(p,1)_i^n &= \int_{I(n,i)} \left(\overline{B}(p)_s^j - \overline{B}(p)_{(i-1)\Delta_n}^j\right) b_s^k\, ds \\
\chi(p,2)_i^n &= \int_{I(n,i)} \left(X(p)_s^j - X(p)_{(i-1)\Delta_n}^j\right)\left(b_s^k - b(p)_s^k\right) ds \\
\chi(p,3)_i^n &= \sum_{l=1}^{d'} \int_{I(n,i)} \left(\overline{B}(p)_s^j - \overline{B}(p)_{(i-1)\Delta_n}^j\right) \sigma_s^{kl}\, dW_s^l,
\end{aligned}$$

and (5.4.15) will follow if we prove that for all $t > 0$ and $r = 1, 2, 3$, and as $p \to \infty$:

$$\sup_n \frac{1}{\sqrt{\Delta_n}} \mathbb{E}\left(\sum_{i=1}^{[t/\Delta_n]} \left|\chi(p,r)_i^n\right|\right) \to 0. \tag{5.4.16}$$

Recalling (5.4.14) and the boundedness of σ, we deduce from (2.1.33) and (2.1.34) that

$$\mathbb{E}\Big(\sup_{t\in I(n,i)} \left|X(p)_t^j - X(p)_{(i-1)\Delta_n}^j\right|^2\Big) \le K\Delta_n.$$

Therefore, with the notation $\alpha(p)_i^n = \mathbb{E}(\int_{I(n,i)} \|b_s - b(p)_s\|^2\, ds)$, further applications of (2.1.33) and (2.1.34) and the Cauchy-Schwarz inequality yield

$$\mathbb{E}\left(\left|\chi(p,r)_i^n\right|\right) \le \begin{cases} K\Delta_n \sqrt{\alpha(p)_i^n} & \text{if } r = 1,2 \\ K\Delta_n\, \alpha(p)_i^n & \text{if } r = 3. \end{cases}$$

Since $\sum_{i\ge 1} \alpha(p)_i^n \le \alpha(p)$ and $\sum_{i=1}^{[t/\Delta_n]} \sqrt{\alpha(p)_i^n} \le \sqrt{t\alpha(p)/\Delta_n}$, we deduce (5.4.16). □

Lemma 5.4.10 *Assume that X has the form*

$$X_t = X_0 + \int_0^t b_s\, ds + \int_0^t \sigma_s\, dW_s + \delta \star \mathfrak{p}_t, \tag{5.4.17}$$

where b and σ are as in the previous lemma, and where δ is as in (5.1.8), *with further*

$$\delta(\omega,t,z)\neq 0 \quad\Rightarrow\quad \Gamma(z)>\frac{1}{p} \tag{5.4.18}$$

for some integer $p\geq 1$. *Then* $(X^{(n)},Z^n(X))\stackrel{\mathcal{L}\text{-}s}{\Longrightarrow}(X,Z(X))$.

Proof 1) Below, we write $X'=\delta * \mathfrak{p}$ and $X''=X-X'$. We also use the notation (5.1.9), including $\mathcal{P}_m$ and the stopping times S_p. Recalling (5.4.10), we also set

$$\begin{aligned}
&\eta(n,p)^{jk}=R_-(n,p)^j\,\Delta X^k_{S_p}+R_+(n,p)^k\,\Delta X^j_{S_p}\\
&\overline{\eta}(n,p)=\big(\Delta X_{S_p},\eta(n,p)\big),\qquad \overline{\eta}_p=(\Delta X,\eta_p)\\
&\Theta^n_t=\textstyle\sum_{p:S_p\leq\Delta_n[t/\Delta_n]}\eta(n,p),\quad \overline{\Theta}^n_t=\sum_{p:S_p\leq\Delta_n[t/\Delta_n]}\overline{\eta}(n,p)\\
&\overline{\Theta}_t=\textstyle\sum_{p:S_p\leq t}\overline{\eta}_p
\end{aligned} \tag{5.4.19}$$

($\overline{\eta}(n,p)=\overline{\eta}_p=0$ if $p\notin\mathcal{P}_m$ because of (5.4.18), so the sums above are finite sums). So $\overline{\Theta}^n_t$ and $\overline{\Theta}_t$ are $(d+d^2)$-dimensional, and the first d components of $\overline{\Theta}^n$ are the discretized process $X'^{(n)}$, and $\overline{\Theta}=(X',Z(X'))$. Finally, $Z^{n,jk}(X'')=\sum_{i=1}^{[t/\Delta_n]}(\zeta(B^j,B^k)^n_i+\zeta(X^{c,j},B^k)^n_i+\zeta(B^j,X^{c,k})^n_i+\zeta(X^{c,j},X^{c,k})^n_i)$, so we deduce from the previous lemma that, with the notation (5.4.11) and $\zeta'^n_i=(\zeta'^{n,jk}_i)_{1\leq j,k\leq d}$,

$$Z^n(X'')-Z'^n\stackrel{\text{u.c.p.}}{\Longrightarrow}0,\quad\text{where } Z'^n_t=\sum_{i=1}^{[t/\Delta_n]}\zeta'^n_i. \tag{5.4.20}$$

2) Let us now consider the setting of Theorem 4.2.1, with $u_n=\Delta_n$ and $q=d^2$ and $q'=d'^2$. We choose a d'^2-dimensional $\mathcal{F}^W_1$-measurable function Φ on the canonical Wiener space Ω^W such that

$$\Phi^{lm}=\int_0^1 W^l_s\,dW^m_s\qquad \mathbb{P}^W\text{-almost surely}$$

(with $\mathbb{P}^W$ the Wiener measure), so (4.2.3) is satisfied. Finally we take θ with components $\theta^{jk,lm}=\sigma^{jl}\sigma^{km}$, and $G(x,y)=y$ for $y\in\mathbb{R}^{d^2}$ (so the process Y does not show). Then, with the notation $\overline{U}^n(G)$ and $\overline{U}(G)$ of (4.2.6) and (4.2.7), Proposition 4.4.10 implies

$$\big(\overline{U}^n(G),(\overline{\eta}(n,p))_{p\geq 1}\big)\stackrel{\mathcal{L}\text{-}s}{\longrightarrow}\big(\overline{U}(G),(\overline{\eta}_p)_{p\geq 1}\big). \tag{5.4.21}$$

The quantities in (4.2.5) are easily computed and write as follows:

$$\gamma^\Phi_\alpha(x,G)=0,\qquad \widehat{\gamma}^\Phi_\alpha(x,G)=0,\qquad \overline{\gamma}^\Phi_\alpha\big(x,G^{jk},G^{j'k'}\big)=\frac{1}{2}\sum_{l,r=1}^{d'}\alpha^{jk,lr}\alpha^{j'k',lr}$$

for any $d^2 \times d'^2$ matrix α. This has two consequences: first, that $\overline{U}^n(G) = Z'^n$, as defined by (5.4.20); second, that $\overline{\gamma}^{\Phi}_{\theta_s}(x, G^{jk}, G^{j'k'}) = \frac{1}{2}\, c_s^{jk}\, c_s^{j'k'}$, which implies that $\overline{U}(G)$ is a version of the process $Z(X^c)$ associated with the continuous process X^c by (5.4.9). Thus (5.4.21) is the same as

$$\bigl(Z'^n, (\overline{\eta}(n,p))_{p\geq 1}\bigr) \stackrel{\mathcal{L}\text{-s}}{\longrightarrow} \bigl(Z(X^c), (\overline{\eta}_p)_{p\geq 1}\bigr), \tag{5.4.22}$$

where the convergence takes place in the space $\mathbb{D}^{d^2} \times (\mathbb{R}^d)^{\mathbb{N}^*}$. Using (5.4.20) and $X''^{(n)} \stackrel{\mathbb{P}}{\Longrightarrow} X''$ and the continuity of X'', we deduce (recall (2.2.5) and (2.2.9)):

$$\bigl(\bigl(X''^{(n)}, Z^n(X'')\bigr), (\overline{\eta}(n,p))_{p\geq 1}\bigr) \stackrel{\mathcal{L}\text{-s}}{\longrightarrow} \bigl(\bigl(X'', Z(X^c)\bigr), (\overline{\eta}_p)_{p\geq 1}\bigr)$$

in $\mathbb{D}^{d+d^2} \times (\mathbb{R}^d)^{\mathbb{N}^*}$. Since the set $\{S_p : p \in \mathcal{P}_m\}$ is locally finite, this yields

$$\bigl(\bigl(X''^{(n)}, Z^n(X'')\bigr), \overline{\Theta}^n\bigr) \stackrel{\mathcal{L}\text{-s}}{\Longrightarrow} \bigl(\bigl(X'', Z(X^c)\bigr), \overline{\Theta}\bigr)$$

for the product topology on $\mathbb{D}^{d+d^2} \times \mathbb{D}^{d+d^2}$. Now, $(X''^{(n)}, Z^n(X'')) + \overline{\Theta}^n = (X^{(n)}, Z^n(X'') + \Theta^n)$ and $\overline{\Theta} = (X', Z(X'))$ and $(X', Z(X^c))$ is continuous, so another application of (2.2.9) allows us to deduce

$$\bigl(X^{(n)}, Z^n(X'') + \Theta^n\bigr) \stackrel{\mathcal{L}\text{-s}}{\Longrightarrow} \bigl(X'' + X', Z(X^c) + Z(X')\bigr) = \bigl(X, Z(X)\bigr). \tag{5.4.23}$$

To finish the proof, we plug $X = X' + X''$ into $\zeta(X^j, X^k)^n_i$ and develop linearly, to obtain the following property: on the set Ω^n_T on which $|S_p - S_q| > \Delta_n$ for all $p, q \in \mathcal{P}_m$ with $S_p \leq T$, we have for all $i \leq [T/\Delta_n]$:

$$\zeta(X^j, X^k)^n_i = \zeta(X''^j, X''^k)^n_i + \sum_{p\geq 1} \eta(n,p)^{jk}\, 1_{\{S_p \in I(n,i)\}}.$$

Hence

$$t \leq T \;\Rightarrow\; Z(X)^n_t = Z(X'')^n_t + \Theta^n_t \quad \text{on the set } \Omega^n_T.$$

Since $\mathbb{P}(\Omega^n_T) \to 1$, we deduce $(X^{(n)}, Z^n(X)) \stackrel{\mathcal{L}\text{-s}}{\Longrightarrow} (X, Z(X))$ from (5.4.23). □

Lemma 5.4.11 *Assume that X is continuous and* (SQ)-(i) *holds. Then* $(X^{(n)}, Z^n(X)) \stackrel{\mathcal{L}\text{-s}}{\Longrightarrow} (X, Z(X))$.

Proof The proof is somewhat similar to the proof of Lemma 5.4.9. One can find a sequence $\sigma(p)$ of adapted bounded and càdlàg processes, such that

$$\mathbb{E}\biggl(\int_0^\infty \|\sigma(p)_s\|^4\, ds\biggr) \leq 2A, \qquad \alpha'(p) := \mathbb{E}\biggl(\int_0^\infty \|\sigma(p)_s - \sigma_s\|^4\, ds\biggr) \to 0 \tag{5.4.24}$$

as $p \to \infty$. Set

$$X(p)^c_t = \int_0^t \sigma(p)_s \, dW_s, \qquad X(p) = X_0 + B + X(p)^c, \qquad \overline{X}(p) = X - X(p).$$

For each p, the processes $(X(p)^{(n)}, Z^n(X(p)))$ converge stably in law to $(X(p), Z(X(p)))$ by the previous lemma. Thus, by Proposition 2.2.4, it suffices to prove the following three properties, as $p \to \infty$:

$$\overline{X}(p) \stackrel{\text{u.c.p.}}{\Longrightarrow} 0, \quad \eta > 0,\ t > 0 \Rightarrow \sup_n \mathbb{P}\Big(\sup_{s \le t} \|\overline{X}^{(n)}_s\| > \eta\Big) \to 0 \tag{5.4.25}$$

$$\eta > 0,\ t > 0 \ \Rightarrow\ \sup_n \mathbb{P}\Big(\sup_{s \le t} \big\|Z^n(X)_s - Z^n\big(X(p)\big)_s\big\| > \eta\Big) \to 0 \tag{5.4.26}$$

$$Z\big(X(p)\big) \stackrel{\mathcal{L}\text{-s}}{\Longrightarrow} Z(X). \tag{5.4.27}$$

The proof of (5.4.27) is exactly the same as for (5.2.20) in Lemma 5.2.8. By Doob's inequality we have $\mathbb{E}(\sup_{s \le t} \|\overline{X}(p)_s\|^2) \le 4\mathbb{E}(\int_0^t \|\sigma(p)_s - \sigma_s\|^2 \, ds)$, which goes to 0 by (5.4.24). This implies the first claim in (5.4.25), which in turn implies the second one.

It remains to prove (5.4.26). For this, observe that

$$Z^n(X)^{jk}_t - Z^n\big(X(p)\big)^{jk}_t = \frac{1}{\sqrt{\Delta_n}} \sum_{r=1}^{4} \sum_{i=1}^{[t/\Delta_n]} \chi(p,r)^n_i, \quad \text{where}$$

$$\chi(p,1)^n_i = \int_{I(n,i)} \big(\overline{X}(p)^j_s - \overline{X}(p)^j_{(i-1)\Delta_n}\big) b^k_s \, ds$$

$$\chi(p,2)^n_i = \sum_{l=1}^{d'} \int_{I(n,i)} \big(B^j_s - B^j_{(i-1)\Delta_n}\big) \big(\sigma^{kl}_s - \sigma(p)^{kl}_s\big) \, dW^l_s$$

$$\chi(p,3)^n_i = \sum_{l=1}^{d'} \int_{I(n,i)} \big(X(p)^{c,j}_s - X(p)^{c,j}_{(i-1)\Delta_n}\big) \big(\sigma^{kl}_s - \sigma(p)^{kl}_s\big) \, dW^l_s$$

$$\chi(p,4)^n_i = \sum_{l=1}^{d'} \int_{I(n,i)} \big(\overline{X}(p)^j_s - \overline{X}(p)^j_{(i-1)\Delta_n}\big) \sigma^{kl}_s \, dW^l_s.$$

Moreover $\chi(p,r)^n_i$ is a martingale increment for $r = 2, 3, 4$. Therefore (5.4.26) will follow if we prove that for all $t > 0$ we have, as $p \to \infty$:

$$\sup_n \frac{1}{\Delta_n^{v/2}} \mathbb{E}\bigg(\sum_{i=1}^{[t/\Delta_n]} \big|\chi(p,r)^n_i\big|^v\bigg) \to 0 \ \text{ when } \begin{cases} \text{either } r = 1, 2 \text{ and } v = 1 \\ \text{or } r = 3, 4 \text{ and } v = 2. \end{cases} \tag{5.4.28}$$

We set

$$a_i^n = \mathbb{E}\bigg(\int_{I(n,i)} \|b_s\|^2\,ds\bigg), \qquad a'(p)_i^n = \mathbb{E}\bigg(\int_{I(n,i)} \|\sigma(p)_s\|^4\,ds\bigg)$$

$$\alpha'(p)_i^n = \mathbb{E}\bigg(\int_{I(n,i)} \|\sigma_s - \sigma(p)_s\|^4\,ds\bigg).$$

We deduce from (2.1.34) and the Cauchy-Schwarz inequality that

$$\mathbb{E}\Big(\sup_{t\in I(n,i)} \big|\overline{X}(p)_t^j - \overline{X}(p)_{(i-1)\Delta_n}^j\big|^4\Big) \le K\,\Delta_n\,\alpha'(p)_i^n$$

$$\mathbb{E}\Big(\sup_{t\in I(n,i)} \big|X(p)_t^{c,j} - X(p)_{(i-1)\Delta_n}^{c,j}\big|^4\Big) \le K\,\Delta_n\,a'(p)_i^n$$

$$\mathbb{E}\Big(\sup_{t\in I(n,i)} \big|B_t^j - B_{(i-1)\Delta_n}^j\big|^2\Big) \le K\,\Delta_n\,a_i^n.$$

Then we obtain, by a repeated use of the Cauchy-Schwarz inequality and (2.1.34) again:

$$\mathbb{E}\big(\big|\chi(p,r)_i^n\big|^v\big) \le \begin{cases} K\,\Delta_n^{3/4}\,(\alpha'(p)_i^n)^{1/4}\,(a_i^n)^{1/2} & \text{if } r=1,2 \text{ and } v=1\\ K\,\Delta_n\,(\alpha'(p)_i^n)^{1/2}\,(a'(p)_i^n)^{1/2} & \text{if } r=3,\ v=2\\ K\,\Delta_n^{3/2}\,\sqrt{\alpha'(p)_i^n} & \text{if } r=4,\ v=2. \end{cases}$$

Hölder's inequality and $\sum_{i\ge1} a_i^n \le A$ and $\sum_{i\ge1} a'(p)_i^n \le 2A$ yield

$$\sum_{i=1}^{[t/\Delta_n]} \big(\alpha'(p)_i^n\big)^{1/4}\big(a_i^n\big)^{1/2} \le \sqrt{A}\,\frac{t^{1/4}}{\Delta_n^{1/4}}\,\alpha'(p)^{1/4}$$

$$\sum_{i=1}^{[t/\Delta_n]} \big(\alpha'(p)_i^n\big)^{1/2}\big(a'(p)_i^n\big)^{1/2} \le \sqrt{2A}\,\alpha'(p)^{1/2}$$

$$\sum_{i=1}^{[t/\Delta_n]} \big(\alpha'(p)_i^n\big)^{1/2} \le \frac{t^{1/2}}{\Delta_n^{1/2}}\,\alpha'(p)^{1/2}.$$

At this stage, we deduce (5.4.28) from the property $\alpha'(p)\to 0$. □

Lemma 5.4.12 *Assume* (SH). *Then* $(X^{(n)}, Z^n(X)) \stackrel{\mathcal{L}\text{-}s}{\Longrightarrow} (X, Z(X))$.

Proof 1) For each $p\in\mathbb{N}^*$ we set

$$X^{\#}(p) = (\delta 1_{\{\Gamma\le 1/p\}}) \star (\mathfrak{p}-\mathfrak{q}), \quad X(p) = X - X^{\#}(p). \tag{5.4.29}$$

We can also write $X(p)$ as (5.4.17), provided we replace δ by $\delta(p) = \delta\, 1_{\{\Gamma>1/p\}}$ and b_t by $b(p)_t = b_t - \int_{\{z:\Gamma(z)>1/p,\|\delta(t,z)\|\le 1\}} \delta(t,z)\,\lambda(dz)$. Observe that $\delta(p)$ satisfies (5.4.18), and $b(p)$ is bounded for any fixed p. Thus $(X(p)^{(n)}, Z^n(X(p)) \stackrel{\mathcal{L}\text{-s}}{\Longrightarrow} (X(p), Z(X(p))$ for each p by Lemma 5.4.10. Therefore, by Proposition 2.2.4, it suffices to prove the three properties (5.4.25), (5.4.26) and (5.4.27), as $p\to\infty$ (with of course the present definition of $X(p)$).

2) We can realize all processes $Z(X(p))$ and also $Z(X)$ on the same extension of the space, with the same W', R_{n-} and R_{n+}, via the formula (5.4.3). Then

$$\begin{aligned} Z(X)^{ij}_t - Z\big(X(p)\big)^{ij}_t &= Z\big(X^\#(p)\big)^{ij}_t \\ &= \sum_{q=1}^{\infty} \big(\Delta X^\#(p)^j_{T_q} R^j_{q-} + \Delta X^\#(p)^i_{T_q} R^j_{p+}\big) 1_{\{T_q\le t\}}. \end{aligned}$$

Hence, since c_t is bounded, and with $a(p) = \int_{\{z:\Gamma(z)\le 1/p\}} \Gamma(z)^2\,\lambda(dz)$, we have

$$\widetilde{\mathbb{E}}\Big(\sup_{s\le t} \big\|Z(X)_s - Z\big(X(p)\big)_s\big\|^2\Big) \le K\mathbb{E}\bigg(\sum_{s\le t} \big\|\Delta X^\#(p)_s\big\|^2\bigg) \le K t\, a(p)$$

because $\|\delta(\omega,t,z)\| \le \Gamma(z)$. Since $a(p)\to 0$ as $p\to\infty$, we have (5.4.27). Moreover, we also have $\mathbb{E}(\sup_{s\le t} \|X^\#(p)_s\|^2) \le Kta(p)$, hence (5.4.25).

3) It remains to prove (5.4.26), and this is done as in the previous lemma. We have

$$\begin{aligned} Z^n(X)^{jk}_t - Z^n\big(X(p)\big)^{jk}_t &= \frac{1}{\sqrt{\Delta_n}} \sum_{r=1}^{4} \sum_{i=1}^{[t/\Delta_n]} \chi(p,r)^n_i, \quad \text{where} \\ \chi(p,1)^n_i &= \int_{I(n,i)} \big(X(p)^{\#,j}_{s-} - X(p)^{\#,j}_{(i-1)\Delta_n}\big)\, b^k_s\,ds \\ \chi(p,2)^n_i &= \sum_{l=1}^{d'} \int_{I(n,i)} \big(X(p)^{\#,j}_{s-} - X(p)^{\#,j}_{(i-1)\Delta_n}\big)\, \sigma^{kl}_s\,dW^l_s \\ \chi(p,3)^n_i &= \int_{I(n,i)} \int \big(X(p)^{\#,j}_{s-} - X(p)^{\#,j}_{(i-1)\Delta_n}\big)\, \delta(s,z)^k\,(\mathfrak{p}-\mathfrak{q})(ds,dz) \\ \chi(p,4)^n_i &= \int_{I(n,i)} \int_{\{z:\Gamma(z)\le 1/p\}} \big(X^j_{s-} - X^j_{(i-1)\Delta_n}\big)\, \delta(s,z)^k\,(\mathfrak{p}-\mathfrak{q})(ds,dz). \end{aligned}$$

Since $\chi(p,r)^n_i$ for $r=2,3,4$ are martingale increments, (5.4.26) will follow if we prove that for all $t>0$, and as $p\to\infty$, we have (5.4.28) for $r=v=1$, and for $r=2,3,4$ and $v=2$.

(2.1.33), (2.1.34) and (2.1.37), plus (SH), yield

$$s \in I(n,i) \Rightarrow \begin{cases} \mathbb{E}\big((X^j_{s-} - X^j_{(i-1)\Delta_n})^2\big) \le K\Delta_n \\ \mathbb{E}\big((X(p)^{\#,j}_{s-} - X(p)^{\#,j}_{(i-1)\Delta_n})^2\big) \le Ka(p)\,\Delta_n. \end{cases}$$

Another application of the same properties and the Cauchy-Schwarz inequality then yield

$$\mathbb{E}\big(|\chi(p,1)^n_i|\big) \le K\Delta_n^2\sqrt{a(p)}$$

$$r = 2,3,4 \quad \Rightarrow \quad \mathbb{E}\big(|\chi(p,r)^n_i|^2\big) \le K\Delta_n^2\, a(p).$$

Thus the left side of (5.4.28) is smaller than $Kt\sqrt{\Delta_n\, a(p)}$ when $r = v = 1$, and than $Kta(p)$ when $r = 2,3,4$ and $v = 2$. The result follows because $a(p) \to 0$. □

Proof of Theorem 5.4.6 As said before, we can assume (SQ). Under this assumption, the convergence $(X^{(n)}, Z^n(X)) \stackrel{\mathcal{L}\text{-}s}{\Longrightarrow} (X, Z(X))$ has been proved in Lemma 5.4.12 when X has jumps and satisfies (SH), and in Lemma 5.4.11 when X is continuous because in this case X has the form (5.4.17) with a vanishing δ.

When X and σ have no common jumps, the discontinuous part of $Z(X)$ is $\mathcal{F}$-conditionally Gaussian by Proposition 5.1.1, and its continuous part is always $\mathcal{F}$-conditionally Gaussian, and those two parts are $\mathcal{F}$-conditionally independent: then $Z(X)$ itself is $\mathcal{F}$-conditionally Gaussian. □

Proof of Theorem 5.4.2 In view of (5.4.6) and since $\overline{Z}^{ij} = Z^{ij} + Z^{ji}$, the first claim follows from Theorem 5.4.6.

It remains to prove the last claims, for which we can again assume (SQ). With the notation $t_n = \Delta_n[t/\Delta_n]$ and $f^{ij}(x) = x^i x^j$, we have

$$[X,X]_t - [X,X]_{\Delta_n[t/\Delta_n]} = \eta^n_t + \eta'^n_t, \qquad \eta^n_t = \int_{t_n}^t c_s\,ds, \qquad \eta'^n_t = \sum_{t_n < s \le t} f(\Delta X_s).$$

On the one hand, we have for all $m \ge 1$:

$$\begin{aligned} \|\eta^n_t\| &\le \int_{t_n}^t (\|c_r\| \wedge m)\,dr + \int_{t_n}^t (\|c_r\| - m)^+ dr \\ &\le m\Delta_n + \left(\Delta_n \int_{t_n}^t ((\|c_r\| - m)^+)^2\,dr\right)^{1/2}, \end{aligned}$$

because $t - t_n \le \Delta_n$, and where the second inequality comes from Cauchy-Schwarz. Then

$$\limsup_n \sup_{s \le t} \frac{1}{\sqrt{\Delta_n}} \|\eta^n_s\| \le \left(\int_0^t ((\|c_r\| - m)^+)^2\,dr\right)^{1/2}$$

for all m, and since $\int_0^t \|c_s\|^2\,ds<\infty$ we deduce $\eta_t^n/\sqrt{\Delta_n}\stackrel{\text{u.c.p.}}{\Longrightarrow}0$. From this and the property $\frac{1}{\sqrt{\Delta_n}}([X,X]_t^n-[X,X]_{\Delta_n[t/\Delta_n]})\stackrel{\mathcal{L}\text{-s}}{\Longrightarrow}\overline{Z}$ it follows that, when X is continuous, we indeed have $\frac{1}{\sqrt{\Delta_n}}([X,X]^n-[X,X])\stackrel{\mathcal{L}\text{-s}}{\Longrightarrow}\overline{Z}$.

Finally when X has jumps, and under (SH), we still have $\eta_t^n/\sqrt{\Delta_n}\stackrel{\text{u.c.p.}}{\Longrightarrow}0$ because c is bounded, and we also have

$$\mathbb{E}\big(\|\eta_t'^n\|\big)\le K\mathbb{E}\Bigg(\sum_{t_n<s\le t}\|\Delta X_s\|^2\Bigg)=K\mathbb{E}\big(\|\delta\|^2\star\mathfrak{q}_t-\|\delta\|^2\star\mathfrak{q}_{t_n}\big)\le K\Delta_n,$$

hence $\eta_t'^n/\sqrt{\Delta_n}\stackrel{\mathbb{P}}{\longrightarrow}0$. Then the stable convergence in law of the variables (5.4.5) towards $\overline{Z}_t$ follows. □

5.5 A Joint Central Limit Theorem

In the previous section, we have obtained a CLT for $\overline{V}^n(f,X)$, another one for $\overline{V}'^n(f,X)$, and a last one for $\overline{Z}^n$ or Z^n, as defined in (5.4.1) and (5.4.7). However, there also exists a joint CLT for the triples $(\overline{V}^n(f,X),\overline{V}'^n(f',X),\overline{Z}^n)$ or $(\overline{V}^n(f,X),\overline{V}'^n(f',X),Z^n)$. Such a joint CLT, besides its own interest, has applications in various problems for high-frequency data: for example it is used (or, rather, an extension of it, given in Chap. 11, is used) for testing whether a process jumps or not; it is also used for estimating the relative importance of the "jump part" (as measured by a suitable functional of type $V^n(f,X)$) and the continuous martingale part (as measured by some $V'^n(f',X)$).

The assumptions for the joint CLT are those under which the CLT for each of the non-vanishing components holds, as they should be. The two test functions f and f' are of dimensions q and q', and of course we suppose that at least one of them is not identically 0, and also that they satisfy the conditions in Theorems 5.1.2 respectively in Theorems 5.3.5 or 5.3.6, in connection with the properties of X.

The limit of the processes $(\overline{V}^n(f,X),\overline{V}'^n(f',X),\overline{Z}^n)$, for example, is expected to be the process $(\overline{V}(f,X),\overline{V}'(f',X),\overline{Z})$, but this triple is *a priori* meaningless as long as we have not specified the "joint" law of this triple. For this, we do as follows:

- With *the same* $R_{p\pm}$ (recall (5.1.2) and (5.1.3)), define $\overline{V}(f,X)$ by (5.1.4) and the process Z'' by

$$Z_t''^{ij}=\sum_{p=1}^{\infty}\big(\Delta X_{T_p}^j R_{p-}^i+\Delta X_{T_p}^i R_{p+}^j\big)1_{\{T_p\le t\}}. \tag{5.5.1}$$

- Let $(\overline{U}'(f',X),Z')$ be a $(q'+d^2)$-dimensional continuous process on the extended space, which conditionally on $\mathcal{F}$ is a centered Gaussian martingale independent of all $R_{p\pm}$ and with variance-covariance given by (5.3.7) with f' instead of f and by

$$\begin{aligned}\mathbb{E}\big(Z_t'^{ij}\,Z_t'^{kl}\mid\mathcal{F}\big) &= \frac{1}{2}\int_0^t c_s^{ik}\,c_s^{jl}\,ds\\ \mathbb{E}\big(\overline{U}'(f'^i,X)_t\,Z_t'^{kl}\mid\mathcal{F}\big) &= \int_0^t \widetilde{\gamma}_{\sigma_s}\big(f'^i\big)^{kl}\,ds,\end{aligned}\qquad(5.5.2)$$

where, for a $d\times d'$ matrix α and a function g on $\mathbb{R}^d$ we have set

$$\widetilde{\gamma}_\alpha(g)^{kl} = \mathbb{E}\big(\big(g(\alpha U)-\widehat{\gamma}_\alpha(g)\big)U\,(\alpha U)^k\,(\alpha U)^l\big)\qquad(5.5.3)$$

(this complements the notation (5.2.4), and U is an $\mathcal{N}(0,I_{d'})$-distributed variable).

- Set $Z=Z'+Z''$, and $\overline{Z}^{ij}=Z^{ij}+Z^{ji}$.
- Set $\overline{V}'(f',X)=\overline{U}'(f',X)+\overline{A}(f',X)+\overline{A}'(f',X)+\overline{U}''(f',X)$, where we use the notation (5.3.8).

The above completely specifies the process $(\overline{V}(f,X),\overline{V}'(f',X),\overline{Z},Z)$ "globally", and separately the processes $\overline{V}(f,X)$, $\overline{V}'(f',X)$, Z, $\overline{Z}$ are the same as in the individual CLTs.

Theorem 5.5.1 *Let f be as in Theorem* 5.1.2, *and assume either that $f'=0$ and* (H) *holds, or that f' and X satisfy any one of the sets of assumptions of Theorems* 5.3.5 *or* 5.3.6. *Then we have the following stable convergence in law*:

$$\begin{aligned}\big(\overline{V}^n(f,X),\overline{V}'^n(f',X),\overline{Z}^n\big) &\stackrel{\mathcal{L}\text{-}s}{\Longrightarrow} \big(\overline{V}(f,X),\overline{V}(f',X),\overline{Z}\big)\\ \big(\overline{V}^n(f,X),\overline{V}'^n(f',X),Z^n\big) &\stackrel{\mathcal{L}\text{-}s}{\Longrightarrow} \big(\overline{V}(f,X),\overline{V}(f',X),Z\big)\end{aligned}\qquad(5.5.4)$$

with $(\overline{V}(f,X),\overline{V}(f',X),\overline{Z})$ and $(\overline{V}(f,X),\overline{V}(f',X),Z)$ as above.

Proof Exactly as for Theorem 5.4.2, it suffices to prove the second convergence. By localization we may assume (SH) in all cases, and also (SK), (SK-r) and (SK') when relevant, depending on the properties of f'.

1) The key ingredient is an extension of (5.4.21). As in the proof of Lemma 5.4.10 we consider the setting of Theorem 4.2.1 with $u_n=\Delta_n$, but now q and q' in this theorem are $d+d^2$ and $d'+d'^2$ here. The function Φ on Ω^W, satisfying (4.2.3), is $(d'+d'^2)$-dimensional with the following components (with obvious notation for the labels of the components):

$$\Phi^i = W_1^i,\qquad \Phi^{lm} = \int_0^1 W_s^l\,dW_s^m\quad \mathbb{P}^W\text{-almost surely.}$$

The process θ is in principle $(d+d^2)(d'+d'^2)$-dimensional, but the only non-vanishing components are

$$\theta^{i,r} = \sigma^{ir},\qquad \theta^{jk,lm} = \sigma^{jl}\sigma^{km}.\qquad(5.5.5)$$

The $(q' + d^2)$-dimensional function G on $\mathbb{R}^w \times (\mathbb{R}^d \times \mathbb{R}^{d^2})$ is given by

$$x \in \mathbb{R}^w,\ y \in \mathbb{R}^d,\ z \in \mathbb{R}^{d^2} \rightsquigarrow G^i\big(x, (y,z)\big) = f'^i(y), \quad G^{jk}\big(x, (y,z)\big) = z^{jk}.$$

With this set of notation, the process $\overline{U}^n(G)$ of (4.2.6) has the components

$$\overline{U}^n\big(G^j\big) = \overline{U}^n\big(f'\big), \qquad \overline{U}^n\big(G^{jk}\big) = Z'^{n,jk}, \tag{5.5.6}$$

where $\overline{U}^n(f')$ is given by (5.2.12) (with f' instead of f) and Z'^n is given by (5.4.20).

Moreover, the process $\overline{U}(G)$ of (4.2.7), with θ again given by (5.5.5), is characterized by the numbers given in (4.2.5) (the argument x does not appear here), in which α is a $(d + d^2) \times (d' + d'^2)$ matrix which will eventually take the values θ_t. Using these numbers, an elementary computation allows us to check that, with $\overline{U}(f', X) = \overline{U}'(f', X) + \overline{U}''(f', X)$ and Z' as described before the statement of the theorem, the pair $(\overline{U}(f', X), Z')$ is a version of $\overline{U}(G)$.

To complete the picture, we also recall that, with the notation (5.1.12) and with $\zeta_p = \sum_{i=1}^d \partial_i f(\Delta X_{S_p})\, R(n,p)^i$, we have $\zeta_p^n - \zeta_p \xrightarrow{\mathbb{P}} 0$ for all p. Then, at this stage, we deduce from Proposition 4.4.10 that as soon as X satisfies (SH), and with $\eta(n,p)$ and η_p as in the proof of Lemma 5.4.10,

$$\Big(\overline{U}^n(f'), Z'^n, \big(\zeta_p^n, \eta(n,p)\big)_{p\ge1}\Big) \xrightarrow{\mathcal{L}\text{-s}} \Big(\overline{U}\big(f', X\big), Z', (\zeta_p, \eta_p)_{p\ge1}\Big) \tag{5.5.7}$$

(we also use (5.5.6) here), for the product topology on $\mathbb{D}^{q'+d^2} \times (\mathbb{R}^{q+d^2})^{\mathbb{N}^*}$.

2) In this step, we fix the integer $m \ge 1$. We use the notation of the previous sections, in addition to the already mentioned Z'^n and ζ_p^n: in particular $X(m)$, $X'(m)$, $Y^n(m)$ are given by (5.1.10) and (5.1.12). As for the processes introduced in (5.4.29), we employ another notation to distinguish them from the previously defined $X(m)$: namely, we set

$$\widetilde{X}(m) = X - (\delta\, 1_{A_m^c}) * (\mathfrak{p} - \mathfrak{q}).$$

We also set

$$\Theta^n(m)_t = \sum_{p\in\mathcal{P}_m:\, S_p\le\Delta_n[t/\Delta_n]} \eta(n,p), \qquad \Theta(m)_t = \sum_{p\in\mathcal{P}_m:\, S_p\le t} \eta_p \tag{5.5.8}$$

which are the process Θ^n and the last d^2 components of the process $\overline{\Theta}$ associated with $\widetilde{X}(m)$ by (5.4.19). In view of the definitions of $\Theta(m)^n$ and $Y^n(m)$ and of the fact that

$$\overline{V}\big(f, X'(m)\big)_t = \sum_{p\in\mathcal{P}_m:\, S_p\le\Delta_n[t/\Delta_n]} \sum_{i=1}^d \partial f_i(\Delta X_{S_p})\, R_p^i,$$

we deduce from (5.5.7) that

$$\big(Y^n(m), \overline{U}^n(f'), Z^{m} + \Theta(m)^n\big) \stackrel{\mathcal{L}\text{-s}}{\Longrightarrow} \big(\overline{V}(f, X'(m)), \overline{U}(f', X), Z' + \Theta(m)\big). \tag{5.5.9}$$

The process $\widetilde{X}(m)$ satisfies the assumptions of Lemma 5.4.10 (with b replaced by the bounded process $b(m)$ of (5.1.10)), and $\widetilde{X}(m)^c = X^c$ for all m. Therefore, by (5.4.20) and the final step of the proof of Lemma 5.4.10, we deduce from (5.5.9) that

$$\big(Y^n(m), \overline{U}^n(f'), Z^n(\widetilde{X}(m))\big) \stackrel{\mathcal{L}\text{-s}}{\Longrightarrow} \big(\overline{V}(f, X'(m), \overline{U}'(f', X)), Z' + \Theta(m)\big). \tag{5.5.10}$$

3) Now we are ready to prove the second convergence in (5.5.4). The convergence (5.5.10) holds for any m, and as $m \to \infty$ we have seen $\overline{V}(f, X(m)) \stackrel{\text{u.c.p.}}{\Longrightarrow} \overline{V}(f, X)$ in (5.1.16), and $\Theta(m) \stackrel{\text{u.c.p.}}{\Longrightarrow} Z''$ in step 2 of the proof of Lemma 5.4.12. Therefore, it remains to prove the following three properties (for all $T, \eta > 0$):

$$\lim_{m\to\infty} \limsup_{n\to\infty} \mathbb{P}\Big(\sup_{t\le T} \big\|\overline{V}^n(f, X)_t - Y^n(m)_t\big\| > \eta\Big) = 0 \tag{5.5.11}$$

$$\lim_{m\to\infty} \limsup_{n\to\infty} \mathbb{P}\Big(\sup_{t\le T} \big\|Z^n(X)_t - Z^n\big(\widetilde{X}(m)\big)_t\big\| > \eta\Big) = 0 \tag{5.5.12}$$

$$\overline{V}'^n(f', X) - \overline{U}^n(f') \stackrel{\text{u.c.p.}}{\Longrightarrow} \overline{A}(f', X) + \overline{A}'(f', X). \tag{5.5.13}$$

(5.5.11) results from the property $\lim_n \mathbb{P}(\Omega_n(T, m)) = 1$ for all m and a combination of (5.1.13) and (5.1.17). (5.5.12) is shown in step 3 of the proof of Lemma 5.4.12. As for (5.5.13), it follows from the next three arguments: first, in Lemma 5.2.5 we have $\overline{U}^n(f') - Y^n(f', X) \stackrel{\text{u.c.p.}}{\Longrightarrow} 0$, under the conditions of Theorem 5.2.2 (on f', instead of f; these conditions are satisfied under the assumptions of the present theorem). Second, by Lemma 5.3.13 it is enough to prove the result when X is continuous. Third, recalling the decomposition (5.3.1), it remains to prove that $A^n(f, X) \stackrel{\text{u.c.p.}}{\Longrightarrow} \overline{A}(f', X) + \overline{A}'(f', X)$, and this is exactly (5.3.24) plus (5.3.25).

Thus all three properties above are satisfied, and the proof is complete. □

5.6 Applications

In this section we pursue the applications which have been outlined in the previous chapter, and give another very important application which is the Euler approximation schemes for a stochastic differential equation.

5.6.1 Estimation of the Volatility

To begin with, we consider the problem of estimating the volatility or the integrated volatility. This has been introduced in Sect. 3.5.1, from which we borrow all notation, and we suppose that the underlying process is *continuous*. That is, we have a one-dimensional Itô semimartingale X of the form

$$X_t = X_0 + \int_0^t b_s\,ds + \int_0^s \sigma_s\,dW_s. \tag{5.6.1}$$

Our concern is to "estimate" the variable $\int_0^t c_s\,ds$ at some terminal time t, where $c_t = \sigma_t^2$.

Let us recall the following first. We set for $p > 0$:

$$D(X, p, \Delta_n)_t = \sum_{i=1}^{[t/\Delta_n]} |\Delta_i^n X|^p.$$

Then when $p \le 2$, or $p > 2$ and (3.4.6) holds, and if σ is càdlàg, we have

$$\Delta_n^{1-p/2}\,D(X, p, \Delta_n)_t \stackrel{\text{u.c.p.}}{\Longrightarrow} m_p A(p)_t, \quad \text{where } A(p)_t = \int_0^t |\sigma_s|^p\,ds \tag{5.6.2}$$

(since X is continuous) and $m_p = \mathbb{E}(|U|^p)$ is the p absolute moment of an $\mathcal{N}(0,1)$ random variable U.

When $p = 2$, then $D(X, 2, \Delta_n) = [X, X]^n$ is the approximate quadratic variation and, if X is continuous, we can rewrite Theorem 5.4.2 as

$$\frac{1}{\sqrt{\Delta_n}}\big(D(X, 2, \Delta_n) - A(2)\big) \stackrel{\mathcal{L}\text{-s}}{\Longrightarrow} Z_t = \sqrt{2}\int_0^t c_s\,dW'_s, \tag{5.6.3}$$

and W' is another Brownian motion, defined on a very good extension of the original filtered probability space, and is independent of $\mathcal{F}$. This holds without the càdlàg property of σ, but we need $\int_0^t b_s^2\,ds < \infty$ for all t.

This results tells us that the approximation of the integrated volatility $A(2)_t$ given by $D(X, 2, \Delta_n)_t$ is accurate with the "rate" $\sqrt{\Delta_n}$. However this is of little help for constructing, for example, a confidence interval for the (unknown) value $A(2)_t$ on the basis of the observations $X_{i\Delta_n}$ at stage n. To this end we need more, namely a *standardized version of the CLT*, which goes as follows (we are not looking here for the minimal hypotheses):

Theorem 5.6.1 *Let X be a continuous Itô semimartingale satisfying Assumption* (H). *Then for each $t > 0$ the random variables*

$$\frac{\sqrt{3}\,(D(X, 2, \Delta_n)_t - A(2)_t)}{\sqrt{2D(X, 4, \Delta_n)_t}} \tag{5.6.4}$$

converge stably in law to a limit which is $\mathcal{N}(0,1)$ and independent of $\mathcal{F}$, in restriction to the set $\Omega_t^W = \{A(2)_t > 0\}$ of (3.5.10).

Proof On the one hand the left side of (5.6.3) at time t, say Z_t^n, converges stably in law to Z_t. On the other hand, $G_t^n = \frac{1}{\Delta_n} D(X, 4, \Delta_n)_t$ converges in probability to $3A(4)_t$ (recall $m_4 = 3$). By (2.2.5), the pair (G^n_t, Z^n_t) converges stably in law to $(3A(4)_t, Z_t)$, and by the continuous mapping theorem $Z^n_t/\sqrt{G^n_t}$ converges stably in law as well to $Z'_t = Z_t/\sqrt{3A(4)_t}$, in restriction to the set $\{A(4)_t > 0\}$ which coincides with Ω^W_t.

Now, conditionally on $\mathcal{F}$, the variable Z_t is centered normal with variance $2A(4)_t$, so Z'_t is $\mathcal{F}$-conditionally centered normal with variance 2/3, in restriction to the $\mathcal{F}$ measurable set Ω^W_t. Upon observing that the left side of (5.6.4) is $\sqrt{3}\, Z^n_t/\sqrt{2G^n_t}$, the result follows. □

The stable convergence in law above seems of little practical importance, only the fact that the limit in law is $\mathcal{N}(0,1)$ really matters, but it is given for free, and in the case Ω^W_t is a proper subset of Ω it is necessary for having a sound statement: indeed, the stable convergence in law restricted to a subset $\Omega' \subset \Omega$ is meaningful, whereas the simple convergence in law cannot be "restricted" to a subset.

Note that the statement of the theorem is "empty" at time $t = 0$, because $\Omega^W_0 = \emptyset$. Also, we can consider (5.6.4) as a process, but there is nothing like a functional convergence for this process.

In practice, when X is continuous we usually have $\Omega^W_t = \Omega$. In this case, we are in good shape for deriving an (asymptotic) confidence interval: if $\alpha \in (0,1)$, denote by z_α the α-symmetric quantile of $\mathcal{N}(0,1)$, that is the number such that $\mathbb{P}(|U| > z_\alpha) = \alpha$ where U is an $\mathcal{N}(0,1)$ variable. Then a *confidence interval with asymptotic significance level α for $A(2)_t$* is given by

$$\left[D(X,2,\Delta_n)_t - z_\alpha\sqrt{\frac{2D(X,4,\Delta_n)_t}{3}},\ D(X,2,\Delta_n)_t + z_\alpha\sqrt{\frac{2D(X,4,\Delta_n)_t}{3}}\right].$$

All the quantities above are, as they should be, known to the statistician when the variables $X_{i\Delta_n}$ are observed.

If we are interested in estimating $A(p)_t$ for $p \neq 0$ we can again use (5.6.2). However the CLTs which now apply are Theorem 5.3.5-(i) when $p > 1$ and Theorem 5.3.6-(i) when $p \leq 1$: in both cases this requires X to be continuous with further (K) when $p > 1$ and (K') when $p \leq 1$. The following result is then proved exactly as the previous theorem:

Theorem 5.6.2 *Let $p > 0$. If X is a continuous Itô semimartingale satisfying* (K) *when $p > 1$ and* (K') *when $p \leq 1$, and for each $t > 0$, the random variables*

$$\frac{\sqrt{m_{2p}}\,(\Delta_n^{1-p/2} D(X,p,\Delta_n)_t - m_p A(p)_t)}{\sqrt{(m_{2p} - m_p^2)\Delta_n^{2-p} D(X,2p,\Delta_n)_t}}$$

converge stably in law to a limit which is $\mathcal{N}(0,1)$ and independent of $\mathcal{F}$, in restriction to the set Ω^W_t (which equals Ω when (K'-1) *holds).*

A confidence interval with asymptotic significance level α for $A(p)_t$ is then given by

$$\left[\frac{\Delta_n^{1-p/2}}{m_p}\left(D(X,p,\Delta_n)_t - z_\alpha\sqrt{\frac{(m_{2p}-m_p^2)D(X,2p,\Delta_n)_t}{m_{2p}}}\right),\right.$$
$$\left.\frac{\Delta_n^{1-p/2}}{m_p}\left(D(X,p,\Delta_n)_t + z_\alpha\sqrt{\frac{(m_{2p}-m_p^2)D(X,2p,\Delta_n)_t}{m_{2p}}}\right)\right].$$

Finally, observe that when $p \in (0,1)$ the same holds even if X has jumps, provided it satisfies (K'-p), but this is wrong if $p \geq 1$.

5.6.2 Detection of Jumps

For the problem of jump detection, as expounded in Sect. 3.5.2, we do not have the tools for going further. Indeed, the choice of the cut-off level $x = x_n$ in (3.5.9) requires a CLT for the statistic S_n of (3.5.6). In this chapter we do have a CLT for $D(X,p,\Delta_n)_t$ or $D(X,p,k\Delta_n)_t$ when X is continuous (or equivalently, in restriction to $\Omega_t^{(c)}$), as seen just above. Theorem 5.1.2 also provide a CLT for $D(X,p,\Delta_n)_t$ or $D(X,p,k\Delta_n)_t$ in restriction to $\Omega_t^{(d)}$. But in both cases we need a joint CLT for the pair $(D(X,p,\Delta_n)_t, D(X,p,k\Delta_n)_t)$, and such results will be given in Chap. 11 only.

5.6.3 Euler Schemes for Stochastic Differential Equations

This subsection is concerned with probably the most useful and widespread example of discretization of processes, outside the realm of statistics of processes. Historically speaking, it is also the oldest example, its roots going back to Euler who introduced the well known method bearing his name for solving differential equations.

This example necessitates some non-trivial background about the general theory of stochastic differential equations (SDE, in short). A comprehensive exposition of the theory would go way beyond the scope of this book, and we will make use of the results of the theory (including the convergence results) without reproving everything. For the most basic results and a general understanding of the setting of SDEs, we refer for example to Protter [83].

1 – The setting. The setting is as follows: we have a filtered probability space $(\Omega, \mathcal{F}, (\mathcal{F}_t)_{t\geq 0}, \mathbb{P})$ endowed with a q-dimensional *driving process* X for the SDE. The most usual case, which gives rise to Itô differential equations, is the continuous case where X is a $(d-1)$-dimensional Brownian motion, plus a last component which is the drift t. However, here we consider a more general situation where the driving process X is a d-dimensional Itô semimartingale satisfying Assumption (H) (the initial value X_0 of X is irrelevant here).

Our SDE will be q-dimensional, and apart from the driving process X it involves two other ingredients: one is the initial condition, which for simplicity we take non-random, as a given point y_0 in $\mathbb{R}^q$. The other is the *coefficient*, which is a function f from $\mathbb{R}^q$ into $\mathbb{R}^q \otimes \mathbb{R}^d$ (the set of $q \times d$ matrices). The equation reads as follows:

$$dY_t = f(Y_{t-})\,dX_t, \qquad Y_0 = y_0 \tag{5.6.5}$$

or, component-wise and in integral form, as

$$i = 1, \ldots, q \;\Rightarrow\; Y_t^i = y_0^i + \sum_{j=1}^{d} \int_0^t f(Y_{s-})^{ij}\,dX_s^j. \tag{5.6.6}$$

Below we consider only the most common case where the function f is C^1 and with linear growth, that is $\|f(y)\| \leq K(1 + \|y\|)$ for some constant K. This implies that the equation has a unique (strong) solution: that is, there is a process Y which is a semimartingale on $(\Omega, \mathcal{F}, (\mathcal{F}_t)_{t \geq 0}, \mathbb{P})$ satisfying (5.6.6), and any other semimartingale satisfying the same is almost surely equal to Y. Furthermore, the solution is a strong Markov process as soon as the driving process X is a Lévy process.

2 – *The Euler scheme.* Although existence and uniqueness are ensured, the "explicit" solution and its law are in general not available. However for practical purposes it is often the case that we need to evaluate the law of Y, at least in the following sense: we want $\mathbb{E}(g(Y_t))$ for some functions g on $\mathbb{R}^q$, or maybe $\mathbb{E}(G(Y))$ for a function G on the Skorokhod space $\mathbb{D}^q$. In the absence of explicit or closed formulas, this evaluation is necessarily done via a numerical approximation.

Now, suppose that X is a Lévy process. If we want $\mathbb{E}(g(Y_t))$, which is a function $h(y_0)$ of the starting point, we can use the integro-differential equation (or pseudo-differential equation) which is satisfied by the function h and try to solve it numerically: there are several ways for doing this when X, hence Y as well, are continuous, since in this case h satisfies a partial differential equation whose coefficients are specified by the coefficient f. However these methods work effectively only when the dimension q is small, and are not really available when X is not continuous. Moreover evaluating $\mathbb{E}(G(Y))$ for a functional G on the Skorokhod space is virtually impossible by analytical methods. And of course, when X is a general semimartingale, no analytical method is available, even for $\mathbb{E}(g(Y_t))$.

Therefore we must rely in general on *Monte-Carlo simulations*. This means that we "simulate" a large number N of independent copies of Y, say $Y^{(1)}, \ldots, Y^{(N)}$. By the ordinary Law of Large Numbers, as $N \to \infty$ we have

$$\frac{1}{N} \sum_{r=1}^{N} G\big(Y^{(r)}\big) \to \mathbb{E}\big(G(Y)\big). \tag{5.6.7}$$

This leads to another problem: how do we simulate the process Y? An exact simulation is in general impossible again, and this is where the Euler schemes come into play. First, simulating the whole driving process X, say on a fixed time interval $[0, T]$, is numerically meaningless. At the best, we can simulate the discrete approximation $X_t^{(n)} = X_{\Delta_n[t/\Delta_n]}$ for some time step $\Delta_n > 0$ which is as small as one wishes (non regular discretization schemes are of course possible here). Simulating $X^{(n)}$ amounts to simulating the increments $\Delta_i^n X$. This is still in general an arduous task, although for quite a few Lévy processes it is indeed possible: if the components of X are Brownian motions, of course, or stable processes, or tempered stable processes, there are easy methods to perform the simulation in a very efficient way.

Now we turn to the simulation of the solution Y itself, assuming that simulating the increments $\Delta_i^n X$ is actually feasible. At stage n, the *Euler approximation* of Y is the process Y^n defined recursively on i as follows:

$$Y_0^n = y_0, \qquad Y_{i\Delta_n}^n = Y_{(i-1)\Delta_n}^n + f\big(Y_{(i-1)\Delta_n}^n\big)\Delta_i^n X \tag{5.6.8}$$

(we use here a matrix notation). This defines Y^n on the grid $(i\Delta_n : i \in \mathbb{N})$, in a feasible way in the sense of simulation. For the mathematical analysis, we need to extend Y^n to all times t. The simplest way consists in setting

$$t \in \big[(i-1)\Delta_n, i\Delta_n\big) \;\Rightarrow\; Y_t^n = Y_{(i-1)\Delta_n}^n, \tag{5.6.9}$$

which is of course consistent with (5.6.8).

At this point, we have to emphasize the fact that Y^n is *not* the discretized version $Y^{(n)}$ of the solution X, in the sense of (5.6.7).

3 – The error process. Our main concern will be an evaluation of the error incurred by replacing Y with its Euler approximation Y^n. There are two sorts of errors: one is the "weak" error, that is $\mathbb{E}(g(Y_t^n)) - \mathbb{E}(g(Y_t))$, or $\mathbb{E}(G(Y^n)) - \mathbb{E}(G(Y))$ for a functional G, but this is in general impossible to evaluate accurately, especially for a functional G, and especially also when X is an arbitrary semimartingale. The other sort of error is the "strong" error, that is $Y^n - Y$. Of course, the only sensible thing to do is to compare Y^n and Y on the grid points, so our *strong error process* will be the piecewise constant process

$$U^n = Y^n - Y^{(n)}. \tag{5.6.10}$$

Our analysis of the behavior of U^n is based upon a result from Kurtz and Protter [68], which will be stated without proof. We first need some notation. For any two real-valued semimartingales V and V', we write

$$\mathcal{Z}\big(V, V'\big)_t^n = \sum_{i=1}^{[t/\Delta_n]} \int_{I(n,i)} (V_{s-} - V_{(i-1)\Delta_n})\, dV'_s.$$

Next, a sequence V^n of processes of the form $V^n_t = \sum_{i=1}^{[t/\Delta_n]} \zeta^n_i$ with ζ^n_i being $\mathcal{F}_{i\Delta_n}$ measurable is said to have the P-UT property if for each t the sequence of variables

$$\sum_{i=1}^{[t/\Delta_n]} \Big(\big|\mathbb{E}\big(\zeta^n_i 1_{\{|\zeta^n_i|\le 1\}} \mid \mathcal{F}_{(i-1)\Delta_n}\big)\big| + \mathbb{E}\big(|\zeta^n_i|^2 1_{\{|\zeta^n_i|\le 1\}} \mid \mathcal{F}_{(i-1)\Delta_n}\big) - \mathbb{E}\big(\zeta^n_i 1_{\{|\zeta^n_i|\le 1\}} \mid \mathcal{F}_{(i-1)\Delta_n}\big)^2 + |\zeta^n_i| 1_{\{|\zeta^n_i|>1\}}\Big) \tag{5.6.11}$$

is bounded in probability. This is the P-UT property stated in Jacod and Shiryaev [57], see Theorem VI.6.15, the acronym meaning "predictably uniformly tight", and restricted here to processes that are constant on each interval $[(i-1)\Delta_n, i\Delta_n]$.

The next theorem gathers the results that are needed later. It says that the Euler approximation X^n always converges to the solution X, and it provides a criterion for obtaining a rate of convergence. The statement below is a mixture of Theorem 3.2 of Jacod and Protter [55] and Theorem 2.2 of Jacod [58].

Theorem 5.6.3 *Let X be a q-dimensional semimartingale and f be a C^1 function from $\mathbb{R}^q$ into $\mathbb{R}^q \otimes \mathbb{R}^d$, with linear growth.*

a) *We have $U^n \stackrel{\text{u.c.p.}}{\Longrightarrow} 0$.*

b) *Suppose that, for some sequence α_n of positive numbers tending to infinity, each sequence $(\alpha_n \mathcal{Z}(X^i, X^j)^n)$ has the P-UT property, and suppose also that $(X^{(n)}, (\alpha_n \mathcal{Z}(X^i, X^j)^n)_{1\le i,j\le q})$ converges stably in law to $(X, \mathcal{Z})$, where $\mathcal{Z} = (\mathcal{Z}^{ij})_{1\le i,j\le q}$ is defined on a very good filtered extension of the space $(\Omega, \mathcal{F}, (\mathcal{F}_t)_{t\ge0}, \mathbb{P})$. Then $\mathcal{Z}$ is a semimartingale on the extended space and $\alpha_n U^n$ converges stably in law to U, where $U = (U^i)_{1\le i\le d}$ is the unique solution on the extended space of the following linear equation*:

$$U^i_t = \sum_{j=1}^d \sum_{k=1}^q \int_0^t \partial_k f^{ij}(Y_{s-}) U^k_{s-}\, dX^j_s - \sum_{j,l=1}^d \sum_{k=1}^q \int_0^t \partial_k f^{ij}(Y_{s-}) f^{kl}(Y_{s-})\, d\mathcal{Z}^{lj}_s. \tag{5.6.12}$$

Taking this theorem for granted, we are now in a position to state (and prove) the CLT-type result which we want for U^n. In this result, the ingredients T_p, R_{p-}, R_{p+}, W' and $\widetilde{c}$ and $\widehat{\sigma}$ are exactly those occurring in (5.4.3), and in particular W' and R_{p-} and R_{p+} are defined on a very good extension $(\widetilde{\Omega}, \widetilde{\mathcal{F}}, (\widetilde{\mathcal{F}}_t)_{t\ge0}, \widetilde{\mathbb{P}})$ of $(\Omega, \mathcal{F}, (\mathcal{F}_t)_{t\ge0}, \mathbb{P})$.

Theorem 5.6.4 *Let X be a d-dimensional Itô semimartingale which satisfies Assumption* (Q). *Let f be a C^1 function from $\mathbb{R}^q$ into $\mathbb{R}^q \otimes \mathbb{R}^d$, with linear growth. Let U^n be the error process* (5.6.10) *of the Euler scheme Z^n associated by* (5.6.9) *with* (5.6.5).

Then $\frac{1}{\sqrt{\Delta_n}} U^n$ converges stably in law to a limiting process U which is the unique solution on the extended space $(\widetilde{\Omega}, \widetilde{\mathcal{F}}, (\widetilde{\mathcal{F}}_t)_{t\ge0}, \widetilde{\mathbb{P}})$ of the following linear equation:

$$
\begin{aligned}
U_t^i = & \sum_{j=1}^{d} \sum_{k=1}^{q} \int_0^t \partial_k f^{ij}(Y_{s-}) U_{s-}^k \, dX_s^j \\
& - \frac{1}{\sqrt{2}} \sum_{j,l,m,r=1}^{d} \sum_{k=1}^{q} \int_0^t \partial_k f^{ij}(Y_{s-}) f^{kl}(Y_{s-}) \widetilde{c}_s^{lj,mr} \, dW_s'^{mr} \\
& - \sum_{j,l=1}^{d} \sum_{k=1}^{q} \sum_{p=1}^{\infty} \partial_k f^{ij}(Y_{T_p-}) f^{kl}(Y_{T_p-}) \\
& \times \left(\Delta X_{T_p}^j R_{p-}^l + \Delta X_{T_n}^l R_{p+}^j\right) 1_{\{T_p \le t\}}.
\end{aligned}
\tag{5.6.13}
$$

This equation is (5.6.12), written with $\mathcal{Z} = Z$, the process given by (5.4.3), in line with Theorem 5.4.6 and the fact that Z^n in that theorem is $(\mathcal{Z}(X^i, X^j)^n/\sqrt{\Delta_n})_{i,j \le d}$ here.

Coming back to the computation of $\mathbb{E}(G(Y))$ for a continuous functional G on the Skorokhod space, we use (5.6.7) with N simulated copies of the Euler approximation Y^n, for some given time step Δ_n. The limit in (5.6.7) (as $N \to \infty$) is then $\mathbb{E}(G(Y^n))$ and, on top of the statistical (Monte-Carlo) error of order $1/\sqrt{N}$, we have the approximation error $\mathbb{E}(G(Y^n)) - \mathbb{E}(G(Y))$. This is where the theorem helps: it basically says that, provided G is well behaved, this approximation error is of order $\sqrt{\Delta_n}$. It also gives (in principle) a way to evaluate this error by using the equation (5.6.13) in the same way as in the "standardization" made in Theorem 5.6.1.

Note however that, although the rate $\sqrt{\Delta_n}$ is sharp "in general", it is not so for the error $\mathbb{E}(g(Y_t^n)) - E(g(Y_t))$, and when X is a Lévy process: in this case, and provided Δ_n is such that t/Δ_n is an integer, this error is of order Δ_n: see Talay and Tubaro [91] or Jacod, Kurtz, Méléard and Protter [59].

Proof By localization, it is enough to prove the result when X satisfies (SQ). In view of Theorems 5.4.6 and 5.6.3, it remains to prove that for all $j, k \le d$ the sequence of processes $Z^{n,jk}$ has the P-UT property.

1) Letting $V_t^n = \sum_{i=1}^{[t/\Delta_n]} \zeta_i^n$ be adapted processes, we give two criteria for the P-UT property. First, the expectation of (5.6.11) is smaller than $2\mathbb{E}(\sum_{i=1}^{[t/\Delta_n]} |\zeta_i^n|)$, so

$$
\sup_n \mathbb{E}\left(\sum_{i=1}^{[t/\Delta_n]} |\zeta_i^n|\right) < \infty \ \forall t \ \Rightarrow \ \left(V^n\right) \text{ has P-UT.} \tag{5.6.14}
$$

Second, suppose that $\mathbb{E}(\zeta_i^n \mid \mathcal{F}_{(i-1)\Delta_n}) = 0$ (we have martingale increments). The first summand in (5.6.11) equals $|\mathbb{E}(\zeta_i^n 1_{\{|\zeta_i^n|>1\}} \mid \mathcal{F}_{(i-1)\Delta_n})|$, and $|\zeta_i^n 1_{\{|\zeta_i^n|>1\}}| \le (\zeta_i^n)^2$. Then the expectation of (5.6.11) is smaller than $2\mathbb{E}(\sum_{i=1}^{[t/\Delta_n]} |\zeta_i^n|^2)$, and we have

$$
\left.\begin{array}{l}
\mathbb{E}\left(\zeta_i^n \mid \mathcal{F}_{(i-1)\Delta_n}\right) = 0 \\
\sup_n \mathbb{E}\left(\sum_{i=1}^{[t/\Delta_n]} |\zeta_i^n|^2\right) < \infty \ \forall t
\end{array}\right\} \Rightarrow \left(V^n\right) \text{ has P-UT.} \tag{5.6.15}
$$

Another evident property is that if, two sequences (V^n) and (V'^n) have the P-UT property, then so does the sum $(V^n + V'^n)$.

2) For any process Y we write for $s \geq (i-1)\Delta_n$:

$$Y^n_{i,s} = Y_s - Y_{(i-1)\Delta_n}. \tag{5.6.16}$$

Under (SH) we set $b'_t = b_t + \int_{\{\|\delta(t,z)>1\}} \delta(t,z)\lambda(dz)$, so we have

$$Z^{n,jk}_t = \sum_{l=1}^{d+2} \sum_{i=1}^{[t/\Delta_n]} \zeta(l)^n_i, \quad \text{where}$$

$$\zeta(l)^n_i = \begin{cases} \frac{1}{\sqrt{\Delta_n}} \int_{I(n,i)} X^{j,n}_{i,s-} \sigma^{kl}_s \, dW^l_s & \text{if } l = 1, \dots, d \\ \frac{1}{\sqrt{\Delta_n}} \int_{I(n,i)} \int X^{j,n}_{i,s-} \delta(s,z)^k (\mathfrak{p} - \mathfrak{q})(ds,dz) & \text{if } l = d+1 \\ \frac{1}{\sqrt{\Delta_n}} \int_{I(n,i)} X^{j,n}_{i,s-} b'^k_s \, ds & \text{if } l = d+2 \end{cases}$$

and it is enough to prove that each sequence $V(l)^n = \sum_{i=1}^{[t/\Delta_n]} \zeta(l)^n_i$ has the P-UT property. (2.1.33), (2.1.34) and (2.1.36) yield $\mathbb{E}(\|X^n_{i,s}\|^2) \leq K\Delta_n$ if $s \in I(n,i)$, and b' and σ are bounded. Therefore by the Cauchy-Schwarz and Doob's inequalities $\mathbb{E}(|\zeta(l)^n_i|^p) \leq K\Delta_n$ if $p = 2$ and $l = 1, \dots, d+1$, and also if $p = 1$ and $l = d+2$. Then the P-UT property for $Y(l)^n$ follows from (5.6.15) for $l = 1, \dots, d+1$ and from (5.6.14) for $l = d+2$.

3) Finally we suppose that X is continuous and $\int_0^t (\|b_s\|^2 + \|c_s\|^2)\, ds \leq K_t$ for all t. Then $X = X_0 + B + X^c$ and $Z^{n,jk}_t = \sum_{l=1}^{2d+2} \sum_{i=1}^{[t/\Delta_n]} \zeta(l)^n_i$, where

$$\zeta(l)^n_i = \begin{cases} \frac{1}{\sqrt{\Delta_n}} \int_{I(n,i)} X^{c,j,n}_{i,s-} \sigma^{kl}_s \, dW^l_s & \text{if } l = 1, \dots, d \\ \frac{1}{\sqrt{\Delta_n}} \int_{I(n,i)} B^{j,n}_{i,s-} \sigma^{kl}_s \, dW^l_s & \text{if } l = d+1, \dots, 2d \\ \frac{1}{\sqrt{\Delta_n}} \int_{I(n,i)} X^{c,j,n}_{i,s-} b^k_s \, ds & \text{if } l = 2d+1 \\ \frac{1}{\sqrt{\Delta_n}} \int_{I(n,i)} B^{j,n}_{i,s-} b^k_s \, ds & \text{if } l = 2d+2. \end{cases}$$

Again it is enough to prove that each sequence $Y(l)^n = \sum_{i=1}^{[t/\Delta_n]} \zeta(l)^n_i$ has the P-UT property. (2.1.33) and (2.1.34) yield for all $p \geq 2$ and $s \in I(n,i)$:

$$\|B^n_{i,s}\|^2 \leq \Delta_n \int_{I(n,i)} \|b_r\|^2 \, dr, \quad \mathbb{E}(\|X^{c,n}_{i,s}\|^p) \leq K_p \Delta_n^{p/2-1} \mathbb{E}\left(\int_{I(n,i)} \|c_r\|^{p/2} \, dr \right).$$

Therefore by the Cauchy-Schwarz and Burkholder-Davis-Gundy inequalities, we get

$$\mathbb{E}\big(|\zeta(l)_i^n|^p\big) \le \begin{cases} \mathbb{E}\big(\int_{I(n,i)} \|c_s\|^2\, ds\big) & \text{if } p=2, l=1,\dots,d \\ \mathbb{E}\big(\int_{I(n,i)} \|b_s\|^2\, ds \int_{I(n,i)} \|c_s\|\, ds\big) & \text{if } p=2, l=d+1,\dots,2d \\ \mathbb{E}\big(\int_{I(n,i)} (\|b_s\|^2 + \|c_s\|)\, ds\big) & \text{if } p=1, l=2d+1 \\ \sqrt{\Delta_n}\, \mathbb{E}\big(\int_{I(n,i)} \|b_s\|^2\, ds\big) & \text{if } p=1, l=2d+2. \end{cases}$$

Since $\int_0^t \|b_s\|^2\, ds$ and $\int_0^t \|c_s\|^2\, ds$ and $\int_0^t \|c_s\|\, ds$ are bounded by hypothesis, it follows that the array $\zeta(l)_i^n$ satisfies the conditions in (5.6.15) when $l = 1,\dots,2d$ and in (5.6.14) when $l = 2d+1, 2d+2$. This completes the proof. □

Bibliographical Notes

Processes with conditionally independent increments have been introduced by Grigelionis [41]. The associated convergence result (Theorem 4.2.1) comes from Jacod [54], and the unpublished paper [53] contains a more general version of the Central Limit Theorem 5.3.6 when the underlying Itô semimartingale is continuous. A CLT of the type of Theorem 5.3.5 was already proved, in a statistical context and for diffusions, by Dohnal [29]. The first CLT for discontinuous Itô semimartingales seems to appear in Jacod and Protter [55], which also contains the analysis of the Euler scheme made in Sect. 5.6.3. One can also mention that, often independently, results about the quadratic variation in the continuous case are scattered through the econometrical literature, as in Andersen, Bollerslev, Diebold and Labys [6] for example. The theorems of Sects. 5.1 and 5.3, are taken from Jacod [60], except Theorem 5.3.6 which was proved in more generality by Barndorff-Nielsen, Graversen, Jacod, Podolskij and Shephard [11].

Chapter 6
Integrated Discretization Error

In contrast with the previous chapters and most of the forthcoming ones, here we study a different kind of functionals, namely $\sum_{i=1}^{N_n(t)} f(X_{(i-1)\Delta_n})$, for a regular discretization scheme with time step $\Delta_n \to 0$.

The law of large numbers for this sort of functionals is evident: as soon as the d-dimensional process X is, say, càdlàg and f is a continuous function on $\mathbb{R}^d$, we have

$$\Delta_n \sum_{i=1}^{[t/\Delta_n]} f(X_{(i-1)\Delta_n}) \to \int_0^t f(X_s)\,ds,$$

the convergence taking place for each ω and being locally uniform in time: this comes from the convergence of Riemann sums. The choice of $X_{(i-1)\Delta_n}$ in the ith summand is arbitrary, we could have taken $X_{t_{n,i}}$ for any $t_{n,i} \in I(n,i) = ((i-1)\Delta_n, i\Delta_n]$ instead, but it fits better for later results.

The associated Central Limit Theorem is more interesting. It concerns the behavior of the following processes (recall the discretized version $X_t^{(n)} = X_{\Delta_n[t/\Delta_n]}$ of the process X):

$$\begin{aligned} \widetilde{V}^n(f,X)_t &= \Delta_n \sum_{i=1}^{[t/\Delta_n]} f(X_{(i-1)\Delta_n}) - \int_0^{\Delta_n[t/\Delta_n]} f(X_s)\,ds \\ &= \int_0^{\Delta_n[t/\Delta_n]} \big(f\big(X_s^{(n)}\big) - f(X_s)\big)\,ds \end{aligned}$$

(taking $\Delta_n[t/\Delta_n]$ instead of t as the upper bound is for convenience: it avoids boundary terms). The variable $\widetilde{V}^n(f,X)$ may be viewed as a kind of "integrated" measure of the error incurred by the discretization, when X is replaced by $X^{(n)}$. Perhaps more to the point, we could take the absolute value of the error, or some power of it, which leads us to introduce the following processes for *a priori* any $p > 0$:

$$\widetilde{V}^n(f,p,X)_t = \int_0^{\Delta_n[t/\Delta_n]} \big|f\big(X_s^{(n)}\big) - f(X_s)\big|^p\,ds.$$

J. Jacod, P. Protter, *Discretization of Processes*,
Stochastic Modelling and Applied Probability 67,
DOI 10.1007/978-3-642-24127-7_6, © Springer-Verlag Berlin Heidelberg 2012

It may appear that this chapter digresses from the main stream of this book. However, consider for a moment the case of the identity function $f(x)=x$. The functional $V^n(f,X)_t=\sum_{i=1}^{[t/\Delta_n]} f(\Delta_i^n X)$ is equal to $X_t^{(n)}-X_0$, which converges pathwise to X_t-X_0 for the Skorokhod topology. In this case the CLT given in Theorem 5.1.2 does not apply, whereas Theorem 5.3.6 amounts simply to $X^{(n)}-X_0\to X-X_0$, because with the notation of this theorem we have in the present case $\overline{V}'^n(f,X)=X^{(n)}-X_0$ and $\overline{U}'(f,X)=\overline{A}'(f,X)=0$ and $\overline{A}(f,X)+\overline{U}''(f,X)=X-X_0$.

In other words, one can view a CLT for $\widetilde{V}^n(f,X)$ or for $\widetilde{V}^n(f,p,X)$ as a substitute of proper CLTs for $V^n(f,X)$ or $V'^n(f,X)$, which do not exist as such when $f(x)=x$.

The content of this chapter is taken from the paper [56] of Jacod, Jakubowski and Mémin, with a few improvements.

6.1 Statements of the Results

The process of interest X is a d-dimensional Itô semimartingale, on the filtered probability space $(\Omega,\mathcal{F},(\mathcal{F}_t)_{t\geq 0},\mathbb{P})$, written in its Grigelionis form as

$$X_t=X_0+\int_0^t b_s\,ds+\int_0^t \sigma_s\,dW_s+(\delta 1_{\{\|\delta\|\leq 1\}})\star(\mathfrak{p}-\mathfrak{q})_t+(\delta 1_{\{\|\delta\|>1\}})\star\mathfrak{p}_t. \quad (6.1.1)$$

Here, W is a d'-dimensional Brownian motion and $\mathfrak{p}$ is a Poisson measure on $\mathbb{R}_+\times E$ with compensator $\mathfrak{q}(dt,dz)=dt\otimes\lambda(dz)$ (λ is a σ-finite measure on the auxiliary space $(E,\mathcal{E})$). Then b and δ are d-dimensional and σ is $d\times d'$-dimensional, and as usual $c=\sigma\sigma^\star$.

Most of the time, we will need an assumption of the same type but stronger than Assumption 4.4.2, or (H). Below, r is a real in $[0,2]$.

Assumption 6.1.1 (or (**H**-r)) X is an Itô semimartingale given by (6.1.1), and we have:

(i) The process b is locally bounded.
(ii) The process σ is càdlàg.
(iii) There is a localizing sequence (τ_n) of stopping times and, for each n, a *deterministic* nonnegative function Γ_n on E such that $\int \Gamma_n(z)^r\lambda(dz)<\infty$ (with the convention $0^0=0$) and $\|\delta(\omega,t,z)\|\wedge 1\leq\Gamma_n(z)$ for all (ω,t,z) with $t\leq\tau_n(\omega)$.

(H-2) is exactly (H), whereas (H-r) for $r\in[0,2)$ implies $r\in\mathcal{I}(X)$ where, according to (3.2.2), $\mathcal{I}(X)$ is the set of all $p\geq 0$ such that $\sum_{s\leq t}\|\Delta X_s\|^p<\infty$ a.s. for all t. In fact, (H-r) when $r<2$ is slightly stronger than (H) plus the property $r\in\mathcal{I}(X)$. Furthermore all (H-r) are the same as (H) when X is continuous. Observe also that, as far as the coefficient δ is concerned, (H-r) is the same as Assumption 5.3.2, that is (K-r), except that we do not require $r\leq 1$. Finally (H-r) implies (H-r') for all $r'\in[r,2]$.

The results are somewhat similar to Theorem 5.4.2, and we start with a description of the limiting processes. As in the previous chapter, we consider an auxiliary space $(\Omega', \mathcal{F}', \mathbb{P}')$ supporting a d'-dimensional Brownian motion W' (the same dimension as for W) and a sequence $(\kappa_n)_{n\geq 1}$ of i.i.d. variables, uniformly distributed on $(0, 1)$ and independent of W'. We also consider an arbitrary weakly exhausting sequence $(T_n)_{n\geq 1}$ for the jumps of X, see, e.g., before (5.1.2). Then we construct the very good filtered extension $(\widetilde{\Omega}, \widetilde{\mathcal{F}}, (\widetilde{\mathcal{F}}_t)_{t\geq 0}, \widetilde{\mathbb{P}})$ of $(\Omega, \mathcal{F}, (\mathcal{F}_t)_{t\geq 0}, \mathbb{P})$ by (4.1.16), that is:

$$\begin{array}{l} \widetilde{\Omega} = \Omega \times \Omega', \qquad \widetilde{\mathcal{F}} = \mathcal{F} \otimes \mathcal{F}', \qquad \widetilde{\mathbb{P}} = \mathbb{P} \otimes \mathbb{P}' \\ (\widetilde{\mathcal{F}}_t) \text{ is the smallest filtration containing } (\mathcal{F}_t) \text{ and such that} \\ \qquad \kappa_n \text{ is } \widetilde{\mathcal{F}}_{T_n} \text{ measurable for all } n \text{ and } W' \text{ is } (\widetilde{\mathcal{F}}_t)\text{-adapted.} \end{array}$$

Then W' is a Brownian motion independent of $(W, \mathfrak{p})$, and $\mathfrak{p}$ is a Poisson measure with compensator $\mathfrak{q}$ and independent of (W, W'), on the extended space.

We start with the processes $\widetilde{V}^n(f, X)$, or rather with $-\widetilde{V}^n(f, X)$ which are somewhat more natural to consider.

Theorem 6.1.2 *Let f be a C^2 function from $\mathbb{R}^d$ into $\mathbb{R}^q$. Assume that X is an Itô semimartingale, which either is continuous or satisfies* (H) *(that is,* (H-2)*). Then the processes $-\frac{1}{\Delta_n}\widetilde{V}^n(f, X)$ converge stably in law to the process*

$$\widetilde{V}(f, X)_t = \frac{1}{2}\big(f(X_t) - f(X_0)\big) + \widetilde{V}'(f, X)_t + \widetilde{V}''(f, X)_t, \tag{6.1.2}$$

where

$$\widetilde{V}'(f, X)_t = \frac{1}{\sqrt{12}} \sum_{j=1}^{d} \sum_{k=1}^{d'} \int_0^t \partial_j f(X_s) \sigma_s^{jk} \, dW_s'^k \tag{6.1.3}$$

$$\widetilde{V}''(f, X)_t = \sum_{n:\, T_n \leq t} \big(f(X_{T_n-} + \Delta X_{T_n}) - f(X_{T_n-})\big)\left(\kappa_n - \frac{1}{2}\right). \tag{6.1.4}$$

The first term in (6.1.2) is a sort of "bias", analogous to the last three terms in (5.3.6), and it is defined on the original space, whereas $\widetilde{V}'(f, X)$ and $\widetilde{V}''(f, X)$ involve extra randomness and necessitates the extension of the space. The process $\widetilde{V}'(f, X)$ is obviously well defined. The process $\widetilde{V}''(f, X)$ is well defined by Proposition 4.1.4, because the variables $\kappa_n - 1/2$ are centered and

$$\sum_{s\leq t} \big\| f(X_{s-} + \Delta X_s) - f(X_{s-}) \big\|^2 < \infty \tag{6.1.5}$$

for all t. Note also that $\widetilde{V}'(f, X)$ and $\widetilde{V}''(f, X)$ are, conditionally on $\mathcal{F}$, two independent processes with independent increments, centered and with covariances

$$\mathbb{E}\big(\widetilde{V}'(f^j, X)_t \, \widetilde{V}'(f^k, X)_t \mid \mathcal{F}\big) = \frac{1}{12} \sum_{l,m=1}^{d} \int_0^t \partial_l f^j(X_s) \, c_s^{lm} \, \partial_m f^k(X_s) \, ds$$

$$\mathbb{E}\big(\widetilde{V}''(f^j,X)_t\,\widetilde{V}''(f^k,X)_t\mid\mathcal{F}\big)=\frac{1}{12}\sum_{s\le t}\big(f^j(X_{s-}+\Delta X_s)-f^j(X_{s-})\big)$$
$$\times\big(f^k(X_{s-}+\Delta X_s)-f^k(X_{s-})\big)$$

(we use here the fact that the variance of κ_n is $1/12$).

Now, (6.1.5) holds as soon as f is C^1, but the C^2 property of f is heavily used in the proof, through Itô's formula. However, we can take advantage of the generalized Itô's formula given in Theorem 3.2.2, to obtain a similar result under weaker conditions on f (but stronger assumptions on X).

For this we recall that a C^p function, when p is not an integer, is a $[p]$-continuously differentiable function whose $[p]$th partial derivatives are Hölder with index $p-[p]$ on every compact subset of $\mathbb{R}^d$. We also need the following process, well defined as soon as $1\in\mathcal{I}(X)$:

$$B'_t = \int_0^t b_s\,ds-(\delta\,1_{\{\|\delta\|\le1\}})*\mathfrak{p}_t. \tag{6.1.6}$$

Theorem 6.1.3 *Let f be a C^r function from $\mathbb{R}^d$ into $\mathbb{R}^q$, for some $r\in(0,2)$, and assume that X satisfies* (H-r), *with moreover $X^c=0$ when $r\in[1,2)$ and $X^c=B'=0$ when $r<1$. Then the processes $-\frac{1}{\Delta_n}\widetilde{V}^n(f,X)$ converge stably in law to the process*

$$\widetilde{V}(f,X)_t = \frac{1}{2}\big(f(X_t)-f(X_0)\big)+\widetilde{V}''(f,X)_t, \tag{6.1.7}$$

where $\widetilde{V}''(f,X)$ is again defined by (6.1.4).

Above, we have $r\in\mathcal{I}(X)$ and (6.1.5) holds when f is $C^{r/2}$, hence *a fortiori* when it is C^r: so again $\widetilde{V}''(f,X)$ is well defined. Since $X^c=0$ we have $\widetilde{V}'(f,X)=0$, and the two formulas (6.1.2) and (6.1.7) agree. Note also that $\widetilde{V}'(f,X)$ is of finite variation when $r\le1$, but in general not when $r>1$. Moreover, again when $r<1$, the limit (6.1.7) can also be written as

$$\widetilde{V}(f,X)_t = \sum_{n:\,T_n\le t}\big(f(X_{T_n-}+\Delta X_{T_n})-f(X_{T_n-})\big)\kappa_n,$$

because then $X_t=X_0+\sum_{n:T_n\le t}\Delta X_{T_n}$ by our assumption $B'=0$.

Now we turn to the processes $\widetilde{V}^n(f,p,X)$. Of course, here, f is one-dimensional, and for simplicity we assume that is C^2, although C^1 would be enough when $r\le1$ below.

Theorem 6.1.4 *Let $p>0$ and f be a C^2 real-valued function on $\mathbb{R}^d$. Assume that X satisfies* (H-r), *where $r=p\wedge2$, with moreover $X^c=0$ when $p\in(1,2)$ and $X^c=B'=0$ when $p\le1$. Then the processes $\frac{1}{\Delta_n}\widetilde{V}^n(f,p,X)$ converge stably in law to the following limit:*

$$\widetilde{V}(f,p,X)_t = \begin{cases} \frac{1}{2}\sum_{j,k=1}^{d}\int_0^t \partial_j f(X_s)\partial_k f(X_s)c_s^{jk}\,ds + \widetilde{V}'(f,p,X)_t & \text{if } p=2 \\ \widetilde{V}'(f,p,X)_t & \text{otherwise,} \end{cases} \tag{6.1.8}$$

where

$$\widetilde{V}'(f,p,X)_t = \sum_{n:\,T_n\le t} \left|f(X_{T_n-}+\Delta X_{T_n}) - f(X_{T_n-})\right|^p \kappa_n. \tag{6.1.9}$$

Note that our assumptions imply

$$\sum_{s\le t}\left|f(X_{s-}+\Delta X_s) - f(X_{s-})\right|^p < \infty,$$

hence $\widetilde{V}'(f,p,X)$ is finite-valued (by Proposition 4.1.3 in this case). The limiting process $\widetilde{V}(f,p,X)$ is increasing in time, as it should be because each $\widetilde{V}^n(f,p,X)$ is also increasing by construction. The forms of the limits in Theorems 6.1.2 and 6.1.4 are thus deeply different.

Remark 6.1.5 The reader will observe that, in contrast with Theorem 6.1.3, we assume $B'=0$ when $p=r\le 1$, and not just when $p<1$. The asymptotic behavior of $\widetilde{V}^n(f,1,X)$ when $X^c=0$ and $B'\neq 0$ is unknown, although it is known that for each t the sequence of variables $\widetilde{V}^n(f,1,X)_t/\Delta_n$ is bounded in probability.

Remark 6.1.6 This result is remarkable in the sense that, since $\widetilde{V}^n(f,p,X)$ is a kind of pth power of the error involved when replacing $f(X_s)$ by $f(X_s^{(n)})$, one would expect the rate to be the pth power of a basic rate. But this is not the case: the rate is *always* $1/\Delta_n$.

This is of course due to the jumps. Indeed, if there is a single jump in the interval $I(n,i)=((i-1)\Delta_n, i\Delta_n]$, say at time T, the integral $\int_{I(n,i)}|f(X_s^{(n)})-f(X_s)|^p\,ds$ is approximately $|f(X_{(i-1)\Delta_n}+\Delta X_T)-f(X_{(i-1)\Delta_n})|^p\,(i\Delta_n-T)$, whose order of magnitude is always Δ_n.

Remark 6.1.7 Assumption (H-r) for any $r\in[0,2]$ entails that σ_t is càdlàg, but this property is not required for the previous result, as will be apparent from the proof. We only need σ_t to be locally bounded. This will *not* be the case in the forthcoming result, for which the càdlàg property is needed when $p\neq 2$.

When X is continuous we have another result, with a "true rate". This result coincides with the previous one when $p=2$ and improves on it when $p>2$. It also shows that, when $p<2$, one cannot have a result like Theorem 6.1.4, unless we assume $X^c=0$.

To describe the limit, we need a notation. For $x=(x^j)_{1\le j\le d'}\in\mathbb{R}^{d'}$ and $p>0$ we set (the dimension d' is implicit in the next formula):

$$\rho(p,x) = \mathbb{E}\left(\int_0^1 \left|\sum_{j=1}^{d'} x^j W_s^j\right|^p ds\right). \tag{6.1.10}$$

These are explicitly calculable, using the multinomial formula, when p is an even integer, and in particular $\rho(2,x) = \|x\|^2/2$, implying that for $p=2$ the limit below is the same as in (6.1.8) in the absence of jumps. When $d'=1$ and for all $p>0$, it is also explicit: we have $\rho(p,x) = 2m_p|x|^p/(p+2)$, where as usual m_p is the p absolute moment of the law $\mathcal{N}(0,1)$. Moreover $\rho(p,x)$ is obviously continuous in x.

Theorem 6.1.8 *Let f be a C^2 function on $\mathbb{R}^d$, and let $p>0$. Assume also that X is a continuous Itô semimartingale satisfying* (H). *Then if ξ_t is the d'-dimensional càdlàg process with components $\xi_t^k = \sum_{j=1}^d \partial_j f(X_t)\sigma_t^{jk}$, we have*

$$\frac{1}{\Delta_n^{p/2}} \widetilde{V}^n(f,p,X)_t \stackrel{\text{u.c.p.}}{\Longrightarrow} \int_0^t \rho(p,\xi_s)\,ds.$$

This result is in fact a law of large numbers, in contrast with Theorem 6.1.4, and one could look for an associated CLT. We will not do that here.

6.2 Preliminaries

The theorems of this chapter encompass two cases:

(a) The process X is a continuous Itô semimartingale, we then set $r=2$.
(b) The process X satisfies (H-r) for some $r \in (0,2]$.

As is now customary for us, we introduce a strengthened version of (H-r):

Assumption 6.2.1 (or (**SH**-r)) We have (H-r), and the processes b and σ are bounded, and $\|\delta(\omega,t,z)\| \wedge 1 \le \Gamma(z)$ with Γ bounded and $\int \Gamma(z)^r \lambda(dz) < \infty$.

An obvious extension of the localization Lemma 4.4.9 yields that, if one of our theorems holds for all X satisfying (SH-r), it also holds for any X satisfying (H-r). So in case (b) above we may replace (H-r) by (SH-r), and in both cases the process X itself is bounded. Then a further (trivial) localization allows us to reduce the problem to the following two cases, still labeled (a) and (b), and where A is some constant:

$$\left.\begin{array}{ll} \bullet \text{ case (a):} & X \text{ is continuous, } r=2, \text{ and} \\ & \int_0^t \|b_s\|\,ds + \int_0^t \|\sigma_s\|^2\,dt + \|X_t\| \le A \\ \bullet \text{ case (b):} & \big\|\delta(t,z)\big\| \le \Gamma(z) \text{ and} \\ & \Gamma(z) + \int \Gamma(z)^r \lambda(dz) + \|b_t\| + \|\sigma_t\| + \|X_t\| \le A. \end{array}\right\} \tag{6.2.1}$$

In case (b) we also use the following notation:

$$\begin{aligned} & b_t'' = b_t + \int_{\{\|\delta(t,z)\|>1\}} \delta(t,z)\lambda(dz) \\ & r \le 1 \Rightarrow b_t' = b_t - \int_{\{\|\delta(t,z)\|\le 1\}} \delta(t,z)\lambda(dz). \end{aligned} \tag{6.2.2}$$

Then $\|b''_t\| \le A + A^2$, and $\|b'_t\| \le 2A$ when $r \le 1$. We can rewrite (6.1.1) as

$$\begin{aligned} X_t &= X_0 + \int_0^t b''_s\,ds + \int_0^t \sigma_s\,dW_s + \delta * (\mathfrak{p} - \mathfrak{q})_t \quad \text{if } r \in (0,2] \\ X_t &= X_0 + \int_0^t b'_s\,ds + \int_0^t \sigma_s\,dW_s + \delta * \mathfrak{p}_t \qquad \text{if } r \in (0,1]. \end{aligned} \tag{6.2.3}$$

Then $B'_t = \int_0^t b'_s\,ds$ is the same as in (6.1.6).

6.2.1 An Application of Itô's Formula

The functionals $\widetilde{V}^n(f,X)$ and $\widetilde{V}^n(f,p,X)$ involve in an essential way the variables $f(X_s) - f(X_{(i-1)\Delta_n})$ and $|f(X_s) - f(X_{(i-1)\Delta_n})|^p$, and to evaluate these variables we use Itô's formula, in its classical version when f is C^2 or $p \ge 2$ and in its extended form of Theorem 3.2.2 when f is C^r or $p < 2$. Note that under (6.2.1) the values that $f(x)$ takes when $\|x\| > A$ are irrelevant, so it is no restriction to assume that f has compact support.

The two types of variables above are quite different, the second one showing an interplay between f and the exponent p. In order to treat both at once, we consider below a (possibly multidimensional) function $g = g(y;x)$ on $\mathbb{R}^d \times \mathbb{R}^d$ (we use this special notation because y plays the role of a parameter). We will assume either one of the following two properties for g:

$$g \text{ is continuous with compact support, and } (y,x) \mapsto g(y;x) \text{ is } C^r. \tag{6.2.4}$$

$$g \text{ is Lipschitz, with compact support.} \tag{6.2.5}$$

So (6.2.5) is slightly weaker than (6.2.4) with $r = 1$, and typically below $r \in (0,2]$ will be the *same* number here and in (6.2.1).

We denote by $\partial_j g$ and $\partial^2_{jk} g$ the partial derivatives of the function $x \mapsto g(y;x)$, when they exist (derivatives with respect to y will not be used).

We associate the function h, and also the function k when $r \ge 1$, on $\mathbb{R}^d \times \mathbb{R}^d \times \mathbb{R}^d$, by

$$h(y;x,w) = g(y;x+w) - g(y;x), \qquad k(y;x,w) = h(y;x,w) - \sum_{j=1}^d \partial_j g(y;x) w^j.$$

Then h, and k when $r \ge 1$, are continuous, and

$$\|h(y;x,w)\| \le K\|w\|^{r\wedge 1}, \qquad r \ge 1 \Rightarrow \|k(y;x,w)\| \le K\|w\|^r. \tag{6.2.6}$$

Our processes of interest in this subsection are, for $s \ge (i-1)\Delta_n$:

$$s \mapsto g(X^n_i; X_s), \quad \text{where } X^n_i = X_{(i-1)\Delta_n}.$$

We give the results in three lemmas, starting with the simplest case $r \le 1$.

Lemma 6.2.2 *Assume* (6.2.1), *case* (b) *with* $r \le 1$ *and* $B' = X^c = 0$. *Then under* (6.2.5) *when* $r = 1$ *and* (6.2.4) *when* $r < 1$, *we have for* $s \ge (i-1)\Delta_n$:

$$g\big(X_i^n; X_s\big) = \int_{(i-1)\Delta_n}^s \int_E h\big(X_i^n; X_{v-}, \delta(v,z)\big)\,\mathfrak{p}(dv,dz). \tag{6.2.7}$$

Proof We have $X_s = X_i^n + \sum_{(i-1)\Delta_n < v \le s} \Delta X_v$ and the result is a simple consequence of (3.2.5) (which is a "pathwise" result) applied with $f(x) = g(X_i^n; x)$. □

Lemma 6.2.3 *Assume* (6.2.1), *case* (b), *with* $1 \le r < 2$ *and* $X^c = 0$. *Then under* (6.2.4) *we have for* $s \ge (i-1)\Delta_n$:

$$\begin{aligned} g\big(X_i^n, X_s\big) = &\int_{(i-1)\Delta_n}^s \gamma\big(X_i^n; X\big)_v\,dv \\ &+ \int_{(i-1)\Delta_n}^s \int_E h\big(X_i^n; X_{v-}, \delta(v,z)\big)(\mathfrak{p}-\mathfrak{q})(dv,dz), \end{aligned} \tag{6.2.8}$$

where

$$\gamma(y;X)_v = \sum_{j=1}^d \partial_j g(y; X_v) b_v''^j + \int_E k\big(y; X_{v-}, \delta(v,z)\big)\,\lambda(dz). \tag{6.2.9}$$

If further X *satisfies* (SH-1) *and* b_t' *is given by* (6.2.2), *we also have*

$$\begin{aligned} g\big(X_i^n; X_s\big) = &\sum_{j=1}^d \int_{(i-1)\Delta_n}^s \partial_j g\big(X_i^n; X\big)_v\, b_v'^j\,dv \\ &+ \int_{(i-1)\Delta_n}^s \int_E h\big(X_i^n; X_{v-}, \delta(v,z)\big)\mathfrak{p}(dv,dz). \end{aligned} \tag{6.2.10}$$

Note that the integrand in the last part of (6.2.9) has a norm smaller than $K\Gamma(z)^r$ by (6.2.1) and (6.2.6), so $\|\gamma(y,X)_v\| \le K$.

Proof Here we use (3.2.4) with $f(x) = g(X_i^n; x)$ as in the previous proof, on the time interval $[(i-1)\Delta_n, \infty)$ instead of $[0,\infty)$. Taking the first formula (6.2.3) with $\sigma \equiv 0$ into consideration, and if $\gamma'(y;X)_v$ and $\gamma''(y;X)_v$ denote the first and the second terms in the right side of (6.2.9), this gives

$$\begin{aligned} g\big(X_i^n; X_s\big) = &\int_{(i-1)\Delta_n}^s \gamma'\big(X_i^n; X\big)_v\,dv \\ &+ \int_{(i-1)\Delta_n}^s \int_E k\big(X_i^n; X_{v-}, \delta(v,z)\big)\mathfrak{p}(dv,dz) \\ &+ \int_{(i-1)\Delta_n}^s \int_E \sum_{j=1}^d \partial_j g\big(X_i^n; X_{v-}\big)\delta(v,z)^j (\mathfrak{p}-\mathfrak{q})(dv,dz). \end{aligned}$$

Since $\|k(X_i^n; X_{v-}, \delta(v,z))\| \leq K\Gamma(z)^r$ by (6.2.6), the second term on the right above has locally integrable variation. Its compensator is the same integral with respect to $\mathfrak{q}$, that is the process $\int_{(i-1)\Delta_n}^s \gamma''(X_i^n; X)_v \, dv$. So this term is in fact

$$\int_{(i-1)\Delta_n}^s \gamma''\big(X_i^n; X\big)_v \, dv + \int_{(i-1)\Delta_n}^s \int_E k\big(X_i^n; X_{v-}, \delta(v,z)\big)(\mathfrak{p} - \mathfrak{q})(dv, dz).$$

Since $k(y; x, w) + \sum_{j=1}^d \partial_j g(y; x) w^j = g(y; x)$, we deduce (6.2.8).

Finally under (SH-1), and by (6.2.6), the last integral in (6.2.8) is an ordinary integral, which splits into two integrals with respect to $\mathfrak{p}$ and $\mathfrak{q}$, respectively. We also have $b_t'' - b_t' = \int_E \delta(t,z)\lambda(dz)$, so (6.2.10) is just another of writing (6.2.8). □

Lemma 6.2.4 *Assume* (6.2.1), *case* (b) *with* $r = 2$. *Then under* (6.2.4) *with* $r = 2$ *also, we have for* $s \geq (i-1)\Delta_n$:

$$\begin{aligned} g\big(X_i^n; X_s\big) = {} & \int_{(i-1)\Delta_n}^s \big(\gamma\big(X_i^n; X\big)_v + \widetilde{\gamma}\big(X_i^n; X\big)_v\big) \, dv \\ & + \sum_{j=1}^{d'} \int_{(i-1)\Delta_n}^s \widetilde{\gamma}'\big(X_i^n; X\big)_v^j \, dW_v^j \\ & + \int_{(i-1)\Delta_n}^s \int_E h\big(X_i^n; X_{v-}, \delta(v,z)\big)(\mathfrak{p} - \mathfrak{q})(dv, dz), \end{aligned} \tag{6.2.11}$$

where γ *is as in* (6.2.9) *and*

$$\widetilde{\gamma}(y; X)_v = \frac{1}{2} \sum_{j,k=1}^d \partial_{jk}^2 g(y; X_v) c_v^{jk}, \qquad \widetilde{\gamma}'(y; X)_v^k = \sum_{j=1}^d \partial_j g(y; X_v) \sigma_v^{jk}.$$

If further X *satisfies* (SH-1) *and* b_t' *is given by* (6.2.2), *we also have*

$$\begin{aligned} g\big(X_i^n; X_s\big) = {} & \sum_{j=1}^d \int_{(i-1)\Delta_n}^s \partial_j g\big(X_i^n; X\big)_v b_v'^j \, dv \\ & + \int_{(i-1)\Delta_n}^s \widetilde{\gamma}\big(X_i^n; X\big)_v \, dv + + \sum_{j=1}^{d'} \int_{(i-1)\Delta_n}^s \widetilde{\gamma}'\big(X_i^n; X\big)_v^j \, dW_v^j \\ & + \int_{(i-1)\Delta_n}^s \int_E h\big(X_i^n; X_{v-}, \delta(v,z)\big)\mathfrak{p}(dv, dz). \end{aligned} \tag{6.2.12}$$

Proof Taking into account the first formula (6.2.3) again, plus the fact that the jumps of X are bounded by A, this is exactly the version (2.1.20) of Itô's formula applied with the same f as in the previous proof and on the time interval $[(i-1)\Delta_n, \infty)$. The only difference is that instead of the truncation level 1 for the jumps, here we take A, so in last term of (2.1.20) vanishes and in the one before the last we can delete the indicator function. We then get exactly (6.2.11), and under the additional assumption (SH-1), we deduce (6.2.12) from (6.2.11) as in the previous lemma. □

6.2.2 Reduction of the Problem

In this subsection we show that, in case (b), we can reduce the problem to the situation where X has only finitely many jumps on finite intervals, exactly as in Chap. 5. For proving this, we need some notation. We set $A_m = \{z : \Gamma(z) > 1/m\}$ for $m \geq 1$ an integer, and

$$X(m)_t = X_0 + \int_0^t b''_s\,ds + \int_0^t \sigma_s\,dW_s + \delta_m * (\mathfrak{p} - \mathfrak{q})_t, \quad \text{where } \delta_m = \delta\,1_{A_m}.$$

Lemma 6.2.5 *Assume* (6.2.1), *case* (b). *On the extended space, we have*:

(i) *If f is C^r with compact support,* $\widetilde{V}(f, X(m)) \overset{\text{u.c.p.}}{\Longrightarrow} \widetilde{V}(f, X)$ *as* $m \to \infty$.
(ii) *If f is C^2 with compact support and $p \geq r$,* $\widetilde{V}(f, p, X(m)) \overset{\text{u.c.p.}}{\Longrightarrow} \widetilde{V}(f, p, X)$ *as* $m \to \infty$.

Proof 1) We may assume f to be one-dimensional here. First, by Doob's inequality,

$$\mathbb{E}\Big(\sup_{s\leq t} \|X(m)_s - X_s\|^2\Big) \leq 4\mathbb{E}\big((\|\delta\|^2\,1_{A_m^c}) * \mathfrak{q}_t\big) \leq a_m t,$$

where $a_m = 4\int_{A_m^c} \Gamma(z)^2 \lambda(dz)$ goes to 0 as $m \to \infty$ by Lebesgue's theorem. Hence

$$X(m) \overset{\text{u.c.p.}}{\Longrightarrow} X \quad \text{as } m \to \infty. \tag{6.2.13}$$

Therefore the dominated convergence theorem for stochastic and ordinary integrals yields

$$\int_0^t \partial_j f\big(X(m)_s\big)\,\sigma_s^{jk}\,dW_s'^k \overset{\text{u.c.p.}}{\Longrightarrow} \int_0^t \partial_j f(X_s)\,\sigma_s^{jk}\,dW_s'^k$$

$$\int_0^t \partial_j f\big(X(m)_s\big)\,\partial_k f\big(X(m)_s\big)\,c_s^{jk}\,ds \overset{\text{u.c.p.}}{\Longrightarrow} \int_0^t \partial_j f(X_s)\,\partial_k f(X_s)\,c_s^{jk}\,ds$$

as soon as f is C^1, because $\partial_j f$ is then continuous and bounded. We also have $f(X(m)) \overset{\text{u.c.p.}}{\Longrightarrow} f(X)$. Hence in view of (6.1.2) and (6.1.3) if $r = 2$, or (6.1.7) if $r < 2$, and of (6.1.8), it remains to prove that $\widetilde{V}''(f, X(m)) \overset{\text{u.c.p.}}{\Longrightarrow} \widetilde{V}''(f, X)$ and $\widetilde{V}'(f, p, X(m)) \overset{\text{u.c.p.}}{\Longrightarrow} \widetilde{V}'(f, p, X)$.

2) The difference $\widetilde{V}''(f, X(m)) - \widetilde{V}''(f, X)$ is a square-integrable martingale on the extended space, and its quadratic variation process is

$$\begin{aligned} F(m)_t = \sum_{n:\,T_n\leq t} \Big(& f\big(X(m)_{T_n-} + \Delta X(m)_{T_n}\big) - f\big(X(m)_{T_n-}\big) \\ & - f\big(X_{T_n-} + \Delta X(m)_{T_n}\big) + f(X_{T_n-})\Big)^2 \Big(\kappa_n - \frac{1}{2}\Big)^2. \end{aligned}$$

Each summand in $F(m)_t$ above goes to 0 by (6.2.13) and is smaller than $K\|\Delta X_{T_n}\|^{2r\wedge 2}$, because $\|\Delta X(m)\|\leq\|\Delta X\|$ and f is C^r with compact support. Since $\mathbb{E}(\sum_{n:T_n\leq t}\|\Delta X_{T_n}\|^{2r\wedge 2})<\infty$ by hypothesis, an application of the Lebesgue theorem yields $\mathbb{E}(F(m)_t)\to 0$. Then we conclude $\widetilde{V}''(f,X(m))-\widetilde{V}''(f,X)\stackrel{\text{u.c.p.}}{\Longrightarrow}0$ from Doob's inequality.

3) Observe that $|\widetilde{V}'(f,p,X(m))_s-\widetilde{V}'(f,p,X)_s|$ is smaller for $s\leq t$ than

$$\sum_{q:T_q\leq t}\Big|\big|f\big(X(m)_{T_q-}+\Delta X(m)_{T_q}\big)-f\big(X(m)_{T_q-}\big)\big|^p \\ -\big|f(X_{T_q-}+\Delta X_{T_q})-f(X_{T_q-})\big|^p\Big|.$$

The qth summand above is smaller than $K\|\Delta X_{T_q}\|^p$ when f is C^2 with compact support, and it goes to 0 in probability, so the dominated convergence theorem and the property $\sum_{s\leq t}\|\Delta X_s\|^p<\infty$ gives the convergence to 0 of the above sum. □

Lemma 6.2.6 *Assume* (6.2.1), *case* (b), *with* $X^c=0$ *if* $r<2$ *and also* $B'=0$ *if further* $r<1$. *Let* f *be* C^r *with compact support. Then*

$$\lim_{m\to\infty}\ \sup_n\ \mathbb{E}\bigg(\frac{1}{\Delta_n}\sup_{s\leq t}\big|\widetilde{V}^n\big(f,X(m)\big)_s-\widetilde{V}^n(f,X)_s\big|\bigg)=0.$$

Proof 1) Again we may assume f to be one-dimensional. The proof is somewhat reminiscent of the proof of Theorem 5.1.2. We apply Lemmas 6.2.2, 6.2.3 or 6.2.4, according to the value of r, with the function $g(y;x)=f(y)-f(x)$ which satisfies (6.2.4). With the notation of those lemmas, and recalling $I(n,i)=((i-1)\Delta_n,i\Delta_n]$, we see that

$$\widetilde{V}^n(f,X)_t-\widetilde{V}^n\big(f,X(m)\big)_t=\sum_{i=1}^{[t/\Delta_n]}\int_{I(n,i)}\big(g\big(X(m)^n_i;X(m)_s\big)-g\big(X^n_i;X_s\big)\big)\,ds. \tag{6.2.14}$$

2) We consider first the case $r\geq 1$, and observe that in view of the form of the function g, the functions $h(y;x,w)$ and $k(y;x,w)$ and the variables $\gamma(y;X)_v$, $\widetilde{\gamma}(y;X)_v$ and $\widetilde{\gamma}'(y;X)_v$ do not depend on y. Then we set

$$\theta(m,1)_v=\begin{cases}\gamma\big(X(m)\big)_v-\gamma(X)_v & \text{if } r<2\\ \gamma\big(X(m)\big)_v-\gamma(X)_v+\widetilde{\gamma}\big(X(m)\big)_v-\gamma(X)_v & \text{if } r=2\end{cases}$$

$$\theta(m,2)_v=\begin{cases}\int_E\big(h\big(X(m)_{v-},\delta_m(v,z)\big)-h\big(X_{v-},\delta(v,z)\big)\big)^2\lambda(dz) & \text{if } r<2\\ \int_E\big(h\big(X(m)_{v-},\delta_m(v,z)\big)-h\big(X_{v-},\delta(v,z)\big)\big)^2\lambda(dz) & \\ \quad+\sum_{j=1}^{d'}\big(\widetilde{\gamma}'\big(X(m)_v\big)^j-\widetilde{\gamma}'(X_v)^j\big)^2 & \text{if } r=2\end{cases}$$

$$\rho(m)^n_{i,s}=\int_{(i-1)\Delta_n}^s\int_E\big(h\big(X(m)_{v-},\delta_m(v,z)\big)-h\big(X_{v-},\delta(v,z)\big)\big)(\mathfrak{p}-\mathfrak{q})(dv,dz).$$

Then (6.2.14) and Lemmas 6.2.3 or 6.2.4, according to the case, yield

$$\widetilde{V}^n(f,X)_t - \widetilde{V}^n\big(f,X(m)\big)_t = \sum_{i=1}^{[t/\Delta_n]} \int_{I(n,i)} \big(\zeta(m,1)^n_{i,s} + \zeta(m,2)^n_{i,s}\big)\,ds, \quad \text{with}$$

$$\zeta(m,1)^n_{i,s} = \int_{(i-1)\Delta_n}^{s} \theta(m,1)_v\,dv,$$

$$\zeta(m,2)^n_{i,s} = \begin{cases} \rho(m)^n_{i,s} & \text{if } r<2 \\ \rho(m)^n_{i,s} + \sum_{j=1}^{d'} \int_{(i-1)\Delta_n}^{s} (\widetilde{\gamma}'(X(m))^j_v - \widetilde{\gamma}'(X)^j_v)\,dW^j_v & \text{if } r=2 \end{cases}$$

We are thus left to prove that, for $l=1,2$,

$$\frac{1}{\Delta_n}\,\mathbb{E}\bigg(\sup_{s\le t}\bigg|\sum_{i=1}^{[s/\Delta_n]} \int_{I(n,i)} \zeta(m,l)^n_{i,s}\,ds\bigg|\bigg) \to 0. \tag{6.2.15}$$

Now we use (6.2.13) and the facts that δ_m converges pointwise to δ (because $\delta(t,z)=0$ if $z\notin \cup_{m\ge1}A_m$) with $\|\delta_m(\omega,t,z)\|\le \Gamma(z)$ and $\int \Gamma(z)^r\lambda(dz)<\infty$, and that $g(y;x)=f(y)-f(x)$ is C^r with compact support and $r\ge1$, and (6.2.6): all these, plus Lebesgue's theorem for the second statement below, give the following properties for $l=1,2$.

$$\big|\theta(m,l)_t\big| \le K, \qquad \theta(m,l)_t \stackrel{\mathbb{P}}{\longrightarrow} 0 \ \forall t \text{ as } m\to\infty. \tag{6.2.16}$$

We are now ready to prove (6.2.15). First, we have $|\int_{I(n,i)} \zeta(m,1)^n_{i,s}\,ds| \le \Delta_n \int_{I(n,i)} |\theta(m,1)_s|\,ds$, therefore the left-hand side of (6.2.15) is smaller than $\mathbb{E}(\int_0^t |\theta(m,1)_s|\,ds)$ and (6.2.15) for $l=1$ follows from (6.2.16). Second, $\int_{I(n,i)} \zeta(m,2)^n_{i,s}\,ds$ is a martingale increment for the discrete-time filtration $(\mathcal{F}_{i\Delta_n})_{i\ge1}$, and a simple calculation shows that

$$\mathbb{E}\bigg(\bigg(\int_{I(n,i)} \zeta(m,2)^n_{i,s}\,ds\bigg)^2\bigg) \le \Delta_n^2\,\mathbb{E}\bigg(\int_{I(n,i)} \theta(m,2)_s\,ds\bigg). \tag{6.2.17}$$

Hence by the Doob and Cauchy-Schwarz inequalities,

$$\mathbb{E}\bigg(\sup_{s\le t}\bigg|\sum_{i=1}^{[s/\Delta_n]} \int_{I(n,i)} \zeta(m,2)^n_{i,s}\,ds\bigg|\bigg) \le 2\Delta_n \sqrt{\mathbb{E}\bigg(\int_0^t \theta(m,2)_s\,ds\bigg)},$$

and we conclude (6.2.15) for $l=2$ using (6.2.16) again.

3) Finally we assume $r<1$, so $X^c=0$ and $B'=0$. In this case we set

$$\theta(m)_t = \int_E \big|h\big(X(m)_{t-},\delta_m(t,z)\big) - h\big(X_{t-},\delta(t,z)\big)\big|\,\lambda(dz)$$

$$\zeta(m)^n_{i,s} = \int_{(i-1)\Delta_n}^{s} \int_E \big(h\big(X(m)_{v-}, \delta_m(v,z)\big) - h\big(X_{v-}, \delta(v,z)\big)\big)\mathfrak{p}(dv,dz).$$

Exactly as for (6.2.16) we have

$$\theta(m)_t \le K, \qquad \theta(m)_t \xrightarrow{\mathbb{P}} 0 \ \forall t \text{ as } m \to \infty. \tag{6.2.18}$$

By Lemma 6.2.2 and (6.2.14) we now obtain

$$\widetilde{V}^n(f,X)_t - \widetilde{V}^n\big(f,X(m)\big)_t = \sum_{i=1}^{[t/\Delta_n]} \int_{I(n,i)} \zeta(m)^n_{i,s}\,ds. \tag{6.2.19}$$

On the other hand we have $\mathbb{E}(|\zeta(m)^n_{i,s}|) \le \mathbb{E}(\int_{(i-1)\Delta_n}^s \theta(m)_v\,dv)$. Therefore

$$\begin{aligned}\mathbb{E}\Big(\sup_{s\le t}\big|\widetilde{V}^n\big(f,X(m)\big)_s - \widetilde{V}^n(f,X)_s\big|\Big) &\le \sum_{i=1}^{[t/\Delta_n]} \mathbb{E}\bigg(\int_{I(n,i)} \big|\zeta(m)^n_{i,s}\big|\,ds\bigg)\\ &\le \Delta_n \mathbb{E}\bigg(\int_0^t \theta(m)_s\,ds\bigg)\end{aligned}$$

and $\mathbb{E}(\int_0^t \theta(m)_s\,ds) \to 0$ by (6.2.18) as $m \to \infty$. This completes the proof. □

Lemma 6.2.7 *Assume* (6.2.1), *case* (b), *with* $X^c = 0$ *if* $r < 2$ *and also* $B' = 0$ *if further* $r \le 1$. *Let* f *be* C^2 *with compact support and* $p \ge r$. *Then*

$$\lim_{m\to\infty} \sup_n \mathbb{E}\bigg(\frac{1}{\Delta_n}\sup_{s\le t}\big|\widetilde{V}^n\big(f,p,X(m)\big)_s - \widetilde{V}^n(f,p,X)_s\big|\bigg) = 0.$$

Proof 1) We apply again Lemmas 6.2.2, 6.2.3 or 6.2.4, with the function $g(y;x) = |f(x) - f(y)|^p$ which satisfies (6.2.4) when $p > r$ or $p = r \ne 1$, and (6.2.5) when $p = r = 1$, because f is C^2 with compact support. We then have

$$\begin{aligned}&\widetilde{V}^n\big(f,p,X(m)\big)_t - \widetilde{V}^n(f,p,X)_t\\ &\quad= \sum_{i=1}^{[t/\Delta_n]} \int_{I(n,i)} \big(g\big(X(m)^n_i; X(m)_s\big) - g\big(X^n_i; X_s\big)\big)\,ds.\end{aligned} \tag{6.2.20}$$

At this point we reproduce the proof of the previous lemma, with the following changes. We recall the notation $X^n_i = X_{(i-1)\Delta_n}$.

2) We consider first the case $r > 1$, and we introduce some notation, in which $s \ge (i-1)\Delta_n$; when $1 < r < 2$ we set

$$\theta(m,1)^n_{i,s} = \gamma\big(X(m)^n_i; X(m)\big)_v - \gamma\big(X^n_i; X\big)_s$$

$$\theta(m,2)^n_{i,s} = \int_E \big(h\big(X(m)^n_i; X(m)_{s-}, \delta_m(s,z)\big) - h\big(X^n_i; X_{s-}, \delta(s,z)\big)\big)^2 \lambda(dz),$$

whereas when $r=2$ we set

$$\theta(m,1)^n_{i,s} = \gamma\big(X(m)^n_i; X(m)\big)_s - \gamma\big(X^n_i; X\big)_s + \widetilde{\gamma}\big(X(m)^n_i; X(m)\big)_s - \gamma(X)_s$$

$$\theta(m,2)^n_{i,s} = \int_E \big(h\big(X(m)^n_i; X(m)_{s-}, \delta_m(s,z)\big) - h\big(X^n_i; X_{s-}, \delta(s,z)\big)\big)^2 \lambda(dz)$$

$$+ \sum_{j=1}^{d'} \big(\widetilde{\gamma}'(X(m)_s)^j - \widetilde{\gamma}'(X^n_i; X_s)^j\big)^2$$

and for all $r>1$,

$$\rho(m)^n_{i,s} = \int_{(i-1)\Delta_n}^{s} \int_E \big(h\big(X(m)^n_i; X(m)_{v-}, \delta_m(v,z)\big)$$

$$- h\big(X^n_i; X_{v-}, \delta(v,z)\big)\big)(\mathfrak{p}-\mathfrak{q})(dv,dz).$$

Then (6.2.20) and Lemmas 6.2.3 or 6.2.4 yield

$$\widetilde{V}^n\big(f,p,X(m)\big)_t - \widetilde{V}^n(f,p,X)_t = \sum_{i=1}^{[t/\Delta_n]} \int_{I(n,i)} \big(\zeta(m,1)^n_{i,s} + \zeta(m,2)^n_{i,s}\big)\,ds,$$

$$\text{where } \zeta(m,1)^n_{i,s} = \int_{(i-1)\Delta_n}^{s} \theta(m,1)^n_{i,v}\,dv \quad \text{and}$$

$$r<2 \Rightarrow \zeta(m,2)^n_{i,s} = \rho(m)^n_{i,s}$$

$$r=2 \Rightarrow \zeta(m,2)^n_{i,s} = \rho(m)^n_{i,s}$$

$$+ \sum_{j=1}^{d'} \int_{(i-1)\Delta_n}^{s} \big(\widetilde{\gamma}'\big(X(m)^n_i; X(m)\big)^j_v - \widetilde{\gamma}'\big(X^n_i; X\big)^j_v\big)\,dW^j_v$$

and as in the previous lemma we are left to prove that (6.2.15) holds for $l=1,2$.

With

$$\overline{\theta}(m,l)^n_t = \theta(m,l)^n_{i,t} \quad \text{if } (i-1)\Delta_n \le t < i\Delta_n, \quad \overline{\theta}'(m,l)_t = \sup_n \overline{\theta}(m,l)^n_t,$$

the same arguments as for (6.2.16) yield here that, for $l=1,2$:

$$\big|\overline{\theta}'(m,l)_t\big| \le K, \qquad \overline{\theta}'(m,l)_t \xrightarrow{\mathbb{P}} 0 \;\forall t \text{ as } m\to\infty. \tag{6.2.21}$$

We have $|\int_{I(n,i)} \zeta(m,1)^n_{i,s}\,ds| \le \Delta_n \int_{I(n,i)} |\theta(m,1)^n_{i,s}|\,ds$, so for $l=1$ the left-hand side of (6.2.15) is smaller than $\mathbb{E}(\int_0^t |\overline{\theta}'(m,1)_s|\,ds)$ and (6.2.15) for $l=1$ follows from (6.2.21). Next, the same argument as in the previous lemma yields that, instead of (6.2.17), we have

$$\mathbb{E}\bigg(\bigg(\int_{I(n,i)} \zeta(m,2)^n_{i,s}\,ds\bigg)^2\bigg) \le \Delta_n^2\,\mathbb{E}\bigg(\int_{I(n,i)} \overline{\theta}'(m,2)_s\,ds\bigg).$$

We conclude (6.2.15) for $l=2$ as in the previous lemma again.

3) Finally we assume $r\le 1$, and we set, again for $s\ge (i-1)\Delta_n$:

$$\begin{aligned}
\theta(m)^n_{i,s} &= \int_E \big|h\big(X(m)^n_i;X(m)_{s-},\delta_m(v,z)\big)-h\big(X^n_i;X_{s-},\delta(s,z)\big)\big|\,\lambda(dz)\\
\zeta(m)^n_{i,s} &= \int_{(i-1)\Delta_n}^s \int_E \big(h\big(X(m)^n_i;X(m)_{s-},\delta_m(s,z)\big)\\
&\quad - h\big(X^n_i;X_{s-},\delta(s,z)\big)\big)\mathfrak{p}(ds,dz)\\
\overline{\theta}(m)^n_t &= \theta(m)^n_{i,t} \quad \text{if } (i-1)\Delta_n\le t<i\Delta_n, \qquad \overline{\theta}'(m)_t = \sup_n \overline{\theta}(m)^n_t.
\end{aligned}$$

Exactly as for (6.2.21) we have

$$\overline{\theta}'(m)_t \le K, \qquad \overline{\theta}'(m)_t \xrightarrow{\mathbb{P}} 0 \ \forall t \text{ as } m\to\infty.$$

Now we have (6.2.19) and $\mathbb{E}(|\zeta(m)^n_{i,s}|)\le \mathbb{E}(\int_{(i-1)\Delta_n}^s \overline{\theta}(m)^n_{i,v}\,dv)$. Hence

$$\mathbb{E}\Big(\sup_{s\le t}\big|\widetilde{V}^n\big(f,X(m)\big)_s-\widetilde{V}^n(f,X)_s\big|\Big) \le \Delta_n\mathbb{E}\bigg(\int_0^t \overline{\theta}'(m)_s\,ds\bigg)$$

again, and $\mathbb{E}(\int_0^t \overline{\theta}'(m)_s\,ds)\to 0$ by (6.2.18). This completes the proof. □

Combining the three previous lemmas and Proposition 2.2.4, we deduce the following:

Corollary 6.2.8 *Assume* (6.2.1), *case* (b).

(i) *Let f be C^r with compact support, and $X^c=0$ when $r<2$ and also $B'=0$ when further $r<1$. If for each $m\ge 1$ we have $-\frac{1}{\Delta_n}\widetilde{V}^n(f,X(m))\stackrel{\mathcal{L}\text{-}s}{\Longrightarrow}\widetilde{V}(f,X(m))$ as $n\to\infty$, then we also have $-\frac{1}{\Delta_n}\widetilde{V}^n(f,X)\stackrel{\mathcal{L}\text{-}s}{\Longrightarrow}\widetilde{V}(f,X)$.*

(ii) *Let f is C^2 with compact support and p be such that $r=p\wedge 2$, and $X^c=0$ when $p<2$ and also $B'=0$ when further $p\le 1$. If for each $m\ge 1$ we have $\frac{1}{\Delta_n}\widetilde{V}^n(f,p,X(m))\stackrel{\mathcal{L}\text{-}s}{\Longrightarrow}\widetilde{V}(f,p,X(m))$ as $n\to\infty$, then we also have $\frac{1}{\Delta_n}\widetilde{V}^n(f,p,X)\stackrel{\mathcal{L}\text{-}s}{\Longrightarrow}\widetilde{V}(f,p,X)$.*

6.3 Proof of the Theorems

On top of (6.2.1), and in the light of Corollary 6.2.8, when X is discontinuous we see that it suffices to prove the results for each process $X(m)$. This process satisfies (6.2.1) and its jump coefficient δ_m is smaller in modulus than $\Gamma(z)1_{A_m}(z)$. Put in another way, we have to prove the result when X satisfies (6.2.1) with, in case (b), a function Γ having the following property:

$$\Gamma(z) \leq A, \qquad \lambda\big(\{z : \Gamma(z) > 0\}\big) < \infty. \tag{6.3.1}$$

We can take for the weakly exhausting sequence (T_n) of stopping times the successive jump times of the Poisson process $1_{\{\Gamma>0\}} * \mathfrak{p}$, whose parameter is $\lambda(\{z : \Gamma(z) > 0\})$. Moreover, (6.3.1) implies that actually (SH-0) holds, so we can use the second form (6.2.3), that is

$$X_t = X_0 + \int_0^t b'_s\, ds + \int_0^t \sigma_s\, dW_s + \delta * \mathfrak{p}_t.$$

Finally, as already said, we can assume f to be with compact support.

6.3.1 Proof of Theorem 6.1.2

Step 1) This step is devoted to the following lemma.

Lemma 6.3.1 *Let a_t be a progressively measurable process which satisfies $\mathbb{E}(\int_0^t |a_s|\, ds) < \infty$ for all t, and set $A_t = \int_0^t a_s\, ds$. Then*

$$\begin{aligned}
&\frac{1}{\Delta_n} \sum_{i=1}^{[t/\Delta_n]} \int_{I(n,i)} ds \int_{(i-1)\Delta_n}^{s} a_v\, dv \stackrel{\text{u.c.p.}}{\Longrightarrow} A_t/2 \\
&\frac{1}{\Delta_n} \sum_{i=1}^{[t/\Delta_n]} \mathbb{E}\bigg(\int_{I(n,i)} ds \int_{(i-1)\Delta_n}^{s} a_v\, dv \mid \mathcal{F}_{(i-1)\Delta_n}\bigg) \stackrel{\text{u.c.p.}}{\Longrightarrow} A_t/2 \\
&\frac{1}{\Delta_n^2} \sum_{i=1}^{[t/\Delta_n]} \int_{I(n,i)} ds \int_{I(n,i)} ds' \int_{(i-1)\Delta_n}^{s\wedge s'} a_v\, dv \stackrel{\text{u.c.p.}}{\Longrightarrow} A_t/3 \\
&\frac{1}{\Delta_n^2} \sum_{i=1}^{[t/\Delta_n]} \mathbb{E}\bigg(\int_{I(n,i)} ds \int_{I(n,i)} ds' \int_{(i-1)\Delta_n}^{s\wedge s'} a_v\, dv \mid \mathcal{F}_{(i-1)\Delta_n}\bigg) \stackrel{\text{u.c.p.}}{\Longrightarrow} A_t/3.
\end{aligned} \tag{6.3.2}$$

Proof We prove only the second claim, the first claim being similar, and even slightly simpler. Write $A(a)_t^n$ and $A'(a)_t^n$ for the two left sides of (6.3.2), and $A(a)_t = \int_0^t a_s\, ds$, to emphasize the dependency on a_t.

1) Here we prove $A(a)^n \stackrel{\text{u.c.p.}}{\Longrightarrow} A(a)/3$. Suppose first that a is continuous. For each t the variables $\theta(n,t) = \sup(|a(s+r) - a(s)| : s \leq t, r \leq \Delta_n)$ go to 0 as $n \to \infty$.

Since $\int_{I(n,i)} ds \int_{I(n,i)} ds' \int_{(i-1)\Delta_n}^{s\wedge s'} dv = \Delta_n^3/3$, we have

$$\sup_{s\le t} \left| A(a)^n_s - \frac{\Delta_n}{3} \sum_{i=1}^{[s/\Delta_n]} a_{(i-1)\Delta_n} \right| \le \frac{t}{3}\,\theta(n,t).$$

Then $A(a)^n \stackrel{\text{u.c.p.}}{\Longrightarrow} A(a)/3$ follows from Riemann integration.

When a is not continuous but satisfies $\mathbb{E}(\int_0^t |a_s|\,ds) < \infty$ for all t, we can find a sequence $a(q)$ of continuous adapted processes satisfying the same integrability condition and such that $\mathbb{E}(\int_0^t |a(q)_s - a_s|\,ds) \to 0$ as $q \to \infty$. We have $A(a(q))^n \stackrel{\text{u.c.p.}}{\Longrightarrow} A(a(q))/3$ as $n \to \infty$ by what precedes, whereas $A(a(q)) \stackrel{\text{u.c.p.}}{\Longrightarrow} A(a)$ as $q \to \infty$, and

$$\left|A\big(a(q)\big)^n_t - A(a)^n_t\right| \le \int_0^t |a(q)_s - a_s|\,ds,$$

hence $A(a)^n \stackrel{\text{u.c.p.}}{\Longrightarrow} A(a)/3$ follows.

2) Next, we prove $A'(a)^n \stackrel{\text{u.c.p.}}{\Longrightarrow} A(a)/3$, which in view of what precedes amounts to showing that $M^n = A(a)^n - A'(a)^n \stackrel{\text{u.c.p.}}{\Longrightarrow} 0$. Set

$$a(m)_t = a_t\, 1_{\{|a_t|\le m\}}, \qquad M(m)^n = A\big(a(m)\big)^n - A'\big(a(m)\big)^n$$
$$a'(m)_t = a_t\, 1_{\{|a_t|> m\}}, \qquad M'(m)^n = A\big(a'(m)\big)^n - A'\big(a'(m)\big)^n.$$

We then have

$$M(m)^n_t = \sum_{i=1}^{[t/\Delta_n]} \zeta(m)^n_i, \quad \text{where } \zeta(m)^n_i = \eta(m)^n_i - \xi(m)^n_i \quad \text{and}$$

$$\eta(m)^n_i = \frac{1}{\Delta_n^2} \int_{I(n,i)} ds \int_{I(n,i)} ds' \int_{(i-1)\Delta_n}^{s\wedge s'} a(m)_v\,dv$$
$$\xi(m)^n_i = \mathbb{E}\big(\eta(m)^n_i \mid \mathcal{F}_{(i-1)\Delta_n}\big),$$

and the same for $M'(m)$ with the variables $\zeta'(m)^n_i$, $\eta'(m)^n_i$ and $\xi'(m)^n_i$.

On the one hand, $|\eta(m)^n_i| \le m\Delta_n$ by the definition of $a(m)_t$, so $|\zeta(m)^n_i| \le 2m\Delta_n$, and thus $\mathbb{E}(\sum_{i=1}^{[t/\Delta_n]} |\zeta(m)^n_i|^2) \le 4m^2 t\Delta_n$. Since $M(m)^n$ is a martingale for the filtration $(\mathcal{F}_{\Delta_n[t/\Delta_n]})$, for any given m we deduce $M(m)^n \stackrel{\text{u.c.p.}}{\Longrightarrow} 0$ as $n \to \infty$ from Doob's inequality.

On the other hand, it is obvious that $|\eta'(m)^n_i| \le \int_{I(n,i)} |a'(m)_s|\,ds$, hence $\mathbb{E}(|\zeta'(m)^n_i|) \le 2\mathbb{E}(\int_{I(n,i)} |a'(m)_s|\,ds)$ by the contraction property of the conditional expectation. Therefore

$$\mathbb{E}\Big(\sup_{s\le t} |M'(m)^n_s|\Big) \le 2\mathbb{E}\bigg(\int_0^t |a_s| 1_{\{|a_s|>m\}}\,ds\bigg),$$

which tends to 0 as $m \to \infty$ by the dominated convergence theorem and our assumption on a_t. Combining this with $M(m)^n \stackrel{\text{u.c.p.}}{\Longrightarrow} 0$ as $n \to \infty$ and Proposition 2.2.1 and gives $M^n = M(m)^n + M'(m)^n \stackrel{\text{u.c.p.}}{\Longrightarrow} 0$, and the result is proved. □

Step 2) For any t we denote by $\Omega_n(t)$ the set on which any interval $I(n,i)$ included into $[0, t+1]$ contains at most one stopping time T_q, so

$$\mathbb{P}\big(\Omega_n(t)\big) \to 1 \quad \text{as } n \to \infty. \tag{6.3.3}$$

We also use the following notation, with the convention $T_0 = 0$:

$$\begin{aligned} i(n,q) &= i, \quad \kappa(n,q) = i - T_q/\Delta_n \quad \text{on the set } \big\{T_q \in I(n,i)\big\} \\ N_t^n &= q \quad \text{on the set } \big\{T_q \le \Delta_n[t/\Delta_n] < T_{q+1}\big\}. \end{aligned} \tag{6.3.4}$$

We can apply the Itô's formulas (6.2.12) in case (a) and (6.2.7) in case (b) (recall then that (SH-0) holds), and with the function $g(y;x) = f(x) - f(y)$ which here is C^2 by the assumptions of the theorem. With the notation

$$w_t = \sum_{j=1}^d \partial_j f(X_t) b_t'^j + \frac{1}{2} \sum_{j,k=1}^d \partial_{jk}^2 f(X_t) c_t^{jk}, \qquad \xi_t^k = \sum_{j=1}^d \partial_j f(X_t) \sigma_t^{jk} \tag{6.3.5}$$

(so ξ is the process showing in Theorem 6.1.8) we deduce that, on the set $\Omega_n(t)$, we have

$$-\frac{1}{\Delta_n} \widetilde{V}^n(f,X)_t = \sum_{i=1}^{[t/\Delta_n]} \big(\zeta_i^n + \zeta_i'^n + \zeta_i''^n\big) + \sum_{q=1}^{N_t^n} \big(\eta(q)^n - \eta'(q)^n\big), \tag{6.3.6}$$

where

$$\zeta_i''^n = \frac{1}{\Delta_n} \int_{I(n,i)} ds \int_{(i-1)\Delta_n}^s w_v\, dv, \qquad \zeta_i'^n = \frac{1}{2} \sum_{k=1}^{d'} \int_{I(n,i)} \xi_s^k\, dW_s^k$$

$$\zeta_i^n = \frac{1}{\Delta_n} \sum_{k=1}^{d'} \int_{I(n,i)} ds \int_{(i-1)\Delta_n}^s \xi_v^k\, dW_v^k - \zeta_i'^n$$

$$\eta(q)^n = \big(f(X_{T_q-} + \Delta X_{T_q}) - f(X_{T_q-})\big)\kappa(n,q)$$

$$\eta'(q)^n = \zeta_{i(n,q)}^n + \zeta_{i(n,q)}'^n + \zeta_{i(n,q)}''^n.$$

On the other hand, we can write the limit (6.1.2) as follows, by Itô's formula again and because the sum over the T_q's below is a finite sum:

$$\begin{aligned} \widetilde{V}(f,X) = {} & \frac{1}{2} \int_0^t w_s\, ds + \frac{1}{2} \sum_{k=1}^{d'} \int_0^t \xi_s^k\, dW_s^k + \frac{1}{12} \sum_{k=1}^{d'} \int_0^t \xi_s^k\, dW_s'^k \\ & + \sum_{q:T_q \le t} \big(f(X_{T_q-} + \Delta X_{T_q}) - f(X_{T_q-})\big)\, \kappa_q. \end{aligned} \tag{6.3.7}$$

Step 3) Our assumptions imply the following:

$$\begin{aligned} &\bullet \text{ in case (a):} \quad \int_0^\infty (\|w_s\| + \|\xi_s^k\|^2)\, ds \le K, \quad \eta(q)^n = \eta'(q)^n = 0 \\ &\bullet \text{ in case (b):} \quad \|w_s\| + \|\xi_s^k\| \le K. \end{aligned} \tag{6.3.8}$$

Then Lemma 6.3.1 yields in both cases:

$$\sum_{i=1}^{[t/\Delta_n]} \zeta_i''^n \stackrel{\text{u.c.p.}}{\Longrightarrow} \frac{1}{2}\int_0^t w_s\, ds.$$

Next, the variable $\zeta_i'^n$ is the increment over the interval $I(n,i)$ of the continuous process $\frac{1}{2}\sum_{k=1}^{d'}\int_0^t \xi_s^k\, dW_s^k$, so we know that

$$\sum_{i=1}^{[t/\Delta_n]} \zeta_i'^n \stackrel{\text{u.c.p.}}{\Longrightarrow} \frac{1}{2}\sum_{k=1}^{d'}\int_0^t \xi_s^k\, dW_s^k.$$

Therefore we deduce from (6.3.3), (6.3.6) and (6.3.7) and the property $\sup_n N_t^n < \infty$ that it is enough to prove the following two convergences:

$$q \ge 1 \quad \Rightarrow \quad \eta'(q)^n \stackrel{\mathbb{P}}{\longrightarrow} 0 \tag{6.3.9}$$

$$\begin{aligned} \sum_{i=1}^{[t/\Delta_n]} \zeta_i^n + \sum_{q=1}^{N_t^n} \eta(q)^n \stackrel{\mathcal{L}\text{-s}}{\Longrightarrow} & \frac{1}{\sqrt{12}}\sum_{k=1}^{d'}\int_0^t \xi_s^k\, dW_s'^k \\ & + \sum_{q:T_q \le t} \big(f(X_{T_q-} + \Delta X_{T_q}) - f(X_{T_q-})\big)\kappa_q. \end{aligned} \tag{6.3.10}$$

Step 4) In this step we prove (6.3.9). In case (a) there is nothing to prove. In case (b), w_t and ξ^k are bounded, hence $\|\zeta_i''^n\| \le K\Delta_n$ and $\mathbb{E}(\|\zeta_{i(n,q)}'^n\|^2) \le K\Delta_n$ and $\mathbb{E}(\|\zeta_{i(n,q)}^n\|^2) \le K\Delta_n$ by an application of Proposition 2.1.10-(a) with $A = E$, because the integers $i(n,q)$ are then $\mathcal{G}_0^A$ measurable. Therefore (6.3.9) holds.

Step 5) In this step we prove a part of (6.3.10), namely

$$\sum_{i=1}^{[t/\Delta_n]} \zeta_i^n \stackrel{\mathcal{L}\text{-s}}{\Longrightarrow} Y_t = \frac{1}{\sqrt{12}}\sum_{k=1}^{d'}\int_0^t \xi_s^k\, dW_s'^k. \tag{6.3.11}$$

For this we will apply Theorem 2.2.15 to the array (ζ_i^n), which is an array of martingale differences for the discrete-time filtrations $(\mathcal{G}_i^n = \mathcal{F}_{i\Delta_n})_{i\ge 0}$. In particular we have (2.2.39) with $(\Omega_n, \mathcal{G}_n, (\overline{\mathcal{F}}_t^n), \mathbb{P}_n) = (\Omega, \mathcal{F}, (\mathcal{F}_t), \mathbb{P})$. Recall that $\zeta_i^n = (\zeta_i^{n,l})$ and $\xi_t^k = (\xi_s^{k,l})$ are q-dimensional. Moreover the limiting process Y above is, conditionally on $\mathcal{F}$, a continuous centered Gaussian martingale with

$$\widetilde{\mathbb{E}}(Y_t^l Y_t^m \mid \mathcal{F}) = \frac{1}{12}\int_0^t \sum_{k=1}^{d'} \xi_s^{k,l}\,\xi_s^{k,m}\,ds.$$

Therefore, to obtain (6.3.11) it is enough to prove the following three properties, where M is either one of the components of W or is a bounded martingale on $(\Omega,\mathcal{F},(\mathcal{F}_t)_{t\geq 0},\mathbb{P})$ which is orthogonal to W, and $t\geq 0$:

$$\sum_{i=1}^{[t/\Delta_n]} \mathbb{E}\big(\|\zeta_i^n\|^4 \mid \mathcal{F}_{(i-1)\Delta_n}\big) \xrightarrow{\mathbb{P}} 0 \tag{6.3.12}$$

$$\sum_{i=1}^{[t/\Delta_n]} \mathbb{E}\big(\zeta_i^{n,l}\,\zeta_i^{n,m} \mid \mathcal{F}_{(i-1)\Delta_n}\big) \xrightarrow{\mathbb{P}} \frac{1}{12}\int_0^t \sum_{k=1}^{d'} \xi_s^{k,l}\,\xi_s^{k,m}\,ds \tag{6.3.13}$$

$$\sum_{i=1}^{[t/\Delta_n]} \mathbb{E}\big(\zeta_i^n\,\Delta_i^n M \mid \mathcal{F}_{(i-1)\Delta_n}\big) \xrightarrow{\mathbb{P}} 0. \tag{6.3.14}$$

An easy calculation using (2.1.34) shows that

$$\mathbb{E}\big(\|\zeta_i^n\|^4 \mid \mathcal{F}_{(i-1)\Delta_n}\big) \leq K\sum_{k=1}^{d'} \mathbb{E}\bigg(\bigg(\int_{I(n,i)} \|\xi_s^k\|^2\,ds\bigg)^2 \mid \mathcal{F}_{(i-1)\Delta_n}\bigg).$$

In case (b), ξ^k is bounded, so the right side above is smaller than $K\Delta_n^2$ and (6.3.12) follows. In case (a), it is smaller than $K\sum_{k=1}^{d'}\mathbb{E}(\theta(t)_n\int_{I(n,i)}\|\xi_s^k\|^2\,ds \mid \mathcal{F}_{(i-1)\Delta_n})$ if $i\leq[t/\Delta_n]$, where $\theta(t)_n=\sup_{i\leq[t/\Delta_n]}\sum_{k=1}^{d'}\int_{I(n,i)}\|\xi_s^k\|^2\,ds$. Hence the left side of (6.3.12) has an expectation smaller than

$$K\mathbb{E}\bigg(\theta(t)_n\sum_{k=1}^{d'}\int_0^t \|\xi_s^k\|^2\bigg),$$

which goes to 0 by the dominated convergence theorem (recall (6.3.8), which in particular yields that $\theta(t)_n$ is bounded and goes to 0 as $n\to\infty$). Hence again (6.3.12) holds.

Next, we have

$$\begin{aligned}\zeta_i^{n,l}\,\zeta_i^{n,m} &= \frac{1}{\Delta_n^2}\sum_{k,k'=1}^{d'}\int_{I(n,i)} ds\int_{I(n,i)} ds'\int_{(i-1)\Delta_n}^{s}\xi_v^{k,l}\,dW_v^k\int_{(i-1)\Delta_n}^{s'}\xi_{v'}^{k',m}\,dW_{v'}^{k'}\\ &\quad-\frac{1}{2\Delta_n}\sum_{k,k'=1}^{d'}\int_{I(n,i)} ds\int_{(i-1)\Delta_n}^{s}\xi_v^{k,l}\,dW_v^k\int_{I(n,i)}\xi_{v'}^{k',m}\,dW_{v'}^{k'}\end{aligned}$$

$$-\frac{1}{2\Delta_n}\sum_{k,k'=1}^{d'}\int_{I(n,i)} ds\int_{(i-1)\Delta_n}^{s}\xi_v^{k,m}\,dW_v^k\int_{I(n,i)}\xi_{v'}^{k',l}\,dW_{v'}^{k'}$$
$$+\frac{1}{4}\sum_{k,k'=1}^{d'}\int_{I(n,i)}\xi_v^{k,m}\,dW_v^k\int_{I(n,i)}\xi_{v'}^{k',l}\,dW_{v'}^{k'}.$$

Then by standard stochastic calculus and Fubini's theorem, we get that $\mathbb{E}(\zeta_i^{n,l}\,\zeta_i^{n,m}\mid\mathcal{F}_{(i-1)\Delta_n})=\mathbb{E}(\theta_i^{n,l,m}\mid\mathcal{F}_{(i-1)\Delta_n})$, where

$$\theta_i^{n,l,m}=\frac{1}{\Delta_n^2}\sum_{k=1}^{d'}\int_{I(n,i)} ds\int_{I(n,i)} ds'\int_{(i-1)\Delta_n}^{s\wedge s'}\xi_v^{k,l}\,\xi_v^{k,m}\,dv$$
$$-\frac{1}{\Delta_n}\sum_{k=1}^{d'}\int_{I(n,i)} ds\int_{(i-1)\Delta_n}^{s}\xi_v^{k,l}\,\xi_v^{k,m}\,dv+\frac{1}{4}\sum_{k=1}^{d'}\int_{I(n,i)}\xi_v^{k,m}\,\xi_v^{k,l}\,dv.$$

At this stage, (6.3.13) follows from (6.3.8) and Lemma 6.3.1, because $\frac{1}{3}-\frac{1}{2}+\frac{1}{4}=\frac{1}{12}$.

When M is a bounded martingale and is orthogonal to W, we have $\mathbb{E}(\zeta_i^n\,\Delta_i^n M\mid\mathcal{F}_{(i-1)\Delta_n})=0$ because ζ_i^n is a stochastic integral with respect to W, and (6.3.14) holds. If $M=W^k$, we see that $\mathbb{E}(\zeta_i^n\,\Delta_i^n M\mid\mathcal{F}_{(i-1)\Delta_n})=\mathbb{E}(\theta_i^n\mid\mathcal{F}_{(i-1)\Delta_n})$, where

$$\theta_i^n=\frac{1}{\Delta_n}\int_{I(n,i)} ds\int_{(i-1)\Delta_n}^{s}\xi_v^k\,dv-\frac{1}{2}\int_{I(n,i)}\xi_s^k\,ds.$$

Then Lemma 6.3.1 yields $\sum_{i=1}^{[t/\Delta_n]}\mathbb{E}(\theta_i^n\mid\mathcal{F}_{(i-1)\Delta_n})\xrightarrow{\mathbb{P}}0$, giving (6.3.14). We have thus finished the proof of (6.3.11).

Step 6) In this final step, we prove (6.3.10), which is equivalent to proving

$$\left(\sum_{i=1}^{[t/\Delta_n]}\zeta_i^n,\ \big(\kappa(n,q)\big)_{q\geq 1}\right)\xrightarrow{\mathcal{L}\text{-s}}\left(\frac{1}{\sqrt{12}}\sum_{k=1}^{d'}\int_0^t\xi_s^k\,dW_s'^k,\ (\kappa_q)_{q\geq 1}\right)\tag{6.3.15}$$

by the same argument which allows us to deduce (5.1.15) from (5.1.14), for example.

In order to obtain (6.3.15), we can use Theorem 4.3.1, with $v_n=\Delta_n$: there is a slight difference here, since $\kappa(n,q)$ in (6.3.4) is the variable $1-\kappa(n,q)$ in that lemma, but since κ_q and $1-\kappa_q$ have the same law it makes no difference. Let us recall that a consequence of that theorem (and with its notation) is that

$$\big(\overline{U}^n(g),\ \big(\kappa(n,q)\big)_{q\geq 1}\big)\xrightarrow{\mathcal{L}\text{-s}}\big(\overline{U}(g),\ (\kappa_q)_{q\geq 1}\big).\tag{6.3.16}$$

Hence here we need to replace the processes $\overline{U}^n(g)$ and $\overline{U}(g)$ by $\sum_{i=1}^{[t/\Delta_n]}\zeta_i^n$ and $\frac{1}{\sqrt{12}}\sum_{k=1}^{d'}\int_0^t\xi_s^k\,dW_s'^k$, respectively. We can then copy the proof of that theorem with these changes, so that the key property (4.3.7), itself implied by (4.3.5), is provided by (6.3.11), which holds under the measure $\mathbb{P}$, and also under the measures Q_ω which are a regular version of the conditional distribution $\mathbb{P}(. \mid \mathcal{G}_0)$ where $(\mathcal{G}_t)$ is as in Step 3. We leave the details to the reader (the proof is slightly simpler, because we do not need the processes $w(n,p)$ and $W''(p)$ here). We end up with (6.3.16), which completes the proof of Theorem 6.1.2.

6.3.2 Proof of Theorem 6.1.3

The proof is the same as in the previous subsection, with a few changes and simplifications. We are in case (b) here, with $r < 2$ (and still (SH-0), according to (6.3.1)).

We still have (6.3.6) (the function g is now C^r only), with $\xi_t^k = 0$ and:

$$w_t = \begin{cases} \sum_{j=1}^d \partial_j f(X_t) b_t'^j & \text{if } r \geq 1 \\ 0 & \text{if } r < 1, \end{cases}$$

hence $\zeta_i^n = \zeta_i'^n = 0$, and also $\zeta_i''^n = 0$ if $r < 1$. Furthermore (6.3.7) still holds, Step 4 is irrelevant, and the rest of the proof is the same.

6.3.3 Proof of Theorem 6.1.4

Step 1) For this theorem, f is one-dimensional. We are in case (b) of (6.2.1), and we also have (SH-0) because of (6.3.1).

We use the notation $\Omega_n(t)$ of (6.3.3), as well as (6.3.4) and (6.3.5), and for any two reals y and $z > 0$ we write $\{y\}^z = |y|^z \operatorname{sign}(y)$.

First, suppose $p \geq 2$. We then apply the version (6.2.12) of Itô's formula (because (SH-0) holds) with the function $g(y;x) = |f(x) - f(y)|^p$, which is C^2 with compact support and satisfies $\partial_j g(y;x) = p\{f(x) - f(y)\}^{p-1} \partial_j f(x)$ and $\partial^2_{jk} g(y;x) = p\{f(x) - f(y)\}^{p-1} \partial^2_{jk} f(x) + p(p-1)|f(x) - f(y)|^{p-2} \partial_j f(x) \partial_k f(x)$. We then obtain on the set $\Omega_n(t)$:

$$\frac{1}{\Delta_n} \widetilde{V}^n(f,p,X)_t = \sum_{i=1}^{[t/\Delta_n]} \left(\zeta_i^n + \zeta_i'^n + \zeta_i''^n\right) + \sum_{q=1}^{N_t^n} \left(\eta(q)^n - \eta'(q)^n + \eta''(q)^n\right), \tag{6.3.17}$$

where, with $Y_{i,s}^n = f(X_s) - f(X_{(i-1)\Delta_n})$:

$$\zeta_i^n = \frac{p}{\Delta_n} \int_{I(n,i)} ds \int_{(i-1)\Delta_n}^s \{Y_{i,v}^n\}^{p-1} w_v \, dv$$

$$\zeta_i'^n = \frac{p}{\Delta_n} \sum_{k=1}^{d'} \int_{I(n,i)} ds \int_{(i-1)\Delta_n}^s \{Y_{i,v}^n\}^{p-1} \xi_v^k \, dW_v^k$$

$$\begin{aligned}
\zeta_i''^n &= \frac{p(p-1)}{2\Delta_n} \sum_{k=1}^{d'} \int_{I(n,i)} ds \int_{(i-1)\Delta_n}^{s} |Y_{i,v}^n|^{p-2} (\xi_v^k)^2 \, dv \\
\eta(q)^n &= |f(X_{T_q-} + \Delta X_{T_q}) - f(X_{T_q-})|^p \kappa(n,q) \\
\eta'(q)^n &= \zeta_{i(n,q)}^n + \zeta_{i(n,q)}'^n + \zeta_{i(n,q)}''^n \\
\eta''(q)^n &= \big(|f(X_{T_q-} + \Delta X_{T_q}) - f(X_{(i(n,q)-1)\Delta_n})|^p \\
&\quad - |f(X_{T_q-}) - f(X_{(i(n,q)-1)\Delta_n})|^p \\
&\quad - |f(X_{T_q-} + \Delta X_{T_q}) - f(X_{T_q-})|^p\big)\kappa(n,q).
\end{aligned}$$

Next if $1 \le p < 2$ we apply (6.2.10) to obtain (6.3.17) on the set $\Omega_n(t)$, with $\zeta_i'^n = \zeta_i''^n = 0$ and $\eta'(q)^n = \zeta_{i(n,q)}^n$, and $w_t = \sum_{j=1}^d \partial_j f(X_t) b_t'^j$. Finally, when $p \le 1$ we apply (6.2.7) to obtain (6.3.17) again on the set $\Omega_n(t)$, with $\zeta_i^n = \zeta_i'^n = \zeta_i''^n = 0$ and $\eta'(q)^n = 0$.

Step 2) In view of (6.2.1), case (b), we deduce from (2.1.44) that if $l > 0$ and $s \in I(n,i)$,

$$\mathbb{E}\big(|Y_{i,v}^n|^l\big) \le K \Delta_n^{1 \wedge (l/2)}.$$

Then since w and ξ^k are bounded, we deduce when $p \ge 1$:

$$\begin{aligned}
&\mathbb{E}\big(|\zeta_i^n|\big) \le K \Delta_n^{1+(2\wedge(p-1))/2}, \qquad \mathbb{E}\big(|\zeta_i''^n|\big) \le K \Delta_n^{1+(2\wedge(p-2))/2}, \\
&\mathbb{E}\big(\zeta_i'^n \mid \mathcal{F}_{(i-1)\Delta_n}\big) = 0, \qquad \mathbb{E}\big(|\zeta_i'^n|^2\big) \le K \Delta_n^{p\wedge 2},
\end{aligned} \tag{6.3.18}$$

from which the following properties readily follow:

$$\left.\begin{aligned}
p > 1 \ &\Rightarrow\ \mathbb{E}\big(\textstyle\sum_{i=1}^{[t/\Delta_n]} |\zeta_i^n|\big) \to 0 \\
p > 1 \ &\Rightarrow\ \mathbb{E}\big(\sup_{s\le t} |\textstyle\sum_{i=1}^{[s/\Delta_n]} \zeta_i'^n|^2\big) \to 0 \\
p > 2 \ &\Rightarrow\ \mathbb{E}\big(\textstyle\sum_{i=1}^{[t/\Delta_n]} |\zeta_i''^n|\big) \to 0.
\end{aligned}\right\} \tag{6.3.19}$$

When $p = 2$ we have $\zeta_i''^n = \frac{1}{\Delta_n} \sum_{k=1}^{d'} \int_{I(n,i)} ds \int_{(i-1)\Delta_n}^{s} (\xi_v^k)^2 \, dv$, so Lemma 6.3.1 yields

$$p = 2 \ \Rightarrow\ \sum_{i=1}^{[t/\Delta_n]} \zeta_i''^n \stackrel{\text{u.c.p.}}{\Longrightarrow} \frac{1}{2} \sum_{j,k=1}^{d} \int_0^t \partial_j f(X_s)\, \partial_k f(X_s)\, c_s^{jk} \, ds. \tag{6.3.20}$$

Moreover $X_{(i(n,r)-1)\Delta_n} \to X_{T_r-}$, hence by (6.3.18) and the same argument as in Step 3 of the proof of Theorem 6.1.2, we get for all $q \ge 1$:

$$\eta'(q)^n \stackrel{\mathbb{P}}{\longrightarrow} 0, \qquad \eta''(q)^n \to 0. \tag{6.3.21}$$

At this stage we put (6.3.17) and (6.3.3) together with (6.3.19), (6.3.20) and (6.3.21), plus the fact that $\zeta_i'^n = \zeta_i''^n = 0$ when $p < 2$ and $\zeta_i^n = 0$ when $p \le 1$: in view of (6.1.8) and (6.1.9), we deduce that for proving the claim it suffices to show that

$$\sum_{q=1}^{N_t^n} \eta(q)^n \stackrel{\mathcal{L}\text{-s}}{\longrightarrow} \sum_{q:T_q \le t} \left|f(X_{T_q-} + \Delta X_{T_q}) - f(X_{T_q-})\right|^p \kappa_n.$$

Now, as in Step 5 of the proof of Theorem 6.1.2, this amounts to proving $(\kappa(n,q))_{q\ge1} \stackrel{\mathcal{L}\text{-s}}{\longrightarrow} (\kappa_q)_{q\ge1}$, and this follows from Theorem 4.3.1.

6.3.4 Proof of Theorem 6.1.8

Here we suppose that X is a continuous Itô semimartingale satisfying (H), and f is a C^2 function, and $p > 0$. For proving Theorem 6.1.8, and by localization, we can again suppose that for some constant A,

$$\left\|b_t(\omega)\right\| \le A, \qquad \left\|\sigma_t(\omega)\right\| \le A, \qquad \left\|X(\omega)\right\| \le A.$$

Then, as before, it is no restriction to suppose that f has compact support. We can also exclude the case $p = 2$, which has been shown in Theorem 6.1.4.

1) Since f is C^2, with the notation w_t and ξ_t^k of (6.3.5), we deduce from (6.2.12) applied to $g(y;x) = f(x) - g(y)$ and from the continuity of X that

$$f(X_s) - f(X_{(i-1)\Delta_n}) = \int_{(i-1)\Delta_n}^s w_v\,dv + \sum_{k=1}^{d'} \int_{(i-1)\Delta_n}^s \xi_v^k\,dW_v^k.$$

Hence with the additional notation

$$\begin{aligned}
\beta_{i,s}^n &= \int_{(i-1)\Delta_n}^s w_v\,dv \\
\beta_{i,s}'^n &= \sum_{j=1}^{d'} \int_{(i-1)\Delta_n}^s \left(\xi_v^k - \xi_{(i-1)\Delta_n}^k\right) dW_v^k \\
\beta_{i,s}''^n &= \sum_{k=1}^{d'} \xi_{(i-1)\Delta_n}^k \left(W_s^k - W_{(i-1)\Delta_n}^k\right) \\
\alpha_{i,s}^n &= \left|\beta_{i,s}^n + \beta_{i,s}'^n + \beta_{i,s}''^n\right|^p - \left|\beta_{i,s}''^n\right|^p \\
\zeta_i^n &= \frac{1}{\Delta_n^{p/2}} \int_{I(n,i)} \left|\beta_{i,s}''^n\right|^p ds, \qquad \zeta_i'^n = \frac{1}{\Delta_n^{p/2}} \int_{I(n,i)} \alpha_{i,s}^n\,ds
\end{aligned}$$

we have

$$\frac{1}{\Delta_n^{p/2}}\,\widetilde{V}^n(f,p,X)_t \;=\; \sum_{i=1}^{[t/\Delta_n]} \big(\zeta_i^n + \zeta_i'^n\big). \tag{6.3.22}$$

2) In this step we show that the array $(\zeta_i'^n)$ is asymptotically negligible, that is $\sum_{i=1}^{[t/\Delta_n]} \zeta_i'^n \stackrel{\text{u.c.p.}}{\Longrightarrow} 0$. For this we recall the elementary estimates, for $x,y\in\mathbb{R}$ and $\varepsilon\in(0,1]$:

$$\big||x+y|^p-|x|^p\big| \;\le\; \begin{cases} |y|^p & \text{if } p\le 1\\ K_p(|x|^{p-1}|y|+|y|^p) \;\le\; \varepsilon|x|^p+K_{p,\varepsilon}|y|^p & \text{if } p\ge 1,\end{cases}$$

and also $|x+y|^p\le K_p(|x|^p+|y|^p)$. Applying this twice, we see that in all cases for p,

$$\big|\alpha_{i,s}^n\big| \;\le\; \varepsilon\big|\beta_{i,s}''^n\big|^p + K_\varepsilon\big|\beta_{i,s}^n\big|^p + K_\varepsilon\big|\beta_{i,s}'^n\big|^p,$$

where K_ε also depends on p (which here is fixed).

Since w_t and $\xi_t^k|$ are bounded, we have $|\beta_{i,s}^n|^p\le K\Delta_n^p$ and $\mathbb{E}(|\beta_{i,s}''^n|^p)\le K\Delta_n^{p/2}$ by the properties of the Brownian motion. If we set $a(n,i,q)=\sum_{k=1}^{d'}\mathbb{E}(\int_{I(n,i)}|\xi_v^k-\xi_{(i-1)\Delta_n}^k|^q\,dv)$, we deduce from the Hölder and Burkholder-Davis-Gundy inequalities that

$$\mathbb{E}\big(\big|\beta_{i,s}'^n\big|^p\big) \;\le\; K\Delta_n^{(p/2-1)^+}\,a(n,i,p\vee 2)^{(p/2)\wedge 1}.$$

Putting together all these estimates results in

$$\mathbb{E}\big(\big|\alpha_{i,s}^n\big|\big) \;\le\; K\varepsilon\Delta_n^{p/2} + K_\varepsilon\Delta_n^{(p/2-1)^+}\,a(n,i,p\vee 2)^{(p/2)\wedge 1} + K_\varepsilon\Delta_n^p,$$

which in turn gives by Hölder's inequality again when $p<2$

$$\sum_{i=1}^{[t/\Delta_n]}\mathbb{E}\big(\big|\zeta_i'^n\big|\big) \;\le\; Kt\varepsilon + K_\varepsilon t\Delta_n^{p/2} + \bigg(\sum_{i=1}^{[t/\Delta_n]} a(n,i,p\vee 2)\bigg)^{(p/2)\wedge 1}. \tag{6.3.23}$$

We have $\sum_{i=1}^{[t/\Delta_n]}a(n,i,q)\le\sum_{k=1}^{d'}\mathbb{E}(\int_0^t|\xi_s^k-\xi_{\Delta_n[s/\Delta_n]}^k|^q\,ds)$, which goes to 0 as $n\to\infty$ for all $q>0$ because the process ξ^k is càdlàg and bounded. Then letting first $n\to\infty$, then $\varepsilon\to 0$ in (6.3.23), we deduce that $\sum_{i=1}^{[t/\Delta_n]}\mathbb{E}(|\zeta_i'^n|)\to 0$, which prove the asymptotic negligibility of the array $(\zeta_i'^n)$.

3) At this stage, and by (6.3.22), it remains to prove that, with the notation (6.1.10),

$$\sum_{i=1}^{[t/\Delta_n]}\zeta_i^n \;\stackrel{\text{u.c.p.}}{\Longrightarrow}\; \int_0^t \rho(p,\xi_s)\,ds.$$

To see this, we introduce the variables $\eta_i^n=\mathbb{E}(\zeta_i^n\mid\mathcal{F}_{(i-1)\Delta_n})$. In view of the form of ζ_i^n, and since $\xi_{(i-1)\Delta_n}=(\xi_{(i-1)\Delta_n}^k)_{1\le k\le d'}$ is measurable with respect to $\mathcal{F}_{(i-1)\Delta_n}$, whereas $W_s - W_{(i-1)\Delta_n}$ is independent of that σ-field when $s \geq (i-1)\Delta_n$, a simple application of the scaling property of the Brownian motion yields that indeed $\eta_i^n = \Delta_n\, \rho(p, d', \xi_{(i-1)\Delta_n})$. Thus

$$\sum_{i=1}^{[t/\Delta_n]} \eta_i^n \stackrel{\text{u.c.p.}}{\Longrightarrow} \int_0^t \rho(p, d', \xi_s)\, ds$$

by Riemann integration, because $x \mapsto \rho(p, d', x)$ is continuous and ξ_t is càdlàg.

Hence it remains to prove that the array $(\zeta_i^n - \eta_i^n)$ is asymptotically negligible. To this end, we observe that $\mathbb{E}((\zeta_i^n)^2 \mid \mathcal{F}_{(i-1)\Delta_n}) \leq K\Delta_n^2$, implying that the array $(\mathbb{E}((\zeta_i^n)^2 \mid \mathcal{F}_{(i-1)\Delta_n}))$ is asymptotically negligible, and we conclude by Lemma 2.2.11.

Part III
More Laws of Large Numbers

The basic setting considered in the previous chapters is far from covering all cases of interest:

- From a theoretical viewpoint, it would be interesting to see what happens when the test function f is not "smooth enough", or in the degenerate case where the limiting processes vanish and thus other normalizing factors are needed.
- From a purely statistical viewpoint, in parametric statistics one needs test functions f which depend not only on the increments $\Delta_i^n X$ of the process X, but also on the value $X_{(i-1)\Delta_n}$, or maybe on the whole "past" path of X before time $(i-1)\Delta_n$. Also, as seen in Sect. 5.6, we need to consider test functions which depend on several successive increments, or on the "truncated" increments to allow us to estimate for example the volatility when the process X has jumps.
- From a practical viewpoint, regular discretization schemes are obviously insufficient. Very often the process X is observed at irregular, and perhaps random, times. Even worse, in the multivariate case, the discretization schemes may be different for each component of the process of interest.

Below, we study some of these situations. This part is concerned with the law of large numbers only, and the associated central limit theorems will be considered in the next part.

The relevant bibliographical notes are not given in this part; they are provided at the end of the chapters of Part IV, together with the historical comments about the corresponding Central Limit Theorems.

Chapter 7
First Extension: Random Weights

In this chapter, we give a first—and rather straightforward—extension of the Laws of Large Numbers of Chap. 3, in the case where the test function f is *random*.

7.1 Introduction

Let X be our basic d-dimensional semimartingale, on the filtered space $(\Omega, \mathcal{F}, (\mathcal{F}_t)_{t\geq 0}, \mathbb{P})$. So far, we have considered the behavior of following two functionals:

$$\begin{aligned} V^n(f,X)_t &= \sum_{i=1}^{N_n(t)} f\big(\Delta_i^n X\big) \\ V'^n(f,X)_t &= \Delta_n \sum_{i=1}^{[t/\Delta_n]} f\big(\Delta_i^n X/\sqrt{\Delta_n}\big), \end{aligned} \tag{7.1.1}$$

where f is a function on $\mathbb{R}^d$, and where $\Delta_i^n X$ is the increment of X over the interval $I(n,i) = (T(n,i-1), T(n,i)]$. The LLN for $V^n(f,X)$ allows for arbitrary random discretization schemes, whereas for $V'^n(f,X)$ we consider only regular schemes $T(n,i) = i\Delta_n$.

Our aim is to replace the summands $f(\Delta_i^n X)$ or $f(\Delta_i^n X/\sqrt{\Delta_n})$ by more general ones, still depending on $\Delta_i^n X$ of course, but also on other random inputs and, why not, on i and n as well. This can be done in many ways, the most general one being simply to replace $f(\Delta_i^n X)$ by an arbitrary variable χ_i^n, in which case obviously anything can happen!

In fact, we want somehow to retain the structure (7.1.1), and the simplest way for extending the first functional $V^n(f,X)$, for example, is to consider weights χ_i^n (random or not), that is the processes

$$\sum_{i=1}^{N_n(t)} \chi_i^n\, f\big(\Delta_i^n X\big).$$

J. Jacod, P. Protter, *Discretization of Processes*,
Stochastic Modelling and Applied Probability 67,
DOI 10.1007/978-3-642-24127-7_7, © Springer-Verlag Berlin Heidelberg 2012

Again, this formulation is so general that it allows for about any possible limiting behavior. However, one can immediately single out two natural special cases:

(1) Take non-random numbers χ_i^n, with some nice limiting behavior, in connection with the sequence Δ_n, for example $\chi_i^n = g(\tau(n,i))$ for some given function g on $\mathbb{R}_+$ and some points $\tau(n,i)$ in $\overline{I}(n,i) = [T(n,i-1), T(n,i)]$. This amounts to affecting the summand $f(\Delta_i^n X)$ with a weight which is the value of the function g, at the time $\tau(n,i)$ at which this summands physically occurs. Such a setting may be aimed to put more emphasis on certain parts of the half line than on others.

(2) Take the weights χ_i^n to be of the form $g(X_{\tau(n,i)})$, where $\tau(n,i) \in \overline{I}(n,i)$, and for some given function g on $\mathbb{R}^d$. This amounts to put more emphasis on certain parts of the "space", that is when X is in the regions of $\mathbb{R}^d$ where g is biggest.

The setting (2) above is heavily used, for example, in statistics of processes in case of high frequency data, and especially in parametric statistics, as we shall see in Sect. 7.4 below.

Summarizing the previous discussion, it appears that the proper extensions of the two functionals $V^n(f,X)$ and $V'^n(f,X)$ are

$$
\begin{aligned}
V^n(F,X)_t &= \sum_{i=1}^{N_n(t)} F\big(., \tau(n,i), \Delta_i^n X\big), \\
V'^n(F,X)_t &= \Delta_n \sum_{i=1}^{[t/\Delta_n]} F\big(., \tau(n,i), \Delta_i^n X/\sqrt{\Delta_n}\big).
\end{aligned}
\tag{7.1.2}
$$

Here, $F = F(\omega,t,x)$ is a function on $\Omega \times \mathbb{R}_+ \times \mathbb{R}^d$, but as usual we omit to mention the sample point ω, and $\tau(n,i)$ is a—possibly random—time in the interval $[T(n,i-1), T(n,i)]$. As for (7.1.1), the $T(n,i)$ are arbitrary stopping times when we consider $V^n(F,X)$, but are $T(n,i) = i\Delta_n$ when we consider $V'^n(F,X)$, and $N_n(t) = \sup(i : T(n,i) \le t)$. The function F may have values in $\mathbb{R}^q$.

As a function of the last argument x, $F(\omega,t,x)$ should of course satisfy the same conditions as in Theorems 3.3.5 and 3.4.1 respectively, more or less uniformly in (ω,t). We also need some regularity in t, and global measurability in (ω,t,x). However, for the LLN's there is no need of any sort of adaptation property to the filtration $(\mathcal{F}_t)_{t\ge0}$. This is because the limiting results in Chap. 3 are "pathwise" results, up to taking subsequences which allow us to replace the convergence in probability by the almost sure convergence. This is most apparent for $V'^n(F,X)$: the argument is straightforward and has indeed nothing to do with the structure of the process X: as soon as we have an LLN for $V'^n(f,X)$ for *all* test functions f, we have the LLN for $V'^n(F,X)$. Therefore we begin with this case.

7.2 The Laws of Large Numbers for $V'^n(F, X)$

Here we extend Theorem 3.4.1, whose notation and assumptions are in force. In particular we have a *regular discretization scheme* with stepsize $\Delta_n \to 0$, and X is a d-dimensional Itô semimartingale, with characteristics (B, C, ν) having the form (2.1.25), that is

$$B_t = \int_0^t b_s\, ds, \qquad C_t = \int_0^t c_s\, ds, \qquad \nu(dt, dx) = dt\, F_t(dx).$$

We consider the processes $V'^n(F, X)$ of (7.1.2), and we suppose the following on F and the times $\tau(n, i)$:

Assumption 7.2.1 (i) Each $\tau(n, i)$ is a measurable variable with values in $[(i-1)\Delta_n, i\Delta_n]$.

(ii) F is a measurable function on $(\Omega \times \mathbb{R}_+ \times \mathbb{R}^d, \mathcal{F} \otimes \mathcal{R}_+ \otimes \mathcal{R}^d)$, and the set D_F of all (ω, t, x) such that $(s, y) \mapsto F(\omega, s, y)$ is *not* continuous at the point (t, x) satisfies:

$$\mathbb{E}\left(\int_0^\infty \rho_{c_s}(D_{F,.,s})\, ds\right) = 0 \tag{7.2.1}$$

where $D_{F,\omega,t} = \{x : (\omega, t, x) \in D_F\}$ (compare with (3.4.10); here, according to (3.4.4), and for any matrix $a \in \mathcal{M}^+_{d\times d}$, the notation ρ_a stands for the normal law $\mathcal{N}0, a)$).

(iii) There is localizing sequence of random times τ_n (not necessarily stopping times), and continuous positive functions f_n on $\mathbb{R}^d$, such that

$$t < \tau_n(\omega) \;\Rightarrow\; \bigl|F(\omega, t, x)\bigr| \le f_n(x). \tag{7.2.2}$$

The same comments as after Corollary 3.4.4 apply here, about the condition (7.2.1). In particular, when c_t is everywhere invertible, this condition amounts to saying that for almost all ω the function $(t, x) \mapsto F(\omega, t, x)$ is Lebesgue-almost everywhere continuous on $\mathbb{R}_+ \times \mathbb{R}^d$.

Theorem 7.2.2 *Assume that X is a d-dimensional Itô semimartingale, and that the discretization scheme is regular with stepsize Δ_n. Let F satisfy Assumption* 7.2.1, *in which each function f_n satisfies one of the following three conditions* (*the same for all n*):

(a) $f_n(x) = \mathrm{o}(\|x\|^2)$ *as* $\|x\| \to \infty$,

(b) $f_n(x) = \mathrm{O}(\|x\|^2)$ *as* $\|x\| \to \infty$, *and X is continuous*,

(c) $f_n(x) = \mathrm{O}(\|x\|^p)$ *as* $\|x\| \to \infty$ *for some* $p > 2$, *and X is continuous and satisfies*

$$\int_0^t \|b_s\|^{2p/(2+p)}\, ds < \infty, \qquad \int_0^t \|c_s\|^{p/2}\, ds < \infty.$$

Then

$$V'^n(F,X)_t \stackrel{\text{u.c.p.}}{\Longrightarrow} V'(F,X)_t := \int_0^t ds \int_{\mathbb{R}^d} F(s,x)\,\rho_{c_s}(dx).$$

Once more, the conditions on X are exactly the same as for Theorem 3.4.1, whereas the limiting process is a straightforward extension of (3.4.7).

Example 7.2.3 The simplest (non-trivial) functions F satisfying Assumption 7.2.1 for all processes X are $F(\omega,t,x)=g(X_t(\omega),x)$ and $F(\omega,t,x)=g(X_{t-}(\omega),x)$, where g is a continuous function on $\mathbb{R}^d\times\mathbb{R}^d$ such that $|g(y,x)|\le f(x)$ for some positive continuous function f. Then under the assumptions of the theorem on $f_n=f$ we have

$$\Delta_n \sum_{i=1}^{[t/\Delta_n]} g\big(X_{\tau(n,i)}, \Delta_i^n X/\sqrt{\Delta_n}\big) \stackrel{\text{u.c.p.}}{\Longrightarrow} \int_0^t ds \int_{\mathbb{R}^d} g(X_s,x)\rho_{c_s}(dx)$$

and the same if we take $X_{\tau(n,i)-}$ instead of $X_{\tau(n,i)}$ above, and whatever $\tau(n,i)$ is inside $\overline{I}(n,i)$: for example one can take for $\tau(n,i)$ the time at which X reaches its maximum (in the one-dimensional case), or its minimum, inside $\overline{I}(n,i)$, if such a thing exists, or the time at which the biggest jump of X occurs within $\overline{I}(n,i)$.

Proof The proof is an extension of the proof of Corollary 3.4.4, and exactly as in that proof it is enough to show that for each fixed t we have

$$V'^n(F,X)_t \stackrel{\mathbb{P}}{\longrightarrow} \int_0^t ds \int_{\mathbb{R}^d} F(s,x)\,\rho_{c_s}(dx). \tag{7.2.3}$$

Below, we fix $t>0$. Since $\tau_n\to\infty$ in (7.2.2), it is enough to show (7.2.3) on the set $\{\tau_m>t\}$, for any given $m\ge 1$. In other words, it is enough to prove the result when $|F(\omega,s,x)|\le f(x)$ for all $s\le t$ and some function $f>0$ satisfying the appropriate condition (a), (b) or (c), according to the case. For each n we introduce the positive finite (random) measure $\overline{m}_n=\overline{m}_n(\omega,ds,dx)$ on $\mathbb{R}_+\times\mathbb{R}^d$:

$$\overline{m}_n(\omega,ds,dx) = \Delta_n \sum_{i=1}^{[t/\Delta_n]} f\big(\Delta_i^n X(\omega)/\sqrt{\Delta_n}\big)\,\varepsilon_{(\tau(n,i)(\omega),\Delta_i^n X(\omega)/\sqrt{\Delta_n})}(ds,dx)$$

(ε_a is the Dirac mass at a). We also consider the measure $\overline{m}$:

$$\overline{m}(\omega,ds,dx) = 1_{[0,t]}(s)\,f(x)\,ds\,\rho_{c_s(\omega)}(dx).$$

Note that (7.2.1) yields that, for all ω outside a $\mathbb{P}$ null set N, the bounded function $F/f=F(\omega,s,x)/f(x)$ is $\overline{m}(\omega,ds,dx)$ almost everywhere continuous.

We set $\theta_n(s)=0$ if $s<\tau(n,1)$ and $\theta_n(s)=t\wedge(i\Delta_n)$ if $\tau(n,i)\le s<\tau(n,i+1)$ for some $i\ge 1$. For any bounded function g on $\mathbb{R}^d$, we have $\overline{m}_n(1_{[0,s]}\otimes g)=V'^n(gf,X)_{\theta_n(s)}$ and $\overline{m}(1_{[0,s]}\otimes g)=V'(gf,X)_{s\wedge t}$ (notation (3.4.7)). Since $\theta_n(s)\to$

$s\wedge t$ uniformly in s, we deduce from Theorem 3.4.1 that for any bounded continuous g we have

$$\overline{m}_n(1_{[0,s]}\otimes g)\ \stackrel{\mathbb{P}}{\longrightarrow}\ \overline{m}(1_{[0,s]}\otimes g).$$

We take a countable family $\mathcal{G}$ of continuous bounded functions on $\mathbb{R}^d$ which is convergence determining, so the family $(1_{[0,s]}\otimes g; s\in\mathbb{Q}_+, g\in\mathcal{G})$ is convergence-determining for the measures on $\mathbb{R}_+\times\mathbb{R}^d$. Hence, as in the proof of Corollary 3.4.4, from any sequence $n_k\to\infty$ one can extract a subsequence $n_{k_l}\to\infty$ such that, for all ω outside a $\mathbb{P}$ null set N' containing the set N, we have $\overline{m}_{n_{k_l}}(\omega,.)\to\overline{m}(\omega,.)$ weakly. Since the function F/f is bounded and $\overline{m}(\omega,.)$ almost everywhere continuous when $\omega\notin N'$, we deduce $\overline{m}_{n_{k_l}}(F/f)\to\overline{m}(F/f)$ outside N', that is almost surely. The left and right sides of (7.2.3) are respectively equal to $\overline{m}_n(F/f)$ and $\overline{m}(F/f)$, so by the subsequence principle (2.2.17) we deduce that (7.2.3) holds. □

7.3 The Laws of Large Numbers for $V^n(F,X)$

In this section, X is an arbitrary semimartingale with characteristics (B,C,ν) and jump measure μ, on $(\Omega,\mathcal{F},(\mathcal{F}_t)_{t\geq0},\mathbb{P})$, and we consider a random discretization scheme $\mathcal{T}=(T(n,i)_{i\geq0}:n\geq1)$ satisfying (3.1.1), recall Definition 3.1.1.

The processes of interest are $V^n(F,X)$, as given by (7.1.2), but we can no longer choose $\tau(n,i)$ arbitrary in $\overline{I}(n,i)$. Basically we have to choose either end point of that interval, thus giving rise to two different functionals ("l" and "r" are for "left" and "right", respectively):

$$\begin{aligned} V^{n,l}(F,X)_t &= \sum_{i=1}^{N_n(t)} F\big(T(n,i-1),\Delta_i^nX\big)\\ V^{n,r}(F,X)_t &= \sum_{i=1}^{N_n(t)} F\big(T(n,i),\Delta_i^nX\big). \end{aligned}\tag{7.3.1}$$

Our aim is to extend Theorem 3.3.1 to this situation. We use all notation of Sect. 3.3 below, and in particular when $1\in\mathcal{I}(X)$ we recall that

$$X_t=X_0+B'_t+X^c_t+\sum_{s\leq t}\Delta X_s,$$

where B' is a predictable process of finite variation.

Actually, we will extend part (A) of Theorem 3.4.1 in full detail, and simply indicate the extensions of parts (B) and (C) without proof, since they will not be used in the sequel and they are a bit messy to prove.

We have two different assumptions on the test function F, regarding the behavior in time, according to which—left or right—functionals we consider. As in Chap. 3, we allow F to be q-dimensional. We denote by D_F^l and D_F^r respectively the set of all (ω, t, x) for which there exist sequences $t_n \to t$ and $x_n \to x$, with further $t_n < t$, resp. $t_n \geq t$, and such that $F(\omega, t_n, x_n)$ does *not* converge to $F(\omega, t, x)$. Then $D^l_{F,\omega,t}$ and $D^r_{F,\omega,t}$ are the (ω, t)-sections of D^l_F and D^r_F, as in Assumption 7.2.1.

Assumption 7.3.1 The function F is $\mathcal{F} \otimes \mathcal{R}_+ \otimes \mathcal{R}^d$ measurable, and

$$\mathbb{P}\left(\left\{\omega : \exists t > 0 :\ \left(\omega, t, \Delta X_t(\omega)\right) \in D^l_F\right\}\right) = 0. \tag{7.3.2}$$

Moreover, for some $\theta > 0$ we have a sequence f_n of nonnegative functions on $\mathbb{R}^d$ and a localizing sequence (τ_n) of random times (not necessarily stopping times), such that $\|F(\omega, t, x)\| \leq f_n(x)$ for $t < \tau_n(\omega)$ and $\|x\| \leq \theta$.

Assumption 7.3.2 The same as above, upon substituting D^l_F with D^r_F.

Note that Assumption 7.3.1 implies that F is essentially left-continuous in time, whereas Assumption 7.3.2 implies that it is essentially right-continuous. We cannot replace (7.3.2) by $\mathbb{E}(1_{D^l_F} * \nu_t) = 0$ for all t, contrary to what happens in Theorem 3.3.5, unless the set D_F is predictable.

Theorem 7.3.3 *Let X be a d-dimensional semimartingale and $\mathcal{T}$ be any random discretization scheme and F be a q-dimensional function on $\Omega \times \mathbb{R}_+ \times \mathbb{R}^d$. Let either $V^n(F, X) = V^{n,l}(F, X)$ and F satisfy Assumption 7.3.1, or $V^n(F, X) = V^{n,r}(F, X)$ and F satisfy Assumption 7.3.2. Assume that each n the function f_n and X satisfy any one of the following four conditions (the same for all n):*

(A-a) $f_n(x) = \mathrm{o}(\|x\|^2)$ *as* $x \to 0$,
(A-b) $X^c = 0$ *and there is a* $p \in \mathcal{I}(X) \cap (1, 2]$ *such that* $f_n(x) = \mathrm{O}(\|x\|^p)$ *as* $x \to 0$,
(A-c) $X^c = 0$ *and* $1 \in \mathcal{I}(X)$ *and* $f_n(x) = \mathrm{o}(\|x\|)$ *as* $x \to 0$,
(A-d) $X^c = 0$ *and there is a* $p \in \mathcal{I}(X) \cap [0, 1]$ *such that* $f_n(x) = \mathrm{O}(\|x\|^p)$ *as* $x \to 0$, *and* $B' = 0$.

Then we have the following Skorokhod convergence in probability:

$$V^n(F, X) \stackrel{\mathbb{P}}{\Longrightarrow} V(F, X) := F \star \mu.$$

The remarks made after Theorem 3.3.1 also hold in this more general setting. In particular, we have the following slightly stronger property:

$$\overline{W}^n(F)_t = V^n(F, X)_t - V(F, X)_{T_n(t)} \stackrel{\text{u.c.p.}}{\Longrightarrow} 0. \tag{7.3.3}$$

We emphasize once more than there is no adaptation to $(\mathcal{F}_t)$ required from F. The integral $F * \mu_t$ in the limit above is an ordinary (Lebesgue) integral, which exists and is finite in all cases because of the specific assumptions on F, in connection with those on X.

Proof We consider only the case $V^n(F, X) = V^{n,l}(F, X)$, the other case being similar. As for Theorem 3.4.1, we prove the stronger statement (7.3.3).

In a first step we extend Lemma 3.3.7 to obtain (7.3.3) when F satisfies (7.3.2) and $F(\omega,t,x)=0$ when $\|x\|\leq\varepsilon$, for some $\varepsilon>0$. Using the notation of that lemma, we can in fact reproduce its proof, the only change being that

$$\begin{aligned}\overline{W}^n(F)_t = \sum_{q=1}^{Q_T}\Big(&F\big(T(n,i(n,q)-1),\Delta^n_{i(n,q)}X'+\Delta X_{S_q}\big)\\ &- F(S_q,\Delta X_{S_q})\Big)1_{\{S_q\leq N_n(t)\}}.\end{aligned}$$

Outside a null set, $\Delta^n_{i(n,q)}X'\to 0$ by (3.3.6), whereas $T(n,i(n,q)-1)$ is strictly smaller than S_q and tends to S_q. Then (7.3.2) yields $F(T(n,i(n,q)-1),\Delta^n_{i(n,q)}X'+\Delta X_{S_q})\to F(S_q,\Delta X_{S_q}))$ a.s. and, since Q_T is finite, we deduce (7.3.3).

Now we turn to the general case. By our usual localization argument, we can suppose that $\|F(\omega,t,x)\|\leq f_1(x)$ for all $\omega\in\Omega$, $t\geq 0$, and all x with $\|x\|\leq\theta$ (that is, $\tau_1=\infty$). With the notation of (3.3.16), we set

$$G_\varepsilon(\omega,t,x) \;=\; F(\omega,t,x)\psi'_\varepsilon(x), \qquad F_\varepsilon \;=\; F-G_\varepsilon.$$

Then, for (7.3.3) it is enough to prove (3.3.17), which we recall below:

$$\begin{aligned}&\varepsilon>0 \;\;\Rightarrow\;\; \overline{W}^n(F_\varepsilon)\stackrel{\text{u.c.p.}}{\Longrightarrow}0\\ &t,\eta>0 \;\Rightarrow\;\; \lim_{\varepsilon\to 0}\limsup_n\mathbb{P}\Big(\sup_{s\leq t}\|\overline{W}^n(G_\varepsilon)_s\|>\eta\Big)=0.\end{aligned} \tag{7.3.4}$$

Since $F_\varepsilon(\omega,t,x)=0$ when $\|x\|\leq\varepsilon$, and since obviously F_ε satisfies (7.3.2), Step 1 yields the first part of (7.3.4). On the other hand $\|G_\varepsilon(\omega,t,x)\|\leq f_1(x)$ as soon as $\varepsilon<\theta$. In this case it follows that we have exactly the same upper bound for $\|\overline{W}^n(G_\varepsilon)_t\|$ here as in Part 2 of the proof of Theorem 3.3.1, with the same control functions h or h_ε on $\mathbb{R}^d$. Hence the same proof yields the second part of (7.3.4), and the proof is complete. □

For extending (B) of Theorem 3.4.1 we need $F(t,x)$ to be close enough to a quadratic function of x near 0, with coefficients which are random processes themselves:

Assumption 7.3.4 The function F is $\mathcal{F}\otimes\mathcal{R}_+\otimes\mathcal{R}^d$ measurable, and satisfies (7.3.2). Moreover, for some $\theta>0$ we have a sequence f_n of nonnegative functions on $\mathbb{R}^d$ and a localizing sequence (τ_n) of random times (not necessarily stopping times), such that $\|F(\omega,t,x)-\sum_{j,k=1}^f\alpha(\omega)^{jk}_t x^jx^k\|\leq f_n(x)$ for $t<\tau_n(\omega)$ and $\|x\|\leq\theta$, where the α^{jk} are q-dimensional measurable processes (not necessarily adapted), whose paths are left continuous with right limits.

Assumption 7.3.5 The same as above, upon substituting D^l_F with D^r_F in (7.3.2), and with α^{jk} being càdlàg.

As said before, we state the result without proof:

Theorem 7.3.6 *Let X be a d-dimensional semimartingale and $\mathcal{T}$ be any random discretization scheme and F be a q-dimensional function on $\Omega \times \mathbb{R}_+ \times \mathbb{R}^d$. Let either $V^n(F,X) = V^{n,l}(F,X)$ and F satisfy Assumption 7.3.4, or $V^n(F,X) = V^{n,r}(F,X)$ and F satisfy Assumption 7.3.5, with $f_n(x) = \mathrm{o}(\|x\|^2)$ as $x \to 0$ for all n. Then*

$$V^n(F,X) \stackrel{\mathbb{P}}{\Longrightarrow} V(F,X)_t := \sum_{j,k=1}^d \int_0^t \alpha_s^{jk}\, dC_s^{jk} + F \star \mu_t.$$

Finally for extending (C) of Theorem 3.4.1 we need the following:

Assumption 7.3.7 The function F is $\mathcal{F} \otimes \mathcal{R}_+ \otimes \mathcal{R}^d$ measurable, and satisfies (7.3.2). Moreover, for some $\theta > 0$ we have a sequence f_n of nonnegative functions on $\mathbb{R}^d$ and a localizing sequence (τ_n) of random times (not necessarily stopping times), such that $\|F(\omega,t,x) - \sum_{j=1}^f \alpha(\omega)_t^j x^j\| \le f_n(x)$ for $t < \tau_n(\omega)$ and $\|x\| \le \theta$, where the α^j are q-dimensional measurable processes (not necessarily adapted), whose paths are left continuous with right limits.

Assumption 7.3.8 The same as above, upon substituting D_F^l with D_F^r in (7.3.2), and with α^j being càdlàg.

Theorem 7.3.9 *Let X be a d-dimensional semimartingale with $1 \in \mathcal{I}(X)$ and $X^c = 0$, and $\mathcal{T}$ be any random discretization scheme and F be a q-dimensional function on $\Omega \times \mathbb{R}_+ \times \mathbb{R}^d$. Let either $V^n(F,X) = V^{n,l}(F,X)$ and F satisfy Assumption 7.3.7, or $V^n(F,X) = V^{n,r}(F,X)$ and F satisfy Assumption 7.3.8, with $f_n(x) = \mathrm{o}(\|x\|)$ as $x \to 0$ for all n. Then*

$$V^n(F,X) \stackrel{\mathbb{P}}{\Longrightarrow} V(F,X)_t := \sum_{j=1}^d \int_0^t \alpha_s^j\, d\mathrm{Var}\big(B'^j\big)_s + F \star \mu_t.$$

7.4 Application to Some Parametric Statistical Problems

Among many other applications, the previous results can be used for estimating a parameter θ in some parametric statistical models for processes observed at discrete times, on a finite time interval.

Although more general situations are amenable to a similar analysis, we consider below the case of a (continuous) diffusion process, with a diffusion coefficient depending on an unknown parameter θ lying in some domain Θ of $\mathbb{R}^q$. That is, we have

$$X_t^\theta = x + \int_0^t b\big(s, X_s^\theta\big)\, ds + \int_0^t \sigma\big(\theta, s, X_s^\theta\big)\, dW_s.$$

Here, $x \in \mathbb{R}^d$ is the (known) starting point, b is a (possibly unknown) function on $\mathbb{R}_+ \times \mathbb{R}^d$, and σ is a *known* function on $\Theta \times \mathbb{R}_+ \times \mathbb{R}^d$. Of course W is a Brownian motion, say d'-dimensional, and the functions b and $\sigma(\theta, .)$ are nice enough for the above equation to have a unique solution. The "true" (unknown) value of the parameter is θ_0, and we observe the process X^{θ_0} at all times $i\Delta_n$ within some fixed interval $[0, T]$.

As is customary in statistics, we use the "weak" formulation. That is, we consider the filtered space $(\Omega, \mathcal{F}, (\mathcal{F}_t))$ which is the canonical space $\mathbb{C}^d$ of all continuous d-dimensional functions on $\mathbb{R}_+$, and X denotes the canonical process $X_t(\omega) = \omega(t)$. Then for each θ the law of X^θ is a probability measure $\mathbb{P}_\theta$ on it. Upon enlarging this canonical space if necessary, so that it accommodates a d'-dimensional Brownian motion, and without changing the notation, the canonical process X satisfies

$$X_t = x + \int_0^t b(s, X_s)\, ds + \int_0^t \sigma(\theta, s, X_s)\, dW_s, \qquad \mathbb{P}_\theta\text{-almost surely.}$$

(Note that the Brownian motion W depends on θ, in this formulation.)

The aim here is to estimate θ in a consistent way, as $\Delta_n \to 0$. This means finding, for each n, a function h_n on $(\mathbb{R}^d)^{[T/\Delta_n]}$ (not depending on θ, of course), which satisfies the *weak consistency* condition

$$\widehat{\theta}_n := h_n(X_{\Delta_n}, X_{2\Delta_n}, \ldots, X_{\Delta_n[T/\Delta_n]}) \xrightarrow{\mathbb{P}_\theta} \theta \quad \forall \theta \in \Theta.$$

Of course, this is only the first step for the solution of the statistical problem. If it can be achieved, a second step consists in finding, among all such $\widehat{\theta}_n$, the "best" ones, for which $\widehat{\theta}_n - \theta$ goes to 0 as fast as possible. The first step requires Laws of Large Numbers in our sense, whereas the second step needs a Central Limit Theorem, and will be considered in Chap. 10 only.

Typically, estimation problems are solved by using the maximum likelihood estimator, when available. Here, the likelihood at stage n is not available if the drift coefficient b is unknown, and even when it is known it is (almost) never explicit. So a lot of work has been done on this topic, using substitutes like "approximated likelihoods" or "quasi-likelihoods". We will not go deep into this, and will only consider simple *contrast functions*, in the spirit of Genon-Catalot and Jacod [37].

As usual, we set $c(\theta, t, x) = \sigma(\theta, t, x)\sigma(\theta, t, x)^*$, which takes its values in $\mathcal{M}^+_{d\times d}$. We consider a function g on $\mathcal{M}^+_{d\times d} \times \mathbb{R}^d$, and the associated contrast functions (they are functions of θ, depending on the observed values of the X at stage n only):

$$\Phi_n(\theta) = \Delta_n \sum_{i=1}^{[T/\Delta_n]} g\left(c\big(\theta, (i-1)\Delta_n, X_{(i-1)\Delta_n}\big), \frac{\Delta_i^n X}{\sqrt{\Delta_n}}\right).$$

The assumptions made below on c and g imply that each $\Phi_n(\theta)$ is continuous in θ, and as customary for this kind of problem we assume that Θ is *compact*. Then we take for $\widehat{\theta}_n$ the *minimum contrast estimator*:

$$\widehat{\theta}_n = \text{any value realizing the minimum of } \theta \mapsto \Phi_n(\theta).$$

Since c is a known function and g is chosen by the statistician, $\widehat{\theta}_n$ is indeed a function h_n of the observations $X_{i\Delta_n}$ at stage n.

In order to achieve consistency, we need two types of conditions:

1 – The function $(\theta,t,x)\mapsto c(\theta,t,x)$ is continuous, and $(t,x)\mapsto b(t,x)$ is locally bounded.

2 – The function g is continuous with at most polynomial growth, and for all $a,a'\in\mathcal{M}_{d\times d}^{+}$ we have:

$$G(a,a')=\int g(a,x)\rho_{a'}(dx) \text{ satisfies } G(a,a') > G(a',a') \text{ if } a\neq a'. \tag{7.4.1}$$

Furthermore, consistent estimators cannot exist, of course, if for different values of θ the processes X^θ are the same. This leads us to introduce the sets

$$\Omega_T^\theta = \left\{\int_0^T \left\|c(\theta',s,X_s)-c(\theta,s,X_s)\right\| ds>0 \quad \forall\theta'\neq\theta\right\}.$$

On the σ-field $\mathcal{F}_T$, the measures $\mathbb{P}_{\theta'}$ for $\theta'\neq\theta$ are all singular with respect to $\mathbb{P}_\theta$ in restriction to the set Ω_T^θ, whereas there exists $\theta'\neq\theta$ such that $\mathbb{P}_\theta$ and $\mathbb{P}_{\theta'}$ are equivalent in restriction to the complement $(\Omega_T^\theta)^c$: therefore, on $(\Omega_T^\theta)^c$ one cannot distinguish between θ and θ' even when the whole path of X_t for $t\in[0,T]$ is observed, hence there are no estimators which are consistent for estimating θ on this set.

Theorem 7.4.1 *In the previous setting and under the previous assumptions, the estimators $\widehat{\theta}_n$ satisfy for all θ_0:*

$$\widehat{\theta}_n \longrightarrow \theta_0 \quad \text{in } \mathbb{P}_{\theta_0}\text{-probability, in restriction to the set } \Omega_T^{\theta_0}.$$

As mentioned before, the convergence above *cannot hold* outside $\Omega_T^{\theta_0}$. Quite fortunately, this set is equal or almost surely equal to Ω itself in many cases, for example when c is continuous and $c(\theta,0,x)\neq c(\theta',0,x)$ for all $\theta\neq\theta'$, where x is the initial condition.

Notice the compactness assumption on Θ. When Θ is closed but unbounded, it may happen that $\widehat{\theta}_n$ is not well defined, or that it drifts away to infinity instead of going to θ_0. To avoid these two problems, one needs additional assumptions on the behavior of $c(\theta,t,x)$ as $\|\theta\|\to\infty$, a very strong one being that for all A large we have $\|c(\theta',t,x)-c(\theta,t,x)\|\geq C_A$ for a positive constant C_A, for all $t\in[0,T]$ and all $\|x\|\leq A$ and all $\|\theta'\|>A$ and $\|\theta\|\leq A/2$. Weaker assumptions are possible but complicated.

It can also happen that Θ is an open set, so $\widehat{\theta}_n$ may drift toward the boundary. The problem is similar to the previous one, but the conditions ensuring consistency are more difficult to state, although in the same spirit. We will not pursue this topic here.

Proof The "true" value θ_0 is fixed, and for simplicity we write $\mathbb{P}=\mathbb{P}_{\theta_0}$. Under $\mathbb{P}$, the process X is a continuous Itô semimartingale satisfying (H).

1) We observe that $\Phi_n(\theta)=V^n(F_\theta,X)_T$, provided $\tau(n,i)=(i-1)\Delta_n$ and F_θ is

$$F_\theta(\omega,t,x) = g\bigl(c\bigl(\theta,s,X_s(\omega)\bigr),x\bigr).$$

Our hypotheses imply Assumption (7.2.1) and (c) of Theorem 7.2.2 (here, $b_t=b(t,X_t)$ and $c_t=c(\theta_0,t,X_t)$). Then this theorem implies, with the notation G of (7.4.1), that

$$\Phi_n(\theta) \xrightarrow{\mathbb{P}} \Phi(\theta) = \int_0^t G\bigl(c(\theta,s,X_s),c(\theta_0,s,X_s)\bigr)\,ds. \tag{7.4.2}$$

This is a convergence for each fixed $\theta\in\Theta$. However, by localization we may assume that $\|X_t\|\leq C$ for some constant C and all t, whereas Θ is compact by hypothesis and thus the assumptions on c and g yield for all $A>1$:

$$\|x\|\leq C,\ \bigl\|\theta-\theta'\bigr\|\leq\varepsilon \Rightarrow \bigl|g\bigl(c(\theta,s,x),y\bigr)-g\bigl(c\bigl(\theta',s,x\bigr),y\bigr)\bigr|\leq\phi_A(\varepsilon)+K\,\frac{\|y\|^p}{A}$$

for some $p\geq 1$ and functions ϕ_A satisfying $\phi_A(\varepsilon)\to 0$ as $\varepsilon\to 0$. It follows that

$$\mathbb{E}\Bigl(\sup_{\theta,\theta':\,\|\theta-\theta'\|\leq\varepsilon}\bigl|\Phi_n(\theta)-\Phi_n\bigl(\theta'\bigr)\bigr|\Bigr)+\sup_{\theta,\theta':\,\|\theta-\theta'\|\leq\varepsilon}\bigl|\Phi(\theta)-\Phi\bigl(\theta'\bigr)\bigr|\leq\phi_A(\varepsilon)+\frac{KT}{A}$$

because $\mathbb{E}(\|\Delta_i^n X\|^p)\leq K\Delta_n^{p/2}$ for all i,n and also $\|c(\theta_0,s,X_s)\|\leq K$, which in turn implies $\int\|y\|^p\,\rho_{c(\theta_0,s,X_s)}(dy)\leq K$. The compact set Θ may be covered by N_ε balls of radius ε, with centers θ_i for $i=1,\dots,N_\varepsilon$, and we deduce from what precedes that for all $\eta>0$:

$$\begin{aligned}\mathbb{P}\Bigl(\sup_\theta\bigl|\Phi_n(\theta)-\Phi(\theta)\bigr|>\eta\Bigr)&\leq\mathbb{P}\Bigl(\sup_{1\leq i\leq N_\varepsilon}\bigl|\Phi_n(\theta_i)-\Phi(\theta_i)\bigr|>\frac{\eta}{3}\Bigr)\\&\quad+\mathbb{P}\Bigl(\sup_{\theta,\theta':\,\|\theta-\theta'\|\leq\varepsilon}\bigl|\Phi_n(\theta)-\Phi_n\bigl(\theta'\bigr)\bigr|>\frac{\eta}{3}\Bigr)\\&\quad+\mathbb{P}\Bigl(\sup_{\theta,\theta':\,\|\theta-\theta'\|\leq\varepsilon}\bigl|\Phi(\theta)-\Phi\bigl(\theta'\bigr)\bigr|>\frac{\eta}{3}\Bigr)\\&\leq\sum_{i=1}^{N_\varepsilon}\mathbb{P}\Bigl(\bigl|\Phi_n(\theta_i)-\Phi(\theta_i)\bigr|>\frac{\eta}{3}\Bigr)+\frac{6}{\eta}\,\phi_A(\varepsilon)+\frac{KT}{A}.\end{aligned}$$

The first term in the right member above goes to 0 as $n\to\infty$ by (7.4.2) for each $\varepsilon>0$ fixed. Since A is arbitrarily large and $\phi_A(\varepsilon)\to 0$ as $\varepsilon\to 0$, we then deduce

$$\sup_\theta\bigl|\Phi_n(\theta)-\Phi(\theta)\bigr| \xrightarrow{\mathbb{P}} 0. \tag{7.4.3}$$

2) By the subsequence principle, we can replace the convergence in probability in (7.4.3) by the convergence for all ω in a full set Ω_0, and below we fix $\omega \in \Omega_0 \cap \Omega_T^{\theta_0}$. On the one hand, Φ_n converges to Φ uniformly on Θ and, being continuous, admits at least a minimum on the compact set Θ at some (not necessarily unique) point $\widehat{\theta}_n$. On the other hand, (7.4.1) implies that Φ has a unique minimum at θ_0 (because we are on $\Omega_T^{\theta_0}$). Hence, necessarily, $\widehat{\theta}_n(\omega) \to \theta_0$ if $\omega \in \Omega_0 \cap \Omega_T^{\theta_0}$. This finishes the proof. □

Example 7.4.2 The simplest choice for the function g is probably the following one:

$$g(a,x) = \sum_{i,j=1}^{d} \left(a^{ij} - x^i x^j\right)^2.$$

The associated function G is

$$G(a,a') = \sum_{i,j=1}^{d} \left(a^{ij} - a'^{ij}\right)^2 + \sum_{i,j=1}^{d} \left(a'^{ii} a'^{jj} + \left(a'^{ij}\right)^2\right),$$

which clearly satisfies (7.4.1).

Example 7.4.3 The previous choice may be the simplest one, but it is not the optimal one, as we will see in Chap. 10 (although it gives rise to the "optimal" rate of convergence). An optimal choice is not available in general, but it is under the additional assumption that the diffusion coefficient $c(\theta,t,x)$ is invertible, that is, takes its values in the set $\mathcal{M}_{d\times d}^{++}$. In this case it is enough to define the function g on $\mathcal{M}_{d\times d}^{++}$ (the argument is exactly the same as in the previous theorem), and one may take

$$g(a,x) = \log\det a + x^* a^{-1} x,$$

which is also $-\log h_a(x)$ if h_a denotes the product of $(2\pi)^{d/2}$ and the density of the measure ρ_a. Then the associated function G is

$$G(a,a') = -\int \log h_a(x)\, \rho_{a'}(dx) = G(a',a') - \int \log \frac{h_a(x)}{h_{a'}(x)}\, \rho_{a'}(dx).$$

Since $\int \frac{h_a(x)}{h_{a'}(x)} \rho_{a'}(dx) = 1$ we deduce from Jensen's inequality that $G(a,a') \geq G(a',a')$, with equality if and only if $h_a/h_{a'} = 1$ almost everywhere, that is if $a = a'$. Hence, here again, we have (7.4.1).

When $b \equiv 0$ and $c(\theta,t,x) = c(\theta)$ does not depend on (t,x) (so X is a Brownian motion with variance $c(\theta)$ under $\mathbb{P}_\theta$), the contrast Φ_n is, up to a constant, minus the likelihood: our estimator is thus the MLE (maximum likelihood estimator), and the optimality properties are not a surprise (but of course we need here the invertibility of c).

Chapter 8
Second Extension: Functions of Several Increments

8.1 Introduction

Now we want to extend the Laws of Large Numbers to functionals in which the test function depends on several successive increments, which may indeed mean several different things.

1) The most natural meaning is the following one. We take an integer $k \geq 2$ and a function F on $(\mathbb{R}^d)^k$, and we substitute $V^n(f,X)$ and $V'^n(f,X)$ with

$$V^n(F,X)_t = \sum_{i=1}^{N_n(t)-k+1} F\big(\Delta_i^n X, \Delta_{i+1}^n X, \dots, \Delta_{i+k-1}^n X\big), \tag{8.1.1}$$

$$V'^n(F,X)_t = \Delta_n \sum_{i=1}^{[t/\Delta_n]-k+1} F\big(\Delta_i^n X/\sqrt{\Delta_n}, \Delta_{i+1}^n X/\sqrt{\Delta_n}, \dots, \Delta_{i+k-1}^n X/\sqrt{\Delta_n}\big). \tag{8.1.2}$$

For $V^n(F,X)$ the discretization scheme is arbitrary, whereas for $V'^n(F,x)$ it is regular. Note that the upper limit $N_n(t)-k+1$ is exactly what is needed to use all increments $\Delta_i^n X$ occurring up to time t.

In these two definitions the same increment $\Delta_i^n X$ occurs into several different summands. This breaks down the "almost independence" of the summands which in the previous chapters was important for proving the LLN, and even more for the CLT.

To overcome these difficulties we can add up summands which involve non-overlapping intervals and discard the others. This leads us to introduce the processes

$$\begin{aligned} \mathcal{V}^n(F,X)_t &= \sum_{i=1}^{[N_n(t)/k]} F\big(\Delta_{ik-k+1}^n X, \dots, \Delta_{ik}^n X\big) \\ \mathcal{V}'^n(F,X)_t &= \Delta_n \sum_{i=1}^{[t/k\Delta_n]} F\big(\Delta_{ik-k+1}^n X/\sqrt{\Delta_n}, \dots, \Delta_{ik}^n X/\sqrt{\Delta_n}\big). \end{aligned} \tag{8.1.3}$$

J. Jacod, P. Protter, *Discretization of Processes*,
Stochastic Modelling and Applied Probability 67,
DOI 10.1007/978-3-642-24127-7_8,

These functionals are much easier to handle than the previous ones, but also with less practical interest unless F is invariant by a permutation of its k arguments (this property is referred to as F being a *symmetrical function*). When F is not symmetrical, they put more emphasis on some increments than on others: for example if $k=2$ and $F(x,y)=f(x)$ does not depend on the second argument, then $\mathcal{V}^n(F,X)$ and $\mathcal{V}'^n(F,X)$ skip all increments with an even index. This fact implies that $\mathcal{V}^n(F,X)$ simply does *not* converge in general when F is not symmetrical.

2) In most applications, the test function F is of one of the following non-exclusive forms:

- *Product form*, that is

$$F(x_1,\dots,x_k) = \prod_{j=1}^k f_j(x_j). \tag{8.1.4}$$

- *Positively homogeneous of degree* $w\geq 0$, that is, it satisfies

$$\lambda\geq 0,\ x_i\in\mathbb{R}^d \quad\Rightarrow\quad F(\lambda x_1,\dots,\lambda x_k) = \lambda^w F(x_1,\dots,x_k),$$

  with the non-standard convention $0^0=1$.
- *Extended multipower*, that is of product form with each f_j being positively homogeneous of some degree $w_j\geq 0$. Then F is positively homogeneous of degree $w=w_1+\cdots+w_k$.
- *Multipower*, that is

$$F(x_1,\dots,x_k) = \prod_{j=1}^k\prod_{i=1}^d |x_j^i|^{w_j^i}, \quad w_j^i\geq 0, \tag{8.1.5}$$

  where again $0^0=1$, so that when $w_j^i=0$ the factor $|x_j^i|^{w_j^i}$ does not show. A multipower is an extended multipower with degree $w=w_1+\cdots+w_k$, where $w_j=w_j^1+\cdots+w_j^d$. Conversely, when $d=1$ any extended multipower is a multipower.

A *multipower variation*, or realized multipower variation, is the functional $V'^n(F,X)$ with F of the form (8.1.5). These functionals have been introduced by Barndorff-Nielsen and Shephard [8]. Observe that, as soon as F is positively homogeneous of degree w, and if the scheme is regular, we have

$$V^n(F,X) = \Delta_n^{w/2-1}\, V'^n(F,X), \qquad \mathcal{V}^n(F,X) = \Delta_n^{w/2-1}\, \mathcal{V}'^n(F,X).$$

3) In some applications we need to consider functionals whose summands depend on a number k_n of increments which increases to infinity as $n\to\infty$. This occurs for example when the process X is corrupted by a noise, as we shall see in Chap. 16, and to eliminate the noise one takes a moving average of k_n successive observations, and typically k_n is then of the order $1/\sqrt{\Delta_n}$ or bigger.

We could plug in k_n in (8.1.1) or (8.1.2), but then at stage n the function $F = F_n$ depends on n since it is a function on $(\mathbb{R}^d)^{k_n}$. Therefore any limit theorem needs some sort of "compatibility" between the F_n's, which is not easy to formulate for functions on different spaces!

This is a rather tricky issue, which so far has no satisfactory answer for general discretization schemes. However, for a *regular scheme*, and among several possibilities, an appealing way to solve this issue is as follows. The objective is to replace $V^n(F, X)$, say, by

$$\sum_{i=1}^{[t/\Delta_n]-k_n+1} F_n\big(\Delta_i^n X, \Delta_{i+1}^n X, \dots, \Delta_{i+k_n-1}^n X\big). \tag{8.1.6}$$

The ith summand involves the increments of X over the interval $[(i-1)\Delta_n, (i-1+k_n)\Delta_n]$, whose length is

$$u_n = k_n \Delta_n.$$

Then we re-scale time in such a way that this interval becomes $[0, 1]$, by putting

$$t \in [0,1] \quad \mapsto \quad X(n,i)_t = X_{(i-1)\Delta_n + tu_n} - X_{(i-1)\Delta_n}. \tag{8.1.7}$$

The increments of X in the ith summand convey exactly the same information as the restriction to the time interval $[0, 1]$ of the discretized version of the process $X(n, i)$, along the regular discretization scheme with time step $1/k_n$, that is (a notation similar to (2.2.12), except that the time interval is restricted to $[0, 1]$):

$$t \in [0,1] \mapsto X(n,i)_t^{(n)} = X(n,i)_{[k_n t]/k_n}. \tag{8.1.8}$$

The paths of the processes $X(n, i)$ and $X(n, i)^{(n)}$ belong to the space $\mathbb{D}_1^d$ of all càdlàg functions from $[0, 1]$ into $\mathbb{R}^d$ (a function in $\mathbb{D}_1^d$, as just defined, may have a jump at time 1, so $\mathbb{D}_1^d$ is different from the "usual" Skorokhod space $\mathbb{D}([0, 1], \mathbb{R}^d)$, whose elements are continuous at time 1). Then it is natural, instead of the sequence of functions F_n on $(\mathbb{R}^d)^{k_n}$, to take a single function Φ on the space $\mathbb{D}_1^d$, and to interpret the sum (8.1.6) as being

$$V^n(\Phi, k_n, X)_t = \sum_{i=1}^{[t/\Delta_n]-k_n+1} \Phi\big(X(n,i)^{(n)}\big). \tag{8.1.9}$$

Put another way, (8.1.6) is the same as (8.1.9), when the functions F_n are given by

$$F_n(x_1, \dots, x_{k_n}) = \Phi\big(y^{(n)}\big), \quad \text{where } y^{(n)}(t) = \sum_{j=1}^{k_n} x_j 1_{\{j/k_n \le t\}}. \tag{8.1.10}$$

This provides us the needed "compatibility relation" between the different F_n's.

The "normalized" version is as follows:

$$V'^n(\Phi, k_n, X)_t = \Delta_n \sum_{i=1}^{[t/\Delta_n]-k_n+1} \Phi\left(\frac{1}{\sqrt{u_n}}\, X(n,i)^{(n)}\right). \tag{8.1.11}$$

The argument of Φ is somewhat similar to the Brownian motion $w(n,i)$ defined in (4.2.4), and is actually the same when $X = W$ is a Brownian motion. This is the reason motivating the choice of the normalizing factor $1/\sqrt{u_n}$. Because of the time change performed in (8.1.8), this is the same as the factor $1/\sqrt{\Delta_n}$ for the increments $\Delta_i^n X$.

We will use a few specific notations below. The space $\mathbb{D}_1^d$ can be identified with the (closed) subspace of $\mathbb{D}^d$ consisting of all functions which are constant after time 1, and we endow it with the relative topology, still called the Skorokhod topology, and with the associated Borel σ-field, which is $\mathcal{D}_1^d = \sigma(x(t) : t \in [0,1])$. We also need the sup norm

$$x^{\#} = \sup\left(\left\|x(t)\right\| : t \in [0,1]\right). \tag{8.1.12}$$

We always assume that the test function Φ on $\mathbb{D}_1^d$ satisfies the following (note that the continuity for the Skorokhod topology implies the continuity for the sup-norm, but does not imply the *uniform* continuity for the sup-norm on bounded sets):

$$\left.\begin{array}{l}\text{For each } v \geq 0 \text{ the function } \Phi \text{ is continuous for the Skorokhod topology,}\\ \quad\text{and also bounded and uniformly continuous for the sup-norm}\\ \quad\text{on the set } \{x : x^{\#} \leq v\}\text{; we then set } \Phi^{\#}(v) = \sup_{x \in \mathbb{D}_1^d : x^{\#} \leq v} |\Phi(x)|.\end{array}\right\} \tag{8.1.13}$$

4) In the whole chapter, X is a d-dimensional semimartingale defined on $(\Omega, \mathcal{F}, (\mathcal{F}_t)_{t \geq 0}, \mathbb{P})$, with characteristics (B, C, ν) and jump measure μ. The LLNs for $V^n(F,X)$ and $\mathcal{V}^n(F,X)$ on the one hand, and for $V^n(\Phi, k_n, X)$ on the other hand, are studied in two different sections, and they do not reduce one to the other. In contrast, the LLNs for $V'^n(F,X)$, $\mathcal{V}'^n(F,X)$ and $V'^n(\Phi, k_n, X)$ are essentially the same and are treated in a single theorem.

8.2 The Law of Large Numbers for $V^n(F,X)$ and $\mathcal{V}^n(F,X)$

Here, we extend Case (A-a) of Theorem 3.3.1. Cases (A-b,c,d) can be dealt with in the same way, but are not treated here because of their limited interest for applications. Cases (B) and (C) are significantly more complicated, and are not treated at all below.

Theorem 8.2.1 *Let X be a d-dimensional semimartingale and $\mathcal{T} = (T_n)$ be any random discretization scheme. Let also F be a continuous function from $(\mathbb{R}^d)^k$ into $\mathbb{R}^q$, which satisfies*

$$F(z) = \mathrm{o}\left(\|z\|^2\right) \quad \text{as } z \to 0 \text{ in } \left(\mathbb{R}^d\right)^k. \tag{8.2.1}$$

Then for each t we have the following convergence in probability, where f_j denotes the function $f_j(x) = F(0,\dots,0,x,0,\dots,0)$ on $\mathbb{R}^d$ with x occurring as the jth argument:

$$V^n(F,X)_t \xrightarrow{\mathbb{P}} V(F,X)_t := \sum_{j=1}^k f_j \star \mu_t, \quad \textit{in restriction to the set } \{\Delta X_t = 0\}. \tag{8.2.2}$$

If further F is a symmetrical function, hence all f_j are equal to the same function f, we have the following Skorokhod convergence in probability:

$$\mathcal{V}^n(F,X) \overset{\mathbb{P}}{\Longrightarrow} V(f,X) := f \star \mu. \tag{8.2.3}$$

Remark 8.2.2 When F is a symmetrical function we have a bit more than (8.2.3), namely

$$\overline{W}_t^n = \mathcal{V}^n(F,X)_t - V(f,X)_{k\Delta_n[t/k\Delta_n]} \overset{\text{u.c.p.}}{\Longrightarrow} 0. \tag{8.2.4}$$

The process $\overline{W}^n$ is the discretized version of a process similar to (3.3.7), which involves a suitably defined interpolation of $V^n(F,X)$ and could be shown to converge to 0 in the $u.c.p.$ sense as well.

When F is not symmetrical, the convergence (8.2.3) fails. For example let $d=1$ and X be a Poisson process with jump times $T_1, T_2, \dots$, and $k=2$ and $F(x,y)=x^4$. Suppose also that we have a regular scheme with $\Delta_n = 1/n$. On the set $A = \{T_1 < 1/4, T_2 > 1\}$ the variables $\mathcal{V}^n(F,X)_s$ equals 1 when $[nT_1]$ is even, and to 0 when $[nT_1]$ is odd, if $n \geq 4$ and for all $s \in (1/2, 1]$. Hence the processes $\mathcal{V}^n(F,X)$ do not converge.

Remark 8.2.3 The convergence in (8.2.2) does *not* take place on the set $\{\Delta X_t \neq 0\}$, usually. The reason is basically the same as for the lack of convergence of $\mathcal{V}(F,X)_t$ when F is not symmetrical (or, for that matter, of $V^n(f,X)_t$ in Theorem 3.3.1): indeed, if t is one of the discretization points $T(n,i)$ for infinitely many n's, and is not a discretization point for infinitely many n's also, the two—usually different—quantities $\sum_{j=1}^{k-1} f_j * \mu_t + f_k * \mu_{t-}$ and $\sum_{j=1}^{k-1} f_j * \mu_t + f_k * \mu_t$ are limit points of the sequence $V^n(F,X)_t$.

Remark 8.2.4 It is important to observe that we *cannot replace* (8.2.2) *by Skorokhod convergence in probability*, even when F is symmetrical, and this is a fundamental difference with Theorem 3.3.1. Indeed, if X has a jump at time S, say, and if $T(n,i-1) < S \leq T(n,i)$, the process $V^n(F,X)$ has a jump of approximate size $f_j(\Delta X_S)$ at time $T(n,i+k-j)$; then as soon as at least two of the variables $f_j(\Delta X_S)$ are not vanishing, the process $V^n(F,X)$ has two "big jumps" at two distinct times lesser apart than $k\Delta_n$, and this prevents the convergence for the Skorokhod topology.

Example 8.2.5 Suppose that $d=1$ and write $g_p(x) = |x|^p$. A useful example is

$$F(x_1,\dots,x_k) = |x_1 + \cdots + x_k|^p, \tag{8.2.5}$$

which is symmetrical, and satisfies (8.2.1) when $p>2$. Then (8.2.3) yields $\mathcal{V}^n(F,X)\stackrel{\mathbb{P}}{\Longrightarrow} g_p*\mu$, which can also be interpreted as a consequence of Theorem 3.3.1 for the function $f=g_p$ and the discretization scheme $T'(n,i)=T(n,ik)$ for $i\geq 0$ and $n\geq 1$. As to (8.2.2), it reads as $V^n(F,X)_t\stackrel{\mathbb{P}}{\longrightarrow} k\,g_p*\mu_t$ on the set $\{\Delta X_t=0\}$.

Another useful example is the multipower function

$$F(x_1,\ldots,x_k) = \prod_{j=1}^k |x_j|^{p_j} \tag{8.2.6}$$

which satisfies (8.2.1) if $p=p_1+\cdots+p_k>2$. Unless all p_j but one vanish, $V^n(F,X)_t\stackrel{\mathbb{P}}{\longrightarrow} 0$ on the set $\{\Delta X_t=0\}$ because the functions f_j are then all identically 0. In other words, the previous theorem gives a trivial result and is not sharp enough to capture the genuine behavior of $V^n(F,X)$. We will see in Chap. 15 how to circumvent this problem.

Proof of Theorem 8.2.1 Although the results here are significantly different from those in Theorem 3.3.1, the proof is basically the same.

Step 1) This step is the counterpart of Lemma 3.3.7, and this is where the difference between (3.3.2) and (8.2.2) really occurs. We wish to prove the results when F is continuous and, for some $\varepsilon>0$, satisfies $F(x_1,\ldots,x_k)=0$ if $\|x_j\|\leq\varepsilon$ for all j.

We use the notation of Lemma 3.3.7: the successive jump times $S_1,\ldots$ of X with size bigger than $\varepsilon/2$ and the number Q_T of such jumps occurring within the time interval $[0,T]$. We also set $X'=X-(x\,1_{\{\|x\|>\varepsilon/2\}})*\mu$. Recalling the intervals $I(n,i)=(T(n,i-1),T(n,i)]$, we also have an integer-valued variable M_T such that for all $n\geq M_T$ and $i\leq N_n(T)$;

$$\left.\begin{array}{l}
\bullet \text{ either the interval } I(n,i) \text{ contains no } S_q \text{ (recall (3.1.2)), and}\\
\qquad s\in I(n,i)\ \Rightarrow\ \|X_s-X_{T(n,i-1)}\|\leq\varepsilon\\
\bullet \text{ or } I(n,i) \text{ contains exactly one } S_q, \text{ and we write } i=i(n,q), \text{ and}\\
\qquad s\in I(n,i)\ \Rightarrow\ \|X_s-X_{T(n,i-1)}-\Delta X_{S_q}1_{\{S_q\leq s\}}\|\leq\varepsilon\\
\bullet\ q\geq 0,\ S_q<T\ \Rightarrow\ S_{q+1}\wedge T>T(n,i(n,q)+k)
\end{array}\right\} \tag{8.2.7}$$

(this is (3.3.10) plus an additional property). We write the (random) integer $i(n,q)$ as $i(n,q)=m(n,q)+l(n,q)$, where $l(n,q)\in\{0,1,\ldots,k-1\}$ and $m(n,q)$ is a multiple of k.

(a) We start by proving (8.2.4), which implies (8.2.3), when F is symmetrical. In this case the process $\overline{W}^n$ of (8.2.4) satisfies, for $n\geq M_T$ and $t\leq T$,

$$\begin{aligned}
\overline{W}^n_t &= \sum_{q=1}^{Q_T}\big(\zeta^n_q-f(\Delta X_{S_q})\big)\,1_{\{S_q\leq k[t/k\Delta_n]\}}, \qquad\text{where}\\
\zeta^n_q &= F\big(\Delta^n_{m(n,q)+1}X,\ldots,\Delta^n_{m(n,q)+k}X\big),
\end{aligned}$$

and $\Delta^n_{m(n,q)+j}X$ equals $\Delta^n_{m(n,q)+j}X'$ if $j=1,\dots,k$, except when $j=l(n,q)$, in which case it equals $\Delta^n_{i(n,q)}X'+\Delta X_{S_q}$. Since X' is continuous at time S_q and F is continuous and symmetrical, we have $\zeta^n_q \to f(\Delta X_{S_q})$ by the definition of f, and (8.2.4) follows.

(b) Next, we prove (8.2.2). By (3.3.10) and (8.2.7) we have for $n\ge M_T$, and on the set $\{\Delta X_T=0\}$:

$$V^n(F,X)_T = \sum_{q=1}^{Q_T}\sum_{j=1}^{k}\zeta(n,j,q), \quad \text{where}$$
$$\zeta(n,j,q) = F\big(\Delta^n_{i(n,q)+1-j}X,\dots,\Delta^n_{i(n,q)+k-j}X'\big),$$

and $\Delta^n_{i(n,q)+l}X$ equals $\Delta^n_{i(n,q)+l}X'$ if $l=1-j,\dots,k-j$, except when $l=0$, in which case it equals $\Delta^n_{i(n,q)}X'+\Delta X_{S_q}$. Here again, $\zeta(n,j,q)\to f_j(\Delta X_{S_q})$, and (8.2.2) follows.

Step 2) We are now ready to prove the claims when F is continuous and satisfies (8.2.1). We consider the functions of (3.3.16), which are $\psi'_\varepsilon=1-\psi_\varepsilon$ and $\psi_\varepsilon(x)=\psi(\|x\|/\varepsilon)$, where ψ is C^∞ on $\mathbb{R}$ with $1_{[1,\infty)}\le\psi\le 1_{[1/2,\infty)}$. The two functions

$$F'_\varepsilon(x_1,\dots,x_k) = F(x_1,\dots,x_k)\prod_{j=1}^{k}\psi'_\varepsilon(x_j), \qquad F_\varepsilon = F-F'_\varepsilon$$

are continuous and F_ε vanishes on a neighborhood of 0, hence Step 1 yields $V^n(F_\varepsilon,X)_t \stackrel{\mathbb{P}}{\longrightarrow} \sum_{j=1}^k (f_j\psi_\varepsilon)*\mu_t$ on the set $\{\Delta X_t=0\}$. Moreover $(f_j\psi_\varepsilon)*\mu \stackrel{\text{u.c.p.}}{\Longrightarrow} f_j*\mu$ as $\varepsilon\to 0$ by the dominated convergence theorem (since $\|(f_j\psi_\varepsilon)(x)\|\le K\|x\|^2$ by (8.2.1) and $\psi_\varepsilon\to 1$ and $\sum_{s\le t}\|\Delta X_s\|^2<\infty$). Therefore, for (8.2.2) it suffices to prove that

$$t,\eta>0 \quad\Rightarrow\quad \lim_{\varepsilon\to 0}\limsup_n \mathbb{P}\Big(\sup_{s\le t}\big\|V^n(F'_\varepsilon)_s\big\|>\eta\Big)=0. \tag{8.2.8}$$

For this, we first deduce from (8.2.1) and the definition of F'_ε that we have $\|F'_\varepsilon(x_1,\dots,x_k)\|\le\theta(\varepsilon)((\|x_1\|^2\wedge 1)+\cdots+(\|x_k\|^2\wedge 1))$, where $\theta(\varepsilon)\to 0$ as $\varepsilon\to 0$. Then if $h(x)=\|x\|^2\wedge 1$, we have

$$\sup_{s\le t}\big\|V^n\big(F'_\varepsilon\big)_s\big\| \le k\theta(\varepsilon)\,V^n(h,X)_t.$$

Since $V^n(h,X)$ converges in probability for the Skorokhod topology by Theorem 3.3.1-(B), we deduce (8.2.8) $\theta(\varepsilon)\to 0$ as $\varepsilon\to 0$.

When F is symmetrical, the proof of (8.2.3) is the same, upon observing that F_ε is also symmetrical and thus $\mathcal{V}^n(F_\varepsilon,X)\stackrel{\mathbb{P}}{\Longrightarrow}(f\psi_\varepsilon)*\mu$ by Step 1. □

8.3 The Law of Large Numbers for $V^n(\Phi, k_n, X)$

We now turn to the process $V^n(\Phi, k_n, X)$, as defined in (8.1.9), and when the discretization scheme is regular. If $z \in \mathbb{R}^d$ and $t \in (0, 1]$ we define the step function $y_{z,t}$ by

$$s \in [0, 1] \quad \mapsto \quad y_{z,t}(s) = z 1_{\{s \geq t\}}.$$

The map $(t, z) \mapsto y_{t,z}$ from $(0, 1] \times \mathbb{R}^d$ into $\mathbb{D}_1^d$ is continuous for the Skorokhod topology, including at the points $(1, z)$ because of our specific definition of this topology. Hence if the test function Φ satisfies (8.1.13), the map $(t, z) \mapsto \Phi(y_{t,z})$ is continuous, and bounded on $(0, 1] \times \{z : \|z\| \leq v\}$ for any $v > 0$. Thus the formula

$$\overline{\Phi}(z) = \int_0^1 \Phi(y_{z,t})\, dt, \tag{8.3.1}$$

defines a continuous and locally bounded function $\overline{\Phi}$ on $\mathbb{R}^d$.

Theorem 8.3.1 *Let X be a d-dimensional semimartingale and $\mathcal{T}$ be a regular discretization scheme, and assume that $k_n \to \infty$ and $u_n = k_n \Delta_n \to 0$. Let also Φ be a q-dimensional function on $\mathbb{D}^d$, which satisfies* (8.1.13) *and*

$$\Phi^{\#}(v) = \mathrm{o}\big(v^2\big) \quad \text{as } v \to 0. \tag{8.3.2}$$

Then for each t we have the following convergence in probability:

$$\frac{1}{k_n} V^n(\Phi, k_n, X)_t \xrightarrow{\mathbb{P}} V(\Phi, X)_t := \overline{\Phi} * \mu_{t-}. \tag{8.3.3}$$

Remark 8.3.2 The limit in (8.3.3) is equal to $\overline{\Phi} * \mu_t$ on the set $\{\Delta X_t = 0\}$, so this result partially extends (8.2.2). More specifically, but *only in the regular scheme setting*, if k is an integer and F is a continuous function on $(\mathbb{R}^d)^k$ satisfying (8.2.1) it is simple to construct a function Φ on $\mathbb{D}^d$ satisfying (8.1.13) and (8.3.2), and such that (8.1.10) holds with $k_n = k$, and also such that $\Phi(y_{z,t}) = f_j(z)$ when $\frac{j-1}{k} \leq t < \frac{j}{k}$: then (8.2.2) is just another way of writing (8.3.3).

However, this does not provide another proof of Theorem 8.2.1 for regular schemes, since we need $k_n \to \infty$ for the above theorem to apply: it simply reflects the "compatibility" of the two results. There are also obvious differences: for instance, the fact that $k_n \to \infty$ "kills" any particular summand in the definition of $V^n(\Phi, k_n, X)_t$, and this is why the limit in (8.3.3) is not restricted to the set $\{\Delta X_t \neq 0\}$.

Remark 8.3.3 Exactly as for Theorem 8.2.1, we *cannot replace* (8.3.3) *by Skorokhod convergence in probability*. The reason is in a sense the opposite: the limit in (8.3.3) is purely discontinuous, whereas the processes which converge have all jumps smaller than U/k_n for some (random) U.

As for Theorem 3.3.1, we begin the proof with a special case.

Lemma 8.3.4 *When Φ is such that $\Phi^{\#}(\varepsilon) = 0$ for some $\varepsilon > 0$, the convergence* (8.3.3) *holds, and it even holds for every ω.*

Proof *Step 1)* We again use the notation of Lemma 3.3.7: the successive jump times $S_1, \dots$ of X with size bigger than $\varepsilon/2$, the process $X'_t = X_t - \sum_{q\ge 1: S_q \le t} \Delta X_{S_q}$, and now Q_t denotes the number of jumps in the *open* interval $(0, t)$ (a slight change of notation). Similar to (8.2.7), and since $u_n \to 0$, we obtain that for some finite (random) integer $M_t(\omega)$ we have for all $i \le [t/\Delta_n]$, and as soon as $n \ge M_t$,

$$\left.\begin{array}{l} \bullet\ X'(n,i)^{\#} \le \varepsilon \\ \bullet\ Q_t \ge 1 \ \Rightarrow\ S_1 > u_n,\ \ S_{Q_t} < t - u_n \\ \bullet\ 2 \le q \le Q_t \ \Rightarrow\ S_q - S_{q-1} > u_n. \end{array}\right\} \tag{8.3.4}$$

Let $i(n,q)$ be the unique integer such that $(i(n,q)-1)\Delta_n < S_q \le i(n,q)\Delta_n$, and set

$$\begin{aligned} Y(n,q,j)_s &= \Delta X_{S_q} 1_{\{s \ge j/k_n\}} = y_{\Delta X_{S_q}, j/k_n}(s) \\ \alpha(n,q,j) &= \Phi\big(X'\big(n, i(n,q)+1-j\big)^{(n)} + Y(n,q,j)\big) \\ \zeta_q^n &= \frac{1}{k_n} \sum_{j=1}^{k_n \wedge i(n,q)} \alpha(n,q,j), \\ A_t^n &= \begin{cases} \frac{1}{k_n} \Phi(X(n, 1-k_n + t/\Delta_n)) & \text{if } t/\Delta_n \text{ is an integer and } S_{Q_t+1} = t \\ 0 & \text{otherwise.} \end{cases} \end{aligned}$$

We have $\Phi(X(n,i)^{(n)}) = \Phi(X'(n,i)^{(n)}) = 0$ when $n \ge M_t$ and i is such that the interval $((i-1)\Delta_n, (i-1)\Delta_n + u_n]$ is in $[0,t]$ and contains no S_q, because $\Phi^{\#}(\varepsilon) = 0$. Hence

$$n \ge M_t \quad \Rightarrow \quad \frac{1}{k_n} V^n(\Phi, k_n, X)_t = \sum_{q=1}^{Q_t} \zeta_q^n + A_t^n. \tag{8.3.5}$$

On the one hand, if $n \ge M_t$ and t/Δ_n is an integer, (8.3.4) yields $X(n, 1-k_n + t/\Delta_n)^{\#} \le \varepsilon + \|\Delta X_t\|$. Hence (8.1.13) yields $A_t^n \to 0$. On the other hand if $\|z\| \le \varepsilon$ we have $y_{z,s}^{\#} \le \varepsilon$, hence $\Phi(y_{z,s}) = 0$ by our assumption on Φ, and thus $\overline{\Phi}(z) = 0$. This implies $\overline{\Phi} * \mu_{t-} = \sum_{q=1}^{Q_t} \overline{\Phi}(\Delta X_{S_q})$. Therefore in view of (8.3.4) we only have to prove that

$$1 \le q \le Q_t \quad \Rightarrow \quad \zeta_q^n \to \overline{\Phi}(\Delta X_{S_q}). \tag{8.3.6}$$

Step 2) Below we fix ω and also q between 1 and Q_t and we always take $n \ge M_t$. Then $i(n,q) > k_n$ and thus $\zeta_q^n = \frac{1}{k_n} \sum_{j=1}^{k_n} \alpha(n,q,j)$. This step is devoted to proving that

$$\zeta_q^n - \zeta_q'^n \to 0, \quad \text{where } \zeta_q'^n = \frac{1}{k_n} \sum_{j=1}^{k_n} \Phi\big(Y(n,q,j)\big). \tag{8.3.7}$$

For this, we first observe that, since X' is continuous at time S_q, we have

$$\chi_n = \sup_{1\le j\le k_n} X'\big(n, i(n,q)+1-j\big)^\# \to 0. \tag{8.3.8}$$

Then there is some $v>0$ such that $\chi_n + \|\Delta X_{S_q}\| \le v$ for all n, and by (8.1.13) there is a function $\phi(\varepsilon)$ such that $\phi(\varepsilon)\to 0$ as $\varepsilon\to 0$ and $|\Phi(x+y)-\Phi(x)|\le \phi(y^\#)$ for all $x,y\in \mathbb{D}_1^d$ such that $x^\#\le v$ and $(x+y)^\#\le v$. It follows that

$$1\le j\le k_n \quad \Rightarrow \quad \big|\alpha(n,q,j) - \Phi\big(Y(n,q,j)\big)\big| \le \phi(\chi_n).$$

Therefore $|\zeta_q^n - \zeta_q'^n| \le \phi(\chi_n)$, and (8.3.7) follows from (8.3.8).

Step 3) Now we are ready to prove (8.3.6), which by (8.3.7) amounts to

$$\zeta_q'^n \to \overline{\Phi}(\Delta X_{s_q}). \tag{8.3.9}$$

Recalling the definition of $\zeta_q'^n$ and the fact that $Y(n,q,j) = y_{z,j/k_n}$ with $z=\Delta X_{S_q}$, we see that $\zeta_q'^n$ is a Riemann approximation of the integral $\overline{\Phi}(z) = \int_0^1 \Phi(y_{z,s})\,ds$. Since $s\mapsto \Phi(y_{z,s})$ is bounded and continuous on $(0,1]$ for any fixed z, (8.3.9) readily follows. □

Proof of Theorem 8.3.1 *Step 1)* For any $\varepsilon\in(0,1)$ we decompose Φ as $\Phi_\varepsilon + \Phi'_\varepsilon$, where

$$\Phi_\varepsilon(x) = \Phi(x)\,\psi_\varepsilon\big(x^\#\big), \qquad \Phi'_\varepsilon(x) = \Phi(x)\,\psi'_\varepsilon\big(x^\#\big)$$

(here, ψ_ε and ψ'_ε are as before (8.2.8) or in (3.3.16)). The function Φ_ε satisfies the assumptions of the previous lemma, hence $\frac{1}{k_n} V^n(\Phi_\varepsilon, k_n, X)_t \to \overline{\Phi}_\varepsilon * \mu_{t-}$ pointwise. Thus it remains to prove the following two properties:

$$\overline{\Phi}_\varepsilon * \mu_{t-} \stackrel{\mathbb{P}}{\longrightarrow} \overline{\Phi} * \mu_{t-} \quad \text{as } \varepsilon\to 0, \tag{8.3.10}$$

$$\eta>0 \quad\Rightarrow\quad \lim_{\varepsilon\to 0}\limsup_n \mathbb{P}\bigg(\frac{1}{k_n}\big|V^n\big(\Phi'_\varepsilon, k_n, X\big)_t\big| > \eta\bigg) = 0. \tag{8.3.11}$$

The first property is easy. Indeed, on the one hand, $(y_{z,s})^\# = \|z\|$ and thus if $\|z\|\ge\varepsilon$ we have $\Phi(y_{z,s}) = \Phi_\varepsilon(y_{z,s})$ for all $s\in(0,1]$, hence $\overline{\Phi}(z) = \overline{\Phi}_\varepsilon(z)$: therefore $\overline{\Phi}_\varepsilon \to \overline{\Phi}$ pointwise. On the other hand (8.3.2) yields $\overline{\Phi}_\varepsilon(z) = o(\|z\|^2)$ as $z\to 0$. Since $\sum_{s\le t}\|\Delta X_s\|^2<\infty$, (8.3.10) follows from the dominated convergence theorem.

Step 2) The proof of (8.3.11) is more involved. We set $T_0=0$ and let $T_1,\dots$ be the successive jump times of X with size bigger than 1, and $X'_t = X_t - (x 1_{\{\|x\|\le 1\}}) * \mu$.

Let Ω_t^n be the set on which, for each $q \geq 1$ with $T_q \leq t$, we have $T_q - T_{q-1} > u_n$ and $\sup_{s \in (0,u_n]}(\|X_{T_q+s} - X_{T_q}\| + \|X_{T_q-} - X_{T_q-s}\|) \leq 1/2$. We have $\Omega_t^n \to \Omega$ as $n \to \infty$, so it is enough to prove that

$$\eta > 0 \quad \Rightarrow \quad \lim_{\varepsilon \to 0} \limsup_n \mathbb{P}\left(\Omega_t^n \cap \left\{\frac{1}{k_n}\left|V^n(\Phi'_\varepsilon, k_n, X)_t\right| > \eta\right\}\right) = 0.$$

Recall that $\psi'_\varepsilon(v) = 0$ if $v \geq \varepsilon$. Then if $\varepsilon < 1/2$ and on the set Ω_t^n we have $(X(n,i)^{(n)})^\# \geq \varepsilon$ for any i such that $T_q \in (i\Delta_n, i\Delta_n + u_n]$ for some q, in which case $\Phi'_\varepsilon(X(n,i)^{(n)}) = 0$. For the other values of i we have $X(n,i) = X'(n,i)$. Therefore $|V^n(\Phi'_\varepsilon, k_n, X)_t| \leq V^n(|\Phi'_\varepsilon, k_n, X')_t$, and it is enough to show that

$$\eta > 0 \quad \Rightarrow \quad \lim_{\varepsilon \to 0} \limsup_n \frac{1}{k_n} \mathbb{E}\left(V^n\left(\left|\Phi'_\varepsilon\right|, k_n, X'\right)_t\right) = 0. \tag{8.3.12}$$

Step 3) Recall the decomposition $X' = X_0 + B + X^c + (x 1_{\{\|x\| \leq 1\}}) * (\mu - \nu)$ and, with $\mathrm{Var}(B^i)$ denoting the variation process of the ith component B^i, set

$$A = \sum_{i=1}^d \left(\mathrm{Var}\left(B^i\right) + C^{ii}\right) + \left(\|x\|^2 1_{\{\|x\| \leq 1\}}\right) * \nu.$$

This increasing process has jumps smaller than 2 because $\|\Delta B\| \leq 1$ identically, so it is locally bounded, and our usual localization procedure shows that it is enough to prove (8.3.12) when the process A is in fact bounded by a constant.

The process $M = X^c + (x 1_{\{\|x\| \leq 1\}}) * (\mu - \nu)$ is a square integrable martingale, the predictable quadratic variation of its ith component being such that $A - \langle M^i, M^i \rangle$ is non-decreasing. Therefore we deduce from the definition of A and its boundedness and from Doob's inequality that

$$\mathbb{E}\left(\sup_{s \in [0,1]} \left\|X'(n,i)_s\right\|^2\right) \leq K \mathbb{E}(A_{(i-1)\Delta_n + u_n} - A_{(i-1)\Delta_n}). \tag{8.3.13}$$

Now, (8.3.2) yields an increasing function θ on $\mathbb{R}_+$ such that $\lim_{v \to 0} \theta(v) = 0$ and $\Phi(y) \leq (y^\#)^2 \theta(y^\#)$ for all $y \in \mathbb{D}^d$. In view of the definition of Φ'_ε, we deduce $|\Phi'_\varepsilon(y)| \leq (y^\#)^2 \theta(\varepsilon)$. Then by (8.3.13) we obtain

$$\begin{aligned}
\frac{1}{k_n} \mathbb{E}\left(V^n\left(\Phi'_\varepsilon, k_n, X'\right)_t\right) &\leq K\theta(2\varepsilon) \frac{1}{k_n} \sum_{i=1}^{[t/\Delta_n]-k_n+1} \mathbb{E}(A_{(i-1)\Delta_n + u_n} - A_{(i-1)\Delta_n}) \\
&\leq K\theta(\varepsilon)\, \mathbb{E}(A_{t+u_n}).
\end{aligned}$$

At this point, and since $\theta(\varepsilon) \to 0$ as $\varepsilon \to 0$, (8.3.12) follows. □

8.4 The LLN for $V'^n(F,X)$, $\mathcal{V}'^n(F,X)$ and $V'^n(\Phi,k_n,X)$

Now we turn to the processes $V'^n(F,X)$ and $\mathcal{V}'^n(F,X)$ of (8.1.2) and (8.1.3), and also $V'^n(\Phi,k_n,X)$ of (8.1.11), and we extend Theorem 3.4.1. In particular, we suppose that the discretization scheme is *regular*, with stepsize Δ_n at stage n.

8.4.1 The Results

Theorem 3.4.1 has been proved when X is an Itô semimartingale, and here we make the same hypothesis. For convenience, we recall that X has a *Grigelionis representation*, possibly defined on an extended space still denoted as $(\Omega,\mathcal{F},(\mathcal{F}_t)_{t\geq 0},\mathbb{P})$, that is

$$X_t = X_0 + \int_0^t b_s\,ds + \int_0^t \sigma_s\,dW_s + (\delta 1_{\{\|\delta\|\leq 1\}})\star(\mathfrak{p}-\mathfrak{q})_t + (\delta 1_{\{\|\delta\|>1\}})\star\mathfrak{p}_t, \quad (8.4.1)$$

where W is a d'-dimensional Wiener process and $\mathfrak{p}$ is a Poisson measure with compensator $\mathfrak{q}(dt,dz) = dt\otimes\lambda(dz)$. Recall that a version of the process C_t is given by

$$C_t = \int_0^t c_s\,ds, \quad \text{where } c_t = \sigma_t\sigma_t^*.$$

In fact, for simplicity we do not do extend Theorem 3.4.1 in the utmost generality, and rather we make the Assumption (H), or Assumption 4.4.2, which we recall here:

Assumption (H) X has the form (4.2.1), with b_t locally bounded and σ_t càdlàg. Moreover $\|\delta(\omega,t,z)\|\wedge 1 \leq \Gamma_n(z)$ for all (ω,t,z) with $t\leq\tau_n(\omega)$, where (τ_n) is a localizing sequence of stopping times and each function Γ_n on E satisfies $\int \Gamma_n(z)^2\lambda(dz) < \infty$.

When $a\in\mathcal{M}^+_{d\times d}$ (a $d\times d$ symmetric nonnegative matrix), we use the notation $\rho_a = \mathcal{N}(0,a)$ of (3.4.4). Then $\rho_a^{k\otimes}$ denotes the kth power of ρ_a and is a Gaussian law on $(\mathbb{R}^d)^k$. We also denote by $\overline{\rho}_a$ the law (on the space $\mathbb{D}^d$ and also, by restriction, on the space $\mathbb{D}^d_1$) of the (non-standard) d-dimensional Brownian motion with covariance matrix a at time 1.

We state the results in two different theorems.

Theorem 8.4.1 *Assume that X satisfies* (H). *Let F be a continuous function on $(\mathbb{R}^d)^k$, which is of polynomial growth when X is continuous and which satisfies the following property when X jumps*:

$$\left|F(x_1,\dots,x_k)\right| \leq \prod_{j=1}^k \Psi\left(\|x_j\|\right)\left(1+\|x_j\|^2\right) \quad (8.4.2)$$

where Ψ is a continuous function on $[0,\infty)$ which goes to 0 at infinity. Then, with $c_s = \sigma_s\sigma_s^*$, we have

$$V'^n(F,X)_t \stackrel{\text{u.c.p.}}{\Longrightarrow} V'(F,X)_t := \int_0^t \rho_{c_s}^{k\otimes}(F)\,ds, \tag{8.4.3}$$

$$\mathcal{V}'^n(F,X) \stackrel{\text{u.c.p.}}{\Longrightarrow} \frac{1}{k}\,V'(F,X). \tag{8.4.4}$$

Theorem 8.4.2 *Assume that X satisfies* (H). *Let Φ be a function on $\mathbb{D}_1^d$ satisfying* (8.1.13) *and with $\Phi^\#$ being of polynomial growth when X is continuous, and satisfying $\Phi^\#(v) = \mathrm{o}(v^2)$ as $v \to \infty$ when X jumps. If moreover $k_n \to \infty$ and $u_n = k_n\Delta_n \to 0$, and with $c_s = \sigma_s\sigma_s^*$, we have*

$$V'^n(\Phi,k_n,X)_t \stackrel{\text{u.c.p.}}{\Longrightarrow} V'(\Phi,X)_t := \int_0^t \overline{\rho}_{c_s}(\Phi)\,ds. \tag{8.4.5}$$

Before giving the proof, we state a corollary similar to Corollary 3.4.3.

Corollary 8.4.3 *Assume that X satisfies* (H). *Let F be a Borel function on $(\mathbb{R}^d)^k$ which satisfies $F(z) \sim H(z)$ as $z \to 0$, where H is a positively homogeneous continuous function of degree $p \in (0,2)$ on $(\mathbb{R}^d)^k$. Then, with $c_s = \sigma_s\sigma_s^*$, we have*

$$\Delta_n^{1-p/2} V^n(F,X)_t \stackrel{\text{u.c.p.}}{\Longrightarrow} \int_0^t \rho_{c_s}^{k\otimes}(H)\,ds.$$

Example 8.4.4 The first theorem takes the following form, when F has the product form (8.1.4) with each f_i continuous and of polynomial growth: If X is continuous, then

$$V'^n(F,X)_t \stackrel{\text{u.c.p.}}{\Longrightarrow} \int_0^t \prod_{i=1}^k \rho_{c_s}(f_i)\,ds.$$

The same holds when X jumps, provided each f_i satisfies $f_i(x) = \mathrm{o}(\|x\|^2)$ as $\|x\| \to \infty$.

When further the f_i are positively homogeneous of order p_i, and again under (H) and either $\sup_{i=1,\dots,k} p_i < 2$ or X continuous, we deduce that

$$\Delta_n^{1-\frac{1}{2}(p_1+\dots+p_k)}\, V^n(F,X)_t \stackrel{\text{u.c.p.}}{\Longrightarrow} \int_0^t \prod_{i=1}^k \rho_{c_s}(f_i)\,ds.$$

8.4.2 The Proofs

Proof of Theorem 8.4.2 *Step 1)* The proof is basically the same as for Theorem 3.4.1, with some simplifications due to the fact that we assume (H). To begin with, by virtue of the localization lemma 4.4.9 we can replace Assumption (H) by the strengthened Assumption (SH) (Assumption 4.4.6, according to which we further have b_t and σ_t and X_t are bounded and also $\|\delta(\omega,t,x)\| \le \Gamma(z)$ for a bounded function Γ satisfying $\int \Gamma(z)^2\lambda(dz) < \infty$). Then, up to modifying the drift term in the Grigelionis decomposition (8.4.1), we can write

$$X = X' + X'', \quad \text{where } X'_t = X_0 + \int_0^t b'_s\, ds + \int_0^t \sigma_s\, dW_s, \quad X''_t = \delta \star (\mathfrak{p} - \mathfrak{q})_t, \tag{8.4.6}$$

and b'_t and σ_t are bounded, σ_t is càdlàg, and $\|\delta(t,z)\| \le \Gamma(z)$ with Γ bounded and with $\int \Gamma(z)^2\lambda(dz) < \infty$.

Step 2) In this step, we deduce from (SH) a number of estimates, after some notation:

$$\left.\begin{array}{ll} \beta(n,i) = \frac{1}{\sqrt{u_n}}\,\sigma_{(i-1)\Delta_n}\, W(n,i), & \overline{X}(n,i) = \frac{1}{\sqrt{u_n}}\, X(n,i), \\ \overline{X}'(n,i) = \frac{1}{\sqrt{u_n}}\, X'(n,i), & \overline{X}''(n,i) = \frac{1}{\sqrt{u_n}}\, X''(n,i) \end{array}\right\} \tag{8.4.7}$$

(recall the notation $Y(n,i)$ of (8.1.7), associated with any process Y, and also $x^{\#}$ as given by (8.1.12)). We also set

$$\begin{aligned} \gamma_i^n &= \mathbb{E}\left(\frac{1}{u_n}\int_{(i-1)\Delta_n}^{(i-1)\Delta_n+u_n} \|\sigma_s - \sigma_{(i-1)\Delta_n}\|^2\, ds\right) \\ \gamma'_n &= \int_{\{z:\,\Gamma(z)\le u_n^{1/4}\}} \Gamma(z)^2\lambda(dz). \end{aligned} \tag{8.4.8}$$

Then (SH) and (2.1.44) yield for $q > 0$:

$$\mathbb{E}\big(\beta(n,i)^{\#q}\big) + \mathbb{E}\big(\overline{X}'(n,i)^{\#q}\big) \le K_q, \qquad \mathbb{E}\big(\overline{X}''(n,i)^{\#2}\big) \le K. \tag{8.4.9}$$

Since $\overline{X}'(n,i) - \beta(n,i) = \frac{1}{\sqrt{u_n}} \int_{(i-1)\Delta_n}^{(i-1)\Delta_n+u_n} (b'_s\, ds + (\sigma_s - \sigma_{(i-1)\Delta_n})\, dW_s)$, we deduce from (2.1.43) that for $q \ge 2$,

$$\mathbb{E}\big(\big(\overline{X}'(n,i) - \beta(n,i)\big)^{\#q}\big) \le K_q\big(u_n^{q/2} + \gamma_i^n\big). \tag{8.4.10}$$

Finally, (2.1.39) applied with $r = 2$ and $q = 1/2$ gives

$$\mathbb{E}\big(\overline{X}''(n,i)^{\#2} \wedge 1\big) \le K\big(u_n^{1/4} + \gamma'_n\big). \tag{8.4.11}$$

Step 3) We associate with the test function Φ on $\mathbb{D}_1^d$ the following variables:

$$\left.\begin{array}{ll} \zeta_i^n = \Delta_n \Phi(\beta(n,i)^{(n)}), & U_t^n = \sum_{i=1}^{[t/\Delta_n]-k_n+1} \zeta_i^n \\ \zeta_i'^n = \mathbb{E}(\zeta_i^n \mid \mathcal{F}_{(i-1)\Delta_n}), & U_t'^n = \sum_{i=1}^{[t/\Delta_n]-k_n+1} \zeta_i'^n \\ \zeta_i''^n = \Delta_n \overline{\rho}_{c_{(i-1)\Delta_n}}(\Phi), & U_t''^n = \sum_{i=1}^{[t/\Delta_n]-k_n+1} \zeta_i''^n. \end{array}\right\} \tag{8.4.12}$$

In this step we prove that

$$U_t^n \stackrel{\text{u.c.p.}}{\Longrightarrow} \int_0^t \overline{\rho}_{c_s}(\Phi)\, ds. \tag{8.4.13}$$

We start by showing that

$$U'^n - U''^n \stackrel{\text{u.c.p.}}{\Longrightarrow} 0. \tag{8.4.14}$$

To see this, observe that $\zeta_i'^n = \Delta_n \overline{\rho}_{c_{(i-1)\Delta_n}}(\Phi^{(n)})$, where $\Phi^{(n)}(x) = \Phi(x^{(n)})$ for all $x \in \mathbb{D}_1^d$ and $x^{(n)}$ denotes the discretized version of x, as defined by (8.1.8). Then (8.4.14) will follow from the property

$$\sup_{a:\|a\| \leq K} \left| \overline{\rho}_a(\Phi^{(n)}) - \overline{\rho}_a(\Phi) \right| \to 0 \tag{8.4.15}$$

for any $K > 0$, because our assumptions yield $\|c_t\| \leq K$. If (8.4.15) fails, we have a sequence (α_n) of $d \times d'$ matrices converging to a limit α, such that $\limsup_n |\overline{\rho}_{a_n \alpha_n^*}(\Phi^{(n)}) - \overline{\rho}_{\alpha_n \alpha_n^*}(\Phi)| > 0$. Then if W' denotes a d'-dimensional standard Brownian motion with time interval $[0, 1]$, we have $\overline{\rho}_{\alpha_n \alpha_n^*}(\Phi^{(n)}) = \mathbb{E}(\Phi(\alpha_n W'^{(n)}))$ and $\overline{\rho}_{\alpha_n \alpha_n^*}(\Phi) = \mathbb{E}(\Phi(\alpha_n W'))$. On the one hand, $\alpha_n W'^{(n)} \to \alpha W'$ uniformly over $[0, 1]$, hence $\Phi(\alpha_n W'^{(n)})) \to \Phi(\alpha_n W')$ because Φ is continuous for the sup-norm. On the other hand the growth assumption on Φ implies $|\Phi(\alpha_n W'^{(n)}))| \leq K(1 + W'^{\#q})$ for some $q \geq 0$, and this variable is integrable. Hence $\overline{\rho}_{\alpha_n \alpha_n^*}(\Phi^{(n)}) \to \overline{\rho}_{\alpha_n \alpha_n^*}(\Phi)$ by the dominated convergence theorem, and we get a contradiction: so (8.4.15) holds, hence (8.4.14) as well.

Next, since Φ is the difference of two nonnegative functions having the same continuity and growth properties, it suffices to consider the case when $\Phi \geq 0$. In this case, by the criterion (2.2.16), proving (8.4.13) amounts to proving $U_y^n \stackrel{\mathbb{P}}{\longrightarrow} \int_0^t \overline{\rho}_{c_s}(\Phi)\, ds$ for any given t. Therefore, in view of (8.4.12) and of the (pathwise) convergence $U_t''^n \to \int_0^t \overline{\rho}_{c_s}(\Phi)\, ds$, which follows from Riemann integration because $t \mapsto \overline{\rho}_{c_t}(\Phi)$ is càdlàg, it is enough to prove that for all t,

$$\mathbb{E}\left(\left(\sum_{i=1}^{[t/\Delta_n]} (\zeta_i^n - \zeta_i'^n)\right)^2\right) \to 0. \tag{8.4.16}$$

The variables $\zeta_i^n - \zeta_i'^n$ are centered, and because of the growth condition on Φ we have $\mathbb{E}((\zeta_i^n - \zeta_i'^n)^2) \leq K \Delta_n^2$. Furthermore, $\mathbb{E}((\zeta_i^n - \zeta_i'^n)(\zeta_j^n - \zeta_j'^n)) = 0$ when $|i - j| > k_n$. Hence the left side of (8.4.16) is smaller than $K t u_n$, and this finishes the proof of (8.4.13).

Step 4) At this stage, it remains to prove that

$$H_t^n := \mathbb{E}\left(\sum_{i=1}^{[t/\Delta_n]-k_n+1} |\chi_i^n|\right) \to 0, \quad \text{where } \chi_i^n = \Delta_n \Phi\big(X(n,i)/\sqrt{u_n}\big) - \zeta_i^n. \tag{8.4.17}$$

Suppose first that X is continuous and that the function $\Phi^\#$ has polynomial growth. Using (8.1.13), we obtain the following property: there are two positive constants K and p, and for any $A>1$ a positive function θ_A satisfying $\theta_A(\varepsilon)\to 0$ as $\varepsilon\to 0$, such that for all $x,y\in\mathbb{D}^d$ all $\varepsilon\in(0,1]$ and all $A>1$, we have:

$$\bigl|\Phi(x+y)-\Phi(x)\bigr| \le \theta_A(\varepsilon)+\frac{A^p\,y^{\#2}}{\varepsilon^2}+K\frac{x^{\#2p}+y^{\#2p}}{A^p}. \tag{8.4.18}$$

This applied to $x=\beta(n,i)$ and $y=\overline{X}(n,i)-\beta(n,i)$ allows us to deduce from (8.4.9) and (8.4.10) that

$$H_t^n\le t\left(\theta_A(\varepsilon)+\frac{K}{A^p}\right)+\frac{KA^p}{\varepsilon^2}\left(t\sqrt{u_n}+\mathbb{E}\left(\int_0^t\sup_{v\in[(s-u_n)^+,s]}\|\sigma_v-\sigma_s\|^2\,ds\right)\right),$$

where K depends on p. The càdlàg and boundedness properties of σ yield that the last term above goes to 0 as $n\to\infty$. Therefore $\limsup_n H_t^n\le\frac{t}{k}(\theta_A(\varepsilon)+\frac{K}{A^p})$, for all $A>1$ and $\varepsilon\in(0,1]$. By taking A large first, and then ε small, we deduce (8.4.17).

Second, suppose that X has jumps, so $\Phi^\#(v)=\mathrm{o}(v^2)$ as $v\to\infty$, and there is a positive function Ψ on $\mathbb{R}_+$ with $\Psi(v)\to 0$ as $v\to\infty$, such that $\Phi^\#(v)\le\Psi(v)v^2$. Then there are functions θ_A as above, such that for all $x,y,z\in\mathbb{D}_1^d$ and $\varepsilon\in(0,1]$ and $A>3$ we have:

$$\begin{aligned}&\bigl|\Phi(x+y+z)-\Phi(x)\bigr|\\&\quad\le\theta_A(\varepsilon)+K\left(\Psi(A)\bigl(x^{\#2}+y^{\#2}+z^{\#2}\bigr)+\frac{A^2y^{\#2}}{\varepsilon^2}+\frac{A^2(z^{\#2}\wedge 1)}{\varepsilon^2}\right).\end{aligned}\tag{8.4.19}$$

This applied with $x=\beta(n,i)$ and $y=\overline{X}'(n,i)-\beta(n,i)$ and $z=\overline{X}''(n,i)$, plus (8.4.9)–(8.4.11), yields

$$\begin{aligned}H_t^n&\le t\bigl(\theta_A(\varepsilon)+K\Psi(A)\bigr)\\&\quad+\frac{KA^2}{\varepsilon^2}\left(tu_n^{1/4}+t\gamma_n'+\mathbb{E}\left(\int_0^t\sup_{v\in[(s-u_n)^+,s]}\|\sigma_s-\sigma_v\|^2\,ds\right)\right),\end{aligned}$$

and we conclude (8.4.17) as above. □

Proof of Theorem 8.4.1 The convergence (8.4.3) looks like a particular case of (8.4.5): take $k_n=k$ for all n, and define the function Φ on $\mathbb{D}^d$ as

$$\Phi(x) = F\left(x\left(\frac{1}{k}\right)-x(0),\,x\left(\frac{2}{k}\right)-x\left(\frac{1}{k}\right),\dots,x(1)-x\left(\frac{k-1}{k}\right)\right).$$

Then $V'^n(F,X)=V'^n(\Phi,k_n,X)$ and also $\overline{\rho}_a(\Phi)=\rho_a^{k\otimes}(F)$.

In fact, we cannot deduce (8.4.3) from (8.4.5) straight away for two reasons: here k_n does not go to infinity (a fact explicitly used for showing (8.4.15)), and when

X jumps the assumption (8.4.2) does not imply the required growth condition on Φ. However here, and with the notation (8.4.12) and Φ as above, we simply have $\zeta_i''^n = \zeta_i'^n$ identically and thus $U'^n = U''^n$. Therefore the previous proof applies without changes in the case X is continuous, and we have (8.4.3).

When X jumps, we have to take care of the weakened growth condition (8.4.2), and this leads us to reproduce the previous proof with some changes which we now explain. We use the notation (8.4.12) and (8.4.17), except that

$$\begin{aligned}
&\beta_{i,j}^n = \sigma_{(i-1)\Delta_n}\Delta_{i+j-1}^n W/\sqrt{\Delta_n}\\
&\zeta_i^n = \Delta_n F\big(\beta_{i,1}^n,\dots,\beta_{i,k}^n\big), \qquad \zeta_i''^n = \Delta_n \rho_{c_{(i-1)\Delta_n}}^{k\otimes}(F)\\
&\chi_i^n \,=\, \Delta_n\big(F\big(\Delta_i^n X/\sqrt{\Delta_n},\dots,\Delta_{i+k-1}^n X/\sqrt{\Delta_n}\big) - \zeta_i^n
\end{aligned}$$

(so here $\zeta_i'^n = \zeta_i''^n$). Then we replace (8.4.9)–(8.4.11) by estimates on conditional expectations, for the second moments only, and we get in exactly the same way that

$$\begin{aligned}
&\mathbb{E}\bigg(\|\beta_{i,j}^n\|^2 + \frac{\|\Delta_{i+j-1}^n X\|^2}{\Delta_n} \mid \mathcal{F}_{(i+j-2)\Delta_n}\bigg) \,\le\, K,\\
&\mathbb{E}\bigg(\bigg\|\frac{\Delta_{i+j-1}^n X}{\sqrt{\Delta_n}} - \beta_{i,j}^n\bigg\|^2 \wedge 1 \mid \mathcal{F}_{(i+j-2)\Delta_n}\bigg) \le K\Delta_n^{1/4} + K\gamma_n' \qquad (8.4.20)\\
&\quad + \frac{K}{\Delta_n}\mathbb{E}\bigg(\int_{(i+j-2)\Delta_n}^{(i+j-1)\Delta_n} \|\sigma_s - \sigma_{(i-1)\Delta_n}\|^2\, ds \mid \mathcal{F}_{(i+j-2)\Delta_n}\bigg).
\end{aligned}$$

The estimate (8.4.19) is no longer valid, but (8.4.2) gives the following: for all $\varepsilon \in (0,1]$ and $A > 3$ and $x_i, y_i \in \mathbb{R}^d$ we have

$$\begin{aligned}
&\big|F(x_1+y_1,\dots,x_k+y_k) - F(x_1,\dots,x_k)\big|\\
&\quad \le \theta_A(\varepsilon) + K\Psi(A)\prod_{j=1}^k\big(1+\|x_j\|^2+\|y_j\|^2\big) + KA^{2k}\sum_{j=1}^k\bigg(\frac{\|y_j\|^2\wedge 1}{\varepsilon^2}\bigg). \qquad (8.4.21)
\end{aligned}$$

Applying this with $x_j = \beta_{i,j}^n$ and $y_j = \Delta_{i+j-1}^n X/\sqrt{\Delta_n} - \beta_{i,j}^n$ and (8.4.20) repeatedly, and by successive conditioning, we see that

$$\begin{aligned}
\mathbb{E}\bigg(\sum_{i=1}^{[t/\Delta_n]-k_n+1} |\chi_i^n|\bigg) &\le Kt\bigg(\theta_A(\varepsilon) + \Psi(A) + \frac{A^{2k}}{\varepsilon^2}\big(\Delta_n^{1/4} + \gamma_n'\big)\bigg)\\
&\quad + \frac{KA^{2k}}{\varepsilon^2}\mathbb{E}\bigg(\int_0^t \sup_{v\in[(s-u_n)^+,s]} \|\sigma_s - \sigma_v\|^2\, ds\bigg), \qquad (8.4.22)
\end{aligned}$$

which goes to 0 by the same argument as in the previous theorem.

Finally, the convergence (8.4.4) is proved in exactly the same way. The only difference is that we now have

$$U_t'^n = \sum_{i=1}^{[t/\Delta_n]-k_n+1} \rho_{c_{(i-1)k\Delta_n}}^{k\otimes}(F),$$

which is a Riemann sum for $\frac{1}{k}\int_0^t \rho_{c_s}^{k\otimes}(F)\,ds$. □

Proof of Corollary 8.4.3 Since H is continuous and positively homogeneous of degree $p \in (0,2)$, (8.4.3) yields, for $j=1,2$, and where $H_1 = H$ and $H_2 = |H|$;

$$\Delta_n^{1-p/2} V^n(H_j, X)_t = V'^n(H_j, X)_t \stackrel{\text{u.c.p.}}{\Longrightarrow} \int_0^t \rho_{c_s}^{k\otimes}(H_j)\,ds. \tag{8.4.23}$$

Using the same functions ψ_ε as in the proof of Theorem 8.2.1 (or (3.3.16)) for $\varepsilon \in (0,1]$, we see that the function $G_\varepsilon(z) = (|F(z)| + H_2(z))\psi_\varepsilon(\|z\|)$ vanishes on a neighborhood of 0, so Theorem 8.2.1 yields that $V^n(G_\varepsilon, X)$ converges in probability in the Skorokhod sense. Since $p<2$ it follows that

$$\Delta_n^{1-p/2} V^n(G_\varepsilon, X) \stackrel{\text{u.c.p.}}{\Longrightarrow} 0. \tag{8.4.24}$$

Moreover the assumption $F \sim H$ near 0 implies the existence of a function θ on $\mathbb{R}_+$ such that $\theta(\varepsilon) \to 0$ as $\varepsilon \to 0$, and that $|F-H| \le \theta(\varepsilon)H_2 + G_\varepsilon$. Therefore

$$\begin{aligned} &\left|\Delta_n^{1-p/2} V^n(F,X) - \Delta_n^{1-p/2} V^n(H_1,X)\right| \\ &\quad \le \theta(\varepsilon)\Delta_n^{1-p/2} V^n(H_2,X) + \Delta_n^{1-p/2} V^n(G_\varepsilon,X), \end{aligned}$$

and the result follows from (8.4.23) and (8.4.24). □

8.5 Applications to Volatility

We continue the first example started in Chap. 3 about the estimation of the integrated (squared) volatility, or other powers of it.

The setting is as in the previous section: an Itô semimartingale X given by (8.4.1), and a regular discretization scheme, and we start with the $d=1$ dimensional case. We look for estimates for

$$A(p)_t = \int_0^t |\sigma_s|^p\,ds.$$

When X is continuous, or when $p<2$, we have already seen that $\Delta_n^{1-p/2} D(X,p,\Delta_n)_t$ converges to $m_p A(p)_t$ in probability, see (3.5.4), and we even have an associated CLT when X is continuous, see Theorem 5.6.1. When X has jumps and $p \ge 2$ these estimators badly fail, since they actually converge to another limit, but an alternative method is provided by Theorem 8.4.1 via the multipower variations which have been introduced by Barndorff-Nielsen and Shephard [8] especially for this reason. We briefly explain this here.

We take the special product form (8.2.6) for F. Then Theorem 8.4.1 yields that

$$D(X, p_1, \dots, p_k, \Delta_n)_t = \sum_{i=1}^{[t/\Delta_n]-k+1} |\Delta_i^n X|^{p_1} |\Delta_{i+1}^n X|^{p_2} \cdots |\Delta_{i+k-1}^n X|^{p_k}$$

(the realized multipower variation with indices $p_1, \dots, p_k$) satisfies under (H):

$$\Delta_n^{1-(p_1+\cdots+p_k)/2} D(X, p_1, \dots, p_k, \Delta_n)_t \overset{\text{u.c.p.}}{\Longrightarrow} \prod_{j=1}^k m_{p_j} \, A(p_1 + \cdots + p_k), \quad (8.5.1)$$

as soon as $p_j < 2$ *for all* j. This provides (many) estimators for $A(p)_t$, for any value of p. For example the *equal-multipowers variation*, with all p_j the same, gives us

$$k > \frac{p}{2} \Rightarrow \Delta_n^{1-p/2} \sum_{i=1}^{[t/\Delta_n]-k+1} \prod_{j=1}^k |\Delta_{i+j-1}^n X|^{p/k} \overset{\text{u.c.p.}}{\Longrightarrow} m_{p/k}^k \, A(p)_t.$$

Next, we turn to the d-dimensional case. The counterpart of $A(p)$ is the following family of processes, indexed by all multi-indices of the form $I = (l; (r_j, m_j, p_j)_{1\le j\le l})$, where $l \in \mathbb{N}^*$ and r_j and m_j range through $\{1, \dots, d\}$, and the powers p_j range through $(0, \infty)$ if $r_j = m_j$, and through N^* otherwise:

$$A(I)_t = \int_0^t \prod_{j=1}^l (c_s^{r_j m_j})^{p_j} \, ds.$$

When $l \ge 2$ and $r_j = m_j$ for all j, we easily extend (8.5.1) to obtain the following (below, all k_j are integers, and we set $K_0 = 0$ and $K_j = k_1 + \cdots + k_j$ and $p = p_1 + \cdots + p_l$):

$$k_j > p_j \ \forall j \Rightarrow \begin{cases} \Delta_n^{1-p} \sum_{i=1}^{[t/\Delta_n]-K_l+1} \prod_{j=1}^l \prod_{u=1}^{k_j} |\Delta_{i+K_{j-1}+u-1}^n X^{r_j}|^{2p_j/k_j} \\ \overset{\text{u.c.p.}}{\Longrightarrow} \prod_{j=1}^l m_{2p_j/k_j}^{k_j} \int_0^t \prod_{j=1}^l (c_s^{r_j r_j})^{p_j} \, ds. \end{cases} \quad (8.5.2)$$

It is more difficult to approximate $A(I)_t$ when the indices r_j and m_j are different and the p_j's are not integers. However when all p_j's are integers (so in fact we may assume that $p_j = 1$ for all j, up to repeating the same indices) we may use the equalities $uv = \frac{1}{2}((u+v)^2 - u^2 - v^2)$ and $uv = \frac{1}{4}((u+v)^2 - (u-v)^2)$ to obtain that, for example,

$$\begin{aligned} \Delta_n^{1-l} \sum_{i=1}^{[t/\Delta_n]-2l+1} \prod_{j=1}^l \Big(& |\Delta_{i+2j-2}^n X^{r_j} + \Delta_{i+2j-2}^n X^{m_j}| \, |\Delta_{i+2j-1}^n X^{r_j} + \Delta_{i+2j-1}^n X^{m_j}| \\ & - |\Delta_{i+2j-2}^n X^{r_j}| \, |\Delta_{i+2j-1}^n X^{r_j}| - |\Delta_{i+2l-2}^n X^{m_j}| \, |\Delta_{i+2j-1}^n X^{m_j}| \Big) \\ \overset{\text{u.c.p.}}{\Longrightarrow} 2^l m_1^{2l} \int_0^t \prod_{j=1}^l c_s^{r_j m_j} \, ds & \end{aligned} \quad (8.5.3)$$

and

$$\Delta_n^{1-l} \sum_{i=1}^{[t/\Delta_n]-2l+1} \prod_{j=1}^{l} \Big(\big|\Delta_{i+2j-2}^n X^{r_j} + \Delta_{i+2j-2}^n X^{m_j}\big| \big|\Delta_{i+2j-1}^n X^{r_j} + \Delta_{i+2j-1}^n X^{m_j}\big|$$
$$- \big|\Delta_{i+2j-2}^n X^{r_j} - \Delta_{i+2l-2}^n X^{m_j}\big| \big|\Delta_{i+2j-1}^n X^{r_j} - \Delta_{i+2j-1}^n X^{m_j}\big| \Big)$$
$$\stackrel{\text{u.c.p.}}{\Longrightarrow} 4^l m_1^{2l} \int_0^t \prod_{j=1}^{l} c_s^{r_j m_j} \, ds. \tag{8.5.4}$$

Chapter 9
Third Extension: Truncated Functionals

As seen before, the results of Chap. 3 do not allow one to approximate, or estimate, such quantities as the second characteristic $C = \langle X^c, X^c \rangle$ of a discontinuous semimartingale X. The realized multipower variations allow one to do so when X is an Itô semimartingale. Here we propose another method, in many respect more natural and easier to understand, and which consists in "truncating" the increments at some level. This method allows us to separate the jumps and the continuous martingale part of X, and it requires less assumptions on the test functions than multipower variations.

The choice of the truncation level should be done in connection with the mesh of the discretization scheme, and should also be related with the process X itself. These two requirements can be fulfilled in a reasonably simple way only when we have a *regular* discretization scheme and X is an Itô semimartingale. So in the whole chapter, X is an Itô semimartingale with the Grigelionis decomposition

$$X_t = X_0 + \int_0^t b_s\, ds + \int_0^t \sigma_s\, dW_s + (\delta 1_{\{\|\delta\| \le 1\}}) \star (\mathfrak{p} - \mathfrak{q})_t + (\delta 1_{\{\|\delta\| > 1\}}) \star \mathfrak{p}_t, \quad (9.0.1)$$

where W is a d'-dimensional Wiener process and $\mathfrak{p}$ is a Poisson measure with compensator $\mathfrak{q}(dt, dz) = dt \otimes \lambda(dz)$. Recall that $c_t = \sigma_t \sigma_t^*$ is a version of the process c_t such that $C_t = \int_0^t c_s\, ds$. We also set

$$X'_t = X_0 + \int_0^t b_s\, ds + \int_0^t \sigma_s\, dW_s, \qquad X''_t = X_t - X'_t. \quad (9.0.2)$$

The basic observation is that, for a regular scheme with time step Δ_n, a typical increment $\Delta_i^n X'$ is of order of magnitude $\sqrt{\Delta_n}$, whereas an increment $\Delta_i^n X''$ is either "big" because there is a big jump in the interval $((i-1)\Delta_n, i\Delta_n]$, or it is negligible with respect to $\sqrt{\Delta_n}$, as illustrated by Lemmas 2.1.6 and 2.1.8 for example. Therefore a solution to our problem is to "truncate" the increments $\Delta_i^n X$ from below or from above, depending on whether we want to approximate the jumps or the continuous part. The truncation level v_n goes to 0, but not as fast as $\sqrt{\Delta_n}$. Do-

J. Jacod, P. Protter, *Discretization of Processes*,
Stochastic Modelling and Applied Probability 67,
DOI 10.1007/978-3-642-24127-7_9,

ing so, we may hope that $\Delta_i^n X\, 1_{\{\|\Delta_i^n X\|\le v_n\}}$ is approximately equal to $\Delta_i^n X'$, and $\Delta_i^n X\, 1_{\{\|\Delta_i^n X\|>v_n\}}$ is approximately equal to $\Delta_i^n X''$.

This idea originates in Mancini [73], who takes $v_n = \sqrt{\Delta_n \log(1/\Delta_n)}$. This particular choice of v_n is good in some cases, and "too close" to $\sqrt{\Delta_n}$ for other results. So it is more convenient to take

$$v_n = \alpha\, \Delta_n^{\varpi} \quad \text{for some } \alpha > 0,\ \varpi \in \left(0, \frac{1}{2}\right) \tag{9.0.3}$$

(when $\varpi = .48$ or $.49$ the difference between the convergence rates toward 0 of this v_n and of $\sqrt{\Delta_n \log(1/\Delta_n)}$ is tiny; moreover, in practice, what is important are the relative sizes of v_n and of a "typical" increment $\Delta_i^n X$, so that the choice of the constant α, in connection with the "average" value of σ_t, is probably more important than the choice of the exponent ϖ).

Then, instead of the functionals $V^n(f, X)$ for example, we have the upwards and downwards truncated functionals, defined as

$$\begin{aligned} V^n(f, v_n+, X)_t &= \sum_{i=1}^{[t/\Delta_n]} f\big(\Delta_i^n X\big)\, 1_{\{\|\Delta_i^n X\|>v_n\}} \\ V^n(f, v_n-, X)_t &= \sum_{i=1}^{[t/\Delta_n]} f\big(\Delta_i^n X\big)\, 1_{\{\|\Delta_i^n X\|\le v_n\}}. \end{aligned} \tag{9.0.4}$$

The last subsections of this chapter are devoted to another important, and closely related, question: we want to approximate the matrix-valued variable c_t for a given time t or maybe c_T for some finite stopping time T. We call these approximations "local approximations".

9.1 Approximation for Jumps

In this first section we consider the simplest question, which is the approximation of $f * \mu$. Although a version for arbitrary discretization schemes is available, we consider only a regular scheme here, and in this case the upwards truncated functionals $V^n(f, v_n+, X)$ answer the question.

The next result looks like (A) of Theorem 3.3.1, but the assumptions on f, in connection with those on X itself, are much weaker when the number p below is smaller than 2. On the other hand, we need X to be an Itô semimartingale and, although it is not necessary when $r = 2$ below, we make the simplifying assumption (H-r), or Assumption 6.1.1, for some $r \in [0, 2]$. This assumption, which is the same as (H) when $r = 2$, is recalled below:

Assumption (H-r) X has the form (9.0.1), with b_t locally bounded and σ_t càdlàg. Moreover $\|\delta(\omega, t, z)\| \wedge 1 \le \Gamma_n(z)$ for all (ω, t, z) with $t \le \tau_n(\omega)$, where (τ_n) is a localizing sequence of stopping times and each function Γ_n on E satisfies $\int \Gamma_n(z)^r \lambda(dz) < \infty$.

Under (H-1) we have the decomposition (3.2.3), that is

$$X = X_0 + B' + X^c + x \star \mu, \qquad B' = B - (x 1_{\{\|x\| \le 1\}}) * \nu. \tag{9.1.1}$$

Since (9.0.1) holds, we also have in this case:

$$B'_t = \int_0^t b'_s \, ds, \qquad b'_t = b_t - \int \delta(t,z)\, 1_{\{\|\delta(t,z)\| \le 1\}}\, \lambda(dz).$$

Theorem 9.1.1 *Let X be an Itô semimartingale satisfying* (H-r) *for some $r \in [0,2]$. If f is a q-dimensional continuous function on $\mathbb{R}^d$, such that $f(x) = \mathrm{O}(\|x\|^r)$ as $x \to 0$, and if v_n satisfy* (9.0.3), *we have the following convergence for the Skorokhod topology*:

$$V^n(f, v_n+, X) \stackrel{\mathbb{P}}{\Longrightarrow} V(f,X) := f \star \mu. \tag{9.1.2}$$

Proof As for Theorem 3.3.1 we prove slightly more, namely

$$\overline{W}^n(f)_t := V^n(f, v_n+, X)_t - f * \mu_{\Delta_n[t/\Delta_n]} \stackrel{\text{u.c.p.}}{\Longrightarrow} 0. \tag{9.1.3}$$

As usual, by localization we may assume the strengthened assumption (SH-r), that is we suppose, additionally to (H-r), that b_t and σ_t and X_t are bounded and $\|\delta(\omega,t,z)\| \le \Gamma(z)$, with Γ bounded and $\int \Gamma(z)^r \lambda(dz) < \infty$. Then b'_t is also bounded when $r \le 1$.

When the function f vanishes on a neighborhood of 0, say $f(x) = 0$ for $\|x\| \le \varepsilon$, we have $V^n(f, v_n+, X) = V^n(f,X)$ as soon as $v_n \le \varepsilon$, and (9.1.3) amounts to Lemma 3.3.7.

Coming back to a general function f, we consider the functions of (3.3.16), which are $\psi'_\varepsilon = 1 - \psi_\varepsilon$ and $\psi_\varepsilon(x) = \psi(\|x\|/\varepsilon)$, where ψ is C^∞ on $\mathbb{R}$ with $1_{[1,\infty)} \le \psi \le 1_{[1/2,\infty)}$. Letting $f_\varepsilon = f\psi_\varepsilon$ and $f'_\varepsilon = f\psi'_\varepsilon$, what precedes shows that $V^n(f_\varepsilon, v_n+, X) \stackrel{\mathbb{P}}{\Longrightarrow} f_\varepsilon * \mu$, whereas $f_\varepsilon * \mu \stackrel{\text{u.c.p.}}{\Longrightarrow} f * \mu$ as $\varepsilon \to 0$ by the dominated convergence theorem (recalling $\sum_{s \le t} \|\Delta X_s\|^r < \infty$ and $\|f_\varepsilon(x)\| \le K\|x\|^r$ when $\|x\| \le 1$). Hence it remains to prove that

$$t, \eta > 0 \quad \Rightarrow \quad \lim_{\varepsilon \to 0} \limsup_n \mathbb{P}\Big(\sup_{s \le t} \big\|\overline{W}^n(f'_\varepsilon)_s\big\| > \eta\Big) = 0.$$

Let $h'_\varepsilon(x) = \|x\|^r \psi'_\varepsilon(x)$, so $\|f'_\varepsilon\| \le K h'_\varepsilon$ and

$$\big\|\overline{W}^n(f'_\varepsilon)_t\big\| \le K V^n(h'_\varepsilon, v_n+, X)_t + K h'_\varepsilon * \mu_t.$$

Since $h'_\varepsilon * \mu_t \to 0$ as $\varepsilon \to 0$ by the same argument as above, we are left to prove

$$t, \eta > 0 \quad \Rightarrow \quad \lim_{\varepsilon \to 0} \limsup_n \mathbb{P}\big(V^n(h_\varepsilon, v_n+, X)_t > \eta\big) = 0. \tag{9.1.4}$$

We first consider the case $1 < r \leq 2$, and we use the decomposition (9.0.2). Then (2.1.33) and (2.1.34) yield $\mathbb{E}(\|\Delta_i^n X'\|^p) \leq K_p \Delta_n^{p/2}$ for all p, and Markov's inequality yields

$$\sum_{i=1}^{[t/\Delta_n]} \mathbb{P}\big(\|\Delta_i^n X'\| > v_n/2\big) \leq \frac{2^{3/(1-2\varpi)}}{v_n^{3/(1-2\varpi)}} \sum_{i=1}^{[t/\Delta_n]} \mathbb{E}\big(\|\Delta_i^n X'\|^{3/(1-2\varpi)}\big) \leq Kt\sqrt{\Delta_n}. \tag{9.1.5}$$

Then the set Ω_t^n on which $\|\Delta_i^n X'\| \leq v_n/2$ for all $i \leq [t/\Delta_n]$ satisfies $\mathbb{P}(\Omega_t^n) \to 1$. On Ω_t^n, and if $\|\Delta_i^n X\| > v_n$, we must have $\|\Delta_i^n X''\| > v_n/2$ and thus $\|\Delta_i^n X\| \leq 2\|\Delta_i^n X''\|$. It follows that

$$V^n\big(h'_\varepsilon, v_n+, X\big)_t \leq 2^r\, V^n\big(h'_\varepsilon, X''\big)_t \quad \text{on } \Omega_t^n. \tag{9.1.6}$$

Theorem 3.3.1-(A-b) yields $V^n(h'_\varepsilon, X'')_t \stackrel{\mathbb{P}}{\longrightarrow} h'_\varepsilon * \mu_t$, hence we conclude (9.1.4) as above.

Next, suppose that $0 \leq r \leq 1$. We use (9.1.1) and put $\overline{X}'_t = X_0 + B'_t + X^c_t$ and $\overline{X}'' = X - \overline{X}'$. The assumptions implies that $\overline{X}'$ satisfies (9.1.5), so as above we have (9.1.6) with $\overline{X}''$ instead of X'', on a set $\Omega_t'^n$ whose probability goes to 1 as $n \to \infty$. Since Theorem 3.3.1-(A-d) yields $V^n(h'_\varepsilon, \overline{X}'')_t \stackrel{\mathbb{P}}{\longrightarrow} h'_\varepsilon * \mu_t$, we conclude (9.1.4) again. □

We can weaken the continuity assumption on f exactly as in Theorem 3.3.5. More interesting is the following version of Corollary 3.3.4, whose proof is omitted because it is exactly the same:

Corollary 9.1.2 *Let X be an Itô semimartingale satisfying* (H), *and let f be a q-dimensional continuous function on $\mathbb{R}^d$ such that $f(x) = \mathrm{O}(\|x\|^r)$ as $x \to 0$, for some $r \in [0, 2]$. Then if T is a finite stopping time with $\mathbb{P}(\Delta X_T \neq 0) = 0$, the stopped processes $V^n(f, v_n+, X)_{t\wedge T}$ converge in probability, for the Skorokhod topology, to the stopped process $f * \mu_{t\wedge T}$, in restriction to the set*

$$\Omega_T = \Big\{\sum_{s\leq T} \|\Delta X_s\|^r < \infty\Big\}.$$

9.2 Approximation for the Continuous Part of X

The second main result of this chapter is a version of the Law of Large Numbers for functionals of normalized increments, similar to Theorem 3.4.1. Now we truncate from above, using cut-off levels v_n satisfying (9.0.3): that is, instead of $V'^n(f, X)$, we consider

$$V'^n(f, v_n-, X)_t = \Delta_n \sum_{i=1}^{[t/\Delta_n]} f\big(\Delta_i^n X/\sqrt{\Delta_n}\big)\, 1_{\{\|\Delta_i^n X\| \leq v_n\}}.$$

More generally, and it is useful for applications, we can consider functionals of successive increments, as in Chap. 8, but subject to truncation. If k is an integer and F is a function on $(\mathbb{R}^d)^k$

$$V'^n(F, v_n-, X)_t = \Delta_n \sum_{i=1}^{[t/\Delta_n]-k+1} F\left(\frac{\Delta_i^n X}{\sqrt{\Delta_n}}, \dots, \frac{\Delta_{i+k-1}^n X}{\sqrt{\Delta_n}}\right) \prod_{l=0}^{k-1} 1_{\{\|\Delta_{i+l}^n X\| \le v_n\}}. \tag{9.2.1}$$

We do not consider the analogue of $\mathcal{V}'^n(F, X)$ here, neither the analogue of $V'^n(\Phi, k_n, X)$ when at stage n we have a functional of k_n successive increments, with $k_n \to \infty$. However, when in (9.2.1) the function F has the form $F(x_1, \dots, x_k) = f(x_1 + \cdots + x_k)$, another truncated functional is probably more natural than $V'^n(F, v_n-, X)$, and given by

$$\begin{aligned} &V'^n(f, k, v_n-, X)_t \\ &\quad = \Delta_n \sum_{i=1}^{[t/\Delta_n]-k+1} f\left(\frac{\Delta_i^n X + \cdots + \Delta_{i+k-1}^n X}{\sqrt{\Delta_n}}\right) 1_{\{\|\Delta_i^n X + \cdots + \Delta_{i+k-1}^n X\| \le v_n\}}. \end{aligned} \tag{9.2.2}$$

When $k = 1$ this is again the same as $V'^n(f, v_n-, X)_t$; it can also be viewed as the sum $\sum_{i=0}^{k-1} V'^n(f, v_n-, X(n,i))_{t-i\Delta_n}$, when Δ_n is replaced by $k\Delta_n$, and $X(n,i)_t = X_{t-i\Delta_n} - X_{i\Delta_n}$.

As in Theorem 8.4.1, $\rho_a^{k\otimes}$ denotes the k-fold product of the law $\rho_a = \mathcal{N}(0, a)$ on $\mathbb{R}^d$.

Theorem 9.2.1 *Assume that X satisfies* (H-r) *for some $r \in [0, 2]$, and let v_n satisfy* (9.0.3). *Let F be a continuous function on $(\mathbb{R}^d)^k$ which satisfies for some $p \ge 0$:*

$$\left|F(x_1, \dots, x_k)\right| \le K \prod_{j=1}^{k} \left(1 + \|x_j\|^p\right). \tag{9.2.3}$$

Then when X is continuous, or when X jumps and either $p \le 2$ or

$$p > 2, \quad 0 < r < 2, \quad \varpi \ge \frac{p-2}{2(p-r)}, \tag{9.2.4}$$

we have, with $c_s = \sigma_s \sigma_s^*$:

$$V'^n(F, v_n-, X)_t \overset{\text{u.c.p.}}{\Longrightarrow} V'(F, X)_t := \int_0^t \rho_{c_s}^{k\otimes}(F)\, ds. \tag{9.2.5}$$

If f is a function on $\mathbb{R}^d$ satisfying $|f(x)| \le K(1 + \|x\|^p)$, and under the same conditions on X and p, we also have for all integers $k \ge 1$:

$$V'^n(f, k, v_n-, X)_t \overset{\text{u.c.p.}}{\Longrightarrow} V'(f, X)_t := \int_0^t \rho_{kc_s}(F)\, ds. \tag{9.2.6}$$

An analogue of Corollary 8.4.3 also holds, and we leave the details to the reader.

When X is continuous, this is simply the version of Theorem 8.4.1 for the truncated functionals, a result indeed of very little interest. When X is discontinuous and under (H-2) = (H) and when $k = 1$, this gives—for truncated functionals—a seemingly very slight improvement upon Theorem 3.4.1, or of Theorem 8.4.1 when $k \geq 2$: namely we replace the assumption $f(x) = \mathrm{o}(\|x\|^2)$ by the assumption $f(x) = \mathrm{O}(\|x\|^2)$, as $\|x\| \to \infty$, or the assumption (8.4.2) by (9.2.3) with $p = 2$. It is however quite significant for applications, because $f(x) = x^j x^l$ satisfies (9.2.3) with $p = 2$ and we deduce that

$$\sum_{i=1}^{[t/\Delta_n]} \Delta_i^n X^j \Delta_i^n X^l \, 1_{\{\|\Delta_i^n X\| \leq v_n\}} \stackrel{\text{u.c.p.}}{\Longrightarrow} C_t^{jl} = \int_0^t c_s^{jl} \, ds.$$

This is simpler than the estimators given in (8.5.3) and (8.5.4) for $l = 1$. This also gives, together with (9.1.2) for the same function f, and with j, l ranging from 1 to d, a way of estimating separately the continuous part and the purely discontinuous part of the quadratic variation-covariation of X.

If (H-r) holds for some $r < 2$ the improvement is more significant: indeed, whatever p in (9.2.3), we can choose ϖ (hence v_n) in such a way that the theorem holds.

Proof We mainly focus on (9.2.5). By our usual localization argument it is enough to prove the result under the strengthened Assumption (SH-r) recalled in the proof of Theorem 9.1.1, instead of (H-r). Without loss of generality we can assume that $p \geq 2$ because the condition (9.2.3) weakens as p increases. Then as in (3.4.15), and instead of the decomposition (9.0.2), we can write

$$X'_t = X_0 + \int_0^t b''_s \, ds + \int_0^t \sigma_s \, dW_s, \qquad X'' = X - X' = \delta * (\mathfrak{p} - \mathfrak{q}), \qquad (9.2.7)$$

where $b''_t = b_t + \int \delta(t,z) 1_{\{\|\delta(t,z)\| > 1\}} \, \lambda(dz)$ is bounded. Below, we also use the functions ψ, ψ_ε, ψ'_ε of the proof of Theorem 9.1.1.

Step 1) Set

$$F_m(x_1, \ldots, x_k) = F(x_1, \ldots, x_k) \prod_{j=1}^k \psi'_m(x_j). \qquad (9.2.8)$$

Each function F_m being continuous and bounded, Theorem 8.4.1 yields, as $n \to \infty$:

$$V'^n(F_m, X)_t \stackrel{\text{u.c.p.}}{\Longrightarrow} \int_0^t \rho_{c_s}^{k\otimes}(F_m) \, ds. \qquad (9.2.9)$$

Since F is of polynomial growth and c_t is bounded, we have $\int_0^t \rho_{c_s}^{k\otimes}(F_m) \, ds \stackrel{\text{u.c.p.}}{\Longrightarrow} \int_0^t \rho_{c_s}^{k\otimes}(F) \, ds$ as $m \to \infty$, by Lebesgue's theorem (we have $F_m \to F$ pointwise and $|F_m| \leq |F|$). Hence by Proposition 2.2.1, and in view of (9.2.9), we are left to prove that for all t,

$$\lim_{m\to\infty} \limsup_{n\to\infty} \mathbb{E}\Big(\sup_{s\le t}\big|V'^n(F,v_n-,X)_s - V'^n(F_m,X)_s\big|\Big) = 0. \tag{9.2.10}$$

Step 2) For all n bigger than some n_m we have $m \le u_n := v_n/\sqrt{\Delta_n}$ because of (9.0.3), and thus $\psi'_m(x) \le 1_{\{\|x\|\le u_n\}}$. Thus (9.2.3) implies

$$\begin{aligned}
&\Big|F(x_1,\dots,x_k)\prod_{j=1}^k 1_{\{\|x_j\|\le u_n\}} - F_m(x_1,\dots,x_k)\Big| \\
&\quad \le K\prod_{j=1}^k \big(1+\|x_j\|^p\,1_{\{\|x_j\|\le u_n\}}\big)\sum_{j=1}^k 1_{\{\|x_j\|>\frac{m}{2}\}}.
\end{aligned}$$

By singling out the two cases $\|y\|\le\|x\|/2$ and $\|y\|>\|x\|/2$, we also get

$$\begin{aligned}
1+\|x+y\|^p\,1_{\{\|x+y\|\le u_n\}} &\le K\big(1+\|x\|^p+\|y\|^p\wedge u_n^p\big) \\
\big(1+\|x+y\|^p\,1_{\{\|x+y\|\le u_n\}}\big)1_{\{\|x+y\|>\frac{m}{2}\}} &\le \frac{K}{m}\,\|x\|^{p+1} + K\,\|y\|^p\wedge u_n^p.
\end{aligned}$$

Therefore

$$n\ge n_m \;\Rightarrow\; \sup_{s\le t}\big|V'^n(F,v_n-,X)_s - V'^n(F_m,X)_s\big| \;\le\; K\sum_{j=1}^k U_t^{n,m,j},$$

where, with the notation (9.2.7),

$$\begin{aligned}
U_t^{n,m,j} &= \sum_{i=1}^{[t/\Delta_n]} \zeta(j,m)_i^n \\
\zeta(j,m)_i^n &= \Delta_n \prod_{l=1}^{j-1} Z_{i+l-1}^n \prod_{l=j+1}^{k} Z_{i+l-1}^n\, Z(m)_{i+j-1}^n \\
Z_i^n &= 1+\Big(\frac{\|\Delta_i^n X'\|}{\sqrt{\Delta_n}}\Big)^p + \Big(\frac{\|\Delta_i^n X''\|}{\sqrt{\Delta_n}}\bigwedge u_n\Big)^p \\
Z(m)_i^n &= \frac{1}{m}\Big(\frac{\|\Delta_i^n X'\|}{\sqrt{\Delta_n}}\Big)^{p+1} + \Big(\frac{\|\Delta_i^n X''\|}{\sqrt{\Delta_n}}\bigwedge u_n\Big)^p.
\end{aligned}$$

Hence it is enough to prove that for all j and t,

$$\lim_{m\to\infty}\limsup_{n\to\infty} \mathbb{E}\big(U_t^{n,m,j}\big) = 0. \tag{9.2.11}$$

Step 3) By (SH-r) and (2.1.33) and (2.1.34) we have, with the notation (9.0.2),

$$\mathbb{E}\big(\big\|\Delta_i^n X'/\sqrt{\Delta_n}\big\|^q \mid \mathcal{F}_{(i-1)\Delta_n}\big) \le K_q \tag{9.2.12}$$

for all $q > 0$. On the other hand, we apply (2.1.45) when $r > 1$ and (2.1.46) when $0 < r \le 1$ with $q = \varpi$ and $p = 2$ to get when $r > 0$:

$$\mathbb{E}\big(\|\Delta_i^n X''/\Delta_n^{\varpi}\|^2 \bigwedge 1 \mid \mathcal{F}_{(i-1)\Delta_n}\big) \le K\Delta_n^{1-r\varpi}\,\phi_n, \tag{9.2.13}$$

where $\phi_n \to 0$ as $n \to \infty$. Since $p \ge 2$ we also have

$$\begin{aligned}\big|\big(\|\Delta_i^n X\|/\sqrt{\Delta_n}\big) \wedge u_n\big|^p &= u_n^p \,\big|\big(\|\Delta_i^n X\|/u_n\sqrt{\Delta_n}\big) \wedge 1\big|^p \\ &\le K u_n^p \big|\big(\|\Delta_i^n X\|/\Delta_n^{\varpi}\big) \wedge 1\big|^2.\end{aligned}$$

Since $p \ge 2$, and with the convention that $\kappa = 0$ when $X = X'$ is continuous and $\kappa = 1$ otherwise, we deduce from (9.2.12) and (9.2.13) that

$$\left.\begin{aligned}&\mathbb{E}\big(Z_i^n \mid \mathcal{F}_{(i-1)\Delta_n}\big) \le K + \kappa\, K\, \Delta_n^w \phi_n \\ &\mathbb{E}\big(Z(m)_i^n \mid \mathcal{F}_{(i-1)\Delta_n}\big) \le \frac{K}{m} + \kappa\, K\, \Delta_n^w \phi_n\end{aligned}\right\} \quad \text{where } w = \varpi(p-r) + 1 - \frac{p}{2}. \tag{9.2.14}$$

Observe that, since $\varpi \in (0, 1/2)$ and $r \in (0, 2]$, we have $w \ge 0$ when $p = 2$ or when (9.2.4) holds. Hence, coming back to the definition of $\zeta(j,m)_i^n$, by successive conditioning we deduce from the previous estimates that, under this condition,

$$\mathbb{E}\big(\zeta(j,m)_i^n\big) \le K\Delta_n\left(\frac{1}{m} + \phi_n\right). \tag{9.2.15}$$

Then $\mathbb{E}(U_t^{n,m,j}) \le Kt(\phi_n + 1/m)$ and since $\phi_n \to 0$ we obtain (9.2.11).

Step 4) It remains to prove (9.2.6), which is of course a particular case of (9.2.5) when $k = 1$. When $k \ge 2$, the proof is exactly similar, and left to the reader, after pointing out that when $F(x_1, \dots, x_k) = f(x_1 + \dots + x_k)$ we have $\rho_{c_s}^{k\otimes}(F) = \rho_{kc_s}(f)$. (One could also argue that this is an application of (9.2.5) with $k = 1$, for each of the k processes $V'^n(f, v_n-, X(n,i))$ corresponding to $k\Delta_n$ instead of Δ_n, as introduced after (9.2.2); although $X(n,i)$ depends on n, it is basically the same as X itself.) □

9.3 Local Approximation for the Continuous Part of X: Part I

Instead of approximating the process C_t, it is may be useful to approximate c_t at some given time t. This is possible, using the methods of the previous section or of Chap. 8.

We could more generally approximate the variable $g(c_t)$, at least for functions g on $\mathcal{M}_{d\times d}^+$ which have the form $g(a) = \rho_a^{k\otimes}(F)$ for some function F on $(\mathbb{R}^d)^k$ to which Theorem 9.2.1 applies, for example. However, below we will only do the approximation for c_t itself. The reason is that, if the sequence $\widehat{c}_t^n$ approximates c_t, then $g(\widehat{c}_t^n)$ approximates $g(c_t)$ as soon as g is continuous; moreover, the performances of the approximation $g(\widehat{c}_t^n)$ are just as good as those obtained by a direct approximation of $g(c_t)$.

The idea is straightforward. Suppose for a moment that X is continuous. An approximation of C_t^{ij} is given by $\sum_{i=1}^{[t/\Delta_n]} \Delta_i^n X^i \Delta_i^n X^j$, hence an approximation of $C_{t+s}^{ij} - C_t^{ij}$ is $\sum_{i=[t/\Delta_n]+1}^{[(t+s)/\Delta_n]} \Delta_i^n X^i \Delta_i^n X^j$, and in turn $\frac{1}{s}(C_{t+s}^{ij} - C_t^{ij})$ converges to c_t^{ij} as $s \to 0$, because c is assumed to be right-continuous. If we mix these two approximations, we may hope that

$$\frac{1}{k_n \Delta_n} \sum_{i=[t/\Delta_n]+1}^{[t/\Delta_n]+k_n} \Delta_i^n X^i \Delta_i^n X^j$$

converges to c_t^{ij}, provided we take a sequence of integers k_n increasing to ∞ and such that $k_n \Delta_n \to 0$. It is indeed the case, and this section is devoted to proving this property.

When X jumps, the same heuristic argument leads first to eliminate the jumps, and this can be done in two ways: we truncate the increments, or we take multipowers. The first method requires choosing cut-off levels v_n as in (9.0.3), the second one needs an appropriate (but relatively arbitrary) choice of the powers, and we will use the same as in (8.5.4) with $l = 1$. This leads us to introduce the following three sequences of variables:

$$\begin{aligned}
\widehat{c}_i^n(k_n)^{jl} &= \frac{1}{k_n \Delta_n} \sum_{m=0}^{k_n-1} \Delta_{i+m}^n X^j \Delta_{i+m}^n X^l \\
\widehat{c}'^n(k_n)_i^{jl} &= \frac{\pi}{8 k_n \Delta_n} \sum_{m=0}^{k_n-1} \Big(\big|\Delta_{i+m}^n X^j + \Delta_{i+m}^n X^l\big| \big|\Delta_{i+m+1}^n X^j + \Delta_{i+m+1}^n X^l\big| \\
&\quad - \big|\Delta_{i+m}^n X^j - \Delta_{i+m}^n X^l\big| \big|\Delta_{i+m+1}^n X^j - \Delta_{i+m+1}^n X^l\big| \Big) \\
\widehat{c}_i^n(k_n, v_n)^{jl} &= \frac{1}{k_n \Delta_n} \sum_{m=0}^{k_n-1} \Delta_{i+m}^n X^j \Delta_{i+m}^n X^l 1_{\{\|\Delta_{i+m}^n X\| \le v_n\}}.
\end{aligned} \tag{9.3.1}$$

Of course we do this for all $j, l = 1, \dots, d$, thus obtaining $\mathcal{M}_{d\times d}^+$-valued variables $\widehat{c}_i^n(k_n)$, $\widehat{c}_i'^n(k_n)$ and $\widehat{c}_i^n(k_n, v_n)$. These variables are *a priori* well defined when $i \ge 1$. However, for convenience we make the following convention:

$$i \in \mathbb{Z},\ i \le 0 \quad \Rightarrow \quad \Delta_i^n Y = 0 \quad \text{for any process } Y. \tag{9.3.2}$$

Then the variables $\widehat{c}_i^n(k_n)$, $\widehat{c}_i'^n(k_n)$ and $\widehat{c}_i^n(k_n, v_n)$ become also defined by (9.3.1) when $i \le 0$, and of course they vanish identically when $i \le -k_n$.

These variables approximate $c_{i\Delta_n}$, but we usually want an approximation of c_t or c_{t-} for some time t, possibly random. For this, recalling $I(n,i) = ((i-1)\Delta_n, i\Delta_n]$, we can take

$$t \in I(n,i) \Rightarrow \begin{cases} \widehat{c}^n(k_n,t-) = \widehat{c}^n_{i-k_n}(k_n), & \widehat{c}^n(k_n,t) = \widehat{c}^n_{i+1}(k_n) \\ \widehat{c}'^n(k_n,t-) = \widehat{c}'^n_{i-k_n-k+1}(k_n), & \widehat{c}'^n(k_n,t) = \widehat{c}'^n_{i+1}(k_n) \\ \widehat{c}^n(k_n,v_n,t-) = \widehat{c}^n_{i-k_n}(k_n,v_n), & \widehat{c}^n(k_n,v_n,t) = \widehat{c}^n_{i+1}(k_n,v_n). \end{cases} \tag{9.3.3}$$

So with the convention (9.3.2), when for example $t > 0$, the variables $\widehat{c}^n(k_n,t-)$ are well defined for all n, and not only when n is big enough to have $t \geq (k_n+2)\Delta_n$. Note also that $\widehat{c}^n(k_n,0) = \widehat{c}^n_1(k_n)$.

Remark 9.3.1 In (9.3.3) the choice of the indices $i - k_n$ and $i+1$ is somewhat arbitrary, but designed in such a way that, for example, $\widehat{c}^n(k_n,t-)$ and $\widehat{c}^n(k_n,t)$ involve the increments of X over the k_n successive intervals which are closest to t on the left and right sides of t respectively, but do not contain t itself.

We could alternatively take the indices $i - l_n$ and $i + l_n - k_n$, for any sequence $l_n \in \mathbb{N}$ such that $l_n/k_n \to 1$: the results would be the same for the LLN below, at the price of a slightly more complicated proof when t is inside the intervals on which the increments are taken, but the associated CLT described in Chap. 13 would then fail unless $(l_n - k_n)/\sqrt{k_n} \to 0$.

When c_s is almost surely continuous at time t, we can take $\widehat{c}^n(k_n,t)$ to be $\widehat{c}^n_{i_n}(k_n)$ with any sequence i_n of integer such that $i_n\Delta_n \to t$, and using a window of size $k_n\Delta_n$ which is approximately symmetrical about t is clearly best.

The same comments apply to the other estimators as well.

For the second claim of the next theorem, we recall that if both X and c are Itô semimartingales, the pair (X,c) is also a $d+d^2$-dimensional Itô semimartingale. Moreover, it has a "joint" Grigelionis representation, and the Poisson random measure appearing in this representation is called a *driving Poisson measure for* (X,c) (recall also that there are many Grigelionis representations, hence many possible driving Poisson measures).

Theorem 9.3.2 *Assume that X satisfies* (H), *and let k_n satisfy $k_n \to \infty$ and $k_n\Delta_n \to 0$, and v_n satisfy* (9.0.4). *Let T be a stopping time.*

a) *In restriction to the set $\{T < \infty\}$ we have*

$$\widehat{c}^n(k_n,T) \xrightarrow{\mathbb{P}} c_T, \qquad \widehat{c}'^n(k_n,T) \xrightarrow{\mathbb{P}} c_T, \qquad \widehat{c}^n(k_n,v_n,T) \xrightarrow{\mathbb{P}} c_T. \tag{9.3.4}$$

b) *In restriction to the set $\{0 < T < \infty\}$ we have*

$$\widehat{c}^n(k_n,T-) \xrightarrow{\mathbb{P}} c_{T-}, \qquad \widehat{c}'^n(k_n,T-) \xrightarrow{\mathbb{P}} c_{T-}, \qquad \widehat{c}^n(k_n,v_n,T-) \xrightarrow{\mathbb{P}} c_{T-}, \tag{9.3.5}$$

provided either one of the following two hypotheses holds:

(b-1) *for some stopping time S we have $T > S$ identically and T is $\mathcal{F}_S$ measurable,*
(b-2) *the process c_t is an Itô semimartingale, and on the set $\{0 < T < \infty\}$ we have $\mathfrak{p}(\{T\} \times E) = 1$, that is T is a "jump time" of $\mathfrak{p}$, for some driving Poisson measure $\mathfrak{p}$ for the pair (X,c).*

The condition (b-1) is satisfied when T is non-random, but otherwise it is very restrictive. When c_t is an Itô semimartingale the condition (b-2) is satisfied in particular by any stopping time T satisfying $\|\Delta X_T\| + \|\Delta c_T\| > 0$ on $\{0 < T < \infty\}$: quite happily, this is a rather mild condition, because on the set $\{\Delta c_T = 0\}$ we have of course $c_{T-} = c_T$, and on this set we can use $\widehat{c}^n(k_n, T+)$ instead of $\widehat{c}^n(k_n, T-)$, for example. Moreover, there are many stopping times T satisfying (b-2), although $\|\Delta X_T\| + \|\Delta c_T\| > 0$ fails.

The convergence of the truncated and bipower approximate quadratic variations towards c_T in (9.3.4), for example, is intuitive, since these quantities are designed to eliminate the jumps of X. In contrast, the convergence $\widehat{c}^n(k_n, T) \xrightarrow{\mathbb{P}} c_T$ is rather counter-intuitive when X has jumps. This convergence is due to the fact that, since T is fixed (albeit random), the discontinuous part of X on the shrinking interval $(T, T + (k_n + 2)\Delta_n]$ is asymptotically negligible in front of its Brownian part, as seen from Corollary 2.1.9, which will play a crucial role in the proof.

So it may seem that considering $\widehat{c}^n(k_n, v_n, T)$ and $\widehat{c}'^n(k_n, T)$ is superfluous. However, from a practical point of view, and when $k_n\Delta_n$ is not very small, it may be wise to use the truncated or bipower versions in case of jumps. Furthermore, as we will see in the next section, the truncated versions are also useful from a mathematical viewpoint.

Proof *Step 1)* We will prove in a unified way (a), and (b) under the hypothesis (b-1), whereas (b) under (b-2) will be deduced at the very end.

We begin with some preliminaries. It is enough to prove the results separately for each component (j, l). Then by the polarization identity (2.1.7) this amounts to proving the result for the one-dimensional processes $X^j + X^l$ and $X^j - X^l$. In other words, it is no restriction to assume that X is one-dimensional. Up to modifying the Brownian motion W and the process σ (but not c), we can also assume that W is one-dimensional.

Next, as usual our localization procedure allows us to assume the strengthened assumption (SH) instead of (H).

Another simplification arises naturally in the context of (b-1): it is enough to prove (9.3.5) in restriction to the set $\{T - S \geq a\}$ for all $a > 0$. Thus, up to replacing T by $T \vee (S + a)$, we can suppose that $T - S \geq a$ identically for some $a > 0$, and of course we can suppose also that n is large enough to have $(k_n + 3)\Delta_n < a$. A last simplification is possible: upon replacing T by $T \wedge N$ (which satisfies (b-1) with $S \wedge (N - a)$ when T satisfies (b-1)), for N arbitrarily large, it is no restriction to assume that T is bounded.

Finally we will vary the process X in the course of the proof, so we mention it in our notation, writing for example $\widehat{c}_i^n(k_n, v_n, X)$ instead of $\widehat{c}_i^n(k_n, v_n)$.

We will unify the proof as follows. When we prove (a), if $T \in I(n, i)$ we set $i_n = i + 1$ and $T_n = (i_n - 1)\Delta_n$, so $c_{T_n} \to c_T$. When we prove (b) under (b-1) and if $T \in I(n, i)$ again, we set $i_n = i - k_n$ for the first and last parts of (9.3.5), and $i_n = i - k_n - 1$ for the second part; in these cases we again set $T_n = (i_n - 1)\Delta_n$, so $c_{T_n} \to c_{T-}$. Therefore in both cases T_n is a stopping time, and our claims amount to the following:

$$\widehat{c}^n_{i_n}(k_n,X)-c_{T_n} \stackrel{\mathbb{P}}{\longrightarrow} 0, \qquad \widehat{c}'^n_{i_n}(k_n,X)-c_{T_n} \stackrel{\mathbb{P}}{\longrightarrow} 0, \qquad \widehat{c}^n_{i_n}(k_n,v_n,X)-c_{T_n} \stackrel{\mathbb{P}}{\longrightarrow} 0.$$

Step 2) We introduce the processes

$$Y^n_t = \sigma_{T_n}(W_t - W_{T_n})\,1_{\{T_n\le t\}}, \qquad Y'^n_t = \int_{T_n\wedge t}^t (\sigma_s - \sigma_{T_n})\,dW_s$$

and in this step we prove that

$$\widehat{c}^n_{i_n}(k_n,Y^n) - c_{T_n} \stackrel{\mathbb{P}}{\longrightarrow} 0, \qquad \widehat{c}'^n_{i_n}(k_n,Y^n) - c_{T_n} \stackrel{\mathbb{P}}{\longrightarrow} 0. \tag{9.3.6}$$

Observe that (9.3.1) and (9.3.3) for Y^n give us (recalling that σ_t and W are one-dimensional, and our definition of T_n) that $\widehat{c}^n_{i_n}(k_n,Y^n) = c_{T_n}\widetilde{c}_n$ and $\widehat{c}'^n_{i_n}(k_n,Y^n) = c_{T_n}\widetilde{c}'_n$, where

$$\widetilde{c}_n = \frac{1}{k_n\Delta_n}\sum_{i=0}^{k_n-1}\left(\Delta^n_{i_n+i}W\right)^2, \qquad \widetilde{c}'_n = \frac{\pi}{2k_n\Delta_n}\sum_{i=0}^{k_n-1}\left|\Delta^n_{i_n+i}W\right|\left|\Delta^n_{i_n+i+1}W\right|.$$

Therefore, (9.3.6) will be a consequence of

$$\widetilde{c}_n \stackrel{\mathbb{P}}{\longrightarrow} 1, \qquad \widetilde{c}'_n \stackrel{\mathbb{P}}{\longrightarrow} 1. \tag{9.3.7}$$

Since T_n is a stopping time, the variables $\Delta^n_{i_n+i}W$ for $i\ge 0$ are i.i.d. with the law $\mathcal{N}(0,\sqrt{\Delta_n})$. Thus (9.3.7) follows from standard calculations (both $\widetilde{c}_n$ and $\widetilde{c}'_n$ have mean 1 and a variance going to 0), and (9.3.6) is proved.

At this stage, it thus remains to prove the following three properties:

$$\left.\begin{array}{lll} Z_n & = \widehat{c}^n_{i_n}(k_n,X) - \widehat{c}^n_{i_n}(k_n,Y^n) \stackrel{\mathbb{P}}{\longrightarrow} 0 \\ Z'_n & = \widehat{c}'^n_{i_n}(k_n,X) - \widehat{c}'^n_{i_n}(k_n,Y^n) \stackrel{\mathbb{P}}{\longrightarrow} 0 \\ Z''_n & = \widehat{c}^n_{i_n}(k_n,v_n,X) - \widehat{c}^n_{i_n}(k_n,Y^n) \stackrel{\mathbb{P}}{\longrightarrow} 0. \end{array}\right\} \tag{9.3.8}$$

Step 3) In this step we prove the last part of (9.3.8). Since $|(x+y)^2 - x^2| \le \varepsilon x^2 + \frac{1+\varepsilon}{\varepsilon}y^2$ for all $x,y\in\mathbb{R}$ and $\varepsilon>0$, we observe that for all $v\ge 1$ and $\varepsilon\in(0,1]$ we have

$$\left|(x+y+z+w)^2\,1_{\{|x+y+z+w|\le v\}} - x^2\right| \le K\frac{|x|^4}{v^2} + \varepsilon x^2 + \frac{K}{\varepsilon}\left((v^2\wedge y^2) + z^2 + w^2\right). \tag{9.3.9}$$

Recalling (9.2.7) and putting $B''_t = \int_0^t b''_s\,ds$, we use the above with $x = \Delta^n_{i_n+i}Y^n/\sqrt{\Delta_n}$ and $y = \Delta^n_{i_n+i}X''/\sqrt{\Delta_n}$ and $z = \Delta^n_{i_n+i}Y'^n/\sqrt{\Delta_n}$ and $w = \Delta^n_{i_n+i}B''/\sqrt{\Delta_n}$ (so

$x+y+z+w=\Delta^n_{i_n+i}X/\sqrt{\Delta_n})$, and $v=v_n/\sqrt{\Delta_n}=\alpha\Delta_n^{\varpi-1/2}$, to get

$$|Z''_n|\le\frac{1}{k_n}\sum_{i=0}^{k_n-1}\Bigg(K\Delta_n^{1-2\varpi}\bigg|\frac{\Delta^n_{i_n+i}Y^n}{\sqrt{\Delta_n}}\bigg|^4+\varepsilon\bigg|\frac{\Delta^n_{i_n+i}Y^n}{\sqrt{\Delta_n}}\bigg|^2$$
$$+\frac{K}{\varepsilon}\Delta_n^{2\varpi-1}\bigg|\frac{\Delta^n_{i_n+i}X''}{\Delta_n^{\varpi}}\bigwedge 1\bigg|^2+\frac{K}{\varepsilon}\bigg|\frac{\Delta^n_{i_n+i}Y'^n}{\sqrt{\Delta_n}}\bigg|^2+\frac{K}{\varepsilon}\bigg|\frac{\Delta^n_{i_n+i}B''}{\sqrt{\Delta_n}}\bigg|^2\Bigg).$$

We set $\gamma_n=\sup_{s\in[T_n,T_n+(k_n+2)\Delta_n)}|\sigma_s-\sigma_{T_n}|^2$, which is bounded and goes to 0 for each ω because of our definition of i_n. In view of (SH), (2.1.33), (2.1.34), and (2.1.45) applied with $p=2$ and $q=\varpi$ and $r=2$, plus the fact that $(i_n+i-1)\Delta_n$ is a stopping time for all $i\ge 0$, we deduce that for some sequence ϕ_n of numbers decreasing to 0 and all $q\ge 2$ we have for $0\le i\le k_n-1$:

$$\begin{aligned}
&\big|\Delta^n_{i_n+i}B''\big|\le K\Delta_n,\qquad \mathbb{E}\big(\big|\Delta^n_{i_n+i}Y^n\big|^q\mid\mathcal{F}_{(i_n+i-1)\Delta_n}\big)\le K_q\Delta_n^{q/2}\\
&\mathbb{E}\big(\big|\Delta^n_{i_n+i}Y'^n\big|^q\mid\mathcal{F}_{(i_n+i-1)\Delta_n}\big)\le K_q\Delta_n^{q/2}\,\mathbb{E}(\gamma_n\mid\mathcal{F}_{(i_n+i-1)\Delta_n})\le K_q\Delta_n^{q/2}\\
&\mathbb{E}\bigg(\frac{|\Delta^n_{i_n+i}X''|}{\Delta_n^{\varpi}}\bigwedge 1\mid\mathcal{F}_{(i_n+i-1)\Delta_n}\bigg)\le K\Delta_n^{1-2\varpi}\,\phi_n.
\end{aligned}\tag{9.3.10}$$

Therefore

$$\mathbb{E}\big(|Z_n|\big)\le K\varepsilon+\frac{K}{\varepsilon}\big(\Delta_n^{1-2\varpi}+\phi_n+\mathbb{E}(\gamma_n)\big).$$

Letting first $n\to\infty$ and then $\varepsilon\to 0$, we deduce the first part of (9.3.8).

Step 4) Next, we prove the second part of (9.3.8). Instead of (9.3.9) we use the following, which holds for all $A\ge 1$:

$$\begin{aligned}
&\big||x+y+z+w|\,|x'+y'+z'+w'|-|x|\,|x'|\big|\\
&\quad\le|x+y+z+w|\bigg(A\big(|y'|\wedge 1\big)+\frac{|y'|^2}{A^2}+|z'|+|w'|\bigg)\\
&\qquad+|x'+y'+z'+w'|\bigg(A\big(|y|\wedge 1\big)+\frac{|y|^2}{A^2}+|z|+|w|\bigg).
\end{aligned}$$

Using this formula with x,y,z,w as in Step 4 and $x'=\Delta^n_{i_n+i+1}Y^n/\sqrt{\Delta_n}$ and $y'=\Delta^n_{i_n+i+1}X''/\sqrt{\Delta_n}$, $z'=\Delta^n_{i_n+i+1}Y'/\sqrt{\Delta_n}$ and $w'=\Delta^n_{i_n+i+1}B''/\sqrt{\Delta_n}$, we get

$$|Z'_n|\le\frac{1}{k_n}\sum_{i=0}^{k_n-1}\Bigg[\frac{|\Delta^n_{i_n+i}X|}{\sqrt{\Delta_n}}\Bigg(A\bigg(\frac{|\Delta^n_{i_n+i+1}X''|}{\sqrt{\Delta_n}}\bigwedge 1\bigg)$$
$$+\frac{|\Delta^n_{i_n+i+1}X''|^2}{A^2\Delta_n}+\frac{|\Delta^n_{i_n+i+1}Y'^n|}{\sqrt{\Delta_n}}+\frac{|\Delta^n_{i_n+i+1}B''|}{\sqrt{\Delta_n}}\Bigg)$$

$$+\left(A\left(\frac{|\Delta_{i_n+i}^n X''|}{\sqrt{\Delta_n}}\bigwedge 1\right)+\frac{|\Delta_{i_n+i}^n X''|^2}{A^2\sqrt{\Delta_n}}+\frac{|\Delta_{i_n+i}^n Y'^n|}{\sqrt{\Delta_n}}+\frac{|\Delta_{i_n+i}^n B''|}{\sqrt{\Delta_n}}\right)$$
$$\times\frac{|\Delta_{i_n+i+1}^n X|}{\sqrt{\Delta_n}}\Bigg].$$

Then, by the same argument (and with the same notation) as in Step 3, and upon using (2.1.45) with $p=1$ and $q=1/2$ and $r=2$ now and also the properties $\mathbb{E}(|\Delta_{i_n+i}^n X''|^2 \mid \mathcal{F}_{(i_n+i-1)\Delta_n}) \leq K\Delta_n$ and $\mathbb{E}(|\Delta_{i_n+i}^n X| \mid \mathcal{F}_{(i_n+i-1)\Delta_n}) \leq K\sqrt{\Delta_n}$, we get by successive conditioning:

$$\mathbb{E}(|Z_n'|) \leq K\left(A\phi_n+\frac{1}{A^2}+\sqrt{\Delta_n}+\mathbb{E}(\sqrt{\gamma_n})\right).$$

Letting first $n\to\infty$ and then $A\to\infty$, we deduce the second part of (9.3.8).

Step 5) Next, we prove the first part of (9.3.8). We complement the notation (9.2.7) by putting $X''(\kappa)=(\delta 1_{\{\Gamma\leq\kappa\}})*(\mathfrak{p}-\mathfrak{q})$ and $B''(\kappa)=B''-(\delta 1_{\{\Gamma>\kappa\}})*\mathfrak{q}$, for any $\kappa\in(0,1)$. The set $\Omega_n(\kappa)$ on which the Poisson process $\mathfrak{p}([0,t]\times\{z:\Gamma(z)>\kappa\})$ has no jump on the interval $(T-(k_n+2)\Delta_n,T)\cup(T,T+(k_n+2)\Delta_n]$ converges to Ω as $n\to\infty$, and on this set we have $\Delta_i^n X=\Delta_i^n B''(\kappa)+\Delta_i^n Y^n+\Delta_i^n Y'^n+\Delta_i^n X''(\kappa)$ for $i=i_n,i_n+1,\dots,i_n+k_n-1$. Then we can apply (9.3.9) with $v=\infty$ (so the first term on the right does not show), with $x=\Delta_{i_n+i}^n Y^n/\sqrt{\Delta_n}$ and $y=\Delta_{i_n+i}^n X''(\kappa)/\sqrt{\Delta_n}$ and $z=\Delta_{i_n+i}^n Y'^n/\sqrt{\Delta_n}$ and $w=\Delta_{i_n+i}^n B''(\kappa)/\sqrt{\Delta_n}$, to get that on the set $\Omega_n(\kappa)$ we have $|Z_n|\leq Z_n(\kappa)$, where

$$Z_n(\kappa)=\frac{1}{k_n}\sum_{i=0}^{k_n-1}\left(\varepsilon^2\left|\frac{\Delta_{i_n+i}^n Y^n}{\sqrt{\Delta_n}}\right|^2+\frac{K}{\varepsilon}\left|\frac{\Delta_{i_n+i}^n X''(\kappa)}{\sqrt{\Delta_n}}\right|^2\right.$$
$$\left.+\frac{K}{\varepsilon}\left|\frac{\Delta_{i_n+i}^n Y'^n}{\sqrt{\Delta_n}}\right|^2+\frac{K}{\varepsilon}\left|\frac{\Delta_{i_n+i}^n B''(\kappa)}{\sqrt{\Delta_n}}\right|^2\right).$$

Then we use (9.3.10) and also $|\Delta_i^n B''(\kappa)|\leq K_\kappa\Delta_n$ and $\mathbb{E}(|\Delta_{i_n+i}^n X''(\kappa)|^2)\leq K\Delta_n\theta(\kappa)$, where $\theta(\kappa)=\int_{\{z:\Gamma(z)\leq\kappa\}}\Gamma(z)^2\lambda(dz)$ (the latter coming from (2.1.36) with $p=2$): we then deduce

$$\mathbb{E}(Z_n(\kappa)) \leq K\varepsilon^2+\frac{K\theta(\kappa)}{\varepsilon}+\frac{K}{\varepsilon}\mathbb{E}(\gamma_n)+\frac{K_\kappa\Delta_n}{\varepsilon}. \tag{9.3.11}$$

On the other hand, for all $\eta>0$ we have $\mathbb{P}(|Z_n|>\eta)\leq\mathbb{P}(\Omega_n(\kappa))+\frac{1}{\eta}\mathbb{E}(Z_n(\kappa))$. Letting $n\to\infty$, we deduce from (9.3.11) and $\Omega_n(\kappa)\to\Omega$ that

$$\limsup_n \mathbb{P}(|Z_n|>\eta) \leq \frac{K\varepsilon^2}{\eta}+\frac{K\theta(\kappa)}{\varepsilon\eta}.$$

This is true for all $\varepsilon,\kappa\in(0,1)$. Letting $\varepsilon=\theta(\kappa)^{1/3}$ and taking advantage of the property $\theta(\kappa)\to 0$ as $\kappa\to 0$, we deduce that $\mathbb{P}(|Z_n|>\eta)\to 0$, and this completes the proof of the first part of (9.3.8).

Step 6) So far, we have proved (a) and (b) under (b-1). We now turn to (b) under (b-2), and we will consider the first convergence in (9.3.5) only, the other two being proved in the same way. We come back to the general case $d \ge 1$. The assumption is that $\mathfrak{p}(\{T\} \times E) = 1$ if $0 < T < \infty$, where $\mathfrak{p}$ is a driving Poisson measure for the pair (X, σ).

Choose any function $\Gamma' : E \to (0, \infty)$ such that $\int \Gamma'(z)\lambda(dz) < \infty$. Let $A = \{z : \Gamma'(z) > 1/m\}$ for some $m \in \mathbb{N}^*$ (so $\lambda(A) < \infty$), and denote by $(\mathcal{G}_t)$ the smallest filtration containing $(\mathcal{F}_t)$ and such that the restriction of $\mathfrak{p}$ to $\mathbb{R}_+ \times A$ is $\mathcal{G}_0$ measurable. Let $R_1, R_2, \dots$ be the successive jump times of the Poisson process $1_A * \mathfrak{p}$, and $X_t^A = X_t - \sum_{p \ge 1} \Delta X_{R_p} 1_{\{R_p \le t\}}$. By Proposition 2.1.10, page 44, W is a $(\mathcal{G}_t)$-Brownian motion, and the restriction $\mathfrak{p}'$ of $\mathfrak{p}$ to $\mathbb{R}_+ \times A^c$ is a Poisson measure whose compensator is the restriction $\mathfrak{q}'$ of $\mathfrak{q}$ to the same set.

Then X^A is an Itô semimartingale, relative to the filtration $(\mathcal{G}_t)$, with the same Grigelionis representation (9.0.1) as X, except that $\mathfrak{p}$ and $\mathfrak{q}$ are replaced by $\mathfrak{p}'$ and $\mathfrak{q}'$, and b_t is replaced by $b_t^A = b_t - \int_{\{\|\delta(t,z)\| \le 1\} \cap A} \delta(t,z)\lambda(dz)$, whereas W, σ and δ are unchanged. In particular, X^A satisfies (H).

Observe that each R_p is positive and $\mathcal{G}_0$ measurable, so it satisfies (b-1) (with $S = 0$), relative to the filtration $(\mathcal{G}_t)$. Therefore we deduce that $\widehat{c}^n(k_n, R_p-, X^A) \xrightarrow{\mathbb{P}} c_{R_p-}$. For $p \ge 1$, the set $\Omega_p^n = \{R_{p-1} < R_p - (k_n + 2)\Delta_n\})$ (with $R_0 = 0$) goes to Ω as $n \to \infty$, because $k_n \Delta_n \to 0$. Since $\Delta_i^n X = \Delta_i^n X^A$ if there is no R_q in $I(n,i)$, we see that on Ω_p^n we have $\widehat{c}^n(k_n, R_p-, X^A) = \widehat{c}^n(k_n, R_p-, X)$, and thus $\widehat{c}^n(k_n, R_p-, X) \xrightarrow{\mathbb{P}} c_{R_p-}$. Then if $G_m = \{R_1, R_2, \dots\}$ (this depends on m, as are the R_p's implicitly), the previous property obviously yields

$$\widehat{c}^n(k_n, T-, X) \xrightarrow{\mathbb{P}} c_{T-} \tag{9.3.12}$$

in restriction to the set $\{T \in G_m\}$, hence also in restriction to the set $\{T \in \bigcup_{m \ge 1} G_m\}$. It remains to observe that $\{0 < T < \infty\} = \{T \in \bigcup_{m \ge 1} G_m\}$. □

9.4 From Local Approximation to Global Approximation

As seen before, and still in the same general setting, one of the main problems is the approximation of the integral $\int_0^t g(c_s)\, ds$ for some test function g on the space $\mathcal{M}_{d \times d}^+$. When g has the form $g(a) = \rho_a^{k\otimes}(F)$ for a suitable function F on $(\mathbb{R}^d)^k$, Theorems 8.4.1 or 9.2.1 provide approximations, under suitable assumptions on F.

Now, we can also use our local estimators: since the variables of (9.3.1) are approximations of $c_{i\Delta_n}$, one may hope that the variables

$$C^n(g, k_n, X)_t = \Delta_n \sum_{i=1}^{[t/\Delta_n]-k_n+1} g\big(\widehat{c}_i^n(k_n)\big)$$

$$C^n(g, k_n, v_n, X)_t = \Delta_n \sum_{i=1}^{[t/\Delta_n]-k_n+1} g\big(\widehat{c}_i^n(k_n, v_n)\big)$$

converge to $\int_0^t g(c_s)\,ds$, under appropriate assumptions on g (we leave out the estimators based on the bipower approximations $\widehat{c}_i'^n(k_n)$). This is the object of the next result:

Theorem 9.4.1 *Assume that X satisfies* (H), *and let g be a continuous function on $\mathcal{M}_{d\times d}^+$, satisfying for some $p \geq 0$:*

$$\big|g(x)\big| \leq K\big(1 + \|x\|^p\big). \tag{9.4.1}$$

a) *If either X is continuous or $p < 1$, we have*

$$C^n(g, k_n, X)_t \overset{\text{u.c.p.}}{\Longrightarrow} \int_0^t g(c_s)\,ds. \tag{9.4.2}$$

b) *If either $p \leq 1$, or* (H-r) *holds for some $r \in [0, 2)$ and*

$$p > 1, \qquad \varpi \geq \frac{p-1}{2p-r}, \tag{9.4.3}$$

we have

$$C^n(g, k_n, v_n, X)_t \overset{\text{u.c.p.}}{\Longrightarrow} \int_0^t g(c_s)\,ds. \tag{9.4.4}$$

The bigger r is, the more stringent is (9.4.3): when r increases we need to truncate more, and the same when p increases.

Remark 9.4.2 We can compare this result with Theorem 9.2.1. We want to approximate $\int_0^t g(c_s)\,ds$. On the one hand, if $g(a) = \rho_a^{k\otimes}(F)$ with some k and some F, the condition (9.2.3) on F with $p = q$ is basically the condition (9.4.1) on g, with $p = q/2$, and the assumptions in Theorem 9.2.1 and in (b) above are thus basically the same: so, as far as the growth of the test function is concerned, the two theorems are very similar.

On the other hand, the existence of k and F such that $g(a) = \rho_a^{k\otimes}(F)$ is a serious restriction on g. For example it implies that g is C^∞ on the restriction of g to the set of invertible matrices, whereas in Theorem 9.4.1 only the continuity of g is needed. This is perhaps of little practical relevance, but mathematically speaking it means that the present theorem quite significantly improves on Theorem 9.2.1.

Finally, in most practical situations g is given and, when it corresponds to some k and F, finding an explicit function F solving this question is often difficult.

Example 9.4.3 When X has jumps, (9.4.2) fails in general when we do not have (9.4.1) with $p < 1$. For example take $g(x) = x$. Then, up to some (negligible) boundary terms, $C^n(g, k_n, X)$ is the approximate quadratic variation, which does not converge to $C_t = \int_0^t c_s\,ds$.

Proof of Theorem 9.4.1 By localization we can and will assume (SH-r) throughout the proof, with $r=2$ when $p\le 1$. Up to proving the desired convergence separately for g^+ and g^-, we can also assume $g\ge 0$, and then it is enough to prove the convergence in probability for each t fixed.

1) We first prove the results when g is bounded. For any $t>0$ we have $\widehat{c}^n(k_n,t)=\widehat{c}_i^n(k_n)$ when $(i-2)\Delta_n\le t<(i-1)\Delta_n$, hence

$$C^n(g,k_n,X)_t \;=\; \Delta_n g\big(\widehat{c}_1^n(k_n)\big)+\int_0^{([t/\Delta_n]-k_n)\Delta_n} g\big(\widehat{c}^n(k_n,s)\big)\,ds.$$

Thus

$$\mathbb{E}\bigg(\bigg|C^n(g,k_n,g)_t-\int_0^t g(c_s)\,ds\bigg|\bigg)\;\le\; Kk_n\Delta_n+\int_0^{([t/\Delta_n]-k_n)\Delta_n} a_n(s)\,ds,$$

where $a_n(s)=\mathbb{E}(|g(\widehat{c}^n(k_n,s))-g(c_s)|)$. Now, (9.3.4) implies that $a_n(s)\to 0$ for each s and stays bounded uniformly in (n,s) because g is bounded. Hence (9.4.2) follows from the dominated convergence theorem, and (9.4.4) is obtained in exactly the same way.

2) With $\psi_\varepsilon,\psi'_\varepsilon$ as given by (3.3.16) or in the proof of Theorem 9.1.1, we write $g_m=g\psi_m$ and $g'_m=g\psi'_m$. The function g'_m is continuous and bounded, so the previous step yields $C^n(g'_m,k_n,X)_t\stackrel{\mathbb{P}}{\longrightarrow}\int_0^t g'_m(c_s)\,ds$ and $C^n(g'_m,k_n,v_n,X)_t\stackrel{\mathbb{P}}{\longrightarrow}\int_0^t g'_m(c_s)\,ds$ for any m fixed, and $\int_0^t g'_m(c_s)\,ds=\int_0^t g(c_s)\,ds$ for all m large enough because c_s is bounded. On the other hand $g'_m(x)\le K\|x\|^p\,1_{\{\|x\|>m\}}$ for all $m\ge 1$. Then it remains to prove that

$$\lim_{m\to\infty}\;\limsup_{n\to\infty}\;\mathbb{E}\Bigg(\Delta_n\sum_{i=1}^{[t/\Delta_n]}\big\|\widehat{c}_i^n(k_n)\big\|^p\,1_{\{\|\widehat{c}_i^n(v_n)\|>m\}}\Bigg)=0 \qquad (9.4.5)$$

$$\lim_{m\to\infty}\;\limsup_{n\to\infty}\;\mathbb{E}\Bigg(\Delta_n\sum_{i=1}^{[t/\Delta_n]}\big\|\widehat{c}_i^n(k_n,v_n)\big\|^p\,1_{\{\|\widehat{c}_i^n(k_n,v_n)\|>m\}}\Bigg)=0, \qquad (9.4.6)$$

under the relevant assumptions in (a) or (b).

We start with (9.4.5). Letting $\kappa=0$ when X is continuous and $\kappa=1$ otherwise, for all $q\ge 2$ we have by (2.1.44) and (SH-2):

$$\mathbb{E}\big(\|\Delta_i^n X\|^q\big)\;\le\;K_q\big(\Delta_n^{q/2}+\kappa\,\Delta_n^{(q/2)\wedge 1}\big).$$

In view of the definition of $\widehat{c}_i^n(k_n)$, we deduce from Hölder's inequality, and for $q\ge 1$:

$$\mathbb{E}\big(\|\widehat{c}_i^n(k_n)\|^q\big)\;\le\;K_q\big(1+\kappa\,\Delta_n^{q\wedge 1-q}\big).$$

Therefore if $q>p$, Markov's inequality yields

$$\mathbb{E}\big(\big\|\widehat{c}_i^n(k_n)\big\|^p\,1_{\{\|\widehat{c}_i^n(v_n)\|>m\}}\big)\;\le\;\frac{K_q}{m^{q-p}}\,\big(1+\kappa\,\Delta_n^{q\wedge 1-q}\big).$$

Taking $q=2p$ in the continuous case and $q=1>p$ otherwise, we deduce (9.4.5).

Now we turn to (9.4.6), with a proof somewhat similar to the proof of (9.2.10). Without loss of generality, we may assume $p\ge 1$. For all n bigger than some n_m we have $2m\le u_n=v_n/\sqrt{\Delta_n}$. With the notation (9.2.7) we have $\|\widehat{c}_i^n(k_n,v_n)\|\le \zeta_i'^n+\zeta_i'^n$, where

$$\zeta_i'^n\;=\;\frac{1}{k_n}\sum_{j=0}^{k_n-1}\bigg(\frac{\|\Delta_{i+j}^n X'\|}{\sqrt{\Delta_n}}\bigg)^2,\qquad \zeta_i''^n\;=\;\frac{u_n^2}{k_n}\sum_{j=0}^{k_n-1}\bigg(\frac{\|\Delta_{i+j}^n X''\|}{\Delta_n^{\varpi}}\bigwedge 1\bigg)^2.$$

Therefore

$$\big\|\widehat{c}_i^n(k_n,v_n)\big\|\,1_{\{\|\widehat{c}_i^n(k_n,v_n)\|>m\}}\;\le\;\frac{2}{m}\,\big(\zeta_i'^n\big)^2+2\zeta_i''^n.$$

By (9.2.12), (9.2.13) and Hölder's inequality, for all $q\ge 1$ we have

$$\mathbb{E}\big(\big|\zeta_i'^n\big|^q\mid\mathcal{F}_{(i-1)\Delta_n}\big)\;\le\;K_q,\qquad \mathbb{E}\big(\big\|\zeta_i''^n\big\|^q\mid\mathcal{F}_{(i-1)\Delta_n}\big)\;\le\;Ku_n^{2q}\,\Delta_n^{1-r\varpi}\,\phi_n$$

for some sequence $\phi_n\to 0$. Setting $w=1-p+\varpi(2p-r)$, we deduce

$$\mathbb{E}\big(\big\|\widehat{c}_i^n(k_n,v_n)\big\|^p\,1_{\{\|\widehat{c}_i^n(k_n,v_n)\|>m\}}\mid\mathcal{F}_{(i-1)\Delta_n}\big)\;\le\;\frac{K}{m^p}+K\Delta_n^w\phi_n. \tag{9.4.7}$$

Since $w\ge 0$ when either $p=1$ or (9.4.3) holds, (9.4.6) follows. □

9.5 Local Approximation for the Continuous Part of X: Part II

Still another family of processes occurs in applications, when the process X jumps. For example we may have to approximate such quantities as

$$\sum_{s\le t} f(\Delta X_s)g(c_{s-})\quad\text{or}\quad\sum_{s\le t} f(\Delta X_s)g(c_s)$$

for suitable functions f and g: we have already seen such expressions in (5.1.5) for example. This sort of process mixes the jumps with the "local" value of c_t, and the approximation for them mixes the approach of Theorem 3.3.1 when $f(x)=\mathrm{o}(\|x\|^2)$ as $x\to 0$ and Theorem 9.1.1 otherwise, with the results of the two previous sections.

More generally, we may have to approximate a process of the form

$$\sum_{s\le t} G(\Delta X_s, c_{s-}, c_s),$$

of which the two previous ones are special cases. This leads us to introduce the following: we consider a function G on $\mathbb{R}^d \times \mathcal{M}^+_{d\times d} \times \mathcal{M}^+_{d\times d}$. We also take a sequence k_n of integers with $k_n \to \infty$ and $k_n \Delta_n \to 0$, and a cut-off sequence v_n as in (9.0.3), with which we associate the $\mathcal{M}^+_{d\times d}$-valued variables $\widehat{c}^n_i(k_n, v_n)$ by (9.3.1). Then we set

$$V^n(G; k_n, v_n, X)_t = \sum_{i=k_n+1}^{[t/\Delta_n]-k_n} G\big(\Delta^n_i X, \widehat{c}^n_{i-k_n}(k_n, v_n), \widehat{c}^n_{i+1}(k_n, v_n)\big)\, 1_{\{\|\Delta^n_i X\| > v_n\}}. \tag{9.5.1}$$

The summation bounds are chosen in such a way that the right side only depends on the increments $\Delta^n_i X$ within the interval $[0, t]$, and that one does not use the convention (9.3.2). If we are willing to use it we can start the summation at $i = 0$. When $G(x, y, y')$ only depends on (x, y), one can extend the summation up to $i = [t/\Delta_n]$: the results below are not modified if we perform those changes.

Theorem 9.5.1 *Assume that X satisfies* (H-r) *for some $r \in [0, 2]$, and choose v_n as in* (9.0.3) *and k_n such that $k_n \to \infty$ and $k_n \Delta_n \to 0$. Let G be a continuous function on $\mathbb{R}^d \times \mathcal{M}^+_{d\times d} \times \mathcal{M}^+_{d\times d}$ and, when $r > 0$, assume*

$$\|x\| \le \eta \Rightarrow \big|G(x, y, y')\big| \le f(x)\big(1 + \|y\|^p\big)\big(1 + \|y'\|^p\big) \tag{9.5.2}$$

for some $\eta > 0$, with f a function on $\mathbb{R}^d$ satisfying $f(x) = \mathrm{o}(\|x\|^r)$ as $x \to 0$, and either $p \in [0, 1]$, or $p > 1$ and (9.4.3) *holds. Then, we have the following Skorokhod convergence in probability*:

$$V^n(G; k_n, v_n, X)_t \stackrel{\mathbb{P}}{\Longrightarrow} \sum_{s\le t} G(\Delta X_s, c_{s-}, c_s). \tag{9.5.3}$$

Under (H-r) we have $\sum_{s\le t} \|\Delta X_s\|^r < \infty$ for all t, hence the assumptions on G made above imply that the limiting process in (9.5.3) is well defined and càdlàg of finite variation. The condition $f(x) = \mathrm{o}(\|x\|^r)$ as $x \to 0$ is not really restrictive for applications, although in view of Theorem 9.1.1 one would rather expect $f(x) = \mathrm{O}(\|x\|^r)$; however, we do not know how to prove the result under this (slightly) weaker condition on f.

Remark 9.5.2 The functionals $V^n(G; k_n, v_n, X)$ involve two truncations.

a) It is possible to use two different truncation levels $v_n = \alpha \Delta_n^{\varpi}$ and $v'_n = \alpha' \Delta_n^{\varpi'}$ with $\alpha, \alpha' > 0$ and $\varpi, \varpi' \in (0, 1/2)$, and replace $\|\Delta^n_i X\| > v_n$ by $\|\Delta^n_i X\| > v'_n$ in (9.5.1). However, since both truncations serve the same purpose of separating the jumps from the "continuous" part of X, there seems to be no need for taking two different levels v_n and v'_n.

b) On the other hand, one could delete one, or even both, truncations: when $f(x) = \mathrm{o}(\|x\|^2)$ as $x \to 0$ one can dispense with the first one (v'_n above), and when $p<1$ in (9.5.2) we can use the non-truncated $\widehat{c}_i^{\prime n}(k_n)$. If for example $f(x)=\mathrm{o}(\|x\|^2)$ as $x\to 0$ in (9.5.2) the processes

$$\sum_{i=k_n+1}^{[t/\Delta_n]-k_n} G\big(\Delta_i^n X, \widehat{c}_{i-k_n}^n(k_n,v_n), \widehat{c}_{i+1}^n(k_n,v_n)\big)$$

also satisfy (9.5.3). In practice, however, it is probably better always to use the truncated versions.

c) One could also replace $\widehat{c}_i^n(k_n,v_n)$ by the bipower version $\widehat{c}_i^{\prime n}(k_n)$, and the results would be the same.

Remark 9.5.3 The reader will notice that no growth condition on $(y,y')\mapsto G(x,y,y')$ is required when $\|x\|>\eta$, whereas $\eta>0$ is also arbitrarily small. This is due to the fact that for any given $\eta>0$ and on any fixed time interval $[0,t]$, there are finitely many jumps of X bigger than η, hence also a (bounded in n) number of increments $\Delta_i^n X$ bigger than η, and thus the continuity of $(x,y,y')\mapsto G(x,y,y')$ when $\|x\|>\varepsilon$ is enough to imply the convergence of the corresponding summands in (9.5.1) to those in (9.5.3).

One could even weaken the continuity assumption, exactly as in Theorem 3.3.5. Namely, if D is a subset of $\mathbb{R}^d$ such that $\mathbb{P}(\exists t>0: \Delta X_t\in D)=0$, or equivalently $1_D * \nu_\infty=0$ a.s., the previous result holds when the function G is continuous outside $D\times\mathcal{M}_{d\times d}^+\times\mathcal{M}_{d\times d}^+$.

Proof *Step 1)* By localization we may assume (SH-r). We use the simplifying notation $V^n(G)=V^n(G;k_n,v_n,X)$, because only G will vary in the proof, and the right side of (9.5.3) is denoted by $V(G)_t$. We have

$$V^n(G)_t=\sum_{i=k_n+1}^{[t/\Delta_n]-k_n}\zeta(G)_i^n, \quad \text{where}$$

$$\zeta(G)_i^n = G\big(\Delta_i^n X, \widehat{c}_{i-k_n}^n(k_n,v_n), \widehat{c}_{i+1}^n(k_n,v_n)\big)\, 1_{\{\|\Delta_i^n X\|>v_n\}}. \tag{9.5.4}$$

Step 2) In this step, and according to the scheme of the proof of Theorem 9.1.1, to which the present theorem reduces when $G(x,y,y')=f(x)$, we prove the result when $G(x,y,y')=0$ if $\|x\|\le\varepsilon$, for some $\varepsilon>0$.

(SH-r) implies $\|\delta(\omega,t,z)\|\le\Gamma(z)$ for some bounded function Γ with $\int\Gamma(z)^r\times\lambda(dz)<\infty$. We let S_q be the successive jumps times of the Poisson process $1_{\{\Gamma>\varepsilon/2\}}*\mathfrak{p}$, with the convention $S_0=0$, and $i(n,q)$ denotes the unique random integer i such that $S_q\in I(n,i)$. The jumps of X outside the set $\cup_{q\ge1}\{S_q\}$ have a size smaller than $\varepsilon/2$, hence the properties of G yields for some (random) integers n_t:

- $s\notin\cup_{q\ge1}\{S_q(\omega)\} \;\Rightarrow\; G(\Delta X_s(\omega), c_{s-}(\omega), c_s(\omega))=0$
- $\Delta_{i(n,q)}^n X(\omega)\to\Delta X_{S_q}(\omega)$ as $n\to\infty$
- $n\ge n_{t,\omega},\ i\in\{j:1\le j\le[\Delta_n/t]\}\backslash\{i(n,q)(\omega):q\ge1\}\;\Rightarrow\;\zeta(G)_i^n(\omega)=0.$

Then, in view of (9.5.4) and of the continuity of G, the convergence (9.5.3) holds as soon as for each $q \geq 1$ we have

$$\widehat{c}^n_{i(n,q)-k_n}(k_n, v_n) \xrightarrow{\mathbb{P}} c_{S_q-}, \qquad \widehat{c}^n_{i(n,q)+1}(k_n, v_n) \xrightarrow{\mathbb{P}} c_{S_q}. \tag{9.5.5}$$

Since S_q is a jump time of the driving Poisson measure $\mathfrak{p}$, these two convergences follow from Theorem 9.3.2.

When $r = 0$ we take for S_q the successive jump times of $1_{\{\Gamma>0\}} * \mathfrak{p}$. We know that $\|\Delta^n_i X\| \leq v_n$, hence $\zeta(G)^n_i = 0$, for all i smaller than $[t/\Delta_n]$ and different from all $i(n,q)$, on a set whose probability goes to 1 as $n \to \infty$. Then the same argument as above shows the convergence (9.5.2), without any growth restriction on G, and the theorem is proved in this case.

Step 3) In this step we provide some estimates. It is understood below that the integer i satisfies $i \geq k_n + 1$, and for simplicity we write $\xi^n_i = 1 + \|\widehat{c}^n_i(k_n, v_n)\|^p$. First, (SH-$r$) and the fact that either $p \leq 1$ or (9.4.3) holds imply (9.4.7) with $w \geq 0$ and $m = 1$, and we deduce

$$\mathbb{E}\big(|\xi^n_i| \mid \mathcal{F}_{(i-1)\Delta_n}\big) \leq K. \tag{9.5.6}$$

Next, we take $\varepsilon \in (0,1)$ and we set

$$A_\varepsilon = \big\{z : \Gamma(z) \leq \varepsilon\big\}, \qquad a(\varepsilon) = \int_{A_\varepsilon} \Gamma(z)^r \,\lambda(dz).$$

We use the decomposition $X = X' + X''$ of (9.0.2) and we also set $X(\varepsilon) = X - (\delta 1_{A^c_\varepsilon}) * \mathfrak{p}$. Thus $X(\varepsilon) = Y(\varepsilon) + Y'(\varepsilon)$, where

$$\begin{aligned} r \in (1,2] &\Rightarrow\ Y(\varepsilon) = (\delta 1_{A_\varepsilon}) * (\mathfrak{p} - \mathfrak{q}), \qquad Y'(\varepsilon) = X' - (\delta 1_{\{\|\delta\|\leq 1\}\cap A^c_\varepsilon}) * \mathfrak{q} \\ r \in (0,1] &\Rightarrow\ Y(\varepsilon) = (\delta 1_{A_\varepsilon}) * \mathfrak{p}, \qquad\quad\ \ Y'(\varepsilon) = X' - (\delta 1_{\{\|\delta\|\leq 1\}}) * \mathfrak{q}. \end{aligned}$$

Since $\|\delta(s,z)\| \leq \Gamma(z)$, we deduce from (2.1.36) when $r > 1$ and (2.1.40) when $r \leq 1$ that

$$\mathbb{E}\big(\big\|\Delta^n_i Y(\varepsilon)\big\|^r \mid \mathcal{F}_{(i-1)\Delta_n}\big) \leq K\Delta_n\, a(\varepsilon), \tag{9.5.7}$$

where $a(\varepsilon) \to 0$ as $\varepsilon \to 0$. Moreover when $r \leq 1$ the process $Y'(\varepsilon)$ is as X' in (9.0.2), except that b_t is substituted with $b(\varepsilon)_t = b_t - \int_{\{\|\delta(t,z)\|\leq 1\}} \delta(t,z)\,\lambda(dz)$, which satisfies $\|b(\varepsilon)_t\| \leq K$ for a constant K which does not depend on ε. When $r > 1$ we have the same, except that now $b(\varepsilon)_t = b_t - \int_{\{\|\delta(t,z)\|\leq 1\}\cap A^c_\varepsilon} \delta(t,z)\,\lambda(dz)$ satisfies $\|b(\varepsilon)_t\| \leq K\varepsilon^{1-r}$. Hence in both cases we deduce from (2.1.33) and (2.1.34) that, for all $q > 0$:

$$\mathbb{E}\big(\big\|\Delta^n_i Y'(\varepsilon)\big\|^q \mid \mathcal{F}_{(i-1)\Delta_n}\big) \leq K_q\big(\Delta_n^{q/2} + \Delta_n^q\, \varepsilon^{-q(r-1)^+}\big). \tag{9.5.8}$$

Now, $\|x+y\|^r 1_{\{\|x+y\|>v_n\}} \leq K_q(\|x\|^{q+r}/v_n^q + \|y\|^r)$ for all $x, y \in \mathbb{R}^d$ and $q > 0$. Taking $q = \frac{4-r}{1-2\varpi}$ and putting together (9.5.7) and (9.5.8), plus $X(\varepsilon) = Y(\varepsilon) +$

$Y'(\varepsilon)$, we obtain

$$\mathbb{E}\big(\big\|\Delta_i^n X(\varepsilon)\big\|^r 1_{\{\|\Delta_i^n X(\varepsilon)\|>v_n\}} \mid \mathcal{F}_{(i-1)\Delta_n}\big) \le K_\varepsilon \Delta_n^2 + K \Delta_n a(\varepsilon). \tag{9.5.9}$$

Step 4) We use once more the functions ψ_ε and ψ'_ε of (3.3.16) and we set

$$G_\varepsilon\big(x,y,y'\big) = G\big(x,y,y'\big)\,\psi_\varepsilon(x), \qquad G'_\varepsilon\big(x,y,y'\big) = G\big(x,y,y'\big)\,\psi'_\varepsilon(x).$$

Step 2 gives us $V^n(G_\varepsilon) \stackrel{\mathbb{P}}{\Longrightarrow} V(G_\varepsilon)$ for each $\varepsilon > 0$. On the other hand, the processes c and ΔX are bounded by (SH-r), so (9.5.2) yields $|G(\Delta X_t, c_{t-}, c_t)| \le K\|\Delta X_t\|^r$. Hence the behavior of f near 0 and the fact that $r \in \mathcal{I}(X)$ yield that $V(G_\varepsilon) \stackrel{\text{u.c.p.}}{\Longrightarrow} V(G)$ as $\varepsilon \to 0$. It thus remains to prove that

$$t > 0 \quad \Rightarrow \quad \lim_{\varepsilon\to 0} \limsup_n \mathbb{E}\Big(\sup_{s\le t}\big|V^n\big(G'_\varepsilon\big)_s\big|\Big) = 0.$$

Our assumptions on G yield that, as soon as $\varepsilon < \eta$, we have $|G'_\varepsilon(x,y,y')| \le \theta(\varepsilon)\,\|x\|^r(1+\|y\|^p)(1+\|y'\|^p)\,1_{\{\|x\|\le\varepsilon\}}$ for some continuous function θ on $[0,1]$ with $\theta(0) = 0$. Then, if $T(\varepsilon)_1, T(\varepsilon)_2, \ldots$ are the successive jump times of the Poisson process $1_{A_\varepsilon^c} * \mathfrak{p}$, and recalling the notation ξ_i^n of Step 3 and also (9.5.4), we see that

$$\big|\zeta\big(G'_\varepsilon\big)_i^n\big| \le \begin{cases} K\,\theta(\varepsilon)\,\varepsilon^r\,\xi_{i-k_n}^n \xi_{i+1}^n & \text{if } T(\varepsilon)_q \in I(n,i) \text{ for some } q \ge 1 \\ K\,\|\Delta_i^n X(\varepsilon)\|^r\,\xi_{i-k_n}^n \xi_{i+1}^n\,1_{\{\|\Delta_i^n X(\varepsilon)\|>v_n\}} & \text{otherwise.} \end{cases}$$

Therefore we see that $|V^n(G'_\varepsilon)_t| \le K Z^n(\varepsilon)_t + K\theta(\varepsilon)\,Z'^n(\varepsilon)_t$, where

$$Z^n(\varepsilon)_t = \sum_{i=k_n+1}^{[t/\Delta_n]-k_n} \big\|\Delta_i^n X(\varepsilon)\big\|^r\,\xi_{i-k_n}^n \xi_{i+1}^n\,1_{\{\|\Delta_i^n X(\varepsilon)\|>v_n\}}$$

$$Z'^n(\varepsilon)_t = \varepsilon^r \sum_{i=k_n+1}^{[t/\Delta_n]-k_n} \xi_{i-k_n}^n \xi_{i+1}^n\,z(\varepsilon)_i^n$$

$$z(\varepsilon)_i^n = 1_{\cup_{q\ge1}\{T(\varepsilon)_q \in I(n,i)\}}$$

and, since $\theta(\varepsilon) \to 0$ as $\varepsilon \to 0$, it remains to prove that

$$\lim_{\varepsilon\to 0}\limsup_{n\to\infty} \mathbb{E}\big(Z^n(\varepsilon)_t\big) = 0, \qquad \sup_{n\ge1,\,\varepsilon\in(0,1)} \mathbb{E}\big(Z'^n(\varepsilon)_t\big) < \infty. \tag{9.5.10}$$

Combining (9.5.6), (9.5.9) and the $\mathcal{F}_{i\Delta_n}$ measurability of ξ_{i+1}^n, we get by successive conditioning that $\mathbb{E}(Z^n(\varepsilon)_t) \le t(K_\varepsilon\Delta_n + K a(\varepsilon))$, and the first part of (9.5.10) follows because $a(\varepsilon) \to 0$ as $\varepsilon \to 0$.

Another application of (9.5.6) and of the $\mathcal{F}_{i\Delta_n}$-measurability of $\xi_{i-k_n}^n\,z(\varepsilon)_i^n$ yields $\mathbb{E}(\xi_{i-k_n}^n \xi_{i+1}^n\,z(\varepsilon)_i^n \mid \mathcal{F}_{i\Delta_n}) \le K\xi_{i-k_n}^n\,z(\varepsilon)_i^n$. By definition of the $T(\varepsilon)_q$'s, the variable

$z(\varepsilon)_i^n$ is independent of $\mathcal{F}_{(i-1)\Delta_n}$ and satisfies, with the notation $a'(\varepsilon) = \lambda(\{\Gamma > \varepsilon\})$:

$$\mathbb{E}\big(z(\varepsilon)_i^n \mid \mathcal{F}_{(i-1)\Delta_n}\big) = \mathbb{P}\big(z(\varepsilon)_i^n = 1\big) = 1 - e^{-\Delta_n a'(\varepsilon)} \le \Delta_n a'(\varepsilon).$$

Therefore by conditioning on $\mathcal{F}_{(i-1)\Delta_n}$ and then taking the expectation and using once more (9.5.6), we obtain $\mathbb{E}(\xi_{i-k_n}^n \xi_{i+1}^n z(\varepsilon)_i^n) \le K \Delta_n a'(\varepsilon)$, from which we deduce $\mathbb{E}(Z'^n(\varepsilon)_t) \le K t \varepsilon^r a'(\varepsilon)$. The assumption $\int \Gamma(z)^r \lambda(dz) < \infty$ yields $\varepsilon^r a'(\varepsilon) \le K$, and the second part of (9.5.10) follows. □

9.6 Applications to Volatility

We continue again the application about the estimation of the volatility. The setting is as in the previous section: an Itô semimartingale X given by (8.4.1) and satisfying (H), and a regular discretization scheme. Below, we will put in use the truncated functional, with v_n as in (9.0.3).

We have seen in (8.5.2), (8.5.3) and (8.5.4) how to estimate the variable $\int_0^t \prod_{j=1}^l (c_s^{r_j m_j})^{p_j} ds$ for all integers r_j and m_j in $\{1, \dots, d\}$ and all integers p_j, or all reals $p_j > 0$ when further $m_j = r_j$ for all j, using the multipower variations. The same quantities can also be approximated with the help of Theorem 9.2.1. Namely, if k_j are integers, and $K_0 = 0$ and $K_j = k_1 + \cdots + k_j$ and $p_j > 0$ and $p = p_1 + \cdots + p_l$, we have

$$\begin{aligned} &k_j > p_j/2 \quad \forall j \quad \Rightarrow \\ &\quad \Delta_n^{1-p} \sum_{i=1}^{[t/\Delta_n]-K_l+1} \prod_{j=1}^{l} \prod_{u=1}^{k_j} \big|\Delta_{i+K_{j-1}+u-1}^n X^{r_j}\big|^{2p_j/k_j} 1_{\{\|\Delta_{i+K_{j-1}+u-1}^n X\| \le v_n\}} \\ &\quad \stackrel{\text{u.c.p.}}{\Longrightarrow} \prod_{j=1}^{l} m_{2p_j/k_j}^{k_j} \int_0^t \prod_{j=1}^{l} \big(c_s^{r_j r_j}\big)^{p_j} ds, \end{aligned} \tag{9.6.1}$$

and also, when m_j does not necessarily agree with r_j,

$$\begin{aligned} &\Delta_n^{1-l} \sum_{i=1}^{[t/\Delta_n]-l+1} \prod_{j=1}^{l} \Delta_{i+j-1}^n X^{r_j} \, \Delta_{i+j-1}^n X^{m_j} \, 1_{\{\|\Delta_{i+j-1}^n X\| \le \alpha \Delta_n^{\varpi}\}} \\ &\quad \stackrel{\text{u.c.p.}}{\Longrightarrow} \int_0^t \prod_{j=1}^{l} c_s^{r_j m_j} ds. \end{aligned} \tag{9.6.2}$$

There is no theoretical reason to prefer either (8.5.2) or (9.6.1), when X jumps. We recall that when X is continuous, we also have

$$\Delta_n^{1-p} \sum_{i=1}^{[t/\Delta_n]-l+1} \prod_{j=1}^{l} \big|\Delta_{i+K_{j-1}+u-1}^n X^{r_j}\big|^{2p_j} \stackrel{\text{u.c.p.}}{\Longrightarrow} \prod_{j=1}^{l} m_{2p_j} \int_0^t \prod_{j=1}^{l} \big(c_s^{r_j r_j}\big)^{p_j} ds \tag{9.6.3}$$

(by Theorem 8.4.1). This last estimator is better behaved from a practical viewpoint than those in both (8.5.2) and (9.6.1), in the continuous case: we will see that, although the rate of convergence of all these estimators is the same, the asymptotic variance of (9.6.3) is less than for the others.

Note that Theorem 9.4.1 provides still another method for approximating the right side of (9.6.2), namely

$$\Delta_n \sum_{i=1}^{[t/\Delta_n]-k_n+1} \prod_{j=1}^{l} \widehat{c}_i^n(k_n, v_n)^{r_j m_j} \stackrel{\text{u.c.p.}}{\Longrightarrow} \int_0^t \prod_{j=1}^{l} c_s^{r_j m_j}\, ds$$

where $k_n \to \infty$ and $k_n \Delta_n \to 0$. However, unless $l = 1$ above, we need to have (H-r) for some $r < 2$ and $\varpi \geq \frac{l-1}{2l-r}$.

Another application of truncated functionals consists in determining the "proportion of jumps" for an observed path of a one-dimensional process X. This is a rather vague notion, often understood as follows: what is the ratio of the part of the quadratic variation due to the jumps (or, to the continuous part), over the total quadratic variation? These ratios are

$$R_t = \frac{[X,X]_t - C_t}{[X,X]_t}, \qquad 1 - R_t = \frac{C_t}{[X,X]_t}.$$

The motivation for looking at those ratios is that, at least when t is large enough, $[X,X]_t$ and C_t are roughly proportional to the "variance of the increments" and the "variance of the increments of the continuous part", respectively, the drift part being treated roughly as a non-random factor. This statement is of course mathematically rather unfounded, but the variable R_t is well defined and does provide some insight on the relative importance of the jumps versus the continuous part.

If we put together Theorems 9.1.1 and 9.2.1, we get an estimator for R_t, on the basis of a discretely (and regularly) observed path. Namely, if we complement the notation $D(X, p, \Delta_n)_t$ of (5.3.23) by putting

$$D(X, p, v_n+, \Delta_n)_t = \sum_{i=1}^{[t/\Delta_n]} |\Delta_i^n X|^p \, 1_{\{|\Delta_i^n X| > v_n\}},$$

$$D(X, p, v_n-, \Delta_n)_t = \sum_{i=1}^{[t/\Delta_n]} |\Delta_i^n X|^p \, 1_{\{|\Delta_i^n X| \leq v_n\}},$$

we have for $t > 0$:

$$\frac{D(X, 2, v_n+, \Delta_n)_t}{D(X, 2, \Delta_n)_t} \stackrel{\mathbb{P}}{\longrightarrow} R_t.$$

Part IV
Extensions of the Central Limit Theorems

The applications sketched at the end of most previous chapters show us that, as useful as the Laws of Large Numbers may be, they are not enough. In all cases we need a way to assert the rate at which our functionals converge. This is the aim of this part: stating and proving Central Limit Theorems associated with the LLNs of the four previous chapters.

Chapter 10
The Central Limit Theorem for Random Weights

In this chapter, the setting and notation are the same as in Chap. 7.

Only regular discretization schemes are considered. The basic d-dimensional process X is an Itô semimartingale on the space $(\Omega, \mathcal{F}, (\mathcal{F}_t)_{t\geq 0}, \mathbb{P})$, with the Grigelionis decomposition

$$X_t = X_0 + \int_0^t b_s\,ds + \int_0^t \sigma_s\,dW_s + (\delta 1_{\{\|\delta\|\leq 1\}}) \star (\mathfrak{p} - \mathfrak{q})_t + (\delta 1_{\{\|\delta\|>1\}}) \star \mathfrak{p}_t, \quad (10.0.1)$$

where W is a d'-dimensional Wiener process and $\mathfrak{p}$ is a Poisson measure with compensator $\mathfrak{q}(dt, dz) = dt \otimes \lambda(dz)$. We set $c = \sigma\sigma^\star$, and $\mu = \mu^X$ is the jump measure of X.

The process X will satisfy various additional conditions, according to the case. But it will at least satisfy Assumption (H), or 4.4.2, which we recall here:

Assumption (H) We have (10.0.1), with b_t locally bounded and σ_t càdlàg, and $\|\delta(\omega, t, z)\| \wedge 1 \leq \Gamma_n(z)$ for all (ω, t, z) with $t \leq \tau_n(\omega)$, where (τ_n) is a localizing sequence of stopping times and each function Γ_n satisfies $\int \Gamma_n(z)^2 \lambda(dz) < \infty$.

10.1 Functionals of Non-normalized Increments—Part I

The test function, here and in the whole chapter, is "random", that is, it is a function F on $\Omega \times \mathbb{R}_+ \times \mathbb{R}^d$. For the non-normalized functional associated with F, we have introduced in (7.3.1) two different versions, according to whether we plug the left or the right endpoint of the discretization interval $I(n, i)$, as the time argument in F. For the Central Limit Theorem it will be important to take the left endpoint. Hence we consider the following, called $V^{n,l}(F, X)$ in Chap. 7:

$$V^n(F, X)_t = \sum_{i=1}^{[t/\Delta_n]} F\big(., (i-1)\Delta_n, \Delta_i^n X\big). \quad (10.1.1)$$

J. Jacod, P. Protter, *Discretization of Processes*,
Stochastic Modelling and Applied Probability 67,
DOI 10.1007/978-3-642-24127-7_10, © Springer-Verlag Berlin Heidelberg 2012

The function F on $\Omega \times \mathbb{R}_+ \times \mathbb{R}^d$ is q-dimensional. Under appropriate assumptions on F we have proved (Theorem 7.3.3) that $V^n(F,X) \stackrel{\mathbb{P}}{\Longrightarrow} F * \mu$ (convergence in probability for the Skorokhod topology).

Our aim is toward an associated CLT and, as in Chap. 5, we will need F to be C^2 in x and $\mathrm{o}(\|x\|^3)$ as $x \to 0$. But this is not enough. Indeed, the key point in the proof of Theorem 5.1.2 is the behavior of the variables ζ_p^n defined in (5.1.12), page 130, and which are of the form

$$\zeta_p^n = \frac{1}{\sqrt{\Delta_n}} \big(f\big(\Delta X_{S_p} + \sqrt{\Delta_n}\, R(n,p)\big) - f(\Delta X_{S_p}) - f\big(\sqrt{\Delta_n}\, R(n,p)\big)\big),$$

where $R(n,p)$ is a suitable sequence of d-dimensional variables which converges (stably in law) as $n \to \infty$, and S_p is a stopping time. Then we use a Taylor's expansion for f, which says that ζ_p^n is approximately equal to $\nabla f(\Delta X_{S_p}) R(n,p)$. In the present setting, and for the left functional $V^{n,l}(F,X)$ for example, the variable ζ_p^n takes the form

$$\begin{aligned} \zeta_p^n = \frac{1}{\sqrt{\Delta_n}} \big(& F\big(S_-(n,p), \Delta X_{S_p} + \sqrt{\Delta_n}\, R(n,p)\big) \\ & - F(S_p, \Delta X_{S_p}) - F\big(S_-(n,p), \sqrt{\Delta_n}\, R(n,p)\big)\big), \end{aligned} \tag{10.1.2}$$

where $S_-(p,n) = (i-1)\Delta_n$ when $(i-1)\Delta_n < S_p \le i\Delta_n$. Therefore we need also some smoothness of $F(\omega,t,x)$, as a function of t, in order to find an equivalent to ζ_p^n: the situation here is akin to the situation in Theorem 5.3.5, regarding the regularity of σ_t as a function of t, see after Example 5.3.1. A way to solve the problem is to assume the following:

Assumption 10.1.1 The q-dimensional function F is such that the maps $\omega \mapsto F(\omega,t,x)$ are $\mathcal{F}_t$ measurable for all (t,x). Each function $x \mapsto F(\omega,t,x)$ is of class C^2 (the partial derivatives are denoted $\partial_i F$ and $\partial_{ij}^2 F$), and the functions $t \mapsto \partial_i F(\omega,t,x)$ are càdlàg, and $F(\omega,t,0) = \partial_i F(\omega,t,0) = 0$.

Moreover there is a real $\gamma \in (1/2, 1]$, a localizing sequence (τ_n) of stopping times and, for each n, two continuous functions f_n and $\overline{f}_n$ on $\mathbb{R}^d$, with $f_n(x) = \mathrm{o}(\|x\|)$ and $\overline{f}_n(x) = \mathrm{O}(\|x\|^2)$ as $x \to 0$, and such that we have identically

$$s,t \le \tau_n(\omega) \;\Rightarrow\; \begin{cases} \|\partial_{ij}^2 F(\omega,t,x)\| \le f_n(x) \\ \|F(\omega,t,x) - F(\omega,s,x)\| \le \overline{f}_n(x)|t-s|^\gamma. \end{cases} \tag{10.1.3}$$

This implies Assumption 7.3.1, because $(t,x) \mapsto F(\omega,t,x)$ is continuous.

We have to describe the limiting process, and for this we basically need the same ingredients as in Sect. 5.1.1: we consider an arbitrary weakly exhausting sequence $(T_n)_{n\ge1}$ for the jumps of X; we also consider an auxiliary space $(\Omega', \mathcal{F}', \mathbb{P}')$ endowed with a triple sequence $(\Psi_{n-}, \Psi_{n+}, \kappa_n)_{n\ge1}$ of variables, all independent, and with the following laws:

$$\Psi_{n\pm} \text{ are } d'\text{-dimensional, } \mathcal{N}(0, I_{d'}), \qquad \kappa_n \text{ is uniform on } [0,1].$$

Then we consider the very good filtered extension $(\widetilde{\Omega}, \widetilde{\mathcal{F}}, (\widetilde{\mathcal{F}}_t)_{t\geq 0}, \widetilde{\mathbb{P}})$ of $(\Omega, \mathcal{F}, (\mathcal{F}_t)_{t\geq 0}, \mathbb{P})$ defined as in (4.1.16), that is

$$\begin{aligned}
&\widetilde{\Omega} = \Omega \times \Omega', \qquad \widetilde{\mathcal{F}} = \mathcal{F} \otimes \mathcal{F}', \qquad \widetilde{\mathbb{P}} = \mathbb{P} \otimes \mathbb{P}' \\
&(\widetilde{\mathcal{F}}_t) \text{ is the smallest filtration containing } (\mathcal{F}_t), \\
&\qquad \text{such that } (\Psi_{n-}, \Psi_{n+}, \kappa_n) \text{ is } \widetilde{\mathcal{F}}_{T_n} \text{ measurable for all } n.
\end{aligned}$$

We define the d-dimensional random variables

$$R_n = \sqrt{\kappa_n}\, \sigma_{T_n -} \Psi_{n-} + \sqrt{1-\kappa_n}\, \sigma_{T_n} \Psi_{n+}$$

where σ is the process occurring in (10.0.1).

The next proposition describes the limiting process, and it is proved exactly as Proposition 5.1.1, using Proposition 4.1.4, page 102.

Proposition 10.1.2 *Suppose that X satisfies Assumption* (H) *and that F satisfies Assumption* 10.1.1. *The formula*

$$\overline{V}(F, X)_t = \sum_{n=1}^{\infty} \left(\sum_{i=1}^{d} \partial_i F(T_n -, \Delta X_{T_n}) R_n^i \right) 1_{\{T_n \leq t\}} \tag{10.1.4}$$

defines a q-dimensional process $Z(F, X)$ on $(\widetilde{\Omega}, \widetilde{\mathcal{F}}, (\widetilde{\mathcal{F}}_t)_{t\geq 0}, \widetilde{\mathbb{P}})$ which is a.s. càdlàg, adapted, and conditionally on $\mathcal{F}$ has centered and independent increments and satisfies

$$\widetilde{\mathbb{E}}\big(\overline{V}(F, X)_t^i\, \overline{V}(F, X)_t^j \mid \mathcal{F}\big) = \frac{1}{2} \sum_{s \leq t} \sum_{k,l=1}^{d} \big(\partial_k F^i\, \partial_l F^j\big)(s-, \Delta X_s)\big(c_{s-}^{kl} + c_s^{kl}\big),$$

and its $\mathcal{F}$-conditional law does not depend on the choice of the exhausting sequence T_n. Moreover, if X and σ have no common jumps, the process $\overline{V}(F, X)$ is $\mathcal{F}$-conditionally Gaussian.

The main result is the exact analogue of Theorem 5.1.2, recall that $X^{(n)}$ is the discretized process $X_t^{(n)} = X_{[\Delta_n [t/\Delta_n]}$.

Theorem 10.1.3 *Suppose that X satisfies Assumption* (H) *and that F satisfies Assumption* 10.1.1. *Then if*

$$\overline{V}^n(F, X)_t = \frac{1}{\sqrt{\Delta_n}} \big(V^n(F, X)_t - F \star \mu_{\Delta_n [t/\Delta_n]}\big),$$

where $V^n(F, X)$ is given by (10.1.1), *the $(d+q)$-dimensional processes $(X^{(n)}, \overline{V}^n(F, X))$ converge stably in law to $(X, \overline{V}(F, X))$, where $\overline{V}(F, X)$ is defined in* (10.1.4). *Moreover, for each fixed t the variables*

$$\frac{1}{\sqrt{\Delta_n}} \big(V^n(F, X)_t - F \star \mu_t\big) \tag{10.1.5}$$

converge stably in law to the variable $\overline{V}(F, X)_t$.

Proof The proof is essentially the same as for Theorem 5.1.2, page 127. To avoid lengthy developments, we constantly refer to the proof of that theorem and use the same notation, and we only point out the necessary changes to be made to that proof.

Step 1) The localization lemma 4.4.9, page 118, applies, so we can suppose that X satisfies the strengthened Assumption (SH), or 4.4.6: in addition to (H) the processes b_t and σ_t and X_t are all bounded by a constant A, and $\|\delta(\omega,t,z)\| \le \Gamma(z)$ with Γ bounded and $\int \Gamma(z)^2\lambda(dz) < \infty$. In particular, $\|\Delta_i^n X\| \le 2A$ always, so it is no restriction to suppose that $F(\omega,t,x) = 0$ whenever $\|x\| \ge 3A$, and we can then choose the functions f_n and $\overline{f}_n$ in Assumption 10.1.1 to vanish as well when $\|x\| \ge 4A$, so that they are bounded.

A further localization procedure can be performed, along the sequence τ_n of Assumption 10.1.1. For each q set $F_q(\omega,t,x) = F(\omega,t\wedge\tau_q(\omega),x)$. Assuming that the result holds for each F_q, and since $\overline{V}^n(F_q,X)_t = \overline{V}^n(F,X)_t$ and $\overline{V}(F_q,X)_t = \overline{V}(F,X)_t$ for all $t \le \tau_q$, we deduce that $(X^{(n)},\overline{V}^n(F,X)) \stackrel{\mathcal{L}\text{-s}}{\Longrightarrow} (X,\overline{V}(F,X))$ in restriction to the time interval $[0,T]$ (as processes indexed by time), and to the set $\{\tau_q \ge T\}$. Since $\mathbb{P}(\tau_q \ge T) \to 1$, we indeed have $(X^{(n)},\overline{V}^n(F,X)) \stackrel{\mathcal{L}\text{-s}}{\Longrightarrow} (X,\overline{V}(F,X))$ in restriction to any time interval $[0,T]$, hence also on the whole half-line $\mathbb{R}_+$. The same argument works for (10.1.5) as well.

Therefore it is enough to prove the results for each F_q, or equivalently for a function F which satisfies Assumption 10.1.1 and vanishes when $\|x\| > A$ for some $A > 0$ and satisfies also

$$\big\|\partial^2_{ij}F(\omega,t,x)\big\| \le \|x\|\,\theta\big(\|x\|\big), \quad \big\|F(\omega,t,x) - F(\omega,s,x)\big\| \le K\big(\|x\|^2 \wedge 1\big)\,|s-t|^\gamma \tag{10.1.6}$$

where $\gamma > \frac{1}{2}$ and the function θ on $\mathbb{R}_+$ is bounded, increasing and continuous with $\theta(0) = 0$. Then, since $F(\omega,t,0) = \partial_i F(\omega,t,0) = 0$, we also have

$$\begin{aligned} \big\|F(\omega,t,x)\big\| &\le K\|x\|^3\,\theta\big(\|x\|\big), \quad \big\|\partial_i F(\omega,t,x)\big\| \le K\|x\|^2\,\theta\big(\|x\|\big) \\ \big\|\partial_i F(\omega,t,x) - \partial_i F(\omega,t,y)\big\| &\le K\|x-y\|. \end{aligned} \tag{10.1.7}$$

Step 2) We use the same specific choice S_p as in Sect. 5.1.2 for the sequence of stopping times that weakly exhausts the jumps of X, and we also use the notation (5.1.9)–(5.1.12), page 129, except that ζ_p^n is now given by (10.1.2). Then the analogue of (5.1.13) becomes

$$\overline{V}^n(F,X)_t = \overline{V}^n\big(F,X(m)\big)_t + Y^n(m)_t \quad \forall t \le T, \quad \text{on the set } \Omega_n(T,m).$$

We can rewrite ζ_p^n as $\zeta_p^n = \zeta_p^{n,1} + \zeta_p^{n,2} + \zeta_p^{n,3}$, where

$$\begin{aligned} \zeta_p^{n,1} = \frac{1}{\sqrt{\Delta_n}}\Big(&F\big(S_-(n,p),\Delta X_{S_p} + \sqrt{\Delta_n}\,R(n,p)\big) \\ &- F\big(S_p,\Delta X_{S_p} + \sqrt{\Delta_n}\,R(n,p)\big)\Big) \end{aligned}$$

$$\zeta_p^{n,2} = \frac{1}{\sqrt{\Delta_n}} \big(F\big(S_p, \Delta X_{S_p} + \sqrt{\Delta_n}\, R(n,p)\big) - F(S_p, \Delta X_{S_p})\big)$$

$$\zeta_p^{n,3} = -\frac{1}{\sqrt{\Delta_n}} F\big(S_-(n,p), \sqrt{\Delta_n}\, R(n,p)\big).$$

The last part of (10.1.6) implies $\|\zeta_p^{n,1}\| \leq K\Delta_n^{\gamma-1/2}$, and since the sequence $R(n,p)$ is bounded in probability the first part of (10.1.7) yields $\zeta_p^{n,3} \stackrel{\mathbb{P}}{\longrightarrow} 0$. Therefore a Taylor expansion for the term $\zeta_p^{n,2}$ gives us that

$$p \geq 1 \quad \Rightarrow \quad \zeta_p^n - \sum_{i=1}^d \partial_i F(S_p, \Delta X_{S_p}) R(n,p)^i \stackrel{\mathbb{P}}{\longrightarrow} 0. \tag{10.1.8}$$

Then, another application of Proposition 4.4.10 allows us to deduce from (10.1.8) that

$$\big(\overline{\zeta}_p^n\big)_{p\geq 1} \stackrel{\mathcal{L}\text{-s}}{\longrightarrow} (\overline{\zeta}_p)_{p\geq 1}, \quad \text{where } \overline{\zeta}_p = \left(\Delta X_{S_p}, \sum_{i=1}^d \partial_i F(S_p, \Delta X_{S_p}) R_p^i\right). \tag{10.1.9}$$

This is the present form for (5.1.14), from which we deduce (5.1.15), page 130, (with $\overline{V}(F, X'(m))$). Then we can go along the proof of Theorem 5.1.2 down to (5.1.17), that is, in order to prove our first claim it is enough to show that:

$$\lim_{m\to\infty} \limsup_{n\to\infty} \mathbb{P}\Big(\Omega_n(t,m) \cap \Big\{\sup_{s\leq t} \big\|\overline{V}^n\big(F, X(m)\big)_s\big\| > \eta\Big\}\Big) = 0. \tag{10.1.10}$$

Step 3) For proving (10.1.10) we proceed again as in Theorem 5.1.2, with F being one-dimensional, with a few changes. Instead of (5.1.18), we set

$$\begin{aligned} k_i^n(\omega; x, y) &= F\big(\omega, (i-1)\Delta_n, x+y\big) - F\big(\omega, (i-1)\Delta_n, x\big) - F\big(\omega, (i-1)\Delta_n, y\big) \\ g_i^n(\omega; x, y) &= k_i^n(\omega; x, y) - \sum_{i=1}^d \partial_i F\big(\omega, (i-1)\Delta_n, x\big) y_i. \end{aligned} \tag{10.1.11}$$

These, as well as $F(\omega, (i-1)\Delta_n, x)$, are $\mathcal{F}_{(i-1)\Delta_n}$ measurable. Therefore we can apply Itô's formula to the process $X(m)_t - X(m)_{(i-1)\Delta_n}$ for $t > (i-1)\Delta_n$, to get

$$\begin{aligned} \xi(m)_i^n &:= F\big((i-1)\Delta_n, \Delta_i^n X(m)\big) - \sum_{s\in I(n,i)} F\big((i-1)\Delta_n, \Delta X(m)_s\big) \\ &= A(n,m,i)_{i\Delta_n} + M(n,m,i)_{i\Delta_n}, \end{aligned}$$

where $M(n,m,i)$ and $A(n,m,i)$ are the same as in Theorem 5.1.2, except that in the definitions of $a(n,m,i)_t$ and $a'(n,m,i)_t$ one substitutes k and g with k_i^n and g_i^n.

Now, (10.1.7) implies that the functions k_i^n and g_i^n satisfy (5.1.22), uniformly in ω, n, i, so the estimates (5.1.23), and of course (5.1.24), are valid here. Then, with the same notation for $T(n,m,i)$, we have (5.1.21), and if $Z^n(m)_t = \frac{1}{\sqrt{\Delta_n}} \sum_{i=1}^{[t/\Delta_n]} \xi(m)_i^n$ we deduce as in Theorem 5.1.2 that

$$\lim_{m\to\infty} \limsup_{n\to\infty} \mathbb{P}\left(\Omega_n(t,m) \cap \left\{\sup_{s\le t} \left|Z^n(m)_s\right| > \frac{\eta}{2}\right\}\right) = 0. \tag{10.1.12}$$

Therefore, proving (10.1.10) amounts to showing that

$$\lim_{m\to\infty} \limsup_{n\to\infty} \mathbb{P}\left(\sup_{s\le t} \left|\overline{V}^n\big(F, X(m)\big)_s - Z^n(m)_s\right| > \frac{\eta}{2}\right) = 0. \tag{10.1.13}$$

For this, we observe that $\overline{V}^n(F, X(m))_t - Z^n(m)_t = \frac{1}{\sqrt{\Delta_n}} \sum_{i=1}^{[t/\Delta_n]} \xi'(m)_i^n$, where

$$\xi'(m)_i^n = \sum_{s\in I(n,i)} \Big(F\big((i-1)\Delta_n, \Delta X(m)_s\big) - F\big(s, \Delta X(m)_s\big)\Big).$$

(10.1.6) yields $|\xi'(m)_i^n| \le K\Delta_n^\gamma \sum_{s\in I(n,i)} \|\Delta X(m)_s\|^2$, and since the variable $\sum_{s\le t} \|\Delta X(m)_s\|^2$ is finite and $\gamma > \frac{1}{2}$, we readily deduce (10.1.13), and thus (10.1.10) follows.

Step 4) It remains to prove the second claim, that is $\frac{1}{\sqrt{\Delta_n}}(V^n(F,X)_t - F \star \mu_t) \xrightarrow{\mathcal{L}\text{-s}} \overline{V}(F,X)_t$ for any fixed t. This is deduced from the first claim exactly as in Theorem 5.1.2, page 133. □

10.2 Functionals of Non-normalized Increments—Part II

Assumption 10.1.1 is reasonably weak, *except* for the second part of (10.1.3). A typical test function F which we want to consider in applications is of the form $F(\omega,t,x) = \overline{F}(Y_t(\omega), x)$ for an auxiliary Itô semimartingale Y and a smooth function $\overline{F}$, but then this assumption fails because $t \mapsto Y_t$ is not Hölder with index $\gamma > \frac{1}{2}$. Note that in such a case we can always consider the Itô semimartingale $Z = (X, Y)$ and of course $V^n(F,X) = V^n(G,Z)$, where $G(\omega,t,(x,y)) = F(\omega,t,x)$. This is why, without any loss of generality, we state the result when $Y = X$.

Therefore we consider the functionals

$$V^n\big(\overline{F}(X), X\big)_t = \sum_{i=1}^{[t/\Delta_n]} \overline{F}\big(X_{(i-1)\Delta_n}, \Delta_i^n X\big).$$

The precise assumptions on $\overline{F}$ are as follows:

Assumption 10.2.1 The q-dimensional function $\overline{F}$ on $\mathbb{R}^d \times \mathbb{R}^d$ is C^2 and satisfies for all $A > 0$:

$$\begin{cases} \overline{F}(y,0) = \frac{\partial}{\partial x^j}\overline{F}(y,0) = 0, \\ \lim_{x\to 0} \sup_{y:\|y\|\le A} \frac{1}{\|x\|}\left\|\frac{\partial^2}{\partial x^j \partial x^k}\overline{F}(y,x)\right\| = 0, \\ \sup_{(y,x):\|y\|\le A,\, 0<\|x\|\le A} \frac{1}{\|x\|^2}\left\|\frac{\partial}{\partial y^j}\overline{F}(y,x)\right\| < \infty. \end{cases}$$

The limiting process takes a different form here. We still use the previous notation T_n, Ψ_{n-}, Ψ_{n+}, κ_n. Below, when writing the partial derivatives of $\overline{F}$, the two arguments are ordered according to the notation $\overline{F}(y,x)$. Note that for all $A > 0$, all first partial derivatives of $\overline{F}$ (with respect to the components of y or those of x) are smaller than $K_A\|x\|$ when $\|y\| \le A$ and $\|x\| \le 1$. Then again the next proposition is proved exactly as Proposition 5.1.1, page 127.

Proposition 10.2.2 *Suppose that X satisfies* (H) *and that $\overline{F}$ satisfies Assumption* 10.2.1. *The formula*

$$\overline{V}\big(\overline{F}(X),X\big)_t = \sum_{n=1}^{\infty}\left(\sum_{i=1}^{d}\left(\frac{\partial\overline{F}}{\partial x^i} - \frac{\partial\overline{F}}{\partial y^i}\right)(X_{T_n-},\Delta X_{T_n})\sqrt{\kappa_n}\,(\sigma_{T_n-}\Psi_{n-})^i \right.$$
$$\left. + \sum_{i=1}^{d}\frac{\partial\overline{F}}{\partial x^i}(X_{T_n-},\Delta X_{T_n})\sqrt{1-\kappa_n}\,(\sigma_{T_n}\Psi_{n+})^i\right)1_{\{T_n\le t\}} \quad (10.2.1)$$

defines a q-dimensional process $\overline{V}(\overline{F}(X),X)$ on $(\widetilde{\Omega},\widetilde{\mathcal{F}},(\widetilde{\mathcal{F}}_t)_{t\ge 0},\widetilde{\mathbb{P}})$ which is a.s. càdlàg, adapted, and conditionally on $\mathcal{F}$ has centered and independent increments and satisfies

$$\widetilde{\mathbb{E}}\big(\overline{V}\big(\overline{F}(X),X\big)^i_t\,\overline{V}\big(\overline{F}(X),X\big)^j_t \mid \mathcal{F}\big)$$
$$= \frac{1}{2}\sum_{s\le t}\sum_{k,l=1}^{d}\left(\left(\frac{\partial\overline{F}^i}{\partial x^k} - \frac{\partial\overline{F}^i}{\partial y^k}\right)\left(\frac{\partial\overline{F}^j}{\partial x^l} - \frac{\partial\overline{F}^j}{\partial y^l}\right)(X_{s-},\Delta X_s)c^{kl}_{s-}\right.$$
$$\left. + \left(\frac{\partial\overline{F}^i}{\partial x^k}\frac{\partial\overline{F}^j}{\partial x^l}\right)(X_{s-},\Delta X_s)c^{kl}_s\right),$$

and its $\mathcal{F}$-conditional law does not depend on the choice of the exhausting sequence T_n. If further the processes X and σ have no common jumps, the process $Z(\overline{F}(X),X)$ is $\mathcal{F}$-conditionally Gaussian.

Our second main theorem is as follows (the notation $\overline{F}(X_-) * \mu$ means the integral process of $(\omega,t,x) \mapsto \overline{F}(X_{t-}(\omega),x)$ with respect to μ):

Theorem 10.2.3 *Suppose that X satisfies* (H) *and that $\overline{F}$ satisfies Assumption* 10.2.1. *Then if*

$$\overline{V}^n\big(\overline{F}(X),X\big)_t = \frac{1}{\sqrt{\Delta_n}}\big(V^n\big(\overline{F}(X),X\big)_t - \overline{F}(X_-)*\mu_{\Delta_n[t/\Delta_n]}\big),$$

the processes $(X^{(n)},\overline{V}^n(\overline{F}(X),X))$ converge stably in law to the process $(X, \overline{V}(\overline{F}(X),X))$, where $\overline{V}(\overline{F}(X),X)$ is defined in (10.2.1). *Moreover for each fixed t the variables*

$$\frac{1}{\sqrt{\Delta_n}}\big(V^n\big(\overline{F}(X),X\big)_t - \overline{F}(X_-)*\mu_t\big)$$

converge stably in law to the variable $\overline{V}(\overline{F}(X),X)_t$.

Remark 10.2.4 In the definition of R_n, hence in the process $\overline{V}(F,X)$ of (10.1.4), there is a complete symmetry between "before" and "after" a jump, because $\sqrt{\kappa_n}$ and $\sqrt{1-\kappa_n}$ have the same distribution. On the other hand, the process $Z(\overline{F},X)$ of (10.2.1) exhibits an essential dissymmetry, due to the fact the ith summand in $V^n(\overline{F}(X),X)$ involves $\Delta_i^n X$, which is "symmetrical" around the jump time if there is one, and also $X_{(i-1)\Delta_n}$ which only involves what happens before the jump.

Remark 10.2.5 One could also take F satisfying a "mixture" of the two Assumptions 10.1.1 and 10.2.1, namely

$$F(\omega,t,x) = G\big(\omega,t,X_t(\omega),x\big)$$

with a function $G(\omega,t,y,x)$ that satisfies Assumption 10.1.1 when considered as a function of (ω,t,x) (with some uniformity in y), and that satisfies Assumption 10.2.1 when considered as a function of (y,x). Details are left to the reader.

Proof *Step 1)* The proof is basically the same as for Theorem 10.1.3, and we briefly indicate the necessary changes.

First, we can again assume (SH) and thus, since X is bounded, the values of $\overline{F}$ outside some compact set do not matter. So we can suppose that $\overline{F}$ and its derivatives are bounded, hence

$$\begin{aligned} &\left\|\frac{\partial^2}{\partial x^i\partial x^j}\overline{F}(y,x)\right\| \le \big(\|x\|\wedge 1\big)\theta\big(\|x\|\big), && \left\|\frac{\partial}{\partial x^i}\overline{F}(y,x)\right\| \le \big(\|x\|^2\wedge 1\big)\theta\big(\|x\|\big), \\ &\big\|\overline{F}(y,x)\big\| \le \big(\|x\|^3\wedge 1\big)\theta\big(\|x\|\big), && \left\|\frac{\partial}{\partial y^i}\overline{F}(y,x)\right\| \le K\big(\|x\|^2\wedge 1\big), \end{aligned} \tag{10.2.2}$$

where θ is as in the proof of Theorem 10.1.3.

Here again, the second claim follows from the first one as in Step 5 of the proof of Theorem 5.1.2, page 133, so we concentrate on the convergence $(X^{(n)},\overline{V}^n(\overline{F}(X),X)) \stackrel{\mathcal{L}\text{-}s}{\Longrightarrow} (X,\overline{V}(\overline{F}(X),X))$.

Step 2) As in the previous proof, we use the notation (5.1.9)–(5.1.12), except that ζ_p^n is now given by

$$\zeta_p^n = \frac{1}{\sqrt{\Delta_n}} \big(\overline{F}\big(X_{S_-(n,p)}, \Delta X_{S_p} + \sqrt{\Delta_n}\, R(n,p)\big) \\ - \overline{F}(X_{S_p-}, \Delta X_{S_p}) - \overline{F}\big(X_{S_-(n,p)}, \sqrt{\Delta_n}\, R(n,p)\big)\big).$$

This can be rewritten as $\zeta_p^n = \zeta_p^{n,1} + \zeta_p^{n,2}$, where

$$\zeta_p^{n,1} = \frac{1}{\sqrt{\Delta_n}} \big(\overline{F}\big(X_{S_p-} - \sqrt{\Delta_n}\, R_-(n,p), \Delta X_{S_p} + \sqrt{\Delta_n}\, R_-(n,p) \\ + \sqrt{\Delta_n}\, R_+(n,p)\big) - \overline{F}(X_{S_p-}, \Delta X_{S_p})\big)$$

$$\zeta_p^{n,2} = -\frac{1}{\sqrt{\Delta_n}} \overline{F}\big(X_{S_-(n,p)}, \sqrt{\Delta_n}\, R(n,p)\big).$$

The sequences $R_-(n,p)$ and $R_+(n,p)$ are bounded in probability, so as in the previous theorem we have $\zeta_p^{n,2} \xrightarrow{\mathbb{P}} 0$ by (10.2.2), and a Taylor's expansion in $\zeta_p^{n,1}$ gives, with the notation $\overline{F}^1 = \partial\overline{F}/\partial x^i$ and $\overline{F}^2 = \overline{F}^1 - \partial\overline{F}/\partial y^i$:

$$\zeta_p^n - \sum_{i=1}^d \big(\overline{F}^2(X_{S_p-}, \Delta X_{S_p}) R_-(n,p)^i + \overline{F}^1(X_{S_p-}, \Delta X_{S_p}) R_+(n,p)^i\big) \xrightarrow{\mathbb{P}} 0. \tag{10.2.3}$$

Then Proposition 4.4.10 and (10.2.3) yield (10.1.9), where now

$$\overline{\zeta}_p^n = \left(\Delta X_{S_p}, \sum_{i=1}^d \big(\overline{F}_i^2(X_{S_p-}, \Delta X_{S_p}) R_{p-}^i + \overline{F}_i^1(X_{S_p-}, \Delta X_{S_p}) R_{p+}^i\big)\right).$$

At this point, the same argument as in the previous theorem shows that we are left to prove the analogue of (10.1.10), that is

$$\lim_{m\to\infty} \limsup_{n\to\infty} \mathbb{P}\Big(\Omega_n(t,m) \cap \Big\{\sup_{s\le t} \big\|\overline{V}^n\big(\overline{F}(X), X(m)\big)_s\big\| > \eta\Big\}\Big) = 0. \tag{10.2.4}$$

Step 3) For proving (10.2.4) we can again assume that $\overline{F}$ is one-dimensional, and we replace (10.1.11) by

$$k_i^n(\omega; x, y) = \overline{F}\big(X_{(i-1)\Delta_n}(\omega), x+y\big) - \overline{F}\big(X_{(i-1)\Delta_n}(\omega), x\big) - \overline{F}\big(X_{(i-1)\Delta_n}(\omega), y\big)$$

$$g_i^n(\omega; x, y) = k_i^n(\omega; x, y) - \sum_{i=1}^d \overline{F}_i^1\big(X_{(i-1)\Delta_n}(\omega), x\big) y_i.$$

By (10.2.2) and $\overline{F}(y,0) = \overline{F}_i^1(y,0) = 0$, we see that k_i^n and g_i^n satisfy (5.1.22), page 132, uniformly in ω, n, i. Hence as in the previous theorem we deduce (10.1.12), where now $Z^n(m)_s = \frac{1}{\sqrt{\Delta_n}} \sum_{i=1}^{[t/\Delta_n]} \xi(m)_i^n$ and

$$\xi(m)_i^n = \overline{F}\big(X_{(i-1)\Delta_n}, \Delta_i^n X(m)\big) - \sum_{s\in I(n,i)} \overline{F}\big(X_{(i-1)\Delta_n}, \Delta X(m)_s\big).$$

Therefore (10.2.4) will follow if we can prove

$$\lim_{m\to\infty} \limsup_{n\to\infty} \mathbb{P}\Big(\sup_{s\le t} \big|\overline{V}^n\big(\overline{F}(X), X(m)\big)_s - Z^n(m)_s\big| > \frac{\eta}{2}\Big) = 0. \tag{10.2.5}$$

We still have $\overline{V}^n(\overline{F}(X), X(m))_t - Z^n(m)_t = \frac{1}{\sqrt{\Delta_n}} \sum_{i=1}^{[t/\Delta_n]} \xi'(m)_i^n$, except that now

$$\xi'(m)_i^n = \sum_{s\in I(n,i)} \big(\overline{F}\big(X_{(i-1)\Delta_n}, \Delta X(m)_s\big) - \overline{F}\big(X_{s-}, \Delta X(m)_s\big)\big).$$

The last property in (10.2.2) yields $|\overline{F}(y',x) - \overline{F}(y,x)| \le K\|x\|^2 \|y'-y\|$. Recalling $\Delta X(m)_s = \int_{\{z:\Gamma(z)\le 1/m\}} \delta(s,z) \mathfrak{p}(\{s\}, dz)$ and $\|\delta(t,z)\| \le \Gamma(z)$, we obtain with $u_m = \int_{\{z:\Gamma(z)\le 1/m\}} \Gamma(z)^2 \lambda(dz)$, and using the fact that $\mathfrak{q}$ is the compensator of $\mathfrak{p}$:

$$\begin{aligned}
\mathbb{E}\big(\big|\xi'(m)_i^n\big|\big) &\le K\mathbb{E}\Big(\sum_{s\in I(n,i)} \|X_{(i-1)\Delta_n} - X_{s-}\| \, \big\|\Delta X(m)_s\big\|^2\Big) \\
&= K\mathbb{E}\Big(\int_{I(n,i)} \int_{\{z:\Gamma(z)\le 1/m\}} \|X_{(i-1)\Delta_n} - X_{s-}\| \, \big\|\delta(s,z)\big\|^2 \mathfrak{p}(ds,dz)\Big) \\
&= K\mathbb{E}\Big(\int_{I(n,i)} \int_{\{z:\Gamma(z)\le 1/m\}} \|X_{(i-1)\Delta_n} - X_{s-}\| \, \big\|\delta(s,z)\big\|^2 \mathfrak{q}(ds,dz)\Big) \\
&\le K u_m \mathbb{E}\Big(\int_{I(n,i)} \|X_{(i-1)\Delta_n} - X_{s-}\| \, ds\Big).
\end{aligned}$$

(SH) implies $\mathbb{E}(\|X_{(i-1)\Delta_n} - X_{s-}\|) \le K\sqrt{\Delta_n}$ if $s \in I(n,i)$, thus $\mathbb{E}(|\xi'(m)_i^n|) \le K u_m \Delta_n^{3/2}$, which yields

$$\mathbb{E}\Big(\sum_{i=1}^{[t/\Delta_n]} \big|\xi'(m)_i^n\big|\Big) \le K t\, u_m \sqrt{\Delta_n},$$

and (10.2.5) follows because $\lim_{m\to\infty} u_m = 0$. This ends the proof. ∎

10.3 Functionals of Normalized Increments

For the Central Limit Theorem for functionals of normalized increments, it is as important as before to take the time $\tau(n,i) = (i-1)\Delta_n$ at the left endpoint of the discretization interval $I(n,i)$. That is, we consider the functionals

$$V'^n(F,X)_t = \Delta_n \sum_{i=1}^{[t/\Delta_n]} F\big(.,(i-1)\Delta_n, \Delta_i^n X/\sqrt{\Delta_n}\big).$$

Exactly as in Chap. 5, there is a CLT like Theorem 5.2.2 when we "center" each summand in $V'^n(F,X)$ around its $\mathcal{F}_{(i-1)\Delta_n}$-conditional expectation: this basically requires Assumption 7.2.1 on F, plus the $\mathcal{F}_t$ measurability of $\omega \mapsto F(\omega,t,x)$. We do not do this here, since it is mainly a tool for proving the CLT when we center $V'^n(F,X)$ around $\int_0^t ds \int_{\mathbb{R}^d} F(s,x)\rho_{c_s}(dx)$, which is then an extension of Theorems 5.3.5 and 5.3.6.

For this, and in addition to the properties of $F(\omega,t,x)$ as a function of x, we also need some regularity in t. This can be the Hölder property (10.1.3), with $\gamma > 1/2$, or we can instead consider $F(\omega,t,x) = \overline{F}(X_t(\omega),x)$ as in the previous section. For the application to statistics which we develop in the next section we really need a mixture of the two approaches. That is, we consider functionals of the form

$$V'^n\big(\overline{F}(X),X\big)_t = \Delta_n \sum_{i=1}^{[t/\Delta_n]} \overline{F}\big((i-1)\Delta_n, X_{(i-1)\Delta_n}, \Delta_i^n X/\sqrt{\Delta_n}\big),$$

where $\overline{F}$ is a (non-random) function on $\mathbb{R}_+ \times \mathbb{R}^d \times \mathbb{R}^d$. In view of Theorem 7.2.2, the CLT will describe the behavior of the following processes (recall that for $a \in \mathcal{M}^+_{d\times d}$, ρ_a denotes the $\mathcal{N}(0,a)$ law):

$$\overline{V}'^n\big(\overline{F}(X),X\big)_t = \frac{1}{\sqrt{\Delta_n}}\left(V'^n\big(\overline{F}(X),X\big)_t - \int_0^t ds \int_{\mathbb{R}^d} \overline{F}(s,X_s,x)\,\rho_{c_s}(dx)\right). \tag{10.3.1}$$

The assumptions on $x \mapsto \overline{F}(t,y,x)$ are the same as in Theorems 5.3.5 or 5.3.6, with some kind of uniformity in (t,y), plus some smoothness in t and y. We gather in a single statement the extensions of both Theorems 5.3.5 and 5.3.6, so we also state the assumptions on $\overline{F}$ as a whole package, although only a part of them is used in any specific situation. Recall that $\overline{F}$ is an $\mathbb{R}^q$-valued function on $\mathbb{R}_+ \times \mathbb{R}^d \times \mathbb{R}^d$. We suppose the existence of an integer $q' \in \{0,\dots,q\}$ and, when $q' < q$, of a nonempty subset B of $\mathbb{R}^d$ which is a finite union of affine hyperplanes, such that

$$\begin{aligned} j \le q' \quad &\Rightarrow \quad x \mapsto \overline{F}^j(t,y,x) \ \text{ is } C^1 \text{ on } \mathbb{R}^d \\ j > q' \quad &\Rightarrow \quad x \mapsto \overline{F}^j(t,y,x) \ \text{ is } C^1 \text{ outside the set } B. \end{aligned} \tag{10.3.2}$$

We denote by $d(x,B)$ the distance between $x \in \mathbb{R}^d$ and B. For each $A > 0$ we have a function g_A on $\mathbb{R}^d$ of polynomial growth, which varies with A of course, but may also be different in each of the conditions under consideration below, whereas $w \in (0,1]$ and $0 < s \leq s'$ always:

$$t \leq A,\ \|y\| \leq A \ \Rightarrow\ \|\overline{F}(t,y,x)\| \ \leq\ g_A(x) \tag{10.3.3}$$

$$t \leq A, \|y\| \leq A \ \Rightarrow\ \left|\frac{\partial}{\partial x^i}\overline{F}^j(t,y,x)\right| \leq \begin{cases} g_A(x) & \text{if } j \leq q' \\ g_A(x)\big(1+\frac{1}{d(x,B)^{1-w}}\big) & \text{if } j > q', x \in B^c \end{cases} \tag{10.3.4}$$

$$\begin{aligned} & t \leq A,\ \|y\| \leq A,\ x \in B^c,\ \|z\| \leq 1 \bigwedge \frac{d(x,B)}{2},\ j > q' \\ & \Rightarrow \left|\frac{\partial}{\partial x^i}\overline{F}^j(t,y,x+z) - \frac{\partial}{\partial x^i}\overline{F}^j(t,y,x)\right| \\ & \quad \leq \big(g_A(x)+g_A(z)\big)\left(1+\frac{1}{d(x,B)^{2-w}}\right)\|z\| \end{aligned} \tag{10.3.5}$$

$$t \leq A,\ \|y\| \leq A \ \Rightarrow\ \|\overline{F}(t,y,x+z) - \overline{F}(t,y,x)\| \leq g_A(x)\big(\|z\|^s + \|z\|^{s'}\big) \tag{10.3.6}$$

$$u,t \leq A,\ \|y\| \leq A \ \Rightarrow\ \|\overline{F}(t,y,x) - \overline{F}(u,y,x)\| \ \leq\ g_A(x)|t-u|^\gamma. \tag{10.3.7}$$

Observe that (10.3.3) with the function g_A and (10.3.6) with the (possibly different) function g'_A imply $\|\overline{F}(t,y,x)\| \leq L_A(1+\|x\|^{r'})$ when $t \leq A$ and $\|y\| \leq A$, with the constant $L_A = g_A(0) + 2g'_A(0)$. In the opposite direction, and exactly as in Remark 5.3.7, if $\|\overline{F}(t,y,x)\| \leq L_A(1+\|x\|^{r'})$ and (10.3.4) hold, then the components $\overline{F}^j$ for $j \leq q'$ satisfy (10.3.6). Note also the two powers s and s' in (10.3.6): the biggest one controls the behavior of the left side as $\|z\| \to \infty$, the smallest one controls its behavior as $z \to 0$.

Remark 10.3.1 Splitting the components of $\overline{F}$ into two distinct families may seem strange at first glance. This is done because we extend both Theorems 5.3.5 or 5.3.6 in a single statement: the components that are everywhere C^1 need no condition like (10.3.5), which is a kind of (local) Lipschitz condition for the derivative, outside B.

As for X, we need the assumptions (K), (K-r), (K') or (K'-r), which we briefly recall (below, r is a number in $[0,1)$):

Assumption (K) We have (H) and

$$\sigma_t \ = \sigma_0 + \int_0^t \widetilde{b}_s\,ds + \int_0^t \widetilde{\sigma}_s\,dW_s + M_t + \sum_{s\leq t} \Delta\sigma_s\,1_{\{\|\Delta\sigma_s\|>1\}}$$

where M is a local martingale with $\|\Delta M_t\| \leq 1$, orthogonal to W, and $\langle M,M\rangle_t = \int_0^t a_s\,ds$ and the compensator of $\sum_{s\leq t} 1_{\{\|\Delta\sigma_s\|>1\}}$ is $\int_0^t \widetilde{a}_s\,ds$, with the following properties: the processes $\widetilde{b}$, $\widetilde{\sigma}$, $\widetilde{a}$ and a are progressively measurable, the processes $\widetilde{b}$, a and $\widetilde{a}$ are locally bounded, and the processes $\widetilde{\sigma}$ and b are càdlàg or càglàd.

Assumption (K-r) (for $r \in [0,1)$) We have (K) except for the càdlàg or càglàd property of b, and $\|\delta(\omega,t,z)\| \wedge 1 \le \Gamma_n(z)$ for all (ω,t,z) with $t \le \tau_n(\omega)$, where (τ_n) is a localizing sequence of stopping times and the Borel functions Γ_n on E satisfy $\int \Gamma_n(z)^r \lambda(dz) < \infty$. Moreover the process $b'_t = b_t - \int_{\{\|\delta(t,z)\| \le 1\}} \delta(t,z)\,\lambda(dz)$ is càdlàg or càglàd.

Assumption (K') We have (K) and both processes c_t and c_{t-} take their values in $\mathcal{M}^{++}_{d\times d}$.

Assumption (K'-r) We have (K-r) and both processes c_t and c_{t-} take their values in $\mathcal{M}^{++}_{d\times d}$.

Finally, in the limiting process we need the notation ρ_a of course, but also the notation of (5.2.4) and (5.3.4) which we state again (below, α is a $d \times d'$ matrix, U is an $\mathcal{N}(0, I_{d'})$ variable, and W is the d'-dimensional standard Brownian motion, and $a = \alpha\alpha^*$, and g, g' are continuous with polynomial growth on $\mathbb{R}^d$):

$$\left.\begin{array}{ll}\widehat{\gamma}_\alpha(g) = \mathbb{E}(g(\alpha U)\,U^*), & \widehat{\gamma}'_\alpha(g)^{jk} = \mathbb{E}(g(\alpha W_1)\int_0^1 W_s^j\,dW_s^k) \\ \multicolumn{2}{l}{\overline{\gamma}_\alpha(g,g') = \mathbb{E}((g(\alpha U) - \widehat{\gamma}_\alpha(g)U)\,(g'(\alpha U) - \widehat{\gamma}_\alpha(g')U)) - \rho_a(g)\,\rho_a(g').}\end{array}\right\} \tag{10.3.8}$$

These functions are one-dimensional, except for $\widehat{\gamma}_\alpha(g)$ which is a d'-dimensional row vector. They all are continuous, as functions of α. They will be applied to the "sections" of $\overline{F}$ or its derivatives, which we denote below as $\overline{F}_{t,y}(x) = \overline{F}(t,y,x)$, and $\partial_j \overline{F}_{t,y}(x) = \frac{\partial}{\partial x^j}\overline{F}(t,y,x)$.

Theorem 10.3.2 *Let X be a d-dimensional Itô semimartingale, and $\overline{F}$ be a q-dimensional function on $\mathbb{R}_+ \times \mathbb{R}^d \times \mathbb{R}^d$ which is continuous and satisfies* (10.3.2). *We also assume* (10.3.3) *and* (10.3.4) *and* (10.3.7) *for some $\gamma > \frac{1}{2}$ (recall that each g_A is of polynomial growth, and $w \in (0,1]$), plus one of the following five sets of hypotheses*:

(a) *We have $q' = q$ and* (K) *and X is continuous.*
(b) *We have* (K') *and* (10.3.5) *for some $w \in (0,1]$ and X is continuous.*
(c) *We have $q' = q$ and* (K-1) *and the functions g_A in* (10.3.3) *and* (10.3.4) *are bounded.*
(d) *We have $q' = q$ and* (K-r) *and* (10.3.6) *with $r \le s \le s' < 1$.*
(e) *We have* (K'-r) *and* (10.3.5) *and* (10.3.6) *with $r \le s \le s' < 1$.*

Then

$$\overline{V}'^n\big(\overline{F}(X), X\big) \stackrel{\mathcal{L}\text{-}s}{\Longrightarrow} \overline{V}'\big(\overline{F}(X), X\big)$$

(*stable convergence in law*), *with the following limiting process*:

(i) *When, for all t,y, the function $x \mapsto \overline{F}(t,y,x)$ is globally even, the process $\overline{V}'(\overline{F}(X), X) = \overline{U}'(\overline{F}'X), X)$ is defined on a very good filtered extension*

$(\widetilde{\Omega}, \widetilde{\mathcal{F}}, (\widetilde{\mathcal{F}}_t)_{t\geq 0}, \widetilde{\mathbb{P}})$ *of the initial space* $(\Omega, \mathcal{F}, (\mathcal{F}_t)_{t\geq 0}, \mathbb{P})$, *and conditionally on* $\mathcal{F}$ *is a continuous centered Gaussian process with independent increments satisfying*

$$\begin{aligned}
&\widetilde{\mathbb{E}}\big(\overline{U}'\big(\overline{F}(X), X\big)_t^j\, \overline{U}'(\overline{F}, X)_t^k \mid \mathcal{F}\big) \\
&\quad = \int_0^t \big(\rho_{c_s}\big(\overline{F}^j_{s,X_s} \overline{F}^k_{s,X_s}\big) - \rho_{c_s}\big(\overline{F}^j_{s,X_s}\big)\,\rho_{c_s}\big(\overline{F}^j_{s,X_s}\big)\big)\, ds. \qquad (10.3.9)
\end{aligned}$$

(ii) *Otherwise, we have*

$$\overline{V}'\big(\overline{F}(X), X\big) = \overline{U}'\big(\overline{F}(X), X\big) + \overline{A}\big(\overline{F}(X), X\big) + \overline{A}'\big(\overline{F}(X), X\big) + \overline{U}\big(\overline{F}(X), X\big),$$

where $\overline{U}'(\overline{F}(X), X)$ *is as in* (i), *except that* (10.3.9) *is replaced with*

$$\widetilde{\mathbb{E}}\big(\overline{U}'\big(\overline{F}(X), X\big)_t^j\, \overline{U}'(\overline{F}, X)_t^k \mid \mathcal{F}\big) = \int_0^t \overline{\gamma}_{\sigma_s}\big(\overline{F}^j_{s,X_s}, \overline{F}^k_{s,X_s}\big)\, ds,$$

and where, with $\widetilde{\sigma}_t$ *and* b'_t *as in* (5.3.2) *and* (5.3.3),

$$\begin{aligned}
\overline{A}\big(\overline{F}(X), X\big)_t^i &= \sum_{j=1}^d \int_0^t b_s^{\prime j}\, \rho_{c_s}\big(\partial_j \overline{F}^i_{s,X_s}\big)\, ds \\
\overline{A}'\big(\overline{F}(X), X\big)_t^i &= \sum_{j=1}^d \sum_{m,k=1}^{d'} \int_0^t \widetilde{\sigma}_s^{jkm}\, \widehat{\gamma}'_{\sigma_s}\big(\partial_j \overline{F}^i_{s,X_s}\big)^{mk}\, ds \\
\overline{U}\big(\overline{F}(X), X\big)_t^i &= \sum_{k=1}^{d'} \int_0^t \widehat{\gamma}_{\sigma_s}\big(\overline{F}^i_{s,X_s}\big)^k\, dW_s^k.
\end{aligned}$$

As for Theorems 5.3.5 and 5.3.6, when $x \mapsto \overline{F}(t, y, x)$ is even, (ii) reduces to (i).

Remark 10.3.3 Suppose that $\overline{F}(t, y, x) = g(y) f(x)$. Then

$$\begin{aligned}
V^n\big(\overline{F}(X), X\big)_t &= \sum_{i=1}^{[t/\Delta_n]} g(X_{(i-1)\Delta_n})\, \Delta_i^n V^n(f, X) \\
V'^n\big(\overline{F}(X), X\big)_t &= \sum_{i=1}^{[t/\Delta_n]} g(X_{(i-1)\Delta_n})\, \Delta_i^n V'^n(f, X).
\end{aligned}$$

Put otherwise, $V^n(\overline{F}(X), X)$ is the integral process, with respect to the process $V^n(f, X)$, of the piecewise constant process taking the value $g(X_{(i-1)\Delta_n})$ on each interval $I(n, i)$, and the same for $V'^n(\overline{F}(X), X)$.

This leads one to guess that the limit of these functionals is of the same type, that is $\int_0^t g(X_{s-})\, dV(g, X)_s$ and $\int_0^t g(X_{s-})\, dV'(g, X)_s$, under appropriate assumptions, of course: this is exactly what Theorems 7.3.3 and 7.2.2 say, in this product situation. Then one would also guess the same for the limit in the associated

CLT. And indeed, this is what happens in Theorem 10.3.2, that is $\overline{V}'(\overline{F}(X), X)_t = \int_0^t g(X_{s-})\, d\overline{V}'(f, X)_s$, as an elementary calculation based on (10.3.8) and on the form of $\overline{V}'(\overline{F}(X), X)$ shows.

On the other hand, for the limits of the CLT for the non-normalized functionals, we *do not have* $Z(\overline{F}(X), X)_t = \int_0^t g(X_{s-})\, dZ(f, X)_s$, because the derivative of g also enters the picture. In other words, the two Theorems 10.2.3 and 10.3.2 are indeed deeply different.

Remark 10.3.4 One could of course suppose that $\overline{F} = \overline{F}(\omega, t, y, x)$ depends on ω as well, in an adapted way. If the conditions (10.3.4)–(10.3.7) are uniform in ω, Theorem 10.3.2 still holds, with the same proof. However, we cannot think of any application for which this generalized setting would be useful.

Proof It is enough to prove (ii), and (i) is a particular case.

As we have seen, Theorems 5.3.5 and 5.3.6 are rather long to prove, and here we need only slight modifications of their proofs. Thus below we use all notation of those proofs and simply point out the changes which are needed for dealing with the present situation.

Step 1) The localization lemma 5.3.12 holds in this context, so we can replace (K), (K'), (K-r) or (K'-r) by the strengthened assumptions (SK), (SK'), (SK-r) or (SK'-r), respectively, and which are

Assumption (SK) We have (K) and the processes b, $\widetilde{b}$, σ, X, $\widetilde{\sigma}$, a and $\widetilde{a}$ are bounded, and $\|\delta(\omega, t, z)\| \le \Gamma(z)$ with Γ bounded and $\int \Gamma(z)^2 \lambda(dz) < \infty$.

Assumption (SK-r) We have (SK) and also $\int \Gamma(z)^r \lambda(dz) < \infty$.

Assumption (SK') We have (SK) and the process c^{-1} is bounded.

Assumption (SK'-r) We have (SK-r) and the process c^{-1} is bounded.

It is of course enough to prove the convergence results in restriction to any finite interval, and X is bounded: so the argument (t, y) in $\overline{F}(t, y, x)$ remains in a compact and the values of $\overline{F}$ outside this compact are irrelevant. In other words, we may assume that in all properties (10.3.3)–(10.3.7) the functions g_A is replaced by a single function g, not depending on A.

Note that if (10.3.6) holds for some $s \le s'$, it also holds with the exponents r, r' if $r \le s$ and $r' \ge s'$. Then, a glance at the proof of Lemma 5.3.13 shows that this proof does not change if we substitute $V'^n(f, X)$ with $V'^n(\overline{F}(X), X)$, implying that we need only to prove the CLT for $V'^n(\overline{F}(X), X')$ instead of $V'^n(\overline{F}(X), X)$. A *warning* should however be issued here: the second argument of $\overline{F}$ should remain X, that is, those processes are

$$V'^n\big(\overline{F}(X), X'\big)_t = \Delta_n \sum_{i=1}^{[t/\Delta_n]} \overline{F}\big((i-1)\Delta_n, X_{(i-1)\Delta_n}, \Delta_i^n X'/\sqrt{\Delta_n}\big).$$

Step 2) As in (5.3.1), we have $\overline{V}'^n(\overline{F}(X), X') = Y^n + A^n$, where

$$Y_t^n = \sqrt{\Delta_n} \sum_{i=1}^{[t/\Delta_n]} \left(\zeta_i^n - \mathbb{E}\left(\zeta_i^n \mid \mathcal{F}_{(i-1)\Delta_n}\right)\right)$$

$$\zeta_i^n = \overline{F}_{(i-1)\Delta_n, X_{(i-1)\Delta_n}}\left(\Delta_i^n X' / \sqrt{\Delta_n}\right)$$

$$A_t^n = \frac{1}{\sqrt{\Delta_n}} \left(\Delta_n \sum_{i=1}^{[t/\Delta_n]} \mathbb{E}\left(\zeta_i^n \mid \mathcal{F}_{(i-1)\Delta_n}\right) - \int_0^t \rho_{c_s}(\overline{F}_{s,X_s})\, ds\right).$$

When $\overline{F}$ is bounded, and since as a function of the third argument x it satisfies (5.2.14), uniformly in (t, y) by Step 1, we can reproduce the proof of Lemma 5.2.5 to obtain that $Y^n - \overline{U}'^n(\overline{F}) \stackrel{\text{u.c.p.}}{\Longrightarrow} 0$, where, with the notation $\beta_i^n = \sigma_{(i-1)\Delta_n} \Delta_i^n W / \sqrt{\Delta_n}$,

$$\overline{U}'^n(\overline{F})_t = \sqrt{\Delta_n} \sum_{i=1}^{[t/\Delta_n]} \left(\overline{F}_{(i-1)\Delta_n, X_{(i-1)\Delta_n}}(\beta_i^n) - \rho_{\sigma_{(i-1)\Delta_n}}\left(\overline{F}_{(i-1)\Delta_n, X_{(i-1)\Delta_n}}\right)\right)$$

(note that (5.2.10) holds here because of (SK-1)). When $\overline{F}$ is not bounded, the same property $Y^n - \overline{U}'^n(\overline{F}) \stackrel{\text{u.c.p.}}{\Longrightarrow} 0$ (under our standing assumptions on $\overline{F}$) is deduced from the bounded case exactly as in Lemma 5.2.8, upon using again the fact that the estimates in (5.3.21) and (5.3.22) are uniform in (t, y).

At this stage, we apply Theorem 4.2.1 with $\theta = \sigma$ and $Y = X$ and $\Phi(y) = y(1)$ and $u_n = \Delta_n$ to obtain that $\overline{U}'^n(\overline{F}) \stackrel{\mathcal{L}\text{-s}}{\Longrightarrow} \overline{U}(\overline{F}(X), X) + \overline{U}'(\overline{F}(X), X)$, as defined in (ii). From what precedes, we deduce

$$Y^n \stackrel{\mathcal{L}\text{-s}}{\Longrightarrow} \overline{U}\left(\overline{F}(X), X\right) + \overline{U}'\left(\overline{F}(X), X\right).$$

Step 3) In view of the decomposition $\overline{V}'^n(\overline{F}(X), X') = Y^n + A^n$, it remains to prove that

$$A^n \stackrel{\text{u.c.p.}}{\Longrightarrow} \overline{A}\left(\overline{F}(X), X\right) + \overline{A}'\left(\overline{F}(X), X\right).$$

It is enough to prove this property for each component, so we can and will assume below $\overline{F}$ to be one-dimensional. In this case, either $q' = 1$ (under the hypotheses (a,c,d) here), or $q' = 0$ (under the hypotheses (b,e) here). We can thus suppose that we are in case (a) of (5.3.20), page 152, for X' (when $q' = 1$) or in case (b) of (5.3.20) for X' (when $q' = 0$), and each function $\overline{F}(t, y, .)$ satisfies (5.3.21) and (5.3.22) uniformly in (t, y).

Step 4) Let us introduce the notation, similar to (5.3.23), page 152:

$$A^n(1)_t = \sqrt{\Delta_n} \sum_{i=1}^{[t/\Delta_n]} \mathbb{E}\left(\overline{F}_{(i-1)\Delta_n, X_{(i-1)\Delta_n}}\left(\frac{\Delta_i^n X'}{\sqrt{\Delta_n}}\right)\right.$$

$$- \overline{F}_{(i-1)\Delta_n, X_{(i-1)\Delta_n}}\big(\beta_i^n\big) \mid \mathcal{F}_{(i-1)\Delta_n}\Big)$$

$$A^n(2)_t = \frac{1}{\sqrt{\Delta_n}} \left(\Delta_n \sum_{i=1}^{[t/\Delta_n]} \rho_{c_{i-1)\Delta_n}}(\overline{F}_{(i-1)\Delta_n, X_{(i-1)\Delta_n}}) - \int_0^t \rho_{c_s}(\overline{F}_{s,X_s})\, ds \right),$$

we are left to prove

$$A^n(1) \stackrel{\text{u.c.p.}}{\Longrightarrow} \overline{A}\big(\overline{F}(X), X\big) + \overline{A}'\big(\overline{F}(X), X\big), \tag{10.3.10}$$

$$A^n(2) \stackrel{\text{u.c.p.}}{\Longrightarrow} 0. \tag{10.3.11}$$

For (10.3.10) one can reproduce the proof given in Part C of Sect. 5.3.3, pages 154–160, with f substituted with $\overline{F}_{(i-1)\Delta_n, X_{(i-1)\Delta_n}}$. We also substitute b_s with b'_s (because we use the process X', instead of X, supposed to be continuous in Sect. 5.3.3 but not necessarily here), and we use again the uniformity in (t, y) of all estimates for $\overline{F}_{t,y}$. The only (small) changes occur in the proofs of Lemmas 5.3.18 and 5.3.19, where the right sides of (5.3.51) and (5.3.52) are respectively

$$\Delta_n \sum_{j=1}^d b'^j_{(i-1)\Delta_n}\, \rho_{c_{(i-1)\Delta_n}}\big(\partial_j \overline{F}^l_{(i-1)\Delta_n, X_{(i-1)\Delta_n}}\big), \quad \text{and}$$

$$\Delta_n \sum_{j=1}^d \sum_{k,m=1}^r \widetilde{\sigma}^{jkm}_{(i-1)\Delta_n}\, \widehat{\gamma}'_{\sigma_{(i-1)\Delta_n}}\big(\partial_j \overline{F}^l_{(i-1)\Delta_n, X_{(i-1)\Delta_n}}\big)^{mk}.$$

In both cases those variables are Riemann approximations for the integrals of the two càdlàg processes

$$\sum_{j=1}^d b'^j_s\, \rho_{c_s}\big(\partial_j \overline{F}^l_{(i-1)\Delta_n, X_{(i-1)\Delta_n}}\big)$$

$$\sum_{j=1}^d \sum_{k,m=1}^r \widetilde{\sigma}^{jkm}_s\, \widehat{\gamma}'_{\sigma_s}\big(\partial_j \overline{F}^l_{(i-1)\Delta_n, X_{(i-1)\Delta_n}}\big)^{mk},$$

and (10.3.10) follows.

Step 5) It remains to prove (10.3.11). With the notation $\overline{\psi}_t(\alpha, y) = \rho_{\alpha\alpha^*}(\overline{F}_{t,y})$ when α is a $d \times d'$ matrix, we have

$$A^n(2)_t = \frac{1}{\sqrt{\Delta_n}} \left(\Delta_n \sum_{i=1}^{[t/\Delta_n]} \overline{\psi}_{(i-1)\Delta_n}(\sigma_{(i-1)\Delta_n}, X_{i-1)\Delta_n}) - \int_0^t \overline{\psi}_s(\sigma_s, X_s)\, ds \right)$$

$$= -\overline{\eta}^n_t - \sum_{i=1}^{[t/\Delta_n]} \zeta^n_i - \sum_{i=1}^{[t/\Delta_n]} \zeta'^n_i,$$

where

$$
\begin{aligned}
\overline{\eta}_t^n &= \frac{1}{\sqrt{\Delta_n}} \int_{[t/\Delta_n]\Delta_n}^t \overline{\psi}_s(\sigma_s, X_s)\, ds \\
\zeta_i^n &= \frac{1}{\sqrt{\Delta_n}} \int_{I(n,i)} \big(\overline{\psi}_{(i-1)\Delta_n}(\sigma_u, X_u) - \overline{\psi}_{(i-1)\Delta_n}(\sigma_{(i-1)\Delta_n}, X_{(i-1)\Delta_n})\big)\, du \\
\zeta_i'^n &= \frac{1}{\sqrt{\Delta_n}} \int_{I(n,i)} \big(\overline{\psi}_u(\sigma_u, X_u) - \overline{\psi}_{(i-1)\Delta_n}(\sigma_u, X_u)\big)\, du.
\end{aligned}
$$

Exactly as for (5.3.27), the function $\overline{\psi}_t$ is C^1 on the set $\mathcal{M}_A \times \mathbb{R}^d$ or $\mathcal{M}'_A \times \mathbb{R}^d$, according to whether we are in case (a) or (b) of (5.3.20), and with $\nabla\overline{\psi}_t$ denoting the family of all first partial derivatives and $Z_t = (\sigma_t, X_t)$ (a $d \times d' + d$-dimensional bounded process), we have

$$
\left.
\begin{array}{l}
\big|\overline{\psi}_s(Z_t)\big| + \big\|\nabla\overline{\psi}_s(Z_t)\big\| \le K \\
\big|\overline{\psi}_s(Z_t) - \overline{\psi}_s(Z_s)\big| \le K\|Z_t - Z_s\| \\
\big|\overline{\psi}_s(Z_t) - \psi_s(Z_s) - \nabla\psi_s(Z_s)(Z_t - Z_s)\big| \le \Psi\big(\|Z_t - Z_s|\big)\,\big(\|Z_t - Z_s\|\big) \\
\big|\overline{\psi}_t(Z_t) - \overline{\psi}_s(Z_t)\big| \le K|t-s|^\gamma
\end{array}
\right\}
\tag{10.3.12}
$$

for some constant K and some increasing function Ψ on $\mathbb{R}_+$, continuous and null at 0.

The first and last properties in (10.3.12) yield $|\overline{\eta}_t^n| \le K\sqrt{\Delta_n}$ and $|\zeta_i'^n| \le K\Delta_n^{1/2+\gamma}$, hence $\overline{\eta}_t^n \stackrel{\text{u.c.p.}}{\Longrightarrow} 0$ and $\sum_{i=1}^{[t/\Delta_n]} \zeta_i'^n \stackrel{\text{u.c.p.}}{\Longrightarrow} 0$, because $\gamma > \frac{1}{2}$ for the latter. It thus remains to prove that $\sum_{i=1}^{[t/\Delta_n]} \zeta_i^n \stackrel{\text{u.c.p.}}{\Longrightarrow} 0$. But for this we observe that, since (SK-1) holds, the process Z is an Itô semimartingale satisfying (SH). Therefore the argument of Part B of Sect. 5.3.3, see page 153, which is performed for the Itô semimartingale σ satisfying (SH) and for the function ψ satisfying (5.3.27), works identically here for Z and the function $\overline{\psi}_t$ which satisfies (10.3.12) (note that ζ_i^n here is the analogue of $\eta_i^n + \eta_i'^n$ there). Thus we get $\sum_{i=1}^{[t/\Delta_n]} \zeta_i^n \stackrel{\text{u.c.p.}}{\Longrightarrow} 0$, hence (10.3.11), and the proof is complete. □

10.4 Application to Parametric Estimation

We continue here our (rather sketchy) description of the application of the functionals "with random weights" to the estimation of a parameter, which was begun in Chap. 7. We recall the setting: the d-dimensional canonical continuous process X is, under the probability measure $\mathbb{P}_\theta$ (where θ belongs to a compact subset Θ of $\mathbb{R}^q$), a diffusion of the form

$$
X_t = x + \int_0^t b(s, X_s)\, ds + \int_0^t \sigma(\theta, s, X_s)\, dW_s.
$$

Here, the initial value $x \in \mathbb{R}^d$ is known, as well as the function $c(\theta, s, x) = \sigma(\theta, s, x)\sigma(\theta, s, x)^*$, whereas the function b is not necessarily known. The process X is observed at times $i\Delta_n$, within a *fixed* time interval $[0, T]$.

We have introduced a minimum contrast estimator for estimating θ as follows. We choose a smooth enough function g on $\mathcal{M}^+_{d\times d} \times \mathbb{R}^d$ and we set

$$\Phi_n(\theta) = \Delta_n \sum_{i=1}^{[T/\Delta_n]} g\left(c\big(\theta, (i-1)\Delta_n, X_{(i-1)\Delta_n}\big), \frac{\Delta_i^n X}{\sqrt{\Delta_n}}\right).$$

Then we set

$$\widehat{\theta}_n = \text{any value realizing the minimum of } \theta \mapsto \Phi_n(\theta).$$

Here we want to study the asymptotic behavior of the estimation error $\widehat{\theta}_n - \theta$ under $\mathbb{P}_\theta$, and for this we need assumptions on the coefficient c and on the chosen functions g which are stronger than in Chap. 7. Namely, we assume the following (the smoothness of g could be slightly relaxed but, since g is at the choice of the statistician, it is innocuous to use assumptions which may be too strong):

1 – The function $(t, x) \mapsto b(t, x)$ is locally bounded, continuous in x and either càdlàg or càglàd in t.

2 – The function $\theta \mapsto \sigma(\theta, t, x)$ is C^2, and the functions $\sigma(\theta, t, x)$ and $\frac{\partial}{\partial\theta^j}\sigma(\theta, t, x)$ for $j = 1, \dots, q$ are locally Hölder in t with some Hölder exponent $\gamma > \frac{1}{2}$, and C^2 in x.

3 – The function g is C^3 in both argument with third derivatives of at most polynomial growth, and $x \mapsto g(y, x)$ is globally even on $\mathbb{R}^d$, and it satisfies slightly more than (7.4.1), namely the C^3 function $a \mapsto G(a, a') = \int g(a, x)\rho_{a'}(dx)$ on $\mathcal{M}^+_{d\times d}$ satisfies

(i) $a = a'$ is the unique minimum of $G(., a')$
(ii) the $d^2 \times d^2$ symmetrical matrix $\partial^2 G(a, a')/\partial a^{ij}\partial a^{i'j'}$ is definite positive and denoted by $H(a)^{ij,i'j'}$, when $a = a'$. (10.4.1)

As seen in Theorem 7.4.1 we have the consistency, that is

$$\widehat{\theta}_n \xrightarrow{\mathbb{P}_{\theta_0}} \theta_0 \text{ on the set } \Omega_T^{\theta_0} = \left\{\int_0^T \big\|c(\theta, s, X_s) - c(\theta_0, s, X_s)\big\|\, ds > 0\ \forall \theta \neq \theta_0\right\}.$$

Saying that $\omega \in \Omega_T^{\theta_0}$ is saying that the observed path satisfies a "global" identifiability condition. This is not enough for having a distributional result about the error $\widehat{\theta}_n - \theta_0$, and for this we also need a "local" identifiability condition, expressed by the following:

$$\Omega_T^{\prime\theta_0} = \left\{\int_0^T \big\|\nabla_\theta\, c(\theta_0, s, X_s)\big\|\, ds > 0\right\}, \tag{10.4.2}$$

where $\nabla_\theta c$ denotes the family of all first partial derivatives with respect to θ (for all components of c as well). Note that if $\omega \in \Omega_T^{\prime\theta_0}$ then the integrals in the definition of $\Omega_T^{\theta_0}$ are positive for $\theta \neq \theta_0$ which are close enough to θ_0. In a sense, the distinction between global and local identifiability looks like the distinction between (i) and (ii) in (10.4.1), corresponding to $a = a'$ being a global or a local minimum of $G(., a')$.

We need some more notation. The time T is fixed, and we introduce the following $q \times q$ (random) symmetric matrices with components

$$\begin{aligned} A(\theta)^{lm} = & \sum_{j,k,j',k'=1}^{d} \int_0^T ds \int \frac{\partial}{\partial a^{jk}}\, g\big(c(\theta,s,X_s),x\big)\, \frac{\partial}{\partial a^{j'k'}}\, g\big(c(\theta,s,X_s),x\big) \\ & \times \frac{\partial}{\partial \theta^l}\, c(\theta,s,X_s)^{jk}\, \frac{\partial}{\partial \theta^m}\, c(\theta,s,X_s)^{j'k'}\, \rho_{c(\theta,s,X_s)}(dx) \end{aligned} \tag{10.4.3}$$

$$\begin{aligned} B(\theta)^{lm} = & \sum_{j,k,j',k'=1}^{d} \int_0^T ds \int \frac{\partial^2}{\partial a^{jk}\,\partial a^{j'k'}}\, g\big(c(\theta,s,X_s),x\big) \\ & \times \frac{\partial}{\partial \theta^l}\, c(\theta,s,X_s)^{jk}\, \frac{\partial}{\partial \theta^m}\, c(\theta,s,X_s)^{j'k'}\, \rho_{c(\theta,s,X_s)}(dx). \end{aligned} \tag{10.4.4}$$

Observe that, by (10.4.1) and (10.4.2), the symmetric matrix $B(\theta_0)$ is nonnegative, and also invertible on the set $\Omega_T^{\prime\theta_0}$. We also consider the variables

$$\begin{aligned} A_n(\theta)^{lm} = \Delta_n & \sum_{i=1}^{[T/\Delta_n]} \sum_{j,k,j',k'=1}^{d} \frac{\partial}{\partial a^{jk}}\, g\Big(c\big(\theta,(i-1)\Delta_n, X_{(i-1)\Delta_n}\big), \frac{\Delta_i^n X}{\sqrt{\Delta_n}}\Big) \\ & \times \frac{\partial}{\partial a^{j'k'}}\, g\Big(c\big(\theta,(i-1)\Delta_n, X_{(i-1)\Delta_n}\big), \frac{\Delta_i^n X}{\sqrt{\Delta_n}}\Big) \\ & \times \frac{\partial}{\partial \theta^l}\, c\big(\theta,(i-1)\Delta_n, X_{(i-1)\Delta_n}\big)^{jk}\, \frac{\partial}{\partial \theta^m}\, c\big(\theta,(i-1)\Delta_n, X_{(i-1)\Delta_n}\big)^{j'k'} \\ B_n(\theta)^{lm} = \Delta_n & \sum_{i=1}^{[T/\Delta_n]} \sum_{j,k,j',k'=1}^{d} \\ & \times \frac{\partial^2}{\partial a^{jk}\,\partial a^{j'k'}}\, g\Big(c\big(\theta,(i-1)\Delta_n, X_{(i-1)\Delta_n}\big), \frac{\Delta_i^n X}{\sqrt{\Delta_n}}\Big) \\ & \times \frac{\partial}{\partial \theta^l}\, c\big(\theta,(i-1)\Delta_n, X_{(i-1)\Delta_n}\big)^{jk}\, \frac{\partial}{\partial \theta^m}\, c\big(\theta,(i-1)\Delta_n, X_{(i-1)\Delta_n}\big)^{j'k'}. \end{aligned}$$

The variable $A_n(\theta)$ takes its values in the set $\mathcal{M}_{q\times q}^{+}$; when it is invertible, that is it belongs to $\mathcal{M}_{q\times q}^{++}$, we need to take below a version of the square-root of its inverse, written $A_n(\theta)^{-1/2}$. We need to do that in a "continuous" way, that is when we write $A_n(\theta)^{-1/2}$ we implicitly use a map $a \mapsto a^{-1/2}$ which is continuous on $\mathcal{M}_{q\times q}^{++}$ (for example if $q = 1$ then $a^{-1/2}$ is always $\sqrt{a}$, or always $-\sqrt{a}$).

Theorem 10.4.1 *In the previous setting and under the previous assumptions, assume further that θ_0 belongs to the interior of Θ. Then, under the measure $\mathbb{P}_{\theta_0}$, we have:*

(i) *The sequence $\frac{1}{\sqrt{\Delta_n}}(\widehat{\theta}_n - \theta_0)$ converges stably in law, in restriction to the set $\Omega_T^{\theta_0} \cap \Omega_T'^{\theta_0}$, to a variable defined on an extension of the space and which, conditionally on $\mathcal{F}$, is centered Gaussian with covariance matrix*

$$B(\theta_0)^{-1}\, A(\theta_0)\, B(\theta_0)^{-1} \tag{10.4.5}$$

where $A(\theta)$ and $B(\theta)$ are defined by (10.4.3) *and* (10.4.4).

(ii) *In restriction to the set $\Omega_T^{\theta_0} \cap \Omega_T'^{\theta_0} \cap \{A(\theta_0)$ is invertible$\}$, the symmetric nonnegative matrix $A_n(\widehat{\theta}_n)$ is invertible for all n large enough, and the standardized sequence*

$$\frac{1}{\sqrt{\Delta_n}}\, A_n(\widehat{\theta}_n)^{-1/2}\, B_n(\widehat{\theta}_n)\,(\widehat{\theta}_n - \theta_0)$$

(where $A_n(\widehat{\theta}_n)^{-1/2}$ is chosen arbitrarily when $A_n(\widehat{\theta}_n)$ is not invertible) converges stably in law to a variable defined on an extension of the space and which, conditionally on $\mathcal{F}$, is centered Gaussian with the identity matrix as its covariance.

Proof For simplicity we write $\mathbb{P} = \mathbb{P}_{\theta_0}$. Our assumptions on b and σ yield that X satisfies Assumption (K).

1) We begin with some preliminaries. We introduce the two following multidimensional functions on $\mathbb{R}_+ \times \mathbb{R}^d \times \mathbb{R}^d$, depending on θ:

$$\overline{F}_\theta(t,y,x)^l = \sum_{j,k=1}^d \frac{\partial}{\partial a^{jk}}\, g\big(c(\theta,t,y),x\big)\, \frac{\partial}{\partial \theta^l}\, c^{jk}(\theta,t,y)$$

$$\begin{aligned}\overline{G}_\theta(t,y,x)^{lm} = &\sum_{j,k,j',k'=1}^d \frac{\partial^2}{\partial a^{jk}\,\partial a^{j'k'}}\, g\big(c(\theta,t,y),x\big)\\ &\times \frac{\partial}{\partial \theta^l}\, c^{jk}(\theta,t,y) \frac{\partial}{\partial \theta^m}\, c^{j',k'}(\theta,t,y)\\ &+ \sum_{j,k=1}^d \frac{\partial}{\partial a^{jk}}\, g\big(c(\theta,t,y),x\big)\, \frac{\partial^2}{\partial \theta^l\,\partial \theta^m}\, c^{jk}(\theta,t,y),\end{aligned}$$

where l,l' range between 1 and q. With this notation we see that the first two partial derivatives of the contrast function $\Phi_n(\theta)$, which is clearly C^2, are

$$\frac{\partial}{\partial \theta^l}\, \Phi_n(\theta) = V'^n\big(\overline{F}^l_\theta(X),X\big)_T, \qquad \frac{\partial^2}{\partial \theta^l\,\partial \theta^m}\, \Phi_n(\theta) = V'^n\big(\overline{G}^{lm}_\theta(X),X\big)_T.$$

In the proof of Theorem 7.4.1, see (7.4.3), we have proved that $\Phi_n(\theta)$ converges in probability, uniformly in θ, to

$$\Phi(\theta) = \int_0^T ds \int g\big(c(\theta, s, X_s), x\big)\, \rho_{c(\theta_0,s,X_s)}(dx).$$

Note that Φ is C^2, with

$$\frac{\partial}{\partial\theta^l}\Phi(\theta) = \int_0^T ds \int \overline{F}_\theta(s, X_s, x)^l\, \rho_{c(\theta_0,s,X_s)}(dx)$$
$$\frac{\partial^2}{\partial\theta^l\,\partial\theta^m}\Phi(\theta) = \int_0^T ds \int \overline{G}_\theta(s, X_s, x)^{lm}\, \rho_{c(\theta_0,s,X_s)}(dx).$$

By (10.4.1), for all (s, y), the function $\theta \mapsto \int g(c(\theta, t, y), x)\rho_{c(\theta_0,t,y)}(dx)$ reaches its minimum at $\theta = \theta_0$, hence with the notation H of (10.4.1) we have for all t, y:

$$\begin{aligned} &\int \overline{F}_{\theta_0}(t, y, x)^l\, \rho_{c(\theta_0,t,y)}(dx) = 0 \\ &\int \overline{G}_{\theta_0}(t, y, x)^{lm}\, \rho_{c(\theta_0,t,y)}(dx) \\ &\quad = \sum_{j,k,j',k'=1}^{d} H\big(c(\theta_0, t, y)\big)^{jk,j'k'} \frac{\partial}{\partial\theta^l} c^{jk}(\theta, t, y) \frac{\partial}{\partial\theta^m} c^{j',k'}(\theta, t, y). \end{aligned} \tag{10.4.6}$$

It follows in particular that, by comparing with (10.4.4),

$$\frac{\partial^2}{\partial\theta^l\,\partial\theta^m}\Phi(\theta_0) = B(\theta_0)^{lm}. \tag{10.4.7}$$

2) Now we give some limit theorems. First, the function $\overline{G}_\theta^{lm}$ satisfies Assumption 7.2.1, and is continuous in θ, so we can reproduce the proof of the uniform convergence (7.4.3) to obtain that

$$\sup_\theta \left| \frac{\partial^2}{\partial\theta^l\,\partial\theta^m}\Phi_n(\theta) - \frac{\partial^2}{\partial\theta^l\,\partial\theta^m}\Phi(\theta)\right| \xrightarrow{\mathbb{P}} 0.$$

Since $\widehat{\theta}_n \xrightarrow{\mathbb{P}} \theta_0$ on $\Omega_T^{\theta_0}$, we deduce that

$$\frac{\partial^2}{\partial\theta^l\,\partial\theta^m}\Phi_n(\widehat{\theta}_n) \xrightarrow{\mathbb{P}} B(\theta_0)^{lm} \quad \text{on } \Omega_T^{\theta_0}. \tag{10.4.8}$$

In exactly the same way, we also obtain

$$A_n(\widehat{\theta}_n)^{lm} \xrightarrow{\mathbb{P}} A(\theta_0)^{lm}, \qquad B_n(\widehat{\theta}_n)^{lm} \xrightarrow{\mathbb{P}} B(\theta_0)^{lm} \quad \text{on } \Omega_T^{\theta_0}. \tag{10.4.9}$$

Second, the function $\overline{F}_{\theta_0}$ satisfies (10.3.4), (10.3.5) with $B=\emptyset$, and (10.3.7) with the Hölder exponent $\gamma > \frac{1}{2}$ of σ and its derivatives in θ, considered as functions of time. We can then apply Theorem 10.3.2, case (a), and with a vanishing centering term in (10.3.1) because of (10.4.6). We thus obtain

$$\left(\frac{1}{\sqrt{\Delta_n}}\,\frac{\partial}{\partial\theta^l}\,\Phi_n(\theta_0)\right)_{1\le j\le q} \stackrel{\mathcal{L}\text{-s}}{\longrightarrow} Z, \tag{10.4.10}$$

where Z is defined on an extension on the space $(\Omega,\mathcal{F},\mathbb{P})$ and, conditionally on $\mathcal{F}$, is centered Gaussian with covariance

$$\begin{aligned}\widetilde{\mathbb{E}}(Z^l\,Z^m\mid\mathcal{F}) &= \int_0^T ds\int \overline{F}_{\theta_0}(s,X_s,x)^l\,\overline{F}_{\theta_0}(s,X_s,x)^m\,\rho_{c(\theta_0,s,X_s)}(dx)\\ &= A(\theta_0)^{lm}.\end{aligned}$$

3) We can now start solving the statistical problem. The—well established—idea consists in putting together two facts, both relying upon the smoothness of Φ_n:

(i) on the set Ω_n where $\widehat{\theta}_n$ is a minimum of Φ_n and belongs to the interior of Θ, we have $\frac{\partial}{\partial\theta^j}\,\Phi_n(\widehat{\theta}_n)=0$.

(ii) we can use a Taylor expansion to evaluate the difference $\frac{\partial}{\partial\theta^j}\,\Phi_n(\widehat{\theta}_n) - \frac{\partial}{\partial\theta^j}\,\Phi_n(\theta_0)$, on the set Ω'_n on which the ball centered at θ_0 and with radius $\|\widehat{\theta}_n-\theta_0\|$ is contained in Θ.

In other words, on the set $\Omega_n\cap\Omega'_n$, and with vector notation, we have

$$0=\frac{\partial}{\partial\theta^j}\,\Phi_n(\widehat{\theta}_n)=\frac{\partial}{\partial\theta^j}\,\Phi_n(\theta_0)+\sum_{l=1}^q\frac{\partial^2}{\partial\theta^j\,\partial\theta^l}\,\Phi_n(\widetilde{\theta}_n)\,\big(\widehat{\theta}^l_n-\theta^l_0\big), \tag{10.4.11}$$

where $\widetilde{\theta}_n$ is random, with $\|\widetilde{\theta}_n-\theta_0\|\le\|\widehat{\theta}_n-\theta_0\|$. Since θ_0 is in the interior of Θ, one readily checks that $\mathbb{P}(\Omega^{\theta_0}_T\backslash(\Omega_n\cap\Omega'_n))\to 0$. Therefore we deduce from (10.4.10) and (10.4.11) that the random vector with components $\frac{1}{\sqrt{\Delta_n}}\sum_{l=1}^q\frac{\partial^2}{\partial\theta^j\,\partial\theta^l}\,\Phi_n(\widetilde{\theta}_n)\times(\widehat{\theta}^l_n-\theta^l_0)$ converges stably in law to Z, in restriction to $\Omega^{\theta_0}_T$.

On the other hand, (10.4.7), (10.4.8) and $\widehat{\theta}_n \stackrel{\mathbb{P}}{\longrightarrow} \theta_0$ on the set $\Omega^{\theta_0}_T$ imply that $\frac{\partial^2}{\partial\theta^j\,\partial\theta^l}\,\Phi_n(\widetilde{\theta}_n) \stackrel{\mathbb{P}}{\longrightarrow} B(\theta_0)^{jl}$ on the set $\Omega^{\theta_0}_T$. Since further $B(\theta_0)$ is invertible on $\Omega'^{\theta_0}_T$, we deduce that

$$\frac{1}{\sqrt{\Delta_n}}\,(\widehat{\theta}_n-\theta_0) \stackrel{\mathcal{L}\text{-s}}{\longrightarrow} B(\theta_0)^{-1}\,Z \quad \text{in restriction to the set } \Omega^{\theta_0}_T\cap\Omega'^{\theta_0}_T. \tag{10.4.12}$$

The variable on the right side above is $\mathcal{F}$-conditionally centered Gaussian with covariance given by (10.4.5), and (i) is proved.

4) It remains to prove (ii), whose first claim readily follows from (10.4.9). In view of (10.4.12), and exactly as in the proof of Theorem 5.6.1 for example, the second claim follows from the fact that $A_n(\widehat{\theta}_n)^{-1/2}\,B_n(\widehat{\theta}_n) \stackrel{\mathbb{P}}{\longrightarrow} A(\theta_0)^{-1/2}\,B(\theta_0)$ on

the set $\Omega_T^{\theta_0} \cap \{A(\theta_0) \text{ is invertible}\}$, which in turn follows from (10.4.9) and from the "continuous" choice made for $a^{-1/2}$ when $a \in \mathcal{M}_{q\times q}^{++}$. □

Example 10.4.2 Let us come back to Example 7.4.2, with the function $g(a,x) = \sum_{i,j=1}^d (a^{ij} - x^i x^j)^2$. Then (10.4.1) is satisfied (the second derivative of G is the identity matrix). Moreover we have

$$\begin{aligned} A(\theta)^{lm} &= 4 \sum_{j,k,j',k'=1}^{d} \int_0^T \big(c^{jj'}c^{kk'} + c^{jk'}c^{j'k}\big)(\theta,s,X_s) \\ &\quad \times \frac{\partial}{\partial\theta^l} c(\theta,s,X_s)^{jk} \frac{\partial}{\partial\theta^m} c(\theta,s,X_s)^{j'k'}\, ds \\ B(\theta)^{lm} &= 2 \sum_{j,k=1}^{d} \int_0^T \frac{\partial}{\partial\theta^l} c(\theta,s,X_s)^{jk} \frac{\partial}{\partial\theta^m} c(\theta,s,X_s)^{jk}\, ds. \end{aligned}$$

Example 10.4.3 Now we turn to Example 7.4.3, assuming that $c(\theta,t,x) \in \mathcal{M}_{d\times d}^{++}$ identically, and the function g is $g(a,x) = \log\det a + x^* a^{-1} x$. In this case it turns out that

$$A(\theta) = 2B(\theta), \qquad B(\theta)^{lm} = \int_0^T \operatorname{trace}\Big(\frac{\partial c}{\partial\theta^l} c^{-1} \frac{\partial c}{\partial\theta^m} c^{-1}\Big)(\theta,s,X_s)\, ds,$$

see Genon-Catalot and Jacod [37] for a proof. Note that this form of $B(\theta)$ implies that g satisfies (10.4.1).

Under the assumption that c is invertible, it can be checked that the value $B(\theta_0)^{-1}A(\theta_0)B(\theta_0)^{-1}$ with A, B as above realizes the "minimum" of all possible asymptotic covariances of minimum contrast estimators, in the sense that if another sequence of estimators $\widehat{\theta'_n}$ is associated with another function g' giving rise to the matrices $A'(\theta_0)$ and $B'(\theta_0)$, then the difference $B'(\theta_0)^{-1}A'(\theta_0)B'(\theta_0)^{-1} - B(\theta_0)^{-1}A(\theta_0)B(\theta_0)^{-1}$ is always nonnegative. Thus the contrast associated with the present function g *is best* in the asymptotic sense, provided of course we deal with a model in which c is everywhere invertible.

Bibliographical Notes

The results of this chapter and the corresponding Laws of Large Numbers of Chap. 7 seem to have appeared for the first time in the statistical literature for parametric inference for Markov diffusions, as described in Sects. 7.4 and 10.4. For example, a special case of the Law of Large Numbers (Theorem 7.3.6) and the Central Limit Theorem 10.3.2 can be found in the paper [29] of Dohnal, and a more general version, in the continuous Markov case again, may be found in the paper [37] by Genon-Catalot and Jacod. A version of the results when X is a continuous Itô semimartingale is in [53].

The present form of these two chapters is essentially taken from the thesis [27] of Diop, see also [28].

Chapter 11
The Central Limit Theorem for Functions of a Finite Number of Increments

In this chapter we give the Central Limit Theorems associated with the Laws of Large Numbers of Chap. 8, when the number of increments in the test function is fixed, the case when the number of increments increases to infinity being postponed to the next chapter. This is again a rather straightforward extension of Chap. 5.

Only regular discretization schemes are considered. The d-dimensional Itô semimartingale X has the Grigelionis decomposition

$$X_t = X_0 + \int_0^t b_s\,ds + \int_0^t \sigma_s\,dW_s + (\delta 1_{\{\|\delta\|\le 1\}}) \star (\mathfrak{p} - \mathfrak{q})_t + (\delta 1_{\{\|\delta\|>1\}}) \star \mathfrak{p}_t, \quad (11.0.1)$$

where W is a d'-dimensional Wiener process and $\mathfrak{p}$ is a Poisson measure with compensator $\mathfrak{q}(dt,dz) = dt \otimes \lambda(dz)$, and $c = \sigma\sigma^\star$. We also assume at least Assumption (H), that is

Assumption (H) In (11.0.1), b_t is locally bounded and σ_t is càdlàg, and $\|\delta(\omega,t,z)\| \wedge 1 \le \Gamma_n(z)$ for all (ω,t,z) with $t \le \tau_n(\omega)$, where (τ_n) is a localizing sequence of stopping times and each function Γ_n satisfies $\int \Gamma_n(z)^2 \lambda(dz) < \infty$.

11.1 Functionals of Non-normalized Increments

In this section we consider the functionals

$$\begin{aligned} V^n(F,X)_t &= \sum_{i=1}^{[t/\Delta_n]-k+1} F\big(\Delta_i^n X, \Delta_{i+1}^n X, \dots, \Delta_{i+k-1}^n X\big), \\ \mathcal{V}^n(F,X)_t &= \sum_{i=1}^{[t/k\Delta_n]} F\big(\Delta_{ik-k+1}^n X, \Delta_{ik-k+2}^n X, \dots, \Delta_{ik}^n X\big), \end{aligned}$$

J. Jacod, P. Protter, *Discretization of Processes*,
Stochastic Modelling and Applied Probability 67,
DOI 10.1007/978-3-642-24127-7_11,

where k is an integer and F is a function on $(\mathbb{R}^d)^k$. We have seen in Theorem 8.2.1 that, if F is continuous and $F(z) = \mathrm{o}(\|z\|^2)$ as $z \to 0$ in $(\mathbb{R}^d)^k$, then

$$\begin{aligned} &\bullet\ t>0 \ \Rightarrow\ V^n(F,X)_t \stackrel{\mathbb{P}}{\longrightarrow} \textstyle\sum_{j=1}^k f_j * \mu_t \\ &\bullet\ F \text{ symmetrical } \Rightarrow\ \mathcal{V}^n(F,X)_t \stackrel{\mathbb{P}}{\Longrightarrow} f * \mu \end{aligned} \tag{11.1.1}$$

where $f_j(x) = F(0,\dots,0,x,0,\dots,0)$ (with x at the jth place), and where $f = f_1 = \dots,\ f_k$ in the symmetrical case (meaning that F is invariant by all the permutations of the k variables in $\mathbb{R}^d$), and $\mu = \mu^X$ is the jump measure of X. The first convergence holds on Ω here, because $\mathbb{P}(\Delta X_t \neq 0) = 0$ for any t under our standing assumption (H).

We wish to provide the CLTs associated with these two convergences. The reader will have noticed the difference in the two statements (11.1.1): functional convergence for $\mathcal{V}^n(F,X)$ (so we may hope for a "functional" CLT), convergence for any fixed time t for $V^n(F,X)$ (so we may only hope for a finite-dimensional CLT).

11.1.1 The Results

The description of the limiting process involves more auxiliary random variables than for Theorem 5.1.2. We set $\mathcal{K}_- = \{-k+1, -k+2, \dots, -1\}$ and $\mathcal{K}_+ = \{1,2,\dots,k-1\}$ and $\mathcal{K} = \mathcal{K}_- \cup \mathcal{K}_+$. We have a family of variables $((\Psi_{n,j})_{j\in\mathcal{K}}, \Psi_{n-}, \Psi_{n+}, \kappa_n, L_n)_{n\geq 1}$, defined on an auxiliary space $(\Omega', \mathcal{F}', \mathbb{P}')$, all independent, and with the following laws:

$$\begin{aligned} &\Psi_{n,j}, \Psi_{n-}, \Psi_{n-} \text{ are } d'\text{-dimensional, } \mathcal{N}(0, I_{d'}), \\ &\kappa_n \text{ is uniform on } [0,1], \\ &L_n \text{ is integer-valued, uniform on } \{0,1,\dots,k-1\}. \end{aligned} \tag{11.1.2}$$

We also consider an arbitrary weakly exhausting sequence $(T_n)_{n\geq 1}$ for the jumps of X (see after (5.1.1), page 126). The very good filtered extension $(\widetilde{\Omega}, \widetilde{\mathcal{F}}, (\widetilde{\mathcal{F}}_t)_{t\geq 0}, \widetilde{\mathbb{P}})$ of $(\Omega, \mathcal{F}, (\mathcal{F}_t)_{t\geq 0}, \mathbb{P})$ is defined by (4.1.16), that is:

$$\begin{aligned} &\widetilde{\Omega} = \Omega \times \Omega', \qquad \widetilde{\mathcal{F}} = \mathcal{F} \otimes \mathcal{F}', \qquad \widetilde{\mathbb{P}} = \mathbb{P} \otimes \mathbb{P}' \\ &(\widetilde{\mathcal{F}}_t) \text{ is the smallest filtration containing } (\mathcal{F}_t) \text{ and such that} \\ &\quad \big((\Psi_{n,j})_{j\in\mathcal{K}}, \Psi_{n-}, \Psi_{n+}, \kappa_n, L_n\big) \text{ is } \widetilde{\mathcal{F}}_{T_n} \text{ measurable for all } n. \end{aligned} \tag{11.1.3}$$

We also define the d-dimensional random variables

$$R_{n,j} = \begin{cases} \sigma_{T_n-}\Psi_{n,j} & \text{if } j \in \mathcal{K}_- \\ \sqrt{\kappa_n}\,\sigma_{T_n-}\Psi_{n-} + \sqrt{1-\kappa_n}\,\sigma_{T_n}\Psi_{n+} & \text{if } j = 0 \\ \sigma_{T_n}\Psi_{n,j} & \text{if } j \in \mathcal{K}_+. \end{cases} \tag{11.1.4}$$

Next, the test function F is a $\mathbb{R}^q$-valued function on $(\mathbb{R}^d)^k$, where $k \geq 2$, and we assume that is it C^2. Exactly as in the case $k = 1$, the limiting process involves the derivatives of F, but here they are taken at points like $(0, \dots, 0, \Delta X_s, 0, \dots, 0)$, and we must establish carefully our notation, which is unfortunately but necessarily cumbersome. The first and second partial derivatives $\partial F/\partial x_l^i$ (where $x_l = (x_l^i)_{1 \leq i \leq d} \in \mathbb{R}^d$) and $\partial^2 F/\partial x_l^i\, \partial x_{l'}^j$ are denoted globally as ∇F and $\nabla^2 F$. We associate the following $\mathbb{R}^q$-valued functions on $\mathbb{R}^d$:

$$\left.\begin{array}{l} f_j(x) = F(0, \dots, 0, x, 0, \dots, 0) \\ \partial_i f_{(l);j}(x) = \frac{\partial F}{\partial x_l^i}(0, \dots, 0, x, 0, \dots, 0) \end{array}\right\} \quad \text{where } x \text{ is at the } j\text{th place.} \tag{11.1.5}$$

Note that $\partial_i f_{(l);l} = \partial_i f_l$, with the usual notation $\partial_i f = \partial f/\partial x^i$. When F is *symmetrical* there are a function f and a family of functions $(\partial_i^* f)_{1 \leq i \leq d}$ on $\mathbb{R}^d$ such that

$$1 \leq i \leq d,\ 1 \leq j, l \leq k \ \Rightarrow\ f_j = f, \qquad \partial_i f_{(l);j} = \begin{cases} \partial_i f & \text{if } l = j \\ \partial_i^* f & \text{if } l \neq j \end{cases} \tag{11.1.6}$$

(as the notation suggests, $\partial_i^* f$ is a partial derivative, but the "$*$" indicates that it is *not* the derivative of f; it is in fact the ith partial derivative of the function $y \mapsto F(x, y, 0, \dots, 0)$, evaluated at $y = 0$).

The next proposition introduces the limiting processes.

Proposition 11.1.1 *Assume* (H), *and suppose that F satisfies $\|\partial_i f_{(l);j}(x)\| \leq K\|x\|$ when $\|x\| \leq 1$ for all $l, j \leq k$ and $i \leq d$. The formulas*

$$\overline{V}(F, X)_t = \sum_{n=1}^{\infty} \Bigg(\sum_{j,l=1}^{k} \sum_{i=1}^{d} \partial_i f_{(l);j}(\Delta X_{T_n})\, R_{n,l-j}^i \Bigg) 1_{\{T_n \leq t\}} \tag{11.1.7}$$

and, when further F is symmetrical,

$$\overline{\mathcal{V}}(F, X)_t = \sum_{n=1}^{\infty} \Bigg(\sum_{j,l=1}^{k} \sum_{i=1}^{d} \partial_i f_{(l);j}(\Delta X_{T_n})\, R_{n,l-j}^i\, 1_{\{L_n = j-1\}} \Bigg) 1_{\{T_n \leq t\}} \tag{11.1.8}$$

define two q-dimensional processes $\overline{V}(F, X)$ and $\overline{\mathcal{V}}(F, X)$ on $(\widetilde{\Omega}, \widetilde{\mathcal{F}}, (\widetilde{\mathcal{F}}_t)_{t \geq 0}, \widetilde{\mathbb{P}})$ which are a.s. càdlàg, adapted, and conditionally on $\mathcal{F}$ have centered and independent increments and satisfy (*recall $c = \sigma\sigma^*$*):

$$\widetilde{\mathbb{E}}\big(\overline{V}(F, X)_t^i\, \overline{V}(F, X)_t^{i'} \mid \mathcal{F}\big)$$
$$= \sum_{s \leq t} \sum_{r,r'=1}^{d} \Bigg(\frac{1}{2} \sum_{j,j'=1}^{k} \big(\partial_r f_{(j);j}^i\, \partial_{r'} f_{(j');j'}^{i'}\big)(\Delta X_s)\, \big(c_{s-}^{rr'} + c_s^{rr'}\big)$$

$$+\sum_{j=2}^{k}\sum_{l=1}^{j-1}\sum_{l'=1}^{k+l-j}\left(\partial_r f^i_{(l);j}\,\partial_{r'} f^{i'}_{(l');j+l'-l}\right)(\Delta X_s)\,c^{rr'}_{s-}$$

$$+\sum_{j=1}^{k-1}\sum_{l=j+1}^{k}\sum_{l'=1+l-j}^{k}\left(\partial_r f^i_{(l);j}\,\partial_{r'} f^{i'}_{(l');j+l'-l}\right)(\Delta X_s)\,c^{rr'}_{s}\Bigg) \tag{11.1.9}$$

$$\widetilde{\mathbb{E}}\big(\overline{\mathcal{V}}(F,X)^i_t\,\overline{\mathcal{V}}(F,X)^{i'}_t \mid \mathcal{F}\big)$$

$$=\sum_{s\le t}\sum_{r,r'=1}^{d}\frac{1}{2}\left(\partial_r f^i\,\partial_{r'} f^{i'}+(k-1)\partial^*_r f^i\,\partial^*_{r'} f^{i'}\right)(\Delta X_s)\left(c^{rr'}_{s-}+c^{rr'}_{s}\right) \tag{11.1.10}$$

and their $\mathcal{F}$-conditional laws do not depend on the choice of the exhausting sequence T_n. If further the processes X and σ have no common jumps, both processes are $\mathcal{F}$-conditionally Gaussian processes.

Proof The process $\overline{V}(F,X)$ is formally of the form (4.1.7), that is the nth summand in (11.1.7) is $V_{T_n}U_n 1_{\{T_n\le t\}}$, where

$$U_n=\left((\Psi_{n,j})_{j\in\mathcal{K}},\sqrt{\kappa_n}\,\Psi_{n-},\sqrt{1-\kappa_n}\,\Psi_{n+}\right)$$

and V_t is a suitable process whose components are linear combinations of the components of σ_t and σ_{t-} times the components of the functions $\partial_i f_{(l);j}$ for all possible values of j,l,i. Moreover the U_n's are i.i.d. centered variables independent of $\mathcal{F}$, with moments of all orders, whereas our assumption on F yields $\|V_t\|\le K(\|\sigma_t\|+\|\sigma_{t-}\|)\|\Delta X_t\|$ as soon as $\|\Delta X_t\|\le 1$. Then (4.1.10) holds, and Proposition 4.1.4 gives that the process $Z(F,X)$ satisfies all the claims, provided we prove that the right sides of (4.1.11) and (11.1.9) agree.

This last property comes from the fact, again, that the U_n are i.i.d. centered and independent of $\mathcal{F}$, from the obvious property

$$\mathbb{E}\left(R^i_{n,j}\,R^{i'}_{n,j'}\mid\mathcal{F}\right)=\begin{cases} c^{ii'}_{T_n-} & \text{if } j=j'\in\mathcal{K}_-\\ \frac{1}{2}\left(c^{ii'}_{T_n-}+c^{ii'}_{T_n}\right) & \text{if } j=j'=0\\ c^{ii'}_{T_n} & \text{if } j=j'\in\mathcal{K}_+\\ 0 & \text{otherwise}\end{cases}$$

and from rather tedious but elementary calculations.

The proof for the process $\mathcal{Z}(F,X)$ is similar, provided we take

$$U_n=\Big((\Psi_{n,j}1_{\{L_n=l\}})_{j\in\mathcal{K},0\le l\le k-1},\left(\sqrt{\kappa_n}\,\Psi_{n-}1_{\{L_n=l\}}\right)_{0\le l\le k-1},$$
$$\left(\sqrt{1-\kappa_n}\,\Psi_{n+}1_{\{L_n=l\}}\right)_{0\le l\le k-1}\Big)$$

which are still i.i.d. centered with all moments and independent of $\mathcal{F}$. The calculation of (11.1.10) is simpler than (11.1.9), due to the simpler structure (11.1.6). □

We are now ready to state the main results of this section.

Theorem 11.1.2 *Assume* (H), *and let F be a C^2 function from* $(\mathbb{R}^d)^k$ *into* $\mathbb{R}^q$, *satisfying*

$$F(0)=0, \qquad \nabla F(0)=0, \qquad \left\|\nabla^2 F(z)\right\| = \mathrm{o}\big(\|z\|\big) \quad \text{as } z\to 0 \text{ in } \big(\mathbb{R}^d\big)^k. \tag{11.1.11}$$

a) *For each t, the q-dimensional variables*

$$\frac{1}{\sqrt{\Delta_n}}\left(V^n(F,X)_t - \sum_{j=1}^{k} f_j \star \mu_t\right) \tag{11.1.12}$$

converge stably in law to the variable $\overline{V}(F,X)_t$ defined by (11.1.7).

b) *If further F is symmetrical, the q-dimensional processes*

$$\frac{1}{\sqrt{\Delta_n}}\big(\mathcal{V}^n(F,X)_t - f \star \mu_{k\Delta_n[t/k\Delta_n]}\big) \tag{11.1.13}$$

converge stably in law to the process $\overline{\mathcal{V}}(F,X)$ defined by (11.1.8). *Moreover, for each fixed t, the variables*

$$\frac{1}{\sqrt{\Delta_n}}\big(\mathcal{V}^n(F,X)_t - f \star \mu_t\big) \tag{11.1.14}$$

converge stably in law to the variable $\overline{\mathcal{V}}(F,X)_t$.

The condition (11.1.11) obviously implies the assumptions of Proposition 11.1.1. When $k=1$, so $F=f$ is a function on $\mathbb{R}^d$, we have $\mathcal{V}^n(F,X)=V^n(F,X)=V^n(f,X)$, and (11.1.11) is exactly what is imposed on the function f in Theorem 5.1.2, which is thus a particular case of the above (except for the joint Skorokhod convergence with the discretized process $X^{(n)}$, which does *not* hold when $k\geq 2$). Indeed, we then have $R_{n,0}=R_n$ (notation (5.1.3)) and $\partial_i f_{(1);1}=\partial_i f$ and further $L_n=0$, so the processes (5.1.4), (11.1.7) and (11.1.8) are all the same.

11.1.2 An Auxiliary Stable Convergence

Before starting the proof, and by the localization lemma 4.4.9, we see that we can replace (H) by the stronger assumption (SH), that is Assumption 4.4.6, under which b_t and σ_t and X_t are bounded and $\|\delta(\omega,t,z)\|\leq \Gamma(z)$ with Γ bounded and $\int \Gamma(z)^2\lambda(dz)<\infty$.

In this subsection we basically extend Proposition 4.4.10 to the present situation, and in the case where the process $\overline{U}^n(G)$ is absent.

We first introduce some notation. The integer k is fixed, and $m \geq 1$ is another integer and $T > 0$. We take $A_m = \{z : \Gamma(z) > 1/m\}$. We choose the weakly exhausting sequence for the jumps of X to be the sequence (S_p) defined in (4.3.1), that is

$$\left.\begin{array}{l} (S_p)_{p\geq 1} \text{ is a reordering of the double sequence } (S(m,j) : m, j \geq 1) \\ \text{where } S(m,1), S(m,2), \dots \text{ are the successive jump times of} \\ \text{the process } 1_{\{A_m \setminus A_{m-1}\}} * \mathfrak{p}. \end{array}\right\}$$

Next, similar to (4.4.20) and (4.4.21), we set

$$\left.\begin{array}{l} b(m)_t = b_t - \int_{A_m \cap \{z:\|\delta(t,z)\|\leq 1\}} \delta(t,z)\,\lambda(dz), \\ X(m)_t = X_0 + \int_0^t b(m)_s\,ds + \int_0^t \sigma_s\,dW_s + (\delta\,1_{A_m^c}) \star (\mathfrak{p} - \mathfrak{q})_t, \\ X'(m) = X - X(m) \;=\; (\delta\,1_{A_m}) \star \mathfrak{p}. \end{array}\right\} \tag{11.1.15}$$

$$\begin{array}{l} \Omega_n(T,m) = \text{the set of all } \omega \text{ such that the jumps of } X'(m) \text{ in} \\ \qquad [0,T] \text{ are spaced by more than } k\Delta_n, \text{ and no such} \\ \qquad \text{jump occurs in } [0,k\Delta_n] \text{ or } [T-k\Delta_n, T], \text{ and} \\ \qquad \text{for all } t \in [0,T],\ s \in [0,k\Delta_n] \\ \qquad \text{we have } \|X(m)_{t+s} - X(m)_t\| \leq 2/m. \end{array} \tag{11.1.16}$$

Note the slight modification in the definition of the sets $\Omega_n(T,m)$, which nevertheless satisfy

$$\mathbb{P}\big(\Omega_n(T,m)\big) \to 1 \quad \text{as } n \to \infty. \tag{11.1.17}$$

Last, for $j \in \mathcal{K}$, and with the convention $\Delta_i^n Y = 0$ when $i \leq 0$, we define the variables

$$\left.\begin{array}{l} R(n,p,j) = \frac{1}{\sqrt{\Delta_n}}\,\Delta_{i+j}^n X \\ R(n,p,0) = \frac{1}{\sqrt{\Delta_n}}\,(\Delta_i^n X - \Delta X_{S_p}) \\ L(n,p) = i - 1 - k\big[\frac{i-1}{k}\big] \end{array}\right\} \quad \text{if } (i-1)\Delta_n < S_p \leq i\Delta_n \tag{11.1.18}$$

(compare with (4.4.16), we have $R(n,p,0) = R_-(n,p) + R_+(n,p)$).

Lemma 11.1.3 *Under* (SH), *and recalling the notation* $R_{p,j}$ *of* (11.1.4), *we have*

$$\big((R(n,p,j))_{-k+1\leq j\leq k-1}, L(n,p)\big)_{p\geq 1} \stackrel{\mathcal{L}\text{-s}}{\longrightarrow} \big((R_{p,j})_{-k+1\leq j\leq k-1}, L_p\big)_{p\geq 1}. \tag{11.1.19}$$

Proof The proof is basically the same as for Proposition 4.4.10. On the set $\{(i-1)\Delta_n < S_p \leq i\Delta_n\}$ we put

$$\alpha(n,p,j) = \frac{1}{\sqrt{\Delta_n}}\,\Delta_{i+j}^n W$$

$$\alpha_-(n,p) = \frac{1}{\sqrt{\Delta_n}}\,(W_{S_p} - W_{(i-1)\Delta_n}), \qquad \alpha_+(n,p) = \frac{1}{\sqrt{\Delta_n}}\,(W_{i\Delta_n} - W_{S_p})$$

$$\overline{w}'(n,p)_s = \frac{1}{\sqrt{k\Delta_n}}\,(W_{([(i-1)/k]-1+s)k\Delta_n} - W_{([(i-1)/k]-1)k\Delta_n})$$

$$\kappa(n,p) = \frac{S_p}{\Delta_n} - (i-1), \qquad \kappa'(n,p) = \frac{S_p}{k\Delta_n} - \left[\frac{i-1}{k}\right]$$

(so $\overline{w}'(n,p)_s$ and $\kappa'(n,p)$ are exactly like $\overline{w}(n,p)$ and $\kappa(n,p)$ in (4.3.2) with $v_n = k\Delta_n$, and further $\kappa'(n,p) = (L(n,p) + \kappa(n,p))/k$). Then a simple calculation shows:

$$\left.\begin{array}{l} \alpha(n,p,j) = \sqrt{k}\,(\overline{w}'(n,p)_{1+(L(n,p)+j+1)/k} - \overline{w}'(n,p)_{1+(L(n,p)+j)/k}), \\ \alpha_-(n,p) = \sqrt{k}\,(\overline{w}'(n,p)_{1+\kappa'(n,p)} - \overline{w}'(n,p)_{1+L(n,p)/k}), \\ \alpha_+(n,p) = \sqrt{k}\,(\overline{w}'(n,p)_{1+(1+L(n,p))k} - \overline{w}'(n,p)_{1+\kappa'(n,p)}), \\ L(n,p) = j-1 \quad \text{on the set } \left\{\frac{j-1}{k} < \kappa'(n,p) \le \frac{j}{k}\right\}, \quad j = 1\ldots,k. \end{array}\right\} \tag{11.1.20}$$

Theorem 4.3.1 implies that $(\overline{w}'(n,p), \kappa'(n,p))_{p\ge1} \stackrel{\mathcal{L}\text{-s}}{\longrightarrow} (W''(p), \kappa'_p)_{p\ge1}$ where the $W''(p)$ and κ'_p are all independent one from another and from $\mathcal{F}$ as well, and each $W''(p)$ is a d'-dimensional Brownian motion and κ'_p is uniform over $[0,1]$. We then put $L_p = j-1$ on the set $\{\frac{j-1}{k} < \kappa'_p \le \frac{j}{k}\}$ for $j = 1\ldots,k$. Taking advantage of (11.1.20), and since the law of κ'_p has no atom, we deduce (as for (5.3.25)) that

$$\begin{aligned} &\big((\alpha(n,p,j))_{j\in\mathcal{K}}, \alpha_-(n,p), \alpha_+(n,p), L(n,p)\big)_{p\ge1} \\ &\quad\stackrel{\mathcal{L}\text{-s}}{\longrightarrow} \big((\sqrt{k}\,(W''(p)_{1+(L'_p+j+1)/k} - W''(p)_{1+(L'_p+j)/k}))_{j\in\mathcal{K}}, \\ &\quad\sqrt{k}\,(W''(p)_{1+\kappa'_p} - W''(p)_{1+L'_p/k}), \\ &\quad\sqrt{k}\,(W''(p)_{1+(1+L'_p)/k} - W''(p)_{1+\kappa'_p}), L'_p\big)_{p\ge1}. \end{aligned} \tag{11.1.21}$$

At this stage we set $\kappa_p = \kappa'_p - L_p/k$. Since κ'_p is uniform over $(0,1]$, it is easy to check that L_p is uniform over $\{0,\ldots,k-1\}$ and independent of κ_p, which itself is uniform over $(0,1]$. Because of the independence between $W''(p)$ and κ'_p, plus the scaling property and the independence of the increments of $W''(p)$, we see that the right side of (11.1.21) has exactly the same law as $((\Psi_{p,j})_{j\in\mathcal{K}}, \sqrt{\kappa_p}\,\Psi_{p-}, \sqrt{1-\kappa_p}\,\Psi_{p-}, L_p)_{p\ge1}$, as given by (11.1.2), and is independent of $\mathcal{F}$. In other words, we have proved that

$$\begin{aligned} &\big(\alpha(n,p,j)_{j\in\mathcal{K}}, \alpha_-(n,p), \alpha_+(n,p), L(n,p)\big)_{p\ge1} \\ &\quad\stackrel{\mathcal{L}\text{-s}}{\longrightarrow} \big((\Psi_{p,j})_{j\in\mathcal{K}}, \sqrt{\kappa_p}\,\Psi_{p-}\,, \sqrt{1-\kappa_p}\,\Psi_{p+}, L_p\big)_{p\ge1}. \end{aligned}$$

Therefore, since σ_t is càdlàg, it is now enough to prove that for any $p \ge 1$ we have, with $S_-(n,p,j) = \sup(i\Delta_n : i\Delta_n < S_p)$ for $j \le 0$ and $S_+(n,p,j) = \inf((i -$

$1)\Delta_n : i\Delta_n \geq S_p)$ when $j \geq 1$:

$$\left.\begin{array}{l} R(n,p,0) - \sigma_{S_-(n,p,0)}\alpha_-(n,p) - \sigma_{S_p}\alpha_+(n,p) \xrightarrow{\mathbb{P}} 0 \\ j \in \mathcal{K}_- \quad \Rightarrow \quad R(n,p,j) - \sigma_{S'_-(n,p,j)}\alpha(n,p,j) \xrightarrow{\mathbb{P}} 0 \\ j \in \mathcal{K}_+ \quad \Rightarrow \quad R(n,p,j) - \sigma_{S'_+(n,p,j)}\alpha(n,p,j) \xrightarrow{\mathbb{P}} 0. \end{array}\right\}$$

The first part above is in fact (4.4.19). The two other parts are shown in exactly the same way. □

11.1.3 Proof of Theorem 11.1.2

Now we proceed to the proof of Theorem 11.1.2. As said before, we can and will assume (SH). We heavily use the notation (11.1.5), as well as those of the previous subsection, and we closely follow the proof of Theorem 5.1.2. We consider the two processes

$$\begin{aligned} \overline{V}^n(F,X)_t &= \frac{1}{\sqrt{\Delta_n}}\left(V^n(F,X)_t - \sum_{j=1}^k f_j * \mu_t^X\right), \\ \overline{\mathcal{V}}^n(F,X)_t &= \frac{1}{\sqrt{\Delta_n}}\left(\mathcal{V}^n(F,X)_t - f \star \mu^X_{k\Delta_n[t/k\Delta_n]}\right), \end{aligned}$$

which replace (5.1.6), the second one being defined when F is symmetrical only.

Step 1) As before, $\mathcal{P}_m$ denotes the set of all indices p such that $S_p = S(m',j)$ for some $j \geq 1$ and some $m' \leq m$, and $i(n,p)$ is the integer such that $(i(n,p)-1)\Delta_n < S_p \leq i(n,p)\Delta_n$. As in (5.1.12), we set

$$Y^n(m)_t = \sum_{p \in \mathcal{P}_m:\, S_p \leq \Delta_n[t/\Delta_n]} \zeta_p^n, \quad \text{where } \zeta_p^n = \frac{1}{\sqrt{\Delta_n}} \sum_{j=1}^k \zeta(j)_p^n \quad \text{and}$$

$$\begin{aligned} \zeta(j)_p^n = {} & F\big(\sqrt{\Delta_n}\, R(n,p,1-j), \dots, \Delta X_{S_p} \\ & + \sqrt{\Delta_n}\, R(n,p,0), \dots, \sqrt{\Delta_n}\, R(n,p,k-j)\big) - f_j(\Delta X_{S_p}) \\ & - F\big(\sqrt{\Delta_n}\, R(n,p,1-j), \dots, \sqrt{\Delta_n}\, R(n,p,0), \dots, \sqrt{\Delta_n}\, R(n,p,k-j)\big). \end{aligned}$$

In view of (11.1.16) and (11.1.18), we have

$$\overline{V}^n(F,X)_T = \overline{V}^n\big(F,X(m)\big)_T + Y^n(m)_T \quad \text{on the set } \Omega_n(T,m). \tag{11.1.22}$$

(In contrast with (5.1.13), this equality holds for T, but not for all $t \leq T$; it holds in fact for any $t \leq T$ such that $|t - S_p| > (k+1)\Delta_n$ for all $p \in \mathcal{P}_m$, but not for the t's which are too close to some S_p.)

For any (p,j) the sequence $R(n,p,j)$ is bounded in probability by (11.1.19). Since F is C^2 and satisfies (11.1.11), the definition of $\zeta(j)^n_p$ and a Taylor expansion of F around $(0,\dots,0,\Delta X_{S_p},0,\dots,0)$ give

$$\frac{1}{\sqrt{\Delta_n}}\zeta(j)^n_p - \sum_{l=1}^{k}\sum_{i=1}^{d}\partial_i f_{(l);j}(\Delta X_{S_p})\,R(n,p,l-j)^i \xrightarrow{\mathbb{P}} 0. \tag{11.1.23}$$

Then another application of Lemma 11.1.3 yields that

$$\left(\zeta^n_p\right)_{p\ge 1} \xrightarrow{\mathcal{L}\text{-s}} (\zeta_p)_{p\ge 1}, \quad \text{where } \zeta_p = \sum_{j,l=1}^{k}\sum_{i=1}^{d}\partial_i f_{(l);j}(\Delta X_{S_p})\,R^i_{p,l-j}.$$

Since the set $\{S_p : p\in\mathcal{P}_m\}\cap[0,t]$ is finite, we deduce that, as $n\to\infty$:

$$Y^n(m) \overset{\mathcal{L}\text{-s}}{\Longrightarrow} \overline{V}\big(F,X'(m)\big), \tag{11.1.24}$$

where $\overline{V}(F,X'(m))$ is associated with the process $X'(m)$ of (11.1.15) by (11.1.7).

Step 2) When F is symmetrical we have a similar result. We set, with $\zeta(j)^n_p$ as above:

$$\mathcal{Y}^n(m)_t = \sum_{p\in\mathcal{P}_m:\, S_p\le t}\zeta'^n_p, \quad \text{where } \zeta'^n_p = \frac{1}{\sqrt{\Delta_n}}\sum_{j=1}^{k}\zeta'(j)^n_p\,1_{\{L(n,p)=j-1\}}$$

and now we have

$$\overline{\mathcal{V}}^n(F,X)_t = \overline{\mathcal{V}}^n\big(F,X(m)\big)_t + \mathcal{Y}^n(m)_t \quad \forall t\le T, \qquad \text{on the set } \Omega_n(T,m) \tag{11.1.25}$$

(note the difference with (11.1.22): this holds for *all* $t\le T$). Using again (11.1.23), we deduce from Lemma 11.1.3 that

$$\left(\zeta'^n_p\right)_{p\ge 1} \xrightarrow{\mathcal{L}\text{-s}} \left(\zeta'_p\right)_{p\ge 1}, \quad \text{where } \zeta'_p = \sum_{j,l=1}^{k}\sum_{i=1}^{d}\partial_i f_{(l);j}(\Delta X_{S_p})\,R^i_{p,l-j}\,1_{\{L_p=j-1\}}.$$

Therefore we have, as $n\to\infty$,

$$\mathcal{Y}^n(m) \overset{\mathcal{L}\text{-s}}{\Longrightarrow} \overline{\mathcal{V}}\big(f,X'(m)\big). \tag{11.1.26}$$

Step 3) All processes $\overline{V}(F,X'(m))$ are defined on the same extension, and $\overline{V}(F,X) = \overline{V}(F,X'(m)) + \overline{V}(F,X(m))$. Then, using the properties (11.1.2), we can reproduce the proof of (5.1.16) to obtain that (with F symmetrical for the second statement below):

$$\overline{V}\big(F,X'(m)\big) \overset{\text{u.c.p.}}{\Longrightarrow} \overline{V}(F,X), \qquad \overline{\mathcal{V}}\big(F,X'(m)\big) \overset{\text{u.c.p.}}{\Longrightarrow} \overline{\mathcal{V}}(F,X). \tag{11.1.27}$$

At this stage, and in view of (11.1.17) and (11.1.22), in order to prove (a) it remains to show that for all $t>0$, $\eta>0$ we have

$$\lim_{m\to\infty}\ \limsup_{n\to\infty}\ \mathbb{P}\big(\big|\overline{V}^n\big(F,X(m)\big)_t\big|>\eta\big)\ =\ 0. \tag{11.1.28}$$

Analogously, for (b) it suffices, by (11.1.25), to show that for all $t>0$, $\eta>0$ we have

$$\lim_{m\to\infty}\ \limsup_{n\to\infty}\ \mathbb{P}\Big(\sup_{s\le t}\big|\overline{\mathcal{V}}^n\big(F,X(m)\big)_s\big|>\eta\Big)\ =\ 0 \tag{11.1.29}$$

(this gives the convergence of (11.1.13), which in turn implies the convergence of (11.1.14) by Step 5 of the proof of Theorem 5.1.2).

Step 4) In this step we begin the proof of (11.1.28), and it is enough to do this when F is one-dimensional. We fix $t>0$. For $j=1,\dots,k$ we set $G^j(x_1,\dots,x_j)=F(x_1,\dots,x_j,0,\dots,0)$, a function on $(\mathbb{R}^d)^j$. Note that $G^1=f_1$ and $G^k=F$. Since $F(0)=0$, an easy computation shows that, as soon as $k\Delta_n<t$,

$$\overline{V}^n(F,X)_t\ =\ \sum_{j=1}^k Y^{n,j}(X)_t-W^n(X)_t-W'^n(X)_t,$$

where

$$\begin{aligned}
Y^{n,j}(X)_t &= \frac{1}{\sqrt{\Delta_n}}\sum_{i=1}^{[t/\Delta_n]-k+1}\zeta_i^{n,j}(X)\\
\zeta_i^{n,j}(X) &= G^j\big(\Delta_i^nX,\dots,\Delta_{i+j-1}^nX\big)-G^j\big(\Delta_i^nX,\dots,\Delta_{i+j-2}^nX,0\big)\\
&\quad-\Delta_{i+j-1}^n\big(f_j*\mu^X\big)\\
W^n(X)_t &= \frac{1}{\sqrt{\Delta_n}}\sum_{j=2}^k f_j*\mu^X_{(j-1)\Delta_n}\\
W'^n(X)_t &= \frac{1}{\sqrt{\Delta_n}}\sum_{j=1}^k\big(f_j*\mu_t^X-f_j*\mu^X_{\Delta_n([t/\Delta_n]+j-k)}\big).
\end{aligned}$$

Therefore, in view of (11.1.17), it is enough to prove the following properties, for all $\eta>0$:

$$\lim_{m\to\infty}\ \limsup_{n\to\infty}\ \mathbb{P}\big(\Omega_n(t,m)\cap\big\{\big|Y^{n,j}\big(X(m)\big)_t\big|>\eta\big\}\big)\ =\ 0. \tag{11.1.30}$$

$$\limsup_{n\to\infty}\ \mathbb{P}\big(\big|W^n\big(X(m)\big)_t\big|>\eta\big)\ =\ 0,\qquad \limsup_{n\to\infty}\ \mathbb{P}\big(\big|W'^n\big(X(m)\big)_t\big|>\eta\big)\ =\ 0. \tag{11.1.31}$$

The proof of (11.1.31) is elementary. Indeed, $|f_j(x)| \le K\|x\|^2$ if $\|x\| \le 1$ by (11.1.11), hence

$$\mathbb{E}\big(\big|f_j * \mu_{s+r}^{X(m)} - f_j * \mu_s^{X(m)}\big|\big) \le \mathbb{E}\big(\|\delta\|^2 * \mathfrak{q}_{s+r} - \|\delta\|^2 * \mathfrak{q}_s\big|\big) \le Kr,$$

where the last inequality comes from (SH). Thus both $\mathbb{E}(|W^n(X(m))|)$ and $\mathbb{E}(|W'^n(X(m))|)$ are smaller than $Kk\sqrt{\Delta_n}$, implying (11.1.31).

Step 5) The next two steps are devoted to proving (11.1.30), so j is fixed. In the present step we generalize the notation (5.1.18) and the estimates (5.1.22).

We denote by $u = (x_1, \dots, x_{j-1})$ the current point in $(\mathbb{R}^d)^{j-1}$, so $G^j(x_1, \dots, x_j)$ is also written as $G^j(u, x_j)$. We set (compare with (5.1.18), page 131):

$$\begin{aligned} k_u(x,y) &= G^j(u, x+y) - G^j(u,x) - G^j(0,y), \\ g_u(x,y) &= k_u(x,y) - \sum_{i=1}^d \frac{\partial G^j}{\partial x^i}(u,x) y^i. \end{aligned}$$

Below, ∇G^j and $\nabla^2 G^j$ denote the families of all first and second order derivatives of G^j, with respect to all variables $x_1, \dots, x_j$. The hypothesis (11.1.11) yields that, for some α_m going to 0 as $m \to \infty$, and recalling that G^j is one-dimensional here,

$$\|u\| \le \frac{2k}{m},\ \|x\| \le \frac{3}{m} \Rightarrow \begin{cases} |G^j(u,x)| \le \alpha_m(\|u\|^3 + \|x\|^3) \\ |\nabla G^j(u,x)| \le \alpha_m(\|u\|^2 + \|x\|^2) \\ |\nabla^2 G^j(u,x)| \le \alpha_m(\|u\| + \|x\|). \end{cases} \tag{11.1.32}$$

This implies

$$\|u\| \le \frac{2k}{m},\ \|x\| \le \frac{3}{m},\ \|y\| \le \frac{1}{m} \Rightarrow \begin{cases} |k_u(x,y)| \le K\alpha_m(\|u\| + \|x\|)\,\|y\| \\ |g_u(x,y)| \le K\alpha_m(\|u\| + \|x\|)\,\|y\|^2, \end{cases} \tag{11.1.33}$$

although the proof is not as straightforward as for (5.1.22). More precisely, under the above condition on x, y, u, a Taylor expansion (with y being the "increment" of the variable) and (11.1.32) yield $|k_u(x,y)| \le K\alpha_m(\|u\|^2 + \|x\|^2 + \|y\|^2)\|y\|$ and $|g_u(x,y)| \le K\alpha_m(\|u\| + \|x\| + \|y\|)\|y\|^2$, which give (11.1.33) when $\|y\| \le \|u\| \vee \|x\|$.

When $\|y\|$ is bigger than both $\|u\|$ and $\|x\|$, we use Taylor's formula for u and (11.1.32) again to get that $|k_u(x,y) - k_0(x,y)|$ and $|g_u(x,y) - g_0(x,y)|$ are smaller than $K\alpha_m\,\|y\|^2\,\|u\|$, and then we can apply (5.1.22), page 132, to the functions k_0 and g_0, to finally obtain (11.1.33) again.

Step 6) Here we prove (11.1.30). We will in fact prove more, namely

$$\lim_{m\to\infty} \limsup_{n\to\infty} \mathbb{P}\Big(\Omega_n(t,m) \cap \Big\{\sup_{s\le t} \big|Y^{n,j}(X(m))_s\big| > \eta\Big\}\Big) = 0. \tag{11.1.34}$$

We will construct a decomposition

$$\frac{1}{\sqrt{\Delta_n}}\,\zeta_i^{n,j}\big(X(m)\big) = \eta(m)_i^n + \eta'(m)_i^n \quad \text{on the set } \Omega_n(t,m), \tag{11.1.35}$$

in such a way that we have

$$\eta'(m)_i^n \text{ is } \mathcal{F}_{(i+j-1)\Delta_n} \text{ measurable and } \mathbb{E}\big(\eta'(m)_i^n \mid \mathcal{F}_{(i+j-2)\Delta_n}\big) = 0, \tag{11.1.36}$$

$$\left.\begin{array}{l} \lim_{m\to\infty} \limsup_{n\to\infty} \mathbb{E}\big(\sum_{i=1}^{[t/\Delta_n]-k+1} |\eta(m)_i^n|\big) = 0 \\ \lim_{m\to\infty} \limsup_{n\to\infty} \mathbb{E}\big(\sum_{i=1}^{[t/\Delta_n]-k+1} (\eta'(m)_i^n)^2\big) = 0. \end{array}\right\} \tag{11.1.37}$$

By (11.1.36) the expectation in the second part of (11.1.37) is also the expectation of the variable $(\sum_{i=1}^{[t/\Delta_n]-k+1} \eta'(m)_i^n)^2$. Hence it is clear that (11.1.35)–(11.1.37) imply (11.1.34).

For each $i \geq 1$ write $U(n,m,i) = (\Delta_i^n X(m), \dots, \Delta_{i+j-2}^n X(m))$. As in Theorem 5.1.2, we set $T(n,m,i) = \inf(s > (i+j-2)\Delta_n : \|X(m)_s - X(m)_{(i-1)\Delta_n}\| > 2/m)$, so $T(n,m,i) > (i+j-1)\Delta_n$ for all $i \leq [t/\Delta_n] - k + 1$ on $\Omega_n(t,m)$. Moreover, on the set $\Omega_n(t,m)$ we have $\|\Delta_i^n X(m)\| \leq 2/m$ for all $i \leq t/\Delta_n$, so $\Omega_n(t,m) \subset B(n,m,i) := \{\|U(n,m,i)\| \leq 2k/m\}$ for all $i \leq t/\Delta_n - k + 1$. Then we can apply Itô's formula to the process $X(m)_t - X(m)_{(i+j-2)\Delta_n}$ for $t \geq (i+j-2)\Delta_n$ and to the C^2 function $x \mapsto G^j(u,x)$: this yields the decomposition (11.1.35), with

$$\eta(m)_i^n = \frac{1}{\sqrt{\Delta_n}}\, 1_{B(n,m,i)} \int_{(i+j-2)\Delta_n}^{(i+j-1)\Delta_n \wedge T(n,m,i)} a\big(U(n,m,i),n,m,i\big)_s\, ds$$

$$\eta'(m)_i^n = \frac{1}{\sqrt{\Delta_n}}\, 1_{B(n,m,i)}\, M\big(U(n,m,i),n,m,i\big)_{(i+j-1)\Delta_n \wedge T(n,m,i)},$$

where, on the one hand,

$$\begin{aligned} a(u,n,m,i)_s = {} & \sum_{r=1}^d \frac{\partial G^j}{\partial x^r}\big(u, X(m)_s - X(m)_{(i+j-2)\Delta_n}\big)\, b(m)_s^r \\ & + \frac{1}{2}\sum_{r,l=1}^d \frac{\partial^2 G^j}{\partial x^r\,\partial x^l}\big(u, X(m)_s - X(m)_{(i+j-2)\Delta_n}\big)\, c_s^{rl} \\ & + \int_{A_m} g_u\big(X(m)_t - X(m)_{(i+j-2)\Delta_n}, \delta(s,z)\big)\, \lambda(dz) \end{aligned}$$

and on the other hand $M(u,n,m,i)_t$ for $t \geq (i+j-2)\Delta_n$ is a square-integrable martingale vanishing at time $(i+j-2)\Delta_n$ and with predictable bracket $\int_{(i-1)\Delta_n}^t a'(u,$

$n,m,i)_s\,ds$, where $a'(u,n,m,i)_s$ is given by

$$a'(u,n,m,i)_s = \sum_{r,l=1}^{d} \Big(\frac{\partial G^j}{\partial x^r}\,\frac{\partial G^j}{\partial x^l}\Big)\big(u, X(m)_s - X(m)_{(i+j-2)\Delta_n}\big)\,c_s^{rl} + \int_{A_m} k_u\big(X(m)_s - X(m)_{(i+j-2)\Delta_n}, \delta(s,z)\big)^2\,\lambda(dz).$$

In particular, we have (11.1.36), and also

$$\mathbb{E}\big((\eta'(m)_i^n)^2 = \frac{1}{\Delta_n}\,\mathbb{E}\bigg(1_{B(n,m,i)} \int_{(i+j-2)\Delta_n}^{(i+j-2)\Delta_n \wedge T(n,m,i)} a'\big(U(n,m,i),n,m,i\big)_s\,ds\bigg).$$

We can now reproduce the end of Step 4 of the proof of Theorem 5.1.2: by (SH) we have $\|\delta(t,z)\| \le \Gamma(z)$ and $\|b(m)_t\| \le Km$. Then (11.1.32) and (11.1.33) yield, as soon as $(i-1)\Delta_n \le t \le T(n,m,i)$ (so $\|X(m)_t - X(m)_{(i-1)\Delta_n}\| \le 3/m$), and on the set $B(n,m,i)$:

$$\begin{aligned} \big|a(n,m,i)_t\big| &\le K\alpha_m \Big(\big\|U(n,m,i)\big\| + \big(\big\|X(m)_t - X(m)_{(i-1)\Delta_n}\big\| \\ &\quad + m\big(\big\|U(n,m,i)\big\|^2 + \big\|X(m)_t - X(m)_{(i-1)\Delta_n}\big\|^2\big)\big)\\ a'(n,m,i)_t &\le K\alpha_m^2 \Big(\big\|U(n,m,i)\big\|^2 + \big\|X(m)_t - X(m)_{(i-1)\Delta_n}\big\|^2\Big). \end{aligned}$$

We also have (5.1.24), page 133, which implies for $p=1,2$:

$$\mathbb{E}\big(\big\|U(m,n,i)\big\|^p\big) \le K\big(s^{(p/2)\wedge 1} + m^p s^p\big).$$

At this stage, we get that the two "lim sup" in (11.1.37) are smaller than $Kt\alpha_m$ and $Kt\alpha_m^2$ respectively. This ends the proof of (11.1.37), hence of (11.1.30), hence also of (a) of Theorem 11.1.2.

Step 7) It remains to prove (11.1.29), when F is symmetrical. For this, we observe that with the same notation $\zeta_i^{n,j}(X)$ as in Step 4 above, except that now $f_j = f$ for all j, we have

$$\overline{V}^n(F,X)_t = \sum_{j=1}^{k} \mathcal{Y}^{n,j}(X)_t, \qquad \mathcal{Y}^{n,j}(X)_t = \frac{1}{\sqrt{\Delta_n}} \sum_{i=1}^{[t/k\Delta_n]} \zeta_{ki-k+1}^{n,j}(X).$$

We have the decomposition (11.1.35) with (11.1.36), and we have proved the limiting results (11.1.37). Those results also hold if, instead of summing over all i between 1 and $[t/\Delta_n]-k+1$, we sum over $i = kl-k+1$ for all l ranging from 1 to $[t/k\Delta_n]$. Hence, similar to (11.1.34), we get for all $\eta > 0$:

$$\lim_{m\to\infty}\ \limsup_{n\to\infty}\ \mathbb{P}\Big(\Omega_n(t,m) \cap \Big\{\sup_{s\le t} \big|\mathcal{Y}^{n,j}\big(X(m)\big)_s\big| > \eta\Big\}\Big) = 0,$$

and we deduce (11.1.29).

11.2 Functionals of Normalized Increments

In this section we study the functionals

$$\begin{aligned} V'^n(F,X)_t &= \Delta_n \sum_{i=1}^{[t/\Delta_n]-k+1} F\big(\Delta_i^n X/\sqrt{\Delta_n},\dots,\Delta_{i+k-1}^n X/\sqrt{\Delta_n}\big) \\ \mathcal{V}'^n(F,X)_t &= \Delta_n \sum_{i=1}^{[t/k\Delta_n]} F\big(\Delta_{ik-k+1}^n X/\sqrt{\Delta_n},\dots,\Delta_{ik}^n X/\sqrt{\Delta_n}\big). \end{aligned} \tag{11.2.1}$$

Under appropriate conditions on F they converge in the *u.c.p.* sense to $\int_0^t \rho_{c_s}^{k\otimes}(F)\,ds$ and $\frac{1}{k}\int_0^t \rho_{c_s}^{k\otimes}(F)\,ds$ respectively, see Theorem 8.4.1. The associated CLTs are thus about the processes:

$$\overline{V}'^n(F,X)_t \;=\; \frac{1}{\sqrt{\Delta_n}}\left(V'^n(F,X)_t - \int_0^t \rho_{c_s}^{k\otimes}(F)\,ds\right) \tag{11.2.2}$$

and

$$\overline{\mathcal{V}}'^n(F,X)_t \;=\; \frac{1}{\sqrt{\Delta_n}}\left(\mathcal{V}'^n(F,X)_t - \frac{1}{k}\int_0^t \rho_{c_s}^{k\otimes}(F)\,ds\right). \tag{11.2.3}$$

11.2.1 The Results

Not surprisingly, we need the same assumptions on X as for Theorems 5.3.5 and 5.3.6, in connection with the assumptions on the test function F. Those assumptions, namely 4.4.3, 5.3.2, 4.4.4 and 5.3.4, are briefly recalled for the reader's convenience below:

Assumption (K) We have (H) and

$$\sigma_t \;=\; \sigma_0 + \int_0^t \widetilde{b}_s\,ds + \int_0^t \widetilde{\sigma}_s\,dW_s + M_t + \sum_{s\le t} \Delta\sigma_s\,1_{\{\|\Delta\sigma_s\|>1\}}$$

where M is a local martingale with $\|\Delta M_t\| \le 1$, orthogonal to W, and $\langle M,M\rangle_t = \int_0^t a_s\,ds$ and the compensator of $\sum_{s\le t} 1_{\{\|\Delta\sigma_s\|>1\}}$ is $\int_0^t \widetilde{a}_s\,ds$, with the following properties: the processes $\widetilde{b}$, $\widetilde{\sigma}$, $\widetilde{a}$ and a are progressively measurable, the processes $\widetilde{b}$, a and $\widetilde{a}$ are locally bounded, and the processes $\widetilde{\sigma}$ and b are càdlàg or càglàd.

Assumption (K-r) (for $r\in[0,1]$) We have (K) except for the càdlàg or càglàd property of b, and $\|\delta(\omega,t,z)\|\wedge 1 \le \Gamma_n(z)$ for all (ω,t,z) with $t\le\tau_n(\omega)$, where (τ_n) is a localizing sequence of stopping times and the Borel functions Γ_n on E satisfy $\int \Gamma_n(z)^r\lambda(dz)<\infty$. Moreover the process $b'_t = b_t - \int_{\{\|\delta(t,z)\|\le 1\}} \delta(t,z)\,\lambda(dz)$ is càdlàg or càglàd.

Assumption (K') We have (K) and both processes c_t and c_{t-} take their values in $\mathcal{M}_{d\times d}^{++}$.

Assumption (K'-r) We have (K-r) and both processes c_t and c_{t-} take their values in $\mathcal{M}_{d\times d}^{++}$.

Under appropriate conditions on F, the limit will be like the process $\overline{U}'(f,X)$ in Theorem 5.3.5, that is an $\mathcal{F}$-conditionally continuous centered Gaussian process with independent increments. The description of the conditional variance is slightly more involved: if $a\in\mathcal{M}_{d\times d}^{+}$, we take independent $\mathcal{N}(0,a)$ distributed variables $U_1,U_2,\dots$, we consider the σ-fields $\mathcal{G}=\sigma(U_1,\dots,U_{k-1})$ and $\mathcal{G}'=\sigma(U_1,\dots,U_k)$ and we set for any two functions F and G on $(\mathbb{R}^d)^k$ with polynomial growth:

$$\begin{aligned}R_a(F,G)=\sum_{j,j'=0}^{k-1}\mathbb{E}\Big(&\mathbb{E}\big(F(U_{k-j},\dots,U_{2k-j-1})\mid\mathcal{G}'\big)\\&\times\mathbb{E}\big(G(U_{k-j'},\dots,U_{2k-j'-1})\mid\mathcal{G}'\big)\\&-\mathbb{E}\big(F(U_{k-j},\dots,U_{2k-j-1})\mid\mathcal{G}\big)\\&\times\mathbb{E}\big(G(U_{k-j'},\dots,U_{2k-j'-1})\mid\mathcal{G}'\big)\Big).\end{aligned}\tag{11.2.4}$$

This has a covariance structure, in the sense that the matrix with entries $R_a(F_i,F_j)$ is symmetric nonnegative for any functions $F_1,\dots,F_q$. When $k=1$ we have the simple expression $R_a(F,G)=\rho_a(FG)-\rho_a(F)\rho_a(G)$.

Below, we combine the extensions of both Theorems 5.3.5 and 5.3.6 into a single statement, as in the previous chapter, and we present all assumptions on F together. However, depending on the hypotheses made on the process X, only a part of them is used for each statement. For simplicity we consider only the "globally even" case, which here means that the function F on $(\mathbb{R}^d)^k$ satisfies for all $x_1,\dots,x_k\in\mathbb{R}^d$:

$$F(-x_1,\dots,-x_k)\;=\;F(x_1,\dots,x_k).\tag{11.2.5}$$

We need a multivariate result, that is, $F=(F^1,\dots,F^q)$ is q-dimensional. We have some integer $q'\in\{0,\dots,q\}$ and, when $q'<q$ we also have a non-empty subset B of $(\mathbb{R}^d)^k$ which is a finite union of affine hyperplanes, and we suppose that

$$\begin{aligned}&j\le q'\quad\Rightarrow\quad x\mapsto F^j(x)\ \text{ is } C^1 \text{ on } \big(\mathbb{R}^d\big)^k\\&j>q'\quad\Rightarrow\quad x\mapsto F^j(x)\ \text{ is continuous on } \big(\mathbb{R}^d\big)^k \text{ and } C^1 \text{ outside } B.\end{aligned}\tag{11.2.6}$$

We denote by $d(z,B)$ the distance between $z\in(\mathbb{R}^d)^k$ and B. Below, x_j and v run through $\mathbb{R}^d$, and z and y run through $(\mathbb{R}^d)^k$. As usual, ∇F is the family of all first partial derivatives of F. In the forthcoming conditions, the numbers w,s,s',p are subject to $0<w\le 1$ and $0<s\le s'$ and $p\ge 0$, but otherwise unspecified, although in the various statements of the theorem some of them may be further restricted:

$$\big\|F(z)\big\| \le K\big(1+\|z\|^p\big) \tag{11.2.7}$$

$$\big|\nabla F^j(z)\big| \le \begin{cases} K(1+\|z\|^p) & \text{if } j\le q' \\ K(1+\|z\|^p)\big(1+\frac{1}{d(z,B)^{1-w}}\big) & \text{if } j>q' \text{ and } z\in B^c \end{cases} \tag{11.2.8}$$

$$z\in B^c,\ \|y\|\le 1\bigwedge \frac{d(z,B)}{2},\ j>q' \ \Rightarrow$$

$$\big|\nabla F^j(z+y)-\nabla F^j(z)\big| \le K\,\|y\|\left(1+\frac{1}{d(z,B)^{2-w}}\right)\big(1+\|z\|^p\big) \tag{11.2.9}$$

$$\big\|F(x_1,\dots,x_{j-1},x_j+v,x_{j+1},\dots,x_k)-F(x_1,\dots,x_k)\big\|$$

$$\le K\left(\|v\|^s+\|v\|^{s'}\right)\prod_{l=1}^k\big(1+\|x_l\|^2\big). \tag{11.2.10}$$

As already mentioned, in the last condition above s controls the behavior as $v\to 0$, and s' controls the behavior as $\|v\|\to\infty$. The last condition implies the first one, with $p=(k-1)p'+s'$. Conversely, and as in Remark 5.3.7, when $q'=q$ the two conditions (11.2.7) and (11.2.8) imply (11.2.10) with $s=1\wedge p$ and $p'=s'=p$. The same comments as in Remark 10.3.1) apply to the present set of assumptions, and the reader will notice that in (11.2.10) we have all $\|x_l\|$ to the power 2, instead of $\|x\|$ to an arbitrary power p in (5.3.11): this is due to the fact that we consider several increments instead of a single one in our functional $V'^n(F,X)$.

Theorem 11.2.1 *Let X be a d-dimensional Itô semimartingale and F be a function from $(\mathbb{R}^d)^k$ into $\mathbb{R}^q$ which is continuous, globally even, and satisfies* (11.2.6). *We also assume* (11.2.7) *and* (11.2.8) *(recall $w\in(0,1]$), plus one of the following five sets of hypotheses*:

(a) *We have $q'=q$ and* (K) *and X is continuous.*
(b) *We have* (K') *and* (11.2.9) *and X is continuous.*
(c) *We have $q'=q$ and* (K-1), *and F and ∇F are bounded.*
(d) *We have $q'=q$ and* (K-*r*) *with some $r\in(0,1)$, and* (11.2.10) *with $r\le s\le s'<1$.*
(e) *We have* (K'-*r*) *with some $r\in(0,1)$, and* (11.2.9) *and* (11.2.10) *with $r\le s\le s'<1$.*

Then the sequence of processes $\overline{V}'^n(F,X)$ in (11.2.2) *converges stably in law to a continuous process $\overline{V}'(F,X)$ which is defined on a very good filtered extension $(\widetilde{\Omega},\widetilde{\mathcal{F}},(\widetilde{\mathcal{F}}_t)_{t\ge0},\widetilde{\mathbb{P}})$ of $(\Omega,\mathcal{F},(\mathcal{F}_t)_{t\ge0},\mathbb{P})$, and conditionally on $\mathcal{F}$ is a centered Gaussian process with independent increments satisfying*

$$\widetilde{\mathbb{E}}\big(\overline{V}'(F^i,X)_t\,\overline{V}'(F^j,X)\mid\mathcal{F}\big)=\int_0^t R_{c_s}\big(F^i,F^j\big)\,ds. \tag{11.2.11}$$

The same holds for the sequence $\overline{\mathcal{V}}'^n(F,X)$ of (11.2.3), *with a limit $\overline{\mathcal{V}}'(F,X)$ which conditionally on $\mathcal{F}$ is a centered Gaussian process with independent increments* *satisfying*

$$\widetilde{\mathbb{E}}\big(\overline{\mathcal{V}}'(F^i,X)_t\,\overline{\mathcal{V}}'(F^j,X)\mid\mathcal{F}\big)=\frac{1}{k}\int_0^t\big(\rho_{c_s}^{k\otimes}(F^iF^j)-\rho_{c_s}^{k\otimes}(F^i)\,\rho_{c_s}^{k\otimes}(F^j)\big)\,ds. \tag{11.2.12}$$

The reader will notice the difference in the covariances in (11.2.11) and (11.2.12), which is due to the fact that in the first case the functional involves overlapping intervals and thus has a more complicated covariance structure. In the discontinuous case the conditions are rather restrictive, as shown in the following example:

Example 11.2.2 We consider the multipower variations, corresponding to the test functions (8.1.5), that is

$$F(x_1,\dots,x_k)=\prod_{j=1}^k\prod_{i=1}^d|x_j^i|^{w_j^i},\quad w_j^i\geq 0 \tag{11.2.13}$$

(recall the non-standard convention $0^0=1$, so that the factor $|x_j^i|^{w_j^i}$ does not show when $w_j^i=0$). Let u be the minimum of all non-vanishing w_j^i, and $w_j=w_j^1+\dots+w_j^d$. The function F is C^1 on $(\mathbb{R}^d)^k$ if and only if $u>1$, in which case it satisfies (11.2.7) and (11.2.8) (case $j\leq q'$). When $u\leq 1$, F is C^1 outside $B=\cup_{j=1}^{kd}\{z\in(\mathbb{R}^d)^k: z^j=0\}$, and it satisfies (11.2.7), (11.2.8) and (11.2.9) (case $j>q'$) with $p=\sum_j w_j$ and $w=u$. As for (11.2.10), and with w as above, it is satisfied with $r=w$ and $r'=p'=\sup_j w_j$.

Therefore in the continuous case the theorem applies under (K), and also (K') if there is at least a w_j^i in the interval (0, 1]. In the discontinuous case, we need $w_j<1$ for all j (hence $w<1$), and also (K'-r) with $r=w$. This is quite restrictive, but Vetter [93] has shown that the result as stated in the above theorem fails in dimension $d=1$, when for example $k=2$ and $w_1=w_2=1$ and (K'-1) holds.

Example 11.2.3 Another interesting example is

$$F(x_1,\dots,x_k)=\prod_{i=1}^d|x_1^i+\dots+x_k^i|^{w^i},\quad w^i\geq 0 \tag{11.2.14}$$

(in Sect. 11.4.3 below we will use a 2-dimensional test function, when $d=1$, whose first component is given by (11.2.13) and second component given by (11.2.14)). Here u is the smallest non-vanishing power w^i. Then F is C^1 on $(\mathbb{R}^d)^k$ if and only if $u>1$, in which case it satisfies (11.2.7) and (11.2.8) (case $j\leq q'$). When $u\leq 1$, F is C^1 outside $B=\cup_{j=1}^d\{z=(x_1,\dots,x_k): x_1^j+\dots+x_k^j=0\}$, which is again a finite union of hyperplanes. It then satisfies (11.2.7), (11.2.8) and (11.2.9) (case $j>q'$) with $p=\sum_i w^i$ and $w=u$, and also (11.2.10) with $r=w$ and $r'=p'=p$.

11.2.2 Elimination of Jumps

The proof of Theorem 11.2.1 goes along the same route as Theorems 5.3.5 and 5.3.6.

First of all, the localization Lemma 5.3.12 holds here without change, so instead of (K), (K-r), (K') or (K'-r) we can and will assume the strengthened versions (SK), (SK-r), (SK') or (SK'-r), that is Assumptions 4.4.7, 5.3.10, 4.4.8 or 5.3.11. In other words, $\|\delta(\omega,t,z)\| \leq \Gamma(z)$ with Γ bounded and $\int \Gamma(z)^r \lambda(dz) < \infty$, and all processes $b, \sigma, \widetilde{b}, \widetilde{\sigma}, a, \widetilde{a}$ are bounded, and furthermore in case of (SK') or (SK'-r) the inverse process c_t^{-1} exists and is also bounded. In particular we can write X as in (5.3.13), that is

$$X = X' + X'' \quad \text{where } X'_t = X_0 + \int_0^t b'_s\, ds + \int_0^t \sigma_s\, dW_s, \qquad X'' = \delta * \mathfrak{p}, \quad (11.2.15)$$

where $b'_t = b_t - \int_{\{z:\|\delta(t,z)\|\leq 1\}} \delta(t,z)\,\lambda(dz)$ is also bounded.

Second, the next lemma, similar to Lemma 5.3.13, shows that we only need to prove the results when $X = X'$ is continuous.

Lemma 11.2.4 *Under* (SK-1) *and the assumptions* (c), *or under* (SK-r) *and the assumptions* (d) *or* (e), *we have*

$$\frac{1}{\sqrt{\Delta_n}}\big(V'^n(F,X) - V'^n(F,X')\big) \stackrel{\text{u.c.p.}}{\Longrightarrow} 0$$

$$\frac{1}{\sqrt{\Delta_n}}\big(\mathcal{V}'^n(F,X) - \mathcal{V}'^n(F,X')\big) \stackrel{\text{u.c.p.}}{\Longrightarrow} 0.$$

Proof The two claims are proved in the same way, and we prove the first one only. We need to show that the array (η_i^n) defined by

$$\eta_i^n = \sqrt{\Delta_n}\left(F\left(\frac{\Delta_i^n X}{\sqrt{\Delta_n}}, \dots, \frac{\Delta_{i+k-1}^n X}{\sqrt{\Delta_n}}\right) - F\left(\frac{\Delta_i^n X'}{\sqrt{\Delta_n}}, \dots, \frac{\Delta_{i+k-1}^n X'}{\sqrt{\Delta_n}}\right)\right)$$

is asymptotically negligible, that is $\sum_{i=1}^{[t/\Delta_n]} \eta_i^n \stackrel{\text{u.c.p.}}{\Longrightarrow} 0$. For $j = 0, \dots, k$ we set

$$\overline{X}_{i,j}^n = \left(\frac{\Delta_i^n X}{\sqrt{\Delta_n}}, \dots, \frac{\Delta_{i+j-1}^n X}{\sqrt{\Delta_n}}, \frac{\Delta_{i+j}^n X'}{\sqrt{\Delta_n}}, \dots, \frac{\Delta_{i+k-1}^n X'}{\sqrt{\Delta_n}}\right), \quad (11.2.16)$$

with obvious conventions when $j = 0$ (there is only X') and when $j = k$ (there is only X). Then

$$\eta_i^n = \sum_{j=0}^{k-1} \eta_{i,j}^n \quad \text{where } \eta_{i,j}^n = \sqrt{\Delta_n}\big(F(\overline{X}_{i,j+1}^n) - F(\overline{X}_{i,j}^n)\big),$$

and it suffices to prove that for each j the array $(\eta_{i,j}^n)$ is asymptotically negligible.

In case (c) the function F is C_b^1 on $(\mathbb{R}^d)^k$, hence $\|\eta^n_{i,j}\| \le K\sqrt{\Delta_n}\,(1 \wedge \|\Delta^n_{i+j}X''/\sqrt{\Delta_n}\|)$ and we conclude as in Lemma 5.3.13, page 150.

In cases (d) and (e) the function F satisfies (11.2.10), with $p' \le 2$ and $r \le s \le s' < 1$, and we can always take $s = r$. Hence

$$\|\eta^n_{i,j}\| \le K\sqrt{\Delta_n}\,\big((\overline{\alpha}^n_{i+j})^r + (\overline{\alpha}^n_{i+j})^{s'}\big)\prod_{l=0}^{j-1}\alpha^n_{i+l}\prod_{l=j}^{k-1}\alpha'^n_{i+l}, \quad \text{where}$$

$$\alpha^n_i = 1 + \big\|\Delta^n_i X/\sqrt{\Delta_n}\big\|^2,$$

$$\alpha'^n_i = 1 + \big\|\Delta^n_i X'/\sqrt{\Delta_n}\big\|^2, \quad \overline{\alpha}^n_i = \big\|\Delta^n_i X''/\sqrt{\Delta_n}\big\|.$$

As in the proof of Lemma 5.3.13, we have $\mathbb{E}(\alpha^n_i \mid \mathcal{F}_{(i-1)\Delta_n}) \le K$ and $\mathbb{E}((\alpha'^n_i)^q \mid \mathcal{F}_{(i-1)\Delta_n}) \le K_q$ for all $q > 0$ and $\mathbb{E}((\overline{\alpha}^n_i)^q \mid \mathcal{F}_{(i-1)\Delta_n}) \le K_q \Delta_n^{1-q/2}$ for all $q \in [r,1]$. By Hölder's inequality and successive conditioning we get $\mathbb{E}(\|\eta^n_{i,j}\|) \le K(\Delta_n^{1+\frac{1-r}{4}} + \Delta_n^{1+\frac{1-s'}{4}})$, again as in Lemma 5.3.13, page 150, and the result follows. □

11.2.3 Preliminaries for the Continuous Case

In view of Lemma 11.2.4, we need only to consider the case where $X = X'$ is continuous, under the strengthened assumptions. In other words we only have to prove (a) and (b) and, as in (5.3.20), page 152 and with the notation (5.3.19), in those two cases we have for some $A \ge 1$:

$$\begin{array}{ll} \text{(a)} & q' = q,\ \text{(SK)},\ F \text{ is } C^1 \text{ and } \nabla F \text{ has polynomial growth},\ \sigma_t \in \mathcal{M}_A \\ \text{(b)} & q' < q,\ \text{(SK')},\ F \text{ satisfies (11.2.7), (11.2.8), (11.2.9)},\ \sigma_t \in \mathcal{M}'_A. \end{array} \tag{11.2.17}$$

As in pages 151–152, we use the notation (5.3.18), that is $\phi_B(z) = 1 + 1/d(z,B)$ with the convention $B = \emptyset$ and $d(z,B) = \infty$ in case (a) (now, ϕ_B is a function on $(\mathbb{R}^d)^k$). Then F is C^1 outside B and our assumptions on F yield for $z, y \in (\mathbb{R}^d)^k$, and some $w \in (0,1]$ and $p \ge 0$:

$$\|F(z)\| \le K\big(1 + \|z\|^p\big), \quad z \notin B \Rightarrow \|\nabla F(z)\| \le K\big(1 + \|z\|^p\big)\phi_B(z)^{1-w} \tag{11.2.18}$$

$$\big\|F(z+y - F(z))\big\| \le \phi'_C(\varepsilon) + \frac{K}{C}\big(\|z\|^{p+1} + \|y\|^{p+1}\big) + \frac{KC^p\|y\|}{\varepsilon} \tag{11.2.19}$$

$$\big\|\nabla F^j(z+y) - \nabla F^j(z)\big\| \le \begin{cases} \phi'_C(\varepsilon) + \frac{K}{C}\,(\|z\|^{p+1} + \|y\|^{p+1}) + \frac{KC^p\|y\|}{\varepsilon} & \text{if } j \le q' \\ K(1 + \|z\|^p + \|y\|^p)\phi_B(z)^{2-w}\|y\| & \\ \qquad \text{if } j > q' \text{ and if } z \notin B,\ \|y\| \le \frac{d(z,B)}{2} & \end{cases} \tag{11.2.20}$$

for some p (which may be bigger than the values p appearing in the conditions (11.2.7)–(11.2.9), because we use here the Euclidean norm of z and y), and for all $C \geq 1$ and $\varepsilon \in (0,1]$, and where $\phi'_C(\varepsilon) \to 0$ as $\varepsilon \to 0$ for all C.

Next we introduce some notation, somewhat similar to (8.4.7), and where $i, j \geq 1$:

$$\beta^n_{i,j} = \sigma_{(i-1)\Delta_n}\, \Delta^n_{i+j-1} W/\sqrt{\Delta_n}, \quad \overline{\beta}^n_i = \bigl(\beta^n_{i,1}, \dots, \beta^n_{i,k}\bigr), \quad \overline{X}^n_i = \overline{X}^n_{i,k} \tag{11.2.21}$$

with the notation (11.2.16). Note that $\overline{\beta}^n_i$ and $\overline{X}^n_i$ are $(\mathbb{R}^d)^k$-valued random variables. We have the following extension of (5.3.30):

$$\begin{aligned}\frac{\Delta^n_{i+j-1} X}{\sqrt{\Delta_n}} - \beta^n_{i,j} = &\frac{1}{\sqrt{\Delta_n}} \int_{I(n,i+j-1)} b_s\, ds \\ &+ \frac{1}{\sqrt{\Delta_n}} \int_{I(n,i+j-1)} (\sigma_s - \sigma_{(i-1)\Delta_n})\, dW_s.\end{aligned}$$

Using this for all $j = 1, \dots, k$, and similar to (5.3.31), we deduce from (SK) the following estimates, where $l \geq 2$ (below, the constants K also depend on k; recall that $X = X'$):

$$\mathbb{E}\bigl(\|\overline{\beta}^n_i\|^l\bigr) + \mathbb{E}\bigl(\|\overline{X}^n_i\|^l\bigr) \leq K_l, \qquad \mathbb{E}\bigl(\|\overline{X}^n_i - \overline{\beta}^n_i\|^l\bigr) \leq K_l\, \Delta_n. \tag{11.2.22}$$

Therefore, since F is of polynomial growth, the variables $F(\overline{X}^n_i)$ and $F(\overline{\beta}^n_i)$ are integrable, and the following q-dimensional variables are well-defined:

$$\begin{aligned}&\chi^n_i = \sqrt{\Delta_n}\,\bigl(F(\overline{X}^n_i)) - F(\overline{\beta}^n_i)\bigr), \quad \chi'^n_i = \mathbb{E}\bigl(\chi^n_i \mid \mathcal{F}_{(i-1)\Delta_n}\bigr), \quad \chi''^n_i = \chi^n_i - \chi'^n_i \\ &\zeta^n_i = \sqrt{\Delta_n}\,\bigl(F(\overline{\beta}^n_i) - \mathbb{E}\bigl(F(\overline{\beta}^n_i) \mid \mathcal{F}_{(i-1)\Delta_n}\bigr)\bigr) = \sqrt{\Delta_n}\,\bigl(F(\overline{\beta}^n_i) - \rho^{k\otimes}_{c_{(i-1)\Delta_n}}(F)\bigr).\end{aligned} \tag{11.2.23}$$

We now write the decomposition $\overline{V}'^n(F,X) = Y^n + A^n(0) + A^n(1) + A^n(2)$, where $Y^n + A^n(0)$ is analogous to $Y^n(F,X)$ in (5.2.3), and $A^n(1)$ and $A^n(2)$ are as in (5.3.23):

$$\left.\begin{aligned}&Y^n_t = \sum_{i=1}^{[t/\Delta_n]-k+1} \zeta^n_i, \qquad A^n(0)_t = \sum_{i=1}^{[t/\Delta_n]-k+1} \chi''^n_i \\ &A^n(1)_t = \sum_{i=1}^{[t/\Delta_n]-k+1} \chi'^n_i \\ &A^n(2)_t = \frac{1}{\sqrt{\Delta_n}} \Bigl(\sum_{i=1}^{[t/\Delta_n]-k+1} \Delta_n\, \rho^{k\otimes}_{c_{(i-1)\Delta_n}}(F) - \int_0^t \rho^{k\otimes}_{c_s}(F)\, ds\Bigr).\end{aligned}\right\} \tag{11.2.24}$$

We also have $\overline{\mathcal{V}}'^n(F,X) = \mathcal{Y}^n + \mathcal{A}^n(0) + \mathcal{A}^n(1) + \mathcal{A}^n(2)$, where

$$\left.\begin{aligned}&\mathcal{Y}^n_t = \sum_{i=1}^{[t/k\Delta_n]} \zeta^n_{ik-k+1}, \qquad \mathcal{A}^n(0)_t = \sum_{i=1}^{[t/k\Delta_n]} \chi''^n_{ik-k+1} \\ &\mathcal{A}^n(1)_t = \sum_{i=1}^{[t/k\Delta_n]} \chi'^n_{ik-k+1} \\ &\mathcal{A}^n(2)_t = \frac{1}{\sqrt{\Delta_n}} \Bigl(\sum_{i=1}^{[t/k\Delta_n]} \Delta_n\, \rho^{k\otimes}_{c_{(ik-k)\Delta_n}}(F) - \frac{1}{k} \int_0^t \rho^{k\otimes}_{c_s}(F)\, ds\Bigr).\end{aligned}\right\} \tag{11.2.25}$$

Therefore the theorem will follow from the next three lemmas:

Lemma 11.2.5 *Under* (11.2.17) *we have* $Y^n \stackrel{\mathcal{L}\text{-s}}{\Longrightarrow} \overline{V}'(F,X)$.

Lemma 11.2.6 *Under* (11.2.17) *we have* $\mathcal{Y}^n \stackrel{\mathcal{L}\text{-s}}{\Longrightarrow} \overline{\mathcal{V}}'(F,X)$.

Lemma 11.2.7 *Under* (11.2.17), *for* $j=0,1,2$ *we have*

$$A^n(j) \stackrel{\text{u.c.p.}}{\Longrightarrow} 0, \qquad \mathcal{A}^n(j) \stackrel{\text{u.c.p.}}{\Longrightarrow} 0. \tag{11.2.26}$$

11.2.4 The Processes Y^n and $\mathcal{Y}^n$

Proof of Lemma 11.2.6 We apply Theorem 4.2.1, page 106, in the following setting: take $u_n = k\Delta_n$ and $q' = kd'$ and $q = kd$, and the function Φ on Ω^W and the process θ with components $\Phi(y)^{jl} = y^l(j/k) - y^l((j-1)/k)$ and $\theta^{j,i,j',l} = \sigma^{il} 1_{\{j=j'\}}$ for $j,j' = 1,\dots,k$ and $l = 1,\dots,d'$ and $i = 1,\dots,d$, and finally $G(y,x) = F(\sqrt{k}\,x)$ for $x \in (\mathbb{R}^d)^k$, so the process Y does not enter the picture. With these conventions, we have $\mathcal{Y}^n = \frac{1}{\sqrt{k}} \overline{U}^n(G)$, where $\overline{U}^n(G)$ is defined by (4.2.6), page 106.

Now, since F satisfies (11.2.5), it is straightforward to check that the functions of (4.2.5) satisfy $\widehat{\gamma}^{\Phi}_{\theta_t}(x, G^j) = 0$ and $\overline{\gamma}^{\Phi}_{\theta_t}(x, G^j, G^l) = \rho^{k\otimes}_{c_t}(F^j F^l) - \rho^{k\otimes}_{c_t}(F^j) \times \rho^{k\otimes}_{c_y}(F^l)$. Therefore the lemma is a special case of Theorem 4.2.1. □

The previous proof is simple because the summands in $\mathcal{Y}^n$ involve non-overlapping intervals. For Y^n this is no longer the case: we do have $\mathbb{E}(\zeta^n_i \mid \mathcal{F}_{(i-1)\Delta_n}) = 0$, but ζ^n_i is *not* $\mathcal{F}_{i\Delta_n}$ measurable but only $\mathcal{F}_{(i+k-1)\Delta_n}$ measurable. So we need to rewrite Y^n as a sum of martingale increments for the discrete-time filtration $(\mathcal{F}_{i\Delta_n})_{i\geq 0}$. Doing so, we significantly complicate the form of the (discrete-time) predictable quadratic variation, and we need a non trivial extension of the convergence $\Delta_n \sum_{i=1}^{[t/\Delta_n]} g(Z_{(i-1)\Delta_n}) \to \int_0^t g(Z_s)\,ds$ when Z is a càdlàg process and g a continuous function.

We begin by solving the second problem. We will use the simplifying notation

$$w^n_i = \Delta^n_i W/\sqrt{\Delta_n}, \qquad \sigma^n_i = \sigma_{i\Delta_n}, \qquad c^n_i = c_{i\Delta_n}, \tag{11.2.27}$$

so in particular $\beta^n_{i,j} = \sigma^n_{i-1} w^n_{i+j-1}$. Next, we let $u, v \geq 1$ be integers and D be a compact subset of $\mathbb{R}^v$, and g be a function on $D^2 \times (\mathbb{R}^{d'})^u$ such that, for some $\gamma \geq 0$,

$$g \text{ is continuous and } \bigl|g(z_1,z_2;y_1,\dots,y_u)\bigr| \leq K \prod_{j=1}^u \bigl(1 + \|y_j\|^\gamma\bigr). \tag{11.2.28}$$

Lemma 11.2.8 *Let g be as above with $1 \leq u \leq k-1$ and Z be a D-valued càdlàg adapted process satisfying*

$$s,t \geq 0 \quad \Rightarrow \quad \mathbb{E}\bigl(\|Z_{t+s} - Z_t\|^2\bigr) \leq K\,s. \tag{11.2.29}$$

Then if $1 \le j, j' \le k$ *we have, with the notation* (11.2.27) *and* $Z_i^n = Z_{i\Delta_n}$,

$$\Delta_n \sum_{i=k}^{[t/\Delta_n]} g\big(Z_{i-j}^n, Z_{i-j'}^n; w_{i-u}^n, \dots, w_{i-1}^n\big)$$
$$\stackrel{\mathbb{P}}{\longrightarrow} \int_0^t ds \int g(Z_s, Z_s; y_1, \dots, y_u)\, \rho(dy_1) \dots \rho(dy_u). \tag{11.2.30}$$

Proof Set

$$\mu_i^n = \Delta_n\, g\big(Z_{i-k}^n, Z_{i-k}^n; w_{i-u}^n, \dots, w_{i-1}^n\big)$$
$$\Gamma_s = \int g(Z_s, Z_s; y_1, \dots, y_u)\, \rho(dy_1) \dots \rho(dy_u).$$

(11.2.28) yields the existence of a family of continuous increasing function θ_C with $\theta_C(0) = 0$ such that, for any $\varepsilon \in (0, 1]$ and $C > 1$, we have

$$\big|g(z_1, z_2; y_1, \dots, y_u) - g(z, z; y_1, \dots, y_u)\big|$$
$$\le K \prod_{j=1}^{u} \big(1 + \|y_j\|^{\gamma+1}\big)\left(\theta_C(\varepsilon) + \frac{1}{C} + \frac{\|z_1 - z\| + \|z_2 - z\|}{\varepsilon}\right).$$

The variables w_i^n have moments of all order (they are $\mathcal{N}(0,1)$), so in view of (11.2.29) we can combine the previous estimate and the Cauchy-Schwarz inequality to get

$$\mathbb{E}\left(\sum_{i=1}^{[t/\Delta_n]} \big|\Delta_n\, g\big(Z_{i-j}^n, Z_{i-j'}^n; w_{i-u}^n\big) - \mu_i^n\big|\right) \le Kt\left(\theta_C(\varepsilon) + \frac{1}{C} + \frac{\sqrt{\Delta_n}}{\varepsilon}\right).$$

By choosing first C large, then ε small, then n large, we deduce that the left side above actually goes to 0 as $n \to \infty$. Therefore, (11.2.30) will hold, provided we have

$$\sum_{i=k}^{[t/\Delta_n]} \mu_i^n \stackrel{\mathbb{P}}{\longrightarrow} \int_0^t \Gamma_s\, ds.$$

For this, we observe that μ_i^n is $\mathcal{F}_{(i-1)\Delta_n}$ measurable, and we set $\mu_i'^n = \mathbb{E}(\mu_i^n \mid \mathcal{F}_{(i-k)\Delta_n})$ and $\mu_i''^n = \mu_i^n - \mu_i'^n$. We deduce from (11.2.28) that

$$\mathbb{E}\big(|\mu_i^n|^2\big) \le K\Delta_n^2, \qquad \mathbb{E}\big(|\mu_i''^n|^2\big) \le K\Delta_n^2, \quad j \ge k \Rightarrow \mathbb{E}\big(\mu_i''^n\, \mu_{i+j}''^n\big) = 0.$$

Hence

$$\mathbb{E}\left(\left(\sum_{i=k}^{[t/\Delta_n]} \mu_i''^n\right)^2\right) \le Kkt\Delta_n \to 0$$

and it remains to show $\sum_{i=k}^{[t/\Delta_n]} \mu_i'^n \stackrel{\mathbb{P}}{\longrightarrow} \int_0^t \Gamma_s\,ds$. Observe that $\mu_i'^n = \Gamma_{(i-k)\Delta_n}$, whereas Γ_s is càdlàg because Z_s is so and g is continuous. Then the result follows because $\sum_{i=k}^{[t/\Delta_n]} \mu_i'^n$ is a Riemann approximation of $\int_0^t \Gamma_s\,ds$. □

Proof of Lemma 11.2.5 1) For $j \in \{0,\dots,k-1\}$, we define by a downward induction on j the following functions F_j on $\mathcal{M}_A \times (\mathbb{R}^{d'})^j$, and with ρ as in the previous lemma:

$$\begin{aligned} F_k(\alpha; y_1,\dots,y_k) &= F(\alpha y_1,\dots,\alpha y_k),\\ F_j(\alpha; y_1,\dots,y_j) &= \int F_{j+1}(\alpha; y_1,\dots,y_j,z)\rho(dz). \end{aligned} \tag{11.2.31}$$

In particular $F_0(\alpha) = \rho_{\alpha\alpha^*}^{k\otimes}(F)$. With this notation, we have the decomposition

$$\zeta_i^n = \sum_{j=0}^{k-1} \zeta_{i,j}^n, \qquad \text{where}$$

$$\zeta_{i,j}^n = \sqrt{\Delta_n}\,\big(F_{j+1}\big(\sigma_{i-1}^n; w_i^n,\dots,w_{i+j}^n\big) - F_j\big(\sigma_{i-1}^n; w_i^n,\dots,w_{i+j-1}^n\big)\big)$$

(recall (11.2.23) and (11.2.27)). Therefore, if we set

$$\begin{aligned} \overline{U}_t'^n &= \sum_{i=k}^{[t/\Delta_n]} \eta_i^n, \quad \text{where} \quad \eta_i^n = \sum_{j=0}^{k-1} \zeta_{i-j,j}^n\\ L_t^n &= \sum_{j=0}^{k-1}\Bigg(\sum_{i=1}^{k-j-1} \zeta_{i,j}^n - \sum_{i=[t/\Delta_n]-k+1}^{[t/\Delta_n]-j} \zeta_{i,j}^n\Bigg), \end{aligned} \tag{11.2.32}$$

we have $Y_t^n = \overline{U}_t'^n + L_t^n$. Observe that $\zeta_{i-j,j}^n$ is $\mathcal{F}_{i\Delta_n}$ measurable, and w_i^n is independent of $\mathcal{F}_{(i-1)\Delta_n}$ and with law ρ, so $\mathbb{E}(\zeta_{i-j,j}^n \mid \mathcal{F}_{(i-1)\Delta_n}) = 0$. Thus η_i^n is $\mathcal{F}_{i\Delta_n}$ measurable and $\mathbb{E}(\eta_i^n \mid \mathcal{F}_{(i-1)\Delta_n}) = 0$, and the process $\overline{U}'^n$ is a discrete sum of martingale increments.

2) In this step, we prove

$$\sup_{s\le t} \big\|L_s^n\big\| \stackrel{\mathbb{P}}{\longrightarrow} 0. \tag{11.2.33}$$

By (11.2.17), as soon as $i+j \le [t/\Delta_n]$ we have $\|\zeta_{i,j}^n\| \le K\sqrt{\Delta_n}\,Z_t^n$, where $Z_t^n = 1 + \sup(\|w_i^n\|^p : 1 \le i \le [t/\Delta_n])$ with p as in (11.2.7). Since $\mathbb{E}(\|\Delta_i^n W\|^{8p}) \le K\Delta_n^{4p}$, Markov's inequality yields $\mathbb{P}(\|\Delta_i^n W\| > \Delta_n^{1/2-1/4p}) \le K_p\Delta_n^2$, and thus $Z_t^n \le \Delta_n^{-1/4}$ when n is bigger than some a.s. finite variable M_t. Since L_t^n contains less than k^2 summands we deduce $\sup_{s\le t}\|L_s^n\| \le Kk^2\sqrt{\Delta_n}\,Z_t^n \le Kk^2\Delta_n^{1/4}$ when $n \ge M_t$, hence the result.

3) Due to the decomposition $Y_t^n = \overline{U}_t^{\prime n} + L_t^n$ and to (11.2.33), it remains to prove that the sequence $\overline{U}^{\prime n}$ converges stably in law to $\overline{V}'(F,X)$. Since η_i^n are martingale increments, by Theorem 2.2.15 it is enough to show the following three properties:

$$\sum_{i=k}^{[t/\Delta_n]} \mathbb{E}\big(\|\eta_i^n\|^4 \mid \mathcal{F}_{(i-1)\Delta_n}\big) \stackrel{\mathbb{P}}{\longrightarrow} 0 \tag{11.2.34}$$

$$\sum_{i=k}^{[t/\Delta_n]} \mathbb{E}\big(\eta_i^{n,l}\,\eta_i^{n,r} \mid \mathcal{F}_{(i-1)\Delta_n}\big) \stackrel{\mathbb{P}}{\longrightarrow} \int_0^t R_{c_s}\big(F^l,F^r\big)\,ds \tag{11.2.35}$$

$$\sum_{i=k}^{[t/\Delta_n]} \mathbb{E}\big(\eta_i^n\,\Delta_i^n N \mid \mathcal{F}_{(i-1)\Delta_n}\big) \stackrel{\mathbb{P}}{\longrightarrow} 0 \tag{11.2.36}$$

for all $t>0$ and for any bounded martingale N orthogonal to W and also for $N = W^l$ for any l.

As in the previous step, $\|\eta_i^n\| \le K\sqrt{\Delta_n}\,(1+\sum_{j=0}^{k-1}\|w_{i-j}^n\|^p)$. Hence the expected value of the left side of (11.2.34) is smaller than $Kt\Delta_n$, thus goes to 0, and (11.2.34) holds.

4) In this step we prove (11.2.36). When N is a bounded martingale orthogonal to W, exactly the same argument as for (4.2.10), page 108, shows that the left side of (11.2.36) vanishes. So it remains to consider the case $N=W^l$. For simplicity, we write

$$\overline{\zeta}_{i,j}^n = \mathbb{E}\big(\zeta_{i,j}^n\,\Delta_{i+j}^n W^l \mid \mathcal{F}_{(i+j-1)\Delta_n}\big), \tag{11.2.37}$$

and we will prove the following:

$$\begin{array}{ll} \text{(i)} & \overline{\zeta}_{i,j}^n \text{ is } \mathcal{F}_{(i+k-1)\Delta_n}\text{-measurable} \\ \text{(ii)} & \mathbb{E}\big(\overline{\zeta}_{i,j}^n \mid \mathcal{F}_{(i-1)\Delta_n}\big) = 0 \\ \text{(iii)} & \sum_{i=1}^{[t/\Delta_n]} \mathbb{E}\big(\|\overline{\zeta}_{i,j}^n\|^2\big) \to 0. \end{array} \tag{11.2.38}$$

Indeed, assuming this, a simple computation shows that

$$\mathbb{E}\Bigg(\Bigg\|\sum_{i=1}^{[t/\Delta_n]} \overline{\zeta}_{i,j}^n\Bigg\|^2\Bigg) \le (2k-1)\sum_{i=1}^{[t/\Delta_n]} \mathbb{E}\big(\|\overline{\zeta}_{i,j}^n\|^2\big) \to 0$$

and we conclude (11.2.36) by the definition (11.2.32) of η_i^n.

Now we proceed to prove (11.2.38). The variables $w_i^n,\dots,w_{i+j-1}^n$ are $\mathcal{F}_{(i+j-1)\Delta_n}$-measurable, and $\Delta_i^n W_{i+j} = \sqrt{\Delta_n}\,w_{i+j}^n$. Therefore

$$\overline{\zeta}_{i,j}^n = \Delta_n \int F_{j+1}\big(\sigma_{i-1}^n;\,w_i^n,\dots,w_{i+j-1}^n,z\big)\,z^l\,\rho(dz)$$

and, since the function F is of polynomial growth, the same is true of each F_j. Thus $\mathbb{E}(\|\overline{\zeta}_{i,j}^n\|^2) \le K\Delta_n^2$, and the part (iii) of (11.2.38) follows, whereas part (i) is obvious since $j \le k-1$. For (ii), we deduce from what precedes and from the definition of the functions F_j that

$$\mathbb{E}\big(\overline{\zeta}^n_{i,j} \mid \mathcal{F}_{(i-1)\Delta_n}\big) = \int F_{j+1}\big(\sigma^n_{i-1}, z_1, \dots, z_{j+1}\big)\, z^l_{j+1}\, \rho(dz_1)\dots\rho(dz_{j+1})$$
$$= \int F_k\big(\sigma^n_{i-1}, z_1, \dots, z_k\big)\, z^l_{j+1}\, \rho(dz_1)\dots\rho(dz_k).$$

Since F is globally even, the function $G(z_1,\dots,z_k) = F_k(\alpha; z_1,\dots,z_k)\, z^r_{j+1}$ is globally odd in the sense that $G(-y) = -G(y)$ for all $y \in (\mathbb{R}^d)^k$. Thus the above integral vanishes, and (i) of (11.2.38) is proved, hence (11.2.36) as well.

5) For (11.2.35) we need to evaluate $\mathbb{E}(\eta^{n,l}_i \eta^{n,r}_i \mid \mathcal{F}_{(i-1)\Delta_n})$. For j, j' between 0 and $k-1$ we introduce some new functions on $\mathcal{M}_A \times \mathcal{M}_A \times (\mathbb{R}^{d'})^{j\vee j'}$:

$$\begin{aligned} F^{lr}_{j,j'}\big(\alpha, \alpha'; y_1, \dots, y_{j\vee j'}\big) = \int & F^l_{j+1}\big(\alpha; y_{1+(j'-j)^+}, \dots, y_{j\vee j'}, y\big) \\ & \times F^r_{j'+1}\big(\alpha'; y_{1+(j-j')^+}, \dots, y_{j\vee j'}, y\big)\, \rho(dy) \\ & - F^l_j\big(\alpha; y_{1+(j'-j)^+}, \dots, y_{j\vee j'}\big) \\ & \times F^r_{j'}\big(\alpha'; y_{1+(j-j')^+}, \dots, y_{j\vee j'}\big) \end{aligned} \tag{11.2.39}$$

A simple calculation shows that

$$\mathbb{E}\big(\zeta^{n,l}_{i-j,j}\, \zeta^{n,r}_{i-j',j'} \mid \mathcal{F}_{(i-1)\Delta_n}\big) => \Delta_n\, F^{lr}_{j,j'}\big(\sigma^n_{i-j-1}, \sigma^n_{i-j'-1}; w^n_{i-j\vee j'}, \dots, w^n_{i-1}\big). \tag{11.2.40}$$

Summing these over j, j' between 0 and $k-1$ gives us the left side of (11.2.35), but it also gives the right side, as seen below. Indeed, a version of the variables $U_1, U_2, \dots$ inside (11.2.4) is given by $U_i = \alpha w_i$ if $a = \alpha\alpha^*$: in this formula, and if one takes $F = F^l$ and $G = F^r$ and the σ-fields $\mathcal{G} = \mathcal{F}_{(k-1)\Delta_n}$ and $\mathcal{G}' = \mathcal{F}_{k\Delta_n}$, the (j, j')th summand is exactly $\mathbb{E}(\zeta^{n,l}_{k-j,j}\, \zeta^{n,r}_{k-j',j'} \mid \mathcal{F}_{(k-1)\Delta_n})$ in the case $\sigma_t = \alpha$ identically. In other words,

$$R_{\alpha\alpha^*}\big(F^l, F^r\big) = \sum_{j,j'=0}^{k-1} \int F^{lr}_{j,j'}(\alpha, \alpha; y_1, \dots, y_{j\vee j'})\, \rho(dy_1)\dots\rho(dy_{j\vee j'}). \tag{11.2.41}$$

Since $\mathbb{E}(\eta^{n,l}_i \eta^{n,r}_i \mid \mathcal{F}_{(i-1)\Delta_n}) = \sum_{j,j'=0}^{k-1} \mathbb{E}(\zeta^{n,l}_{i-j,j}\, \zeta^{n,r}_{i-j',j'} \mid \mathcal{F}_{(i-1)\Delta_n})$, by (11.2.40) and (11.2.41) we see that (11.2.35) amounts to the convergence (11.2.30) for the functions $g = F^{lr}_{j,j'}$ and the process $Z_t = \sigma_t$. That is, we have to check the conditions of Lemma 11.2.8.

For this, we first observe that under (SK) the process $Z_t = \sigma_t$ satisfies (11.2.29) and takes its values in the compact sets $D = \mathcal{M}_A$ or $D = \mathcal{M}'_A$, according to the case. Next, in view of the definitions of F_j and $F^{lr}_{j,j'}$, plus the continuity of the test function F and (11.2.7), by repeated applications of the dominated convergence theorem we can check that the function $g = F^{lr}_{j,j'}$ satisfies (11.2.28). This completes the proof. □

11.2.5 Proof of Lemma 11.2.7

For proving (11.2.26) we can argue component by component, so we assume here that F is one-dimensional. That is, $q=1$ and in case (a) $q'=1$, whereas in case (b) $q'=0$. We only prove the results for $A^n(j)$, since for $\mathcal{A}^n(j)$ it is the same.

Step 1) The process $A^n(2)$ is the same as in (5.3.23) except that $[t/\Delta_n]$ and ρ_{c_s} are substituted with $[t/\Delta_n]-k+1$ and $\rho_{c_s}^{k\otimes}$. So the proof of Part B of Sect. 5.3.3 holds here as well, giving $A^n(2) \stackrel{\text{u.c.p.}}{\Longrightarrow} 0$.

Step 2) We use the notation (11.2.23). The variable $\chi_i^{\prime\prime n}$ is $\mathcal{F}_{(i+k-1)\Delta_n}$-measurable and $\mathbb{E}(\chi_i^{\prime\prime n} \mid \mathcal{F}_{(i-1)\Delta_n})=0$ by construction. Hence, exactly as in (11.2.38) and right after this, for proving $A^n(0) \stackrel{\text{u.c.p.}}{\Longrightarrow} 0$ it suffices to show that

$$\sum_{i=1}^{[t/\Delta_n]} \mathbb{E}\big(|\chi_i^n|^2\big) \to 0. \tag{11.2.42}$$

Combining (11.2.19) and the estimates (11.2.22), we deduce from the definition of χ_i^n that

$$\mathbb{E}\bigg(\sum_{i=1}^{[t/\Delta_n]} |\chi_i^n|^2\bigg) \le Kt\bigg(\phi_C'(\varepsilon)^2 + \frac{1}{C^2} + \frac{\Delta_n C^{2p}}{\varepsilon^2}\bigg). \tag{11.2.43}$$

Letting $n\to\infty$, then $\varepsilon\to 0$, then $C\to\infty$, we deduce (11.2.42), and $A^n(0) \stackrel{\text{u.c.p.}}{\Longrightarrow} 0$ is proved.

Step 3) We now start the proof of $A^n(1) \stackrel{\text{u.c.p.}}{\Longrightarrow} 0$. The argument is the same as for the proof of (5.3.25), pages 152–160, except that we argue with the function F on $(\mathbb{R}^d)^k$ and thus replace β_i^n and $\Delta_i^n X/\sqrt{\Delta_n}$ by $\overline{\beta}_i^n$ and $\overline{X}_i^n$, hence θ_i^n by $\overline{\theta}_i^n = \overline{X}_i^n - \overline{\beta}_i^n$.

The variable $\overline{\theta}_i^n$ is indeed kd-dimensional, with the components

$$\overline{\theta}_i^{n,jl} = \frac{\Delta_{i+j-1}^n X^l}{\sqrt{\Delta_n}} - \frac{(\sigma_{(i-1)\Delta_n}\Delta_{i+j-1}^n W)^l}{\sqrt{\Delta_n}},$$

for $j=1\dots,k$ and $l=1,\dots,d$. Then, exactly as after (5.3.31), we have a decomposition $\overline{\theta}_i^n = \frac{1}{\sqrt{\Delta_n}}\sum_{r=1}^4 \overline{\zeta}(r)_i^n$, where the d-dimensional variables $\overline{\zeta}(r)_i^{n,j\cdot} = (\overline{\zeta}(r)_i^{n,jl})_{1\le l\le d}$ are given by

$$\overline{\zeta}(1)_i^{n,j\cdot} = \Delta_n\, b_{(i-1)\Delta_n}$$

$$\overline{\zeta}(2)_i^{n,j\cdot} = \int_{I(n,i+j-1)} \big(\widetilde{\sigma}_{(i-1)\Delta_n}(W_s - W_{(i-1)\Delta_n})\big)\, dW_s$$

$$\overline{\zeta}(3)_i^{n,j\cdot} = \int_{I(n,i+j-1)} \big(M_s' - M_{(i-1)\Delta_n}'\big)\, dW_s$$

$$\overline{\zeta}(4)_i^{n,j.} = \int_{I(n,i+j-1)} (b_s - b_{(i-1)\Delta_n})\,ds + \int_{I(n,i+j-1)} \left(\int_{(i-1)\Delta_n}^{s} \widetilde{b}'_u du \right) dW_s$$
$$+ \int_{I(n,i+j-1)} \left(\int_{(i-1)\Delta_n}^{s} (\widetilde{\sigma}_u - \widetilde{\sigma}_{(i-1)\Delta_n})\,dW_u \right) dW_s. \qquad (11.2.44)$$

We also set

$$A_i^n = \left\{ \|\overline{\theta}_i^n\| > d\left(\overline{\beta}_{i,j}^n, B\right)/2 \right\}.$$

Then, with obvious vector notation (the gradient ∇F is kd-dimensional, and below $\nabla F(.)\overline{\theta}_i^n$ stands for the usual scalar product), (5.3.32) becomes

$$F(\overline{Y}_i^n) - F(\overline{\beta}_i^n) = \nabla F(\overline{\beta}_i^n)\overline{\theta}_i^n + (F(\overline{\beta}_i^n + \overline{\theta}_i^n) - F(\overline{\beta}_i^n))1_{A_i^n}$$
$$-\nabla_j F(\overline{\beta}_i^n)\overline{\theta}_i^n\, 1_{A_i^n} + (\nabla_j F(\overline{\beta}_i^n + u_i^n \overline{\theta}_i^n) - \nabla_j F(\overline{\beta}_i^n))\,\overline{\theta}_i^n\, 1_{(A_i^n)^c}$$

where u_i^n is a random number between 0 and 1. Exactly as for (5.3.32), all variables in the right side above are almost surely well defined.

We still have the decomposition (5.3.33), that is $A^n(1) = \sum_{r=1}^{7} D^n(r)$, where

$$D^n(r)_t = \sum_{i=1}^{[t/\Delta_n]-k+1} \delta(r)_i^n, \qquad \delta(r)_i^n = \mathbb{E}\left(\delta'(r)_i^n \mid \mathcal{F}_{(i-1)\Delta_n}\right) \quad \text{and}$$
$$\delta'(r)_i^n = \nabla F\left(\overline{\beta}_i^n\right) \overline{\zeta}(r)_i^n \quad \text{for } r = 1,2,3,4$$
$$\delta'(5)_i^n = -\sqrt{\Delta_n}\, \nabla F\left(\overline{\beta}_i^n\right)\overline{\theta}_i^n\, 1_{A_i^n}$$
$$\delta'(6)_i^n = \sqrt{\Delta_n}\left(\nabla F\left(\overline{\beta}_i^n + u_i^n \overline{\theta}_i^n\right) - \nabla F\left(\overline{\beta}_i^n\right)\right)\overline{\theta}_i^n\, 1_{(A_i^n)^c}$$
$$\delta'(7)_i^n = \sqrt{\Delta_n}\left(F\left(\overline{\beta}_i^n + \overline{\theta}_i^n\right) - F\left(\overline{\beta}_i^n\right)\right) 1_{A_i^n}.$$

Step 4) We replace the definition (5.3.35) of α_i^n by

$$\alpha_i^n = \Delta_n^{3/2} + \mathbb{E}\left(\int_{(i-1)\Delta_n}^{(i+k-1)\Delta_n} \left(\|b_s - b_{(i-1)\Delta_n}\|^2 + \|\widetilde{\sigma}_s - \widetilde{\sigma}_{(i-1)\Delta_n}\|^2 \right) ds \right). \qquad (11.2.45)$$

Then (5.3.36) and (5.3.37) become for any $l > 0$ (here and below the constants depend on k, without special mention):

$$\mathbb{E}\left(\left\|\zeta(1)_i^n\right\|^l\right) + \mathbb{E}\left(\left\|\zeta(2)_i^n\right\|^l\right) \le K_l \Delta_n^l$$
$$\mathbb{E}\left(\left\|\zeta(4)_i^n\right\|^l\right) \le K_l\, \Delta_n^{l-1}\, \alpha_i^n, \qquad \mathbb{E}\left(\left\|\zeta(3)_i^n\right\|^l\right) \le K_l \Delta_n^{l/2+(1\wedge(l/2))}. \qquad (11.2.46)$$

Next, we set

$$\gamma_i^n = \begin{cases} 1 & \text{if } w = 1 \\ \phi_B(\overline{\beta}_i^n) & \text{if } w < 1 \end{cases}$$

(in particular in case (a) we have $\gamma_i^n = 1$). When $w < 1$, we have $\sigma_t \in \mathcal{M}'_A$, hence conditionally on $\mathcal{F}_{(i-1)\Delta_n}$ the kd-dimensional vector $\overline{\beta}_i^n$ is centered Gaussian with a covariance which is bounded, as well as its inverse (by a non-random bound). Then Lemma 5.3.14 applies in this situation, giving us $\mathbb{E}((\gamma_i^n)^t) \le K_t$ for all $t \in (0,1)$. In view of this, the proof of Lemma 5.3.15 applies without changes other than notational, apart from Step 2 of that proof. For this step, for any fixed j one should replace (5.3.44) by

$$\begin{aligned}\overline{\zeta}(3)_i^{n,j\cdot} &= \big(M'_{(i+j-1)\Delta_n} - M'_{(i-1)\Delta_n}\big)\Delta_{i+j-1}^n W \\ &\quad - \int_{I(n,i+j-1)} \sigma'_s(W_s - W_{(i+j-2)\Delta_n})\, dW'_s \\ &\quad - \int_{I(n,i+j-1)} \int_{E'} \delta'(s,z)(W_s - W_{(i+j-2)\Delta_n})\big(\mathfrak{p}' - \mathfrak{q}'\big)(ds,dz),\end{aligned}$$

and take the augmented filtration defined by $\mathcal{F}'_t = \mathcal{F}_t$ if $t < (i+j-1)\Delta_n$ and $\mathcal{F}'_t = \mathcal{F}_t \wedge \sigma(W_s : s \ge 0)$ otherwise. Then with

$$\widetilde{W}_i^n = \sup_{s \in ((i-1)\Delta_n, (i+k-1)\Delta_n]} \|W_s - W_{(i-1)\Delta_n}\|$$

we still have the estimate (5.3.45) for $\overline{\zeta}(3)_i^{n,j\cdot}$, conditionally on $\mathcal{F}_{(i+j-2)\Delta_n}$. If we do this for all j, we end up with

$$l \le 2 \quad \Rightarrow \quad \mathbb{E}\big(\big\|\overline{\zeta}(3)_i^n\big\|^l \mid \mathcal{F}'_{(i-1)\Delta_n}\big) \le K_l\, \Delta_n^{l/2} \big(\widetilde{W}_i^n\big)^l, \tag{11.2.47}$$

and the rest of the proof of Lemma 5.3.15 is unchanged.

The proof of Lemma 5.3.16, page 158, also goes true without major changes, due to the properties (11.2.18)–(11.2.20) of F: since we have here

$$\mathbb{E}\big(\big\|\overline{\beta}_i^n\big\|^l\big) \le K_l, \qquad \mathbb{E}\big(\big\|\overline{\theta}_i^n\big\|^l\big) \le K_n\, \Delta_n^{(l/2)\wedge 1}$$

for all $l \ge 0$, the arguments which all rely upon a repeated use of Hölder's inequality and Lemma 5.3.15 are still true. Therefore we obtain that $D^n(r) \overset{\text{u.c.p.}}{\Longrightarrow} 0$ for $r = 4,5,6,7$.

Step 5) At this stage, it remains to prove only $D^n(r) \overset{\text{u.c.p.}}{\Longrightarrow} 0$ for $r = 1,2,3$, and this property is clearly true if we have

$$r = 1,2,3 \quad \Rightarrow \quad \mathbb{E}\big(\nabla F\big(\overline{\beta}_i^n\big)\overline{\zeta}(r)_i^n \mid \mathcal{F}_{(i-1)\Delta_n}\big) = 0.$$

For this, it is enough to prove that, for all $j = 1,\dots,k$ and $j = 1,\dots,d$ and $r = 1,2,3$, we have with $G = \partial F/\partial x_j^l$:

$$\mathbb{E}\big(G\big(\overline{\beta}_i^n\big)\overline{\zeta}(r)_i^{n,jl} \mid \mathcal{F}_{(i-1)\Delta_n}\big) = 0. \tag{11.2.48}$$

When $r=3$, this property is proved exactly as in Lemma 5.3.17, page 159, upon taking the augmented filtration $(\mathcal{F}'_t)$ described in Step 4 above: then $\overline{\beta}_i^{n,j}$ is $\mathcal{F}'_{(i+j-1)\Delta_n}$ measurable, whereas as in Step 3 of the proof of Lemma 5.3.15, page 156, one sees that $\mathbb{E}(\overline{\zeta}(3)_i^{n,jl} \mid \mathcal{F}'_{(i+j-1)\Delta_n}) = 0$, hence (11.2.48) follows.

Finally, suppose that $r=1$ or $r=2$. Let $Y_s = W_{(i-1)\Delta_n+s} - W_{(i-1)\Delta_n}$ which, as a process, takes its values in the space $\mathbb{C}^{d'}$ of all continuous functions from $\mathbb{R}_+$ into $\mathbb{R}^{d'}$. We endow this space with the σ-field $\mathcal{C}^{d'}$ generated by all coordinates $y \mapsto y(s)$ and the Wiener measure $\mathbb{P}_W$, which is the law of W, and also the $\mathcal{F}_{(i-1)\Delta_n}$-conditional law of Y. From the definition of $\overline{\zeta}_i^{n,jl}$, we see that this variable is almost surely equal (and even everywhere equal when $r=1$) to $L(\omega, Y)$ for an appropriate function L on $\Omega \times \mathbb{C}^{d'}$, which is $\mathcal{F}_{(i-1)\Delta_n} \otimes \mathcal{C}^{d'}$-measurable and is "even" in the sense that $L(\omega, -y) = L(\omega, y)$. As for $G(\overline{\beta}_i^n)$, it is also (everywhere) equal to some function $L'(\omega, Y)$, with the same property as L except that now it is "odd", that is $L'(\omega, -y) = -L(\omega, y)$: this is because F is globally even, so its first partial derivative is globally odd. Hence the product $L'' = LL'$ is odd in this sense.

Now, a version of the conditional expectation in (11.2.48) is given by $\int L''(\omega, y) \times \mathbb{P}_W(dy)$, and since $\mathbb{P}_W$ is invariant by the map $y \mapsto -y$ and $y \mapsto L''(\omega, y)$ is odd, we obtain (11.2.48). This completes the proof.

11.3 Joint Central Limit Theorems

So far, we have "separate" CLTs for the non-normalized functionals $V^n(F,X)$ and the normalized functionals $V'^n(F,X)$. In some applications, we need a "joint" CLT for these two functionals, with different test functions and perhaps different numbers of increments k. And even, in some cases, it is necessary to have a joint CLT for these, together with the approximate quadratic variation $[X,X]^n$ studied in Chap. 5. In other words, we need an analogue of Theorem 5.5.1, see page 174, in the setting of the present chapter.

The two test functions are F and F', respectively with dimensions q and q', and it is no restriction to suppose that both are defined on $(\mathbb{R}^d)^k$ with the same k. We set

$$\begin{aligned}
\overline{V}^{n\#}(F,X)_t &= \frac{1}{\sqrt{\Delta_n}} \left(V^n(F,X)_t - \sum_{j=1}^k f_j * \mu_t \right) \\
\overline{V}'^n\big(F',X\big)_t &= \frac{1}{\sqrt{\Delta_n}} \left(V'^n(F,X)_t - \int_0^t \rho_{c_s}^{k\otimes}(F)\,ds \right) \\
\overline{Z}_t^{n\#,jl} &= \frac{1}{\sqrt{\Delta_n}} \left(\sum_{i=1}^{[t/\Delta_n]-k+1} \Delta_i^n X^j \Delta_i^n X^l - \big[X^j, X^l\big]_t \right) \\
\overline{Z}_t^{n,jl} &= \frac{1}{\sqrt{\Delta_n}} \left(\sum_{i=1}^{[t/\Delta_n]} \Delta_i^n X^j \Delta_i^n X^l - \big[X^j, X^l\big]_{\Delta_n[t/\Delta_n]} \right).
\end{aligned} \tag{11.3.1}$$

In particular $\overline{V}^{n\#}(F,X)$ is as in (11.1.12), and $\overline{V}'(F',X)$ as in (11.2.2), and $\overline{Z}^n$ as in (5.4.1).

The joint CLT will be considered in two cases:

case (A): for the variables $(\overline{V}^{n\#}(F,X)_t, \overline{V}'^n(F',X)_t, \overline{Z}_t^{n\#})$, with t fixed
case (B): for the processes $(\overline{V}'^n(F',X), \overline{Z}^n)$.

Basically, the joint CLT combines the CLTs for the components, and needs a description of the dependency between the components of the limit. For describing the "joint" limit we do as in Sect. 5.5, see page 173:

- We define $\overline{V}(F,X)$ by (11.1.7) and, with *the same* $R_{n,j}$, we set

$$\overline{Z}_t''^{ij} = \sum_{p=1}^{\infty}\sum_{n=1}^{\infty}\left(\Delta X_{T_p}^i R_{p,0}^j + \Delta X_{T_p}^j R_{p,0}^i\right) 1_{\{T_p \le t\}}. \tag{11.3.2}$$

- We consider the $\mathbb{R}^{d^2}$-valued function on $(\mathbb{R}^d)^k$ whose components are $\overline{F}^{ij}(x_1,\dots,x_k) = x_1^i x_1^j$. Then we use the notation $R_a(F,G)$ of (11.2.4) with F and G being the components of F' or those of $\overline{F}$; note that, due to the special form of $\overline{F}$, we have

$$R_a\left(\overline{F}^{ij}, \overline{F}^{ml}\right) = a^{im}a^{jl} + a^{il}a^{jm}. \tag{11.3.3}$$

- Let $(\overline{V}'(F',X), \overline{Z}')$ be a $(q'+d^2)$-dimensional continuous process on the extended space, which conditionally on $\mathcal{F}$ is a centered Gaussian martingale independent of all $R_{p,j}$ and with variance-covariance given by

$$\begin{aligned}
&\widetilde{\mathbb{E}}\left(\overline{V}'(F^i,X)_t \, \overline{V}'(F^j,X) \mid \mathcal{F}\right) = \int_0^t R_{c_s}\left(F^i,F^j\right) ds \\
&\widetilde{\mathbb{E}}\left(\overline{V}'(F^i,X)_t \, \overline{Z}_t'^{lm} \mid \mathcal{F}\right) = \int_0^t R_{c_s}\left(F^i,\overline{F}^{lm}\right) ds \\
&\widetilde{\mathbb{E}}\left(\overline{Z}_t'^{ij} \, \overline{Z}_t'^{lm} \mid \mathcal{F}\right) = \int_0^t R_{c_s}\left(\overline{F}^{ij},\overline{F}^{lm}\right) ds.
\end{aligned} \tag{11.3.4}$$

- Set $\overline{Z} = \overline{Z}' + \overline{Z}''$.

The results are as follows:

Theorem 11.3.1 *Let F be a C^2 function from $(\mathbb{R}^d)^k$ into $\mathbb{R}^q$, satisfying* (11.1.11), *and let F' be a function from $(\mathbb{R}^d)^k$ into $\mathbb{R}^{q'}$. Assume*

- *either X satisfies* (H) *and F' is identically* 0,
- *or F' and X satisfy the conditions of Theorem* 11.2.1 (*in particular, F' is globally even, and* (H) *still holds*).

Then for each t we have

$$\left(\overline{V}^{n\#}(F,X)_t,\overline{V}'^n\left(F',X\right)_t,\overline{Z}_t^{n\#}\right)\stackrel{\mathcal{L}\text{-s}}{\longrightarrow}\left(\overline{V}(F,X)_t,\overline{V}'\left(F',X\right)_t,\overline{Z}_t\right). \qquad (11.3.5)$$

Theorem 11.3.2 *If F' and X satisfy the conditions of Theorem* 11.2.1, *we have the (functional) stable convergence in law*

$$\left(\overline{V}'^n\left(F',X\right),\overline{Z}^n\right)\stackrel{\mathcal{L}\text{-s}}{\Longrightarrow}\left(\overline{V}'\left(F',X\right),\overline{Z}\right). \qquad (11.3.6)$$

Proof of Theorems 11.3.1 and 11.3.2 The proof of the two theorems is conducted together, and it is an extension of the proof of Theorem 5.5.1. We freely use the notation of the previous sections. It as usual, by localization we assume the strengthened assumptions (SH), (SK), (SK'), (SK-r), according to the case.

1) Set

$$\overline{Z}_t'^{n,jl}=\frac{1}{\sqrt{\Delta_n}}\sum_{i=1}^{[t/\Delta_n]-k+1}\left(\sum_{u,v=1}^{d'}\sigma_{(i-1)\Delta_n}^{ju}\,\sigma_{(i-1)\Delta_n}^{lv}\,\Delta_i^n W^u\,\Delta_i^n W^v-c_{(i-1)\Delta_n}^{jl}\right).$$

We let G be the $(q'+d^2)$-dimensional function on $(\mathbb{R}^d)^k$ whose components are (with obvious notation for the labels of components):

$$G^i(x_1,\dots,x_k)=F'^i(x_1,\dots,x_k),\qquad G^{lm}(x_1,\dots,x_k)=\overline{F}^{lm}(x_1).$$

Lemma 11.2.5 is valid with the function G instead of F', and it then takes the form

$$\left(Y^n,\overline{Z}'^n\right)\stackrel{\mathcal{L}\text{-s}}{\Longrightarrow}\left(\overline{U}'(G,X),\overline{Z}'\right). \qquad (11.3.7)$$

This result is indeed based on Theorem 4.2.1, but we also have a joint convergence with the "jumps". Namely, as in Theorem 4.3.1 (and with the same proof), Lemma 11.1.3 can be improved to give

$$\begin{aligned}&\left(Y^n,\overline{Z}'^n,\left(\left(R(n,p,j)\right)_{-k+1\le j\le k-1},L(n,p)\right)_{p\ge1}\right)\\&\quad\stackrel{\mathcal{L}\text{-s}}{\longrightarrow}\left(\overline{U}'\left(F',X\right),\overline{Z}',\left((R_{p,j})_{-k+1\le j\le k-1},L_p\right)_{p\ge1}\right).\end{aligned} \qquad (11.3.8)$$

2) We fix the integer $m\ge1$ for this step of the proof. Recall $Y^n(m)$, as given before (11.1.22), and set accordingly

$$\widehat{\Theta}^n(m)_t^{jl}=\sum_{p\in\mathcal{P}_m:\,S_p\le\Delta_n([t/\Delta_n]-k+1)}\left(R(n,p,0)^j\,\Delta X_{S_p}^l+R(n,p,0)^l\,\Delta X_{S_p}^j\right)$$

$$\widehat{\Theta}(m)_t^{jl}=\sum_{p\in\mathcal{P}_m:\,S_p\le t}\left(R_{p,0}^j\,\Delta X_{S_p}^l+R_{p,0}^l\,\Delta X_{S_p}^j\right).$$

Then (11.3.8) and the same argument as for deriving (11.1.24) from (11.1.19) give the following functional convergence, as $n \to \infty$:

$$\left(Y^n(m), Y^n, \overline{Z}'^n + \widehat{\Theta}^n(m)\right) \stackrel{\mathcal{L}\text{-}s}{\Longrightarrow} \left(\overline{V}\left(F, X'(m)\right), \overline{U}'\left(F', X\right), \overline{Z}' + \widehat{\Theta}(m)\right). \quad (11.3.9)$$

3) We are now in a position to prove (11.3.6). We have (11.3.9) for each m, and $\widehat{\Theta}(m) \stackrel{\text{u.c.p.}}{\Longrightarrow} \overline{Z}''$ as $m \to \infty$ is proved as in Step 2 of the proof of Lemma 5.4.12. Then, assuming that

$$\overline{V}'^n\left(F', X\right) - Y^n \stackrel{\text{u.c.p.}}{\Longrightarrow} 0 \qquad \forall T, \eta > 0 \quad (11.3.10)$$

and

$$\lim_{m\to\infty} \limsup_n \mathbb{P}\Big(\sup_{t\le T} \left\|\overline{Z}^n_{(t-(k-1)\Delta_n)^+} - \overline{Z}'^n_t - \widehat{\Theta}^n(m)_t\right\| > \eta\Big) = 0, \quad (11.3.11)$$

we deduce that

$$\left(\overline{V}'^n\left(F', X\right)_t, \overline{Z}^n_{(t-(k-1)\Delta_n)^+}\right) \stackrel{\mathcal{L}\text{-}s}{\Longrightarrow} \left(\overline{U}'\left(F', X\right), \overline{Z}\right).$$

Since $\overline{U}'(F', X)$ is continuous we also have the convergence, toward the same limit, of the sequence $(\overline{V}'^n(F', X)_{(t-(k-1)\Delta_n)^+}, \overline{Z}^n_{(t-(k-1)\Delta_n)^+})$. This in turn yields (11.3.6).

So it remains to prove (11.3.10) and (11.3.11). By Lemma 11.2.4 it suffices to prove (11.3.10) when X is continuous, and then this property follows from Lemma 11.2.7. As for (11.3.11), we observe that with the notation $Z'^n, Z', \Theta^n(m), \Theta(m)$ used in (5.5.9), page 176, we have

$$\overline{Z}'^{n,ij}_t = Z'^{n,ij}_{(t-(k-1)\Delta_n)^+} + Z'^{n,ji}_{(t-(k-1)\Delta_n)^+}, \qquad \overline{Z}''^{ij}_t = Z''^{ij}_t + Z'^{ji}_t$$

$$\widehat{\Theta}^n(m)^{ij}_t = \Theta^n(m)^{ij}_{(t-(k-1)\Delta_n)^+} + \Theta^n(m)^{ij}_{(t-(k-1)\Delta_n)^+},$$

$$\widehat{\Theta}(m)^{ij}_t = \Theta(m)^{ij}_t + \Theta(m)^{ij}_t.$$

Therefore (5.5.12) and the same argument as for deducing (5.5.10) from (5.5.9) give us (11.3.11).

4) Finally we turn to the proof of (11.3.5). We have seen in (11.1.27) that $\overline{V}(F, X(m)) \stackrel{\text{u.c.p.}}{\Longrightarrow} \overline{V}(F, X)$ as $m \to \infty$. Moreover, if t is fixed, we have

$$\overline{Z}^{n\#}_t - \overline{Z}^n_{(t-(k-1)\Delta_n)^+} = -\frac{1}{\sqrt{\Delta_n}}\left([X,X]_t - [X,X]_{\Delta_n([t/\Delta_n]-k+1)^+}\right) \stackrel{\mathbb{P}}{\longrightarrow} 0,$$

where the last convergence is shown in the proof of Theorem 5.4.2, page 172 (this is proved when $k = 1$, but the proof is valid for any k). Therefore, in view of (11.3.9), (11.3.10) and (11.3.11), it remains to show that

$$\lim_{m\to\infty} \limsup_{n\to\infty} \mathbb{P}\left(\overline{V}^n(F, X)_T - Y^n(m)_T| > \eta\right) = 0$$

for all t. This is a consequence of (11.1.22) and $\lim_n \mathbb{P}(\Omega_n(T, m)) = 1$ for all m and (11.1.28), and therefore the proof is complete. □

11.4 Applications

We pursue here the two applications started in Chap. 3. For simplicity we consider the one-dimensional case ($d = d' = 1$) only. We observe the semimartingale X at all times $i\Delta_n$ within a fixed time interval $[0, t]$. We also assume that X satisfies, at least, the following:

$$\text{We have (K) and } \int_0^t c_s \, ds > 0 \text{ a.s.} \tag{11.4.1}$$

11.4.1 Multipower Variations and Volatility

Again, our aim is to estimate the quantity

$$A(p)_t = \int_0^t |\sigma_s|^p \, ds, \tag{11.4.2}$$

mainly when $p = 2$, on the basis of the discrete observations of $X_{i\Delta_n}$ for all $i = 0, \ldots, [t/\Delta_n]$. When the process X is continuous, the problem was solved in Sect. 5.6.

When X jumps, though, these methods do not work, especially in the most interesting case $p = 2$. However we have seen in Sect. 8.5 that, for example, if $p > 0$ and k is an integer and

$$D(X; p, k; \Delta_n)_t = \sum_{i=1}^{[t/\Delta_n]-k+1} \prod_{j=1}^{k} \left|\Delta_{i+j-1}^n X\right|^{p/k}, \tag{11.4.3}$$

(the "equal-multipower" variation), then

$$k > \frac{p}{2} \Rightarrow \Delta_n^{1-p/2} D(X; p, k; \Delta_n) \overset{\text{u.c.p.}}{\Longrightarrow} (m_{p/k})^k \, A(p). \tag{11.4.4}$$

Now, Theorem 11.2.1 allows us to obtain a rate of convergence in (11.4.4), under appropriate conditions. More precisely, the left side of (11.4.4) is $\Delta_n V'^n(F, X)_t$ for the function $F(x_1, \ldots, x_k) = \prod_{j=1}^k |x_j|^{p/k}$. As mentioned in Example 11.2.2, this function is C^1 (and even C^∞) on $(\mathbb{R}\backslash\{0\})^k$ and satisfies (11.2.7), and also (11.2.8) and (11.2.9) with $w = 1 \wedge (p/k)$, and also (11.2.10) with $r = r' = p/k$. So we require $k > p$.

In this case, the asymptotic variance for the CLT is easily computed. Indeed, one has an explicit form for $R_a(F, F)$ in (11.2.4), which is (after some tedious calculations):

$$R_a(F, F) = M(p, k) a^p \quad \text{where}$$
$$M(p, k) = \frac{(m_{2p/k})^{k+1} + (m_{2p/k})^k (m_{p/k})^2 - (2k+1) m_{2p/k} (m_{p/k})^{2k} + (2k-1)(m_{p/k})^{2k+2}}{m_{2p/k} - (m_{p/k})^2}. \tag{11.4.5}$$

If we want $\Delta_n^{1-p/2} D(X; p,k; \Delta_n)_t$ to be a feasible estimator for the variable $(m_{p/k})^k A(p)_t$, we also need a consistent estimator for the asymptotic variance, which is $M(p,k) A(2p)_t$. For this, we can use again (11.4.4), and the same proof as for Theorem 5.6.1, page 177, gives us the following, by taking advantage of Theorem 11.2.1 (note that under our standing assumption (11.4.1) the set Ω_t^W of (3.5.10) has probability 1, and thus $A(p)_t > 0$ a.s. for all $p > 0$):

Theorem 11.4.1 *Let $p > 0$. If X is a (possibly discontinuous) Itô semimartingale satisfying* (11.4.1) *and* (K'-r) *for some $r < 1$ and if k is an integer satisfying*

$$p < k \le \frac{p}{r},$$

then for each $t > 0$ the random variables

$$\frac{(m_{2p/k})^{k/2}\,(\Delta_n^{1-p/2} D(X; p,k; \Delta_n)_t - (m_{p/k})^k A(p)_t)}{\sqrt{M(p,k)\,\Delta_n^{2-p} D(X; 2p,k,\Delta_n)_t}}$$

converge stably in law to a limit which is $\mathcal{N}(0,1)$ and independent of $\mathcal{F}$.

11.4.2 Sums of Powers of Jumps

Here we consider the problem of estimating the quantity

$$\overline{A}(p)_t = \sum_{s\le t} |\Delta X_s|^p$$

when $p > 3$. This as such is of doubtful interest for applications, but it plays a crucial role for the detection of jumps, as we will see in the next subsection. The condition $p > 3$ is not needed for consistency, but is crucial for obtaining rates of convergence.

The previous results provide us with a whole family of possible estimators. Indeed, fix an integer k and consider the function F on $\mathbb{R}^k$ given by

$$F(x_1,\dots,x_k) = |x_1+\dots+x_k|^p,$$

that is (8.2.5). It is symmetric and satisfies (11.1.11) because $p > 3$. The associated function f is $f(x) = |x|^p$, whose first derivative is $\partial f(x) = p\{x\}^p$, with the notation $\{y\}^p = y^p$ when $y \ge 0$ and $\{y\}^p = -|y|^p$ when $y < 0$. More generally with the notation (11.1.5) we have $\partial f_{(l);j}(x) = p\{x\}^{p-1}$ for all j,l.

By Theorem 8.2.1, both variables $\frac{1}{k} V^n(F,X)_t$ and $\mathcal{V}^n(F,X)_t$ are estimators for $\overline{A}(p)_t$, as is also the variable $V^n(f,X)_t$ (which is $V^n(f,X)_t = V^n(F,X)_t =$

$\mathcal{V}^n(F,X)_t$ if we take $k=1$). The normalized "estimation errors" are then

$$\begin{aligned} E(p,k)_n &= \frac{1}{\sqrt{\Delta_n}}\left(\frac{1}{k}\,V^n(F,X)_t - \overline{A}(p)_t\right) \\ \mathcal{E}(p,k)_n &= \frac{1}{\sqrt{\Delta_n}}\big(\mathcal{V}^n(F,X)_t - \overline{A}(p)_t\big). \end{aligned} \tag{11.4.6}$$

Theorem 11.1.2 says that

$$E(p,k)_n \xrightarrow{\mathcal{L}\text{-s}}$$

$$E(p,k) = \frac{p}{k}\sum_{n=1}^{\infty}\{\Delta X_{T_n}\}^{p-1}\Big(kR_{n,0} + \sum_{j=1}^{k-1}(k-j)(R_{n,j}+R_{n,-j}\Big)1_{\{T_n\le t\}}$$

$$\mathcal{E}(p,k)_n \xrightarrow{\mathcal{L}\text{-s}}$$

$$\mathcal{E}(p,k) = p\sum_{n=1}^{\infty}\{\Delta X_{T_n}\}^{p-1}\left(\sum_{j=1}^{k}\sum_{l=1}^{k} R^i_{n,l-j}\,1_{\{L_n=j-1\}}\right)1_{\{T_n\le t\}}$$

respectively, where $\{x\}^q = |x|^q\,\mathrm{sign}(x)$. The associated $\mathcal{F}$-conditional variances are

$$\begin{aligned} \widetilde{\mathbb{E}}\big(E(p,k)^2 \mid \mathcal{F}\big) &= \frac{(2k^2+1)p^2}{6k}\sum_{s\le t}|\Delta X_s|^{2p-2}(c_{s-}+c_s), \\ \widetilde{\mathbb{E}}\big(\mathcal{E}(p,k)^2 \mid \mathcal{F}\big) &= \frac{kp^2}{2}\sum_{s\le t}|\Delta X_s|^{2p-2}(c_{s-}+c_s). \end{aligned} \tag{11.4.7}$$

It is interesting to see that (11.4.8) follows from (5.1.5), without resorting to Theorem 11.1.2 at all: indeed $\mathcal{V}^n(F,X)$ is nothing else than $V^n(f,X)$, but for the regular discretization scheme with stepsize $k\Delta_n$, so, taking into account the fact that one normalizes by $\frac{1}{\sqrt{\Delta_n}}$ to get the limit, we see that (11.4.8) is exactly the same as (5.1.5) in this case.

We can compare the variances. The following inequalities are simple:

$$\widetilde{\mathbb{E}}\big(E(p,1)^2 \mid \mathcal{F}\big) \le \widetilde{\mathbb{E}}\big(E(p,k)^2 \mid \mathcal{F}\big) \le \widetilde{\mathbb{E}}\big(\mathcal{E}(p,k)^2 \mid \mathcal{F}\big).$$

Therefore it is always best (asymptotically) to use the simple estimator $V^n(f,X)_t$. As to the ratio of the variance of $E(p,k)$ over the variance of $\mathcal{E}(p,k)$, it decreases from 1 to 2/3 as k increases from 1 to ∞: this means that if we need to use an estimator with $k\ge 2$, it is always best to use $\frac{1}{k}\,V^n(F,X)_t$, and the advantage over $\mathcal{V}^n(F,X)_t$ is bigger when k is large.

11.4.3 Detection of Jumps

Now we turn to detection of jumps, as started in Sect. 3.5.2, page 93. With $k \geq 2$ an integer, $p > 3$, and F and f as above, we introduce two different test statistics:

$$S(p,k)_n = \frac{V^n(F,X)_t}{kV^n(f,X)_t}, \qquad \mathcal{S}(p,k)_n = \frac{\mathcal{V}^n(F,X)_t}{V^n(f,X)_t}, \tag{11.4.8}$$

Note that $\mathcal{S}(p,k)_n$ is the same as S_n in (3.5.6). Recalling the two subsets $\Omega_t^{(c)}$ and $\Omega_t^{(d)}$ on which the path $s \mapsto X_s$ on $[0,t]$ is continuous or not, and recalling (11.4.1), we have

$$\mathcal{S}(p,k)_n \xrightarrow{\mathbb{P}} \begin{cases} 1 & \text{on the set } \Omega_t^{(d)} \\ k^{p/2-1} & \text{on the set } \Omega_t^{(c)}, \end{cases}$$

according to Theorem 3.5.1. The very same proof as for this theorem, based upon Theorems 8.2.1 and 8.4.1, shows that, under the same assumptions, we have the same result for $S(p,k)_n$:

$$S(p,k)_n \xrightarrow{\mathbb{P}} \begin{cases} 1 & \text{on the set } \Omega_t^{(d)} \\ k^{p/2-1} & \text{on the set } \Omega_t^{(c)}. \end{cases}$$

To establish the distributional asymptotic behavior of these test statistics, we need a joint Central Limit Theorem for the pairs $(V^n(f,X)_t, V^n(F,X)_t)$ or $(V^n(f,X)_t, \mathcal{V}^n(F,X)_t)$. This amounts to a CLT for the pairs of variables $(E(p,1)_n, E(p,k)_n)$ and $(E(p,1)_n, \mathcal{E}(p,k)_n)$, with the notation (11.4.6).

Proposition 11.4.2 *In the above setting, the following holds*:

a) *The pair $(E(p,1)_n, E(p,k)_n)$ converges stably in law to a 2-dimensional variable $(Z, \overline{Z})$ which, conditionally on $\mathcal{F}$, is centered with covariance*

$$\begin{aligned} \widetilde{\mathbb{E}}\big(Z^2 \mid \mathcal{F}\big) &= \widetilde{\mathbb{E}}(Z\,\overline{Z} \mid \mathcal{F}) = \frac{p^2}{2} \sum_{s \leq t} |\Delta X_s|^{2p-2}\,(c_{s-} + c_s) \\ \widetilde{\mathbb{E}}\big(\overline{Z}^2 \mid \mathcal{F}\big) &= \frac{(2k^2+1)p^2}{6k} \sum_{s \leq t} |\Delta X_s|^{2p-2}(c_{s-} + c_s). \end{aligned} \tag{11.4.9}$$

b) *The pair $(E(p,1)_n, \mathcal{E}(p,k)_n)$ converges stably in law to a 2-dimensional variable $(Z, \widetilde{Z})$ which, conditionally on $\mathcal{F}$, is centered, with*

$$\begin{aligned} \widetilde{\mathbb{E}}\big(Z^2 \mid \mathcal{F}\big) &= \widetilde{\mathbb{E}}(Z\,\widetilde{Z} \mid \mathcal{F}) = \frac{p^2}{2} \sum_{s \leq t} |\Delta X_s|^{2p-2}\,(c_{s-} + c_s) \\ \widetilde{\mathbb{E}}\big(\widetilde{Z}^2 \mid \mathcal{F}\big) &= \frac{kp^2}{2} \sum_{s \leq t} |\Delta X_s|^{2p-2}(c_{s-} + c_s). \end{aligned} \tag{11.4.10}$$

Moreover when the processes X and σ have no common jumps, the variables $(Z,\overline{Z})$ and $(Z,\widetilde{Z})$ are $\mathcal{F}$-conditionally Gaussian.

Proof a) Consider the two-dimensional function G on $(\mathbb{R}^d)^k$ whose first component is $G^1=F$ and second component is $G^2(x_1,\dots,x_k)=|x_1|^p$. We have $V^n(G^1,X)_t=V^n(F,X)_t$ and $V^n(G^2,X)_t=V^n(f,X)_{([t/\Delta_n]-k+1)\Delta_n}$.

Set $t_n=([t/\Delta_n]-k+1)\Delta_n$. We have $t_n\to t$ and the limiting process $Z(f,X)$ of (5.1.4) is a.s. continuous at time t. Then we deduce from Theorem 5.1.2 that

$$\frac{1}{\sqrt{\Delta_n}}\big(V^n(f,X)_{t_n}-\overline{A}(p)_{\Delta_n[t_n/\Delta_n]}\big)-\frac{1}{\sqrt{\Delta_n}}\big(V^n(f,X)_t-\overline{A}(p)_{\Delta_n[t/\Delta_n]}\big)$$

converges in law to 0, hence also in probability. Moreover, we have (5.1.25) with t, and also with t_n (same proof), so $(V^n(G^2,X)_t-V^n(f,X)_t)/\sqrt{\Delta_n}\xrightarrow{\mathbb{P}}0$. In other words, proving (a) amounts to proving

$$\left(\frac{1}{\sqrt{\Delta_n}}\big(V^n\big(G^2,X\big)_t-\overline{A}(p)_t\big),\frac{1}{\sqrt{\Delta_n}}\big(V^n\big(G^1,X\big)_t-k\overline{A}(p)_t\big)\right)\xrightarrow{\mathcal{L}\text{-s}}(Z,\overline{Z}).\tag{11.4.11}$$

Now, the function G satisfies (11.1.11), and $g_j^1=f$ for all j, and $g_1^2=f$ and $g_j^2=0$ when $j=2,\dots,k$. Moreover $g'^1_{j(l)}(x)=g'^2_{1(1)}(x)=p\{x\}^{p-1}$ for all j,l and also $g'^2_{j(l)}=0$ when j and l are not both equal to 1. Therefore (a) of Theorem 11.1.2 yields that the pair of variables (11.4.11) converge stably in law to $\overline{V}(G,X)_t$, which is $\mathcal{F}$-conditionally centered, and whose components and $Z=\overline{V}(G^2,X)_t$ and $\overline{Z}=\frac{1}{k}\overline{V}(G^1,X)_t$ satisfy (11.4.9) by a simple computation based on (11.1.9).

b) Here we consider the two-dimensional function G on $(\mathbb{R}^d)^k$ with components $G^1=F$ and $G^2(x_1,\dots,x_k)=|x_1|^p+\cdots+|x_k|^p$, so G is symmetrical. We have $\mathcal{V}^n(G^1,X)_t=\mathcal{V}^n(F,X)_t$ and $\mathcal{V}^n(G^2,X)_t=V^n(f,X)_{(k\Delta_n[t/k\Delta_n]}$. Exactly as above, we have $(\mathcal{V}^n(G^2,X)_t-V^n(f,X)_t)/\sqrt{\Delta_n}\xrightarrow{\mathbb{P}}0$, so we need to prove

$$\left(\frac{1}{\sqrt{\Delta_n}}\big(\mathcal{V}^n\big(G^2,X\big)_t-\overline{A}(p)_t\big),\frac{1}{\sqrt{\Delta_n}}\big(\mathcal{V}^n\big(G^1,X\big)_t-\overline{A}(p)_t\big)\right)\xrightarrow{\mathcal{L}\text{-s}}(Z,\widetilde{Z}).\tag{11.4.12}$$

The function G satisfies (11.1.11), and $g_j^1=g_j^2=f$ for all j, and $g'^1_{j(l)}(x)=g'^2_{j(j)}(x)=p\{x\}^{p-1}$ for all j,l and also $g'^2_{j(l)}=0$ when $j\neq l$. Therefore (b) of Theorem 11.1.2 yields that the variables (11.4.12) converge stably in law to $(Z,\widetilde{Z})=\overline{\mathcal{V}}(G,X)_t$, which is $\mathcal{F}$-conditionally centered and satisfies (11.4.10) because of (11.1.10). □

This result becomes degenerate (all limits vanish) when X is continuous. In this case, we should replace $E(p,k)_n$ and $\mathcal{E}(p,k)_n$ by

$$E'(p,k)_n=\frac{1}{\sqrt{\Delta_n}}\left(\frac{\Delta_n^{1-p/2}}{k}V^n(F,X)_t-k^{p/2-1}m_pA(p)_t\right)$$

$$\mathcal{E}'(p,k)_n = \frac{1}{\sqrt{\Delta_n}}\big(\Delta_n^{1-p/2} \mathcal{V}^n(F,X)_t - k^{p/2-1} m_p A(p)_t\big)$$

(we use here the notation (11.4.2)). To express the result, we again need some notation. Recall that $d = 1$ here. We then set

$$\begin{aligned} &\overline{M}_p(k,l) = R_1(F,G), \quad \text{where} \\ &\quad F(x_1,\dots,x_k) = |x_1+\dots+x_k|^p, \quad G(x_1,\dots,x_k) = |x_1+\dots+x_l|^p \\ &\overline{M}'_p(k) = \mathbb{E}\big(|U|^p\,|U+\sqrt{k-1}\,V|^p\big), \quad \text{where} \\ &\quad U \text{ and } V \text{ are independent } \mathcal{N}(0,1). \end{aligned} \tag{11.4.13}$$

There is no explicit form for these quantities, unless p is an even integer or when $k = l = 1$, in which case $\overline{M}_p(1,1) = m_{2p} - m_p^2$ and $\overline{M}'_p(1) = m_{2p}$.

Proposition 11.4.3 *In the above setting, and if further X is continuous, we have:*
a) *The pair $(E'(p,1)_n, E'(p,k)_n)$ converges stably in law to a 2-dimensional variable $(Z', \overline{Z}')$ which, conditionally on $\mathcal{F}$, is centered Gaussian with covariance*

$$\begin{aligned} \widetilde{\mathbb{E}}\big(Z'^2 \mid \mathcal{F}\big) &= \overline{M}_p(1,1)\,A(2p)_t = \big(m_{2p} - m_p^2\big)A(2p)_t \\ \widetilde{\mathbb{E}}\big(Z'\overline{Z}' \mid \mathcal{F}\big) &= \frac{1}{k}\overline{M}_p(k,1)\,A(2p)_t \\ \widetilde{\mathbb{E}}\big(\overline{Z}'^2 \mid \mathcal{F}\big) &= \frac{1}{k^2}\overline{M}_p(k,k)\,A(2p)_t. \end{aligned}$$

b) *The pair $(E'(p,1)_n, \mathcal{E}'(p,k)_n)$ converges stably in law to a 2-dimensional variable $(Z', \widetilde{Z}')$ which, conditionally on $\mathcal{F}$, is centered Gaussian with covariance*

$$\begin{aligned} &\widetilde{\mathbb{E}}\big(Z'^2 \mid \mathcal{F}\big) = \big(m_{2p} - m_p^2\big)A(2p)_t \\ &\widetilde{\mathbb{E}}\big(Z'\widetilde{Z}' \mid \mathcal{F}\big) = \big(\overline{M}'_p(k) - k^{p/2}m_p^2\big)\,A(2p)_t \\ &\widetilde{\mathbb{E}}\big(\widetilde{Z}'^2 \mid \mathcal{F}\big) = k^{p-1}\big(m_{2p} - m_p^2\big)\,A(2p)_t. \end{aligned}$$

(This proposition does not require $p > 3$, it holds for all $p > 1$, and even for $p \in (0,1]$ when σ_s and σ_{s-} never vanish).

Proof a) We use the same function G and times t_n as in (a) of the previous proof. We have $E'(p,k)_n = \frac{1}{k}\overline{V}'^n(G^1,X)_t$, where we use the notation (11.2.2). We also have $E'(p,1)_n = \overline{V}'^n(G^2,X)_{t_n} + \alpha_n$, where $\alpha_n = \frac{1}{\sqrt{\Delta_n}} m_p \int_{t_n}^t c_s^{p/2}\,ds$. As in the previous proof again, $\overline{V}'^n(G^2,X)_{t_n} - \overline{V}'^n(G^2,X)_t \xrightarrow{\mathbb{P}} 0$, and obviously $\alpha_n \to 0$ because $t - t_n \le k\Delta_n$. Therefore $(k\,E'(p,k)_n, E'(p,1)_n)$ has the same stable limit in law as $\overline{V}'^n(G,X)_t$. The assumptions of (a) of Theorem 11.2.1 being fulfilled, the result follows from this theorem and the definition (11.4.13).

b) Now we use G as in (b) of the previous proof, and again the result is a consequence of the same Theorem 11.2.1, part (b). □

Coming back to the variables $S(p,k)_n$ and $\mathcal{S}(p,k)_n$ of (11.4.8), we have the following behavior.

Theorem 11.4.4 *Assume* (11.4.1). *With the notation* $u_t = \lambda(\{z : \delta(t,z) \neq 0\})$ *and* $T = \inf(t : \int_0^t u_s\,ds = \infty)$, *assume also that the process*

$$w_t = \begin{cases} \int_{\{|\delta(t,z)|\leq 1\}} \delta(t,z)\,\lambda(dz) & \text{if } u_t < \infty \\ +\infty & \text{otherwise.} \end{cases} \tag{11.4.14}$$

is Lebesgue-almost everywhere equal on the interval $[0,T]$ *to an* $\mathbb{R}_+$*-valued càdlàg (resp. càglàd) process if* b *is càdlàg (resp. càglàd). Then we have*:

a) *The variables* $\frac{1}{\sqrt{\Delta_n}}(S(p,k)_n - 1)$ *and* $\frac{1}{\sqrt{\Delta_n}}(\mathcal{S}(p,k)_n - 1)$ *converge stably in law, in restriction to the set* $\Omega_t^{(d)}$ *on which* X *has at least one jump on the interval* $[0,t]$, *to two variables* $S(p,k)$ *and* $\mathcal{S}(p,k)$ *which are* $\mathcal{F}$*-conditionally centered with conditional variances*

$$\left.\begin{aligned} \widetilde{\mathbb{E}}\big(S(p,k)^2 \mid \mathcal{F}\big) &= \frac{(k-1)(2k-1)}{6k}\,\frac{\sum_{s\leq t}|\Delta X_s|^{2p-2}(c_{s-}+c_s)}{(\sum_{s\leq t}|\Delta X_s|^p)^2} \\ \widetilde{\mathbb{E}}\big(\mathcal{S}(p,k)^2 \mid \mathcal{F}\big) &= \frac{k-1}{2}\,\frac{\sum_{s\leq t}|\Delta X_s|^{2p-2}(c_{s-}+c_s)}{(\sum_{s\leq t}|\Delta X_s|^p)^2}. \end{aligned}\right\} \tag{11.4.15}$$

Moreover when the processes X *and* σ *have no common jumps, the variables* $S(p,k)$ *and* $\mathcal{S}(p,k)$ *are* $\mathcal{F}$*-conditionally Gaussian.*

b) *The variables* $\frac{1}{\sqrt{\Delta_n}}(S(p,k)_n - k^{p/2-1})$ *and* $\frac{1}{\sqrt{\Delta_n}}(\mathcal{S}(p,k)_n - k^{p/2-1})$ *converge stably in law, in restriction to the set* $\Omega_t^{(c)}$ *on which* X *is continuous on the interval* $[0,t]$, *to two variables* $S'(p,k)$ *and* $\mathcal{S}'(p,k)$, *which are* $\mathcal{F}$*-conditionally centered Gaussian with conditional variances*

$$\left.\begin{aligned} \widetilde{\mathbb{E}}\big(S'(p,k)^2 \mid \mathcal{F}\big) &= \frac{\overline{M}_p(k,k) - 2k^{p/2}\overline{M}_p(k,1) + k^p\overline{M}_p(1,1)}{k^2 m_p^2}\,\frac{A(2p)_t}{(A(p)_t)^2} \\ \widetilde{\mathbb{E}}\big(\mathcal{S}'(p,k)^2 \mid \mathcal{F}\big) &= \frac{k^{p-2}((k+1)m_{2p} + (k-1)m_p^2) - 2k^{p/2-1}\overline{M}'_p(k)}{m_p^2}\,\frac{A(2p)_t}{(A(p)_t)^2}. \end{aligned}\right\} \tag{11.4.16}$$

Proof The first claim (a) is an application of Proposition 11.4.2, upon using the so-called "delta method" in statistics. The second claim (b) follows analogously from Proposition 11.4.3, at least when X is continuous, that is when $\Omega_t^{(c)} = \Omega$. To accommodate the case $\Omega_t^{(c)} \neq \Omega$ (which is the real case of interest), we do exactly as in the proof of Theorem 3.5.1, whose notation is used, see page 94. The only change is that we need the semimartingales $X(q)$ of that proof to satisfy (11.4.1), and for this the only thing to prove is that the process $b - w$ is either càdlàg or càglàd on $[0,T]$.

For this we observe that we can modify the function $z \mapsto \delta(\omega,t,z)$ on a set of times t having a vanishing Lebesgue measure, without altering (11.0.1). Therefore our assumption implies that, upon choosing a "good" version of δ, one may suppose that w itself is càdlàg (resp. càglàd) when b is càdlàg (resp. càglàd). This completes the proof. □

We also have a standardized version for these results, exactly as in Theorem 11.4.1. That is, we can find positive variables Γ_n, depending only on the values $X_{i\Delta_n}$ for $0 \le i \le [t/\Delta_n]$ at stage n, and such that

$$\begin{aligned} &\frac{S(p,k)_n - 1}{\Gamma_n} \stackrel{\mathcal{L}\text{-s}}{\longrightarrow} S \quad \text{on } \Omega_t^{(c)}, \quad \text{where} \\ &S \text{ is } \mathcal{F}\text{-conditionally centered with variance } 1 \\ &\frac{S(p,k)_n - k^{p/2-1}}{\Gamma_n} \stackrel{\mathcal{L}\text{-s}}{\longrightarrow} S \quad \text{on } \Omega_t^{(d)}, \quad \text{where} \\ &S \text{ is } \mathcal{F}\text{-conditionally } \mathcal{N}(0,1) \end{aligned} \tag{11.4.17}$$

and the same for $\mathcal{S}(p,k)_n$. Of course Γ_n is not the same in the two statements above, and it depends on (p,k). More precisely, for the first case above, $(\Gamma_n)^2$ should converge in probability to the first conditional variance in (11.4.15), in restriction to the set $\Omega_t^{(d)}$, and for the second case it should converge in probability to the first conditional variance in (11.4.16), in restriction to the set $\Omega_t^{(c)}$.

There are in fact many choices for Γ_n, all relying upon "estimators" for the random quantities appearing in the numerator and the denominator of the right sides of (11.4.15) and (11.4.16). For $A(p)_t$ and $A(2p)_t$ we can use the variables $D(X;q,k';\Delta_n)_t$ of (11.4.3) (with $q = p$ and $q = 2p$ and a suitable choice of k'), or those given in Chap. 9. For $\overline{A}(p)_t$, appearing in the denominator of (11.4.15) we may use the square of $\sum_{i=1}^{[t/\Delta_n]} |\Delta_i^n X|^p$. For the numerator of (11.4.15) we may use the estimators provided in Theorem 9.5.1.

We are now ready to construct the test. At each stage n, and based upon the statistics $S_n = S(p,k)_n$ or $S_n = \mathcal{S}(p,k)_n$, the critical (rejection) region for the null hypotheses $\Omega_t^{(d)}$ and $\Omega_t^{(c)}$ respectively will be

$$C_n^d = \{S_n > 1 + \gamma_n\}, \qquad C_n^c = \{S_n < k^{p/2-1} - \gamma_n'\}.$$

Here γ_n and γ_n' are positive "observable" variables (that is, depending on $X_{i\Delta_n}$ for $i = 0, 1, \dots$ only), and chosen so that the asymptotic level of either family of tests is some prescribed value $\alpha \in (0,1)$.

Consider for example the second test, for which the normalized variables $(S_n - k^{p/2-1})/\Gamma_n$ are asymptotically $\mathcal{N}(0,1)$ (conditionally under the null hypothesis $\Omega_t^{(c)}$), where Γ_n is as in (11.4.17). We can then take $\gamma_n = \Gamma_n z_\alpha$, where z_α is the α-quantile of $\mathcal{N}(0,1)$, that is the number such that $\mathbb{P}(U > z_\alpha) = \alpha$, where U is $\mathcal{N}(0,1)$. For the first test, the suitably normalized $(S_n - 1)/\Gamma_n$ is asymptotically centered with variance 1, but no longer necessarily Gaussian, so we can replace z_α by a quantity constructed using the Markov inequality, which is of course too large, or we may do "conditional simulations" to get a hand on the correct value of the α-quantile.

In any case, the precise definition of the asymptotic level, and even more of the asymptotic power, would take us too far here, and leads to rather intricate problems. For more information, the reader can consult Aït-Sahalia and Jacod [2] for tests based on $\mathcal{S}(p,k)_n$, whereas Fan and Fan [32] have introduced the (more accurate) tests based on $S(p,k)_n$.

Comments: We have in fact a whole family of tests, according to whether we choose $S(p,k)$ or $\mathcal{S}(p,k)$, and according to the values of k and p. Comparing two tests with different k's and/or p's is a delicate matter, and comparing the asymptotic variances of the test statistics is of course not enough: if $k'^{p'/2-1} > k^{p/2-1}$ the two possible limit values of $S(p',k')_n$ are easier to distinguish for $S(p',k')_n$ than for $S(p,k)_n$, but the asymptotic variance may also be bigger, for one of the two tests, or for both.

Let us just say that, and in restriction to $\Omega_t^{(d)}$, the ratio of the asymptotic variances of $S(p,k)_n$ and $\mathcal{S}(p,k)_n$ is $\frac{2k-1}{3k}$, which increases from $1/2$ when $k=2$ to $2/3$ when k increases to infinity. This suggests that one should preferably use $S(p,k)$ if one wants to construct a test on the basis of this theorem.

Bibliographical Notes

This chapter and its companion Chap. 8 are motivated by the multipower variations and also by the problem of jump detection. Multipower variations were introduced by Barndorff-Nielsen and Shephard [8] for the estimation of the integrated volatility when there are jumps, and the Law of Large Numbers of Theorem 8.4.1 was proved by these authors in the case of multipower variations. The literature, essentially in econometrics, is vast, and we can quote for example Barndorff-Nielsen and Shephard [10], and Woerner [94, 95]; this topic includes estimation based on the range swept by the observed process between two successive observation times (this is a functional of the form (8.1.9) or (8.1.11)), see Garman and Klass [36] and Parkinson [77] for early works, Christensen and Podolskij [22] for a more recent one.

The first Central Limit Theorem 11.1.2 is taken from Jacod [60], the second one (Theorem 11.2.1) is a simple extension of similar results when X is continuous, due to Barndorff-Nielsen, Graversen, Jacod, Podolskij and Shephard [11], in the spirit of Kinnebrock and Podolskij [66]; however, an earlier version of this result, when further the stochastic volatility is independent of the process X, may be found in Barndorff-Nielsen and Shephard [10]. When there are jumps, it is new, although Barndorff-Nielsen, Shephard and Winkel [12] have a version for Lévy jumps. The applications to volatility estimation, to the relative importance of jumps, or to the detection of jumps, may be found in the above mentioned papers and in other papers by Huang and Tauchen [48], Aït-Sahalia and Jacod [2], and Fan and Fan [32] for the comparison given in Theorem 11.4.4.

Chapter 12
The Central Limit Theorem for Functions of an Increasing Number of Increments

As the name suggests, the topic of this chapter is the same as the previous one, except that the number of successive increments involved in each summand of the functionals is no longer fixed but goes to infinity as the time step goes to 0.

Only regular discretization schemes are considered. The d-dimensional Itô semimartingale X has the Grigelionis decomposition

$$X_t = X_0 + \int_0^t b_s\,ds + \int_0^t \sigma_s\,dW_s + (\delta 1_{\{\|\delta\|\leq 1\}}) \star (\mathfrak{p} - \mathfrak{q})_t + (\delta 1_{\{\|\delta\|>1\}}) \star \mathfrak{p}_t, \quad (12.0.1)$$

where W is a d'-dimensional Wiener process and $\mathfrak{p}$ is a Poisson measure with compensator $\mathfrak{q}(dt, dz) = dt \otimes \lambda(dz)$, and $c = \sigma\sigma^\star$. We also assume at least Assumption (H), that is

Assumption (H) In (12.0.1), b_t is locally bounded and σ_t is càdlàg, and $\|\delta(\omega, t, z)\| \wedge 1 \leq \Gamma_n(z)$ for all (ω, t, z) with $t \leq \tau_n(\omega)$, where (τ_n) is a localizing sequence of stopping times and each function Γ_n satisfies $\int \Gamma_n(z)^2 \lambda(dz) < \infty$.

Before starting, we recall the general setting of Chap. 8. We have a sequence k_n of integers satisfying

$$k_n \to \infty, \qquad u_n = k_n \Delta_n \to 0. \qquad (12.0.2)$$

We define the re-scaled processes $X(n, i)$, with time ranging through $[0, 1]$, and the "discretized" versions $X(n, i)^{(n)}$ of those, along the time step $1/k_n$, according to

$$X(n, i)_t = X_{(i-1)\Delta_n + tu_n} - X_{(i-1)\Delta_n}, \qquad X(n, i)_t^{(n)} = X(n, i)_{[k_n t]/k_n}.$$

Then $X(n, i)^{(n)}$ only involves the increments $\Delta_j^n X$ for $j = 0, \ldots, k_n - 1$. The functionals of interest here are

J. Jacod, P. Protter, *Discretization of Processes*,
Stochastic Modelling and Applied Probability 67,
DOI 10.1007/978-3-642-24127-7_12, © Springer-Verlag Berlin Heidelberg 2012

$$V^n(\Phi,k_n,X)_t = \sum_{i=1}^{[t/\Delta_n]-k_n+1} \Phi\big(X(n,i)^{(n)}\big)$$

$$V'^n(\Phi,k_n,X)_t = \Delta_n \sum_{i=1}^{[t/\Delta_n]-k_n+1} \Phi\big(X(n,i)^{(n)}/\sqrt{u_n}\big),$$

where Φ is a function on the Skorokhod space $\mathbb{D}_1^d$ (recall the comments after (8.1.8), about the Skorokhod topology on this space).

As for functionals depending on a single increment, the central limit theorems require some kind of differentiability of the function Φ. Differentiability for a function on $\mathbb{D}_1^d$ is not a trivial notion, and to avoid many complications we will restrict ourselves to a very special situation. Namely, we say that Φ is a *moving average function* if it has the form

$$\left.\begin{array}{l} \Phi(x) = f\big(\int x(s)\,G(ds)\big), \quad \text{where} \\ \quad\bullet\ f \text{ is a continuous function on } \mathbb{R}^d \\ \quad\bullet\ G \text{ is a signed finite measure on } [0,1] \text{ with } G(\{0\})=0. \end{array}\right\} \tag{12.0.3}$$

Such a function Φ on $\mathbb{D}_1^d$ satisfies the basic assumption (8.1.13), page 230. For a better understanding of this class of functions, let us consider the left-continuous function

$$g(s) = G\big([s,1]\big). \tag{12.0.4}$$

A simple computation shows that the ith summand in $V^n(\Phi,k,X)_t$ is

$$\Phi\big(X(n,i)^{(n)}\big) = f\left(\sum_{j=1}^{k_n} g(j/k_n)\,\Delta_{i+j-1}^n X\right).$$

For example, if G is the Dirac mass at 1, then $\Phi(X(n,i)^{(n)})$ is equal to $f(X_{(i-1+k_n)\Delta_n} - X_{(i-1)\Delta_n})$, which reduces to $\Phi(X(n,i)^{(n)}) = f(\Delta_i^n X)$ when further $k_n = 1$ (a case excluded in this chapter, however).

Although very restrictive in a sense, this setting covers many statistical applications, especially in the case when there is an observation noise: see Chap. 16.

The content of this chapter is formally new, although partly contained in the paper [63] of Jacod, Podolskij and Vetter, and its main motivation is to serve as an introduction to Chap. 16 on noise.

12.1 Functionals of Non-normalized Increments

12.1.1 The Results

Below, the test function Φ is *always* a moving average function. As soon as the function f in (12.0.3) satisfies $f(z) = o(\|z\|^2)$ as $z \to 0$ in $\mathbb{R}^d$ (the same as in (A-a) of Theorem 3.3.1), for each $t>0$ we have by Theorem 8.3.1:

$$\begin{aligned}\frac{1}{k_n}V^n(\Phi,k_n,X)_t &= \frac{1}{k_n}\sum_{i=1}^{[t/\Delta_n]-k_n+1} f\left(\sum_{j=1}^{k_n} g(j/k_n)\,\Delta_{i+j-1}^n X\right)\\ &\stackrel{\mathbb{P}}{\longrightarrow}\sum_{s<t}\int_0^1 f\big(g(u)\,\Delta X_s\big)\,du \qquad (12.1.1)\end{aligned}$$

(we do not have Skorokhod convergence here). In this formulation the measure G has disappeared and only the function g shows up, but it is important that g be left-continuous with bounded variation.

In particular, in the one-dimensional case $d=1$, we get when $p>2$:

$$\frac{1}{k_n}\sum_{i=1}^{[t/\Delta_n]-k_n+1}\left|\sum_{j=1}^{k_n} g(j/k_n)\,\Delta_{i+j-1}^n X\right|^p \stackrel{\mathbb{P}}{\longrightarrow}\left(\int_0^1 |g(u)|^p\,du\right)\sum_{s<t}|\Delta X_s|^p.$$

Here we aim to a CLT associated with (12.1.1), and as usual we give a multidimensional result. That is, we consider an $\mathbb{R}^q$-valued function Φ on $\mathbb{D}^d$, whose coordinates Φ^l are of the form (12.0.3) with the associated measures G^l and functions f^l and g^l. Moreover, the functions f^l should have a very special form, namely they should be linear combinations of positively homogeneous functions, as introduced before (3.4.3): recall that h is a positively homogeneous function with degree $p>0$ if it satisfies

$$x\in\mathbb{R}^d,\ \lambda\ge 0\ \Rightarrow\ h(\lambda x)=\lambda^p h(x),\qquad\text{or equivalently}\quad h(x)=\|x\|^p\,h\big(x/\|x\|\big).$$

This is of course very restrictive; for example if $d=1$ this is equivalent to saying that $h(x)=\alpha_+x^p$ for $x\ge 0$ and $h(x)=\alpha_-(-x)^p$ if $x<0$, for $\alpha_\pm$ arbitrary reals. In any dimension, it implies $h(0)=0$.

The "rate of convergence" will be $\sqrt{u_n}$, so we are interested in the convergence of the q-dimensional process $Y^n(X)$ with components

$$\begin{aligned}Y^n(X)_t^l = \frac{1}{\sqrt{u_n}}\Bigg(&\frac{1}{k_n}\sum_{i=1}^{[t/\Delta_n]-k_n+1} f^l\left(\sum_{j=1}^{k_n} g^l(j/k_n)\,\Delta_{i+j-1}^n X\right)\\ &-\sum_{s\le t}\int_0^1 f^l\big(g^l(u)\,\Delta X_s\big)\,du\Bigg) \qquad (12.1.2)\end{aligned}$$

when each f^l is a linear combination of positively homogeneous functions, hence in particular $f^l(0)=0$.

We have to describe the limiting process, which again is rather tedious. We begin with some notation. We assume that each function f^l is C^1 on $\mathbb{R}^d$, and we set for

$x \in \mathbb{R}^d$ and $t \in [0,1]$ and $j, j' = 1, \dots, q$ and $i, i' = 1, \dots, d$:

$$\left.\begin{aligned} h_-(x,t)_i^j &= \int_0^t \partial_i f^j(g^j(s+1-t)x)\, g^j(s)\, ds \\ h_+(x,t)_i^j &= \int_t^1 \partial_i f^j(g^j(s-t)x)\, g^j(s)\, ds \\ H_-(x)_{ii'}^{jj'} &= \int_0^1 h_-(x,t)_i^j\, h_-(x,t)_{i'}^{j'}\, dt \\ H_+(x)_{ii'}^{jj'} &= \int_0^1 h_+(x,t)_i^j\, h_+(x,t)_{i'}^{j'}\, dt. \end{aligned}\right\} \tag{12.1.3}$$

For i, i' fixed, the $\mathbb{R}^q \otimes \mathbb{R}^q$-valued function $H_{\pm ii'}$ on $\mathbb{R}^d$ is locally bounded and takes its values in the set $\mathcal{M}^+_{q\times q}$ of all $q \times q$ symmetric nonnegative matrices. Moreover, the assumptions made on f below will imply that $|H_t(x)_{ii'}^{jj'}| = o(\|x\|^4)$ as $x \to 0$. In this case, and recalling $c = \sigma\sigma^*$, the following formulas define two $\mathcal{M}^+_{q\times q}$-valued processes $\xi_s = (\xi_s^{jj'})$ and $\Xi_t = (\Xi_t^{jj'})$, the second one being non-decreasing for the strong order in $\mathcal{M}^+_{q\times q}$:

$$\xi_s^{jj'} = \sum_{i,i'=1}^d \left(c_{s-}^{ii'} H_-(\Delta X_s)_{ii'}^{jj'} + c_s^{ii'} H_+(\Delta X_s)_{ii'}^{jj'}\right), \qquad \Xi_t = \sum_{s\le t} \xi_s. \tag{12.1.4}$$

This is enough to characterize the limiting process, which will be as follows:

$$\left.\begin{array}{l} Y(X) \text{ is defined on a very good filtered extension } (\widetilde{\Omega}, \widetilde{\mathcal{F}}, (\widetilde{\mathcal{F}}_t)_{t\ge0}, \widetilde{\mathbb{P}}) \text{ of} \\ (\Omega, \mathcal{F}, (\mathcal{F}_t)_{t\ge0}, \mathbb{P}) \text{ and, conditionally on } \mathcal{F}, \text{ is a purely discontinuous} \\ \text{centered Gaussian martingale with } \widetilde{\mathbb{E}}(Y(X)_t^j\, Y(X)_t^{i'} \mid \mathcal{F}) = \Xi_t^{jj'}. \end{array}\right\} \tag{12.1.5}$$

As usual, it is convenient to give a "concrete" realization of this limit. To this end, we consider an auxiliary space $(\Omega', \mathcal{F}', \mathbb{P}')$ endowed with a sequence $(\Psi_n)_{n\ge1}$ of independent q-dimensional variables, standard centered normal. We also consider an arbitrary weakly exhausting sequence $(T_n)_{n\ge1}$ for the jumps of X (see after (5.1.1), page 126). The very good filtered extension $(\widetilde{\Omega}, \widetilde{\mathcal{F}}, (\widetilde{\mathcal{F}}_t)_{t\ge0}, \widetilde{\mathbb{P}})$ of $(\Omega, \mathcal{F}, (\mathcal{F}_t)_{t\ge0}, \mathbb{P})$ is defined by (4.1.16), page 104, as

$$\begin{array}{l} \widetilde{\Omega} = \Omega \times \Omega', \qquad \widetilde{\mathcal{F}} = \mathcal{F} \otimes \mathcal{F}', \qquad \widetilde{\mathbb{P}} = \mathbb{P} \otimes \mathbb{P}' \\ (\widetilde{\mathcal{F}}_t) \text{ is the smallest filtration containing } (\mathcal{F}_t) \text{ and such that} \\ \qquad \Psi_n \text{ is } \widetilde{\mathcal{F}}_{T_n} \text{ measurable for all } n. \end{array}$$

We also need a "square-root" of the matrix-valued process ξ, that is a $\mathbb{R}^q \otimes \mathbb{R}^q$-valued process $\alpha_t = (\alpha_t^{jl})_{1\le j,l\le q}$ which is optional and satisfies $\sum_{l=1}^q \alpha_t^{jl} \alpha_t^{j'l} = \xi_t^{jj'}$. Since Ξ_t above is finite-valued, Proposition 4.1.3, page 101, implies that the following process

$$Y(X)_t = \sum_{n\ge1:\ T_n\le t} \alpha_{T_n} \Psi_n \tag{12.1.6}$$

is well defined on the extended space and satisfies (12.1.5).

Remark 12.1.1 We can make another, more concrete but "high-dimensional", choice for α_t. The matrices $H_\pm(x)$ admit a measurable (in x) $(qd) \times (qd)$-dimensional square-root $\widetilde{H}_\pm(x)$ in the sense that

$$H_\pm(x)^{jj'}_{ii'} = \sum_{l=1}^{q} \sum_{r=1}^{d} \widetilde{H}_\pm(x)^{jl}_{ir} \, \widetilde{H}_\pm(x)^{j'l}_{i'r}.$$

This procedure is purely non-random. Then in (12.1.6) we take independent standard centered Gaussian variables Ψ_n which are qdd'-dimensional, instead of being q-dimensional, and then we use (12.1.6) with the following choice of α_t (which now is $\mathbb{R}^q \otimes \mathbb{R}^{qdd'}$-valued):

$$\alpha_t^{j,lrv} = \sum_{i=1}^{q} \left(\widetilde{H}_-(\Delta X_t)^{jl}_{ir} \sigma^{iv}_{t-} + \widetilde{H}_+(\Delta X_t)^{jl}_{ir} \sigma^{iv}_t \right).$$

Theorem 12.1.2 *Assume* (H), *and let f be a q-dimensional function on $\mathbb{R}^d$, whose components are linear combinations of positively homogeneous C^2 functions with degree strictly bigger than* 3. *For $j = 1, \dots, q$, let G^j be finite signed measures supported by* $(0,1]$, *the associated functions g^j by* (12.0.4) *being all Hölder with some index $\theta \in (0,1]$. Finally, assume that*

$$u_n = k_n \Delta_n \to 0, \qquad k_n^{2\theta+1} \Delta_n \to \infty. \tag{12.1.7}$$

Then for each $t > 0$ the q-dimensional variables $Y^n(X)_t$ defined by (12.1.2) *converge stably in law to the variable $Y(X)_t$ which is the value at time t of the process characterized by* (12.1.5), *or equivalently defined by* (12.1.6).

Remark 12.1.3 When f is a linear combination of r positively homogeneous C^2 functions with degrees $p_1, \dots, p_r$, saying that $p_i > 3$ for all i (that is, the assumption in the theorem) is equivalent to saying that

$$f(0) = \partial_i f(0) = 0, \qquad \partial^2_{ij} f(x) = o\left(\|x\|\right) \quad \text{as } \|x\| \to 0. \tag{12.1.8}$$

This is exactly the same condition as for the CLT for $V^n(f, X)$.

The fact that f is a linear combination of r positively homogeneous functions will be needed, in a crucial way, in the proof of Lemma 12.1.8 below.

Remark 12.1.4 The second condition in (12.1.7) may look strange. It comes from the last—centering—term in (12.1.2): we have to approximate the integral $\int_0^1 f^l(g^l(u)\Delta X_s)\,du$ by Riemann sums with stepsize $1/k_n$, at a rate faster than $\sqrt{u_n}$. Now, under our assumptions the function $u \mapsto f^l(g^l(u)\Delta X_s)$ is Hölder with index θ, so the necessary rate is achieved only under (12.1.7).

Upon replacing the last term in (12.1.2) by

$$\sum_{s \le t} \frac{1}{k_n} \sum_{r=1}^{k_n} f^l\left(g^l(r/k_n)\Delta X_s\right), \tag{12.1.9}$$

we would obtain the above CLT with no conditions on k_n except (12.0.2).

Remark 12.1.5 When all measures G^j equal the Dirac mass at 1, and with $k_n = 1$, the process $Y^n(X)$ is exactly the process $\frac{1}{\sqrt{\Delta_n}}(V^n(f,X) - f * \mu)$ of (5.1.7), and this is the case even if we replace the last centering term in (12.1.2) by (12.1.9). However, the limit obtained in Theorem 5.1.2, page 127, is not the same than the limit we would obtain if Theorem 12.1.2 were true for $k_n = 1$, and for example the latter is always $\mathcal{F}$-conditionally Gaussian, whereas the former is so only when the processes X and σ do not jump at the same times. The conclusion is that, even with the version using (12.1.9), the assumption $k_n \to \infty$ is crucial.

12.1.2 An Auxiliary Stable Convergence Result

The localization lemma 4.4.9 applies here, so instead of (H) we assume (SH), that is, the processes b and σ and X are bounded, and $\|\delta(\omega,t,z)\| \le \Gamma(z)$ with Γ bounded and $\int \Gamma(z)^2 \lambda(dz) < \infty$.

For any $\mathbb{R}^d$-valued process U we write

$$\overline{U}(g^j)_i^n = \sum_{r=1}^{k_n} g^j\left(\frac{r}{k_n}\right) \Delta_{i+r-1}^n U. \tag{12.1.10}$$

The components of $\overline{U}(g^j)_i^n$ are denoted by $\overline{U}(g^j)_i^{n,v}$ for $v = 1,\dots,d$. There is another way to write these variables, when U is a semimartingale. Namely, if we set

$$g_n^j(t) = \sum_{r=1}^{k_n} g^j(r/k_n)\, 1_{((r-1)\Delta_n, r\Delta_n]}(t) \tag{12.1.11}$$

(those are step functions on $\mathbb{R}$), we have

$$\overline{U}(g^j)_i^n = \int_{(i-1)\Delta_n}^{(i-1)\Delta_n + u_n} g_n^j\left(\frac{s-(i-1)\Delta_n}{u_n}\right) dU_s. \tag{12.1.12}$$

Below we make use of the notation of Sect. 4.4.2: we set $A_m = \{z : \Gamma(z) > 1/m\}$. We choose the weakly exhausting sequence for the jumps of X to be the sequence (S_p) defined in (4.3.1), that is

$$\left.\begin{array}{l} (S_p)_{p\ge 1} \text{ is a reordering of the double sequence } (S(m,j) : m, j \ge 1), \\ \text{where } S(m,1), S(m,2), \dots \text{ are the successive jump times} \\ \text{of the process } 1_{\{A_m \setminus A_{m-1}\}} * \mathfrak{p}. \end{array}\right\}$$

Next, similar to (4.4.20), we set

$$\begin{aligned} b(m)_t &= b_t - \int_{A_m \cap \{z : \|\delta(t,z)\| \le 1\}} \delta(t,z)\,\lambda(dz) \\ X(m)_t &= X_0 + \int_0^t b(m)_s\,ds + \int_0^t \sigma_s\,dW_s + (\delta 1_{A_m^c}) \star (\mathfrak{p} - \mathfrak{q})_t \end{aligned} \tag{12.1.13}$$

$$X'(m) = X - X(m) \;=\; (\delta\, 1_{A_m}) \star \mathfrak{p},$$

whereas we slightly modify (4.4.21) by setting

$$\begin{aligned}\Omega_n(T,m) = {} & \text{the set on which the jumps of } X'(m) \text{ in } [0,T] \text{ are more} \\ & \text{than } u_n \text{ apart, and no such jump occurs in } [0,u_n] \text{ nor,} \\ & \text{in } [T-u_n,T], \text{ and } \|X(m)_{t+s} - X(m)_t\| \le 2/m \\ & \text{for all } t \in [0,T],\ s \in [0,u_n],\end{aligned} \tag{12.1.14}$$

so we still have for all $m \ge 1$:

$$\mathbb{P}\big(\Omega_n(T,m)\big) \;\to\; 1 \quad \text{as } n \to \infty. \tag{12.1.15}$$

Finally, we define the q-dimensional variables $R(n,p)$ as follows: with $i(n,p)$ being the unique integer with $i(n,p)\Delta_n < S_p \le (i(n,p)+1)\Delta_n$, we set (recall that $\overline{X}(g)$ is defined by (12.1.10) with $U = X$)

$$\begin{aligned}R(n,p)^j = {} & \frac{1}{k_n\sqrt{u_n}} \sum_{l=(i(n,p)-k_n+1)\vee 1}^{i(n,p)} \bigg(f^j\big(\overline{X}(g^j)^n_l\big) \\ & - f^j\bigg(g^j\bigg(\frac{i(n,p)-l+1}{k_n}\bigg)\Delta X_{S_p}\bigg)\bigg).\end{aligned} \tag{12.1.16}$$

The aim of this subsection is to prove the following key proposition:

Proposition 12.1.6 *Assume* (SH) *and the hypotheses of Theorem* 12.1.2 *about* f *and the* g^j*'s. Then, with* α_t *and* Ψ_p *as in* (12.1.6), *we have*

$$\big(R(n,p)\big)_{p\ge 1} \stackrel{\mathcal{L}\text{-s}}{\longrightarrow} (R_p)_{p\ge 1}, \quad \textit{where } R_p \;=\; \alpha_{S_p}\,\Psi_p. \tag{12.1.17}$$

The proof follows the same scheme as for Proposition 4.4.10, but it is technically more demanding, and we divide it into a series of lemmas. Before starting, we recall that (as in Theorem 4.3.1 for example) it is enough to prove

$$\big(R(n,p)\big)_{1\le p\le P} \;\stackrel{\mathcal{L}\text{-s}}{\longrightarrow}\; (R_p)_{1\le p\le P} \tag{12.1.18}$$

for any finite P. Below we fix P, as well as the smallest integer m such that for any $p = 1,\dots,P$ we have $S_p = S(m',j)$ for some $m' \le m$ and some $j \ge 1$. The process $X(m)$ associated with m by (12.1.13) will play a crucial role.

To begin with, we put for $x \in \mathbb{R}^d$, $1 \le j \le q$ and $1 \le l \le d$ and $1 \le r \le k_n$:

$$\begin{aligned}h_-(x,j,l)^n_r \;&=\; \frac{1}{k_n}\sum_{u=1}^{r} \partial_l f^j\bigg(g^j\bigg(\frac{u+k_n-r}{k_n}\bigg)x\bigg)\, g^j\bigg(\frac{u}{k_n}\bigg) \\ h_+(x,j,l)^n_r \;&=\; \frac{1}{k_n}\sum_{u=1+r}^{k_n} \partial_l f^j\bigg(g^j\bigg(\frac{u-r}{k_n}\bigg)x\bigg)\, g^j\bigg(\frac{u}{k_n}\bigg).\end{aligned} \tag{12.1.19}$$

Next, for $1 \le v \le d'$, and with the convention $\Delta_i^n Y = 0$ when $i \le 0$, we set

$$\begin{aligned} z_{p-}^{n,jlv}(x) &= \frac{1}{\sqrt{u_n}} \sum_{r=1}^{k_n-1} h_-(x,j,l)_r^n \, \Delta_{i(n,p)-k_n+r}^n W^v, \\ z_{p+}^{n,jlv}(x) &= \frac{1}{\sqrt{u_n}} \sum_{r=1}^{k_n-1} h_+(x,j,l)_r^n \, \Delta_{i(n,p)+r}^n W^v, \end{aligned} \tag{12.1.20}$$

$$\begin{aligned} \overline{z}_{p-}^{n,j}(x) &= \frac{1}{\sqrt{u_n}} \sum_{r=1}^{k_n-1} \sum_{l=1}^{d} h_-(x,j,l)_r^n \, \Delta_{i(n,p)-k_n+r}^n X(m)^l, \\ \overline{z}_{p+}^{n,j}(x) &= \frac{1}{\sqrt{u_n}} \sum_{r=1}^{k_n-1} \sum_{l=1}^{d} h_+(x,j,l)_r^n \, \Delta_{i(n,p)+r}^n X(m)^l, \end{aligned} \tag{12.1.21}$$

We end this series of notation by introducing the smallest filtration $(\mathcal{G}_t)$ on Ω which contains $(\mathcal{F}_t)$ and such that S_p is $\mathcal{G}_0$ measurable for all $p \le P$: this is the same as in Theorem 4.3.1 (with P instead of l), and we know that W is a $(\mathcal{G}_t)$-Brownian motion, and the representation (12.1.13) for $X(m)$ is the same relative to $(\mathcal{F}_t)$ and to $(\mathcal{G}_t)$.

Lemma 12.1.7 *Assume* (SH) *and the hypotheses of Theorem* 12.1.2 *about f and the g^j's. Then for each $p \le P$ we have*

$$R(n,p)^j - \left(\overline{z}_{p-}^{n,j}(\Delta X_{S_p}) + \overline{z}_{p+}^{n,j}(\Delta X_{S_p})\right) \stackrel{\mathbb{P}}{\longrightarrow} 0. \tag{12.1.22}$$

Proof It is enough to prove (12.1.22) on the set $\{u_n < S_p < T\}$ for T arbitrary large. So below T is fixed, as well as j, and we may suppose that $i(n,p) \ge k_n$.

On the set $\Omega_n(T,m) \cap \{S_p < T\}$ we have $\Delta_i^n X = \Delta_i^n X(m)$ if $i(n,p) - k_n + 1 \le i < i(n,p)$ and $\Delta_i^n X = \Delta_i^n X(m) + \Delta X_{S_p}$ when $i = i(n,p)$, where $X(m)$ is given by (12.1.13). Hence on $\Omega_n(T,m) \cap \{S_p < T\}$, and by virtue of (12.1.10), the lth summand in (12.1.16) is

$$f^j\left(g^j\left(\frac{i(n,p)-l+1}{k_n}\right)\Delta X_{S_p} + \overline{X(m)}(g^j)_l^n\right) - f^j\left(g^j\left(\frac{i(n,p)-l+1}{k_n}\right)\Delta X_{S_p}\right).$$

A Taylor expansion shows that this difference is $B_l^n + B_l'^n$, where the leading term is

$$B_l^n = \sum_{v=1}^{d} \partial_v f^j\left(g^j\left(\frac{i(n,p)-l+1}{k_n}\right)\Delta X_{S_p}\right)\overline{X(m)}(g^j)_l^{n,v} \tag{12.1.23}$$

and the remainder term satisfies

$$\left|B_l'^n\right| \le K\left(\left\|\overline{X(m)}(g^j)_l^n\right\|^2 + \left\|\overline{X(m)}(g^j)_l^n\right\|^{2+w}\right) \tag{12.1.24}$$

for some $w \geq 0$, because $\nabla^2 f$ has polynomial growth and g^j is bounded, as well as ΔX_{S_p}. Moreover, by a change of order of summation and a relabeling, it is tedious but straightforward to deduce from (12.1.10), (12.1.19) and (12.1.21) that

$$\begin{aligned} &\frac{1}{k_n\sqrt{u_n}}\sum_{l=i(n,p)-k_n+1}^{i(n,p)} B_l^n = \overline{z}_{p-}^{n,j}(\Delta X_{S_p}) + \overline{z}_{p+}^{n,j}(\Delta X_{S_p}) + B''^n, \\ &\quad \text{with } B''^n = \frac{1}{\sqrt{u_n}}\sum_{v=1}^{d} h_+(\Delta X_{S_p}, j, v)_0^n\, \Delta_{i(n,p)}^n X(m)^v. \end{aligned} \tag{12.1.25}$$

Since ΔX_{S_p} is bounded and $\sup_n |h_+(x,j,v)_0^n|$ is locally bounded in x, we have $|h_+(\Delta X_{S_p}, j, v)_0^n| \leq K$. Hence, putting these facts together, we see that, on the set $\{S_p < T\} \cap \Omega_n(T,m)$, the left side of (12.1.22) is smaller in absolute value than

$$\begin{aligned} &\frac{K}{k_n\sqrt{u_n}}\sum_{l=i(n,p)-k_n+1}^{i(n,p)} \left(\|\overline{X(m)}(g^j)_l^n\|^2 + \|\overline{X(m)}(g^j)_l^n\|^{2+w}\right) \\ &\quad + \frac{K}{\sqrt{u_n}}\|\Delta_{i(n,r)}^n X(m)\|. \end{aligned}$$

Taking advantage of (12.1.15), we see that the result will hold if we can prove

$$\begin{aligned} &\mathbb{E}\left(\|\Delta_{i(n,r)}^n X(m)\|\right) \leq K\sqrt{\Delta_n} \\ &\mathbb{E}\left(\sum_{l=(i(n,p)-k_n+1)\vee 1}^{i(n,p)} \|\overline{X(m)}(g^j)_l^n\|^{w'}\right) \leq K k_n u_n \end{aligned} \tag{12.1.26}$$

for any $w' \geq 2$; Since $X(m)$ has the representation (12.1.13) relative to the filtration $(\mathcal{G}_t)$ and since $i(n,p)$ is $\mathcal{G}_0$ measurable, the first part of (12.1.26) comes from (SH) and (2.1.44). The second part holds for the same reason, because of (12.1.12) and $|g_n^j(t)| \leq K$. □

Lemma 12.1.8 *Assume* (SH) *and the hypotheses of Theorem* 12.1.2 *about* f *and the* g^j*'s. Then for each* $p \leq P$ *we have*

$$\begin{aligned} &\overline{z}_{p-}^{n,j}(\Delta X_{S_p}) - \sum_{l=1}^{d}\sum_{v=1}^{d'} \sigma_{S_p-}^{lv} z_{p-}^{n,jlv}(\Delta X_{S_p}) \stackrel{\mathbb{P}}{\longrightarrow} 0 \\ &\overline{z}_{p+}^{n,j}(\Delta X_{S_p}) - \sum_{l=1}^{d}\sum_{v=1}^{d'} \sigma_{S_p}^{lv} z_{p+}^{n,jlv}(\Delta X_{S_p}) \stackrel{\mathbb{P}}{\longrightarrow} 0. \end{aligned}$$

Proof The proofs of the two claims are similar, and we only prove the first one. We have $|h_-(\Delta X_{S_p}, j, l)_r^n| \leq K$ because $\|\Delta X_{S_p}\| \leq K$. Therefore $\mathbb{E}(|z_{p-}^{n,jlv}(\Delta X_{S_p})|) \leq$

K and, since σ is càdlàg, the first claim is equivalent to proving that

$$B_n := \left(\overline{z}_{p-}^{n,j}(\Delta X_{S_p}) - \sum_{l=1}^{d}\sum_{v=1}^{d'} \sigma_{S_p-u_n}^{lv}\, z_{p-}^{n,jlv}(\Delta X_{S_p})\right) 1_{\{S_p>u_n\}} \xrightarrow{\mathbb{P}} 0. \quad (12.1.27)$$

Now we apply the property $f^j = \sum_{r=1}^{R} \alpha_r f_r^j$, where f_r^j is positively homogeneous of degree $w(r,j) > 3$. By construction, the maps $f \mapsto \overline{z}_{p-}^{n,j}$ and $f \mapsto z_{p-}^{n,jlv}$ are linear, so it is obviously enough to prove the result with each f_r^j instead of f^j, or equivalently to suppose that f^j itself is positively homogeneous of degree $w > 3$. Therefore each $\partial_l f^j$ is also positively homogeneous of degree $w-1$, hence $\partial_l f^j(zx) = |z|^{w-1}(\partial_l f^j(x) 1_{\{z>0\}} + \partial_l f^j(-x) 1_{\{z<0\}})$ for all $z \in \mathbb{R}$ and $x \in \mathbb{R}^d$. Plugging this into (12.1.19) gives us

$$\begin{aligned} h_-(x,j,l)_r^n &= \partial_l f^j(x)\, h_{-+}(j,l)_r^n + \partial_l f^j(-x)\, h_{--}(j,l)_r^n, \quad \text{where} \\ h_{-+}(j,l)_r^n &= \frac{1}{k_n}\sum_{u=1}^{r}\left|g^j\left(\frac{u+k_n-r}{k_n}\right)\right|^{w-1} g^j\left(\frac{u}{k_n}\right) 1_{\{g^j(\frac{u+k_n-r}{k_n})>0\}} \\ h_{--}(j,l)_r^n &= \frac{1}{k_n}\sum_{u=1}^{r}\left|g^j\left(\frac{u+k_n-r}{k_n}\right)\right|^{w-1} g^j\left(\frac{u}{k_n}\right) 1_{\{g^j(\frac{u+k_n-r}{k_n})<0\}}. \end{aligned}$$

Comparing (12.1.20) and (12.1.21), and using the above, we see that, if $S_p > u_n$,

$$B_n = \sum_{l=1}^{d}\left(\partial_l f^j(\Delta X_{S_p})\, B_{n+}^l + \partial_l f^j(-\Delta X_{S_p})\, B_{n-}^l\right) \quad (12.1.28)$$

where, with the notation $\phi_{n\pm}^l(t) = \sum_{r=1}^{k_n-1} h_{-\pm}(j,l)_r^n\, 1_{((r-1)\Delta_n, r\Delta_n]}(t)$,

$$\begin{aligned} B_{n\pm}^l = &\frac{1}{\sqrt{u_n}} \int_{(i(n,p)-k_n)^+\Delta_n}^{(i(n,p)-1)\Delta_n} \phi_{n\pm}^l\big(s - (i(n,p)-k_n)\Delta_n\big) \\ &\times \left(dX(m)_s^l - \sum_{u=1}^{d'} \sigma_{(S_p-u_n)^+}^{lu}\, dW_s^u\right). \end{aligned}$$

In view of (12.1.28), in order to get (12.1.27) it suffices to prove that

$$B_{n+}^l \xrightarrow{\mathbb{P}} 0, \qquad B_{n-}^l \xrightarrow{\mathbb{P}} 0. \quad (12.1.29)$$

For this, we proceed as in the proof of Proposition 4.4.10: recalling (12.1.13), we apply (2.1.33), (2.1.34) and (2.1.39) (with $r=2$ and $q=1/2$ and $s=u_n$), all relative to the filtration $(\mathcal{G}_t)$ (so $i(n,p)$ is $\mathcal{G}_0$ measurable), plus the fact that $|\phi_{n\pm}^l| \le K$, to obtain

$$\mathbb{E}\big((B^l_{n\pm})^2 \wedge 1\big) \leq K\sqrt{u_n} + K\int_{\{\Gamma(z)\leq u_n^{1/4}\}} \Gamma(z)^2 \,\lambda(dz)$$
$$+ K\mathbb{E}\Big(\sup_{0\leq s\leq u_n} \|\sigma_{(S_p-u_n)^+ + s} - \sigma_{(S_p-u_n)^+}\|^2\Big)$$

which goes to 0. Hence (12.1.29) holds. □

We can consider $x \mapsto z^{n,jlv}_{p\pm}(x)$ as two processes $z^n_{p\pm}$ indexed by $\mathbb{R}^d$ and with values in $\mathbb{R}^{qdd'}$, and as such they have continuous paths because $x \mapsto h_\pm(x,t)$ are Lipschitz, uniformly in t, on every compact subset of $\mathbb{R}^d$ (recall that f is C^2 and all g^j are bounded). In other words z^n_{p-} and z^n_{p+} are random variables taking their values in the set $\mathbf{C} = \mathbf{C}(\mathbb{R}^d, \mathbb{R}^{qdd'})$ of continuous functions, a set which we endow with the local uniform topology.

We are now going to prove that these processes converge stably in law, hence we need to describe the limits. Those are defined as follows, on an auxiliary space $(\Omega', \mathcal{F}', \mathbb{P}')$ which supports two independent i.i.d. sequences $(W(p-) : p \geq 1)$ and $(W(p+) : p \geq 1)$ of standard d'-dimensional Brownian motions. Then, with x ranging through $\mathbb{R}^d$ and j, l, v ranging through the appropriate sets of indices, we set

$$z^{jlv}_{p-}(x) = \int_0^1 h_-(x,t)^j_l \, dW(p-)^v_t, \qquad z^{jlv}_{p+}(x) = \int_0^1 h_+(x,t)^j_l \, dW(p+)^v_t. \tag{12.1.30}$$

Those are independent centered Gaussian processes, the laws of all z_{p-}, resp. all z_{p+}, being the same, and it turns out that they have versions with paths in $\mathbf{C}$ (this will be a consequence of the forthcoming proof). As is usual for stable convergence, we take the product extension (4.1.16), page 104. The following lemma is stated for p between 1 to P, but there is a version for all $p \geq 1$. It could be deduced from Theorem 4.3.1, but a direct proof is in fact simpler.

Lemma 12.1.9 *Assume* (SH) *and the hypotheses of Theorem* 12.1.2 *about f and the g^j's. Then*

$$\big(z^n_{p-}, z^n_{p+}\big)_{1\leq p\leq P} \stackrel{\mathcal{L}\text{-}s}{\longrightarrow} (z_{p-}, z_{p+})_{1\leq p\leq P}$$

for the product topology on $(\mathbf{C}(\mathbb{R}^d, \mathbb{R}^{qdd'}) \times \mathbf{C}(\mathbb{R}^d, \mathbb{R}^{qdd'}))^P$, each $\mathbf{C}(\mathbb{R}^d, \mathbb{R}^{qdd'})$ being endowed with the local uniform topology.

Proof 1) We need to prove that if $F_{1\pm}, \dots, F_{P\pm}$ are bounded Lipschitz functions on $\mathbf{C}$ and Z is a bounded variable on $(\Omega, \mathcal{F})$, and if $V_n = \prod_{p=1}^P F_{p-}(z^n_{p-})\, F_{p+}(z^n_{p+})$ and z_- and z_+ are two independent processes on $(\Omega', \mathcal{F}', \mathbb{P}')$ with the same laws as z_{p-} and z_{p+}, then

$$\mathbb{E}(Z\, V_n) \to \mathbb{E}(Z) \prod_{p=1}^P \mathbb{E}'\big(F_{p-}(z_-)\big)\, \mathbb{E}'\big(F_{p+}(z_+)\big). \tag{12.1.31}$$

This is the analogue of (4.3.4), where the processes $\overline{U}^n(g)$ and the variables $\kappa(n,p)$ do not show up, and $\overline{w}(n,p)$ and $W''(p)$ are replaced by (z^n_{p-}, z^n_{p+}) and (z_{p-}, z_{p+}); moreover (z^n_{p-}, z^n_{p+}) has exactly the same measurability properties as $\overline{w}(n,p)$, namely it depends only on S_p and on the increments of W over $((S_p - u_n)^+, S_p + u_n]$. Hence the proof of Theorem 4.3.1, page 109, goes through without change down to the end of Step 6, showing that it is enough to prove (12.1.31) when Z is $\mathcal{G}_0$ measurable, where $\mathcal{G}_0$ is as before Lemma 12.1.7.

Step 7 of that proof is not valid here. However, put

$$z^{n,jlv}_-(x) = \frac{1}{\sqrt{u_n}} \sum_{r=1}^{k_n-1} h_-(x,j,l)^n_r \, \Delta^n_r W^v$$

$$z^{n,jlv}_+(x) = \frac{1}{\sqrt{u_n}} \sum_{r=1}^{k_n-1} h_+(x,j,l)^n_r \, \Delta^n_r W^v.$$

Then in restriction to the set Ω_n on which $S_p > u_n$ and $|S_p - S_{p'}| > 2u_n$ for all $p, p' \in \{1, \dots, P\}$, the processes $(z^n_{p-}, z^n_{p+} : 1 \le p \le P)$ are all independent, and independent of $\mathcal{G}_0$, with the same laws as z^n_- and z^n_+ above. Therefore

$$\mathbb{E}(Z \, 1_{\Omega_n} V_n) = \mathbb{E}(Z \, 1_{\Omega_n}) \prod_{p=1}^{P} \mathbb{E}\big(F_{p-}(z^n_-)\big) \, \mathbb{E}\big(F_{p+}(z^n_+)\big).$$

Since $\mathbb{P}(\Omega_n) \to 1$, (12.1.31) follows from the convergence in law (we no longer need stable convergence here) of z^n_- and z^n_+ to z_- and z_+, respectively.

2) We prove for example $z^n_- \xrightarrow{\mathcal{L}} z_-$ (the proof of $z^n_+ \xrightarrow{\mathcal{L}} z_+$ is similar). This can be done in two independent steps: one consists in proving the finite-dimensional convergence in law, the other consists in proving that the processes z^n_- are "C-tight", which together with the first step also implies that there is a continuous version for the limit $x \mapsto z_-(x)$.

The first step amounts to $(z^n_-(x_1), \dots, z^n_-(x_I)) \xrightarrow{\mathcal{L}} (z_-(x_1), \dots, z_-(x_I))$ for arbitrary $x_1, \dots, x_I$ in $\mathbb{R}^d$. The $(qdd'I)$-dimensional random vector with components $z^{n,jlv}_-(x_i)$ is centered Gaussian with covariance

$$\mathbb{E}\big(z^{n,jlv}_-(x_i) \, z^{n,j'l'v'}_-(x_{i'})\big) = \begin{cases} \frac{1}{k_n} \sum_{r=1}^{k_n-1} h_-(x_i,j,l)^n_r \, h_-(x_{i'},j',l')^n_r & \text{if } v = v' \\ 0 & \text{if } v \neq v'. \end{cases}$$

Recalling (12.1.19), when $v = v'$ the right side above is a Riemann approximation for the triple integral

$$\int_0^1 \left(\int_0^t \partial_l f^j\big(g^j(s)x_i\big) \, ds \int_0^t \partial_{l'} f^{j'}\big(g^{j'}(s')x_{i'}\big) \, ds' \right) dt.$$

Hence, recalling (12.1.3), we get

$$\mathbb{E}\big(z_-^{n,jlv}(x_i)\, z_-^{n,j'l'v'}(x_{i'})\big) \to c\big(i,i',j,j',l,l'\big) := \delta_{vv'} \int_0^1 h_-(x_i,t)_l^j\, h_-(x_{i'},t)_{l'}^{j'}\, dt\,.$$

In view of (12.1.30), we have $c(i,i',j,j',l,l') = \mathbb{E}(z_-^{jlv}(x_i)\, z_-^{j'l'v'}(x_{i'}))$, which implies the desired convergence in law because the processes are Gaussian.

The second step, according to a C-tightness criterion which extends Kolmogorov's continuity criterion and may be found for example in Ibragimov and Has'minski [49], is accomplished if one can prove that for any $A > 0$, we have

$$\|x\|,\ \|x'\| \le A \;\Rightarrow\; \mathbb{E}\big(\big|z_-^{n,jlv}(x) - z_-^{n,jlv}(x')\big|^w\big) \le K_A \|x - x'\|^{w'} \tag{12.1.32}$$

for some $w > 0$ and $w' > d$ (since d is the dimension of the "parameter" x here). Proving (12.1.32) is simple: our assumptions on f and g^j yield $|h_-(x,j,l)_r^n - h_-(x',j;l)_r^n| \le K_A\|x - x'\|$ if $\|x\|,\ \|x'\| \le A$, and clearly the law of the process $(z_-^n(x) : x \in \mathbb{R}^d)$ does not depend on $i(n,p)$, so we may assume that $i(n,p) = k_n$. Then an application of the Burkholder-Davis-Gundy inequality yields for $w \ge 1$:

$$\begin{aligned}
&\mathbb{E}\big(\big|z_-^{n,jlv}(x) - z_-^{n,jlv}(x')\big|^w\big)\\
&\quad\le \frac{K}{u_n^{w/2}}\, \mathbb{E}\Bigg(\bigg(\sum_{r=1}^{k_n-1} \big|h_-(x,j,l)_r^n - h_-\big(x',j,l\big)_r^n\big|^2\, \big|\Delta_r^n W^v\big|^2\bigg)^{w/2}\Bigg)\\
&\quad\le \frac{K_{A,w}\|x - x'\|^w}{u_n^{w/2}}\, \mathbb{E}\Bigg(\bigg(\sum_{r=1}^{k_n-1} \big|\Delta_r^n W^v\big|^2\bigg)^{w/2}\Bigg) \le K_{A,w}\|x - x'\|^w,
\end{aligned}$$

the last equality above coming from Hölder's inequality and $\mathbb{E}(|\Delta_i^n W|^w) \le K_w \Delta_n^{w/2}$. Then we get (12.1.32), upon choosing $w = w' > d$. □

Proof of Proposition 12.1.6 First, by (2.2.5) and the previous lemma,

$$\Bigg(\sum_{l=1}^{d}\sum_{v=1}^{d'} \big(\sigma_{S_p-}^{lv} z_{p-}^{n,jlv} + \sigma_{S_p}^{lv} z_{p+}^{n,jlv}\big)\Bigg)_{p,j} \overset{\mathcal{L}\text{-}s}{\Longrightarrow} \Bigg(\sum_{l=1}^{d}\sum_{v=1}^{d'} \big(\sigma_{S_p-}^{lv} z_{p-}^{jlv} + \sigma_{S_p}^{lv} z_{p+}^{jlv}\big)\Bigg)_{p,j} \tag{12.1.33}$$

where p and j range through $\{1,\dots,P\}$ and $\{1,\dots,q\}$ respectively. The above are processes with paths in $\mathbf{C}' = \mathbf{C}(\mathbb{R}^d,\mathbb{R})$, and the convergence takes place for the local uniform topology.

At this stage, we give a useful property of the stable convergence. Let $Z^n = (Z_x^n)_{x\in\mathbb{R}^a}$ be a sequence of processes on $(\Omega,\mathcal{F},\mathbb{P})$, with paths in the set $\mathbf{C}'' = \mathbf{C}(\mathbb{R}^a,\mathbb{R}^b)$ of continuous functions, for some $a, b \in \mathbb{N}^*$, and which converges stably in law to a limiting process $Z=(Z_x)_{x\in\mathbb{R}^a}$, given on an extension $(\widetilde{\Omega},\widetilde{\mathcal{F}},\widetilde{\mathbb{P}})$ of the original space. Then, for any $\mathbb{R}^a$-valued variable V on $(\Omega,\mathcal{F})$, we have

$$Z_V^n \stackrel{\mathcal{L}\text{-s}}{\longrightarrow} Z_V. \tag{12.1.34}$$

This is a simple consequence of the second half of (2.2.6), page 47: indeed, letting Y be bounded measurable on $(\Omega,\mathcal{F})$ and ϕ be bounded continuous on $\mathbb{R}^b$, we see that the function $G(\omega,z)=Y(\omega)\,\phi(z(V(\omega)))$ on $\Omega\times\mathbf{C}''$ is bounded measurable, and continuous in $z\in\mathbf{C}''$. Then we deduce from (2.2.6) that $\mathbb{E}(Y\,\phi(Z_V^n))\to\widetilde{\mathbb{E}}(Y\,\phi(Z_V))$, which is (12.1.34).

We deduce from (12.1.33) and (12.1.34) that

$$\begin{aligned}&\Bigg(\sum_{l=1}^{d}\sum_{v=1}^{d'}\big(\sigma_{S_p-}^{lv}z_{p-}^{n,jlv}(\Delta X_{S_p})+\sigma_{S_p}^{lv}z_{p+}^{n,jlv}(\Delta X_{S_p})\big)\Bigg)_{p,j}\\&\quad\stackrel{\mathcal{L}\text{-s}}{\longrightarrow}\Bigg(\sum_{l=1}^{d}\sum_{v=1}^{d'}\big(\sigma_{S_p-}^{lv}z_{p-}^{jlv}(\Delta X_{S_p})+\sigma_{S_p}^{lv}z_{p+}^{jlv}(\Delta X_{S_p})\big)\Bigg)_{p,j}\end{aligned}$$

which, combined with Lemmas 12.1.7 and 12.1.8, yields

$$\big(R(n,p)^j\big)_{p,j}\stackrel{\mathcal{L}\text{-s}}{\longrightarrow}\Bigg(\sum_{l=1}^{d}\sum_{v=1}^{d'}\big(\sigma_{S_p-}^{lv}z_{p-}^{jlv}(\Delta X_{S_p})+\sigma_{S_p}^{lv}z_{p+}^{jlv}(\Delta X_{S_p})\big)\Bigg)_{p,j}.$$

Then in view of (12.1.22), and in order to obtain (12.1.18), it remains to observe that the right side above is the same as $(R_p^j)_{1\le p\le P,1\le j\le q}$, as given by (12.1.17), in the sense that both Pq-dimensional random vectors have the same $\mathcal{F}$-conditional distribution. This property is an easy consequence of (12.1.3), (12.1.4), (12.1.6) and (12.1.30). □

12.1.3 Proof of Theorem 12.1.2

As said before, we may assume (SH). The time t is fixed throughout. The proof is basically the same as for Theorems 5.1.2 and 11.1.2.

Step 1) We use the notation A_m, $X(m)$, $X'(m)$, $\Omega_n(t,m)$, $i(n,p)$ of Sect. 12.1.2, see page 344, and $\mathcal{P}_m$ is the set of all indices p such that $S_p=S(m',j)$ for some $j\ge1$ and some $m'\le m$.

With the notation (12.1.2) and (12.1.10), we get

$$Y^n(X)_t\;=\;Y^n\big(X(m)\big)_t+Z^n(m)_t\quad\text{on the set }\;\Omega_n(t,m), \tag{12.1.35}$$

where $Z^n(m)_t = \sum_{p\in\mathcal{P}_m:\, S_p\le t} \zeta_p^n$, and $\zeta_p^n = (\zeta_p^{n,j})_{1\le j\le q}$ is given by

$$\zeta_p^{n,j} = \frac{1}{\sqrt{u_n}}\left(\frac{1}{k_n}\sum_{l=(i(n,p)-k_n+1)\vee 1}^{i(n,p)} \left(f^j\big(\overline{X}(g^j)_l^n\big) - f^j\big(\overline{X(m)}(g^j)_l^n\big)\right) - \int_0^1 f^j\big(g^j(u)\Delta X_{S_p}\big)\,du\right).$$

Moreover, again on the set $\Omega_n(t,m)$, we have with the notation (12.1.16):

$$\begin{aligned}
\zeta_p^{n,j} &= R(n,p)^j + \gamma(p,j)_n + \gamma'(j,p)_n, \quad \text{where}\\
\gamma(j,p)_n &= -\frac{1}{k_n\sqrt{u_n}} \sum_{l=i(n,p)-k_n+1}^{i(n,p)} f^j\big(\overline{X(m)}(g^j)_l^n\big)\\
\gamma'(j,p)_n &= -\frac{1}{\sqrt{u_n}}\int_0^1 \left(f^j\big(g^j(u)\Delta X_{S_p}\big) - f^j\big(g^j\big((1+[uk_n])/k_n\big)\Delta X_{S_p}\big)\right) du.
\end{aligned}$$

Step 2) By hypothesis, $u\mapsto f^j(g^j(u)\Delta X_{S_p})$ is Hölder with index θ (pathwise in ω), hence (12.1.7) implies $\gamma'(j,p)_n \xrightarrow{\mathbb{P}} 0$. On the other hand, (12.1.8) yields $|f^j(\overline{X(m)}(g^j)_l^n)| \le K\|\overline{X(m)}(g^j)_l^n)\|^2$ on the set $\Omega_n(t,m)\cap\{S_p\le t\}$ (recall the end of the definition (12.1.14) of $\Omega_n(T,m)$). This, in view of (12.1.15) and (12.1.26) and $u_n\to 0$, implies $\gamma(j,p)_n \xrightarrow{\mathbb{P}} 0$. Therefore $\zeta_p^{n,j} - R(n,p)^j \xrightarrow{\mathbb{P}} 0$ and, by Proposition 12.1.6, we obtain exactly as for (5.1.15):

$$Z^n(m)_t \xrightarrow{\mathcal{L}\text{-s}} Y\big(X'(m)\big)_t, \qquad \text{as } n\to\infty,$$

where $Y(X'(m))$ is associated with the process $X'(m)$ by (12.1.6), that is with the sum extended over the stopping times S_p (instead of T_p) for $p\in\mathcal{P}_m$ only.

The convergence $Y(X'(m))_t \xrightarrow{\mathbb{P}} Y(X)_t$ follows from (12.1.6) and the property $\sum_{s\le t}\|\xi_s\| < \infty$. Hence we are left to prove that, for all $t>0$ and $\eta>0$,

$$\lim_{m\to\infty}\ \limsup_{n\to\infty}\ \mathbb{P}\big(\big|Y^n\big(X(m)\big)_t\big| > \eta\big) = 0. \tag{12.1.36}$$

Step 3) For (12.1.36) it suffices to consider the one-dimensional case $q=1$. So we drop the index j everywhere. With g_n associated to $g=g^j$ by (12.1.11), we write

$$\overline{X}(n,m,i)_t = \int_{(i-1)\Delta_n}^{(i-1)\Delta_n+t} g_n\left(\frac{s-(i-1)\Delta_n}{u_n}\right) dX(m)_s.$$

We thus have $\overline{X(m)}(g)_i^n = \overline{X}(n,m,i)_{u_n}$ by (12.1.12). We also set

$$Y'^n\big(X(m)\big)_t = \frac{1}{k_n\sqrt{u_n}} \sum_{i=1}^{[t/\Delta_n]-k_n+1} \bigg(f\big(\overline{X}(n,m,i)_{u_n}\big) - \sum_{s\le u_n} f\big(\Delta\overline{X}(n,m,i)_s\big)\bigg). \tag{12.1.37}$$

Observe that when $(r-1)\Delta_n < t \le r\Delta_n$ for some $r = 1,\dots,k_n$, we have $\Delta\overline{X}(n,m,i)_t = g(\frac{r}{k_n})\,\Delta X(m)_{(i-1)\Delta_n+t}$. Then, comparing (12.1.37) with (12.1.2), we see that

$$Y^n\big(X(m)\big)_t - Y'^n\big(X(m)\big)_t = W(n,m)_t + W'(n,m)_t + W''(n,m)_t,$$

where, with $J(n,r,t)$ being the union of the two intervals $(0,(r-1)\Delta_n]$ and $(([\frac{t}{\Delta_n}]-k_n+r)\Delta_n, [\frac{t}{\Delta_n}]\Delta_n]$,

$$W(n,m)_t = -\frac{1}{\sqrt{u_n}} \sum_{[t/\Delta_n]\Delta_n < s \le t} \int_0^1 f\big(g(u)\,\Delta X(m)_s\big)\,du$$

$$W'(n,m)_t = \frac{1}{\sqrt{u_n}} \sum_{s\le[t/\Delta_n]\Delta_n} \bigg(\frac{1}{k_n}\sum_{r=1}^{k_n} f\bigg(g\bigg(\frac{r}{k_n}\bigg)\Delta X(m)_s\bigg) - \int_0^1 f\big(g(u)\,\Delta X(m)_s\big)\,du\bigg)$$

$$W''(n,m)_t = -\frac{1}{k_n\sqrt{u_n}} \sum_{r=1}^{k_n} \sum_{s\in J(n,r,t)} f\bigg(g\bigg(\frac{r}{k_n}\bigg)\Delta X(m)_s\bigg).$$

We have $\|\Delta X(m)\| \le 1$, and $\mathbb{E}(\sum_{v\in(t,t+s]} \|\Delta X(m)_v\|^2) \le Ks$ by (SH), and our assumptions on f and g yield

$$\|x\| \le 1 \;\Rightarrow\; \big|f\big(g(u)x\big)\big| \le K\|x\|^2, \quad \big|f\big(g(u)x\big) - f\big(g(u')x\big)\big| \le K\|x\|^2\,\big|u-u'\big|^\theta.$$

Hence we have $\mathbb{E}(|W(n,m)_t|) \le \frac{K\Delta_n}{\sqrt{u_n}}$ and $\mathbb{E}(|W'(n,m)_t|) \le \frac{Kt}{k_n^\theta\sqrt{u_n}}$, and also $\mathbb{E}(|W''(n,m)_t|) \le K\sqrt{u_n}$, which all go to 0 as $n\to\infty$. Therefore instead of (12.1.36) it is enough to prove

$$\lim_{m\to\infty} \limsup_{n\to\infty} \mathbb{P}\big(\Omega_n(t,m) \cap \big\{\big|Y'^n\big(X(m)\big)_t\big| > \eta\big\}\big) = 0.$$

At this stage, we reproduce Step 4 of the proof of Theorem 5.1.2, with the process $\overline{X}(n,m,i)$ instead of $X(m)_t - X(m)_{(i-1)\Delta_n}$. The only difference is that instead of (5.1.21) we now have to prove

$$\lim_{m\to\infty} \limsup_n \frac{1}{k_n\sqrt{u_n}}\, \mathbb{E}\bigg(\sum_{i=1}^{[t/\Delta_n]-k_n+1} \big|A(n,m,i)_{((i-1)\Delta_n+u_n)\wedge T(n,m,i)}\big|\bigg) = 0 \tag{12.1.38}$$

$$\lim_{m\to\infty} \limsup_n \frac{1}{k_n^2 u_n} \mathbb{E}\left(\sum_{i=1}^{[t/\Delta_n]-k_n+1} A'(n,m,i)_{((i-1)\Delta_n+u_n)\wedge T(n,m,i)} \right) = 0, \tag{12.1.39}$$

where $A(n,m,i)$ and $A'(n,m,i)$ are still given by (5.1.20), and with the formula (5.1.23), page 133, replaced by

$$\begin{aligned} \left|a(n,m,i)_t\right| &\leq K\alpha_m \left(\left\|\overline{X}(n,m,i)_t\right\| + m\left\|\overline{X}(n,m,i)_t\right\|^2\right) \\ a'(n,m,i)_t &\leq K\alpha_m^2 \left\|\overline{X}(n,m,i)_t\right\|^2 \end{aligned}$$

(recall that α_m is some sequence going to 0 as $m \to \infty$). Now, since $|g_n| \leq K$ we obtain $\mathbb{E}(\|\overline{X}(n,m,i)_{u_n}\|^2) \leq K(u_n + m^2 u_n^2)$, and both (12.1.38) and (12.1.39) follow.

12.2 Functionals of Normalized Increments

We keep the setting of the previous section, but now we are interested in the functionals $V'^n(\Phi, k_n, X)$. Recall the notation $\overline{\rho}_a$ for the law (on $\mathbb{D}^d$ or on $\mathbb{D}_1^d$) of a d-dimensional Brownian motion with covariance matrix a at time 1. Theorem 8.4.2 yields that, when X satisfies (H),

$$V'^n(\Phi, k_n, X)_t \stackrel{\text{u.c.p.}}{\Longrightarrow} \int_0^t \overline{\rho}_{c_s}(\Phi)\, ds$$

as soon as the function f in (12.0.3) is of polynomial growth when X is continuous, and $f(x) = \mathrm{o}(\|x\|^2)$ as $\|x\| \to \infty$ otherwise. Moreover one has an explicit form for $\overline{\rho}_{\alpha\alpha^*}(\Phi)$ when α is a $d \times d'$ matrix and if Φ satisfies (12.0.3): indeed, the d-dimensional variable $\int_0^1 \alpha W_s\, G(ds)$ is centered Gaussian with covariance $\Lambda(g)\,\alpha\alpha^*$, where

$$\Lambda(g) = \int_{(0,1]^2} (s \wedge t)\, G(ds)\, G(dt) = \int_{(0,1]^2} G(ds)\, G(dt) \int_0^{s\wedge t} du = \int_0^1 g(s)^2\, ds \tag{12.2.1}$$

(the third equality comes from Fubini's theorem). Therefore $\overline{\rho}_a(\Phi) = \rho_{\Lambda(g)a}(f)$ for any $d \times d$ covariance matrix a.

We give an associated CLT, in the q-dimensional case. For $j = 1, \dots, q$, we have a test function Φ^j of the form (12.0.3), with the associated measure G^j and the functions f^j and g^j (see (12.0.4)) and the (strictly) positive numbers $\Lambda(g^j)$ defined above. We will derive the asymptotic behavior of the q-dimensional processes $Y'^n(X)$ with components:

$$\begin{aligned} Y'^n(X)_t^j = \frac{1}{\sqrt{u_n}} \Bigg(&\sum_{i=1}^{[t/\Delta_n]-k_n+1} f^j\left(\frac{1}{\sqrt{u_n}} \sum_{r=1}^{k_n} g^j(r/k_n)\, \Delta_{i+r-1}^n X\right) \\ &- \int_0^t \rho_{\Lambda(g^j)c_s}\left(f^j\right) ds \Bigg). \end{aligned} \tag{12.2.2}$$

In Theorem 11.2.1 we were faced with two additional difficulties, in comparison with the CLT for functionals depending on a single increment: one was the description of the limit, the other was the lack of (conditional) independence of the summands of the functional, because of the overlapping intervals.

Here the two same problems arise. The solution of the first one—the description of the limit—is a simple extension of the description in Sect. 11.2.1. The solution of the second one is significantly more difficult here, because the (conditional) independence of summands is ensured only if they are more than k_n indices apart, whereas $k_n \to \infty$. The solution will be achieved by splitting the sum into big blocks which will be independent when we "separate" them by "small" blocks of k_n successive summands. Then the independence ensures a CLT for the sum of the big blocks, and the sum of the small blocks is seen to be asymptotically negligible when the relative sizes of big versus small blocks goes to infinity.

12.2.1 The Results

Once more, we extend both Theorems 5.3.5 and 5.3.6 at once, but only in the case of an "even" function. According to the case, we need one of the assumptions 4.4.3, 5.3.2, 4.4.4 and 5.3.4, which we recall below:

Assumption (K) We have (H) and

$$\sigma_t = \sigma_0 + \int_0^t \widetilde{b}_s\,ds + \int_0^t \widetilde{\sigma}_s\,dW_s + M_t + \sum_{s\le t} \Delta\sigma_s\,1_{\{\|\Delta\sigma_s\|>1\}}$$

where M is a local martingale with $\|\Delta M_t\| \le 1$, orthogonal to W, and $\langle M, M\rangle_t = \int_0^t a_s\,ds$ and the compensator of $\sum_{s\le t} 1_{\{\|\Delta\sigma_s\|>1\}}$ is $\int_0^t \widetilde{a}_s\,ds$, with the following properties: the processes $\widetilde{b}$, $\widetilde{\sigma}$, $\widetilde{a}$ and a are progressively measurable, the processes $\widetilde{b}$, a and $\widetilde{a}$ are locally bounded, and the processes $\widetilde{\sigma}$ and b are càdlàg or càglàd.

Assumption (K-r) (for $r \in [0,1]$) We have (K) except for the càdlàg or càglàd property of b, and $\|\delta(\omega,t,z)\| \wedge 1 \le \Gamma_n(z)$ for all (ω,t,z) with $t \le \tau_n(\omega)$, where (τ_n) is a localizing sequence of stopping times and the Borel functions Γ_n on E satisfy $\int \Gamma_n(z)^r \lambda(dz) < \infty$. Moreover the process $b'_t = b_t - \int_{\{\|\delta(t,z)\|\le 1\}} \delta(t,z)\,\lambda(dz)$ is càdlàg or càglàd.

Assumption (K') We have (K) and both processes c_t and c_{t-} take their values in $\mathcal{M}^{++}_{d\times d}$.

Assumption (K'-r) We have (K-r) and both processes c_t and c_{t-} take their values in $\mathcal{M}^{++}_{d\times d}$.

As for the test function, the assumptions on $f = (f^j)_{1\le j\le q}$ will vary, exactly as in the afore-mentioned theorems. For the same reasons (see e.g. Remark 10.3.1) we split the components into two parts, and suppose that for some integer $q' \in \{0, \dots, q\}$ we have

$$\begin{array}{lll} j \le q' & \Rightarrow & x \mapsto f^j(x) \text{ is } C^1 \text{ on } \mathbb{R}^d \\ j > q' & \Rightarrow & x \mapsto F^j(x) \text{ is continuous on } \mathbb{R}^d \text{ and } C^1 \text{ outside } B, \end{array} \tag{12.2.3}$$

where, when $q' < q$, B is non empty and a finite union of affine hyperplanes of $\mathbb{R}^d$. We denote by $d(x, B)$ the distance between $x \in \mathbb{R}^d$ and B. Below, the numbers w, s, s', p are subject to $0 < w \le 1$ and $0 < s \le s'$ and $p \ge 0$. Then we set

$$\|f(x)\| \le K(1 + \|x\|^p) \tag{12.2.4}$$

$$|\nabla f^j(x)| \le \begin{cases} K(1 + \|x\|^p) & \text{if } j \le q' \\ K(1 + \|x\|^p)(1 + \frac{1}{d(x,B)^{1-w}}) & \text{if } j > q' \text{ and } x \in B^c \end{cases} \tag{12.2.5}$$

$$x \in B^c,\ \|y\| \le 1 \bigwedge \frac{d(x,B)}{2},\ j > q' \Rightarrow$$

$$|\nabla f^j(x+y) - \nabla f^j(x)| \le K\,\|y\| \left(1 + \frac{1}{d(x,B)^{2-w}}\right)(1 + \|x\|^p) \tag{12.2.6}$$

$$\|f(x+y) - f(x)\| \le K(1 + \|x\|^p)(\|y\|^r + \|y\|^{r'}). \tag{12.2.7}$$

Next, we describe the limit. To this end, we observe that the formula

$$L(g^j)_t = \int_t^{t+1} g^j(s-t)\,dW_s^l \tag{12.2.8}$$

defines a family $(L(g^j) : 1 \le j \le q)$ of d'-dimensional centered processes which is globally Gaussian and stationary, with covariance structure

$$\mathbb{E}\big(L(g^j)_t^i\, L(g^l)_s^m\big) = \delta^{im} \int_{t \vee s}^{(t+1)\wedge(s+1)} g^j(u-t)\, g^l(u-s)\, du.$$

If α is a $d \times d'$ matrix, we write $\alpha L(g^j)_t$ for the d-dimensional variable with components $\sum_{m=1}^{d'} \alpha^{im} L(g^j)_t^m$. The law of the family $(\alpha L(g^j) : 1 \le j \le q)$ only depends on the g^j's and on $a = \alpha\alpha^*$. This is reflected in the following notation, in which we use the notation (12.2.1), and which makes sense because all f^j have polynomial growth:

$$R_a^{jl} = \int_0^2 \mathbb{E}\big(f^j(\alpha L(g^j)_1)\, f^l(\alpha L(g^l)_t)\big)\,dt - 2\rho_{\Lambda(g^j)a}(f^j)\,\rho_{\Lambda(g^l)a}(f^l). \tag{12.2.9}$$

Observing that $\rho_{\Lambda(g^j)a}(f^j) = \mathbb{E}(f^j(\alpha L(g^j)_t)$ for all t, we easily check that R_a is a covariance matrix (which depends on the families f^j and g^j, although it does not show in the notation).

We are now ready to state the result.

Theorem 12.2.1 *Let X be a d-dimensional Itô semimartingale, and $f=(f^j)$ be a globally even q-dimensional function on $\mathbb{R}^d$ which satisfies* (12.2.3). *We also assume* (12.2.4) *and* (12.2.5) (*recall $w\in(0,1]$*), *plus one of the following five sets of hypotheses*:

(a) *We have $q'=q$ and* (K) *and X is continuous.*
(b) *We have* (K') *and* (12.2.6) *and X is continuous.*
(c) *We have $q'=q$ and* (K-1), *and f and ∇f are bounded.*
(d) *We have $q'=q$ and* (K-r) *for some $r\in[0,1)$, and* (12.2.8) *with $r\le s\le s'<1$.*
(e) *We have* (K'-r) *with some $r\in(0,1)$, and* (12.2.7) *and* (12.2.8) *with $r\le s\le s'<1$.*

Finally, we let G^j be finite signed measures supported by $(0,1]$, *such that the functions g^j associated by* (12.0.4) *are Hölder with some index $\theta\in(0,1]$, and we assume* (12.1.7).

Then the processes $Y'^n(X)$ of (12.2.2) *converge stably in law to a process $Y'(X)$ which is defined on a very good filtered extension $(\widetilde{\Omega},\widetilde{\mathcal{F}},(\widetilde{\mathcal{F}}_t)_{t\ge0},\widetilde{\mathbb{P}})$ of the space $(\Omega,\mathcal{F},(\mathcal{F}_t)_{t\ge0},\mathbb{P})$, and conditionally on $\mathcal{F}$ is a continuous centered Gaussian process with independent increments satisfying*

$$\widetilde{\mathbb{E}}\big(Y'(X)^j_t\,Y'(X)^l_t\mid\mathcal{F}\big)=\int_0^t R^{jl}_{c_s}\,ds. \tag{12.2.10}$$

12.2.2 Preliminaries for the Proof

We follow the scheme of the proof of Theorem 11.2.1.

1) First, by the localization lemma 5.3.12, instead of (K), (K-r), (K') or (K'-r) we can and will assume the strengthened versions (SK), (SK-r), (SK') or (SK'-r), that is Assumptions 4.4.7, 5.3.10, 4.4.8 or 5.3.11. In other words, $\|\delta(\omega,t,z)\|\le\Gamma(z)$ with Γ bounded and $\int\Gamma(z)^r\lambda(dz)<\infty$, and all processes $b,\sigma,X,\widetilde{b},\widetilde{\sigma},a,\widetilde{a}$ are bounded, and furthermore in case of (SK') or (SK'-r) the inverse process c_t^{-1} is also bounded. In particular we can write X as in (5.3.13), that is

$$X=X'+X''\quad\text{where } X'_t=X_0+\int_0^t b'_s\,ds+\int_0^t\sigma_s\,dW_s,\qquad X''=\delta*\mathfrak{p},$$

where $b'_t=b_t-\int_{\{z:\|\delta(t,z)\|\le1\}}\delta(t,z)\,\lambda(dz)$ is also bounded. As in (5.3.14), we can also write σ as

$$\sigma_t=\sigma_0+\int_0^t\widetilde{b}'_s\,ds+\int_0^t\widetilde{\sigma}_s\,dW_s+M'_t, \tag{12.2.11}$$

where $\widetilde{b}'_t=\widetilde{b}_t+\widetilde{a}'_t$ is bounded and M' is a martingale orthogonal to W, with bounded jumps, and predictable bracket $\int_0^t a'_s\,ds$ where $\|a'_t\|$ is bounded.

2) The aim of the next lemma is to eliminate the jumps.

Lemma 12.2.2 *Under* (SK-1) *and the assumption* (c), *or under* (SK-r) *and the assumptions* (d) *or* (e), *we have*

$$Y'^n(X) - Y'^n(X') \stackrel{\text{u.c.p.}}{\Longrightarrow} 0.$$

Proof By (12.2.2), and with the notation $\overline{X}(g^j)$ of (12.1.10), the result is equivalent to asymptotic negligibility, for each $j = 1, \dots, q$, of the array

$$\zeta_i^n = \frac{\Delta_n}{\sqrt{u_n}} \left(f^j\left(\overline{X}(g^j)_i^n/\sqrt{u_n}\right) - f^j\left(\overline{X'}(g^j)_i^n/\sqrt{u_n}\right)\right).$$

In case (c), $|\zeta_i^n| \le \frac{K\Delta_n}{\sqrt{u_n}} (1 \wedge (\|\overline{X''}(g^j)_i^n\|/\sqrt{u_n}))$. Then (12.1.12), (SK-1) and (2.1.47) yield $\mathbb{E}(|\zeta_i^n|) \le K\Delta_n \phi_n$, with $\phi_n \to 0$. The asymptotic negligibility follows.

In cases (d) and (e) the function f satisfies (12.2.8), with $p' \le 2$ and $r \le s \le s' < 1$, and we can always take $s = r$. Hence

$$\|\zeta_i^n\| \le K \frac{\Delta_n}{\sqrt{u_n}} \left((\alpha_i''^n)^r + (\alpha_i''^n)^{s'}\right) \alpha_i'^n, \quad \text{where}$$

$$\alpha_i'^n = 1 + \left(\|\overline{X'}(g^j)_i^n\|/\sqrt{u_n}\right)^2, \qquad \alpha_i''^n = \|\overline{X''}(g^j)_i^n\|/\sqrt{u_n}.$$

(SK-r) together with (12.1.12), (2.1.33), (2.1.34) and Lemma 2.1.7 imply $\mathbb{E}((\alpha_i'^n)^q) \le K_q$ for all $q \ge 0$ and $\mathbb{E}((\alpha_i''^n)^q) \le K_q u_n^{1-q/2}$ for all $q \ge r$. Applying twice Hölder's inequality with the first exponent equal to $\frac{4}{3+r}$ and $\frac{4}{3+r'}$, we obtain $\mathbb{E}(|\zeta_i^n|) \le K\Delta_n (u_n^{\frac{1-r}{4}} + u_n^{\frac{1-r'}{4}})$, and the asymptotic negligibility follows from $u_n \to 0$. □

3) In view of this lemma, it remains to consider the case when $X = X'$ is continuous, and we have the same two cases as in (5.3.20):

(a) $q' = q$, (SK), f is C^1 and ∇f has polynomial growth, $\sigma_t \in \mathcal{M}_A$
(b) $q' < qt$, (SK'), f satisfies (12.2.4), (12.2.5) and (12.2.6), $\sigma_t \in \mathcal{M}'_A$

for some $A > 0$. These assumptions will be in force throughout the rest of the section.

Now we introduce some notation. Recalling (12.1.11), we set

$$\left.\begin{aligned} \overline{\beta}_i^{n,j} &= \frac{1}{\sqrt{u_n}} \sigma_{(i-1)\Delta_n} \int_{(i-1)\Delta_n}^{(i-1)\Delta_n + u_n} g^j\left(\tfrac{s-(i-1)\Delta_n}{u_n}\right) dW_s \\ \overline{\beta}_i'^{n,j} &= \frac{1}{\sqrt{u_n}} \sigma_{(i-1)\Delta_n} \int_{(i-1)\Delta_n}^{(i-1)\Delta_n + u_n} g_n^j\left(\tfrac{s-(i-1)\Delta_n}{u_n}\right) dW_s. \end{aligned}\right\} \tag{12.2.12}$$

The d-dimensional random vectors $\overline{\beta}_i^{n,j}$ and $\overline{\beta}_i'^{n,j}$ are, conditionally on $\mathcal{F}_{(i-1)\Delta_n}$, centered Gaussian with respective covariances $\Lambda(g^j)c_{(i-1)\Delta_n}$ and $\Lambda(g_n^j)c_{(i-1)\Delta_n}$, where for any function h on $[0,1]$ we set $\Lambda(h) = \int_0^1 h(s)^2 ds$ (an extension of

(12.2.1)). Then, exactly as for (8.4.9)–(8.4.11) and upon using (12.1.12), we have the following estimates for all $v \geq 2$ (recall that X is continuous and $|g_n^j(t)| \leq K$):

$$\mathbb{E}\big(\|\overline{\beta}_i^{n,j}\|^v\big) + \mathbb{E}\big(\|\overline{\beta}_i^{\prime n,j}\|^v\big) + \mathbb{E}\big(\|\overline{X}(g^j)_i^n/\sqrt{u_n}\|^v\big) \leq K_v, \tag{12.2.13}$$

$$\left.\begin{array}{l} \mathbb{E}\big(\|\frac{1}{\sqrt{u_n}}\overline{X}(g^j)_i^n - \overline{\beta}_i^{\prime n,j}\|^v\big) \leq K_v\big(u_n^{v/2} + \frac{1}{u_n}\overline{\gamma}_i^n\big), \\ \text{where} \quad \overline{\gamma}_i^n = \mathbb{E}\big(\int_{(i-1)\Delta_n}^{(i-1)\Delta_n+u_n} \|\sigma_s - \sigma_{(i-1)\Delta_n}\|^2 ds\big) \leq K u_n^2. \end{array}\right\} \tag{12.2.14}$$

Since f is of polynomial growth, the variables $f^j(\overline{X}(g^j)_i^n/\sqrt{u_n})$, $f^j(\overline{\beta}_i^{n,j})$ and $f^j(\overline{\beta}_i^{\prime n,j})$ are integrable, so we can set

$$\begin{array}{ll} \chi_i^{n,j} = \frac{\Delta_n}{\sqrt{u_n}}\big(f^j\big(\overline{X}(g^j)_i^n/\sqrt{u_n}\big) - f^j\big(\overline{\beta}_i^{\prime n,j}\big)\big) & \\ \widehat{\chi}_i^{n,j} = \frac{\Delta_n}{\sqrt{u_n}}\big(f^j\big(\overline{\beta}_i^{\prime n,j}\big) - f^j\big(\overline{\beta}_i^{n,j}\big)\big) & \\ \chi_i^{\prime n,j} = \mathbb{E}\big(\chi_i^{n,j} \mid \mathcal{F}_{(i-1)\Delta_n}\big), & \widehat{\chi}_i^{\prime n,j} = \mathbb{E}\big(\widehat{\chi}_i^{n,j} \mid \mathcal{F}_{(i-1)\Delta_n}\big) \\ \chi_i^{\prime\prime n,j} = \chi_i^{n,j} - \chi_i^{\prime n,j}, & \widehat{\chi}_i^{\prime\prime n,j} = \widehat{\chi}_i^{n,j} - \widehat{\chi}_i^{\prime n,j}. \end{array}$$

Next, we define a family of processes:

$$\left.\begin{array}{ll} H^n(1)_t^j = \frac{\Delta_n}{\sqrt{u_n}}\sum_{i=1}^{[t/\Delta_n]-k_n+1}\big(f^j\big(\overline{\beta}_i^{n,j}\big) - \rho_{A(g^j)c_{(i-1)\Delta_n}}(f^j)\big) & \\ H^n(2)_t^j = \sum_{i=1}^{[t/\Delta_n]-k_n+1}\chi_i^{\prime\prime n,j}, & H^n(3)_t^j = \sum_{i=1}^{[t/\Delta_n]-k_n+1}\chi_i^{\prime n,j}, \\ H^n(4)_t^j = \sum_{i=1}^{[t/\Delta_n]-k_n+1}\widehat{\chi}_i^{\prime\prime n,j}, & H^n(5)_t^j = \sum_{i=1}^{[t/\Delta_n]-k_n+1}\widehat{\chi}_i^{\prime n,j}, \\ H^n(6)_t^j = \frac{1}{\sqrt{u_n}}\big(\Delta_n\sum_{i=1}^{[t/\Delta_n]-k_n+1}\rho_{A(g^j)c_{(i-1)\Delta_n}}(f^j) - \int_0^t \rho_{A(g^j)c_s}(f^j)\,ds\big). & \end{array}\right\} \tag{12.2.15}$$

Then we have

$$Y'^n(X)^j = \sum_{l=1}^{6} H^n(l)^j,$$

and the theorem will follow from the next two lemmas:

Lemma 12.2.3 *The processes $(H^n(1)^j)_{1\leq j\leq q}$ converge stably in law to the process $\overline{U}'(F,X)$.*

Lemma 12.2.4 *For $l = 2,3,4,5,6$ and $j = 1,\dots,q$ we have*

$$H^n(l)^j \stackrel{\text{u.c.p.}}{\Longrightarrow} 0. \tag{12.2.16}$$

12.2.3 Proof of Lemma 12.2.4

We want to prove (12.2.16), and for this we can argue component by component: that is, we can assume $q = 1$ and the index j does not show up: we simply write f and g, and χ_i^n, $\chi_i^{\prime n}$, and so on.

Step 1) We first study $H^n(6)$. With the notation (5.3.23), page 152, in which we substitute c_s with $\Lambda(g)c_s$, we have

$$H^n(6)_t = \frac{1}{\sqrt{k_n}} A^n(2)_{(t-u_n)^+} - \frac{1}{\sqrt{u_n}} \int_{(t-u_n)^+}^t \rho_{\Lambda(g)c_s}(f)\,ds.$$

The absolute value of the last term in the right side above is smaller than $K\sqrt{u_n}$, and the first term goes to 0 locally uniformly in t by (5.3.24): these two facts obviously imply (12.2.16) for $l=6$.

Step 2) Next, we study $H^n(5)$, and this is where the second condition in (12.1.7) comes into play. With the same function $\psi(\alpha)=\rho_{\alpha\alpha^*}(f)$ as in (5.3.26), we have $H^n(5)_t=\sum_{i=1}^{[t/\Delta_n]-k_n+1}\zeta_i^n$, where

$$\zeta_i^n = \frac{\Delta_n}{\sqrt{u_n}}\big(\psi\big(\sqrt{\Lambda(g_n)}\,\sigma_{(i-1)\Delta_n}\big) - \psi\big(\sqrt{\Lambda(g)}\,\sigma_{(i-1)\Delta_n}\big)\big).$$

The function ψ is C_b^1 on the set $\mathcal{M}_A$ in case (a), and on the set $\mathcal{M}'_A$ in case (b). Moreover, since $\Lambda(g)>0$, we have $\frac{1}{C}\le\Lambda(g_n)\le C$ for some $C>0$ and all n large enough, and thus

$$|\zeta_i^n| \le \frac{K\Delta_n}{\sqrt{u_n}}\big|\sqrt{\Lambda(g_n)}-\sqrt{\Lambda(g)}\big| \le \frac{K\Delta_n}{\sqrt{u_n}}\big|\Lambda(g_n)-\Lambda(g)\big|$$

by the C_b^1 property of ψ and the boundedness of σ_t.

The Hölder property of g and (12.1.11) yield $|g_n-g|\le Kk_n^{-\theta}$, hence (12.2.1) gives $|\Lambda(g_n)-\Lambda(g)|\le Kk_n^{-\theta}$ as well. Then $|\zeta_i^n|\le K\Delta_n/(k_n^{2\theta+1}\Delta_n)^{1/2}$ and (12.1.18) for $l=5$ follows from (12.1.7).

Step 3) Next, we study $H^n(2)$. The conditions on f yield, for some $p\ge0$ and all $\varepsilon\in(0,1)$:

$$\big|f(x+y)-f(x)\big| \le K\big(1+\|x\|^p+\|y\|^p\big)\bigg(\frac{\|y\|}{\varepsilon}+1_{\{d(x,B)\le\varepsilon\}}\bigg) \tag{12.2.17}$$

in both cases (a) and (b) (in case (a), $d(x,B)=\infty$ for all x). Observe that when $B\neq\emptyset$, the proof of Lemma 5.3.14, page 156, written for $\gamma_i^n=\phi_B(\beta_i^n)$, works equally well if we replace β_i^n by $\overline{\beta}_i^{\prime n}=\overline{\beta}_i^{\prime n,j}$, because $\sqrt{\Lambda(g_n)}$ stays in an interval $[a,a']$ away from 0. We deduce that in case (b), we have $\mathbb{E}(d(\overline{\beta}_i^{\prime n},B)^{-1/2})\le K$, and thus

$$\mathbb{P}\big(d\big(\overline{\beta}_i^{\prime n},B\big)\le\varepsilon\big) \le K\sqrt{\varepsilon}$$

by Markov's inequality. This fact, plus (12.2.13) and (12.2.14), plus the Cauchy-Schwarz inequality, allows us to deduce from (12.2.17) that

$$\mathbb{E}\big(|\chi_i^n|^2\big) \le K\frac{\Delta_n^2}{u_n}\bigg(\frac{\sqrt{u_n}}{\varepsilon^2}+\varepsilon^{1/4}\bigg). \tag{12.2.18}$$

We have $H^n(2) = \sum_{r=0}^{k_n-1} H^n(2,r)$, with $H^n(2,r)_t = \sum_{i=1}^{l_n(r,t)} \chi''^n_{(i-1)k_n+r+1}$ and $l_n(r,t) = [([t/\Delta_n] - r)/k_n]$. The variables $\chi''^n_{(i-1)k_n+r+1}$ are martingale increments, relative to the discrete time filtration $(\mathcal{F}_{(ik_n+r)\Delta_n})_{i\geq 0}$. Therefore by Doob's inequality and (12.2.18):

$$\mathbb{E}\Big(\sup_{s\leq t} \big|H^n(2,r)_s\big|^2\Big) \leq 4\mathbb{E}\bigg(\sum_{i=1}^{l_n(r,t)} \big|\chi^n_{(i-1)k_n+r+1}\big|^2\bigg) \leq Kt\,\frac{1}{k_n^2}\bigg(\frac{\sqrt{u_n}}{\varepsilon^2} + \varepsilon^{1/4}\bigg).$$

Then, taking the square-root and summing up over r yields for all $\varepsilon \in (0,1)$:

$$\mathbb{E}\Big(\sup_{s\leq t} \big|H^n(2)_s\big|\Big) \;\leq\; K\sqrt{t}\,\bigg(\frac{u_n^{1/4}}{\varepsilon} + \varepsilon^{1/8}\bigg).$$

and we conclude (12.2.16) for $l=2$ because $u_n \to 0$ and $\varepsilon > 0$ is arbitrarily small.

Step 4) Here we study $H^n(4)$. The random vector $\overline{\beta}'^n_i - \overline{\beta}^n_i$ is, conditionally on $\mathcal{F}_{(i-1)\Delta_n}$, centered Gaussian with covariance $a_n\, c_{(i-1)\Delta_n}$, where $a_n = \int_0^1 (g_n(s) - g(s))^2 ds$. The Hölder property of g implies $a_n \leq K k_n^{-2\theta}$, and thus $\mathbb{E}(\|\overline{\beta}'^n_i - \overline{\beta}^n_i\|^r) \leq K_r k_n^{-r\theta}$ for any $r>0$. Using (12.2.17), and exactly as for (12.2.18), we deduce

$$\mathbb{E}\big(|\widehat{\chi}^n_i|^2\big) \;\leq\; K\,\frac{\Delta_n^2}{u_n}\bigg(\frac{1}{\varepsilon^2 k_n^{2\theta}} + \varepsilon^{1/4}\bigg).$$

We can then repeat the end of the previous step to conclude (12.2.16) for $l=4$.

Step 5) Now we start the proof of (12.2.16) for $l=3$, and this essentially reproduces Part C of Sect. 5.3.3, pages 154–160, except that Δ_n is substituted with u_n. Instead of (5.3.30), and by virtue of (12.1.12), we have

$$\begin{aligned}\theta^n_i := \frac{\overline{X}(g)^n_i}{\sqrt{u_n}} - \overline{\beta}'^n_i \;&=\; \frac{1}{\sqrt{u_n}}\int_{(i-1)\Delta_n}^{(i-1)\Delta_n+u_n} g_n\big(s-(i-1)\Delta_n\big)\, b_s\, ds \\ &+ \frac{1}{\sqrt{u_n}}\int_{(i-1)\Delta_n}^{(i-1)\Delta_n+u_n} g_n\big(s-(i-1)\Delta_n\big)\,(\sigma_s - \sigma_{(i-1)\Delta_n})\, dW_s.\end{aligned}$$

Then as for (5.3.31),

$$l>0 \quad\Rightarrow\quad \mathbb{E}\big(\|\theta^n_i\|^l\big) \;\leq\; K_l u_n^{(l/2)\wedge 1}.$$

Next, since X is continuous, and recalling (12.2.11), we have the decomposition $\theta^n_i = \frac{1}{\sqrt{u_n}}\sum_{j=1}^4 \zeta(j)^n_i$, where

$$\zeta(1)^n_i = b_{(i-1)\Delta_n}\int_{(i-1)\Delta_n}^{(i-1)\Delta_n+u_n} g_n\bigg(\frac{s-(i-1)\Delta_n}{u_n}\bigg)\, ds$$

$$\zeta(2)^n_i = \int_{(i-1)\Delta_n}^{(i-1)\Delta_n+u_n} g_n\bigg(\frac{s-(i-1)\Delta_n}{u_n}\bigg)\,\big(\widetilde{\sigma}_{(i-1)\Delta_n}(W_s - W_{(i-1)\Delta_n})\big)\, dW_s$$

$$\zeta(3)_i^n = \int_{(i-1)\Delta_n}^{(i-1)\Delta_n+u_n} g_n\left(\frac{s-(i-1)\Delta_n}{u_n}\right)(M'_s - M'_{(i-1)\Delta_n})\, dW_s$$

$$\zeta(4)_i^n = \int_{(i-1)\Delta_n}^{(i-1)\Delta_n+u_n} g_n\left(\frac{s-(i-1)\Delta_n}{u_n}\right)(b_s - b_{(i-1)\Delta_n})\, ds$$

$$+ \int_{(i-1)\Delta_n}^{(i-1)\Delta_n+u_n} g_n\left(\frac{s-(i-1)\Delta_n}{u_n}\right)\left(\int_{(i-1)\Delta_n}^{s} \widetilde{b}'_u du\right) dW_s$$

$$+ \int_{(i-1)\Delta_n}^{(i-1)\Delta_n+u_n} g_n\left(\frac{s-(i-1)\Delta_n}{u_n}\right)\left(\int_{(i-1)\Delta_n}^{s} (\widetilde{\sigma}_u - \widetilde{\sigma}_{(i-1)\Delta_n})\, dW_u\right) dW_s.$$

The other notation of Part C of Sect. 5.3.3 are unchanged, except for β_i^n which is replaced by $\overline{\beta}_i^n$ everywhere. Then

$$H^n(3) = \sum_{j=1}^{7} D^n(j) \tag{12.2.19}$$

where, with u_i^n a random number with values in $[0,1]$ and $A_i^n = \{\|\theta_i^n\| > d(\overline{\beta}_i^n, B)/2\}$ and, with vector notation,

$$D^n(j)_t = \sum_{i=1}^{[t/\Delta_n]-k_n+1} \delta(j)_i^n, \quad \text{where } \delta(j)_i^n = \mathbb{E}(\delta'(j)_i^n \mid \mathcal{F}_{(i-1)\Delta_n}), \qquad \text{and}$$

$$\delta'(j)_i^n = \frac{1}{k_n} \nabla f(\overline{\beta}_i^n)\, \zeta(j)_i^n \quad \text{for } j = 1,2,3,4$$

$$\delta'(5)_i^n = -\frac{\sqrt{u_n}}{k_n} \nabla f(\overline{\beta}_i^n)\, \theta_i^n\, 1_{A_i^n}$$

$$\delta'(6)_i^n = \frac{\sqrt{u_n}}{k_n} \left(\nabla f(\overline{\beta}_i^n + u_i^n \theta_i^n) - \nabla f(\overline{\beta}_i^n)\right) \theta_i^n\, 1_{(A_i^n)^c}$$

$$\delta'(7)_i^n = \frac{\sqrt{u_n}}{k_n} \left(f(\overline{\beta}_i^n + \theta_i^n) - f(\overline{\beta}_i^n)\right) 1_{A_i^n}.$$

Next, we replace (5.3.35) by

$$\alpha_i^n = u_n^{3/2} + \mathbb{E}\left(\int_{(i-1)\Delta_n}^{(i-1)\Delta_n+u_n} \left(\|b_s - b_{(i-1)\Delta_n}\|^2 + \|\widetilde{\sigma}_s - \widetilde{\sigma}_{(i-1)\Delta_n}\|^2\right) ds\right).$$

Step 6) As seen in Step 3, (5.3.39) holds true here, with γ_i^n associated by (5.3.38) with $\overline{\beta}_i^{\prime n}$ instead of β_i^n. Then (5.3.36), (5.3.37) and Lemma 5.3.15 holds with u_n instead of Δ_n.

Now we proceed to Lemma 5.3.16, page 158, which we prove in the present setting. First, because $\frac{1}{k_n}\sum_{i=1}^{[t/\Delta_n]}\alpha_i^n \to 0$, we have instead of (5.3.46):

$$0 < v \le 1 \quad \Rightarrow \quad \frac{\Delta_n}{u_n^v}\sum_{i=1}^{[t/\Delta_n]}\left(\alpha_i^n\right)^v \to 0. \tag{12.2.20}$$

Then it remains to prove (5.3.47) for $j = 4, 5, 6, 7$. (5.3.48) holds with the additional factor $1/k_n$, hence

$$\sum_{i=1}^{[t/\Delta_n]}\mathbb{E}\big(\big|\delta'(4)_i^n\big|\big) \le \frac{K\Delta_n}{u_n^v}\sum_{i=1}^{[t/\Delta_n]}\left(\alpha_i^n\right)^v,$$

for some $v \in (0,1)$, and (5.3.47) for $j = 4$ follows from (12.2.20).

In case (a) the result for $j = 5, 7$ is again obvious, and for $j = 6$ we see that, instead of (5.3.49), we have

$$\mathbb{E}\big(\big|\delta'(6)_i^n\big|\big) \le \frac{Ku_n}{k_n}\left(\phi'_C(\varepsilon) + \frac{1}{C} + \frac{K}{\varepsilon}\sqrt{u_n}\right).$$

By $u_n/k_n = \Delta_n$ and $u_n \to 0$, we conclude (5.3.47) for $j = 6$ as in Lemma 5.3.16.

In case (b), we get (5.3.50) with $\sqrt{u_n}/k_n$ instead of $\sqrt{\Delta_n}$. We deduce $\mathbb{E}(|\delta'(j)_i^n|) \le Ku_n^{1+w/4}/k_n = K\Delta_n u_n^{w/4}$ for some $w > 0$, and again we conclude (5.3.47) for $j = 5, 6, 7$.

Step 7) The proof of Lemma 5.3.17 remains valid here, to give $D^n(3) = 0$. Finally we can reproduce the proof of Step 5 of Sect. 11.2.5 to obtain that, since f is globally even on $\mathbb{R}^d$, we also have $D^n(1) = D^n(2) = 0$ identically. In view of (12.2.19), this ends the proof of Lemma 12.2.4.

12.2.4 Block Splitting

Now we start the proof of Lemma 12.2.3. We are concerned with the process $H^n(1)_t = \sum_{i=1}^{[t/\Delta_n]-k_n+1}\zeta_i^n$, where the components of ζ_i^n are

$$\zeta_i^{n,j} = \frac{\Delta_n}{\sqrt{u_n}}\left(f^j\left(\overline{\beta}_i^{n,j}\right) - \rho_{A(g^j)c_{(i-1)\Delta_n}}\left(f^j\right)\right).$$

Although the variable ζ_i^n has a vanishing $\mathcal{F}_{(i-1)\Delta_n}$-conditional expectation, it is not $\mathcal{F}_{i\Delta_n}$ measurable. To tackle this problem, we split the sum over i into blocks of size mk_n, separated by blocks of size k_n, in order to ensure some "conditional independence" of the successive summands, and it remains a residual sum for the summands occurring just before time t.

More specifically, we fix an integer $m \geq 2$ (which will eventually go to infinity). The lth block of size mk_n contains ζ_i^n for all i between $I(m,n,l) = (l-1) \times (m+1)k_n + 1$ and $I(m,n,l) + mk_n - 1$. The number of such blocks which can be accommodated without using data after time t is then $l_n(m,t) = [\frac{[t/\Delta_n]-1}{(m+1)k_n}]$. The "real" times corresponding to the beginning of the lth big block is then $t(m,n,l) = (I(m,n,l)-1)\Delta_n$.

At this stage, we set

$$\zeta(m)_i^n = \sum_{r=0}^{mk_n-1} \zeta_{I(m,n,i)+r}^n, \qquad Z^n(m)_t = \sum_{i=1}^{l_n(m,t)} \zeta(m)_i^n.$$

The process $Z^n(m)$ is still difficult to analyze because any one summand $\zeta(m)_i^n$ in it involves the process σ_s evaluated at many different times. It is convenient to modify the definition by freezing the volatility at the beginning of each large block, and this leads to set (below, $i+r \geq 1$ always):

$$\left.\begin{array}{l} \widehat{\beta}_{i,r}^{n,j} = \frac{1}{\sqrt{u_n}} \sigma_{(i-1-r)\Delta_n} \int_{(i-1)\Delta_n}^{(i-1)\Delta_n+u_n} g^j\big(\frac{s-(i-1)\Delta_n}{u_n}\big)\, dW_s \\ \eta_{i,r}^{n,j} = \frac{\Delta_n}{\sqrt{u_n}} \big(f^j(\widehat{\beta}_{i,r}^{n,j}) - \rho_{\Lambda(g^j)c_{(i-1-r)\Delta_n}}(f^j)\big) \\ \eta(m)_i^n = \sum_{r=0}^{mk_n-1} \eta_{I(m,n,i)+r,r}^n \\ M^n(m)_t = \sum_{i=1}^{l_n(m,t)} \eta(m)_i^n. \end{array}\right\} \tag{12.2.21}$$

The variables $\eta_{i,r}^{n,j}$ are one-dimensional, but the variables $\widehat{\beta}_{i,r}^{n,j}$ are d-dimensional, and we write its components as $\widehat{\beta}_{i,r}^{n,lj}$ for $l = 1, \ldots, d$.

The rest of this subsection is devoted to proving that $H^n(1)$ has the "same behavior" as $M^n(m)$, asymptotically as $n \to \infty$ and also $m \to \infty$.

Lemma 12.2.5 *We have for all $t > 0$:*

$$\lim_{m\to\infty} \limsup_{n\to\infty} \mathbb{E}\Big(\sup_{s\leq t} \big\| H^n(1)_s - Z^n(m)_s \big\|\Big) = 0. \tag{12.2.22}$$

Proof Denote by $J(n,m,t)$ the set of all integers j between 1 and $[t/\Delta_n] - k_n + 1$, which are *not* in the big blocks, that is not of the form $I(m,n,i) + l$ for some $i \geq 1$ and $0 \leq l \leq mk_n - 1$. We further divide $J(n,m,t)$ into k_n disjoint subsets $J(n,m,t,r)$ for $r = 1, \ldots, k_n$, where $J(n,m,t,r)$ is the set of all $j \in J(n,m,t)$ equal to r modulo k_n. Then

$$H^n(1)_t - Z^n(m)_t = \sum_{r=1}^{k_n} \overline{Z}^n(m,r)_t, \qquad \overline{Z}^n(m,r)_t = \sum_{j\in J(n,m,t,r)} \zeta_j^n. \tag{12.2.23}$$

Observe that $\mathbb{E}(\zeta_j^n \mid \mathcal{F}_{(j-1)\Delta_n}) = 0$ and ζ_j^n is $\mathcal{F}_{(j+k_n)\Delta_n}$ measurable. Then $\overline{Z}^n(m,r)_t$ is a sum of martingale increments, because any two distinct indices in

$J(n,m,t,r)$ are more than k_n apart. Furthermore (12.2.13) and the polynomial growth of f imply that $\mathbb{E}(\|\zeta_j^n\|^2) \le K\Delta_n^2/u_n$. Therefore by Doob's inequality,

$$\mathbb{E}\Big(\sup_{s\le t} \big\|\overline{Z}^n(m,r)_s\big\|^2\Big) \le 4\,\mathbb{E}\Big(\sum_{j\in J(n,m,t,r)} \|\zeta_j^n\|^2\Big) \le Kt\,\frac{\Delta_n^2}{mu_n^2},$$

where the last inequality comes from the fact that the number of points in $J(n,m,t,r)$ is not more than $l_n(m,t)$, which is less than t/mu_n. Therefore

$$\mathbb{E}\Big(\sup_{s\le t} \big\|\overline{Z}^n(m,r)_s\big\|\Big) \le K\sqrt{t}\,\frac{\Delta_n}{\sqrt{m}\,u_n} = K\sqrt{t}\,\frac{1}{\sqrt{m}\,k_n}.$$

By virtue of (12.2.23), we deduce that the expectation in (12.2.22) is less than $K\sqrt{t}/\sqrt{m}$, and the result follows. □

Lemma 12.2.6 *We have for all $t>0$ and all $m\ge 1$:*

$$\mathbb{E}\Big(\sup_{s\le t} \big\|Z^n(m)_s - M^n(m)_s\big\|\Big) \to 0.$$

Proof 1) We start as in the previous lemma: $J'(n,m,t)$ the set of all integers j between 1 and $[t/\Delta_n]-k_n+1$, which are *inside* the big blocks, that is of the form $j = I(m,n,i)+l$ for some $i\ge 1$ and $l\in\{0,\dots,mk_n-1\}$. Then $J'(n,m,t,r)$ is the set of all $j\in J(n,m,t)$ equal to r modulo k_n. Hence

$$Z^n(m)_t - M^n(m)_t = \sum_{r=1}^{k_n} M^n(m,r)_t, \qquad M^n(m,r)_t = \sum_{j\in J(n,m,t,r)} \theta_j^n, \tag{12.2.24}$$

where $\theta_j^n = \zeta_j^n - \eta_{j,l}^n$ when $j = I(m,n,i)+l$. Then, as in the previous lemma,

$$\mathbb{E}\Big(\sup_{s\le t} \big\|M'^n(m,r)_s\big\|^2\Big) \le 4\,\mathbb{E}\Big(\sum_{i\in J'(n,m,t,r)} \|\theta_i^n\|^2\Big). \tag{12.2.25}$$

2) Now we give an estimate for $\mathbb{E}(\|\theta_j^n\|^2)$ for $j\in J(n,m,t,r)$. When $j = I(m,n,i)+l$ again, we can write

$$\begin{aligned}\theta_j^n = \frac{\sqrt{u_n}}{k_n}\Big(& f\big(\sqrt{\Lambda(g)}\,\sigma_{(j-1)\Delta_n}U\big) - f\big(\sqrt{\Lambda(g)}\,\sigma_{(j-1-l)\Delta_n}U\big) \\ & - \big(\psi\big(\sqrt{\Lambda(g)}\,\sigma_{(j-1)\Delta_n}\big) - \psi\big(\sqrt{\Lambda(g)}\,\sigma_{(j-1-l)\Delta_n}\big)\big)\Big),\end{aligned} \tag{12.2.26}$$

where $\psi(\alpha) = \mathbb{E}(f(\alpha U))$ is as in (5.3.26) once more and U is $\mathcal{N}(0,I_{d'})$.

Now (SK) implies, as in (5.3.28) and because $l\Delta_n \le mu_n$,

$$\mathbb{E}\big(\|\sigma_{(j-1)\Delta_n} - \sigma_{(j-1-l)\Delta_n}\|^2\big) \le Kmu_n. \tag{12.2.27}$$

On the other hand we have (12.2.17), which, exactly as for (12.2.18), yields for all $\varepsilon \in (0,1)$, all $A' > 1$ and $\alpha, \alpha' \in \mathcal{M}_{A'}$ in case (a) or $\alpha, \alpha' \in \mathcal{M}'_{A'}$ in case (b):

$$\mathbb{E}\big(\|f(\alpha U) - f(\alpha' U)\|^2\big) \leq K_{A'}\left(\frac{\|\alpha - \alpha'\|^2}{\varepsilon^2} + \varepsilon^{1/4}\right). \tag{12.2.28}$$

Combining (12.2.26), (12.2.27) and (12.2.28), and using that $\sqrt{\Lambda(g)}\,\sigma_t$ takes its values in $\mathcal{M}_{A'}$ in case (a) and in $\mathcal{M}'_{A'}$ in case (b), where $A' = A(\Lambda(g) \vee \frac{1}{\Lambda(g)})$, and by successive conditioning, we obtain for all $\varepsilon \in (0,1)$:

$$\mathbb{E}\big(\|\theta_j^n\|^2\big) \leq \frac{K u_n}{k_n^2}\left(\frac{m u_n}{\varepsilon^2} + \varepsilon^{1/4}\right). \tag{12.2.29}$$

Then by (12.2.25), and since $J'(n,m,k,r)$ contains at most $[t/u_n]$ points, we get

$$\mathbb{E}\Big(\sup_{s \leq t} \|M^n(m,r)_s\|^2\Big) \leq \frac{Kt}{k_n^2}\left(\frac{m u_n}{\varepsilon^2} + \varepsilon^{1/4}\right).$$

It remains to use (12.2.24), which yields

$$\mathbb{E}\Big(\sup_{s \leq t} \|Z^n(m)_s - M^n(m)_s\|\Big) \leq K\sqrt{t}\left(\frac{\sqrt{m u_n}}{\varepsilon} + \varepsilon^{1/8}\right)$$

and the result follows because ε is arbitrarily small. □

12.2.5 *Proof of Lemma 12.2.3*

We are now almost ready to prove Lemma 12.2.3. The key step is a CLT for the processes $M^n(m)$ of (12.2.21). Recalling the processes $L(g^j)$ of (12.2.8), we can set

$$A(\alpha; s, t)^{jl} = \mathbb{E}\big(f^j\big(\alpha L(g^j)_s\big) f^l\big(\alpha L(g^l)_t\big)\big) - \rho_{\Lambda(g^j)\alpha\alpha^*}(f^j)\,\rho_{\Lambda(g^l)\alpha\alpha^*}(f^l),$$

and we have:

Lemma 12.2.7 *For each $m \geq 1$ the processes $M^n(m)$ converge stably in law to a limit $M(m)$ defined on a very good extension $(\widetilde{\Omega}, \widetilde{\mathcal{F}}, (\widetilde{\mathcal{F}}_t)_{t\geq 0}, \widetilde{\mathbb{P}})$ of $(\Omega, \mathcal{F}, (\mathcal{F}_t)_{t\geq 0}, \mathbb{P})$, and which conditionally on $\mathcal{F}$ is a continuous centered Gaussian process with independent increments with*

$$\widetilde{\mathbb{E}}\big(M(m)_t^j\, M(m)_t^l \mid \mathcal{F}\big) = \frac{1}{m}\int_0^t \left(\int_{[0,m]^2} A(\sigma_s; u, v)^{jl}\, du\, dv\right) ds. \tag{12.2.30}$$

Proof 1) We will apply Theorem 2.2.15, page 58, to the array $(\eta(m)_i^n)$, with $N_n(t) = [t/u_n(m+1)]$ and $T(n,i) = t(m,n,i+1)$ and $(\Omega_n, \mathcal{G}^n, (\overline{\mathcal{F}}_t^n), \mathbb{P}_n) = (\Omega, \mathcal{F}, (\mathcal{F}_t), \mathbb{P})$ and $\mathcal{G}_i^n = \mathcal{F}_{t(m,n,i+1)}$, so we have (2.2.29) and (2.2.39).

First, $\mathbb{E}(\eta(m)_i^n \mid \mathcal{G}_{i-1}^n) = 0$ is clear. Since f has polynomial growth and c_t is bounded, we have $\mathbb{E}(\|\eta_{i,r}^n\|^4 \mid \mathcal{F}_{(i-1-r)\Delta_n}) \leq K\Delta_n^4/u_n^2$, hence $\mathbb{E}(\|\eta(m)_i^n\|^4 \mid \mathcal{G}_{i-1}^n) \leq Km^4u_n^2$ and it follows that

$$\sum_{i=1}^{l_n(m,t)} \mathbb{E}\big(\big\|\eta(m)_i^n\big\|^4 \mid \mathcal{G}_{i-1}^n\big) \stackrel{\mathbb{P}}{\longrightarrow} 0 \tag{12.2.31}$$

for any $t>0$ because $l_n(m,t) \leq t/mu_n$.

Next, we show that

$$\mathbb{E}\big(\eta(m)_i^n \,(N_{t(m,n,i+1)} - N_{t(m,n,i)}) \mid \mathcal{G}_{i-1}^n\big) = 0 \tag{12.2.32}$$

for any martingale N which either is bounded and orthogonal to the Brownian motion W, or is one of the components of W. In the former case, (12.2.32) is proved exactly as at the very end of the proof of Theorem 4.2.1, because $\zeta(m)_i^{\prime n}$ is a function of the process $W(m,n,i)_s = W_{t(m,n,i)+s} - W_{t(m,n,i)}$. In the latter case, $\eta(m)_i^n\,(N_{t(m,n,i+1)} - N_{t(m,n,i)})$ is again a function of the process $W(m,n,i)$, say $G(W(m,n,i))$, which satisfies $G(-W(m,n,i)) = -G(W(m,n,i))$ because the function f is globally even; since the $\mathcal{G}_{i-1}^n$-conditional laws of $W(m,n,i)$ and $-W(m,n,i)$ are the same, we have (12.2.32).

2) At this stage, by Theorem 2.2.15 it remains to prove the following convergence, for all $t>0$, and where $C(m)_t^{jl}$ denotes the right side of (12.2.30):

$$\sum_{i=1}^{l_n(m,t)} \mathbb{E}\big(\eta(m)_i^{n,j}\,\eta(m)_i^{n,l} \mid \mathcal{G}_{i-1}^n\big) \stackrel{\mathbb{P}}{\longrightarrow} C(m)_t^{jl}. \tag{12.2.33}$$

We fix for a moment i and n, and we set $I = I(m,n,i)$. Then from the first formula in (12.2.21) one easily sees that the $\mathcal{G}_{i-1}^n$-conditional distribution of the family of variables $(\widehat{\beta}_{I+r,r}^{n,jw} : r \geq 0, 1 \leq w \leq d, 1 \leq j \leq q)$ is the same as the law of the variables $(\alpha L(g^j)_{r/k_n}^w : r \geq 0, 1 \leq w \leq d, 1 \leq j \leq q)$ taken at $\alpha = \sigma_{(I-1)\Delta_n}$. Thus

$$\mathbb{E}\big(\eta_{I+r,r}^{n,j}\,\eta_{I+r',r'}^{n,l} \mid \mathcal{G}_{i-1}^n\big) = \frac{\Delta_n^2}{u_n}\,A\big(\sigma_{(I-1)\Delta_n}; r/k_n, r'/k_n\big)^{jl},$$

and a simple calculation yields

$$\mathbb{E}\big(\eta(m)_i^{n,j}\,\eta(m)_i^{n,l} \mid \mathcal{G}_{i-1}^n\big) = \frac{u_n}{k_n^2}\sum_{r,r'=0}^{mk_n-1} A\big(\sigma_{(I(m,n,i)-1)\Delta_n}; r/k_n, r'/k_n\big)^{jl}. \tag{12.2.34}$$

The functions g^j being Hölder on $[0,\infty)$, it is clear from (12.2.8) that the process $L(g^j)_t$ is continuous in probability in t, and is a Gaussian process, and the f^j have polynomial growth and are continuous. So it readily follows that the functions $(\alpha,s,t)\mapsto A(\alpha;s,t)^{jl}$ are continuous. Now, the right side of (12.2.34) is u_n times a Riemann sum for the integral $\int_{[0,m]^2} A(\sigma_{(I(m,n,i)-1)\Delta_n};u,v)^{jl}\,du\,dv$, and the left side of (12.2.33) is thus a Riemann sum for the triple integral (times $1/m$) defining $C(m)^{jl}_t$. Therefore (12.2.33) follows from the continuity of the function $A(.)^{jl}$. □

Proof of Lemma 12.2.3 In view of Lemmas 12.2.5, 12.2.6 and 12.2.7, and because of Proposition 2.2.4, the only property which remains to be proved is

$$M(m) \stackrel{\mathcal{L}\text{-}s}{\Longrightarrow} Y'(X) \quad \text{as } m\to\infty, \tag{12.2.35}$$

where $Y'(X)$ is as in Theorem 12.2.1. Recall that $M(m)$ and $Y'(X)$ are, conditionally on $\mathcal{F}$, centered, Gaussian, continuous and with independent increments and with covariances at time t respectively $C(m)^{jl}_t$ and C^{jl}_t, where C^{jl}_t is the right side of (12.2.10). Therefore, (12.2.35) follows from the fact that the $\mathcal{F}$-conditional distributions of $M(m)$ converge to the $\mathcal{F}$-conditional distribution of $Y'(x)$, and this is implied by

$$C(m)^{jl}_t \stackrel{\mathbb{P}}{\longrightarrow} C^{jl}_t \quad \text{as } m\to\infty$$

for all t. In turn, and since the function $A(.)^{jl}$ is locally bounded, this is implied by

$$\frac{1}{m}\int_{[0,m]^2} A(\alpha;u,v)^{jl}\,du\,dv \;\to\; R^{jl}_{\alpha\alpha^*} \tag{12.2.36}$$

for each $d\times d'$ matrix α.

To see this, we first observe that (we drop the indices j,l below)

$$R_{\alpha\alpha^*} \;=\; \int_0^2 A(\alpha;1,t)\,dt. \tag{12.2.37}$$

Next, since the process $L(\alpha)$ is stationary, we have $A(\alpha;u+t,v+t)=A(\alpha;u,v)$ for all $u,v,t\ge 0$. Moreover the variables $L(\alpha)_t$ and $L(\alpha)_s$ are independent if $|s-t|>1$, and thus $A(\alpha;t,s)=0$ in this case. Hence

$$\begin{aligned}\frac{1}{m}\int_{[0,m]^2} A(\alpha;u,v)\,du\,dv &= \frac{1}{m}\int_0^m du\int_{(u-1)\vee 0}^{(u+1)\wedge m} A(\alpha;1,v-u+1)\,dv\\ &= \frac{1}{m}\int_0^m du\int_{(u-1)^+}^{2\wedge(m+1-u)} A(\alpha;1,t)\,dt\end{aligned}$$

and, in view of (12.2.37), the property (12.2.36) is straightforward by the dominated convergence theorem. □

Chapter 13
The Central Limit Theorem for Truncated Functionals

In this chapter we prove the Central Limit Theorems associated with the Laws of Large Numbers of Chap. 9. The proofs use techniques which we have established and employed earlier in this book, with one exception: new techniques are needed for studying the local estimators of the volatility c_t.

Only regular discretization schemes are considered. The d-dimensional Itô semimartingale X has the Grigelionis decomposition

$$X_t = X_0 + \int_0^t b_s\,ds + \int_0^t \sigma_s\,dW_s + (\delta 1_{\{\|\delta\|\leq 1\}}) \star (\mathfrak{p} - \mathfrak{q})_t + (\delta 1_{\{\|\delta\|>1\}}) \star \mathfrak{p}_t, \quad (13.0.1)$$

where W is a d'-dimensional Wiener process and $\mathfrak{p}$ is a Poisson measure with compensator $\mathfrak{q}(dt,dz) = dt \otimes \lambda(dz)$, and $c = \sigma\sigma^\star$. We also assume at least Assumption (H-r) for some $r \in [0,2]$ (recall that (H-2) = (H)), that is Assumption 6.1.1, recalled below:

Assumption (H-r) We have (13.0.1) with b_t locally bounded and σ_t càdlàg and $\|\delta(\omega,t,z)\| \wedge 1 \leq \Gamma_n(z)$ for all (ω,t,z) with $t \leq \tau_n(\omega)$, where (τ_n) is a localizing sequence of stopping times and each function Γ_n satisfies $\int \Gamma_n(z)^r \lambda(dz) < \infty$.

We also consider a sequence of truncation levels v_n satisfying

$$v_n = \alpha\,\Delta_n^{\varpi} \quad \text{for some } \alpha > 0,\ \varpi \in \left(0, \frac{1}{2}\right). \quad (13.0.2)$$

13.1 A Central Limit Theorem for Approximating the Jumps

Here we consider the functionals

$$V^n(f, v_n+, X)_t = \sum_{i=1}^{[t/\Delta_n]} f(\Delta_i^n X)\, 1_{\{\|\Delta_i^n X\| > v_n\}}.$$

J. Jacod, P. Protter, *Discretization of Processes*,
Stochastic Modelling and Applied Probability 67,
DOI 10.1007/978-3-642-24127-7_13,

Assuming (H-r), we know by Theorem 9.1.1, page 249, that $V^n(f, v_n+, X) \overset{\mathbb{P}}{\Longrightarrow} f * \mu$ if $f(x) = \mathrm{O}(\|x\|^r)$ as $x \to 0$, and we are now looking for the associated central limit theorem, which describes the behavior of the processes

$$\overline{V}^n(f, v_n+, X)_t = \frac{1}{\sqrt{\Delta_n}} \left(V^n(f, v_n+, X)_t - f \star \mu_{\Delta_n[t/\Delta_n]}\right).$$

Although the assumptions on f are different, the results look very much like Theorem 5.1.2, and in particular the limiting process is the same. We briefly recall how this limiting process is constructed. We have an auxiliary space $(\Omega', \mathcal{F}', \mathbb{P}')$ endowed with a triple sequence $(\Psi_{n-}, \Psi_{n+}, \kappa_n)_{n\geq 1}$ of variables, all independent, and with the following laws:

$$\Psi_{n\pm} \text{ are } d'\text{-dimensional, } \mathcal{N}(0, I_{d'}), \qquad \kappa_n \text{ is uniform on } [0,1].$$

We take an arbitrary weakly exhausting sequence $(T_n)_{n\geq 1}$ of stopping times for the jumps of X. The *very good filtered extension* $(\widetilde{\Omega}, \widetilde{\mathcal{F}}, (\widetilde{\mathcal{F}}_t)_{t\geq 0}, \widetilde{\mathbb{P}})$ of $(\Omega, \mathcal{F}, (\mathcal{F}_t)_{t\geq 0}, \mathbb{P})$ is

$$\left.\begin{array}{l} \widetilde{\Omega} = \Omega \times \Omega', \qquad \widetilde{\mathcal{F}} = \mathcal{F} \otimes \mathcal{F}', \qquad \widetilde{\mathbb{P}} = \mathbb{P} \otimes \mathbb{P}' \\ (\widetilde{\mathcal{F}}_t) \text{ is the smallest filtration containing } (\mathcal{F}_t) \text{ and such that} \\ \qquad\qquad (\Psi_{n-}, \Psi_{n+}, \kappa_n) \text{ is } \widetilde{\mathcal{F}}_{T_n} \text{ measurable for all } n, \end{array}\right\}$$

and finally we set

$$R_n = R_{n-} + R_{n+}, \quad \text{where } R_{n-} = \sqrt{\kappa_n}\, \sigma_{T_n-} \Psi_{n-}, \quad R_{n+} = \sqrt{1-\kappa_n}\, \sigma_{T_n} \Psi_{n+}. \tag{13.1.1}$$

Under (H-r), the proof of Proposition 5.1.1 goes through for a q-dimensional C^1 test function f satisfying $\partial_i f(x) = \mathrm{O}(\|x\|^{r/2})$, in which case we can define the process $Z(f, X)$ by (5.1.4), that is

$$\overline{V}(f, X)_t = \sum_{n=1}^{\infty} \left(\sum_{i=1}^{d} \partial_i f(\Delta X_{T_n}) R_n^i \right) 1_{\{T_n \leq t\}}.$$

This process has a version which is càdlàg, adapted, and conditionally on $\mathcal{F}$ has centered and independent increments and satisfies

$$\widetilde{\mathbb{E}}\left(\overline{V}(f, X)_t^i \, \overline{V}(f, X)_t^j \mid \mathcal{F}\right) = \frac{1}{2} \sum_{s\leq t} \sum_{k,l=1}^{d} \left(\partial_k f^i \, \partial_l f^j\right)(\Delta X_s)\left(c_{s-}^{kl} + c_s^{kl}\right). \tag{13.1.2}$$

As in Theorem 5.1.2, we give a joint convergence with the discretized processes $X_t^{(n)} = X_{\Delta_n[t/\Delta_n]}$:

Theorem 13.1.1 *Assume* (H-r) *for some* $r \in [0,2]$, *and let* f *be a* C^1 *function from* $\mathbb{R}^d$ *into* $\mathbb{R}^q$, *with* $f(0) = 0$ *and* $\partial_i f(x) = \mathrm{O}(\|x\|^{p-1})$ *for all* $i = 1, \ldots, d$ *as* $x \to 0$, *for some* $p>1$. *If further*

$$p>r+1\vee r, \qquad \frac{1}{2(p-r)}<\varpi<\frac{1}{2}\bigwedge\frac{1}{2r}, \tag{13.1.3}$$

then

$$\left(X^{(n)},\overline{V}^n(f,v_n+,X)\right)\stackrel{\mathcal{L}\text{-s}}{\Longrightarrow}\left(X,\overline{V}(f,X)\right) \tag{13.1.4}$$

and also, for each fixed t,

$$\frac{1}{\sqrt{\Delta_n}}\left(V^n(f,v_n+,X)_t-f\star\mu_t\right)\stackrel{\mathcal{L}\text{-s}}{\longrightarrow}\overline{V}(f,X)_t.$$

The comments made after Theorem 5.1.2, page 128, hold here as well.

Remark 13.1.2 It is enlightening to compare the conditions on f and X in the two theorems, and for this we consider the case $f(x)=g_p(x)=\|x\|^p$ for some $p>1$, so f is C^1 and $\partial_i f(x)=\mathrm{O}(\|x\|^{p-1})$ as $x\to 0$. Under (H-r), Theorem 5.1.2 applies if $p>3$, and Theorem 13.1.1 applies (with a proper choice of ϖ) if $p>r+1\vee r$, and the latter is smaller than 3 if and only if $r<\frac{3}{2}$.

So the conditions on the power p for the function g_p may be more or less stringent in these two theorems, according to the properties of X.

Remark 13.1.3 The previous remark leads one to inquire why those two theorems have such different assumptions. For Theorem 5.1.2 the reason for which p should be bigger than 3 when $f=g_p$ comes from the interference with the "Brownian part". In Theorem 13.1.1 the Brownian part is eliminated, but the "correct" centering for the processes $V^n(f,v_n,X)$ should be $(f(x)1_{\{\|x\|>v_n\}})*\mu$ rather than $f*\mu$. Then, although $\eta(f)^n_t=(f(x)1_{\{\|x\|>v_n\}})*\mu-f*\mu$ goes to 0 under (H-r) when $f=g_p$ and $p\geq r$, the rate of convergence may be smaller than $\sqrt{\Delta_n}$. Indeed, in the proof below, we prove the CLT with the "correctly centered" processes, and then deduce the CLT as stated in the theorem.

In fact, the second condition in (13.1.3) is designed for a rate faster than $\sqrt{\Delta_n}$ in the above convergence. Of course this condition is not sharp in the literal sense: for example when $r=0$, that is when X has locally finitely many jumps, the result holds (in a straightforward way) with no condition on f except differentiability, and no condition on ϖ except $0<\varpi<\frac{1}{2}$. On the other hand, if the jumps of X are those of a stable process with index $\beta\in(0,2)$, then $\frac{1}{\sqrt{\Delta_n}}\eta(g_p)^n_t\stackrel{\mathbb{P}}{\longrightarrow}0$ if and only if $p>\beta+1/2\varpi$, whereas (H-r) is satisfied if and only if $r>\beta$: so the second condition (13.1.3) is indeed sharp.

Proof By localization we can assume (SH-r), that is Assumption 6.2.1 according to which we have (H-r) and the processes b, σ and X are bounded, and $\|\delta(\omega,t,z)\|\leq\Gamma(z)$ with Γ bounded and $\int\Gamma(z)^r\lambda(dz)<\infty$. Note also that the last claim is deduced from the first one, as in Theorem 5.1.2, so we only have to prove (13.1.4).

Step 1) We introduce the auxiliary processes

$$\overline{Y}^n(f, v_n+, X)_t = \frac{1}{\sqrt{\Delta_n}} \left(V^n(f, v_n+, X)_t - \left(f(x)\, 1_{\{\|x\|>v_n\}} \right) * \mu_{\Delta_n[t/\Delta_n]} \right).$$

In this step we show that it is enough to prove

$$\left(X^{(n)}, \overline{Y}^n(f, v_n+, X) \right) \stackrel{\mathcal{L}\text{-s}}{\Longrightarrow} \left(X, \overline{V}(f, X) \right). \tag{13.1.5}$$

To show this, it clearly suffices to prove that

$$t > 0 \quad \Rightarrow \quad U_t^n = \frac{1}{\sqrt{\Delta_n}} \left(\| f(x) \|\, 1_{\{\|x\| \le v_n\}} \right) * \mu_t \stackrel{\mathbb{P}}{\longrightarrow} 0. \tag{13.1.6}$$

The assumption of f implies $\| f(x) \| \le K \|x\|^p$ if $\|x\| \le 1$. Hence (SH-r) yields

$$\mathbb{E}\left(U_t^n\right) \le \frac{K}{\sqrt{\Delta_n}}\, \mathbb{E}\left((\Gamma \wedge v_n)^p * \mathfrak{p}_t \right) \le \frac{K\, t\, v_n^{p-r}}{\sqrt{\Delta_n}} \int \Gamma(z)^r\, \lambda(dz) \le \frac{K\, t\, v_n^{p-r}}{\sqrt{\Delta_n}},$$

and $\frac{1}{2(p-r)} < \varpi$ in (13.1.3) yields (13.1.6).

Step 2) In this step we suppose that $f(x) = 0$ when $\|x\| \le \varepsilon$ for some $\varepsilon > 0$. In this case $\overline{Y}^n(f, v_n+, X) = \overline{V}^n(f, X)$, as given by (5.1.6), as soon as $v_n \le \varepsilon$. Then our result is nothing else than Theorem 5.1.2, except that f is C^1 instead of C^2.

However, coming back to the proof of Theorem 5.1.2 and using its notation, we see that when $m > 2/\varepsilon$ we have $\overline{V}^n(f, X(m))_t = 0$ and $\overline{V}(f, X(m))_t = 0$ for all $t \le T$, on the set $\Omega_n(T, m)$. Hence, only Step 1 of that proof is needed for a function f as above, and only the C^1 property is used there. In other words, when f is C^1 and vanishes on a neighborhood of 0 we have Theorem 5.1.2 and our present result as well.

Step 3) Here we use the function ψ of (3.3.16) (a C^∞ function on $\mathbb{R}_+$ with $1_{[1,\infty)} \le \psi \le 1_{[1/2,\infty)}$), and $\psi_\varepsilon = \psi(\|x\|/\varepsilon)$ and $\psi'_\varepsilon = 1 - \psi_\varepsilon$. The function $f_\varepsilon = f\psi_\varepsilon$ is C^1 and vanishes on a neighborhood of 0. Then by the previous step and Proposition 2.2.4, it remains to prove the following two properties, where $f'_\varepsilon = f\psi'_\varepsilon$:

$$\overline{V}\left(f'_\varepsilon, X\right) \stackrel{\text{u.c.p.}}{\Longrightarrow} 0 \quad \text{as } \varepsilon \to 0 \tag{13.1.7}$$

$$t > 0,\ \eta > 0 \ \Rightarrow \ \lim_{\varepsilon \to 0}\ \limsup_{n \to \infty}\ \mathbb{P}\left(\sup_{s \le t} \left\| \overline{Y}^n\left(f'_\varepsilon, v_n+, X\right)_s \right\| > \eta \right) = 0. \tag{13.1.8}$$

It is enough to prove these for each component, so below we assume that $q = 1$.

For (13.1.7), $\overline{V}(f'_\varepsilon, X)$ is $\mathcal{F}$-conditionally a martingale satisfying (13.1.2), whereas our assumption on f and the facts that $\psi'_\varepsilon(x) \le 1_{\{\|x\| \le \varepsilon\}}$ and $|\partial_i \psi'_\varepsilon(x)| \le \frac{K}{\varepsilon}\, 1_{\{\|x\| \le \varepsilon\}}$ imply

$$\varepsilon \in (0,1] \ \Rightarrow \ \left| f'_\varepsilon(x) \right| \le K \|x\|^p\, 1_{\{\|x\| \le \varepsilon\}}, \qquad \left| \partial_i f'_\varepsilon(x) \right| \le K \|x\|^{p-1}\, 1_{\{\|x\| \le \varepsilon\}}. \tag{13.1.9}$$

Then, since (13.1.3) implies $p-1\geq r$, Doob's inequality and the boundedness of c_t yield

$$\widetilde{\mathbb{E}}\Big(\sup_{s\leq t}\big|\overline{V}(f'_\varepsilon,X)_s\big|^2\Big)\leq K\mathbb{E}\bigg(\sum_{s\leq t}\|\Delta X_s\|^{2r}\,1_{\{\|\Delta X_s\|\leq\varepsilon\}}\bigg)$$
$$=K\int_0^t\mathbb{E}\bigg(\int_{\{\|\delta(s,z)\|\leq\varepsilon\}}\big\|\delta(s,z)\big\|^{2r}\,\lambda(dz)\bigg)\,ds.$$

Using $\|\delta(s,z)\|\leq\Gamma(z)$ and $\int\Gamma(z)^{2r}\,\lambda(dz)<\infty$, we then deduce (13.1.7) from Lebesgue's theorem.

Step 4) We start proving (13.1.8). Set $H(\varepsilon)^n_t=\sum_{s\leq t}f'_\varepsilon(\Delta X_s)\,1_{\{\|\Delta X_s\|>v_n\}}$. Then

$$\overline{Y}^n(f_\varepsilon,v_n+,X)_t=\sum_{i=1}^{[t/\Delta_n]}\zeta(\varepsilon)^n_i,\quad\text{where}$$
$$\zeta(\varepsilon)^n_i=\frac{1}{\sqrt{\Delta_n}}\big(f'_\varepsilon\big(\Delta^n_i X\big)\,1_{\{\|\Delta^n_i X\|>v_n\}}-\Delta^n_i H(\varepsilon)^n\big)$$

and in this step we show that it is enough to exhibit subsets A^n_i of Ω and variables $\zeta'(\varepsilon)^n_i\geq 0$ such that

$$\left.\begin{array}{l}|\zeta(\varepsilon)^n_i|\;\leq\;\zeta'(\varepsilon)^n_i\quad\text{on the set }A^n_i\\ t>0\;\Rightarrow\;\sum_{i=1}^{[t/\Delta_n]}\mathbb{P}((A^n_i)^c)\;\to\;0\quad\text{as }n\to\infty\\ t>0\;\Rightarrow\;\lim_{\varepsilon\to 0}\;\limsup_{n\to\infty}\;\mathbb{E}\big(\sum_{i=1}^{[t/\Delta_n]}\zeta'(\varepsilon)^n_i\big)\;=\;0.\end{array}\right\}\tag{13.1.10}$$

Indeed, assuming the above properties, we see that the second one yields $\mathbb{P}(\Omega^n_t)\to 1$, where $\Omega^n_t=\cap_{i\leq[t/\Delta_n]}A^n_i$, and the first one implies that on the set Ω^n_t we have $|\overline{Y}^n(f_\varepsilon,v_n+,X)_s|\leq\sum_{i=1}^{[s/\Delta_n]}\zeta'(\varepsilon)^n_i$ for all $s\leq t$: so the third property and $\mathbb{P}(\Omega^n_t)\to 1$ yield (13.1.8).

Step 5) The conditions in (13.1.3) imply that one can choose a real l with

$$1<l<\frac{1}{2r\varpi},\qquad r>1\;\Rightarrow\;l<\frac{2(p-1)\varpi-1}{2(r-1)\varpi}.\tag{13.1.11}$$

We set $u_n=(v_n)^l$ and $F_n=\{z:\Gamma(z)>u_n\}$ and

$$X''^n=(\delta\,1_{F_n})*\mathfrak{p},\quad X''(\varepsilon)^n=(\delta\,1_{F_n\cap\{\|\delta\|\leq 2\varepsilon\}})*\mathfrak{p},\quad N^n_t=1_{F_n}*\mathfrak{p}$$
$$X'^n_t=X_t-X''^n_t$$
$$=X_0+\int_0^t b_s\,ds+\int_0^t\sigma_s\,dW_s+(\delta 1_{(F_n)^c})*(\mathfrak{p}-\mathfrak{q})_t-(\delta 1_{\{\|\delta\|\leq 1\}\cap F_n})*\mathfrak{q}_t$$
$$A^n_i=\big\{\|\Delta^n_i X'^n\|\leq v_n/2\big\}\cap\big\{\Delta^n_i N^n\leq 1\big\}.$$

Below we use an argument already encountered several times, see for example the proof of Theorem 4.3.1, Step 3: Let $(\mathcal{G}_t^n)$ denote the smallest filtration containing $(\mathcal{F}_t)$ and such that the restriction of the measure $\mathfrak{p}$ to $\mathbb{R}_+ \times F_n$ is $\mathcal{G}_0^n$ measurable. Then W and the restriction $\mathfrak{p}_n$ of $\mathfrak{p}$ to $\mathbb{R}_+ \times (F_n)^c$ are still a Brownian motion and a Poisson measure relative to $(\mathcal{G}_t^n)$, and the stochastic integrals defining X'^n, which are with respect to W and $\mathfrak{p}_n$, can be taken relative to $(\mathcal{F}_t)$ or to $(\mathcal{G}_t^n)$ without difference. Hence, taking advantage of the strengthened assumption (SH-r), we can use the estimate (2.1.44) and the properties $\int_{F_n^c} \Gamma(z)^m \,\lambda(dz) \le K u_n^{m-r}$ and $\int_{F_n} \Gamma(z)\,\lambda(dz) \le K u_n^{-(r-1)^+}$ to get for $m \ge 2$:

$$\mathbb{E}\big(\|\Delta_i^n X'^n\|^m \mid \mathcal{G}_0^n\big) \le K_m \Delta_n \big(\Delta_n^{m/2-1} + u_n^{m-r} + \Delta_n^{m-1} u_n^{-m(r-1)^+}\big). \tag{13.1.12}$$

We have $r \le 2$ and (13.1.11) implies $2(r-1)^+ l\varpi \le 1$, so taking $m=2$ yields

$$\mathbb{E}\big(\|\Delta_i^n X'^n\|^2 \mid \mathcal{G}_0^n\big) \le K\Delta_n. \tag{13.1.13}$$

On the other hand N^n is a Poisson process with parameter $\lambda(F_n)$, which is smaller than K/u_n^r, and we deduce

$$\mathbb{P}(\Delta_i^n N^n \ge 2) \le K \Delta_n^2 u_n^{-2r}. \tag{13.1.14}$$

Then (13.1.12), Markov's inequality, $u_n = \alpha^l \Delta_n^{l\varpi}$ and (13.1.14) yield for $m \ge 2$:

$$\begin{aligned}\sum_{i=1}^{[t/\Delta_n]} \mathbb{P}\big((A_i^n)^c\big) \le K_m t \big(&\Delta_n^{m(1/2-\varpi)-1} + \Delta_n^{m\varpi(l-1)-rl\varpi} \\ &+ \Delta_n^{m(1-\varpi-l\varpi(r-1)^+)-1} + \Delta_n^{1-2rl\varpi}\big).\end{aligned}$$

We have $1-2rl\varpi > 0$ and $1/2-\varpi > 0$ and $l > 1$ and $1-\varpi-l\varpi(r-1)^+ > 0$, so by taking m large enough we conclude the second part of (13.1.10).

On the set A_i^n, and if $v_n \le \varepsilon$, we have four mutually exclusive possibilities:

1. $\Delta_i^n N^n = 0$, which implies $\Delta_i^n X = \Delta_i^n X'^n$, hence $f'_\varepsilon(\Delta_i^n X) = 0$, and $\Delta_i^n H(\varepsilon)^n = 0$.
2. $\Delta_i^n N^n = 1$ and $\Delta_i^n X''^n = \Delta_i^n X''(\varepsilon)_i^n = 0$, hence again $f'_\varepsilon(\Delta_i^n X) = 0 = \Delta_i^n H(\varepsilon)^n = 0$.
3. $\Delta_i^n N^n = 1$ and $\Delta_i^n X''^n \ne 0$ and $\Delta_i^n X''(\varepsilon)_i^n = 0$; so the only jump of X inside the interval $((i-1)\Delta_n, i\Delta_n]$ with size bigger than v_n has in fact a size bigger than 2ε and then $\Delta_i^n H(\varepsilon)^n = 0$, and also $\|\Delta_i^n X''^n\| > 2\varepsilon$, implying $\|\Delta_i^n X\| > \varepsilon$ and thus $f'_\varepsilon(\Delta_i^n X) = 0$.
4. $\Delta_i^n N^n = 1$ and $\Delta_i^n X''^n = \Delta_i^n X''(\varepsilon)_i^n \ne 0$; then $\Delta_i^n X = \Delta_i^n X'^n + \Delta_i^n X''(\varepsilon)^n$ and $\Delta_i^n H(\varepsilon)_i^n = f'_\varepsilon(\Delta_i^n X''(\varepsilon)_i^n)\, 1_{\{\|\Delta_i^n X''(\varepsilon)_i^n\| > v_n\}}$.

Therefore the first part of (13.1.10) is satisfied as soon as $v_n \le \varepsilon$, if we take

$$\begin{aligned}\zeta'(\varepsilon)_i^n = \frac{1}{\sqrt{\Delta_n}} \Big| &f'_\varepsilon\big(\Delta_i^n X'^n + \Delta_i^n X''(\varepsilon)^n\big)\, 1_{\{\|\Delta_i^n X'^n + \Delta_i^n X''(\varepsilon)^n\| > v_n\}} \\ &- f'_\varepsilon\big(\Delta_i^n X''(\varepsilon)^n\big) 1_{\{\|\Delta_i^n X''(\varepsilon)^n\| > v_n\}}\Big|\, 1_{\{\|\Delta_i^n X'^n\| \le v_n/2\}}.\end{aligned}$$

Step 6) In this step we prove the last part of (13.1.10), with the above choice for $\zeta'(\varepsilon)_i^n$. One deduces from (13.1.9) that

$$\begin{aligned}&\big|f'_\varepsilon(x+z)1_{\{\|x+z\|>v\}}-f'_\varepsilon(x)1_{\{\|x\|>v\}}\big|\,1_{\{\|z\|\le v/2\}}\\&\quad\le K\big(\|x\|^{p-1}\,\|z\|+\|x\|^p\,1_{\{\|x\|\le 2v\}}\big).\end{aligned}$$

It is then enough to prove the following two properties:

$$\frac{1}{\sqrt{\Delta_n}}\sum_{i=1}^{[t/\Delta_n]}\mathbb{E}\big(\big\|\Delta_i^nX''(\varepsilon)^n\big\|^p\,1_{\{\|\Delta_i^nX''(\varepsilon)^n\|\le 2v_n\}}\big)\to 0\quad\text{as } n\to\infty \tag{13.1.15}$$

$$\lim_{\varepsilon\to0}\ \limsup_{n\to\infty}\ \frac{1}{\sqrt{\Delta_n}}\sum_{i=1}^{[t/\Delta_n]}\mathbb{E}\big(\big\|\Delta_i^nX''(\varepsilon)^n\big\|^{p-1}\,\big\|\Delta_i^nX'''^n\big\|\big)=0. \tag{13.1.16}$$

For (13.1.15) we use $\|x\|^p\,1_{\{\|x\|\le 2v_n\}}\le K\,v_n^{p-r\wedge1}\,\|x\|^{r\wedge1}$. The variables $\widehat{\delta}_n(q)_{s,t}$ associated with $\delta(t,z)1_{F_n}(z)$ by (2.1.35) satisfy $\widehat{\delta}_n(r)_{s,t}\le K$ and also $\widehat{\delta}_n(1)_{s,t}\le K/u_n^{r-1}$ when $r>1$, so (2.1.40) yields

$$\mathbb{E}\big(\big\|\Delta_i^nX''(\varepsilon)^n\big\|^p\,1_{\{\|\Delta_i^nX''(\varepsilon)^n\|\le 2v_n\}}\big)\le K\Delta_n^{1+(p-r\wedge1-l(r-1)^+)\varpi}.$$

By (13.1.3) and (13.1.11) we have $(p-r\wedge1-l(r-1)^+)\varpi>\frac12$, hence (13.1.15) holds.

For (13.1.16), observe that $\|\Delta_i^nX''(\varepsilon)_i^n\|\le\Delta_i^nG(\varepsilon)^n$, where $G(\varepsilon)_t^n=((\Gamma\wedge(2\varepsilon))\,1_{F_n})*\mathfrak{p}$. When $p\le2$ we use $p\ge r+1$ and (2.1.40), and when $p>2$ we use (2.1.41) and also $(\Delta_n/u_n^{r-1})^{p-1}\le K\Delta_n$ if further $r>1$ (because in this case $(p-1)(1-(r-1)l\varpi)\ge1$ by $2rl\varpi<1$), to get that $\mathbb{E}((\Delta_i^nG(\varepsilon)^n)^{p-1})\le K\Delta_n(\phi(\varepsilon)+\Delta_n^{1/(p-1)})$, where $\phi(\varepsilon)\to0$ as $\varepsilon\to0$. Thus (13.1.13) and the Cauchy-Schwarz inequality, plus the $\mathcal{G}_0^n$ measurability of $\Delta_i^nG(\varepsilon)^n$, yield

$$\begin{aligned}\mathbb{E}\big(\big\|\Delta_i^nX''(\varepsilon)^n\big\|^{p-1}\,\big\|\Delta_i^nX'''^n\big\|\big)&=\mathbb{E}\big(\big(\Delta_i^nG(\varepsilon)^n\big)^{p-1}\,\mathbb{E}\big(\big\|\Delta_i^nX'''^n\big\|\mid\mathcal{G}_0^n\big)\big)\\&\le K\sqrt{\Delta_n}\,\mathbb{E}\big(\big(\Delta_i^nG(\varepsilon)^n\big)^{p-1}\big)\le K\Delta_n^{3/2}\big(\phi(\varepsilon)+\Delta_n^{1/(p-1)}\big).\end{aligned}$$

At this stage, (13.1.16) follows from $\phi(\varepsilon)\to0$ as $\varepsilon\to0$. □

13.2 Central Limit Theorem for Approximating the Continuous Part

Now we turn to the functionals

$$V'^n(F,v_n-,X)_t=\Delta_n\sum_{i=1}^{[t/\Delta_n]-k+1}F\left(\frac{\Delta_i^nX}{\sqrt{\Delta_n}},\dots,\frac{\Delta_{i+k-1}^nX}{\sqrt{\Delta_n}}\right)\prod_{l=0}^{k-1}1_{\{\|\Delta_{i+l}^nX\|\le v_n\}}.$$

which converge to $\int_0^t \rho_{c_s}^{k\otimes}(F)\,ds$. Recall that k is an integer and F a continuous function on $(\mathbb{R}^d)^k$ growing at a rate less than $\|x\|^p$ in each of the k arguments. This convergence holds under (H) when X is continuous, and also when it jumps and $p \le 2$, whereas we need (H-r) with $r < 2$ and $\varpi \ge (p-2)/2(p-r)$ when $p > 2$ and X jumps. Here we are looking for the associated Central Limit Theorem, that is we want to find the behavior of the processes

$$\overline{V}'^n(F, v_n-, X)_t = \frac{1}{\sqrt{\Delta_n}}\left(V'^n(F, v_n-, X)_t - \int_0^t \rho_{c_s}^{k\otimes}(F)\,ds\right). \tag{13.2.1}$$

When f is a function on $\mathbb{R}^d$, the following functionals are also of interest, see (9.2.2):

$$V'^n(f, k, v_n-, X)_t = \Delta_n \sum_{i=1}^{[t/\Delta_n]-k+1} f\left(\frac{\Delta_i^n X + \cdots + \Delta_{i+k-1}^n X}{\sqrt{\Delta_n}}\right) 1_{\{\|\Delta_i^n X + \cdots + \Delta_{i+k-1}^n X\| \le v_n\}}. \tag{13.2.2}$$

13.2.1 The Results

The problem is almost the same as in Sect. 11.2, and the result similar to Theorem 11.2.1. More precisely, when X is continuous, we have the following property: for any t, outside a set whose probability goes to 0 as $n \to \infty$, and for all $s \le t$ we have $\overline{V}'^n(F, v_n-, X)_s = \overline{V}'^n(F, X)_s$, as given by (11.2.2) (see the proof below); then the asymptotic results are exactly the same, and no further proof is needed. When X has jumps, the truncation basically eliminates the jumps and consequently the assumptions on the test function F are much weaker for $\overline{V}'^n(F, v_n-, X)$ than for $\overline{V}'^n(F, X)$.

The assumptions on X will be 4.4.3, 5.3.2, 4.4.4 or 5.3.4, which we briefly recall:

Assumption (K) We have (H) and

$$\sigma_t = \sigma_0 + \int_0^t \widetilde{b}_s\,ds + \int_0^t \widetilde{\sigma}_s\,dW_s + M_t + \sum_{s\le t} \Delta\sigma_s\,1_{\{\|\Delta\sigma_s\|>1\}}$$

where M is a local martingale with $\|\Delta M_t\| \le 1$, orthogonal to W, and $\langle M, M\rangle_t = \int_0^t a_s\,ds$ and the compensator of $\sum_{s\le t} 1_{\{\|\Delta\sigma_s\|>1\}}$ is $\int_0^t \widetilde{a}_s\,ds$, with the following properties: the processes $\widetilde{b}$, $\widetilde{\sigma}$, $\widetilde{a}$ and a are progressively measurable, the processes $\widetilde{b}$, a and $\widetilde{a}$ are locally bounded, and the processes $\widetilde{\sigma}$ and b are càdlàg or càglàd.

Assumption (K-r) (for $r \in [0,1]$) We have (K) except for the càdlàg or càglàd property of b, and $\|\delta(\omega,t,z)\| \wedge 1 \le \Gamma_n(z)$ for all (ω,t,z) with $t \le \tau_n(\omega)$, where

(τ_n) is a localizing sequence of stopping times and the Borel functions Γ_n on E satisfy $\int \Gamma_n(z)^r \lambda(dz) < \infty$. Moreover the process $b'_t = b_t - \int_{\{\|\delta(t,z)\| \leq 1\}} \delta(t,z)\,\lambda(dz)$ is càdlàg or càglàd.

Assumption (K') We have (K) and both processes c_t and c_{t-} take their values in $\mathcal{M}_{d\times d}^{++}$.

Assumption (K'-r) We have (K-r) and both processes c_t and c_{t-} take their values in $\mathcal{M}_{d\times d}^{++}$.

We restate the conditions on F given in (11.2.7)–(11.2.10), with slight but important modifications in the exponents. The function F is globally even on $(\mathbb{R}^d)^k$ and q-dimensional. For some integer $q' \in \{0,\dots,q\}$ and, when $q' < q$, for some subset B of $(\mathbb{R}^d)^k$ which is a finite union of affine hyperplanes, we have

$$\begin{aligned} &j \leq q' \quad\Rightarrow\quad x \mapsto F^j(x) \;\text{ is } C^1 \text{ on } \big(\mathbb{R}^d\big)^k \\ &j > q' \quad\Rightarrow\quad x \mapsto F^j(x) \;\text{ is continuous on } \big(\mathbb{R}^d\big)^k \text{ and } C^1 \text{ outside } B. \end{aligned} \tag{13.2.3}$$

We denote by $d(z,B)$ the distance between $z \in (\mathbb{R}^d)^k$ and B. In the conditions below we always have $w \in (0,1]$, and z and y run through $(\mathbb{R}^d)^k$, and x_j and v run through $\mathbb{R}^d$. The numbers p, p', s, s', w are subject to $0 < w \leq 1$ and $0 < s \leq s'$ and $p, p' \geq 0$, but otherwise arbitrary, and ∇F stands for the family of all first partial derivatives of F:

$$\big\|F(z)\big\| \;\leq\; K\big(1 + \|z\|^p\big) \tag{13.2.4}$$

$$\big|\nabla F^j(z)\big| \leq \begin{cases} K\big(1+\|z\|^p\big) & \text{if } j \leq q' \\ K\big(1+\|z\|^p\big)\big(1+\frac{1}{d(z,B)^{1-w}}\big) & \text{if } j > q' \quad \text{and} \quad z \in B^c \end{cases} \tag{13.2.5}$$

$$\begin{aligned} &z \in B^c,\; \|y\| \leq 1 \bigwedge \frac{d(z,B)}{2},\; j > q' \\ &\quad\Rightarrow\; \big|\nabla F^j(z+y) - \nabla F^j(z)\big| \leq K\,\|y\| \left(1 + \frac{1}{d(z,B)^{2-w}}\right)\big(1+\|z\|^p\big) \end{aligned} \tag{13.2.6}$$

$$\begin{aligned} &\big\|F(x_1,\dots,x_{j-1},x_j+v,x_{j+1},\dots,x_k) - F(x_1,\dots,x_k)\big\| \\ &\quad\leq K\,\big(\|v\|^s + \|v\|^{s'}\big) \prod_{l=1}^k \big(1+\|x_l\|^{p'}\big). \end{aligned} \tag{13.2.7}$$

These are exactly the same as (11.2.7)–(11.2.10) (see after (11.2.10), page 312, for comments about the connections between those conditions), except for the last one in which the exponent 2 for $\|x_l\|$ is replaced by p'.

The limiting process is the same as in Theorem 11.2.1, page 312. It is based on the following quantities, where F and G are two functions on $(\mathbb{R}^d)^k$ with polynomial growth and $a \in \mathcal{M}_{d\times d}^{+}$ and the variables $U_1, U_2, \dots$ are independent $\mathcal{N}(0,a)$ distributed, and we use the two σ-fields $\mathcal{G} = \sigma(U_1, \dots, U_{k-1}$ and $\mathcal{G}' = \sigma(U_1, \dots, U_k)$:

$$\begin{aligned} R_a(F,G) = \sum_{j,j'=0}^{k-1} \mathbb{E}\Big(&\mathbb{E}\big(F(U_{k-j}, \dots, U_{2k-j-1}) \mid \mathcal{G}'\big) \\ &\times \mathbb{E}\big(G(U_{k-j'}, \dots, U_{2k-j'-1}) \mid \mathcal{G}'\big) \\ &- \mathbb{E}\big(F(U_{k-j}, \dots, U_{2k-j-1}) \mid \mathcal{G}\big) \\ &\times \mathbb{E}\big(G(U_{k-j'}, \dots, U_{2k-j'-1}) \mid \mathcal{G}'\big)\Big). \end{aligned} \tag{13.2.8}$$

Theorem 13.2.1 *Let X be a d-dimensional Itô semimartingale and F be a function from $(\mathbb{R}^d)^k$ into $\mathbb{R}^q$ which is continuous, globally even, and satisfies* (13.2.3), (13.2.4) *and* (13.2.5). *We also assume* (13.0.2) *for the truncation levels v_n, and one of the following four sets of hypotheses*:

(a) *We have $q' = q$ and* (K) *and X is continuous.*
(b) *We have* (K') *and* (13.2.6) *and X is continuous.*
(c) *We have $q' = q$ and* (K-r) *for some $r \in (0,1]$, and* (13.2.7) *with $r \le s \le 1 \le s'$, and*

$$\varpi \ge \frac{p'-2}{2(p'-r)} \quad \text{if } k \ge 2 \text{ and } p' > 2, \text{ and} \quad \begin{cases} r = 1 \Rightarrow s = s' = 1 \\ r < 1 \Rightarrow s \ge \frac{r}{2-r},\ \varpi \ge \frac{s'-1}{2(s'-r)}. \end{cases} \tag{13.2.9}$$

(d) *We have* (K'-r) *with some $r \in (0,1]$, and* (13.2.6) *and* (13.2.7) *with $r \le s \le 1 \le s'$, as well as* (13.2.9).

Then the processes $\overline{V}'^n(F, v_n-, X)$ of (13.2.1) *converge stably in law to a continuous process $\overline{V}'(F,X)$ which is defined on a very good filtered extension $(\widetilde{\Omega}, \widetilde{\mathcal{F}}, (\widetilde{\mathcal{F}}_t)_{t\ge0}, \widetilde{\mathbb{P}})$ of $(\Omega, \mathcal{F}, (\mathcal{F}_t)_{t\ge0}, \mathbb{P})$, and conditionally on $\mathcal{F}$ is a centered Gaussian process with independent increments satisfying, with the notation* (13.2.8):

$$\widetilde{\mathbb{E}}\big(\overline{V}'(F^i, X)_t\, \overline{V}'(F^j, X) \mid \mathcal{F}\big) = \int_0^t R_{c_s}\big(F^i, F^j\big)\, ds.$$

Moreover, under the same conditions, and for any real $\gamma > 0$, we have

$$\frac{1}{\sqrt{\Delta_n}} \big(V'^n(F, \gamma v_n-, X) - V'^n(F, v_n-, X)\big) \stackrel{\text{u.c.p.}}{\Longrightarrow} 0, \tag{13.2.10}$$

and also, recalling (13.2.2),

$$\frac{1}{\sqrt{\Delta_n}} \big(V'^n(f, k, \gamma v_n-, X) - V'^n(F, v_n-, X)\big) \stackrel{\text{u.c.p.}}{\Longrightarrow} 0 \tag{13.2.11}$$

when F takes the form $F(x_1, \dots, x_k) = f(x_1 + \dots + x_k)$ for a function f on $\mathbb{R}^d$.

Observe that, when $k=1$, (13.2.7) implies (13.2.4), or equivalently (9.2.3) (see page 251) with $p=s'$, whereas when $k\geq 2$ it implies (9.2.3) with $p=p'$. Thus (9.2.4) is weaker than (13.2.9), as it should be because the CLT implies the LLN.

The condition (13.2.4) agrees with (9.2.3), page 251, so (9.2.4) is weaker than (13.2.9), as it should be because the CLT implies the LLN.

This theorem is the same as Theorem 11.2.1 when X is continuous, and in that case there is no reason to look at truncated functionals anyway: so (a) and (b) above are here just for completeness. The main improvement upon Theorem 11.2.1, when X has jumps, is due to the way Condition (13.2.7) is used: here we allow $s'\geq 1$ and $p'>2$, and those two numbers can actually be arbitrarily large, upon choosing ϖ appropriately. This improvement is illustrated, in a more concrete way, by the following example, which deals with the same test function F as Example 11.2.2, page 313.

Example 13.2.2 Consider the truncated multipower variations, corresponding to the one-dimensional test function (11.2.13):

$$F(x)=\prod_{j=1}^{k}\prod_{i=1}^{d}\left|x_j^i\right|^{w_j^i},\quad w_j^i\geq 0,$$

and suppose that X has jumps. Let $w_j=w_j^1+\cdots+w_j^d$ and $u=\min(w_j^i: w_j^i>0, 1\leq i\leq d, 1\leq j\leq k)$. (13.2.4)–(13.2.7) hold with $p=\sum_j w_j$ and $p'=s'=1\vee\max(w_1,\dots,w_d)$, and $s=1\wedge u$, and with $B=\cup_{j=1}^{kd}\{z\in(\mathbb{R}^d)^k: z^j=0\}$ when $u\leq 1$ (otherwise $B=\emptyset$, or equivalently we are in the case $q'=q=1$). Then if (K'-r) holds for some $r\in[0,1)$, the above theorem applies as soon as $s\geq r$ and $\varpi\in[\frac{p'-1}{2(p'-r)},\frac{1}{2})$.

Another interesting case is

$$F(x)=\prod_{j=1}^{k}\prod_{i=1}^{d}\left(x_j^i\right)^{w_j^i},\quad w_j^i\in\mathbb{N}. \tag{13.2.12}$$

(There are no absolute values, which is why the w_j^i's are integers.) This function is differentiable, so we have $q'=q=1$ in (13.2.3). Then (13.2.4), (13.2.5) and (13.2.7) hold with p,p',s' as above and $s=1$. It is globally even if and only if the sum of all w_j^i is even. In this case the theorem applies under (K-1) if $s'=1$, which amounts to having $F(x)=\prod_{j=1}^k x_j^{m_j}$ for indices m_j in $\{1,\dots,d\}$: this is of course an uninteresting case where $\rho_{c_s}^{k\otimes}(F)=0$. It also applies under (K-r) when $r<1$ for any function like (13.2.12), provided $\varpi\in[\frac{p'-1}{2(p'-r)},\frac{1}{2})$.

Remark 13.2.3 If (K-1) holds but not (K-r) for any $r<1$, the present theorem needs F to have linear growth in each variable x_j, because (13.2.7) with $s'=s=1$ implies for example $\|F(x_1,\dots,x_k)\|\leq G(x_1,\dots,x_{k-1})(1+\|x_k\|)$ for some function

G, and then there is no restriction on the truncation levels v_n other than (13.0.2). This contrasts with Theorem 11.2.1, which needs F and ∇F to be bounded in this situation.

When (K-r) holds for some $r < 1$, Theorem 13.2.1 applies without restriction on the growth of F and ∇F (say, when F is C^1 everywhere) apart from being polynomial, again in contrast with Theorem 11.2.1 which needs a sub-linear growth for F. On the other hand, there is a restriction on the truncation levels, which should be "small enough" (that is, ϖ big enough). Note at this juncture that, when X and F satisfy the conditions of the theorem, it is always possible to find an $\varpi < 1/2$ which satisfies (13.2.1).

When $k = 1$ the theorem above is similar to Theorems 5.3.5 and 5.3.6, pages 147–148, and in (K-r) the property that the process σ_t is an Itô semimartingale is crucial. However we have also proved a CLT for the quadratic variation, when X jumps, under the hypothesis (H) only. We may wonder whether the "truncated realized quadratic variation", defined by

$$\widehat{C}^n(v_n-, X)_t^{jl} = \sum_{i=1}^{[t/\Delta_n]} \Delta_i^n X^j \, \Delta_i^n X^l \, 1_{\{\|\Delta_i^n X\| \le v_n\}}$$

enjoys the same property (note that $\widehat{C}^n(v_n-, X)^{jl} = \Delta_n V'^n(F, v_n-, X)$ with $k = 1$ and $F(x) = x^j x^l$). This is indeed true, provided we have (H-r) with $r < 1$. When only (H-1) holds one does not know whether or not this result holds, and Mancini [75] has shown that it fails when the discontinuous part of X is a Cauchy process, which satisfies (H-r) for all $r > 1$ but not for $r = 1$.

For some applications, we need a joint CLT for the above process and the approximate quadratic variation $[X, X]^n$ itself. That is, we consider the two processes

$$\begin{aligned} \overline{Z}_t^n &= \frac{1}{\sqrt{\Delta_n}} \big([X, X]_t^n - [X, X]_{\Delta_n [t/\Delta_n]} \big) \\ \overline{C}^n(v_n-, X)_t &= \frac{1}{\sqrt{\Delta_n}} \big(\widehat{C}^n(v_n-, X)_t - C_t \big) \end{aligned} \tag{13.2.13}$$

($\overline{Z}^n$ is as in (5.4.1)). The limit is, as usual, defined on a very good extension $(\widetilde{\Omega}, \widetilde{\mathcal{F}}, (\widetilde{\mathcal{F}}_t)_{t\ge 0}, \widetilde{\mathbb{P}})$ of $(\Omega, \mathcal{F}, (\mathcal{F}_t)_{t\ge 0}, \mathbb{P})$: recalling the process $\overline{Z}$ defined by (5.4.3), page 161, we define two $d \times d$-dimensional processes by

$$\begin{aligned} \overline{Z}_t'^{ij} &= \frac{1}{\sqrt{2}} \sum_{k,l=1}^{d} \int_0^t \big(\widehat{\sigma}_s^{ij,kl} + \widehat{\sigma}_s^{ji,kl} \big) \, dW_s'^{kl} \\ \overline{Z}_t''^{ij} &= \sum_{p=1}^{\infty} \big(\Delta X_{T_p}^j R_p^i + \Delta X_{T_p}^i R_p^j \big) 1_{\{T_p \le t\}}, \end{aligned} \tag{13.2.14}$$

so that $\overline{Z} = \overline{Z}' + \overline{Z}''$; here, R_p is as in (13.1.1) and W' is a $d' \times d'$-dimensional Brownian motion independent of the R_p's and of $\mathcal{F}$. Note that $\overline{Z}'$ is, conditionally on $\mathcal{F}$, a continuous centered Gaussian martingale with variance-covariance given by

$$\mathbb{E}(\overline{Z}_t^{\prime ij}\,\overline{Z}_t^{\prime kl}\mid\mathcal{F}) = \int_0^t \left(c_s^{ik}c_s^{jl} + c_s^{il}c_s^{jk}\right) ds. \tag{13.2.15}$$

Theorem 13.2.4 *Assume* (H-r) *for some* $r\in(0,1)$, *and suppose that in* (13.0.2) *we have* $\varpi\in[\frac{1}{4-2r},\frac{1}{2})$. *Then with the notation* (13.2.13) *we have the following stable (functional) convergence in law*:

$$\left(\overline{Z}^n, \overline{C}^n(v_n-,X)\right) \stackrel{\mathcal{L}\text{-}s}{\Longrightarrow} \left(\overline{Z}' + \overline{Z}'', \overline{Z}'\right),$$

where $\overline{Z}'$ *and* $\overline{Z}''$ *are defined by* (13.2.14).

Therefore in this case, not only does $V^n(2;v_n-,X)^{jl}$ converge to the "continuous limit" C_t^{jl}, but the CLT for $\overline{C}^n(v_n-,X)$ is the same as if X were continuous.

Finally, we state a theorem which extends Theorem 11.3.2 (page 327), and "almost" includes the previous one as a particular case (take $F=0$ below, but the assumptions on X are stronger than in the previous theorem). The process $\overline{Z}''$ is as in (13.2.14); the function F on $(\mathbb{R}^d)^k$ is q-dimensional, and the pair $(\overline{V}'(F,X),\overline{Z}')$ denotes (as on page 326) a $(q+d\times d)$-dimensional process which, conditionally on $\mathcal{F}$, is continuous, centered, Gaussian, independent of $\overline{Z}''$, and with variance-covariance given by

$$\begin{aligned}
&\widetilde{\mathbb{E}}\left(\overline{V}'(F^i,X)_t\,\overline{V}'(F^j,X)\mid\mathcal{F}\right) = \int_0^t R_{c_s}(F^i,F^j)\,ds\\
&\widetilde{\mathbb{E}}\left(\overline{V}'(F^i,X)_t\,\overline{Z}_t^{\prime lm}\mid\mathcal{F}\right) = \int_0^t R_{c_s}(F^i,\overline{F}^{lm})\,ds\\
&\widetilde{\mathbb{E}}\left(\overline{Z}_t^{\prime ij}\,\overline{Z}_t^{\prime lm}\mid\mathcal{F}\right) = \int_0^t R_{c_s}(\overline{F}^{ij},\overline{F}^{lm})\,ds,
\end{aligned} \tag{13.2.16}$$

where $\overline{F}^{ij}(x_1,\dots,x_k) = x_1^i x_1^j$. Finally we consider the normalized (non-truncated) functionals $V'^n(F,X)$, and the associated processes (the same as in (11.2.2)):

$$\overline{V}'^n(F,X)_t = \frac{1}{\sqrt{\Delta_n}}\left(V'^n(F,X)_t - \int_0^t \rho_{c_s}^{k\otimes}(F)\,ds\right).$$

Theorem 13.2.5 *Let X be a d-dimensional Itô semimartingale and F be a function from* $(\mathbb{R}^d)^k$ *into* $\mathbb{R}^q$ *which is continuous, globally even, and satisfies* (13.2.3). *We also assume* (13.2.4) *and* (13.2.5) (*recall* $w\in(0,1]$), *plus one of the following four sets of hypotheses*:

(a) *We have* $q'=q$ *and* (K) *and X is continuous*; *then we set* $r=0$.
(b) *We have* (K') *and* (13.2.6) *and X is continuous*; *then we set* $r=0$.
(d) *We have* $q'=q$ *and* (K-r) *for some* $r<1$, *and* (13.2.7) *with* $r\le s\le s'<1$ *and* $p'\le 2$.

(e) *We have* (K'-r) *with some* $r \in (0,1)$, *and* (13.2.6), *and* (13.2.7) *with* $r \le s \le s' < 1$ *and* $p' \le 2$.

Let also v_n *satisfy* (13.0.2) *with* $\varpi \in [\frac{1}{4-2r}, \frac{1}{2})$. *Then we have the following stable (functional) convergence in law*:

$$\big(\overline{V}'^n(F,X), \overline{Z}^n, \overline{C}^n(v_n-,X)\big) \stackrel{\mathcal{L}\text{-}s}{\Longrightarrow} \big(\overline{V}'(F,X), \overline{Z}' + \overline{Z}'', \overline{Z}'\big).$$

The cases considered here are the same as in Theorem 11.2.1, page 312, and the reader will notice that there is no (c) here! This is because we need $r < 1$ for the CLT for $\widehat{C}^n(v_n-,X)$ to hold, hence no joint CLT is available in case (c) of Theorem 11.2.1.

This theorem is useful in some statistical applications, but one could as well have a joint CLT for a sequence of functionals whose components are of the form $\overline{Z}^{n,ij}$ or $V^n(F,X)$ or $V^n(F,v_n+,X)$ or $V'^n(F,X)$ or $V'^n(F,v_n-,X)$, that is, all processes considered so far.

13.2.2 Proofs

We begin with a lemma which requires only (SH-r) for some $r \in [0,2]$, and will be used several times in the sequel. According to the value of r, we set

$$X' = X - X'', \qquad X'' = \begin{cases} \delta * \mathfrak{p} & \text{if } r \le 1 \\ \delta * (\mathfrak{p} - \mathfrak{q}) & \text{if } r > 1. \end{cases} \tag{13.2.17}$$

Although this differs from (11.2.15) when $r > 1$, we use the notation $\overline{X}^n_{i,j}$ of (11.2.16), that is

$$\overline{X}^n_{i,j} = \left(\frac{\Delta^n_i X}{\sqrt{\Delta_n}}, \ldots, \frac{\Delta^n_{i+j-1} X}{\sqrt{\Delta_n}}, \frac{\Delta^n_{i+j} X'}{\sqrt{\Delta_n}}, \ldots, \frac{\Delta^n_{i+k-1} X'}{\sqrt{\Delta_n}}\right), \tag{13.2.18}$$

with the process X' as defined here. Then, recalling $u_n = \alpha \Delta_n^{\varpi}$, we set

$$F_u(x_1,\ldots,x_k) = F(x_1,\ldots,x_k) \prod_{j=1}^k 1_{\{\|x_j\| \le u\}} \quad \text{for } u > 0, \tag{13.2.19}$$

$$\eta^n_{ij} = F_{v_n/\sqrt{\Delta_n}}\big(\overline{X}^n_{i,j+1}\big) - F_{v_n/\sqrt{\Delta_n}}\big(\overline{X}^n_{i,j}\big) \quad \text{for } j = 0,1,\ldots,k-1. \tag{13.2.20}$$

Lemma 13.2.6 *Assume* (SH-r) *for some* $r \in (0,2]$, *and suppose that* F *satisfies* (13.2.7) *for some* $p' \ge 0$ *and* $s' \ge 1 \ge s > 0$. *Let* $m \ge 1$ *and suppose that* $k = 1$ *or* $\varpi \ge \frac{m(p' \vee 2) - 2}{2(m(p' \vee 2) - r)}$. *Then, with* $\theta > 0$ *arbitrarily fixed when* $r > 1$ *and* $\theta = 0$ *when* $r \le 1$, *there is a sequence* ϕ_n *(depending on* m, s, s', θ*) of positive numbers going* *to* 0 *as* $n \to \infty$, *such that*

$$\mathbb{E}\big(\|\eta^n_{i,j}\|^m \mid \mathcal{F}_{(i-1)\Delta_n}\big) \le \big(\Delta_n^{\frac{2-r}{2}(1\wedge\frac{ms}{r})-\theta} + \Delta_n^{(1-r\varpi)(1\wedge\frac{ms'}{r})-ms'\frac{1-2\varpi}{2}-\theta}\big)\,\phi_n. \tag{13.2.21}$$

Proof The assumption on F weakens when p' increases, so we can assume $p' \ge 2$. For simplicity we write $u_n = v_n/\sqrt{\Delta_n} = \alpha\Delta_n^{\varpi-1/2}$, which goes to ∞.

Recall $\|\delta(t,z)\| \le \Gamma(z)$ with Γ bounded and $\int \Gamma(z)^r\,\lambda(dz) < \infty$. We set $Y = \Gamma * \mathfrak{p}$ when $r \le 1$ and $Y = X''$ otherwise, so in all cases $\|\Delta^n_i X''\| \le \|\Delta^n_i Y\|$. Then we set

$$U^n_i = \frac{\|\Delta^n_i X'\|}{\sqrt{\Delta_n}}, \qquad V^n_i = \frac{\|\Delta^n_i Y\|}{\Delta_n^{\varpi}}\bigwedge 1, \qquad W^n_i = \frac{\|\Delta^n_i Y\|}{\sqrt{\Delta_n}}\bigwedge 1,$$

$$Z^n_{i,j} = u_n^{-2/(1-2\varpi)}\big(U^n_{i+j}\big)^{p'+2/(1-2\varpi)} + \big(1+(U^n_{i+j})^{p'}\big)\big(W^n_{i+j}\big)^s + \big(1+(U^n_{i+j})^{p'}\big)\,u_n^{s'}\big(V^n_{i+j}\big)^{s'}.$$

Below we suppose that n is large enough to have $u_n > 1$. Singling out the case $\|x_j + y\| \le u_n$ and $\|x_j\| \le u_n$ (implying $\|y\| \le 2u_n$) for which (13.2.7) is used, and the three cases $\|x_j + y\| \le u_n < \|x_j\|$, and $\frac{u_n}{2} < \|x_j\| \le u_n < \|x_j + y\|$, and $2\|x_j\| \le u_n < \|x_j + y\|$ (implying $\|y\| > u_n/2$), for which (13.2.4) is used, we see that for $j = 0,\dots,k-1$ we have

$$\begin{aligned}
&\big\|F_{u_n}(x_1,\dots,x_j,x_{j+1}+y,x_{j+2},\dots,x_k) - F_{u_n}(x_1,\dots,x_k)\big\| \\
&\quad\le K\prod_{l=1}^{j}\big(1+\|x_l\|^{p'}\wedge u_n^{p'}\big)\prod_{l=j+2}^{k}\big(1+\|x_l\|^{p'}\big) \\
&\qquad\times\big(u_n^{-2/(1-2\varpi)}\|x_{j+1}\|^{p'+2/(1-2\varpi)} + \big(1+\|x_{j+1}\|^{p'}\big)\big(\|y\|^s\wedge 1 + \|y\|^{s'}\wedge u_n^{s'}\big)\big)
\end{aligned}$$

(an empty product is set to 1). This is applied with $x_l = \Delta^n_{i+l-1}X/\sqrt{\Delta_n}$ when $l \le j$ and $x_l = \Delta^n_{i+l-1}X'/\sqrt{\Delta_n}$ when $l > j$ and $y = \Delta^n_{i+j}X''/\sqrt{\Delta_n}$. Then, observing that $\|\Delta^n_i Y/\sqrt{\Delta_n}\| \wedge u_n \le \frac{u_n}{1\wedge\alpha}V^n_i$, we see that $\|x_l\|^{p'}\wedge u_n^{p'} \le K((U^n_{i+l-1})^{p'} + u_n^{p'}(V^n_{i+l-1})^2)$ (recall $p' \ge 2$) if $l \le j$, and $\|y\|^s \wedge 1 = (W^n_{i+j})^s$, and $\|y\|^{s'}\wedge u_n^{s'} \le u_n^{s'}(V^n_{i+l-1})^{s'}$, and thus

$$\|\eta^n_{i,j}\| \le K\prod_{l=1}^{j}\big(1+u_n^{p'}(V^n_{i+l-1})^2 + (U^n_{i+l-1})^{p'}\big)\prod_{l=j+2}^{k}\big(1+(U^n_{i+l-1})^{p'}\big)\,Z^n_{i,j}.$$

Now we give some estimates based on (SH-r). We apply Proposition 2.1.10 (page 44) with $\mathcal{G}_t = \mathcal{G}^E_t$ given by (2.1.48) with the set $A = E$, and we deduce from (2.1.33), and (2.1.34) (pages 40–40) that

$$m > 0 \;\Rightarrow\; \mathbb{E}\big((U^n_i)^m \mid \mathcal{G}_{(i-1)\Delta_n}\big) \le K_m. \tag{13.2.22}$$

Next, Corollary 2.1.9 (use (a) or (c), according to whether $r>1$ or not) yields that, for a suitable sequence ϕ_n going to 0 (this sequence varies from line to line below, for example it depends on m in the next formula):

$$\begin{aligned}
&\mathbb{E}\big((V_i^n)^m \mid \mathcal{F}_{(i-1)\Delta_n}\big) \le \Delta_n^{(1-r\varpi)(1\wedge\frac{m}{r})}\phi_n \\
&\mathbb{E}\big((W_i^n)^m \mid \mathcal{F}_{(i-1)\Delta_n}\big) \le \Delta_n^{\frac{2-r}{2}(1\wedge\frac{m}{r})}\phi_n.
\end{aligned} \tag{13.2.23}$$

Assuming $r \le 1$ first, we observe that the variables W_{i+j}^n and V_{i+j}^n are $\mathcal{G}_{(i+j-1)\Delta_n}$ measurable. Then, recalling $ms' \ge 1$, we deduce from (13.2.22) and (13.2.23) and successive conditioning that we have, with $\theta=0$:

$$\begin{aligned}
&\mathbb{E}\big(\big|Z_{i,j}^n\big|^m \mid \mathcal{F}_{(i+j-1)\Delta_n}\big) \\
&\quad \le K\Delta_n^m + \big(\Delta_n^{\frac{2-r}{2}(1\wedge\frac{ms}{r})-\theta} + \Delta_n^{(1-r\varpi)(1\wedge\frac{ms'}{r})-ms'\frac{1-2\varpi}{2}-\theta}\big)\phi_n,
\end{aligned} \tag{13.2.24}$$

and we can indeed dispense with the first term on the right when $m \ge 1$, because the last exponent of Δ_n is strictly smaller than 1.

When $r>1$ the successive conditioning argument no longer works, but we may instead apply Hölder's inequality, to get (13.2.24) with an arbitrarily small $\theta>0$ (and then of course $K=K_\theta$ depends on θ).

(13.2.22) and (13.2.23) also yield

$$\begin{aligned}
&\mathbb{E}\big(\big|1+u_n^{p'}\big(V_{i+l-1}^n\big)^2 + \big(U_{i+l-1}^n\big)^{p'}\big|^m \mid \mathcal{F}_{(i+l-2)\Delta_n}\big) \le K_m\big(1+\Delta_n^{1-p'm/2+\varpi(p'm-r)}\big) \\
&\mathbb{E}\big(\big|1+\big(U_{i+l-1}^n\big)^{p'}\big|^m \mid \mathcal{F}_{(i+l-2)\Delta_n}\big) \le K_m.
\end{aligned}$$

The first expression above is smaller than K if $\varpi \ge \frac{p'm-2}{2(p'm-r)}$, and is needed when $j \ge 1$, that is when $k \ge 2$. Then by successive conditioning again, we get (13.2.21). □

Proof of Theorem 13.2.1 By our usual localization procedure we can assume the strengthened assumption (SH-r), in addition to (K), (K'), (K-r) or (K'-r), according to the case.

Suppose first that X is continuous. As seen in the proof of Theorem 13.1.1, the sets $A_i^n = \{\|\Delta_i^n X\| \le v_n\}$ satisfy $\sum_{i=1}^{[t/\Delta_n]} \mathbb{P}((A_i^n)^c) < \infty$. Then on the set $\Omega_t^n = \bigcap_{i\le[t/\Delta_n]} A_i^n$, whose probability goes to 1 as $n\to\infty$, we have $\overline{V}'^n(F,v_n-,X)_s = \overline{V}'^n(F,X)_s$ for all $s \le t$, and thus the result amounts to Theorem 11.2.1, page 312. Note that the left side of (13.2.10) vanishes in this case (for all times $t \le T$) when n is large enough (depending on T). When $F(x_1,\dots,x_k) = f(x_1+\dots+x_k)$, the left side of (13.2.11) also vanishes for all times $t \le T$ when n is large enough, by the same argument.

It remains to consider the cases (c) and (d). The process X' satisfies (K) in case (c) and (K') in case (d), so $\overline{V}'^n(F,v_n-,X')$ converges stably in law to $\overline{V}'(F,X) =$

$\overline{V}'(F, X')$ by what precedes. Hence it remains to prove that

$$\frac{1}{\sqrt{\Delta_n}}\big(V'^n(F, v_n-, X) - V'^n(F, v_n-, X')\big) \stackrel{\text{u.c.p.}}{\Longrightarrow} 0. \tag{13.2.25}$$

With the notation (13.2.20), at time t the left side above is equal to $\sqrt{\Delta_n}\sum_{j=0}^{k-1} \times \sum_{i=1}^{[t/\Delta_n]-k+1} \eta_{i,j}^n$, hence it is enough to prove that $\mathbb{E}(\|\eta_{i,j}^n\|) \leq K\sqrt{\Delta_n}\,\phi_n$ for a sequence of numbers ϕ_n going to 0. By virtue of the previous lemma, applied with $m = 1$ and with $\theta = 0$ because $r \leq 1$, it is a simple matter to check that this holds under the condition (13.2.9), which in fact is exactly designed for this purpose. This completes the proof of the stable convergence result in cases (c) and (d).

Finally, (13.2.25) holds as well with γv_n instead of v_n, and the same argument (based in fact on Lemma 13.2.6 for $k = 1$) shows that when $F(x_1, \ldots, x_k) = f(x_1 + \cdots + x_k)$ we also have

$$\frac{1}{\sqrt{\Delta_n}}\big(V'^n(f, k, \gamma v_n-, X) - V'^n(f, k, \gamma v_n-, X')\big) \stackrel{\text{u.c.p.}}{\Longrightarrow} 0.$$

Then we deduce (13.2.10) and (13.2.11) in cases (c) and (d) from the same for the continuous process X'. □

Proof of Theorems 13.2.4 and 13.2.5 In all cases of the two theorems we have at least (H-r) for some $r < 1$, and by localization we may and will assume the strengthened assumption (SH-r). We write $\overline{Z}^n(X)$ and $\overline{Z}(X)$, $\overline{Z}'(X)$, $\overline{Z}''(X)$ instead of $\overline{Z}^n$ and $\overline{Z}$, $\overline{Z}'$, $\overline{Z}''$, to emphasize the dependency upon the process X. We also set (without discretization of the process C):

$$\overline{C}^n(X') = \frac{1}{\sqrt{\Delta_n}}\big([X', X']^n - C\big).$$

We consider the $2d$-dimensional process $Y = (X', X)$: it satisfies (SH-r) and the volatility process of Y, say $\overline{\sigma}$, is $\overline{\sigma}^{i,j} = \overline{\sigma}^{i+d,j} = \sigma^{ij}$ for $i = 1, \ldots, d$ and $j = 1, \ldots, d'$, and X' is continuous. Therefore a version of the $(2d) \times (2d)$-dimensional process $\overline{Z}(Y) = \overline{Z}'(Y) + \overline{Z}''(Y)$, see (13.2.14), is given, component by component for $i, j = 1, \ldots, d$, by

$$\begin{aligned}
\overline{Z}(Y)^{ij} &= \overline{Z}(X)^{i,j+d} = \overline{Z}'(X)^{i+d,j} = \overline{Z}'(X)^{ij} \\
\overline{Z}(Y)^{i+d,j+d} &= \overline{Z}'(X)^{ij} + \overline{Z}''(X)^{ij}.
\end{aligned}$$

Moreover, if we extend F so that it becomes a function on $(\mathbb{R}^d \times \mathbb{R}^d)^k$ by $F((x_1, y_1), \ldots) = F(x_1, \ldots)$, the pair (Y, F) satisfies the assumptions of Theorem 11.3.2, page 327, as soon as (X, F) satisfies those of Theorem 13.2.5, and of course $\overline{V}'^n(F, X') = \overline{V}'^n(F, Y)$. Therefore

$$\big(\overline{V}'^n(F, X'), \overline{Z}^n(X'), \overline{Z}^n(X)\big) \stackrel{\mathcal{L}\text{-}s}{\Longrightarrow} \big(\overline{V}'(F, X'), \overline{Z}'(X), \overline{Z}'(X) + \overline{Z}''(X)\big), \tag{13.2.26}$$

where $\overline{Z}'(X)$ is given by (13.2.14) and $(\overline{V}'(F,X'),\overline{Z}'(X))$ is as in (13.2.16). The convergence (13.2.26) is used for proving Theorem 13.2.5, but, when we prove Theorem 13.2.4, we can take $F=0$ and this convergence is implied by Theorem 5.4.2, page 162, and thus the assumption (H) is enough.

In view of (13.2.26), and for proving both theorems, it thus remains to show that

$$\overline{V}'^n(F,X')-\overline{V}'^n(F,X)\stackrel{\text{u.c.p.}}{\Longrightarrow}0,\qquad \overline{Z}^n(X')-\overline{C}^n(v_n-,X)\stackrel{\text{u.c.p.}}{\Longrightarrow}0.$$

The first property above is given by Lemma 11.2.4, page 314. For the second one, we observe that $\widehat{C}^n(v_n-,X')_s=[X',X']^n_s$ for all $s\le t$ on a set Ω^n_t whose probability goes to 1, whereas $\frac{1}{\sqrt{\Delta_n}}(C_t-C_{\Delta_n[t/\Delta_n]})\stackrel{\text{u.c.p.}}{\Longrightarrow}0$ because c_t is locally bounded. Therefore this second property amounts to

$$\frac{1}{\sqrt{\Delta_n}}\big(\widehat{C}^n(v_n-,X)-\widehat{C}^n(v_n-,X')\big)\stackrel{\text{u.c.p.}}{\Longrightarrow}0. \tag{13.2.27}$$

The left side above, evaluated at time t, is $\sqrt{\Delta_n}\sum_{i=1}^{[t/\Delta_n]}\eta^n_{i,0}$, where $\eta^n_{i,0}$ is given by (13.2.20) with $k=1$ and the function $F(x)^{jk}=x^jx^k$. This function satisfies (13.2.4) and (13.2.7) with $p=s'=2$ and $s=1$, so when $r<1$ and $\varpi\ge\frac{1}{4-2r}$ we deduce from (13.2.21) with $\theta=0$ and $m=1$ that $\mathbb{E}(\|\eta^n_{i,0}\|)\le\sqrt{\Delta_n}\,\phi_n$, and (13.2.27) follows. □

13.3 Central Limit Theorem for the Local Approximation of the Continuous Part of X

This section is devoted to establishing a Central Limit Theorem for the approximations of the "spot volatility" c_t given in Sect. 9.3. To do this we need an assumption which is almost the same as (K) plus (H-r), and goes as follows, for some $r\in[0,2]$:

Assumption 13.3.1 (or ($\overline{\mathbf{K}}$-r)) The process X satisfies (H-r), and the process σ satisfies (H) = (H-2).

Although not obvious at first glance, this assumption implies (K), except that it does not require the process b_t to be càdlàg or càglàd. Conversely, (K) "almost" implies ($\overline{\mathbf{K}}$-2). To see this, under (K) we can write a "global" Grigelionis representation for the pair (X,σ): we have a Wiener process W and a Poisson random measure $\mathfrak{p}$, and X is given by (13.0.1) and σ by

$$\sigma_t=\sigma_0+\int_0^t\widetilde{b}_s\,ds+\int_0^t\widetilde{\sigma}_s\,dW_s+(\widetilde{\delta}1_{\{\|\widetilde{\delta}\|\le1\}})\star(\mathfrak{p}-\mathfrak{q})_t+(\widetilde{\delta}1_{\{\|\widetilde{\delta}\|>1\}})\star\mathfrak{p}_t, \tag{13.3.1}$$

for some predictable function $\widetilde{\delta}(\omega,t,z)$ on $\Omega\times\mathbb{R}_+\times E$. (5.3.2) holds with the same $\widetilde{b}$ and $\widetilde{\sigma}$, and with $M=(\widetilde{\delta}1_{\{\|\widetilde{\delta}\|\le1\}})\star(\mathfrak{p}-\mathfrak{q})$ and, as soon as further $\widetilde{\delta}$ satisfies a condition similar to the condition on δ in (H-2), we have ($\overline{\mathbf{K}}$-2).

Warning: One has to be aware of the following fact. Under (K) we have (5.3.2) for some local martingale M which may have a non-vanishing continuous martingale part $M_t^c = \int_0^t \widetilde{\sigma}'_s \, dW'_s$ (again in vector notation), where W' is a q-dimensional Wiener process orthogonal to W. This means that we have *not* considered a global Grigelionis representation for the pair (X, σ), and the Brownian motion W drives X but is not enough to drive the continuous part of σ. In this case we can write (13.3.1) with (W, W') instead of W, and with a process $\widetilde{\sigma}$ which indeed is a mix of the two processes $\widetilde{\sigma}$ and $\widetilde{\sigma}'$ (associated with M as above) showing in (5.3.2).

Apart from some necessary conditions on the jumps of σ, the reason for introducing ($\overline{\mathbf{K}}$-2) instead of simply (K) is precisely that we want the process $\widetilde{\sigma}$ showing in the Grigelionis representation of σ to be càdl'àg: as we will see, its left limit appears in the limits obtained below.

Note also that ($\overline{\mathbf{K}}$-r) is very close to (K-r) as well, except that (K-r) is meaningful for $r \le 1$ only.

13.3.1 Statements of Results

We first recall the estimators for the processes c_t and c_{t-}, as introduced in Sect. 9.3. Only the non-truncated and truncated estimators are studied below, because the analysis of the estimators based on multipowers is more difficult to do. The truncation levels v_n are of the form (13.0.2), and we choose a sequence k_n of integers with $k_n \to \infty$ and $k_n \Delta_n \to 0$. We again use the convention (9.3.2), page 255, according to which $\Delta_i^n Y = 0$ for any process Y when i is a nonpositive integer. Then if $i \in \mathbb{Z}$ we define the $\mathcal{M}_{d\times d}^+$-valued variables $\widehat{c}_i^n(k_n)$ and $\widehat{c}_i^n(k_n, v_n)$ by

$$\widehat{c}_i^n(k_n)^{jl} = \frac{1}{k_n \Delta_n} \sum_{m=0}^{k_n-1} \Delta_{i+m}^n X^j \, \Delta_{i+m}^n X^l$$

$$\widehat{c}_i^n(k_n, v_n)^{jl} = \frac{1}{k_n \Delta_n} \sum_{m=0}^{k_n-1} \Delta_{i+m}^n X^j \, \Delta_{i+m}^n X^l \, 1_{\{\|\Delta_{i+m}^n X\| \le v_n\}}.$$

Recalling $I(n,i) = ((i-1)\Delta_n, i\Delta_n]$, the estimators of c_{t-} and c_t are respectively

$$\left.\begin{array}{ll} \widehat{c}^n(k_n, t-) = \widehat{c}_{i-k_n}^n(k_n), & \widehat{c}^n(k_n, t) = \widehat{c}_{i+1}^n(k_n) \\ \widehat{c}^n(k_n, v_n, t-) = \widehat{c}_{i-k_n}^n(k_n, v_n), & \widehat{c}^n(k_n, v_n, t) = \widehat{c}_{i+1}^n(k_n, v_n) \end{array}\right\} \text{ if } t \in I(n,i). \tag{13.3.2}$$

(Note that, with the convention (9.3.2), $\widehat{c}_i^n(k_n)$ for example is defined for all $i \in \mathbb{Z}$, and vanishes when $i \le -k_n$; hence $\widehat{c}^n(k_n, t-)$ is defined even when $t \le k_n \Delta_n$, and vanishes when $t \le 0$; also, $t = 0$ is considered as belonging to $I(n, 0)$ above.)

According to Theorem 9.3.2, and since we assume at least ($\overline{\mathbf{K}}$-2) below, for any stopping time T we have

$$\begin{array}{lll} \widehat{c}^n(k_n,T) \stackrel{\mathbb{P}}{\longrightarrow} c_T, & \widehat{c}^n(k_n,v_n,T-) \stackrel{\mathbb{P}}{\longrightarrow} c_T & \text{on } \{T<\infty\} \\ \widehat{c}^n(k_n,T-) \stackrel{\mathbb{P}}{\longrightarrow} c_{T-}, & \widehat{c}^n(k_n,v_n,T-) \stackrel{\mathbb{P}}{\longrightarrow} c_{T-} & \text{on } \{0<T<\infty\} \end{array} \tag{13.3.3}$$

without restriction for the first line, and under either one of the next two conditions for the second one (where $\mathfrak{p}$ is a driving Poisson measure for (X,σ)):

$$\begin{array}{ll} \text{(b-1)} & T>S \text{ and } T \text{ is } \mathcal{F}_S \text{ measurable for some stopping time } S \\ \text{(b-2)} & \mathfrak{p}(\{T\}\times E)=1 \text{ almost surely on } \{0<T<\infty\}. \end{array} \tag{13.3.4}$$

Under ($\overline{\mathbf{K}}$-2), if T satisfies (b-1) it is predictable, hence $c_{T-}=c_T$ a.s. So if we want to have access to the left limits of the process c_t we need stopping times satisfying (b-2). Of course if $T=T_1\wedge T_2$ where T_1 and T_2 satisfy (b-1) and (b-2), respectively, we also have the second line of (13.3.3).

Our aim is to prove an associated central limit theorem. We give a multivariate version, so we have a set L which indexes the times T_l and T'_l at which the "pre" and "post" estimators are taken. There is no special reason for taking the same index set for both, but it is also not a restriction. There is no restriction on the size of L, which may be infinite and even uncountable, although, since we consider the product topology below, this is essentially irrelevant and using only finite sets L would be the same. We wish to determine the asymptotic behavior of the "estimation errors" in the truncated case

$$Z_l^{n-}=\widehat{c}^n(k_n,v_n,T_l-)-c_{T_l-}, \quad Z_l^{n+}=\widehat{c}^n\big(k_n,v_n,T'_l\big)-c_{T'_l}, \tag{13.3.5}$$

or in the non-truncated case

$$Z_l^{n-}=\widehat{c}^n(k_n,T_l-)-c_{T_l-}, \quad Z_l^{n+}=\widehat{c}^n\big(k_n,T'_l\big)-c_{T'_l}. \tag{13.3.6}$$

Below we need quite strong restrictions on the stopping times T_l, because those should satisfy at least (13.3.4), and also unfortunately on T'_l. Heuristically, because we want a joint convergence of the $Z_l^{n\pm}$ above for all l, we need that the random times T_l and T'_l be in some sense independent of the Wiener process W. The precise structure of the family $(T_l,T'_l)_{l\in L}$ is as follows, in connection with the driving Poisson measure $\mathfrak{p}$ showing in (13.0.1) and (13.3.1):

$$\begin{array}{l} \bullet\ T_l=t_l\wedge S_l \text{ and } T'_l=t'_l\wedge S'_l, \text{ where} \\ \quad -\ t_l \text{ and } t'_l \text{ are non-random, in } (0,\infty) \text{ and } [0,\infty) \text{ respectively,} \\ \quad -\ S_l \text{ and } S'_l \text{ are stopping times,} \\ \quad -\ \text{if } S=S_l \text{ or } S=S'_l, \text{ then } \mathfrak{p}(\{S\}\times E)=1 \text{ on } \{S<\infty\}. \\ \bullet\ \text{for all } l\neq m \text{ we have } T_l\neq T_m \text{ and } T'_l\neq T'_m \text{ almost surely.} \end{array} \tag{13.3.7}$$

The condition on S is satisfied if $\|\Delta X_S\|+\|\Delta\sigma_S\|>0$ on $\{S<\infty\}$. The last condition above is *essential*: as we will see, the (suitably normalized) errors Z_l^{n-} and Z_m^{n-} for example are, asymptotically, $\mathcal{F}$-conditionally independent when $l\neq m$, and this is clearly wrong on the set $\{T_l=T_m\}$ because they are equal on this set. However, we do *not* impose or exclude any relation between the two families (T_l) and (T'_l): we may for example have $\mathbb{P}(T'_l=T_m)>0$ for some l,m.

Remark 13.3.2 One could very slightly weaken this assumption: if there is a T'_l, say T'_1, which is bigger than all others (that is $T'_1 > T'_l$ for all $l \neq 1$), we can relax the condition on this T'_1 to be an arbitrary finite stopping time. This "improvement" looks ridiculous, but it means that the forthcoming theorem also gives the asymptotic behavior of the "post" estimators $\widehat{c}^n(k_n, v_n, T)$ and $\widehat{c}^n(k_n, T)$ for *any* finite stopping time T.

For example if $L = \{1, 2\}$ and $T'_1 > T'_2$ there is no restriction on T'_1 (other than being a finite stopping time), but the result does *not* extend to the case where T'_2 is an arbitrary stopping time.

3) Next, we describe the limit. We assume ($\overline{\text{K-}r}$), and the càdlàg process $\widetilde{\sigma}_t$ of (13.3.1) has components $(\widetilde{\sigma}_t^{ij,k})_{1\leq i\leq d, 1\leq j,k\leq d'}$. Let $(\widetilde{\Omega}, \widetilde{\mathcal{F}}, (\widetilde{\mathcal{F}}_t)_{t\geq 0}, \widetilde{\mathbb{P}})$ be a very good filtered extension of $(\Omega, \mathcal{F}, (\mathcal{F}_t)_{t\geq 0}, \mathbb{P})$ which supports two independent families of i.i.d. variables $(Y(-,l), Y'(-,l))$ and $(Y(+,l), Y'(+,l))$ indexed by L, independent of $\mathcal{F}$, all with the same law as (Y, Y'), as described below (δ^{ij} is the Kronecker symbol):

$$\begin{aligned} &\bullet\ Y = (Y^{jk} : 1 \leq j,k \leq d') \text{ and } Y' = (Y'^j : 1 \leq j \leq d') \\ &\qquad \text{are independent centered Gaussian vectors,} \\ &\bullet\ \mathbb{E}(Y^{jk} Y^{uv}) = \delta^{ju}\delta^{kv} + \delta^{jv}\delta^{ku}, \qquad \mathbb{E}(Y'^j Y'^k) = \delta^{jk}/3. \end{aligned} \tag{13.3.8}$$

With the notation

$$A_t^{ij,kw} = \sigma_t^{ik}\sigma_t^{jw}, \qquad \widetilde{A}_t^{ij,w} = \sum_{k=1}^{d'} \big(\sigma_t^{ik}\widetilde{\sigma}_t^{jk,w} + \sigma_t^{jk}\widetilde{\sigma}_t^{ik,w}\big), \tag{13.3.9}$$

we associate the d^2-dimensional variables with components

$$\begin{aligned} Z(-,l)^{ij} &= \sum_{k,w=1}^{d'} A_{T_l-}^{ij,kw}\, Y(-,l)^{kw}, & Z(+,l)^{ij} &= \sum_{k,w=1}^{d'} A_{T'_l}^{ij,kw}\, Y(+,l)^{kw} \\ Z'(-,l)^{ij} &= \sum_{w=1}^{d'} \widetilde{A}_{T_l-}^{ij,w}\, Y'(-,l)^{w}, & Z'(+,l)^{ij} &= \sum_{w=1}^{d'} \widetilde{A}_{T'_l}^{ij,w}\, Y'(+,l)^{w}. \end{aligned} \tag{13.3.10}$$

The variables $Z(-,l)$, $Z(+,l)$, $Z'(-,l)$ and $Z'(+,l)$ are also characterized by their $\mathcal{F}$-conditional (global) law: they are $\mathcal{F}$-conditionally independent centered Gaussian, with conditional covariances (compare with (13.2.15)):

$$\begin{aligned} \widetilde{\mathbb{E}}\big(Z(-,l)^{ij}\, Z(-,l)^{kw} \mid \mathcal{F}\big) &= c_{T_l-}^{ik}\, c_{T_l-}^{jw} + c_{T_l-}^{iw}\, c_{T_l-}^{jk} \\ \widetilde{\mathbb{E}}\big(Z(+,l)^{ij}\, Z(+,l)^{kw} \mid \mathcal{F}\big) &= c_{T'_l}^{ik}\, c_{T'_l}^{jw} + c_{T'_l}^{iw}\, c_{T'_l}^{jk} \\ \widetilde{\mathbb{E}}\big(Z'(-,l)^{ij}\, Z'(-,l)^{kw} \mid \mathcal{F}\big) &= \sum_{v=1}^{d'} \widetilde{A}_{T_l-}^{ij,v}\, \widetilde{A}_{T_l-}^{kw,v} \\ \widetilde{\mathbb{E}}\big(Z'(+,l)^{ij}\, Z'(+,l)^{kw} \mid \mathcal{F}\big) &= \sum_{v=1}^{d'} \widetilde{A}_{T'_l}^{ij,v}\, \widetilde{A}_{T'_l}^{kw,v}. \end{aligned} \tag{13.3.11}$$

Theorem 13.3.3 *Let X satisfy $(\overline{\text{K-}r})$ for some $r \in [0,2)$, and let $(T_l, T'_l)_{l\in L}$ satisfy* (13.3.7). *Let v_n be as in* (13.0.2) *and let k_n satisfy, for some $\beta \in [0,\infty]$:*

$$k_n \to \infty, \qquad k_n \Delta_n \to 0, \qquad k_n \sqrt{\Delta_n} \to \beta. \tag{13.3.12}$$

a) *If X is continuous, both the truncated versions* (13.3.5) *and the non-truncated versions* (13.3.6) *satisfy*

$$\left.\begin{array}{l} \beta = 0 \Rightarrow (\sqrt{k_n}\, Z_l^{n-}, \sqrt{k_n}\, Z_l^{n+})_{l\in L} \stackrel{\mathcal{L}\text{-s}}{\longrightarrow} (Z(-,l), Z(+,l))_{l\in L} \\ 0 < \beta < \infty \Rightarrow (\sqrt{k_n}\, Z_l^{n-}, \sqrt{k_n}\, Z_l^{n+})_{l\in L} \\ \qquad \stackrel{\mathcal{L}\text{-s}}{\longrightarrow} (Z(-,l) + \beta\, Z'(-,l), Z(+,l) + \beta\, Z'(+,l))_{l\in p} \\ \beta = \infty \Rightarrow \big(\frac{1}{\sqrt{k_n \Delta_n}}\, Z_l^{n-}, \frac{1}{\sqrt{k_n \Delta_n}}\, Z_l^{n+}\big)_{l\in L} \stackrel{\mathcal{L}\text{-s}}{\longrightarrow} (Z'(-,l), Z'(+,l))_{l\in L}. \end{array}\right\} \tag{13.3.13}$$

b) *If*

$$k_n \Delta_n^\tau \to \beta' \in (0,\infty), \quad \text{where } \tau \in (0,1) \tag{13.3.14}$$

(*so* (13.3.12) *holds, with $\beta = 0$ when $\tau < \frac{1}{2}$, and $\beta = \beta'$ if $\tau = \frac{1}{2}$, and $\beta = \infty$ when $\tau > \frac{1}{2}$*), *we have* (13.3.13) *for the non-truncated versions* (13.3.6), *as soon as*

$$\text{either } r < \frac{4}{3}, \quad \text{or } \frac{4}{3} \le r < \frac{2}{1+\tau} \quad \left(\text{and then } \tau < \frac{1}{2}\right). \tag{13.3.15}$$

c) *Under* (13.3.14) *we have* (13.3.13) *for the truncated versions* (13.3.5), *as soon as*

$$r < \frac{2}{1+\tau\wedge(1-\tau)}, \qquad \varpi > \frac{\tau\wedge(1-\tau)}{2(2-r)}, \tag{13.3.16}$$

The convergence in (13.3.13) is for the product topology, so when L is infinite it really amounts to the convergence for any finite subset of L. We also see why the last condition in (13.3.7) is necessary: the limits of, say, Z_l^{n-} and Z_m^{n-} are $\mathcal{F}$-conditionally independent if $l \neq m$, and this cannot be true in restriction to the set $\{T_l = T_m\}$, because $Z_l^{n-} = Z_m^{n-}$ on this set.

This result is in deep contrast with Theorem 13.2.4 when X is not continuous, and a few comments are in order:

- Instead of (H-r), this theorem requires the stronger assumption $(\overline{\text{K-}r})$: from its very formulation, which involves the process $\widetilde{\sigma}$ (at least in some cases), we see that it *cannot* hold if we only assume (H-r).
- On the other hand, there is virtually no restriction on the jumps of X, that is on the value of r in $(\overline{\text{K-}r})$, except that $r < 2$ is required: for a given r, there is always a choice of τ in (13.3.14) and ϖ satisfying (13.3.16) (since the first condition in (13.3.16) implies $\frac{\tau\wedge(1-\tau)}{2(2-r)} < \frac{1}{2}$).
- Even more, in all cases we can use the non-truncated version of the estimators (with a suitable "rate" for k_n, as expressed by the number τ and depending on the number r such that Assumption $(\overline{\text{K-}r})$ holds). However, when X jumps, it is probably wiser from a practical standpoint to use the truncated versions.

- In practice, the discretization step Δ_n is fixed by the structure of the data, but we can choose the window length k_n at will. In the case of (13.3.14), the "rate of convergence" of the estimator is $\Delta_n^{(\tau\wedge(1-\tau))/2}$ and thus the "optimal" choice, leading to the asymptotically smallest estimation variance, is $\tau = 1/2$ (the rate is then $\Delta_n^{1/4}$), with β small. However in the discontinuous case this works only when $r < \frac{4}{3}$, both for the truncated and the non-truncated versions.
- For practical purposes again, we also need an estimator for the asymptotic variance, as given by (13.3.11). There is no reasonable way to estimate $\widetilde{\sigma}_T$, hence in practice choosing $\beta > 0$ should be avoided, and this is why we have chosen to single out the case $\beta = 0$ in (13.3.13), although it can be viewed as the second convergence taken with $\beta = 0$. In the light of the previous comment, under (13.3.14) we should choose τ smaller than, but as close as possible to $\frac{1}{2}$ if $r \leq \frac{4}{3}$, and $\tau = \frac{2-r}{r}$ if $r > \frac{4}{3}$ (if, by chance, the smallest value of r for which (K-r) holds is known).

Remark 13.3.4 From the comments above, only the case $\beta = 0$ in (13.3.13) has practical relevance. However the three cases are interesting from a mathematical standpoint, and are easily understood. If c_t were constant the error would be of order $1/\sqrt{k_n}$ because we basically use k_n approximately i.i.d. variables (the squared increments of X) to estimate a variance. Now, c_t varies over an interval of length $k_n\Delta_n$, and by a quantity of order of magnitude $\sqrt{k_n\Delta_n}$ because of (K-r). So the two rates $\sqrt{k_n}$ and $1/\sqrt{k_n\Delta_n}$ compete: the first one is smaller when $\beta = 0$ and bigger when $\beta = \infty$, and both are of the same order when $0 < \beta < \infty$.

Remark 13.3.5 We have chosen to assume $(\overline{\text{K-}r})$, that is basically (H-r) (which is "necessary" here) plus the fact that σ_t is an Itô semimartingale. This is a natural choice in the general setting of this book, but other types of requirements could be imposed on σ_t, such as assuming for example that $\mathbb{E}(\|c_{t+s} - c_t\| \wedge 1) \leq K\, s^\theta$ for some constants K and θ. Then we would have the first convergence in (13.3.13) as soon as $k_n\Delta_n^{2\theta/(1+2\theta)} \to 0$.

Remark 13.3.6 In Chap. 9 other estimators $\widehat{c}'^n(k_n, T\pm)$ have been introduced. They are based on bipower variations, and we recall their definition. We set

$$\widehat{c}'^n(k_n)^{jl} = \frac{\pi}{8k_n\Delta_n} \sum_{m=0}^{k_n-1} \big(\big|\Delta_{i+m}^n X^j + \Delta_{i+m}^n X^l\big|\big|\Delta_{i+m+1}^n X^j + \Delta_{i+m+1}^n X^l\big|$$
$$- \big|\Delta_{i+m}^n X^j - \Delta_{i+m}^n X^l\big|\big|\Delta_{i+m+1}^n X^j - \Delta_{i+m+1}^n X^l\big|\big)$$

and the estimators are

$$\widehat{c}'^n(k_n, t-) = \widehat{c}^n_{[t/\Delta_n]-k_n-1}(k_n), \qquad \widehat{c}'^n(k_n, t) = \widehat{c}^n_{[t/\Delta_n]+2}(k_n).$$

These also satisfy (13.3.13) (under appropriate conditions), but with different asymptotic variances.

For example suppose that we are in the one-dimensional case $d = 1$, and we have a single stopping time $T_1 = T$, and we are in the case (13.3.14) with $\tau < \frac{1}{2}$. Then $\sqrt{k_n}\,(\widehat{c}^n(k_n,T) - c_T)$ converges stably in law to $Z = \sqrt{2}\,c_T Y(+,1)$ by our theorem, and one can show that $\sqrt{k_n}\,(\widehat{c}'^n(k_n,T) - c_T)$ converges stably in law to $\overline{Z} = \frac{\pi}{2}\sqrt{M(2,2)}\,c_T Y(+,1)$, where $M(p,k)$ is given by (11.4.5): this is of course not a surprise, in view of Theorem 11.2.1 applied with $k = 2$ and the function $F(x,y) = |xy|$. However, after some elementary computation, we see that

$$\widetilde{\mathbb{E}}(\overline{Z}^2 \mid \mathcal{F}) = \left(\frac{\pi^2}{2} + \frac{\pi}{2} - 3\right) c_T^2 \approx 3.5\,c_T^2,$$

to be compared with $\widetilde{\mathbb{E}}(Z^2 \mid \mathcal{F}) = 2c_T^2$. In other words, from an asymptotic viewpoint, it is always better to use $\widehat{c}^n(k_n,T)$ rather than $\widehat{c}'^n(k_n,T)$ for estimating c_T.

Next, we state a trivial but interesting consequence of that theorem. We can consider $(\widetilde{c}^n(k_n,t))_{t\geq 0}$ as a process, which by construction is piecewise constant in time. For any fixed t we have $\widetilde{c}^n(k_n,t) \stackrel{\mathbb{P}}{\longrightarrow} c_t$. Do we also have stable convergence in law of $\widetilde{c}^n(k_n,t) - c_t$ as processes, after a suitable normalization ?

The candidate for the limit is as follows: we have a process $(Y_t, Y'_t)_{t\geq 0}$ on a very good filtered extension $(\widetilde{\Omega}, \widetilde{\mathcal{F}}, (\widetilde{\mathcal{F}}_t)_{t\geq 0}, \widetilde{\mathbb{P}})$ of $(\Omega, \mathcal{F}, (\mathcal{F}_t)_{t\geq 0}, \mathbb{P})$, such that the variables (Y_t, Y'_t) are i.i.d. with the same law as (Y, Y') in (13.3.8) when t varies. Then, as in (13.3.10), set

$$Z_t^{ij} = \sum_{k,w=1}^{d'} A_t^{ij,kw}\, Y_t^{kw}, \qquad Z_t'^{ij} = \sum_{w=1}^{d'} \widetilde{A}_t^{ij,w}\, Y_t'^{w}.$$

Then Theorem 13.3.3 readily gives the following:

Theorem 13.3.7 *Let X satisfy* $(\overline{\text{K-}r})$ *for some* $r \in [0,2)$. *If either X is continuous and* (13.3.12) *holds, or* (13.3.14) *and* (13.3.15) *hold, we have the following finite-dimensional stable convergence in law (denoted as* $\stackrel{\mathcal{L}_f\text{-s}}{\longrightarrow}$ *below, see* (2.2.15), *page* 50):

$$\beta = 0 \Rightarrow \left(\sqrt{k_n}\,\big(\widetilde{c}^n(k_n,t) - c_t\big)\right)_{t\geq 0} \stackrel{\mathcal{L}_f\text{-s}}{\longrightarrow} (Z_t)_{t\geq 0}$$

$$0 < \beta < \infty \Rightarrow \left(\sqrt{k_n}\,\big(\widetilde{c}^n(k_n,t) - c_t\big)\right)_{t\geq 0} \stackrel{\mathcal{L}_f\text{-s}}{\longrightarrow} \left(Z_t + \beta\, Z'_t\right)_{t\geq 0}$$

$$\beta = \infty \Rightarrow \left(\frac{1}{\sqrt{k_n \Delta_n}}\big(\widetilde{c}^n(k_n,t) - c_t\big)\right)_{t\geq 0} \stackrel{\mathcal{L}_f\text{-s}}{\longrightarrow} \left(Z'_t\right)_{t\geq 0}.$$

There is no way to obtain a functional convergence here. Indeed, the processes (Y_t) and (Y'_t) are *white noise* processes, and thus (Z_t) and (Z'_t) as well have the path structure of white noises, and for example their paths are almost surely not Borel, as functions of t. Note that a similar result would hold for the "pre-t" estimators $\widetilde{c}^n(k_n,t-)$, and even for the joint pre- and post-t estimators, and also for the truncated estimators under (13.3.16).

We end this set of results with an extension of Theorem 13.3.3 in the case where one consider simultaneously the estimators $\widehat{c}_i^n(k_n, v_n)$ for several values of k_n. This is in view of some applications, and to stay relatively simple we consider two of them only, with k_n and with mk_n respectively, where $m \geq 2$ is an integer. So, in the setting of this theorem, we consider $Z_l^{n\pm}$ as defined by (13.3.5) or 13.3.6), and also

$$\overline{Z}_l^{n-} = \widehat{c}^n(mk_n, v_n, T_l-) - c_{T_l-}, \qquad \overline{Z}_l^{n+} = \widehat{c}^n\big(mk_n, v_n, T_l'\big) - c_{T_l'},$$

or in the non-truncated case

$$\overline{Z}_l^{n-} = \widehat{c}^n(mk_n, T_l-) - c_{T_l-}, \qquad \overline{Z}_l^{n+} = \widehat{c}^n\big(mk_n, T_l'\big) - c_{T_l'}.$$

We also suppose that the extension $(\widetilde{\Omega}, \widetilde{\mathcal{F}}, (\widetilde{\mathcal{F}}_t)_{t\geq 0}, \widetilde{\mathbb{P}})$ supports, in addition to the variables $(Y(\pm, l), Y'(\pm, l))$, another family $(\widehat{Y}(\pm, l), \widehat{Y}'(\pm, l))$ independent of the first one and of $\mathcal{F}$, and with the same law, characterized by (13.3.8). Finally, we complement (13.3.10) with the variables

$$\begin{aligned}
\overline{Z}(-,l)^{ij} &= \frac{1}{m}\sum_{k,w=1}^{d'} A_{T_l-}^{ij,kw}\big(Y(-,l)^{kw} + \sqrt{m-1}\,\widehat{Y}(-,l)^{kw}\big)\\
\overline{Z}(+,l)^{ij} &= \frac{1}{m}\sum_{k,w=1}^{d'} A_{T_l'}^{ij,kw}\big(Y(+,l)^{kw} + \sqrt{m-1}\,\widehat{Y}(+,l)^{kw}\big)\\
\overline{Z}'(-,l)^{ij} &= \frac{1}{2m}\sum_{w=1}^{d'} \widetilde{A}_{T_l-}^{ij,w}\big((3m-1)Y'(-,l)^{kw} + (m-1)\sqrt{4m-1}\,\widehat{Y}'(-,l)^{kw}\big)\\
\overline{Z}'(+,l)^{ij} &= \frac{1}{2m}\sum_{w=1}^{d'} \widetilde{A}_{T_l'}^{ij,w}\big((3m-1)Y'(+,l)^{kw} + (m-1)\sqrt{4m-1}\,\widehat{Y}'(+,l)^{kw}\big).
\end{aligned} \tag{13.3.17}$$

Theorem 13.3.8 *Under the assumptions of Theorem* 13.3.3, *and in the three cases* (a), (b), (c) *of this theorem, we have the following stable convergence in law*:

$$\begin{aligned}
&\beta = 0 \Rightarrow \sqrt{k_n}\,\big(Z_l^{n-}, Z_l^{n+}, \overline{Z}_l^{n-}, \overline{Z}_l^{n+}\big)_{l\in L}\\
&\qquad\stackrel{\mathcal{L}\text{-s}}{\longrightarrow} \big(Z(-,l), Z(+,l), \overline{Z}(-,l), \overline{Z}(+,l)\big)_{l\in L}\\
&0 < \beta < \infty \Rightarrow \sqrt{k_n}\,\big(Z_l^{n-}, Z_l^{n+}, \overline{Z}_l^{n-}, \overline{Z}_l^{n+}\big)_{l\in L}\\
&\qquad\stackrel{\mathcal{L}\text{-s}}{\longrightarrow} \big(Z(-,l) + \beta Z'(-,l), Z(+,l) + \beta Z'(+,l),\\
&\qquad\qquad \overline{Z}(-,l) + \beta\overline{Z}'(-,l), \overline{Z}(+,l) + \beta\overline{Z}'(+,l)\big)_{l\in p}\\
&\beta = \infty \Rightarrow \frac{1}{\sqrt{k_n\Delta_n}}\,\big(Z_l^{n-}, Z_l^{n+}, \overline{Z}_l^{n-}, \overline{Z}_l^{n+}\big)_{l\in L}\\
&\qquad\stackrel{\mathcal{L}\text{-s}}{\longrightarrow} \big(Z'(-,l), Z'(+,l), \overline{Z}'(-,l), \overline{Z}'(+,l)\big)_{l\in L}.
\end{aligned}$$

The next three subsections are devoted to the proof of Theorem 13.3.3. In the last subsection below we briefly show how the proof has to be modified in order to obtain Theorem 13.3.8.

13.3.2 Elimination of the Jumps and of the Truncation

We begin with some notation. The rate of convergence in (13.3.13) can be written as

$$z_n = \begin{cases} \sqrt{k_n} & \text{if } \beta < \infty \\ \frac{1}{\sqrt{k_n \Delta_n}} & \text{if } \beta = \infty \end{cases}, \quad \text{hence } z_n \le K\Big(\sqrt{k_n} \bigwedge \frac{1}{\sqrt{k_n \Delta_n}}\Big). \tag{13.3.18}$$

Recalling the Poisson random measure $\mathfrak{p}$ of (13.0.1) and (13.3.1), we choose a positive Borel function Γ' on E which is λ-integrable and with $\lambda(\{z : \Gamma'(z) > 1\}) > 0$, and we define the double sequence of (finite) stopping times $R(m, p)$ for $m, p \ge 1$ as follows:

$$\begin{array}{l}(R(m,p) : p \ge 1) \text{ are the successive jump times} \\ \text{of the Poisson process } 1_{\{\Gamma' > 1/m\}} * \mathfrak{p}.\end{array}$$

The next lemma shows that it is enough to prove the result when the double sequence (T_l, T_l') has a special form: the index set L is finite, with a partition $L = L_1 \cup L_2$, and

$$\left.\begin{array}{l}\bullet \text{ if } l \in L_1 \text{ then } T_l = t_l \in (0,\infty) \text{ and } T_l' = t_l' \in [0,\infty) \ (t_l, t_l' \text{ non-random}) \\ \bullet \text{ if } l \in L_2 \text{ then } T_l = T_l' = R(m_l, p_l) \text{ for some } p_l, m_l \ge 1 \\ \bullet\ l \neq m \ \Rightarrow\ T_l \neq T_m,\ T_l' \neq T_m' \text{ almost surely.}\end{array}\right. \tag{13.3.19}$$

Lemma 13.3.9 *If any one of the claims of Theorem* 13.3.3 *holds for all families* $(T_l, T_l')_{l\in L}$ *satisfying* (13.3.19), *it also holds when* L *is arbitrary and* $(T_l, T_l')_{l\in L}$ *satisfies* (13.3.7) *only.*

Proof In this proof the times T_l, T_l' vary, so to emphasize the dependency we write $Z^{n\pm}(R)$ instead of $Z_l^{n\pm}$ when $T_l = R$ or $T_l' = R$, and for either the truncated or the non-truncated version, according to the case. We also consider the case $\beta = 0$ in (13.3.13), the other cases being proved in the same way.

We are given a family $(T_l, T_l')_{l\in L}$ satisfying (13.3.7), with the associated (S_l, S_l', t_l, t_l'), and we want to prove (13.3.13). As said before, it is enough to consider a finite L, say $L = \{1, \dots, Q\}$, and we can assume also that $t_l \neq t_{l'}$ and $t_l' \neq t_{l'}'$ when $l' \neq l$.

For any $m \ge 1$ we set $T(m)_l = t_l$ and $T'(m)_l = t_l'$ when $1 \le l \le Q$, and $T(m)_l = T'(m)_l = R(m, l - Q)$ when $l > Q$. Any finite sub-family $(T(m)_l, T'(m)_l)_{1\le l\le M}$ satisfies (13.3.19) because the t_l's, resp. the t_l''s, are pairwise distinct, and because

$\mathbb{P}(R(m,p)=t)=0$ for any $t\geq 0$. Therefore we can apply the hypothesis: for any finite sub-family, hence also for the whole family, we have the convergence (13.3.13) for $(T(m)_l, T'(m)_l)_{l\geq 1}$, that is

$$\left(\sqrt{k_n}\, Z^{n-}\big(T(m)_l\big), \sqrt{k_n}\, Z^{n+}\big(T'(m)_l\big)\right)_{l\geq 1} \stackrel{\mathcal{L}\text{-s}}{\longrightarrow} \big(Z(m-,l), Z(m+,l)\big)_{l\geq 1}, \tag{13.3.20}$$

where the limits $Z(m\pm,l)$ are associated with $T(m)_l$ and $T'(m)_l$ by (13.3.10).

With any two sets $\mathcal{L}=\{p_1<\cdots<p_Q\}$ and $\mathcal{L}'=\{p'_1<\cdots<p'_Q\}$ of positive integers we put $\Omega_{m,\mathcal{L},\mathcal{L}'}=\cap_{l=1}^{Q}(\{T_l=T(m)_{p_l}\}\cap\{T'_l=T'(m)_{p'_l}\})$. On $\Omega_{m,\mathcal{L},\mathcal{L}'}$ we have $Z^{n-}(T_l)=Z^n(T(m)_{p_l})$ and $Z^{n+}(T'_l)=Z^{n+}(T'(m)_{p'_l})$, whereas, by virtue of the definition (13.3.10), the $\mathcal{F}$-conditional law of the variable $(Z(m-,p_l),Z(m+,p'_l))_{l\in L}$ is the same, in restriction to $\Omega_{\mathcal{L}}$, as the $\mathcal{F}$-conditional law of $(Z(-,l)Z(+,l))_{l\in L}$. Then, due to a basic property of stable convergence in law, we deduce from (13.3.20) that

$$\left(\sqrt{k_n}\, Z^{n-}(T_l), \sqrt{k_n}\, Z^{n+}(T'_l)\right)_{l\in L} \stackrel{\mathcal{L}\text{-s}}{\longrightarrow} \big(Z(-,l), Z(+,l)\big)_{l\in L}$$
$$\text{in restriction to } \Omega_{m,\mathcal{L},\mathcal{L}'}.$$

Therefore, another basic property of the stable convergence in law yields the above convergence holds on $\Omega'=\cup_{m\geq 1}\Omega_m$, where Ω_m is the union of all $\Omega_{m,\mathcal{L},\mathcal{L}'}$ over all families $\mathcal{L}$ and $\mathcal{L}'$ of Q distinct positive integers.

At this stage, it thus remains to check that $\mathbb{P}(\Omega')=1$. We put $B_m=\{R(m,p): p\geq 1\}$. Observe that Ω_m is the set on which, for all $l\in L$, we have either $T_l=t_l$ or $T_l\in B_m$, together with either $T'_l=t'_l$ or $T'_l\in B_m$. The sets B_m increase to the set $B=\{t>0:\mathfrak{p}(\{t\}\times E)=1\}$, hence Ω' is the set on which, for all $l\in L$, we have either $T_l=t_l$ or $T_l\in B$, together with either $T'_l=t'_l$ or $T'_l\in B$. Since by hypothesis $S_l\in B$ and $S'_l\in B$ almost surely and $T_l=t_l\wedge S_l$ and $T'_l=t'_l\wedge S'_l$, we see that $\mathbb{P}(\Omega')=1$, and the proof is complete. □

In view of this lemma, below we restrict our attention to families $(T_l,T'_l)_{l\in L}$ of the form (13.3.19). We associate with (T_l,T'_l) the following notation:

$$\begin{aligned}
&T_l\in I(n,i) \Rightarrow i_n(-,l)=(i-k_n-1)^+\\
&T'_l\in I(n,i) \Rightarrow i_n(+,l)=i\\
&S_n(\pm,l)=i_n(\pm,l)\Delta_n.
\end{aligned} \tag{13.3.21}$$

Note that $i_n(\pm,l)\geq 0$ always, and as soon as $i_n(-,l)\geq 1$ (that is, for all n large enough) we have

$$\begin{aligned}
\widetilde{c}^n(k_n,v_n,T_l-) &= \widetilde{c}^n_{i_n(-,l)+1}(k_n,v_n), &\quad \widetilde{c}^n(k_n,T_l-) &= \widetilde{c}^n_{i_n(-,l)+1}(k_n),\\
\widetilde{c}^n(k_n,v_n,T'_l) &= \widetilde{c}^n_{i_n(+,l)+1}(k_n,v_n), &\quad \widetilde{c}^n(k_n,T'_l) &= \widetilde{c}^n_{i_n(+,l)+1}(k_n).
\end{aligned} \tag{13.3.22}$$

In the next lemma we compare the above estimators when we vary the process X, so we explicitly mention this process in our notation, writing for example $\widehat{c}^n(k_n, v_n, T_l-, Y)$ when we use a given process Y.

Lemma 13.3.10 *Assume* (SH-r) *for some* $r \in (0, 2)$, *and let* $X = X' + X''$ *be the decomposition introduced in* (13.2.17).

a) *When* $X = X'$, *or when* (13.3.14) *and* $\varpi \leq \frac{1-\tau}{r}$ *hold, we have*

$$\begin{aligned} &\mathbb{P}\big(\widehat{c}^n(k_n, v_n, T_l-, X) \neq \widehat{c}^n(k_n, T_l-, X)\big) \to 0 \\ &\mathbb{P}\big(\widehat{c}^n(k_n, v_n, T'_l, X) \neq \widehat{c}^n(k_n, T'_l, X)\big) \to 0. \end{aligned} \tag{13.3.23}$$

b) *Under* (13.3.14) *and* (13.3.16) *we have (recall* (13.3.18) *for* z_n*)*

$$\begin{aligned} &z_n\big(\widehat{c}^n(k_n, v_n, T_l-, X) - \widehat{c}^n(k_n, v_n, T_l-, X')\big) \stackrel{\mathbb{P}}{\longrightarrow} 0 \\ &z_n\big(\widehat{c}^n(k_n, v_n, T'_l, X) - \widehat{c}^n(k_n, v_n, T'_l, X')\big) \stackrel{\mathbb{P}}{\longrightarrow} 0. \end{aligned} \tag{13.3.24}$$

Proof 1) We start with the statements relative to T'_l in both (13.3.23) and (13.3.24), and we set $i_n = i_n(+, l)$. We have $\mathbb{P}(\widehat{c}^n(k_n, v_n, T'_l, X) \neq \widehat{c}^n(k_n, T'_l, X)) \leq \sum_{i=1}^{k_n} a(n, j)$, where

$$\begin{aligned} a(n, j) &= \mathbb{P}\big(\big\|\Delta^n_{i_n+j} X\big\| > v_n\big) \\ &\leq \mathbb{P}\big(\big\|\Delta^n_{i_n+j} X'\big\| > v_n/2\big) + \mathbb{P}\big(\big\|\Delta^n_{i_n+j} X''\big\| > v_n/2\big). \end{aligned}$$

If $j \geq 0$ we have $\{i_n + j = i\} \in \mathcal{F}_{(i-1)\Delta_n}$ because $S_n(+, l)$ is a stopping time. Therefore we can apply (13.2.22) and (13.2.23) on each of these sets and then sum up over i to obtain

$$\begin{aligned} \mathbb{P}\big(\big\|\Delta^n_{i_n+j} X'\big\| > v_n/2\big) &\leq K_m \Delta_n^{m(1/2-\varpi)} \sum_{i\geq 1} \mathbb{E}\big(\big(U^n_{i_n+j}\big)^m 1_{\{i_n+j=i\}}\big) \\ &\leq K_m \Delta_n^{m(1/2-\varpi)} \\ \mathbb{P}\big(\big\|\Delta^n_{i_n+j} X''\big\| > v_n/2\big) &\leq K \sum_{i\geq 1} \mathbb{E}\big(\big(V^n_{i_n+j}\big)^r 1_{\{i_n+j=i\}}\big) \leq \Delta_n^{1-r\varpi} \phi_n, \end{aligned}$$

where $\phi_n \to 0$ as $n \to \infty$, and $m > 0$ is arbitrary. Upon taking m big enough, and since $\varpi < \frac{1}{2}$, we deduce that $a(n, j) \leq \Delta_n$ when $X = X'$, and this readily gives (13.3.23) for T'_l when $X = X'$ because $k_n \Delta_n \to 0$. Otherwise, we have $a(n, j) \leq 2\Delta_n^{1-r\varpi} \phi_n$, which again implies (13.3.23) when (13.3.14) and $\varpi \leq \frac{1-\tau}{r}$ hold, because we then have $k_n \Delta_n^{1-r\varpi} \phi_n \leq K\phi_n \to 0$.

Next, assume (13.3.14) and (13.3.16). With the notation (13.2.20), we have

$$\widehat{c}^n\big(k_n, v_n, T'_l, X\big)^{jl} - \widehat{c}^n\big(k_n, v_n, T'_l, X'\big)^{jl} = \frac{1}{k_n} \sum_{w=1}^{k_n} \eta^n_{i_n+w,0},$$

provided we take $k=1$ and $F(x)=x^j x^l$. This function satisfies (13.2.4) and (13.2.7) with $p=s'=2$ and $s=1$. Applying (13.2.21) with $m=1$, in restriction to the sets $\{i_n=i-1\}$, and summing over $i\geq 1$, we deduce

$$\mathbb{E}\big(\big\|\widehat{c}^n(k_n,v_n,T'_l,X)-\widehat{c}^n(k_n,v_n,T'_l,X')\big\|\big) \le \big(\Delta_n^{\frac{2-r}{2}(1\wedge\frac{1}{r})-\theta}+\Delta_n^{(2-r)\varpi-\theta}\big)\phi_n,$$

where $\theta>0$ is arbitrarily small and $\phi_n\to 0$ again. Under (13.3.14) we have $z_n\leq K/\Delta_n^{(\tau\wedge(1-\tau))/2}$. Then, as soon as (13.3.16) holds, we deduce (13.3.24) for T'_l.

2) Observe that the previous argument proves the results for T'_l because we have (13.3.22) and $S_n(+,l)$ is a stopping time. It works as well for T_l-, as long as $S_n(-,l)$ is a stopping time, which is the case when $l\in L_1$. When $l\in L_2$, however, $S_n(-,l)$ is no longer a stopping time, so the previous proof fails.

We thus resort to the same trick as in Step 7 of the proof of Theorem 9.3.2. We have $T_l=R(m,p)$ for some $p,m\geq 1$, and we denote by $(\mathcal{G}_t)$ the smallest filtration containing $(\mathcal{F}_t)$ and such that $(R(m,q):q\geq 1)$ is $\mathcal{G}_0$ measurable. By Proposition 2.1.10, page 44, W is a $(\mathcal{G}_t)$-Brownian motion, and the restriction $\mathfrak{p}'$ of $\mathfrak{p}$ to $\mathbb{R}_+\times\{z:\Gamma'(s)\leq\frac{1}{m}\}$ is a $(\mathcal{G}_t)$-Poisson measure whose compensator is the restriction $\mathfrak{q}'$ of $\mathfrak{q}$ to the same set. Then $\overline{X}_t=X_t-\sum_{q\geq 1}\Delta X_{R(m,q)}1_{\{R(m,q)\leq t\}}$ is an Itô semimartingale, relative to the filtration $(\mathcal{G}_t)$, with the same Grigelionis representation (13.0.1) as X, except that $\mathfrak{p}$ and $\mathfrak{q}$ are replaced by $\mathfrak{p}'$ and $\mathfrak{q}'$, and the process b_t is replaced by another bounded process $\overline{b}'_t$, and W, σ and δ are unchanged. In particular, $\overline{X}$ satisfies (SH-r) relative to $(\mathcal{G}_t)$. Moreover, by a simple calculation, the decomposition $\overline{X}=\overline{X}'+\overline{X}''$ similar to $X=X'+X''$ is

$$\overline{X}'_t-X'_t=\int_0^t\alpha_s\,ds,\quad\text{where }\alpha_s=\begin{cases}0 & \text{if } r\leq 1\\ -\int_{\{z:\Gamma'(z)>1/m\}}\delta(t,z)\lambda(dz) & \text{if } r>1.\end{cases}$$

Now we can prove the results, using the fact that $S_n(-,l)$ is a $(\mathcal{G}_t)$-stopping time:

(i) (13.3.23) with T_l- holds for X', and for $\overline{X}$ under (13.3.14) and $\varpi\leq\frac{1-\tau}{r}$. Since $R(m,p-1)<T_l-(k_n+2)\Delta_n$ for all large n, hence $\widehat{c}^n(k_n,v_n,T_l-,X)=\widehat{c}^n(k_n,v_n,T_l-,\overline{X})$ and the same for the non-truncated versions: thus (13.3.23) with T_l- holds for X itself, under the same conditions.

(ii) (13.3.24) with T_l- holds for the pair $(\overline{X},\overline{X}')$. As seen above, we have $\widehat{c}^n(k_n,v_n,T_l-,X)=\widehat{c}^n(k_n,v_n,T_l-,\overline{X})$ for all n large enough, whereas we have (13.3.23) for both X' and $\overline{X}'$, so to get (13.3.24) with T_l- for the pair (X,X') it remains to prove that

$$z_n\big(\widehat{c}^n(k_n,T_l-,X')-\widehat{c}^n(k_n,T_l-,\overline{X}')\big)\stackrel{\mathbb{P}}{\longrightarrow}0. \tag{13.3.25}$$

To see this, we set $A_t=\int_0^t\alpha_s\,ds$, so $\Delta_i^nX'=\Delta_i^n\overline{X}'+\Delta_i^nA$ and $\|\Delta_i^nA\|\leq K\Delta_n$ (because α is bounded), hence $|\Delta_i^nX'^j\Delta_i^nX'^l-\Delta_i^n\overline{X}'^j\Delta_i^n\overline{X}'^l|\leq K\Delta_n^2+K\Delta_n\|\Delta_i^n\overline{X}'\|$. Since $\mathbb{E}(\|\Delta_i^n\overline{X}'\|\mid\mathcal{G}_{(i-1)\Delta_n})\leq K\sqrt{\Delta_n}$ (by (13.2.22) for example), we deduce

$$\mathbb{E}\big(\big\|\widehat{c}_i^n(k_n,X')-\widehat{c}_i^n(k_n,\overline{X}')\big\|\mid\mathcal{G}_{(i-1)\Delta_n}\big)\leq K\sqrt{\Delta_n}.$$

Applying this on each set $\{i_n(-,l) = i-1\}$ (which belongs to $\mathcal{G}_{(i-1)\Delta_n}$), summing over $i \geq 1$, and since $z_n\sqrt{\Delta_n} \to 0$, we deduce (13.3.25). □

We are now ready to state the last result of this subsection:

Lemma 13.3.11 *Assume that* (a) *of Theorem* 13.3.3 *holds for the non-truncated versions* (13.3.6). *Then* (b) *of this theorem also holds, as well as* (a) *and* (c) *for the truncated versions.*

Proof We consider a family $(T_l, T'_l)_{l\in L}$ which, by virtue of Lemma 13.3.9, may be assumed to satisfy (13.3.19). Since $(\overline{\text{K-}r})$ implies (H-r), by localization we may assume (SH-r), so the previous lemma applies.

Our hypothesis is that (13.3.13) holds when X is continuous and for the non-truncated versions (13.3.6) of the estimators. A first application of (13.3.23) (with $X = X'$) gives that (13.3.13) holds as well for the truncated versions, when X is continuous.

Next, assume X discontinuous, and (13.3.14) and (13.3.16). Then the fact that (13.3.13) holds for X' and the truncated version, together with (13.3.24), immediately yield that (13.3.13) holds for X and the truncated versions.

Finally, assume X discontinuous, and (13.3.14) and (13.3.15). We have the first part of (13.3.16), and we can also find a $\varpi \in (0, \frac{1}{2})$ which satisfies the second part of (13.3.16), together with $\varpi \leq \frac{1-\tau}{r}$. Then, with $v_n = \Delta_n^\varpi$ for this particular value of ϖ, we have (13.3.13) for X and the truncated versions, and that it also holds for the non-truncated versions then readily follows from (13.3.23) again. □

13.3.3 The Scheme of the Proof in the Continuous Case

In view of Lemma 13.3.9 and 13.3.11, and upon localizing, it remains to prove the convergence (13.3.13) under the following strengthened assumption:

Assumption $(\overline{\text{SK}})$ We have $(\overline{\text{K-}r})$ with X continuous (so the value of r here is irrelevant), and the processes $b, \widetilde{b}, \sigma$ and $\widetilde{\sigma}$ are bounded, and we have $\|\widetilde{\delta}(\omega,t,z)\| \wedge 1 \leq \Gamma(z)$ with a function Γ on E which is bounded and with $\int_E \Gamma(z)^2 \lambda(dz)$.

It is also enough to consider the non-truncated estimators, so we use the version (13.3.6) for $Z_l^{n\pm}$, and a family $(T_l, T'_l)_{l\in L}$ which satisfies (13.3.19) with a finite set L. Below, and until the end of the section, all these hypotheses are in force.

We again introduce some notation, complementing (13.3.9) and (13.3.21). Below we write the various components of all these variables: the indices j, m are between 1 and d, whereas u, w are between 1 and d':

$$Y^n(\pm, l)^{uw} = \frac{1}{k_n\Delta_n}\sum_{i=1}^{k_n} \Delta^n_{i_n(\pm,l)+i} W^u\, \Delta^n_{i_n(\pm,l)+i} W^w - \delta^{uw} \tag{13.3.26}$$

$$\begin{aligned} Y'^n(+,l)^u &= \frac{1}{k_n}\sum_{i=1}^{k_n}\big(W^u_{(i_n(+,l)+i-1)\Delta_n} - W^u_{i_n(+,l)\Delta_n}\big) \\ Y'^n(-,l)^u &= \frac{1}{k_n}\sum_{i=1}^{k_n}\big(W^u_{(i_n(-,l)+i-1)\Delta_n} - W^u_{(i_n(-,l)+k_n-1)\Delta_n}\big) \end{aligned}$$

$$\begin{aligned} Z^n(\pm,l)^{jm} &= \sum_{u,w=1}^{d'} A^{jm,uw}_{S_n(\pm,l)}\, Y^n(\pm,l)^{uw} \\ Z'^n(\pm,l)^{jm} &= \sum_{w=1}^{d'} \widetilde{A}^{jm,w}_{S_n(\pm,l)}\, Y'^n(\pm,l)^{w}. \end{aligned} \tag{13.3.27}$$

The reader will observe the differences in the definitions of $Y'^n(+,l)$ and $Y'^n(-,l)$. The main terms are $Z^n(\pm,l)$ and $Z'^n(\pm,l)$, and the remainder terms are naturally the differences

$$\begin{aligned} Z''^n(+,l) &= Z^{n+}_l - Z^n(+,l) - Z'^n(+,l) \\ Z''^n(-,l) &= Z^{n-}_l - Z^n(-,l) - Z'^n(-,l). \end{aligned} \tag{13.3.28}$$

Upon multiplying both members of these two equalities by z_n, as given by (13.3.18), and observing that $z_n/\sqrt{k_n} = 1$ and $z_n\sqrt{k_n\Delta_n} \to \beta$ when $\beta < \infty$, whereas $z_n/\sqrt{k_n} \to 0$ and $z_n\sqrt{k_n\Delta_n} = 1$ when $\beta = \infty$, we see that all three cases of (13.3.13) follow from the next two lemmas:

Lemma 13.3.12 *Assuming* ($\overline{\text{SK}}$), *we have the following stable convergence*:

$$\begin{aligned} &\Big(\sqrt{k_n}\, Z^n(-,l), \sqrt{k_n}\, Z^n(+,l), \frac{1}{\sqrt{k_n\Delta_n}}\, Z'^n(-,l), \frac{1}{\sqrt{k_n\Delta_n}}\, Z'^n(+,l)\Big)_{l\in L} \\ &\quad \stackrel{\mathcal{L}\text{-s}}{\longrightarrow} \big(Z(-,l), Z(+,l), Z'(-,l), Z'(+,l)\big)_{l\in L}. \end{aligned} \tag{13.3.29}$$

Lemma 13.3.13 *Under* ($\overline{\text{SK}}$) *we have for all* $l \in L$:

$$z_n\, Z''^n(-,l) \stackrel{\mathbb{P}}{\longrightarrow} 0, \qquad z_n\, Z''^n(+,l) \stackrel{\mathbb{P}}{\longrightarrow} 0.$$

The remainder of this section is thus devoted to proving these two lemmas.

13.3.4 Proof of Lemma 13.3.12

We start with two preliminary results. First, let $(U_n)_{n\geq 1}$ be an i.i.d. sequence of $\mathcal{N}(0, I_{d'})$-distributed variables on some space $(\Omega', \mathcal{F}', \mathbb{P}')$, and set for $1 \leq u, v \leq d'$:

$$Y^{uv}_n = k_n^{-1/2}\sum_{i=1}^{k_n}\big(U^u_i U^v_i - \delta^{uv}\big), \qquad Y'^u_n = k_n^{-3/2}\sum_{i=1}^{k_n}(k_n - i)U^u_i.$$

Lemma 13.3.14 *The variables* (Y_n, Y_n') *converge in law to* (Y, Y'), *as defined in* (13.3.8).

Proof The pair (Y_n, Y_n') is equal to $\sum_{i=1}^{k_n}(\zeta_i^n, \zeta_i'^n)$, where ζ_i^n and $\zeta_i'^n$ are d'^2 and d' dimensional variable with components:

$$\zeta_i^{n,uv} = k_n^{-1/2}\big(U_i^u U_i^v - \delta^{uv}\big), \qquad \zeta_i'^{n,u} = k_n^{-3/2}(k_n - i)U_i^u. \tag{13.3.30}$$

The variables $(\zeta_i^n, \zeta_i'^n)$ are centered, and independent when i varies, for each n. Moreover

$$\begin{aligned}
&\sum_{i=1}^{k_n} \mathbb{E}'\big(\zeta_i^{n,uv}\,\zeta_i^{n,ws}\big) = \delta^{uw}\delta^{vs} + \delta^{us}\delta^{vw} \\
&\sum_{i=1}^{k_n} \mathbb{E}'\big(\zeta_i^{n,uv}\,\zeta_i'^{n,w}\big) = 0 \\
&\sum_{i=1}^{k_n} \mathbb{E}'\big(\zeta_i'^{n,u}\,\zeta_i'^{n,v}\big) = \sum_{i=1}^{k_n} \frac{(k_n-i)^2}{k_n^3}\,\delta^{uv} \to \frac{1}{3}\,\delta^{uv} \\
&\sum_{i=1}^{k_n} \mathbb{E}'\big(\|\zeta_i^n\|^4\big) + \sum_{i=1}^{k_n} \mathbb{E}'\big(\|\zeta_i'^n\|^4\big) \le \frac{K}{k_n} \to 0.
\end{aligned} \tag{13.3.31}$$

The result then follows from Theorem 2.2.14. □

The next stable convergence result does not formally follows from the convergence (4.3.4) in Theorem 4.3.1, but it is very similar, although simpler to prove.

Lemma 13.3.15 *With* $Y(\pm,l)$ *and* $Y'(\pm,l)$ *as defined before* (13.3.8), *we have the following stable convergence*:

$$\left(\sqrt{k_n}\,Y^n(\pm,l), \frac{1}{\sqrt{k_n\Delta_n}}\,Y'^n(\pm,l)\right)_{l\in L} \stackrel{\mathcal{L}\text{-s}}{\longrightarrow} \big(Y(\pm,l), Y'(\pm,l)\big)_{l\in L}.$$

Proof Put $\overline{Y}^n(\pm,l) = \sqrt{k_n}\,Y^n(\pm,l)$ and $\overline{Y}'^n(\pm,l) = \frac{1}{\sqrt{k_n\Delta_n}}\,Y'^n(\pm,l)$, and let (Y, Y') be as in (13.3.8) again. We need to prove that

$$\begin{aligned}
&\mathbb{E}(Z)\prod_{l\in L} f_l\big(\overline{Y}^n(-,l)\big)\,g_l\big(\overline{Y}^n(+,l)\big)\,f_l'\big(\overline{Y}'^n(-,l)\big)\,g_q'\big(\overline{Y}'^n(+,l)\big) \\
&\quad\to \mathbb{E}(Z)\prod_{l\in L}\mathbb{E}'\big(f_l(Y)\big)\,\mathbb{E}'\big(g_l(Y)\big)\,\mathbb{E}'\big(f_l'(Y')\big)\,\mathbb{E}'\big(g_l'(Y')\big),
\end{aligned} \tag{13.3.32}$$

for any continuous bounded functions f_l and g_l on $\mathcal{M}_{d'\times d'}$ (the set of all $d'\times d'$ matrices) and f_l' and g_l' on $\mathbb{R}^{d'}$, and any $\mathcal{F}$ measurable bounded variable Z on $(\Omega, \mathcal{F}, \mathbb{P})$.

Recalling (13.3.19), we set $\mathcal{G}=\sigma(T_l, T'_l : l\in L)$ and $\mathcal{H}=\sigma(W_t : t\geq 0)\bigvee \mathcal{G}$. Since the variables $Y_l^{n\pm}$ are $\mathcal{H}$ measurable, we can replace Z by $Z=\mathbb{E}(Z\mid \mathcal{H})$. In other words it is enough to prove (13.3.32) when Z is $\mathcal{H}$ measurable. Moreover the process W is independent of $\mathcal{G}$ by (13.3.19), hence is an $(\mathcal{H}_t)$-Wiener process where $(\mathcal{H}_t)$ is the smallest filtration such that $\mathcal{G}\subset\mathcal{H}_0$ and to which W is adapted. For each $j\geq 1$ the set

$$B_j = \cup_{l\in L}\big(\big((T_l-1/j)^+, T_l\big]\cup\big(T'_l, T'_l+1/j\big]\big)$$

is $(\mathcal{H}_t)$-predictable. We can thus define $W(j)$ and $\overline{W}(j)$ as in Theorem 4.3.1, and set $\mathcal{H}^j=\sigma(\overline{W}(j)_t : t\geq 0)\bigvee\mathcal{G}$. The σ-fields $\mathcal{H}^j$ increase with j, and $\bigvee_j\mathcal{H}^j=\mathcal{H}$. Hence if Z is bounded $\mathcal{H}$ measurable, we have $Z_j=\mathbb{E}(Z\mid\mathcal{H}^j)\to Z$ in $\mathbb{L}^1(\mathbb{P})$. Thus if (13.3.32) holds for each Z_j, it also holds for Z. In other words it is enough to prove (13.3.32) when Z is $\mathcal{H}^j$-measurable for any given $j\geq 1$.

Second, we use the simplifying notation $u_n=(k_n+1)\Delta_n$. We let Ω_n be the set on which $T_l>u_n$ and $|T_l-T_{l'}|>2u_n$ and $|T'_l-T'_{l'}|>2u_n$ for all $l,l'\in L$ with $l\neq l'$, and also either $T_l=T'_{l'}$ or $|T_l-T'_{l'}|>2u_n$ for all $l,l'\in L$. Since $u_n\to 0$, we have $\Omega_n\to\Omega$. Therefore, instead of (13.3.32) it is enough to prove that

$$\begin{aligned}\mathbb{E}(Z)\,1_{\Omega_n}\prod_{l\in L} f_l\big(\overline{Y}^n(-,l)\big)\,g_l\big(\overline{Y}^n(+,l)\big)\,f'_l\big(\overline{Y}'^n(-,l)\big)\,g'_q\big(\overline{Y}'^n(+,l)\big)\\ \to\ \mathbb{E}(Z)\prod_{l\in L}\mathbb{E}'\big(f_l(Y)\big)\,\mathbb{E}'\big(g_l(Y)\big)\,\mathbb{E}'\big(f'_l(Y')\big)\,\mathbb{E}'\big(g'_l(Y')\big), \qquad (13.3.33)\end{aligned}$$

when Z is bounded and $\mathcal{H}^j$ measurable. Below we take n sufficiently large to have $u_n<\frac{1}{j}$. Then we observe that, because of the independence of W and $\mathcal{G}$, and in view of (13.3.26) and of the scaling and symmetry properties of the $(\mathcal{H}_t)$-Wiener process W, we have the following: in restriction to the $\mathcal{G}$ measurable set Ω_n, for any $l\in L$ the $\mathcal{G}$-conditional law of $(\overline{Y}^n(\pm,l),\overline{Y}'^n(\pm+,l))$ is the same as the law of (Y_n,Y'_n), as given by (13.3.29); moreover, again conditionally on $\mathcal{G}$ and in restriction to Ω_n, the variables $(\overline{Y}^n(-,l),\overline{Y}'^n(-,l))$ and $(\overline{Y}^n(+,l'),\overline{Y}'^n(+,l'))$ are independent when l,l' range through L, and they are globally independent of the σ-field $\mathcal{H}^j$. Therefore when Z is $\mathcal{H}^j$ measurable, the left side of (13.3.33) takes the form

$$\mathbb{E}(Z\,1_{\Omega_n})\prod_{l\in L}\mathbb{E}'\big(f_l(Y_n)\,f'_l(Y'_n)\big)\,\mathbb{E}'\big(g_l(Y_n)\,g'_l(Y'_n)\big).$$

Since $\mathbb{E}(Z1_{\Omega_n})\to\mathbb{E}(Z)$, (13.3.32) amounts to $(Y_n,Y'_n)\stackrel{\mathcal{L}}{\longrightarrow}(Y,Y')$, which is the previous lemma. □

We are now ready to prove Lemma 13.3.12. We have (13.3.27), which is analogous to (13.3.10) except that we replace T_l- and T'_l by $S_n(-,l)$ and $S_n(+,l)$, and of course $Y(\pm,l)$ and $Y'(\pm,l)$ by $Y^n(\pm,l)$ and $Y'^n(\pm,l)$. Now, from the definition (13.3.21) of $S_n(\pm,l)$ and the càdlàg property of σ and $\widehat{\sigma}$, we see that $A_{S_n(-,l)}\to A_{T_l-}$ and $A_{S_n(+,l)}\to A_{T'_l}$, and analogously for $\widetilde{A}_{S_n(\pm,l)}$. Then (13.3.29) follows from the previous lemma and the property (2.2.5) of stable convergence.

13.3.5 Proof of Lemma 13.3.13

For the proof of Theorem 13.3.3 it thus remains to prove Lemma 13.3.13, which we do in a number of steps.

Step 1) We assume ($\overline{\mathrm{SK}}$), so we can rewrite (13.3.1) as

$$\sigma_t = \sigma_0 + \int_0^t \widetilde{b}'_s\, ds + \int_0^t \widetilde{\sigma}_s\, dW_s + M_t, \tag{13.3.34}$$

where $\widetilde{b}'$ is again bounded, and $M = \widetilde{\delta} * (\mathfrak{p} - \mathfrak{q})$. The index l is fixed in the statement of the lemma, and we omit it in our notation: for example we write $i_n(\pm)$, T and T' instead of $i_n(\pm, l)$, T_l and T'_l. Our first task is to express the remainder term $Z''^n(\pm) = Z''^n(\pm, l)$ into tractable pieces.

For this, for i between 1 and k_n, we first define the following variables (we use vector notation there, $\beta_i^n(\pm)$ is d-dimensional, the others are $d \times d'$-dimensional):

$$\begin{aligned}
&\xi_i^n(\pm) = \widetilde{\sigma}_{S_n(\pm)}\big(W_{S_n(\pm)+(i-1)\Delta_n} - W_{S_n(\pm)}\big),\\
&\xi_i'^n(\pm) = \sigma_{S_n(\pm)+(i-1)\Delta_n} - \sigma_{S_n(\pm)}\\
&\widetilde{\xi}_i^n(\pm) = \xi_i'^n(\pm) - \xi_i^n(\pm), \qquad \beta_i^n(\pm) = \sigma_{S_n(\pm)+(i-1)\Delta_n}\Delta_{i_n(\pm)+i}^n W.
\end{aligned}$$

A rather tedious, but otherwise elementary, computation based on (13.3.27) and (13.3.28), and of course on the basic formulas (13.3.1), (13.3.2) and (13.3.6), allows us to check that

$$Z''^n(\pm) = \sum_{j=1}^6 \zeta^n(j, \pm), \tag{13.3.35}$$

where

$$\begin{aligned}
&\zeta^n(1,+) = c_{S_n(+)} - c_{T'}, \qquad \zeta^n(1,-) = c_{S_n(-)+k_n\Delta_n} - c_{T-}\\
&\zeta^n(2,+) = 0\\
&\zeta^n(2,-)^{jm} = c^{jm}_{S_n(-)} - c^{jm}_{S_n(-)+(k_n-1)\Delta_n} + \sum_{w=1}^{d'} \widetilde{A}^{jm,w}_{S_n(-)}\big(W^w_{S_n(-)+(k_n-1)\Delta_n} - W^v_{S_n(-)}\big)
\end{aligned}$$

and further for $j = 3, 4, 5, 6$ we have

$$\zeta^n(j, \pm) = \frac{1}{k_n \Delta_n} \sum_{i=1}^{k_n} \eta_i^n(j, \pm) \tag{13.3.36}$$

where

$$\eta_i^n(3,\pm)^{jm} = \Delta^n_{i_n(\pm)+i} X^j\, \Delta^n_{i_n(\pm)+i} X^m - \beta_i^n(\pm)^j\, \beta_i^n(\pm)^m$$

$$\eta_i^n(4,\pm)^{jm} = \sum_{u,v,w=1}^{d'} \big(\sigma_{S_n(\pm)}^{ju}\,\widetilde{\sigma}_{S_n(\pm)}^{mv,w} + \sigma_{S_n(\pm)}^{mu}\,\widetilde{\sigma}_{S_n(\pm)}^{jv,w}\big)$$
$$\times \big(\Delta_{i_n(\pm)+i}^n W^u\,\Delta_{i_n(\pm)+i}^n W^v - \Delta_n\delta^{uv}\big)\big(W_{S_n(\pm,l)+(i-1)\Delta_n}^w - W_{S_n(\pm)}^w\big)$$

$$\eta_i^n(5,\pm)^{jm} = \sum_{u,w=1}^{d'} \xi_i'^n(\pm)^{ju}\,\xi_i'^n(\pm)^{mw}\,\Delta_{i_n(\pm)+i}^n W^u\,\Delta_{i_n(\pm)+i}^n W^w$$

$$\eta_i^n(6,\pm)^{jm} = \sum_{u,w=1}^{d'} \big(\sigma_{S_n(\pm)}^{ju}\,\widetilde{\xi}_i^n(\pm)^{mw} + \sigma_{S_n(\pm)}^{mu}\,\widetilde{\xi}_i^n(\pm)^{jw}\big)\Delta_{i_n(\pm)+i}^n W^u\,\Delta_{i_n(\pm)+i}^n W^w.$$

In view of the decomposition (13.3.35), it is thus enough to prove that, for $j = 1,2,3,4,5,6$, we have

$$z_n\,\zeta^n(j,\pm) \stackrel{\mathbb{P}}{\longrightarrow} 0. \tag{13.3.37}$$

Step 2) In this step we prove (13.3.37) when $S_n(\pm)$ is a stopping time, that is for $S_n(+)$, and for $S_n(-)$ when $T = T_l$ with $l \in L_1$ (as in (13.3.19)). To simplify the notation, we write $i_n = i_n(\pm)$ and $S_n = S_n(\pm)$.

Proof of (13.3.37) for $j=1$ Recall that $(\overline{\text{SK}})$ implies $\mathbb{E}(\|\sigma_R - \sigma_{R'}\|^2) \le K\Delta_n$ for any two stopping times R, R' with $R \le R' \le R + \Delta_n$, and since σ is bounded we deduce $\mathbb{E}(\|c_R - c_{R'}\|^2) \le K\Delta_n$ as well.

Applying this with $R = T'$ and $R' = S_n(+)$ gives the result for $\zeta^n(+,1)$, because $z_n\Delta_n \to 0$. Applying this with $R = S_n(-) + (k_n - 1)\Delta_n$ and $R' = T$, plus the fact that $c_{T-} = c_T$ a.s. because T is non-random here, we get the result for $\zeta^n(1,-)$. □

Proof of (13.3.37) for $j=3$ Let $\rho_i^n(\pm) = \Delta_{i_n+i}^n X - \beta_i^n(\pm)$. We deduce from $(\overline{\text{SK}})$ and (2.1.44), page 43, and the fact that $(i_n + i - 1)\Delta_n$ is a stopping time that, for all $i \ge 1$ and $q \ge 2$,

$$\mathbb{E}\big(\big\|\rho_i^n(\pm)\big\|^q\big) \le K_q\Delta_n^{1+q/2}, \quad \mathbb{E}\big(\big\|\beta_i^n(\pm)\big\|^q\big) \le K\Delta_n^{q/2}. \tag{13.3.38}$$

We have $\|\eta_i^n(3,\pm)\| \le 2\|\rho_i^n(\pm)\|^2 + \|\rho_i^n(\pm)\|\,\|\beta_i^n(\pm)\|$, hence (13.3.38) and the Cauchy-Schwarz inequality yield $\mathbb{E}(\|\zeta_i^n(3,\pm)\|) \le K\sqrt{\Delta_n}$ and the result follows from $z_n\sqrt{\Delta_n} \to 0$. □

Proof of (13.3.37) for $j=4$ Observe that

$$\zeta^n(4,\pm)^{jm} = \sum_{u,v,w=1}^{d'} A(j,m,u,v,w)_n\,\Phi(j,m,u,v,w)_n,$$

where each $A(j,m,u,v,w)_n$ is bounded $\mathcal{F}_{S_n(\pm)}$ measurable and

$$\Phi(j,m,u,v,w)_n = \frac{1}{k_n\Delta_n}\sum_{i=1}^{k_n}\big(W^w_{(i_n+i-1)\Delta_n} - W^w_{S_n}\big)\big(\Delta^n_{i_n+i}W^u\,\Delta^n_{i_n+i}W^v - \Delta_n\delta^{uv}\big).$$

A straightforward computation shows that the first two $\mathcal{F}_{S_n}$-conditional moments of the variables $\Phi(j,m,u,v,w)_n$ are respectively 0 and smaller than $K\Delta_n$. Then $\frac{1}{\sqrt{\Delta_n}}\Phi(j,m,u,v,w)_n$ is bounded in probability as n varies, and the result follows, again because $z_n\sqrt{\Delta_n}\to 0$. □

Proof of (13.3.37) for $j=5$ As in the proof for $j=1$, we have $\mathbb{E}(\|\xi'^n_i(\pm)\|^2)\le Kk_n\Delta_n$. Since $\xi'^n_i(\pm)$ is $\mathcal{F}_{S_n+(i-1)\Delta_n}$ measurable, successive conditioning yields $\mathbb{E}(\|\eta^n_i(5,\pm)\|)\le Kk_n\Delta_n^2$, hence $\mathbb{E}(\|\zeta(5,\pm)\|)\le Kk_n\Delta_n$. The result follows because $z_nk_n\Delta_n\to 0$. □

Proof of (13.3.37) for $j=6$ Recalling (13.3.34) and ($\overline{\text{SK}}$), we have

$$\widetilde{\xi}^n_i(\pm) = \int_{S_n}^{S_n+(i-1)\Delta_n}\widetilde{b}'_s\,ds + \int_{S_n}^{S_n+(i-1)\Delta_n}(\widetilde{\sigma}_s-\widetilde{\sigma}_{S_n})\,dW_s + M_{S_n+(i-1)\Delta_n} - M_{S_n}, \tag{13.3.39}$$

where $M=\widetilde{\delta}*(\mathfrak{p}-\mathfrak{q})$ and $\widetilde{b}'_t=\widetilde{b}_t+\int_{\{\|\widetilde{\delta}(t,z)\|>1\}}\widetilde{\delta}(t,z)\,\lambda(dz)$ is bounded. Then if $\varepsilon\in(0,1]$, on the set $\Omega(n,i,\varepsilon)=\{\|\Delta M_s\|\le\varepsilon\ \forall s\in(S_n,S_n+(i-1)\Delta_n]\}$ we have

$$\widetilde{\xi}^n_i(\pm) = \int_{S_n}^{S_n+(i-1)\Delta_n}\widetilde{b}(\varepsilon)_s\,ds + \int_{S_n}^{S_n+(i-1)\Delta_n}(\widetilde{\sigma}_s-\widetilde{\sigma}_{S_n})\,dW_s + M(\varepsilon)_{S_n+(i-1)\Delta_n} - M(\varepsilon)_{S_n},$$

where $M(\varepsilon)=(\widetilde{\delta}\,1_{\{\Gamma\le\varepsilon\}})*(\mathfrak{p}-\mathfrak{q})$ and $\widetilde{b}(\varepsilon)_t=\widetilde{b}'_t-\int_{\{\Gamma(z)>\varepsilon\}}\widetilde{\delta}(t,z)\,\lambda(dz)$ (here Γ is the function appearing in ($\overline{\text{SK}}$)). We have $\int_{\{\Gamma(z)>\varepsilon\}}\|\widetilde{\delta}(t,z)\|^2\,\lambda(dz)\le\phi(\varepsilon):=\int_{\{\Gamma(z)>\varepsilon\}}\Gamma(z)^2\,\lambda(dz)$, which goes to 0 as $\varepsilon\to 0$, hence also $\|b(\varepsilon)_t\|\le K/\varepsilon$. Then, with the notation

$$\gamma_n = \frac{1}{k_n\Delta_n}\mathbb{E}\bigg(\int_{S_n}^{S_n+k_n\Delta_n}\|\widetilde{\sigma}_{S_n+s}-\widetilde{\sigma}_{S_n}\|^2\,ds\bigg),$$

we deduce that

$$\mathbb{E}\big(\big\|\widetilde{\xi}^n_i(\pm)\big\|^2\,1_{\Omega(n,i,\varepsilon)}\big)\le Kk_n\Delta_n\rho(n,\varepsilon),\quad \text{where } \rho(n,\varepsilon)=\frac{k_n\Delta_n}{\varepsilon}+\gamma_n+\phi(\varepsilon).$$

Observing that $\widetilde{\xi}_i^n(\pm)$ and $\Omega(n,i,\varepsilon)$ are $\mathcal{F}_{(i_n+i-1)\Delta_n}$ measurable, we deduce by successive conditioning and the above, plus the boundedness of σ and the Cauchy-Schwarz inequality, that

$$\mathbb{E}\big(\|\eta_i^n(6,\pm)\|\, 1_{\Omega(n,i,\varepsilon)}\big) \leq K\Delta_n\sqrt{k_n\Delta_n\rho(n,\varepsilon)}.$$

Since $\Omega(n,k_n,\varepsilon)\subset\Omega(n,i,\varepsilon)$ if $1\leq i\leq k_n$, the previous estimate and (13.3.36) yield

$$\mathbb{E}\big(\|\zeta_i^n(6,\pm)\|\, 1_{\Omega(n,k_n,\varepsilon)}\big) \leq K\sqrt{k_n\Delta_n\rho(n,\varepsilon)}. \tag{13.3.40}$$

Now, since $\widetilde{\sigma}$ is càdlàg and either $S_n+k_n\Delta_n<T$ and $s_n\to T$, or $S_n>T'$ and $S_n+k_n\Delta_n\to T'$, we see that $\gamma_n\to 0$. This and $\lim_{\varepsilon\to 0}\phi(\varepsilon)=0$ yield $\lim_{\varepsilon\to 0}\limsup_n\rho(n,\varepsilon)=0$. Moreover the sequence $z_n\sqrt{k_n\Delta_n}$ is bounded, so we deduce

$$\lim_{\varepsilon\to 0}\ \limsup_n\ z_n\,\mathbb{E}\big(\|\zeta^n(6,\pm)\|\, 1_{\Omega(n,k_n,\varepsilon)}\big) = 0.$$

Finally, $\mathbb{P}(\Omega(n,k_n,\varepsilon))\to 1$ as $n\to\infty$ for each $\varepsilon>0$ because the interval $(S_n,S_n+(k_n-1)\Delta_n]$ tends to the empty set. Then what precedes immediately yields (13.3.37) for $j=6$. □

Proof of (13.3.37) for $j=2$ Only $\zeta^n(2,-)$ needs to be considered. Using (K), we can apply Itô's formula to $c_t=\sigma_t\sigma_t^*$ to get, with $S_n'=S_n+(k_n-1)\Delta_n$:

$$\zeta^n(2,-) = \int_{S_n(-)}^{S_n'}\widehat{b}_s\,ds - \int_{S_n(-)}^{S_n'}(\widetilde{A}_s-\widetilde{A}_{S_n})\,dW_s + \overline{M}_{S_n'} - \overline{M}_{S_n(-)},$$

where $\|\widehat{b}_s\|\leq K$ and $\widetilde{A}$ is given by (13.3.9) and $\overline{M}=G*(\mathfrak{p}-\mathfrak{q})$ with

$$G^{jm}(t,z) = -\sum_{u=1}^{d'}\big(\widetilde{\delta}^{ju}(t,z)\widetilde{\delta}^{mu}(t,z)+\sigma_{t-}^{ju}\widetilde{\delta}^{mu}(t,z)+\sigma_{t-}^{mu}\widetilde{\delta}^{ju}(t,z)\big).$$

Observe that (13.3.37) is similar to (13.3.39), and $\widetilde{A}$ is càdlàg and $\|G(t,z)\|\leq K\Gamma(z)$. Then the same proof as in the previous step yields that $\zeta^n(2,-)$ satisfies (13.3.40) for all $\varepsilon\in(0,1]$, where now $\Omega(n,k_n,\varepsilon)=\{\|\Delta\overline{M}_s\|\leq\varepsilon\ \forall s\in(S_n,S_n+(k_n-1)\Delta_n\}$. Then we conclude (13.3.37) for $j=2$ in the same way as for $j=6$. □

Step 3) In this short step we extend the previous result to a slightly more general situation. Above we have $S_n(\pm)=i_n(\pm)\Delta_n$; however, we could replace $S_n(\pm)$ by a more general stopping time, which is not necessarily a multiple of Δ_n. Namely we could take any $S_n'(\pm)$ satisfying

$$\begin{array}{l} S_n'(\pm) \text{ is a stopping time and, according to the case,}\\ S_n'(-)+k_n\Delta_n<T\leq S_n(-)+(k_n+2)\Delta_n,\quad T'\leq S_n'(+)\leq T'+2\Delta_n. \end{array} \tag{13.3.41}$$

We thus replace $i_n(\pm)$ by $i'_n(\pm) = S'_n(\pm)/\Delta_n$, which is no longer an integer in general. However, if for any process Y we naturally define the increment $\Delta^n_{i'_n(\pm)+i} Y$ as being

$$\Delta^n_{i'_n(\pm)+i} Y = Y_{S'_n(\pm)+i\Delta_n} - Y_{S'_n(\pm)} \tag{13.3.42}$$

then we can define the processes $\zeta'^n(j,\pm)$ as $\zeta^n(j,\pm)$, with $S'_n(\pm)$ and $i'_n(\pm)$ instead of $S_n(\pm)$ and $i_n(\pm)$ everywhere. Moreover, the property (13.3.41) is the only property (in addition to $(\overline{\mathrm{SK}})$ of course) which is used in the above proof. Therefore we have

$$(13.3.41) \quad \Rightarrow \quad z_n \zeta'^n(j,\pm) \xrightarrow{\mathbb{P}} 0 \quad \text{for } j = 1,2,3,4,5,6. \tag{13.3.43}$$

Step 4) It remains to prove (13.3.37) for $\zeta^n(j,-)$ when $S_n(-)$ is *not* a stopping time, that is when $T = T_l = R(m,p)$ for some $p, m \geq 1$.

We use the same trick as in Lemma 13.3.10, slightly modified as follows: with the convention $R(m,0) = 0$ when $p = 1$, we transfer the initial time at $R = R(m,p-1)$, that is we consider the filtration $\mathcal{F}'_t = \mathcal{F}_{R+t}$ and the enlarged filtration $(\mathcal{G}'_t)$ which is the smallest one containing $(\mathcal{F}'_t)$ and such that all $(R(m,q) : q \geq 1)$ are $\mathcal{G}'_0$ measurable. Then we consider the process $W'_t = W_{R+t} - W_R$ and, recalling that $R(m,q)$ are the successive jump times of $1_G * \mathfrak{p}$ where $G = \{z : \Gamma'(z) > \frac{1}{m}\}$, the measure $\mathfrak{p}'$ defined by $\mathfrak{p}'((0,t] \times A) = \mathfrak{p}((R, R+t] \times (A \cap G^c))$. By Proposition 2.1.10, page 44, W' is a $(\mathcal{G}'_t)$-Wiener process and $\mathfrak{p}'$ is a $(\mathcal{G}'_t)$-Poisson random measure with compensator $\mathfrak{q}'(dt,dz) = dt \otimes (1_{G^c}(z)\lambda(dz))$. Then, with the notation

$$\widetilde{\delta}(t,z) = \widetilde{\delta}(R+t,z), \quad \widetilde{\sigma}'_t = \widehat{\sigma}_{R+t}, \quad \widetilde{b}'_t = \widetilde{b}_{R+t} - \int_{\{\|\widetilde{\delta}'(t,z)\| \leq 1\} \cap G} \widetilde{\delta}'(t,z)\,\lambda(dz)$$

(note that $\widetilde{\delta}'$ is a $(\mathcal{G}'_t)$-predictable function), we can define the processes (X',σ') by

$$\sigma'_t = \sigma_R + \int_0^t \widetilde{b}_{R+s}ds + \int_0^t \widetilde{\sigma}'_s dW'_s + \big(\widetilde{\delta}' 1_{\{\|\widetilde{\delta}'\| \leq 1\}}\big) \star (\mathfrak{p}' - \mathfrak{q}')_t + \big(\widetilde{\delta}' 1_{\{\|\widetilde{\delta}'\| > 1\}}\big) \star \mathfrak{p}'_t$$

$$X'_t = X_R + \int_0^t b_{R+s}ds + \int_0^t \sigma'_s dW'_s.$$

The process X' satisfies $(\overline{\mathrm{SK}})$, because X does. Moreover if $i'_n(-) = (i_n(-) - R/\Delta_n)^+$, then $S'_n(-) = i'_n(-)\Delta_n$ is a $(\mathcal{G}'_t)$-stopping time because R and $T = R(m,p)$ are $\mathcal{G}'_0$ measurable,. In other words, we are in the situation of (13.3.41) with T replaced by $R(m,p) - R$, at least as soon as n is large enough for having $i'_n(-) > 0$ Then if we use the notation (13.3.42) and associate $\zeta'^n(j,-)$ with $(i'_n(-), S'_n(-), X', W', \sigma', \widetilde{\sigma}')$ instead of $(i_n(-), S_n(-), X, W, \sigma, \widetilde{\sigma})$ we thus obtain (13.3.43) by virtue of Step 3.

Now, the definition of σ' implies $\sigma'_t = \sigma_t - \sum_{q \geq p} \Delta\sigma_{R(m,q)} 1_{\{R(m,q) - R \leq t\}}$. Therefore $\sigma'_t = \sigma_{R+t}$, hence also $X'_t = X_{R+t}$, for all $t < R(m,p) - R = T_l - R$. Using (13.3.42), we deduce that $\Delta^n_{i'_n(-)+i} X' = \Delta^n_{i_n(-)+i} X$ for all $i = 1, \dots, k_n$, when

$i'_n(-) > 0$, and the same for W' and W. It follows that for all n large enough we have $\zeta'^n(j,-) = \zeta^n(j,-)$, and we thus deduce (13.3.37) from (13.3.43).

This ends the proof of Lemma 13.3.13, hence of Theorem 13.3.3 as well.

13.3.6 Proof of Theorem 13.3.8

Now we consider Theorem 13.3.8. First, Lemma 13.3.9 holds for this theorem as well as for Theorem 13.3.3, with exactly the same proof, whereas the claims of Lemma 13.3.10 are clearly true if we replace k_n by mk_n. Hence Lemma 13.3.11 also holds in the setting of Theorem 13.3.8.

Next, we define $\overline{Y}^n(\pm,l)$ and $\overline{Y}'^n(\pm,l)$ as in the previous subsection, except that k_n is substituted with mk_n. This allows us to define $\overline{Z}^n(\pm,l)$ and $\overline{Z}'^n(\pm,l)$ by (13.3.27) with $\overline{Y}^n(\pm,l)$ and $\overline{Y}'^n(\pm,l)$ instead of $Y^n(\pm,l)$ and $Y'^n(\pm,l)$, and $\overline{Z}''^n(\pm,l) = \overline{Z}_l^{n\pm} - \overline{Z}^n(\pm,l) - \overline{Z}'^n(\pm,l)$, as in (13.3.28). Upon replacing k_n by mk_n all the way through, the proof of Lemma 13.3.13 gives

$$z_n\,\overline{Z}''^n(-,l) \stackrel{\mathbb{P}}{\longrightarrow} 0, \qquad z_n\,\overline{Z}''^n(+,l) \stackrel{\mathbb{P}}{\longrightarrow} 0.$$

At this stage, it thus remains to prove that

$$\begin{aligned}
&\Big(\sqrt{k_n}\,Z^n(-,l), \sqrt{k_n}\,Z^n(+,l), \sqrt{k_n}\,\overline{Z}^n(-,l), \sqrt{k_n}\,\overline{Z}^n(+,l),\\
&\qquad \frac{1}{\sqrt{k_n\Delta_n}}\,Z'^n(-,l), \frac{1}{\sqrt{k_n\Delta_n}}\,Z'^n(+,l), \frac{1}{\sqrt{k_n\Delta_n}}\,\overline{Z}'^n(-,l), \frac{1}{\sqrt{k_n\Delta_n}}\,\overline{Z}'^n(+,l)\Big)_{l\in L}\\
&\quad \stackrel{\mathcal{L}\text{-s}}{\longrightarrow} \big(Z(-,l), Z(+,l), \overline{Z}(-,l), \overline{Z}(+,l),\\
&\qquad\quad Z'(-,l), Z'(+,l), \overline{Z}'(-,l), \overline{Z}'(+,l)\big)_{l\in L}.
\end{aligned} \tag{13.3.44}$$

To this aim, we complement the notation Y_n and Y'_n of Lemma 13.3.14 by setting

$$\overline{Y}_n^{uv} = \frac{1}{m\sqrt{k_n}}\sum_{i=1}^{mk_n}\big(U_i^u U_i^v - \delta^{uv}\big), \qquad \overline{Y}_n'^u = \frac{1}{mk_n^{3/2}}\sum_{i=1}^{mk_n}(mk_n - i)U_i^u.$$

Then, if $(\widehat{Y}, \widehat{Y}')$ is another pair of variables independent from and with the same law as (Y, Y'), we obtain the following extension of Lemma 13.3.14:

Lemma 13.3.16 *The variables $(Y_n, \overline{Y}_n, Y'_n, \overline{Y}'_n)$ converge in law to*

$$\Big(Y, \frac{1}{m}\big(Y + \sqrt{m-1}\,\widehat{Y}\big), Y', \frac{1}{2m}\big((3m-1)Y' + (m-1)\sqrt{4m-1}\,\widehat{Y}'\big)\Big).$$

Proof We can write $(Y_n, \overline{Y}_n, Y'_n, \overline{Y}'_n)$ as is $\sum_{i=1}^{mk_n}(\zeta_i^n, \overline{\zeta}_i^n, \zeta_i'^n, \overline{\zeta}_i'^n)$, where ζ_i^n and $\zeta_i'^n$ are given by (13.3.30) when $i \le k_n$ and vanish otherwise, and

$$\overline{\zeta}_i^{n,uv} = \frac{1}{m\sqrt{k_n}}\big(U_i^u U_i^v - \delta^{uv}\big), \qquad \overline{\zeta}_i'^{n,u} = \frac{1}{mk_n^{3/2}}(mk_n - i)U_i^u.$$

All those variables are centered and independent when i varies, for each n. We still have (13.3.31), and also

$$\begin{aligned}
&\sum_{i=1}^{mk_n} \mathbb{E}'\big(\overline{\zeta}_i^{n,uv}\,\overline{\zeta}_i^{n,ws}\big) = \frac{1}{m}\big(\delta^{uw}\delta^{vs} + \delta^{us}\delta^{vw}\big)\\
&\sum_{i=1}^{mk_n} \mathbb{E}'\big(\zeta_i^{n,uv}\,\overline{\zeta}_i^{n,ws}\big) = \frac{1}{m}\big(\delta^{uw}\delta^{vs} + \delta^{us}\delta^{vw}\big)\\
&\sum_{i=1}^{mk_n} \mathbb{E}'\big(\overline{\zeta}_i^{n,uv}\,\overline{\zeta}_i'^{n,w}\big) = \sum_{i=1}^{mk_n} \mathbb{E}'\big(\overline{\zeta}_i^{n,uv}\,\zeta_i'^{n,w}\big) = \sum_{i=1}^{mk_n} \mathbb{E}'\big(\zeta_i^{n,uv}\,\overline{\zeta}_i'^{n,w}\big) = 0\\
&\sum_{i=1}^{mk_n} \mathbb{E}'\big(\overline{\zeta}_i'^{n,u}\,\overline{\zeta}_i'^{n,v}\big) = \sum_{i=1}^{mk_n} \frac{(mk_n - i)^2}{m^2k_n^3}\,\delta^{uv} \;\to\; \frac{m}{3}\,\delta^{uv}\\
&\sum_{i=1}^{mk_n} \mathbb{E}'\big(\zeta_i'^{n,u}\,\overline{\zeta}_i'^{n,v}\big) = \sum_{i=1}^{k_n} \frac{(mk_n - i)(k_n - i)}{mk_n^3}\,\delta^{uv} \;\to\; \frac{3m-1}{6m}\,\delta^{uv}\\
&\sum_{i=1}^{mk_n} \mathbb{E}'\big(\|\overline{\zeta}_i^n\|^4\big) + \sum_{i=1}^{mk_n} \mathbb{E}'\big(\|\overline{\zeta}_i'^n\|^4\big) \le \frac{K}{k_n} \to 0.
\end{aligned}$$

Then again the result follows from Theorem 2.2.14, plus a simple calculation of the variance-covariance. □

Finally, we can reproduce the proof of Lemma 13.3.15, to deduce (13.3.44) from the previous lemma, and the proof of Theorem 13.3.8 is complete.

13.4 Another Central Limit Theorem Using Approximations of the Spot Volatility

In this section we consider another Central Limit Theorem related to the local approximations of the volatility, and which about the processes

$$V^n(G; k_n, v_n, X)_t = \sum_{i=k_n+1}^{[t/\Delta_n]-k_n} G\big(\Delta_i^n X, \widehat{c}_{i-k_n}^{\,n}(k_n, v_n), \widehat{c}_{i+1}^{\,n}(k_n, v_n)\big)\, 1_{\{\|\Delta_i^n X\| > v_n\}}, \tag{13.4.1}$$

where G is a function on $\mathbb{R}^d \times \mathcal{M}^+_{d\times d} \times \mathcal{M}^+_{d\times d}$. This was introduced in (9.5.1), and under appropriate conditions it converges to the following limit;

$$V(G,X)_t = \sum_{s\le t} G(\Delta X_s, c_{s-}, c_s). \tag{13.4.2}$$

Here we have two approximations; one is for the jumps of X, and the typical rate of convergence is $\sqrt{\Delta_n}$; the other is for the spot volatility, with a typical rate $1/\sqrt{k_n}$. Since we must have $k_n\Delta_n \to 0$, the rate for approximating the spot volatility is always slower than the rate for approximating the jumps. Therefore the rate at which $V^n(G;k_n,v_n,X)$ converges to $V(G)$ will be the slowest of the two, that is $1/\sqrt{k_n}$.

For practical purposes we need to consider two functionals $V^n(G;k_n,v_n,X)$ and $V^n(G;mk_n,v_n,X)$ simultaneously, where $m\ge 2$ is a fixed integer. On the other hand, for simplicity we consider only the one-dimensional case $d=1$, so we may also take $d'=1$ here, and the test function F is also one-dimensional.

13.4.1 Statements of Results

We will have two different CLTs here, depending on the properties of the test function G. In both cases, for the applications we need to somehow relax the continuity assumption on G, in the spirit of Remark 9.5.3. Toward this aim, we consider a subset A of $\mathbb{R}$ (recall that X is one-dimensional here) which satisfies

$$\begin{array}{l} A \text{ is open, with a finite complement, and} \\ D := \{x : \mathbb{P}(\exists s>0 : \Delta X_s = x) > 0\} \subset A. \end{array} \tag{13.4.3}$$

Then if G satisfies the hypotheses of Theorem 9.5.1 except that the continuity holds only on $A\times\mathbb{R}_+\times\mathbb{R}_+$, then we still have $V^n(G;k_n,v_n,X) \stackrel{\mathbb{P}}{\Longrightarrow} V(G)$ (since $d=1$, we can identify $\mathcal{M}^+_{d\times d}$ with $\mathbb{R}_+$ here). Note that $0\in A$ necessarily here.

For the first result, we make the following assumptions on G, which is a function of three variables (x,y,y'); when it is differentiable we write ∂G_i for $\partial_x G(x,y,y')$ and $\partial_y G(x,y,y')$ and $\partial_{y'} G(x,y,y')$, when $i=1,2,3$ respectively, and similarly for higher order derivatives when they exist.

$$\begin{array}{l} \bullet\ A \text{ satisfies (13.4.3) and } \varepsilon>0 \text{ satisfies } [-\varepsilon,\varepsilon]\subset A \\ \bullet\ G \text{ is } C^1 \text{ on } A\times\mathbb{R}^2_+, \quad G(0,y,y')=0 \\ \bullet\ \partial_1 G(x,y,y')/x^2 \text{ is locally bounded on } A\times\mathbb{R}^2_+ \\ \bullet\ |\partial_j G(x,y,y')| \le Kx^2 \text{ on } [-\varepsilon,\varepsilon]\times\mathbb{R}^2_+ \text{ for } j=2,3. \end{array} \tag{13.4.4}$$

The third condition above implies $\partial_j G(0,y,y')=0$ for $j=2,3$.

In order to describe the limiting process we take a weakly exhausting sequence (T_n) of stopping times for the jumps of X, and a very good filtered extension $(\widetilde{\Omega},\widetilde{\mathcal{F}},(\widetilde{\mathcal{F}}_t)_{t\ge0},\widetilde{\mathbb{P}})$ of $(\Omega,\mathcal{F},(\mathcal{F}_t)_{t\ge0},\mathbb{P})$ of the same type as the one in Sect. 13.1, except that it supports four sequence $(\Psi_{n-}, \Psi_{n+}, \Psi'_{n-}, \Psi'_{n+})$ of independent standard normal variables, independent of $\mathcal{F}$. The limiting processes will be combinations of the following two processes

$$\begin{aligned}\mathcal{U}_t &= \sqrt{2}\sum_{q\geq 1}\big(\partial_2 G(\Delta X_{T_q}, c_{T_q-}, c_{T_q})c_{T_q-}\Psi_{q-} \\ &\quad + \partial_3 G(\Delta X_{T_q}, c_{T_q-}, c_{T_q})c_{T_q}\Psi_{q-}\big)\,1_{\{T_q\leq t\}} \\ \mathcal{U}'_t &= \sqrt{2}\sum_{q\geq 1}\big(\partial_2 G(\Delta X_{T_q}, c_{T_q-}, c_{T_q})c_{T_q-}\Psi'_{q-} \\ &\quad + \partial_3 G(\Delta X_{T_q}, c_{T_q-}, c_{T_q})c_{T_p}\Psi_{q-}\big)\,1_{\{T_q\leq t\}}.\end{aligned} \tag{13.4.5}$$

These processes are well defined, and $\mathcal{F}$-conditionally they are purely discontinuous Gaussian martingales, as soon as X satisfies (H-r) and (13.4.4) holds with the same r, by an application of Proposition 4.1.4, and further they are $\mathcal{F}$-conditionally independent, with the following (conditional) variances:

$$\begin{aligned}\widetilde{\mathbb{E}}\big((\mathcal{U}_t)^2\mid\mathcal{F}\big) &= \widetilde{\mathbb{E}}\big((\mathcal{U}'_t)^2\mid\mathcal{F}\big) \\ &= 2\sum_{s\leq t}\big(c_{s-}^2\,\partial_2 G(\Delta X_s, c_{s-}, c_s)^2 + c_s^2\,\partial_3 G(\Delta X_s, c_{s-}, c_s)^2\big).\end{aligned} \tag{13.4.6}$$

Our first result goes as follows.

Theorem 13.4.1 *Assume* ($\overline{\textbf{K}\text{-}r}$) *for some* $r\in[0,2)$, *and let* G *satisfy* (13.4.4) *for some* $\varepsilon\geq 0$. *We take* $v_n \asymp \Delta_n^{\varpi}$ *for some* $\varpi\in(0,\frac{1}{2})$ *and* k_n *such that* $k_n\Delta_n^{\tau}\to\beta'$ *for some* $\beta'\in(0,\infty)$ *and some* $\tau\in(0,\frac{1}{2})$. *We also assume one of the following two conditions*:

(a) *either* $G(x,y,y')=0$ *when* $|x|\leq\eta$, *for some* $\eta>0$,

(c) *or* (*in addition to* $\varpi,\tau\in(0,\frac{1}{2})$)

$$r<\frac{4}{3},\qquad \frac{1}{4-r}\leq\varpi<\frac{1}{2r},\qquad \tau<2\varpi(2-r). \tag{13.4.7}$$

Then, for any t *and any integer* $m\geq 2$ *we have the following stable convergence in law of* 2-*dimensional variables*:

$$\begin{aligned}&\Big(\sqrt{k_n}\,\big(V^n(G;k_n,v_n,X)_t - V(G,X)_t\big),\ \sqrt{k_n}\,\big(V^n(G;mk_n,v_n,X)_t - V(G,X)_t\big)\Big) \\ &\quad\stackrel{\mathcal{L}\text{-}s}{\longrightarrow}\ \Big(\mathcal{U}_t,\ \frac{1}{m}\big(\mathcal{U}_t+\sqrt{m-1}\,\mathcal{U}'_t\big)\Big).\end{aligned} \tag{13.4.8}$$

Remark 13.4.2 In this situation we typically do not have functional convergence for the Skorokhod topology, even when the process $V(G,X)$ is replaced by its discretized version. This is because, if X has a jump at time T, then $V^n(G;k_n,u_n,X)$

and $V^n(G;mk_n,u_n,X)$ have a jump of approximately the same size, at times $([T/\Delta_n]-k_n)\Delta_n$ and $([T/\Delta_n]-mk_n)\Delta_n$. However, we would obtain the functional convergence, were we taking $V^n(G;k_n,v_n,X)_t - V(G,X)_{([t/\Delta_n]-k_n)\Delta_n}$ for the first component, $V^n(G;k_n,v_n,X)_{t+(m-1)k_n\Delta_n} - V(G,X)_{([t/\Delta_n]-k_n)\Delta_n}$ for the second one.

An analogous joint convergence holds, with any number of integers m. There are also (much more complicated) versions in the multidimensional case $d\geq 2$, or when $\tau\geq\frac{1}{2}$, or when G is multidimensional.

The second result is relative to the case where, in the previous theorem, the limit vanishes identically (this is a "degenerate case", in the spirit of the forthcoming Chap. 15, but it is more convenient to study this case here). We then need a different normalization, and also stronger assumptions on the test function G. Namely, on top of (13.4.4), we also assume:

$$\left.\begin{array}{l}\bullet\ G(x,y,y') \text{ is } C^2 \text{ in } (y,y') \text{ on } A\times\mathbb{R}_+^2\\ \bullet\ \left|\partial^2_{j,i}G(x,y,y')\right|\leq K|x|^2 \text{ on } [-\varepsilon,\varepsilon]\times\mathbb{R}_+^2 \text{ for } j,i=2,3\\ \bullet\ x\in A,\ y>0\ \Rightarrow\ G(x,y,y)=\partial_2 F(x,y,y)=\partial_3 G(x,y,y)=0.\end{array}\right. \tag{13.4.9}$$

The limiting processes will now be combinations of the following processes:

$$\begin{aligned}\overline{\mathcal{U}}_t &= \sum_{q\geq 1} c^2_{T_q}\big(\partial^2_{22}G(\Delta X_{T_q},c_{T_q},c_{T_q})\Psi^2_{q-}+2\partial^2_{23}G(\Delta X_{T_q},c_{T_q},c_{T_q})\Psi_{q-}\Psi_{q+}\\ &\quad +\partial^2_{33}G(\Delta X_{T_q},c_{T_q},c_{T_q})\Psi^2_{q+}\big)\,1_{\{T_q\leq t\}}\\ \overline{\mathcal{U}}'_t &= \frac{1}{m^2}\sum_{q\geq 1} c^2_{T_q}\big(\partial^2_{22}G(\Delta X_{T_q},c_{T_q},c_{T_q})(\Psi_{q-}+\sqrt{m-1}\,\Psi'_{q-})^2\\ &\quad +2\partial^2_{23}G(\Delta X_{T_q},c_{T_q},c_{T_q})\big(\Psi_{q-}+\sqrt{m-1}\,\Psi'_{q-}\big)\big(\Psi_{q+}+\sqrt{m-1}\,\Psi'_{q+}\big)\\ &\quad +\partial^2_{33}G(\Delta X_{T_q},c_{T_q},c_{T_q})\big(\Psi_{q+}+\sqrt{m-1}\,\Psi'_{q+}\big)^2\big)\,1_{\{T_q\leq t\}}.\end{aligned} \tag{13.4.10}$$

These are again well defined, by Proposition 4.1.3 this time. Moreover, although no longer $\mathcal{F}$-conditional martingale, they have $\mathcal{F}$-conditionally independent increments, with finite variation, and their (conditional) means are

$$\left.\begin{array}{l}\widetilde{\mathbb{E}}(\overline{\mathcal{U}}_t\mid\mathcal{F})=\overline{B}_t,\qquad \widetilde{\mathbb{E}}(\overline{\mathcal{U}}'_t\mid\mathcal{F})=\frac{1}{m}\,\overline{B}_t,\quad\text{where}\\ \overline{B}_t=\sum_{s\leq t}c^2_s\big(\partial^2_{22}G(\Delta X_s,c_s,c_s)+\partial^2_{33}(\Delta X_s,c_s,c_s)\big).\end{array}\right. \tag{13.4.11}$$

Theorem 13.4.3 *Assume* ($\mathbf{K}$-r) *for some* $r\in[0,2)$, *and let* G *satisfy* (13.4.4) *and* (13.4.9) *for some* $\varepsilon\geq 0$. *We take* $v_n\asymp\Delta_n^{\varpi}$ *for some* $\varpi\in(0,\frac{1}{2})$ *and* k_n *such that* $k_n\Delta_n^{\tau}\to\beta'$ *for some* $\beta'\in(0,\infty)$ *and some* $\tau\in(0,\frac{1}{2})$. *We also assume the following condition on* X, *with* A *as in* (13.4.4):

$$\forall t>0,\ \ \Delta X_t\in A\backslash\{0\}\ \Rightarrow\ \Delta c_t=0, \tag{13.4.12}$$

and also one of the following two conditions:

(a) *either* $G(x,y,y')=0$ *when* $|x|\le\eta$, *for some* $\eta>0$,
(c) *or (in addition to* $\varpi,\tau\in(0,\frac{1}{2})$)

$$r<\frac{4}{3},\quad \frac{1}{4-r}\le\varpi<\frac{1}{2r},\quad \tau<\big(2\varpi(2-r)\big)\wedge\big((3-r)\varpi\big). \tag{13.4.13}$$

Then, for any t and any integer $m\ge 2$ *we have the following stable convergence in law of* 2*-dimensional variables*:

$$\big(k_nV^n(G;k_n,v_n,X)_t,\,k_nV^n(G;mk_n,v_n,X)_t\big)\stackrel{\mathcal{L}\text{-s}}{\longrightarrow}\big(\overline{\mathcal{U}}_t,\overline{\mathcal{U}}'_t\big). \tag{13.4.14}$$

For the sake of comparison with the previous theorem, it is worth mentioning that under (13.4.12) we have $V(G,X)=0$ identically, so (13.4.14) does not contradicts (13.4.8).

13.4.2 Proofs

The proofs of both theorems are conducted together and necessitate quite a few steps.

Step 1. This step is devoted to some preliminaries. First, by localization, we can and will assume the strengthened assumptions according to which X satisfies (SH-r), and σ satisfies (SH).

Next, we use the notation (4.3.1): the function Γ is as in (SH-r), and $A_l=\{z:\Gamma(z)>1/l\}$, and $(S(l,q):q\ge 1)$ is the sequence of successive jump times of the process $1_{A_l\setminus A_{l-1}}*\mathfrak{p}$, and $(S_q)_{q\ge1}$ is a reordering of the double sequence $(S(l,j):l,j\ge1)$, and $i(n,q)$ is the (random) integer such that $S_q\in I(n,i(n,q))$. We let $\mathcal{P}_l$ be the set of all q such that $S_q=S(l',j)$ for some $j\ge1$ and some $l'\le l$.

Similar with (5.1.10), we also write

$$b(l)_t=b_t-\int_{A_l\cap\{z:\|\delta(t,z)\|\le1\}}\delta(t,z)\,\lambda(dz)$$

$$\begin{aligned}
X(l)_t &= X_0+\int_0^t b(l)_s\,ds+\int_0^t\sigma_s\,dW_s+(\delta\,1_{A_l^c})\star(\mathfrak{p}-\mathfrak{q})_t\\
X'(l) &= X-X(l) = (\delta\,1_{A_l})\star\mathfrak{p}\\
\overline{X}(l)_t &= X_0+\int_0^t b(l)_s\,ds+\int_0^t\sigma_s\,dW_s.
\end{aligned}$$

$\Omega_n(t,l)=$ the set on which each interval $[0,t]\cap I(n,i)$
contains at most one jump of $X'(l)$, and that
$|X(l)_{t+s}-X(l)_t|\le 2/l$ for all $t\in[0,t]$, $s\in[0,\Delta_n]$

and we complement this by setting (the fixed integer m is as in our theorems):

$$\Omega'_n(t,l) = \Omega_n(t,l) \cap \big(\cap_{q \in \mathcal{P}_l} \{S_q < t - mk_n\Delta_n \text{ or } S_q > t\}\big). \tag{13.4.15}$$

We also denote by $(\mathcal{G}_t^{(l)})$ the smallest filtration containing $(\mathcal{F}_t)$ and such that the restriction of $\mathfrak{p}$ to $A_l \times \mathbb{R}_+$ is $\mathcal{G}_0^{(l)}$-measurable. We recall that, by our standard estimates, $\mathbb{E}(|\Delta_i^n X(l)|^2 \mid \mathcal{G}_{(i-1)\Delta_n}^{(l)}) \le K_l \Delta_n$ for any i, including random indices, provided they are $\mathcal{G}_0^{(l)}$-measurable. In particular, this yields

$$q \in \mathcal{P}_l \;\Rightarrow\; \Delta_{i(n,q)}^n X(l) \;=\; \mathrm{O_P}(\sqrt{\Delta_n}). \tag{13.4.16}$$

We can apply Theorem 13.3.8 with the sequence of stopping times $T_q = T'_q = S_q$. Since $\tau < \frac{1}{2}$ we are in the case $\beta = 0$, and $d = d' = 1$ yields that the process A_t of (13.3.9) is $A_t = c_t$. We then obtain the following stable convergence in law (in principle for any finite sub-family, but since this convergence is for the product topology it also holds for the whole sequence):

$$\begin{aligned}
&\sqrt{k_n}\big(\widehat{c}^n_{i(n,q)-k_n}(k_n,v_n) - c_{S_q-}, \widehat{c}^n_{i(n,q)+1}(k_n,v_n) - c_{S_q},\\
&\quad \widehat{c}^n_{i(n,q)-mk_n}(mk_n,v_n) - c_{S_q-}, \widehat{c}^n_{i(n,q)+1}(mk_n,v_n) - c_{S_q}\big)_{q\ge 1}\\
&\quad \stackrel{\mathcal{L}\text{-s}}{\longrightarrow} \sqrt{2}\big(c_{T_p-}\Psi_{p-}, c_{T_p}\Psi_{p+}, \tfrac{1}{m}\, c_{T_p-}\big(\Psi_{p-} + \sqrt{m-1}\,\Psi'_{q-}\big),\\
&\quad \tfrac{1}{m}\, c_{T_p}\big(\Psi_{p+} + \sqrt{m-1}\,\Psi'_{q+}\big)\big)_{q\ge 1}
\end{aligned} \tag{13.4.17}$$

Step 2. This step is devoted to proving an auxiliary result which is somehow similar to Lemma 13.3.10, with a precise estimate.

Lemma 13.4.4 *Under* (SH-r) *for X and* (SH) *for σ, and as soon as*

$$\varpi \ge \frac{1}{4-r}, \qquad \tau \le 2(2-r)\varpi, \tag{13.4.18}$$

we have for $j = 1, \dots, k_n + 1$:

$$\mathbb{E}\big(\big|\widehat{c}_i^n(k_n,v_n) - c_{(i-j)\Delta_n}\big|^2 \mid \mathcal{F}_{(i-1)\Delta_n}\big) \le \frac{K}{k_n}. \tag{13.4.19}$$

Proof We will write below $\widehat{c}_i^n(k_n,v_n;X)$ to emphasize the dependency on the process X, and recall that $\widehat{c}_i^n(k_n;X))$ is the estimator with no truncation (or, $v_n = \infty$). We use the decomposition (13.2.17) for X, and we recall the following estimates, following from Corollary 2.1.9 (as (13.2.23)), and for $q \ge 2$ and any $j = 1, \dots, k_n + 1$ and $v \in (0, \frac{1}{2})$:

$$\begin{aligned}
&\mathbb{E}\big(\big|\Delta_i^n X\big|^q \mid \mathcal{F}_{(i-1)\Delta_n}\big) \le K_q\Delta_n, \quad \mathbb{E}\big(\big|\Delta_i^n X'\big|^q \mid \mathcal{F}_{(i-1)\Delta_n}\big) \le K_q\Delta_n^{q/2}\\
&\mathbb{E}\big(\big|\Delta_i^n X' - \sigma_{(i-j}\Delta_i^n W\big|^q \mid \mathcal{F}_{(i-1)\Delta_n}\big) \le K_q\Delta_n^{q/2+1}\, k_n
\end{aligned} \tag{13.4.20}$$

$$\mathbb{E}\left(\left|\frac{\Delta_i^n X''}{\Delta_n^v}\bigwedge 1\right|^q \mid \mathcal{F}_{(i-1)\Delta_n}\right) \leq K_q \Delta_n^{(1-rv)}.$$

First, $\widehat{c}_i^n(k_n; W) = \frac{1}{k_n \Delta_n} \sum_{i=0}^{k_n-1} (\Delta_i^n W)^2$, so a simple calculation shows $\mathbb{E}((\widehat{c}_i^n(k_n; W) - 1)^2 \mid \mathcal{F}_{(i-1)\Delta_n}) = 2/k_n$. Next, the second part of (13.4.20) and Hólder's inequality yield, for any $\theta > 0$:

$$\mathbb{E}\left(\left(\widehat{c}_i^n(k_n; X') - c_{(i-j)\Delta_n} \widehat{c}_i^n(k_n; W)\right)^2 \mid \mathcal{F}_{(i-1)\Delta_n}\right) \leq K_\theta \Delta_n^{1-\theta}.$$

Third, for all $x, y \in \mathbb{R}$, $v > 0$, $q, w > 0$ we have

$$\left||x+y|^2 1_{\{|x+y|\leq v\}} - x^2\right|^q \leq K\left(\left(|y| \wedge v\right)^{2q} + |x|^q \left(|y| \wedge v\right)^q + \frac{|x|^{q(2+w)}}{v^{qw}}\right).$$

We apply this with $x = \Delta_i^n X'$ and $y = \Delta_i^n X''$ and $v = v_n$ and w such that $w(1-2\varpi) \geq 2$: using Hölder's inequality and (13.4.20) again, we deduce

$$\mathbb{E}\left(\left|\left(\Delta_i^n X\right)^2 1_{\{|\Delta_i^n X|\leq u_n\}} - \left(\Delta_i^n X'\right)^2\right|^q \mid \mathcal{F}_{(i-1)\Delta_n}\right) \leq K_q \Delta_n^{1+(2q-r)\varpi},$$

if $q = 1, 2$. By successive conditioning, we deduce

$$\mathbb{E}\left(\left(\widehat{c}_i^n(k_n, v_n; X) - \widehat{c}_i^n(k_n; X')\right)^2 \mid \mathcal{F}_{(i-1)\Delta_n}\right) \leq K\left(\Delta_n^{2(2-r)\varpi} + \frac{\Delta_n^{(4-r)\varpi-1}}{k_n}\right).$$

Putting all these estimates together, and taking advantage of (13.4.18) for the choice of θ and of the boundedness of c, we obtain (13.4.19). □

Step 3. In this step we fix l. We prove the result for the process $X'(l)$, whereas *we still keep the spot volatility estimators to be those constructed on the basis of* X. That is, in this step $V^n(G; k_n, v_n, X'(l))$ denotes the process defined by (13.4.1) with $\widehat{c}_i^n(k_n, v_n)$ as above, but with $\Delta_i^n X$ substituted with $\Delta_i^n X'(l)$. The limiting processes, when associated with $X'(l)$, are denoted as $\mathcal{U}(l)$, $\mathcal{U}'(l)$, $\overline{\mathcal{U}}(l)$ and $\overline{\mathcal{U}}'(l)$: they are given by (13.4.5) and (13.4.10), except that the sum on q is taken for $q \in \mathcal{P}_l$ instead of $q \geq 1$.

Observing that $\Delta_i^n X'(l) \neq 0$ only when $i = i(n, q)$ for some $q \in \mathcal{P}_l$, we see that on the set $\Omega_n'(t, l)$, and for $w = 1$ or $w = m$, we have

$$\begin{aligned} V^n\left(G; wk_n, v_n, X'(l)\right)_t - V\left(G, X'(l)\right)_t &= \sum_{q \in \mathcal{P}_l} \zeta(l, w)_q^n 1_{\{S_q \leq t\}}, \\ \zeta(l, w)_q^n &= \left(G\left(\Delta X_{S_q}, \widehat{c}_{i(n,q)-k_n}^n(wk_n, v_n), \widehat{c}_{i(n,q)+1}^n(wk_n, v_n)\right)\right. \\ &\quad \left. - G(\Delta X_{S_q}, c_{S_q-}, c_{S_q})\right) 1_{\{|\Delta X_{S_q}| > v_n\}} - G(\Delta X_{S_q}, c_{S_q-}, c_{S_q}) 1_{\{|\Delta X_{S_q}| \leq v_n\}}. \end{aligned} \tag{13.4.21}$$

With the same notation as in (13.4.5) or (13.4.10), we also write

$$\zeta_q = \sqrt{2}\big(\partial_2 G(\Delta X_{S_q}, c_{S_q-}, c_{S_q}) c_{S_q-} \Psi_{q-} + \partial_3 G(\Delta X_{T_q}, c_{S_q-}, c_{S_q}) c_{S_q} \Psi_{q-}\big)$$

$$\zeta'_q = \sqrt{2}\big(\partial_2 G(\Delta X_{S_q}, c_{S_q-}, c_{S_q}) c_{S_q-} \Psi'_{q-} + \partial_3 G(\Delta X_{S_q}, c_{S_q-}, c_{S_q}) c_{S_q} \Psi_{q-}\big)$$

$$\begin{aligned}\overline{\zeta}_q = c_{S_q}^2 \big(&\partial^2_{22} G(\Delta X_{S_q}, c_{S_q}, c_{S_q}) \Psi^2_{q-} + 2\partial^2_{23} G(\Delta X_{S_q}, c_{S_q}, c_{S_q}) \Psi_{q-}\Psi_{q+}\big) \\ &+ \partial^2_{33} G(\Delta X_{S_q}, c_{S_q}, c_{S_q}) \Psi^2_{q+} \end{aligned} \tag{13.4.22}$$

$$\begin{aligned}\overline{\zeta}'_q = \frac{1}{m^2}\big(&\partial^2_{22} G(\Delta X_{S_q}, c_{S_q}, c_{S_q}) \big(\Psi_{q-} + \sqrt{m-1}\,\Psi'_{q-}\big)^2 \\ &+ 2\partial^2_{23} G(\Delta X_{S_q}, c_{S_q}, c_{S_q}) \big(\Psi_{q-} + \sqrt{m-1}\,\Psi'_{q-}\big)\big(\Psi_{q+} + \sqrt{m-1}\,\Psi'_{q+}\big) \\ &+ \partial^2_{33} G(\Delta X_{S_q}, c_{S_q}, c_{S_q}) \big(\Psi_{q+} + \sqrt{m-1}\,\Psi'_{q+}\big)^2\big).\end{aligned}$$

We have (13.4.16) and $\mathbb{P}(\Omega'_n(t,l)) \to 1$ as $n \to \infty$, hence for proving Theorem 13.4.1 it is enough to show the following stable convergence

$$\big(\sqrt{k_n}\,\zeta(l,1)^n_q, \sqrt{k_n}\,\zeta(l,m)^n_q\big)_{q \in \mathcal{P}_l} \xrightarrow{\mathcal{L}\text{-s}} \left(\zeta_q, \frac{1}{m}\big(\zeta_q + \sqrt{m-1}\,\zeta'_q\big)\right)_{q \in \mathcal{P}_l}, \tag{13.4.23}$$

whereas for Theorem 13.4.3 it is enough to show

$$\big(k_n \zeta(l,1)^n_q, k_n \zeta(l,m)^n_q\big)_{q \in \mathcal{P}_l} \xrightarrow{\mathcal{L}\text{-s}} \big(\overline{\zeta}_q, \overline{\zeta}'_q\big)_{q \in \mathcal{P}_l}, \tag{13.4.24}$$

For the first of these two claims, recalling (13.4.3) and (13.4.4) and using a Taylor expansion for the function G, we obtain for any q, and as $n \to \infty$:

$$\begin{aligned}\zeta(l,w)^n_i = {} & \mathrm{O}\big(v_n^2\big) + \big(\partial_2 G(\Delta X_{S_q}, c_{S_q-}, c_{S_q})\big(\widehat{c}^n_{i(n,q)-k_n}(wk_n, v_n) - c_{S_q-}\big) \\ & + \partial_3 G(\Delta X_{S_q}, c_{S_q}, c_{S_q})\big(\widehat{c}^n_{i(n,q)+1}(wk_n, v_n) - c_{S_q}\big)\big) 1_{\{|\Delta X_{S_p}| > v_n\}} \\ & + \mathrm{o}\big(\big|\widehat{c}^n_{i(n,q)(wk_n,v_n)-k_n} - c_{S_q-}\big| + \big|\widehat{c}^n_{i(n,q)+1}(wk_n, v_n) - c_{S_q}\big|\big).\end{aligned}$$

Since $\tau < 4\varpi$, with the help of (13.4.16) and (13.4.17) one deduces (13.4.23).

For the second claim, by (13.4.9) and (13.4.12), we have $V(G,X) \equiv 0$ and $c_{T_q-} = c_{T_q}$, hence $\partial_i G(\Delta X_{T_q}, c_{S_q-}, c_{S_q}) = 0$ for $i = 2,3$ and all q. Then we have to resort upon a second order Taylor's expansion, which gives

$$\begin{aligned}\zeta(w)^n_i = {} & \mathrm{O}\big(v_n^2\big) + \frac{1}{2}\big(\partial^2_{22} G(\Delta X_{S_q}, c_{S_q-}, c_{S_q})\big(\widehat{c}^n_{i(n,q)-k_n}(wk_n, v_n) - c_{S_q-}\big)^2 \\ & + 2\partial^2_{23} G(\Delta X_{S_q}, c_{S_q-}, c_{S_q})\big(\widehat{c}^n_{i(n,q)-k_n}(wk_n, v_n) - c_{S_q-}\big) \\ & \times \big(\widehat{c}^n_{i(n,q)+1}(wk_n, v_n) - c_{S_q}\big)\end{aligned}$$

$$+\partial_{33}^2 G(\Delta X_{S_q}, c_{S_q}, c_{S_q})\big(\widehat{c}^n_{i(n,q)+1}(wk_n, v_n) - c_{S_q})^2\big)1_{\{|\Delta X_{S_p}|>v_n\}}$$
$$+\,\mathrm{o}\big(|\widehat{c}^n_{i(n,q)(wk_n,v_n)-k_n} - c_{S_q-}|^2 + |\widehat{c}^n_{i(n,q)+1}(wk_n, v_n) - c_{S_q}|^2\big)$$

By $\tau < 2\varpi$ and (13.4.16) and (13.4.17) we now deduce (13.4.24). This completes the proof of both theorems, for the process $X'(l)$.

Step 4. The previous step yields the convergences (13.4.8) or (13.4.14) when we substitute X with $X'(l)$, without modifying $\widehat{c}^n_i(k_n, v_n)$. Moreover, by the assumptions (13.4.4) or (13.4.9) on G it is straightforward to check that $\mathcal{U}(l) \stackrel{\text{u.c.p.}}{\Longrightarrow} \mathcal{U}$ and $\mathcal{U}'(l) \stackrel{\text{u.c.p.}}{\Longrightarrow} \mathcal{U}'$ in case of Theorem 13.4.1 and $\overline{\mathcal{U}}(l) \stackrel{\text{u.c.p.}}{\Longrightarrow} \overline{\mathcal{U}}$ and $\overline{\mathcal{U}}'(l) \stackrel{\text{u.c.p.}}{\Longrightarrow} \overline{\mathcal{U}}'$ in case of Theorem 13.4.3, on the extended space, and as $l \to \infty$. Thus, observing that $V(G,X) - V(G,X'(l)) = V(G,X(l))$ (which is identically 0 in case of Theorem 13.4.3), and with the notation $V^n(w,l) = V(G; wk_n, v_n, X) - V(G; wk_n, v_n, X'(l)) - V(G, X(l))$ and $\kappa = \frac{1}{2}$ in case of Theorem 13.4.1 and $\kappa = 1$ in case of Theorem 13.4.3, by Proposition 2.2.2 it remains to show the following property

$$\lim_{l\to\infty} \limsup_{n\to\infty} \mathbb{P}\big(k_n^\kappa \big|V^n(w,l)_t\big| > \rho\big) = 0$$

for all $t, \rho > 0$ and for $w = 1$ and $w = m$. The proof is of course the same when $w = 1$ and when $w = m$, so we will prove this for $w = 1$ only.

Below we fix t. We will specify, in restriction to the set $\Omega'_n(t,l)$, a decomposition of $V^n(1,l)_t$ as a sum $\sum_{j=1}^6 A(j,l)_n$ for suitable variables $A(j,l)_n$. Then, since $\mathbb{P}(\Omega'_n(t,l)) \to 1$ as $n \to \infty$, it will be enough to prove that, for all $\rho > 0$ and $j = 1,2,3,4,5,6$,

$$\lim_{l\to\infty} \limsup_{n\to\infty} \mathbb{P}\big(k_n^\kappa \,|A(j,l)_n| > \rho\big) = 0. \tag{13.4.25}$$

Step 5. For simplicity, we write $\widehat{c}^n_i$ instead of $\widehat{c}^n_i(k_n, v_n)$. A simple calculation shows that indeed we have $V^n(1,l)_t = \sum_{j=1}^6 A(j,l)_n$ on $\Omega'_n(t,l)$, provided we define $A(j,l)_n$ as follows: First, for $j = 1,2$ we set

$$A_n(1,l) = -\sum_{s\in J(n,t)} G\big(\Delta X(l)_s, c_{s-}, c_s\big), \quad \text{where}$$
$$J(n,y) = (0, k_n\Delta_n] \cup \big(([t/\Delta_n] - k_n)\Delta_n, t\big]$$
$$A_n(2,l) = \sum_{q\in\mathcal{P}_l:\, S_q\le t} \eta(l)^n_q, \quad \text{where}$$
$$\eta(l)^n_q = G\big(\Delta X_{S_q} + \Delta^n_{i(n,q)}X(l), \widehat{c}^n_{i(n,q)-k_n}, \widehat{c}^n_{i(n,q)+1}\big)1_{\{|\Delta X_{S_q}+\Delta^n_{i(n,q)}X(l)|>v_n\}}$$
$$- G\big(\Delta^n_{i(n,q)}X(l), \widehat{c}^n_{i(n,q)-k_n}, \widehat{c}^n_{i(n,q)+1}\big)1_{\{|\Delta^n_{i(n,q)}X(l)|>v_n\}}$$
$$- G\big(\Delta X_{S_q}, \widehat{c}^n_{i(n,q)-k_n}, \widehat{c}^n_{i(n,q)+1}\big)1_{\{|\Delta X_{S_q}|>v_n\}}.$$

Secondly, for $j=3,4,5,6,7$ we set $A_n(j,l)=\sum_{i=k_n+1}^{[t/\Delta_n]-k_n}\zeta(j,l)_i^n$, where

$$\zeta(3,l)_i^n = -\sum_{s\in I(n,i)} G\big(\Delta X(l)_s, c_{(i-1)\Delta_n}, c_{i\Delta_n}\big)\,1_{\{\Delta X(l)_s|\le v_n\}}$$

$$\zeta(4,l)_i^n = \sum_{s\in I(n,i)} \big(G\big(\Delta X(l)_s, c_{(i-1)\Delta_n}, c_{i\Delta_n}\big) - G\big(\Delta X(l)_s, c_{s-}, c_s\big)\big)$$

$$\begin{aligned}\zeta(5,l)_i^n &= G\big(\Delta_i^n X(l), c_{(i-1)\Delta_n}, c_{i\Delta_n}\big)1_{\{|\Delta_{i(n,q)}^n X(l)|>v_n\}}\\ &\quad - \sum_{s\in I(n,i)} G\big(\Delta X(l)_s, c_{(i-1)\Delta_n}, c_{i\Delta_n}\big)\,1_{\{\Delta X(l)_s|>v_n\}})\end{aligned}$$

$$\begin{aligned}\zeta(6,l)_i^n &= \big(G\big(\Delta_i^n X(l), \widehat{c}_{i-k_n}^n, \widehat{c}_{i+1}^n\big) - G\big(\Delta_i^n X(l), c_{(i-1)\Delta_n}, c_{i\Delta_n}\big)\big)\\ &\quad \times 1_{\{v_n<|\Delta_{i(n,q)}^n X(l)|\le 2/l\}}.\end{aligned}$$

Step 6. In this step we treat the easy cases $j=1,2,3$. Since eventually l goes to ∞ in (13.4.25), it is no restriction below to assume that $\frac{2}{l}<\varepsilon$, where $\varepsilon>0$ is the number occurring in (13.4.4).

We have $|G(x,y,y')|\le K_A|x|^3$ when $|y|,|y'|\le A$, whereas (SH) for X yields $\mathbb{E}(|\Delta_i^n X(l)|^q)\le K_{l,q}\Delta_n$ for all $q\ge 2$, thus $\mathbb{E}(|A(1,l)_t|)\le Kk_n\Delta_n$. Since $k_n\sqrt{\Delta_n}\to 0$ (because $\tau<\frac{1}{2}$), we deduce $k_nA(1,l)_n\stackrel{\mathbb{P}}{\longrightarrow}0$ as $n\to\infty$, for any fixed l: hence (13.4.25) holds for $j=1$ and $\kappa=1$.

(13.4.4) and the boundedness in probability of both sequences $\widehat{c}_{i(n,q)-k_n}^n$ and $\widehat{c}_{i(n,q)+1}^n$ and $\Delta_{i(n,q)}^n X(l)\stackrel{\mathbb{P}}{\longrightarrow}0$ yield that, for each q, we have $\eta(l,q)^n=\mathrm{O}_\mathbb{P}(|\Delta_{i(n,q)}^n X(l)|)$ (argue separately in the two cases $\Delta X_{S_q}=0$ and $\Delta X_{S_q}\ne 0$) Hence (13.4.16) gives us $k_nA(2,l)_n\stackrel{\mathbb{P}}{\longrightarrow}0$ as $n\to\infty$, for any fixed l: hence (13.4.25) holds for $j=2$ and $\kappa=1$.

Finally, using again $|G(x,y,y')|\le K_A|x|^3$ when $|y|,|y'|\le A$, we can write

$$\mathbb{E}\big(\big|A(3,l)_n\big|\big)\le Kv_n^{3-r}\mathbb{E}\bigg(\sum_{s\le t}|\Delta X_s|^r\bigg)\le Ktv_n^{3-r}.$$

Hence (13.4.25) for $j=3$ holds when $\kappa=\frac{1}{2}$ as soon as $\tau<2(3-r)\varpi$ (hence under (13.4.7)), and when $\kappa=1$ as soon as $\tau<(3-r)\varpi$ (hence under (13.4.13)).

Step 7. This step is devoted to proving (13.4.25) for $j=4$ and $\kappa=1$. We have $|\zeta(4,l)_i^n|\le K(\zeta(-,l)_i^n+\zeta(+,l)_i^n$, where

$$\zeta(-,l)_i^n = \sum_{s\in I(n,i)} \big|\Delta X(l)_s\big|^2\,|c_{s-}-c_{(i-1)\Delta_n}|$$

$$\zeta(+,l)_i^n = \sum_{s\in I(n,i)} \big|\Delta X(l)_s\big|^2|c_{i\Delta_n}-c_c|.$$

Recall that $\mathbb{E}(|c_{T+u} - c_T| \mid \mathcal{F}_T) \leq K\sqrt{u}$ for all $u > 0$ and all finite stopping times T. On the one hand, we may write

$$\mathbb{E}\big(|\zeta(-,l)_i^n|\big) = \mathbb{E}\bigg(\int_{I(n,i)} \int_{A_l^c} \delta(s,z)^2 |c_{s-} - c_{(i-1)\Delta_n}| \, \mathfrak{p}(ds,dz)$$

$$\leq \mathbb{E}\bigg(\int_{I(n,i)} |c_{s-} - c_{(i-1)\Delta_n}| \, ds \int_{A_l^c} \Gamma(z)^2 \, \lambda(dz) \;\leq\; K\Delta_n^{3/2}.$$

On the other hand, by successive conditioning we obtain

$$\mathbb{E}\big(|\zeta(+,l)_i^n|\big) = \mathbb{E}\bigg(\sum_{q\geq 1} |\Delta X_{S_q}|^2 \, (c_{i\Delta_n} - c_{S_q}| \, 1_{\{S_q \in I(n,i)\}}\bigg)$$

$$\leq K\sqrt{\Delta_n}\,\mathbb{E}\bigg(\sum_{s\in I(n,i)} |\Delta X_s|^2\bigg) \;\leq\; K\Delta_n^{3/2}.$$

Putting these two estimates together, we get $\mathbb{E}(|A(4,l)_n|) \leq Kt\sqrt{\Delta_n}$ and, since $k_n\sqrt{\Delta_n} \to 0$, we conclude (13.4.25) when $j = 4$ and $\kappa = 1$.

Step 8. Here we consider the case $j = 5$. We choose a number $u \in (1, \frac{1}{2r\varpi} \wedge \frac{1}{2\varpi})$, which is possible because $\varpi < 1/2r$. We write $l_n = [1/v_n^u]$ (the integer part), so for all n large enough we have $1/l_n < v_n < 1/l$. We then set

$$A'_n = A_{l_n} \cap (A_l)^c, \qquad Y^n = \big(\delta(s,z) 1_{A'_n}\big) * \mathfrak{p}$$

$$b(l,n)_t = b(l)_t - \int_{A'_n} \delta(t,z)\lambda(dz)$$

$$\overline{Y}_t^n = X(l)_t - Y_t^n = X_0 + \int_0^t b(l,n)_s \, ds + \int_0^t \sigma_s \, dW_s + (\delta 1_{(A'_n)^c}) * (\mathfrak{p} - \mathfrak{q})_t$$

$$N_t^n = \mathfrak{p}\big([0,t] \times A'_n\big), \quad H(n,i) = \Big\{ \big|\Delta_i^n \overline{Y}^n\big| \leq \frac{u_n}{2} \Big\} \cap \big\{\Delta_i^n N^n \leq 1\big\}.$$

First, N^n is a Poisson process with parameter $\lambda(A'_n) \leq K l_n^r$. Hence

$$\mathbb{P}\big(\Delta_i^n N^n \geq 2\big) \;\leq\; K\Delta_n^{2-2ru\varpi} \;\leq\; K\Delta_n \, \rho_n, \quad \text{where } \rho_n \to 0. \tag{13.4.26}$$

Second, $\Delta_n l_n^r \leq K$ and $|b(l,n)_t| \leq K l_n^{(r-1)^+}$ and Lemma 2.1.5 and 2.1.7 yield

$$w \geq r \;\Rightarrow\; \mathbb{E}\big(|\Delta_i^n \overline{Y}^n|^w \mid \mathcal{G}_{(i-1)\Delta_n}^{(l_n)}\big) \leq K_w\big(\Delta_n^{w/2} + \Delta_n^{1+u\varpi(w-r)}\big). \tag{13.4.27}$$

We apply this with w large enough and use Markov's inequality to get $\mathbb{P}(|\Delta_i^n \overline{Y}^n| > v_n/2) \leq K\Delta_n^2$. Therefore if $\Omega''_n(t) = \cap_{1\leq i\leq [t/\Delta_n]} H(n,i)$, this property and (13.4.27) imply $\mathbb{P}(\Omega''_n(t)) \to 1$.

Next, on the set $H(n,i)$, we have $|\Delta_i^n \overline{Y}^n| \le v_n/2$ and $|\Delta_i^n Y^n| \le 1/l$, and also $|\Delta X(l)_s| \le v_n$ for all $s \in I(n,i)$ except, when $\Delta_i^n N^n = 1$, for a single value of s for which $\Delta X(l)_s = \Delta_i^n Y^n$ (whose absolute value may be smaller or greater than v_n). In other words, on $H(n,i)$ we have

$$\begin{aligned}\zeta(5,l)_i^n = &\big(G\big(\Delta_i^n Y^n + \Delta_i^n \overline{Y}^n, c_{(i-1)\Delta_n}, c_{i\Delta_n}\big) 1_{\{|\Delta_i^n Y^n + \Delta_i^n \overline{Y}^n| > v_n\}} \\ &- G\big(\Delta_i^n Y^n, c_{(i-1)\Delta_n}, c_{i\Delta_n}\big) 1_{\{|\Delta_i^n Y^n| > v_n\}}\big) 1_{\{|\Delta_i^n Y^n| \le 1/m,\, |\Delta_i^n \overline{Y}^n| \le v_n/2\}}.\end{aligned}$$

The following estimate, for $v \in (0,1)$ and $y, z \in (0, A]$ for some A, and also $|x| \le 1$ and $|x'| \le v/2$, follows from (13.4.4):

$$\big|G\big(x+x', y, z\big)1_{\{|x+x'|>v\}} - G(x,y,z)1_{\{|x|>v\}}\big| \le K\big(x^2|x'| + \big(|x| \wedge v\big)^3\big).$$

Therefore, on the set $H(n,i)$ again, we have

$$|\zeta(5,l)_i^n| \le K\big(|\Delta_i^n Y^n|^2 |\Delta_i^n \overline{Y}^n| + \big(|\Delta_i^n Y^n| \wedge v_n\big)^3\big)$$

and by (2.1.42) applied to $\delta\, 1_{A'_n}$ and $r=0$ we get $\mathbb{E}((|\Delta_i^n Y^n| \wedge u_n)^3) \le K \Delta_n^{1+(3-r)\varpi}$. Applying (13.4.27) with $w = 2$ and the Cauchy-Schwarz inequality to obtain $\mathbb{E}(|\Delta_i^n \overline{Y}^n| \mid \mathcal{G}^{(l_n)}_{(i-1)\Delta_n}) \le K\sqrt{\Delta_n}$. We also have $|\Delta_i^n Y^n| \le \Delta_i^n((\Gamma\, 1_{A'_n}) * \mathfrak{p})$, which is $\mathcal{G}_0^{(l_n)}$-measurable. Then by successive conditioning and (2.1.41), we obtain

$$\mathbb{E}\big(|\Delta_i^n Y^n|^2 \Delta_i^n \overline{Y}^n|\big) \le K\sqrt{\Delta_n}\,\mathbb{E}\big(\big|((\Gamma\, 1_{A'_n}) * \mathfrak{p})\big|^2\big) \le K\Delta_n\big(\sqrt{\Delta_n} + \Delta_n^{1-2u\varpi}\big).$$

We then deduce from the previous estimates that

$$\mathbb{E}\big(\big|A(5,l)_n\big|\, 1_{\Omega''_n(t)}\big) \;\le\; Kt\big(\sqrt{\Delta_n} + \Delta_n^{1-2u\varpi} + \Delta_n^{(3-r)\varpi}\big).$$

Since $\mathbb{P}(\Omega''_n(t)) \to 1$, we deduce that (13.4.25) holds for $j = 6$ and $\kappa = 1$.

Step 9. In this step we consider $j = 6$. For $q = 1, 2$ we introduce the notation

$$\eta(l,q)_i^n = \big|\Delta_i^n X(l)\big|^2\, 1_{\{|\Delta_i^n X(l)| > v_n\}}\big(\big|\widehat{c}_{i+1}^{\,n} - c_{i\Delta_n}\big|^q + \big|\widehat{c}_{i-k_n}^{\,n} - c_{(i-1)\Delta_n}\big|^q\big).$$

(SH) and the inequality $|x+y|^2\, 1_{\{|x+y|>v_n\}} \le K(x^2 + |y|^{2+w}/v_n^w)$ for all $w > 0$ imply, by our usual estimates, that

$$\mathbb{E}\big(\Delta_i^n X(l)^2\, 1_{\{|\Delta_i^n X(l)| > v_n\}} \mid \mathcal{F}_{(i-1)\Delta_n}\big) \le \Delta_n(K_l \Delta_n + K\phi_l),$$

where $\phi_l = \int_{A_l^c} \Gamma(z)^2 \lambda(dz) \to 0$ as $l \to \infty$. Then, since $\Delta_i^n X(l)$ is $\mathcal{F}_{i\Delta_n}$-measurable and $\widehat{c}_{i-k_n}^{\,n} - c_{(i-1)\Delta_n}$ is $\mathcal{F}_{(i-1)\Delta_n}$-measurable, by successive conditioning and Lemma 13.4.4 and summing up on i, we obtain

$$\mathbb{E}\big(\eta(l,q)_i^n\big) \le \frac{\Delta_n}{k_n^{q/2}}\,(K_l\Delta_n + K\phi_l). \tag{13.4.28}$$

We now single out the two situations of Theorems 13.4.1 and 13.4.3. For the first theorem, and since $2/l < \varepsilon$, we have $|\zeta(6,l)_i^n| \le K\eta(l,1)_i^n$. Hence (13.4.28) yields

$$\mathbb{E}\big(|A(6,l)_n|\big) \le \frac{t}{\sqrt{k_n}}\,(K_l\Delta_n + K\phi_l),$$

which gives (13.4.25) for $j=6$ and $\kappa = 1/2$.

In the situation of Theorem 13.4.3 this estimate is not sufficient. However, we can now use (13.4.9) and (13.4.9) and do a Taylor expansion for G around the point $(\Delta_i^n X(l), c_{i\Delta_n}, c_{i\Delta_n})$, to get

$$\big|\zeta(6,l)_i^n\big| \le K\big(\eta(l,2)_i^n + \eta(l)_i^n\big), \quad \text{with} \quad \eta(l)_i^n = \big|\Delta_i^n X(l)\big|^2 \,|c_{i\Delta_n} - c_{(i-1)\Delta_n}|^2.$$

In order to evaluate the expectation of $\eta(l)_i^n$ we prove an elementary auxiliary result. Let U and V be two Itô semimartingales satisfying (SH), and with no common jumps and $U_0 = V_0 = 0$. Then Itô's formula yields

$$U_t^2 V_t^2 = M_t + \int_0^t \big(a_s U_s^2 + a_s' U_s V_s + a_s'' V_s^2\big)\,ds,$$

where M is a martingale and a, a', a'' are suitable processes, and also $\mathbb{E}(U_t^2) + \mathbb{E}(V_t^2) \le \alpha t$, where the number α and the processes a, a', a'' are bounded by a constant depending only on the bounds on the characteristics of U and V. Taking the expectation above, we get $\mathbb{E}(U_t^2 V_t^2) \le \alpha' t^2$, where again α' only depends on the bounds on the characteristics of U and V. Applying this to $U_t = X(l)_t - X(l)_{(i-1)\Delta_n}$ and $V = c_t - c_{(i-1)\Delta_n}$ for $t \ge (i-1)\Delta_n$, we deduce that $\mathbb{E}(\eta(l)_i^n) \le K_l \Delta_n^2$. This and (13.4.28) with $q=2$ yield

$$\mathbb{E}\big(\big|A(6,l)_n\big|\big) \le \frac{t}{k_n}\,(K_l\Delta_n + K_l\Delta_n k_n + K\phi_l),$$

which gives (13.4.25) for $j=6$ and $\kappa = 1$.

Step 10. So far, we have proved Theorems 13.4.1 and 13.4.3 in case (b). If we consider case (a) of these theorems, as soon as l is large enough for having $G(x,y,y') = 0$ whenever $|x| \le 2/l$, we see that $A(j,l)_n = 0$ for $j = 1,3,4,5,6$ on the set $\Omega_n'(t,l)$, so we only need to show (13.4.25) when $j=2$, and this was done in Step 6, without requiring (13.4.7) or (13.4.13). So the proof of both theorems is complete.

13.5 Application to Volatility

1) First, we come back to the problem of estimating, in the $d=1$ dimensional case, the quantity

$$A(p)_t \;=\; \int_0^t |\sigma_s|^p\,ds.$$

In Chap. 11 we used the multipower variations given by (11.4.3), but instead we can use the *truncated multipower variations*, with v_n is as in (13.0.2):

$$D(X;p,k;v_n-,\Delta_n)_t \;=\; \sum_{i=1}^{[t/\Delta_n]-k+1}\prod_{j=1}^{k}\left|\Delta_{i+j-1}^n X\right|^{p/k} 1_{\{|\Delta_{i+j-1}^n X|\le v_n\}},$$

which of course includes the truncated power variation

$$D(X;p;v_n-,\Delta_n)_t := D(X;p,1;v_n-,\Delta_n)_t = \sum_{i=1}^{[t/\Delta_n]}\left|\Delta_i^n X\right|^p 1_{\{|\Delta_i^n X|\le v_n\}}. \tag{13.5.1}$$

Then, referring to Example 13.2.2, we obtain exactly the same result as in Theorem 11.4.1, with different requirements: in both cases we assume (K-r) for some $r\in[0,1)$, and the CLT holds for the multipower variations if $k>p/r$, whereas for the truncated multipower variations the value of k is arbitrary (and in particular may be $k=1$), but we need $\varpi\ge(p-2)/2(p-r)$. More precisely, the analogue of Theorem 11.4.1 becomes (with the same proof, note that in Theorem 11.4.1 the property (11.4.1), implying $\Omega_t^W=\Omega$ a.s., was assumed):

Theorem 13.5.1 *Let $p\ge 1$ and let X is a (possibly discontinuous) one-dimensional Itô semimartingale satisfying* (K'-p) *when $p=1$ and* (K-r) *for some $r\in[0,1]$ when $p>1$. Then for each $t>0$ the random variables*

$$\frac{\sqrt{m_{2p}}\,\bigl(\Delta_n^{1-p/2}D(X;p;v_n-,\Delta_n)_t-m_p\,A(p)_t\bigr)}{\sqrt{(m_{2p}-m_p^2)\,\Delta_n^{2-p}D(X;2p;v_n-,\Delta_n)_t}} \tag{13.5.2}$$

converge stably in law to a limit which is $\mathcal{N}(0,1)$ and independent of $\mathcal{F}$, in restriction to the set $\Omega_t^W=\{A(p)_t>0\}$ (which is Ω when (K'-1) *holds), provided either X is continuous, or*

$$r=1\;\Rightarrow\;p=1,\qquad r<1\;\Rightarrow\;p\ge\frac{r}{2-r},\quad\varpi\ge\frac{p-1}{2(p-r)}.$$

We state the result only for the truncated uni-power variations, but a result for the truncated multi-powers is available, and in fact formally takes *exactly* the same form as in Theorem 11.4.1, with exactly the same assumptions (with everywhere p substituted with p/k).

A common feature with (non-truncated) multipower variations is that *we need the jumps of X to be summable*, and even a bit more, as expressed by (K-r) for some $r\le 1$. This seems unavoidable, although quite restrictive, and if this assumption fails the results are simply not true in general, see the already mentioned papers by Vetter [93] for multipowers and Mancini [75] for truncated powers.

So far and, say, for $p \geq 2$ (the relevant case in practice), we have a whole family of estimators for $A(p)_t$ enjoying an associated Central Limit Theorem:

$$\widehat{A}(p,k)_n = \frac{1}{(m_{p/k})^k} \Delta_n^{1-p/2} D(X;p,k;v_n-,\Delta_n)_t, \quad k \geq 1, \ \frac{p-1}{2(p-r)} \leq \varpi < \tfrac{1}{2}$$

$$\widehat{A}'(p,k)_n = \frac{1}{(m_{p/k})^k} \Delta_n^{1-p/2} D(X;p,k,\Delta_n)_t, \quad k > \frac{p}{2}$$

(and of course $v_n = \alpha \Delta_n^{\varpi}$). Then we can compare these estimators on several grounds:

From the viewpoint of the assumptions needed for the CLT: When $p > 2$, all estimators need (K-r) (for $\widehat{A}(p,k)_n$) or (K'-r) (for $\widehat{A}(p,k)_n$), for some $r < 1$. So the truncated uni-powers is what requires less assumptions, but only marginally.

On the other hand, when $p = 2$, we still need (K-r) or (K'-r), for $\widehat{A}(2,k)_n$ or $\widehat{A}'(2,k)_n$, *except* for $\widehat{A}(2,1)_n$; by virtue of Theorem 13.2.4, in this case we only need (H-r), again for some $r < 1$. This is of course significantly less demanding.

From the "feasibility" viewpoint: $\widehat{A}(p,k)_n$ needs to specify the truncation level v_n, with ϖ in $(\frac{p-k}{2(p-kr)}, \frac{1}{2})$: so a more precise prior knowledge of the value of r is required. Furthermore, these are asymptotic results, but for finite samples the chosen value of v_n, hence of the constant α in (13.0.2) as well, play a fundamental role. Therefore in practice the usage of truncated powers needs some preliminary "estimation" of the proper cut-off level v_n, in connection with the "average" value of the volatility σ_t on the interval of interest. These drawbacks do not exist for non-truncated multi-powers.

From the viewpoint of the asymptotic variance of the estimator: here the asymptotic variance is the same for $\widehat{A}(p,k)_n$ and for $\widehat{A}'(p,k)_n$, for any fixed $k \geq 1$. Now, this asymptotic behavior is hidden in a result like the convergence of (13.5.2). However we also know that, after normalization by the same factor $1/\sqrt{\Delta_n}$, the asymptotic variances of $\widehat{A}(p,k)_n$ and $\widehat{A}'(p,k)_n$ are $v(p,k)\,A(2p)_t$, where

$$v(p,k) = \frac{M(p,k)}{(m_{p/k})^{2k}},$$

where $M(p,k)$ is given by (11.4.5), with of course $M(p,1) = m_{2p} - m_p^2$, hence $v(p,1) = (m_{2p} - m_p^2)/m_p^2$. So the "best" from this viewpoint consists in taking the integer k which minimizes $v(p,k)$, with of course p fixed. In this direction, one may show for example that $v(2,k)$ increases from 2 to π^2, as k increases from 1 to ∞. Note that $v(2,1) = 2$ is the Cramer-Rao bound for the variance of unbiased estimators of a for an i.i.d. sample of $\mathcal{N}(0,a)$-distributed variables.

A consequence of these facts is that one should choose k as small as possible, and thus *the truncated power* (13.5.1) *is better (asymptotically) than the multipower.* However the gain is relatively slight: for example when $p = 2$ we have $v(2,1) = 2$ and $v(2,2) = \frac{(\pi-2)(\pi+6)}{4}$, which is approximately 2.61, so the ratio of the two asymptotic standard deviations is approximately 1.27, bigger than but close to 1. In practice, this has to be weighted against the feasibility issues mentioned above.

2) Next, we turn to the multidimensional case $d \geq 2$, and suppose that we want to estimate the following integral

$$C_t^{jk} = \int_0^t c_s^{jk}\, ds.$$

According to Theorem 13.2.4, we may use the estimator $\widehat{C}^n(v_n-, X)_t^{jk}$. The associated standardized CLT again easily follows, upon taking the following estimator for the $\mathcal{F}$-conditional variance

$$H_t^{n,jk} = \sum_{i=1}^{[t/\Delta_n]-1} \left((\Delta_i^n X^j)^2 (\Delta_{i+1}^n X^k)^2 + \Delta_i^n X^j \Delta_i^n X^k \Delta_{i+1}^n X^j \Delta_{i+1}^n X^k \right)$$
$$1_{\{\|\Delta_i^n X\| \leq v_n, \|\Delta_{i+1}^n X\| \leq v_n\}},$$

Theorem 13.5.2 *If X is a (possibly discontinuous) Itô semimartingale satisfying* (H-r) *for some $r \in [0,1)$, for each $t > 0$ the random variables*

$$\frac{\widehat{C}^n(v_n-, X)_t^{jk} - C_t^{jk}}{\sqrt{H_t^{n,jk}}}$$

converge stably in law to a limit which is $\mathcal{N}(0,1)$ and independent of $\mathcal{F}$, in restriction to the set $\{C_t^{11} > 0, C_t^{22} > 0\}$, provided either X is continuous or $\varpi \in [\frac{1}{4-2r}, \frac{1}{2})$.

The estimators $\widehat{C}^n(v_n-, X)_t^{jk}$ have to be compared with the two distinct bipower estimators constructed in (8.5.3) and (8.5.4) (with $l = 1$), which take the following forms with the present notation:

$$\frac{4}{\pi} \sum_{i=1}^{[t/\Delta_n]-1} \left(\left|\Delta_i^n X^j + \Delta_i^n X^k\right| \left|\Delta_{i+1}^n X^j + \Delta_{i+1}^n X^k\right| \right.$$
$$\left. - \left|\Delta_i^n X^j\right| \left|\Delta_{i+1}^n X^j\right| - \left|\Delta_i^n X^k\right| \left|\Delta_{i+1}^n X^k\right| \right)$$

$$\frac{8}{\pi} \sum_{i=1}^{[t/\Delta_n]-1} \left(\left|\Delta_i^n X^j + \Delta_i^n X^k\right| \left|\Delta_{i+1}^n X^j + \Delta_{i+1}^n X^k\right| \right.$$
$$\left. - \left|\Delta_i^n X^j - \Delta_i^n X^k\right| \left|\Delta_{i+1}^n X^j - \Delta_{i+1}^n X^k\right| \right).$$

Here again, one may show that these two estimators satisfy a CLT similar to the above. Their asymptotic variances are bigger than the asymptotic variance of $\widehat{C}^n(v_n-, X)_t^{jk}$, but again they do not suffer the drawback of having to choose the truncation level v_n. However one needs (K-r) for some $r < 1$, instead of (H-r), for a CLT to hold for those bipower estimators.

Bibliographical Notes

Truncated functionals have been introduced by Mancini [73, 74] for estimating the volatility in the presence of jumps (with a threshold v_n equivalent to $\alpha\sqrt{\Delta_n \log(1/\Delta_n)}$, and also by Shimizu and Yoshida [88] (with a threshold of the form (13.0.2)) in a parametric context for diffusion processes with jumps. Thresholds are also underlying in the paper [1] of Aït-Sahalia, and truncated functionals have been heavily used by Aït-Sahalia and Jacod for constructing various tests connected with the jumps of the process, see e.g. [2].

The results referred to as "approximations for jumps" are new. The other results also have a new formulation in both this chapter and Chap. 9. However the Law of Large Numbers of Theorem 9.2.1 when $k = 1$ and for power variations is due to Mancini [73]. In the same setting, various versions of the Central Limit Theorem 13.2.1 can be found in the papers of Mancini [74, 75], Cont and Mancini [23] or Jacod [60].

The spot estimation of the volatility and the associated Central Limit Theorem is mostly new, but related estimators can be found in Lee and Mykland [72] for detecting jumps, see also Alvarez, Panloup, Pontier and Savy [5]. The content of Sect. 9.5 is basically taken from Aït-Sahalia and Jacod [2]. The content of Sect. 9.4 is also original, but the reader will notice that no associated CLT is given: this is an open question which, once solved, could perhaps provide a solution to the problem of estimating the integrated volatility when there are jumps with index activity bigger than 1 (that is, when (H-1) is not satisfied).

Finally, Sect. 13.4 is taken from Jacod and Todorov [64], with a few simplifications, and the aim of this CLT is to provide tests for common arrival of jumps in the underlying process X and its volatility process.

Part V
Various Extensions

This last part is concerned with partial answers to some of the problems which have been omitted so far. First we consider the situation where the discretization scheme is not regular. Second, and back to regular discretization, we study some degenerate situations where the rate of convergence is not the standard one of $1/\sqrt{\Delta_n}$.

Finally we study a case which is clearly motivated by applications, and in particular financial applications, even more than the previous results: this is when the process is subject to some kind of measurement error. The error can be "additive", which typically occurs in many physical applications, or it can have a more complex structure, as occurring in finance under the name "microstructure noise", and it is more difficult to analyze.

Chapter 14
Irregular Discretization Schemes

In practice, the observation times at stage n are quite often not regularly spaced. A reason can be that some data are missing. More important, the observations may be more frequent during some periods of time, or when the observed process is in some regions of space, or when it is "more active". Or, the observations may occur randomly, for example according to the arrival times of a Poisson or Poisson-like process, with an intensity which in turn may depend on time or on the process itself.

The aim of this chapter is to study what happens to our laws of large numbers and central limit theorems when we relax the assumption of regularly spaced observations.

We know by Theorem 3.3.1 that the law of large numbers for non-normalized functionals holds for completely general discretization schemes. On the other hand, the associated central limit theorem is so far unproved, except when the discretization scheme is irregular but not random, and in addition under very strong assumptions on the asymptotic behavior of the sampling times. So we omit this topic here.

Henceforth we consider only the normalized functionals for one or several increments, truncated or not. Consequently, we suppose in the whole chapter that X is an Itô semimartingale on the filtered probability space $(\Omega, \mathcal{F}, (\mathcal{F}_t)_{t\geq 0}, \mathbb{P})$, with the Grigelionis decomposition

$$X_t = X_0 + \int_0^t b_s\,ds + \int_0^t \sigma_s\,dW_s + (\delta 1_{\{\|\delta\|\leq 1\}}) \star (\mathfrak{p} - \mathfrak{q})_t + (\delta 1_{\{\|\delta\|>1\}}) \star \mathfrak{p}_t\,. \tag{14.0.1}$$

The first question which arises is how to normalize the increments in order to obtain some limit theorems. As we will see, in addition to finding the proper normalization, solving this problem requires some structure on the discretization scheme, and explaining these additional assumptions is the object of the first section below.

An important comment should be made here: the underlying process X can be, and will be, multidimensional. *But*, at each stage n all its components are observed at the *same times* $T(n,i)$. This is natural in some applications, when for example one observes the position of a randomly moving object, so all 3 components are measured simultaneously. For other applications it is an over-simplifying hypothesis: for example if X is the vector of the prices of 3 (or 300) different assets, then

J. Jacod, P. Protter, *Discretization of Processes*,
Stochastic Modelling and Applied Probability 67,
DOI 10.1007/978-3-642-24127-7_14,

typically the observation times are not equidistant, but they are also different for different assets. The treatment of a vector-valued semimartingale whose components are observed at different times is *much more* difficult than when the observations are simultaneous, see for example Hayashi and Yoshida [44, 46], and this topic will not be considered at all in the sequel.

14.1 Restricted Discretization Schemes

Let us recall that a random discretization scheme is a double sequence $(T(n,i): i \geq 0, n \geq 1)$ of finite stopping times, increasing in i, with $T(n,0)=0$ and

$$\begin{aligned} &\forall n \geq 1: \quad T(n,i) < \infty \Rightarrow T(n,i) < T(n,i+1), \qquad \lim_{i \to \infty} T(n,i) = \infty \\ &\forall t > 0: \quad \pi_t^n := \sup_{i \geq 1} \big(T(n,i) \wedge t - T(n,i-1) \wedge t \big) \stackrel{\mathbb{P}}{\longrightarrow} 0. \end{aligned} \tag{14.1.1}$$

We use the notation (3.1.1), that is

$$\begin{aligned} \Delta(n,i) &= T(n,i) - T(n,i-1), \qquad & N_n(t) &= \sum_{i \geq 1} 1_{\{T(n,i) \leq t\}} \\ T_n(t) &= T\big(n, N_n(t)\big), & I(n,i) &= \big(T(n,i-1), T(n,i)\big]. \end{aligned}$$

In the regular case we have seen before that $V'^n(f,X)$ converges. The expression for $V'^n(f,X)$ exhibits two normalizing factors: One inside the test function because we use $f(\Delta_i^n X/\sqrt{\Delta_n})$, and one outside, which is Δ_n. The inside normalization is here to ensure that $\Delta_i^n X/\sqrt{\Delta_n}$ is approximately $\sigma_{(i-1)\Delta_n} U_i^n$ for some $\mathcal{N}(0, I_{d'})$ random variable U_i^n independent of $\mathcal{F}_{(i-1)\Delta_n}$. In particular a summand is typically neither vanishing nor exploding, and thus the external normalization is the inverse of the number of summands, up to the factor t. In the irregular case, neither one of these two kinds of normalization stays valid, and here we discuss the first—inside—one.

A natural substitute of $f(\Delta_i^n/\sqrt{\Delta_n})$ is $f(\Delta_i^n X/\sqrt{\Delta(n,i)})$. However, the variable $\Delta_i^n X/\sqrt{\Delta(n,i)}$ is close to U_i^n as before only if $\Delta(n,i)$, or equivalently $T(n,i)$, is $\mathcal{F}_{T(n,i-1)}$ measurable. This property is called the *strong predictability condition* of the sampling scheme.

The strong predictability condition accommodates all deterministic schemes, of course, but otherwise is very restrictive. In particular it rules out two interesting situations:

(i) When $T(n,i) = \inf(t > T(n,i-1) : X_t - X_{T(n,i-1)} \in A_n)$ for some Borel set A_n: for example when $d=1$ with $A_n = \{-1/n, 1/n\}$.
(ii) When for each n the $T(n,i)$'s are random and independent of X; for example they form a Poisson process with parameter n, or a renewal process.

Although case (i) describes what happens in some specific applied situations, we will not treat it at all: An interested reader can consult the papers [33] or [34] of Fukasawa for example. Rather, we will consider an extension of case (ii), in which the observation time $T(n,i)$ may depend on $\mathcal{F}_{T(n,i-1)}$ and also on some extraneous random input.

The setting is as follows: we have the space $(\Omega, \mathcal{F}, (\mathcal{F}_t)_{t\geq 0}, \mathbb{P})$ on which the process X is defined and satisfies (14.0.1), and to accommodate the extra randomness without changing too much the notation we suppose that $\mathbb{P}$ is defined on a σ-field $\mathcal{G}$ which is bigger than $\mathcal{F}$.

Definition 14.1.1 a) A *restricted pre-discretization scheme* is a double sequence $(T(n,i) : i \geq 0, n \geq 1)$ with $T(n,0) = 0$ and

$$T(n,i) \;=\; T(n,i-1) + \theta^n_{T(n,i-1)}\,\varepsilon(n,i) \tag{14.1.2}$$

where

1) The variables $\varepsilon(n,i)$ are $(0,\infty)$-valued, and they generate a σ-field $\mathcal{H}$ which is contained in $\mathcal{G}$ and *independent* of $\mathcal{F}$.

2) For each $n \geq 1$ the sequence $(\varepsilon(n,i) : i \geq 1)$ is i.i.d. with

$$q > 0 \;\Rightarrow\; m'_q(n) \;=\; \mathbb{E}\big(\varepsilon(n,i)^q\big) \;<\; \infty, \qquad m'_1(n) \;=\; 1, \tag{14.1.3}$$

3) Each $\theta^n = (\theta^n_t)_{t\geq 0}$ is an $(\mathcal{F}^n_t)$-adapted càdlàg process, where $(\mathcal{F}^n_t)$ is the smallest filtration containing $(\mathcal{F}_t)$ and for which all $T(n,i)$ for $i \geq 1$ are stopping times, and further neither θ^n_t nor θ^n_{t-} vanish.

b) A *restricted* (*random*) *discretization scheme* is a restricted pre-discretization scheme which satisfies (14.1.1).

The last condition in (14.1.3) is innocuous: if it were not satisfied we could replace θ^n by $\theta^n/m'_1(n)$. The condition that all moments of $\varepsilon(n,i)$ are finite is here for convenience and could be relaxed, but it seems a mild condition from a practical viewpoint. Likewise, the càdlàg property of θ^n could be replaced by progressive measurability, although the assumption $\inf_{s\leq t}\theta^n_s > 0$ for all t and n (which is implied by (2) above) is important, as we shall see.

Note that no assumption is made as to relations between the sequences $(\varepsilon(n,i) : i \geq 1)$ for different values of n: this is irrelevant to our analysis.

Every deterministic scheme is a restricted scheme: take $\varepsilon(n,i) \equiv 1$ and $\theta^n_t = \Delta(n,i)$ for $T(n,i-1) \leq t < T(n,i)$ for example. Any scheme such that each sequence $(T(n,i))_{i\geq 1}$ forms a renewal process independent of $(X, W, \mathfrak{p})$ and with inter-arrival times having all moments, is also a restricted scheme. But restricted schemes are indeed much more general than that, since by (14.1.3) $\Delta(n,i)$ may depend on the whole past $\mathcal{F}^n_{T(n,i-1)}$ before time $T(n,i-1)$. Moreover the strong predictability condition, as stated before Definition 14.1.1, is not necessarily satisfied by a restricted scheme. However the general schemes described in (i) above are still not restricted schemes in general.

Remark 14.1.2 We have seen in the previous chapters that, for some applications, it is useful to consider several time steps at once: for example, in the regular setting, one compares variables like $V^n(f,X)_t$ computed for the time step Δ_n, with the same computed with the time step $2\Delta_n$.

Here we have to be careful: doing this in an irregular scheme setting amounts to comparing the scheme $(T(n,i): i \geq 0)$ with the scheme $(T(n,2i): i \geq 1)$. However, if $(T(n,i))$ is a restricted scheme, the new scheme $(T(n,2i))$ is usually no longer a restricted scheme in the mathematical sense stated above. That is, the notion defined above is not preserved by a thinning of data.

Restricted pre-discretization schemes are easy to construct, via (14.1.2), and starting with the processes θ^n and the variables $\varepsilon(n,i)$. The first part of (14.1.1) is automatically satisfied because $T(n,k) \geq (\inf_{s\leq t}\theta^n_s)\, S^n_k$ on the set $\{T(n,k) \leq t\}$, where $S^n_k = \sum_{i=1}^k \varepsilon(n,i)$, and because $S^n_k \to \infty$ as $k \to \infty$ for any fixed n. In contrast, the property $\pi^n_t \xrightarrow{\mathbb{P}} 0$ is not a completely trivial matter, and no necessary and sufficient condition for this is known.

In the sequel, we need to assume $\pi^n_t \xrightarrow{\mathbb{P}} 0$, of course, and also to prescribe a rate of convergence to 0 for the mesh π^n_t: for this, we suppose that this rate is deterministic, and the same for all times t. This is of course a further serious restriction, which is mathematically expressed in the following assumption:

Assumption 14.1.3 (or **(D)**) There is a sequence of positive numbers $r_n \to \infty$ and a càdlàg $(\mathcal{F}_t)$-adapted process θ such that neither θ_t nor θ_{t-} vanish and

$$r_n\,\theta^n \overset{\text{u.c.p.}}{\Longrightarrow} \theta. \tag{14.1.4}$$

Moreover for all $p > 0$ we have $m'_p(n) \to m'_p < \infty$ as $n \to \infty$ (this is of course always true for $p = 1$ with $m'_1 = 1$, and also for $p = 0$ with $m'_0 = 1$).

A regular discretization scheme obviously satisfies (D) with $r_n = 1/\Delta_n$ and $\theta_t = 1$ (take $\varepsilon(n,i) = 1$).

Remark 14.1.4 The fact that all moments of $\varepsilon(n,i)$ converge as $n \to \infty$ could be weakened: for each result below the convergence of moments of suitable orders would be enough. Assuming the convergence for all moments seems, however, innocuous for applications.

A key property of restricted schemes satisfying (D) concerns the following increasing processes, where $p \geq 0$:

$$H(p)^n_t = \sum_{i=1}^{N_n(t)} \Delta(n,i)^p. \tag{14.1.5}$$

These are, as are many other processes to be seen later, partial sums of a triangular array for the stopping rules $(N_n(t): t \geq 0)$, exactly as in (2.2.28), and with $\zeta^n_i =$

$\Delta(n,i)^p$ in the present case. Of course, we have the basic relation (2.2.30), that is $T(n,i) = \inf(t : N_n(t) \geq i)$. However in order to obtain limit theorems for such triangular arrays, quite often we also need the property (2.2.29): namely there is a discrete-time filtration $(\mathcal{G}_i^n)_{i\geq 0}$, such that ζ_i^n is $\mathcal{G}_i^n$ measurable and

$$n \geq 1,\ t \geq 0 \ \Rightarrow\ N_n(t) \text{ is a } \big(\mathcal{G}_i^n\big)\text{-stopping time.} \tag{14.1.6}$$

If this holds, we also recall that each $T(n,i)$ is a predictable stopping time with respect to the continuous-time filtration $(\overline{\mathcal{F}}_t^n)_{t\geq 0}$ defined by $\overline{\mathcal{F}}_t^n = \mathcal{G}_{N_n(t)}^n$, and that $\overline{\mathcal{F}}_{T(n,i)}^n = \mathcal{G}_i^n$ and $\overline{\mathcal{F}}_{T(n,i)-}^n = \mathcal{G}_{i-1}^n$ when $i \geq 1$. As we see just below, (14.1.6) holds if we take

$$\mathcal{G}_i^n \;=\; \mathcal{F}_{T(n,i)}^n \bigvee \sigma\big(\varepsilon(n,i+1)\big) \qquad \big(\text{hence } \mathcal{F}_{T(n,i)}^n \subset \mathcal{G}_i^n \subset \mathcal{F}_{T(n,i+1)}^n\big). \tag{14.1.7}$$

The next lemma provides some insight about the "average" behavior of restricted discretization schemes, as defined above, over any time interval $[0,t]$. It turns out also to be a key technical result for the proofs of the LLN and CLT given later.

Lemma 14.1.5 a) *For any restricted pre-discretization scheme the choice* (14.1.7) *for* $\mathcal{G}_i^n$ *yields* (14.1.6), *and in this case we have* $\mathcal{F}_t^n \subset \overline{\mathcal{F}}_t^n$, *and for each n the space* $(\Omega, \mathcal{G}, (\overline{\mathcal{F}}_t^n)_{t\geq 0}, \mathbb{P})$ *is a very good filtered extension of* $(\Omega, \mathcal{F}, (\mathcal{F}_t)_{t\geq 0}, \mathbb{P})$.

b) *Any restricted pre-discretization scheme satisfying* (D) *is a discretization scheme, that is, it satisfies* (14.1.1). *Moreover for all* $t \geq 0$ *and* $p \geq 0$ *we have*

$$r_n^{p-1}\, H(p)_t^n \ \stackrel{\mathbb{P}}{\longrightarrow}\ m'_p \int_0^t (\theta_s)^{p-1}\, ds \tag{14.1.8}$$

and also

$$\eta > 0 \ \Rightarrow\ r_n^{1-\eta}\, \pi_t^n \ \stackrel{\mathbb{P}}{\longrightarrow}\ 0, \tag{14.1.9}$$

$$\frac{1}{r_n}\, N_n(t) \ \stackrel{\mathbb{P}}{\longrightarrow}\ \int_0^t \frac{1}{\theta_s}\, ds. \tag{14.1.10}$$

Remark 14.1.6 Although (14.1.10) is simply (14.1.8) for $p = 0$, we state it separately because it gives some insight to the meaning of the "rate" r_n: this is approximately the number of observations up to time t, divided by a positive (random) quantity (increasing with t, of course), namely $\int_0^t \frac{1}{\theta_s}\, ds$. According to (14.1.10), $1/r_n$ is also "almost" the rate at which π_t^n goes to 0. The property of $(\Omega, \mathcal{G}, (\overline{\mathcal{F}}_t^n)_{t\geq 0}, \mathbb{P})$ being a very good filtered extension of $(\Omega, \mathcal{F}, (\mathcal{F}_t)_{t\geq 0}, \mathbb{P})$ is not used for the LLN, but is a crucial property for the CLT.

Proof 1) We begin with (a). Since θ^n is adapted to the filtration $(\mathcal{F}_t^n)$, it follows from (14.1.2) that $T(n,i+1)$ is $\mathcal{G}_i^n$ measurable, as well as $T(n,i)$. Since $\{N_n(t) = i\} = \{T(n,i) \leq t < T(n,i+1)\}$, we deduce that $N_n(t)$ is a $(\mathcal{G}_i^n)$-stopping time, hence (14.1.6) holds. The inclusion $\mathcal{F}_t^n \subset \overline{\mathcal{F}}_t^n$ follows.

It remains to prove the "very good" property of the extension. Using the σ-field $\mathcal{H}$ of Definition 14.1.1, we define a new filtration by setting $\mathcal{F}'_t = \cap_{s>t}(\mathcal{H} \bigvee \mathcal{F}_s)$. Each $T(n,i)$ is an $(\mathcal{F}'_t)$-stopping time (immediate, by induction on i), so $\mathcal{F}^n_t \subset \mathcal{F}'_t$. Moreover $\mathcal{G}^n_i \subset \mathcal{F}'_{T(n,i)}$ by (14.1.7) and what precedes, so since $T(n, N_n(t)) \leq t$ we further deduce that $\overline{\mathcal{F}}^n_t \subset \mathcal{F}'_t$.

It is then enough to show that any bounded $(\mathcal{F}_t)$-martingale M is also a martingale relative to $(\mathcal{F}'_t)$. Since M is right-continuous and bounded, this amounts to showing that

$$\mathbb{E}\big((M_{r+t+s} - M_{r+t})\, 1_A\big) = 0 \tag{14.1.11}$$

for all $r, s, t > 0$ and $A \in \mathcal{F}_r \otimes \mathcal{H}$. By a monotone class argument it is even enough to show (14.1.11) when $A = A' \cap A''$, with $A' \in \mathcal{F}_r$ and $A'' \in \mathcal{H}$. But then, due to the independence of $\mathcal{F}$ and $\mathcal{H}$, the left side of (14.1.11) is $\mathbb{P}(A'')\,\mathbb{E}((M_{r+t+s} - M_{r+t}) 1_{A'})$, which vanishes because M is an $(\mathcal{F}_t)$-martingale. So the result is proved.

2) Next, we start proving (b) by showing that (14.1.8) for all $p > 0$ implies (14.1.9). Under (14.1.8), $M(p)^n_{t+1} = \sup_{s \leq t+1} r_n^{p-1} \Delta H(p)^n_s$ tends to 0 in probability as $n \to \infty$, and $r_n^{p-1}(H(p)^n_{t+1} - H(p)^n_t) \stackrel{\mathbb{P}}{\longrightarrow} m'_p \int_t^{t+1} (\theta_s)^{p-1} ds > 0$. This last property yields $\mathbb{P}(T(n, N_n(t)+1) \leq t+1) \to 1$, whereas if $T(n, N_n(t)+1) \leq t+1$ we have $r_n^{p-1} (\pi^n_t)^p \leq M(p)^n_{t+1}$. Then (14.1.9) with $\eta = 1/p$ follows.

3) In this step we show that by localization, and for proving (14.1.8), we may assume

$$\frac{1}{C} \leq \theta_t \leq C, \qquad \frac{1}{C} \leq r^n \theta^n_t \leq C \tag{14.1.12}$$

identically, for some constant $C > 1$. This is different from our usual localization procedure, but may be easily seen as follows. For $C > 1$ set $S_C = \inf(t : \theta_t \leq \frac{1}{C}$ or $\theta_t \geq C)$ and $S^n_C = \inf(t : r_n \theta^n_t \leq \frac{1}{C}$ or $r_n \theta^n_t \geq C)$, and also

$$\theta(C)_t = \begin{cases} \theta_t & \text{if } t < S_C \\ \theta_{S_C-} & \text{if } t \geq S_C \end{cases} \qquad \theta(C)^n_t = \begin{cases} \theta^n_t & \text{if } t < S^n_C \\ \theta^n_{S^n_C-} & \text{if } t \geq S^n_C. \end{cases}$$

Consider the restricted discretization scheme associated by (14.1.2) with $\theta(C)^n$ and the same $\varepsilon(n,i)$ as before. This new scheme satisfies (14.1.12) and (D) with the same sequence r_n. Moreover the associated processes $H(C,p)^n$ (by (14.1.5)) satisfy $H(C,p)^n_t = H(p)^n_t$ if $t < S^n_C$, whereas the limiting variables in (14.1.8) are the same for θ and $\theta(C)$ if $t \leq S_S$.

Therefore if (14.1.8) holds for the schemes associated with all C, it will then hold for the original scheme, provided $\mathbb{P}(S_C < t) \to 0$ and $\limsup_n \mathbb{P}(S^n_C \leq t) \to 0$ as $C \to \infty$. These properties are satisfied because θ_t and θ_{t-} do not vanish (hence $S_C \to \infty$ as $C \to \infty$), and because of (14.1.4).

4) In all the sequel we suppose that (14.1.12) holds, and in this step we undertake a further reduction of the problem. Namely we show that, for any *a priori* given

number $\gamma > 0$, we can assume

$$\varepsilon(n,i) \leq r_n^\gamma. \tag{14.1.13}$$

To see this we consider a new scheme with the same processes θ^n and the new auxiliary variables $\varepsilon'(n,i) = \varepsilon(n,i) \wedge r_n^\gamma$, which satisfy (14.1.13). This new scheme also satisfies (D) because

$$m'_p(n) - \mathbb{E}\big(\varepsilon'(n,i)^p\big) = \mathbb{E}\big(\big(\varepsilon(n,i)^p - r_n^{p\gamma}\big) 1_{\{\varepsilon(n,i)>r_n^{p\gamma}\}}\big) \leq \frac{m'_{2p}(n)}{r_n^{2p\gamma}} \to 0$$

for all p (note that $\mathbb{E}(\varepsilon'(n,i)) < 1$ in general, but this will have no consequences). We also set $S_k^n = \sum_{j=1}^k \varepsilon(n,j)$ and $R_t^n = \inf(k : S_k^n \geq t r_n)$. In view of the second set of inequalities in (14.1.12), and for any sequence k_n of integers, it is clear that the variables $H(p)_t^n$ for the original scheme and for the new scheme are equal outside the set $\Omega_{n,t}$ on which either $R_{Ct}^n > k_n$ or $\sup_{1\leq i\leq k_n} \varepsilon(n,i) > r_n^\gamma$. Since the variables $(\varepsilon(n,i) : i \geq 1)$ are independent with $\mathbb{E}(\varepsilon(n,i)) = 1$ and $\mathbb{E}(\varepsilon(n,i)^2) \leq K$, we have for all $k \in \mathbb{N}^*$ and $x > 0$:

$$\mathbb{E}\big(\big(S_k^n - k\big)^2\big) \leq Kk, \qquad \mathbb{P}\Big(\sup_{1\leq i\leq k} \varepsilon(n,i) > x\Big) \leq k\,\frac{m'_{2/\gamma}(n)}{x^{2/\gamma}} \leq K\,\frac{k}{x^{2/\gamma}}.$$

We take $k_n = [Ctr_n] + [r_n] + 2$, so $R_{Ct}^n > k_n$ implies $S_{k_n}^n - k_n \leq -r_n$. Thus, upon taking $x = r_n^\gamma$ above, we get

$$\mathbb{P}(\Omega_{n,t}) \leq \frac{Kk_n}{r_n^2} \to 0.$$

Therefore if (14.1.8) holds for the new scheme it also holds for the original scheme. In other words, we can and will assume both (14.1.12) and (14.1.13) in what follows.

5) Now we apply the results of Sect. 2.2.4 to the arrays and sums

$$\zeta(p,L)_i^n = r_n^{p-1} L_{T(n,i-1)} \Delta(n,i)^p, \qquad Z(p,L)_t^n = \sum_{i=1}^{N_n(t)+1} \zeta(p,L)_i^n$$

$$\zeta'(p,L)_i^n = \mathbb{E}\big(\zeta(p,L)_i^n \mid \mathcal{F}_{T(n,i-1)}^n\big), \qquad Z'(p,L)_t^n = \sum_{i=1}^{N_n(t)+1} \zeta'(p,L)_i^n,$$

where L is an arbitrary bounded and nonnegative càdlàg process, adapted to the filtration $(\mathcal{F}_t)$. The sums above are of the form (2.2.28), with the summand $\zeta(p,L)_i^n$ being $\mathcal{F}_{T(n,i)}^n$-measurable and stopping rules $N_n(t)+1$ which are stopping times for the discrete filtrations $(\mathcal{F}_{T(n,i)})_{i\geq 0}$. Moreover if $p \leq 1/\gamma$, and in view of (14.1.12) and (14.1.13), we have $|\zeta(p,L)_i^n| \leq K$. Therefore, for all $p \leq 1/\gamma$ Lemma 2.2.11

yields

$$Z''(p,L)^n_t := \sum_{i=1}^{N_n(t)+1} \big(\zeta(p,L)^n_i\big)^2 \stackrel{\text{u.c.p.}}{\Longrightarrow} 0 \Rightarrow Z(p,L)^n - Z'(p,L)^n \stackrel{\text{u.c.p.}}{\Longrightarrow} 0. \tag{14.1.14}$$

Note also that, since $\pi^n_s \stackrel{\mathbb{P}}{\longrightarrow} 0$ for all s, we have $\pi'^n_t = \sup(\Delta(n,i) : 1 \le i \le N_n(t)+1) \stackrel{\mathbb{P}}{\longrightarrow} 0$ as well.

First take $p=1$. Since $\pi'^n_t \stackrel{\mathbb{P}}{\longrightarrow} 0$ we see that $Z(1,L)^n_t$ is a Riemann sum approximation of the integral $\int_0^t L_s\,ds$. Since L is càdlàg, we deduce that $Z(1,L)^n_t \stackrel{\text{u.c.p.}}{\Longrightarrow} \int_0^t L_s\,ds$. Furthermore $Z''(1,L)^n_t \le Z(1,L^2)^n_t\,\pi'^n_t$, so $Z''(1,L)^n \stackrel{\text{u.c.p.}}{\Longrightarrow} 0$ and (14.1.14) yields

$$\frac{1}{r_n} \sum_{i=1}^{N_n(t)+1} L_{T(n,i-1)}\, r_n \theta^n_{T(n,i-1)} \stackrel{\text{u.c.p.}}{\Longrightarrow} \int_0^t L_s\,ds$$

because $\zeta'(1,L)^n_i = L_{T(n,i-1)}\theta^n_{T(n,i-1)}$. Since $r_n\theta^n \ge 1/C$ we deduce that the sequence of variables $\frac{1}{r_n}\sum_{i=1}^{N'_n(t)} L_{T(n,i-1)}$ is bounded in probability, and using (14.1.4) we end up with

$$\frac{1}{r_n} \sum_{i=1}^{N_n(t)+1} L_{T(n,i-1)}\, \theta_{T(n,i-1)} \stackrel{\text{u.c.p.}}{\Longrightarrow} \int_0^t L_s\,ds. \tag{14.1.15}$$

Next we take $p < 1/\gamma$ and observe that $\zeta'(p,1)^n_i = r_n^{p-1}(\theta^n_{T(n,i-1)})^p m'_p(n)$. Then, by (14.1.4) and $m'_p(n) \to m'_p$, and applying (14.1.15) with $L = \theta^{p-1}$, we obtain

$$Z'(p,1)^n_t \stackrel{\text{u.c.p.}}{\Longrightarrow} m'_p \int_0^t (\theta_s)^{p-1}\,ds.$$

Now (14.1.12) and (14.1.13) yield $\Delta(n,i) \le K/r_n^{1-\gamma}$, hence $Z''(p,1)^n_t \le \frac{K}{r_n^{1+\gamma-2p\gamma}} \times Z(1,1)^n_t$, which goes to 0 as soon as $p < \frac{1+\gamma}{2\gamma}$. In this case, another application of (14.1.14) gives

$$Z(p,1)^n_t \stackrel{\text{u.c.p.}}{\Longrightarrow} m'_p \int_0^t (\theta_s)^{p-1}\,ds.$$

Observing that $Z(p,1)^n_t$ is in fact $r_n^{p-1}H(p)^n_s$ taken at time $s = T(n,N_n(t)+1)$, which converges in probability to t, we deduce (14.1.8).

In other words, when $p < \frac{1+\gamma}{2\gamma}$ we have proved (14.1.8) when (14.1.13) holds, hence also for the original scheme. Since γ is arbitrarily small, we deduce (14.1.8) for all $p \ge 0$. □

14.2 Law of Large Numbers for Normalized Functionals

As said before, the inside normalization of the increment $\Delta_i^n X$ should be $\sqrt{\Delta(n,i)}$. We still have a problem with the outside normalization. A natural approach consists in normalizing each summand with the length $\Delta(n,i)$ of the relevant interval, that is to consider functionals of the form

$$V'^n(f,X)_t = \sum_{i=1}^{N_n(t)} \Delta(n,i)\, f\big(\Delta_i^n X/\sqrt{\Delta(n,i)}\big).$$

In a sense, this is the most straightforward generalization of the formula (3.4.2), and this is the normalization introduced by Barndorff-Nielsen and Shephard in [9]. Likewise, one can consider a function F on $(\mathbb{R}^d)^k$ and set, as in (8.1.2):

$$V'^n(F,X)_t = \sum_{i=1}^{N_n(t)-k+1} \Delta(n,i)\, F\bigg(\frac{\Delta_i^n X}{\sqrt{\Delta(n,i)}},\dots,\frac{\Delta_{i+k-1}^n X}{\sqrt{\Delta(n,i+k-1)}}\bigg). \tag{14.2.1}$$

Another possibility of outside normalization, when (D) holds, would be

$$V'^n(F,p,X)_t = r_n^{p-1} \sum_{i=1}^{N_n(t)-k+1} \Delta(n,i)^p\, F\bigg(\frac{\Delta_i^n X}{\sqrt{\Delta(n,i)}},\dots,\frac{\Delta_{i+k-1}^n X}{\sqrt{\Delta(n,i+k-1)}}\bigg) \tag{14.2.2}$$

and clearly $V'^n(F,X) = V'^n(F,1,X)$.

Finally, it is also possible to consider truncated functionals. The truncation is as in (9.0.3), page 248, so it naturally depends on the length of the interval. That is, the truncation level for the ith interval is

$$v(n,i) = \alpha\Delta(n,i)^{\varpi} \quad \text{for some } \alpha>0,\ \varpi\in\bigg(0,\frac{1}{2}\bigg). \tag{14.2.3}$$

With this notation, the extension of (9.2.1), page 251, takes the following form

$$\begin{aligned} V'^n(F,v_n-,X)_t = & \sum_{i=1}^{N_n(t)-k+1} \Delta(n,i) F\bigg(\frac{\Delta_i^n X}{\sqrt{\Delta(n,i)}},\dots,\frac{\Delta_{i+k-1}^n X}{\sqrt{\Delta(n,i+k-1)}}\bigg) \\ & \times \prod_{l=0}^{k-1} 1_{\{\|\Delta_{i+l}^n X\|\le v_n(n,i+l)\}} \end{aligned} \tag{14.2.4}$$

and also the analogue of (14.2.2):

$$\begin{aligned} V'^n(F,p,v_n-,X)_t = r_n^{p-1} & \sum_{i=1}^{N_n(t)-k+1} \Delta(n,i)^p F\bigg(\frac{\Delta_i^n X}{\sqrt{\Delta(n,i)}},\dots,\frac{\Delta_{i+k-1}^n X}{\sqrt{\Delta(n,i+k-1)}}\bigg) \\ & \times \prod_{l=0}^{k-1} 1_{\{\|\Delta_{i+l}^n X\|\le v_n(n,i+l)\}}. \end{aligned} \tag{14.2.5}$$

The results are analogous to Theorems 8.4.1 and 9.2.1, and need (H-r), or Assumption 6.1.1, which we recall:

Assumption (H-r) X has the form (14.0.1), with b_t locally bounded and σ_t càdlàg. Moreover $\|\delta(\omega,t,z)\| \wedge 1 \leq \Gamma_n(z)$ for all (ω,t,z) with $t \leq \tau_n(\omega)$, where (τ_n) is a localizing sequence of stopping times and each function Γ_n on E satisfies $\int \Gamma_n(z)^r \lambda(dz) < \infty$.

Theorem 14.2.1 *Assume that X satisfies* (H-r) *for some $r \in (0,2]$. Let $p \geq 0$, and assume that we have a restricted discretization scheme satisfying* (D).

a) *Let F be a continuous function on $(\mathbb{R}^d)^k$, which is of polynomial growth when X is continuous and which satisfies the following property when X jumps*:

$$\bigl|F(x_1,\dots,x_k)\bigr| \leq \prod_{j=1}^{k} \Psi\bigl(\|x_j\|\bigr)\bigl(1+\|x_j\|^2\bigr) \tag{14.2.6}$$

where Ψ is a continuous function on $[0,\infty)$ that goes to 0 at infinity. Then

$$V'^n(F,p,X)_t \overset{\text{u.c.p.}}{\Longrightarrow} V'(F,p,X)_t := m'_p \int_0^t \rho_{c_s}^{k\otimes}(F)\,(\theta_s)^{p-1}\,ds. \tag{14.2.7}$$

b) *Let F be a continuous function on $(\mathbb{R}^d)^k$ which satisfies for some $q \geq 0$*:

$$\bigl|F(x_1,\dots,x_k)\bigr| \leq K\prod_{j=1}^{k}\bigl(1+\|x_j\|^q\bigr). \tag{14.2.8}$$

Then

$$V'^n(F,p,v_n-,X) \overset{\text{u.c.p.}}{\Longrightarrow} V'(F,p,X) \tag{14.2.9}$$

when X is continuous, and when X jumps and either $q \leq 2$ or

$$q>2, \quad r<2, \quad \varpi \geq \frac{q-2}{2(q-r)}.$$

Proof 1) The proof is as for Theorems 8.4.1 and 9.2.1. By localization, we can assume (SH-r) (that is, (H-r) with further b_t and σ_t and X_t bounded, and $\|\delta(\omega,t,z)\| \leq \Gamma(z)$ with Γ bounded and $\int \Gamma(z)^r\lambda(dz) < \infty$). Also, exactly as in the proof of Lemma 14.1.5, we can assume that the properties (14.1.12) and (14.1.13) hold, the latter for a $\gamma \in (0,\frac{1}{2})$ as small as needed (the choice depends on k and on the test function F). Note that (14.1.12) and (14.1.13) imply that for all n large enough we have identically

$$\pi_t^n \leq r_n^{\gamma-1} \leq \frac{1}{\sqrt{r_n}}. \tag{14.2.10}$$

A further localization based on the boundedness of θ and on (14.1.8) allows us to assume that for some constants C_p depending on p we have identically

$$r_n^{p-1}\, H(p)_t^n \leq C_p t. \tag{14.2.11}$$

In the course of the proof, we heavily use the following property, deduced from (14.1.8) and which holds for every càdlàg process L:

$$r_n^{p-1} \sum_{i=1}^{N_n(t)} L_{T(n,i-1)}\, \Delta(n,i)^p \stackrel{\text{u.c.p.}}{\Longrightarrow} m'_p \int_0^t L_s\, \theta_s^{1-p}\, ds. \tag{14.2.12}$$

2) A second simplification of the problem arises as follows. We slightly modify the processes of interest, by setting

$$\mathcal{V}'^n(F,p,X)_t = r_n^{p-1} \sum_{i=1}^{N_n(t)} \Delta(n,i)^p\, F\left(\frac{\Delta_i^n X}{\sqrt{\Delta(n,i)}}, \dots, \frac{\Delta_{i+k-1}^n X}{\sqrt{\Delta(n,i+k-1)}}\right), \tag{14.2.13}$$

that is $\mathcal{V}'^n(F,p,X)_t = V'^n(F,p,X)_{T(n,N_n(t)+k-1)}$, and analogously we set $\mathcal{V}'^n(F, p, v_n-, X)_t = V'^n(F,p,v_n-,X)_{T(n,N_n(t)+k-1)}$. The reason for this is that the stopping rule $N_n(t)$ in this new formula becomes a $(\mathcal{G}_i^n)_{i\geq 0}$-stopping time, whereas $N_n(t)-k+1$ is not, unless $k=1$. Then, we claim that it is enough to prove that, for each t,

$$\mathcal{V}'^n(F,p,X)_t \stackrel{\mathbb{P}}{\longrightarrow} V'(F,p,X)_t, \quad \text{or} \quad \mathcal{V}'^n(F,p,v_n-,X)_t \stackrel{\mathbb{P}}{\longrightarrow} V'(F,p,X)_t, \tag{14.2.14}$$

under the appropriate assumptions.

Indeed, when $F \geq 0$, these properties imply the convergence in the *u.c.p.* sense, by the criterion (2.2.16). Since F is the difference of two nonnegative functions having the same continuity and growth properties as F itself, we deduce that (14.2.14) implies the *u.c.p.* convergence without restriction on the sign of F.

Moreover, the properties of our sampling scheme imply that the difference $T(n, N_n(t)+k-1)-t$ goes to 0 in probability, locally uniformly in t. Therefore $\mathcal{V}'^n(F,p,X) \stackrel{\text{u.c.p.}}{\Longrightarrow} V'(F,p,X)$, for example, implies $V'^n(F,p,X) \stackrel{\text{u.c.p.}}{\Longrightarrow} V'(F,p,X)$ as well. This shows that it is enough to prove the first or the second part of (14.2.14), respectively, for any given t, instead of (14.2.7) or (14.2.9).

3) We introduce some notation, somewhat analogous to those of the proof of Theorem 8.4.1, see pages 243–244:

$$\beta_{i,j}^n = \sigma_{T(n,i-1)} \frac{\Delta_{i+j-1}^n W}{\sqrt{\Delta(n,i+j-1)}},$$

$$\zeta_i^n = F\left(\beta_{i,1}^n, \dots, \beta_{i,k}^n\right), \qquad \zeta_i'^n = \rho_{cT(n,i-1)}^{k\otimes}(F),$$

$$\chi_i^n = F\left(\frac{\Delta_i^n X}{\sqrt{\Delta(n,i)}}, \dots, \frac{\Delta_{i+k-1}^n X}{\sqrt{\Delta(n,i+k-1)}}\right) - \zeta_i^n.$$

(Warning: unlike in Theorem 8.4.1, there is no outside normalization in ζ_i^n, $\zeta_i''^n$ and χ_i^n), and also

$$a_n(t) = r_n^{p-1}\,\mathbb{E}\left(\sum_{i=1}^{N_n(t)} \Delta(n,i)^p\,|\chi_i^n|\right).$$

Under our assumptions, the process $\rho_{c_t}^{k\otimes}(F)$ is càdlàg. Thus (14.2.12) yields

$$r_n^{p-1}\sum_{i=1}^{N_n(t)} \Delta(n,i)^p\,\zeta_i''^n \stackrel{\text{u.c.p.}}{\Longrightarrow} V'(F,p,X).$$

Moreover the properties of restricted discretization schemes imply that $\mathbb{E}(\zeta_i^n|\mathcal{G}_{i-1}^n) = \zeta_i''^n$, with $\mathcal{G}_i^n$ given by (14.1.7). Then the first part of (14.2.14) will follow if we prove the next two properties:

$$a_n(t) \to 0 \tag{14.2.15}$$

$$G_t^n := r_n^{p-1}\sum_{i=1}^{N_n(t)} \Delta(n,i)^p\,\zeta_i'^n \stackrel{\mathbb{P}}{\longrightarrow} 0, \quad \text{where } \zeta_i'^n = \zeta_i^n - \zeta_i''^n. \tag{14.2.16}$$

4) In this step we prove (14.2.15) when X is discontinuous. To this end, we first use the fact that $N_n(t)$ is a $(\mathcal{G}_i^n)_{i\geq 0}$-stopping time, to get

$$a_n(t) \leq r_n^{p-1}\,\mathbb{E}\left(\sum_{i=1}^{N_n(t)} \Delta(n,i)^p\,\mathbb{E}\big(|\chi_i^n| \mid \mathcal{G}_{i-1}^n\big)\right). \tag{14.2.17}$$

We replace the variables γ_i^n and the numbers γ_n' of (8.4.8) by

$$\gamma_i^n = \sup_{s\in[T(n,i-1),T(n,i+k)]} \|\sigma_s - \sigma_{T(n,i-1)}\|^2$$

$$\gamma_n' = \int_{\{z:\,\Gamma(z)\leq r_n^{-1/8}\}} \Gamma(z)^2\,\lambda(dz).$$

Once more, the properties of restricted schemes and our usual estimates yield that (8.4.20), page 243, can be replaced by the following:

$$\begin{aligned}
&\mathbb{E}\left(\|\beta_{i,j}^n\|^2 + \frac{\|\Delta_{i+j-1}^n X\|^2}{\Delta(n,i+j-1)} \mid \mathcal{G}_{i+j-2}^n\right) \leq K \\
&\mathbb{E}\left(\left\|\frac{\Delta_{i+j-1}^n X}{\sqrt{\Delta(n,i+j-1)}} - \beta_{i,j}^n\right\|^2 \bigwedge 1 \mid \mathcal{G}_{i+j-2}^n\right) \\
&\quad \leq K\,\mathbb{E}\big(r_n^{-1/8} + \gamma_n' + \gamma_i^n \mid \mathcal{G}_{i+j-2}^n\big) \quad \text{on the set } \big\{T(n,i+j-1)\leq t\big\}
\end{aligned} \tag{14.2.18}$$

(we use here (14.2.10) and the fact that $\Delta(n,i) \leq 1/\sqrt{r_n}$ if $T(n,i-1) \leq t$). Then, in view of (14.2.17) and since F satisfies (14.2.6), we can use (8.4.21), page 243, to deduce, as for (8.4.22), page 243), that for all $\varepsilon \in (0,1)$ and $A > 3$ we have

$$a_n(t) \leq K\,\mathbb{E}\big(L(A,\varepsilon)^n_{t+1}\big), \tag{14.2.19}$$

as soon as $k/\sqrt{r_n} < 1$ (so $T(n,i+k-1) \leq t+1$ if $i \leq N_n(t)$, recall (14.2.10)), and where

$$L(A,\varepsilon)^n_{t+1} = r_n^{p-1}\left(H(p)^n_{t+1}\left(\theta_A(\varepsilon) + \Psi(A) + \frac{A^{2k}}{\varepsilon^2}\left(r_n^{-1/8} + \gamma'_n\right)\right) + \frac{A^{2k}}{\varepsilon^2}\sum_{i=1}^{N_n(t+1)} \Delta(n,i)^p\,\gamma^n_i\right).$$

If $\eta > 0$ and if $(S_q)_{q\geq 1}$ denote the successive jump times of σ_t with size bigger than η, for all n large enough we have $\gamma^n_i \leq 2\eta$ for all $i \leq N_n(t+1)$ such that $|T(n,i) - S_q| > (k+1)\pi^n_t$ for all $q \geq 1$, and otherwise $\gamma^n_i \leq K$. Hence for all n large enough $\sum_{i=1}^{N_n(t+1)} \Delta(n,i)^p \gamma^n_i \leq 2\eta H(p)^n_{t+1} + K Q_t (\pi^n_t)^p$, where Q_t denotes the number of S_q smaller than $t+1$. Therefore, using (14.2.10), we get

$$L(A,\varepsilon)^n_{t+1} \leq \left(\theta_A(\varepsilon) + \Psi(A) + \frac{A^{2k}}{\varepsilon^2}\left(2\eta + r_n^{-1/8} + \gamma'_n\right)\right) r_n^{p-1} H(p)^n_t + \frac{K A^{2k}}{\varepsilon^2}\, Q_t r_n^{p\gamma/2-1}.$$

In view of (14.2.11) and if γ is such that $p\gamma/2 < 1$, the $\limsup$ (as $n \to \infty$) of the right side above is smaller than $a(A,\varepsilon,\eta) = K(\theta_A(\varepsilon) + \Psi(A) + \frac{2\eta A^{2k}}{\varepsilon^2})$. Moreover, again by (14.2.11) and the boundedness of σ and the very definition of $L(A,\varepsilon)^n_{t+1}$, we have $L(A,\varepsilon)^n_t \leq K_{A,\varepsilon}(t+1)$. Hence, by virtue of (14.2.19) we deduce that $\limsup_n a_n(t) \leq K a(A,\varepsilon,\eta)$. Since $\lim_{\varepsilon\to 0}\lim_{A\to\infty}\lim_{\eta\to 0} a(A,\varepsilon,\eta) = 0$, (14.2.15) follows.

5) Next, we prove (14.2.15) when X is continuous. In this case (14.2.18) can be strengthened as follows, for any $q \geq 2$:

$$\begin{aligned} &\mathbb{E}\left(\|\beta^n_{i,j}\|^q + \frac{\|\Delta^n_{i+j-1}X\|^q}{\Delta(n,i+j-1)^{q:2}} \mid \mathcal{G}^n_{i+j-2}\right) \leq K_q \\ &\mathbb{E}\left(\left\|\frac{\Delta^n_{i+j-1}X}{\sqrt{\Delta(n,i+j-1)}} - \beta^n_{i,j}\right\|^2 \mid \mathcal{G}^n_{i+j-2}\right) \\ &\leq K_q\left(\frac{1}{\sqrt{r_n}} + \mathbb{E}\big(\gamma^n_i \mid \mathcal{G}^n_{i+j-2}\big)\right) \quad \text{on the set } \big\{T(n,i+j-1) \leq t\big\}. \end{aligned} \tag{14.2.20}$$

Since now F has polynomial growth, we have, similar to (8.4.18),

$$\begin{aligned}&\big|F(x_1+y_1,\ldots,x_k+y_k)-F(x_1,\ldots,x_k)\big|\\&\quad\leq\theta_A(\varepsilon)+\frac{KA^q}{\varepsilon^2}\sum_{j=1}^k\|y_j\|^2+\frac{K}{A^q}\prod_{j=1}^k\big(\|x_j\|^{2q}+\|y_j\|^{2q}\big)\end{aligned}$$

for some $q\geq 2$. Then (14.2.19) holds with

$$L(A,\varepsilon)^n_{t+1}=r_n^{p-1}\left(H(p)^n_{t+1}\left(\theta_A(\varepsilon)+\frac{A^q}{\varepsilon^2\sqrt{r_n}}+\frac{1}{A^q}\right)+\frac{A^q}{\varepsilon^2}\sum_{i=1}^{N_n(t+1)}\Delta(n,i)^p\,\gamma_i'^n\right)$$

when $k/\sqrt{r_n}<1$. The proof of (14.2.15) is then analogous to the discontinuous case.

6) For the claim (a) it remains to prove (14.2.16). Here again we have to be careful: $\zeta_i'^n$ satisfies $\mathbb{E}(\zeta_i'^n\mid\mathcal{G}^n_{i-1})=0$, but is only $\mathcal{G}^n_{i+k-1}$-measurable. With t fixed, we write $u^n_i=r_n^{p-1}\Delta(n,i)^p\,\zeta_i'^n\,1_{\{T(n,i)\leq t\}}$. Then u^n_i is $\mathcal{G}^n_{i+k-1}$-measurable and $\mathbb{E}(u^n_j\mid\mathcal{G}^n_{i+k-1})=0$ if $j\geq i+k$. Therefore, since $G^n_t=\sum_{i=1}^{N_n(t)}u^n_i$, we have

$$\begin{aligned}\mathbb{E}\big((G^n_t)^2\big)&\leq\mathbb{E}\left(\sum_{1\leq i,j\leq N_n(t),|i-j|<k}u^n_iu^n_j\right)\\&\quad+2\mathbb{E}\left(\sum_{1\leq i<i+k\leq j\leq N_n(t)}u^n_i\,\mathbb{E}\big(u^n_j\mid\mathcal{G}^n_{i+k-1}\big)\right)\\&\leq k\mathbb{E}\left(\sum_{i=1}^{N_n(t)}\big(u^n_i\big)^2\right)=k\mathbb{E}\left(\sum_{i=1}^{N_n(t)}\mathbb{E}\big(\big(u^n_i\big)^2\mid\mathcal{G}^n_{i-1}\big)\right).\end{aligned}$$

In view of the definition of u^n_i, we deduce

$$\begin{aligned}\mathbb{E}\big((G^n_t)^2\big)&\leq k\mathbb{E}\left(r_n^{2p-2}\sum_{i\geq 1}\Delta(n,i)^{2p}\,\mathbb{E}\big(\big|\zeta_i'^n\big|^2\mid\mathcal{G}^n_{i-1}\big)\,1_{\{T(n,i)\leq t\}}\right)\\&\leq K\mathbb{E}\big(r_n^{2p-2}H(2p)^n_t\big)\end{aligned}$$

where we have used the property $\mathbb{E}(|\zeta_i'^n|^2\mid\mathcal{G}^n_{i-1})\leq K$ for the last inequality. By (14.2.11) applied with $2p$, the above goes to 0 as $n\to\infty$. This shows (14.2.16).

7) Finally we prove (b), which amounts to the second part of (14.2.14). For this, we basically reproduce the proof of Theorem 9.2.1, pages 252–254. Using (a) for

the bounded continuous function F_m given by

$$F_m(x_1,\dots,x_k)=F(x_1,\dots,x_k)\prod_{j=1}^k \psi'_m(x_j),$$

(that is, (9.2.8)) we see that it is enough to show

$$\lim_{m\to\infty}\ \limsup_{n\to\infty}\ \mathbb{E}\big(\big|\mathcal{V}'^n(F,v_n-,X)_t-\mathcal{V}'^n(F_m,X)_t\big|\big)\ =\ 0. \tag{14.2.21}$$

In the proof of Theorem 9.2.1, we replace everywhere v_n by $v(n,i)$, hence u_n by $u(n,i)=1/\alpha\Delta(n,i)^{1/2-\varpi}$. Thus for any $m\geq 1$ and $t>0$ we have $m\leq u(n,i)$ for all i such that $T(n,i)\leq t$, as soon as n is bigger than some (random) integer $n_{m,t}$. It follows that (14.2.21) amounts to

$$\lim_{m\to\infty}\ \limsup_{n\to\infty}\ \mathbb{E}\big(U_t^{n,m,j}\big)\ =\ 0, \tag{14.2.22}$$

where now (with q as in (14.2.8), we can assume $q\geq 2$):

$$\begin{aligned}
U_t^{n,m,j}\ &=\ r_n^{p-1}\sum_{i=1}^{N_n(t)}\Delta(n,i)^p\,\zeta(j,m)_i^n\\
\zeta(j,m)_i^n&=\prod_{l=1}^{j-1}Z_{i+l-1}^n\ \prod_{l=j+1}^{k}Z_{i+l-1}^n\,Z(m)_{i+j-1}^n,\\
Z_i^n&=1+\bigg(\frac{\|\Delta_i^nX'\|}{\sqrt{\Delta(n,i)}}\bigg)^q+\bigg(\frac{\|\Delta_i^nX''\|}{\sqrt{\Delta(n,i)}}\bigwedge u(n,i)\bigg)^q\\
Z(m)_i^n&=\frac{1}{m}\bigg(\frac{\|\Delta_i^nX'\|}{\sqrt{\Delta(n,i)}}\bigg)^{q+1}+\bigg(\frac{\|\Delta_i^nX''\|}{\sqrt{\Delta(n,i)}}\bigwedge u(n,i)\bigg)^q.
\end{aligned}$$

Applying (14.2.20) and (2.1.45) we get, as in (9.2.12) and (9.2.13), page 253, and for all $s>0$:

$$\begin{aligned}
&\mathbb{E}\bigg(\frac{\|\Delta_i^nX'\|^s}{\Delta(n,i)^{s/2}}\mid\mathcal{G}_{i-1}^n\bigg)\ \leq\ K_s\\
&\mathbb{E}\bigg(\frac{\|\Delta_i^nX''\|^2}{\Delta(n,i)^{2\varpi}}\bigwedge 1\mid\mathcal{G}_{i-1}^n\bigg)\leq\ K\Delta(n,i)^{1-r\varpi}\,\phi_n
\end{aligned}$$

where $\phi_n\to 0$ as $n\to\infty$. Then, as for (9.2.14), we get

$$\left.\begin{aligned}
&\mathbb{E}\big(Z_i^n\mid\mathcal{G}_{i-1}^n\big)\ \leq\ K\big(1+\kappa\Delta(n,i)^w\phi_n\big),\\
&\mathbb{E}\big(Z(m)_i^n\mid\mathcal{G}_{i-1}^n\big)\ \leq\ \tfrac{K}{m}+\kappa K\Delta(n,i)^w\phi_n.
\end{aligned}\right.\qquad \text{where } w=\varpi(q-r)+1-\frac{q}{2}.$$

Then under the conditions of (b), we end up with the following, instead of (9.2.15):

$$\mathbb{E}\big(\zeta(j,m)_i^n \mid \mathcal{G}_{i-1}^n\big) \leq K\Delta(n,i)\left(\frac{1}{m} + \phi_n\right).$$

Then $\mathbb{E}(U_t^{n,m,j}) \leq K(\phi_n + 1/m)\,\mathbb{E}(r_n^{p-1}\,H(p)_t^n)$, which by (14.2.11) is smaller than $K_t(\phi_n + 1/m)$. Then we obtain (14.2.22), and this finishes the proof. □

14.3 Central Limit Theorem for Normalized Functionals

14.3.1 The Results

Here we give a central limit theorem associated with Theorem 14.2.1. Although it is possible to give a CLT when the test function F depends on k successive increments, we restrict our attention to the case where a single increment is involved, the result for $k \geq 2$ being much more difficult to prove. That is, we have a function f on $\mathbb{R}^d$ and a real $p \geq 0$, and we are looking for the limiting behavior of the processes

$$\begin{aligned}\overline{V}'^n(f,p,X) &= \sqrt{r_n}\,\big(V'^n(f,p,X) - V'(f,p,X)\big)\\ \overline{V}'^n(f,p,\upsilon_n-,X) &= \sqrt{r_n}\,\big(V'^n(f,p,\upsilon_n-,X) - V'(f,p,X)\big),\end{aligned}$$

where $V'(f,p,X)$ is given by (14.2.7). The assumptions on X will be the same as in Theorems 11.2.1 or 13.2.1; that is, one of these:

Assumption (K) We have (H) and

$$\sigma_t = \sigma_0 + \int_0^t \widetilde{b}_s\,ds + \int_0^t \widetilde{\sigma}_s\,dW_s + M_t + \sum_{s\leq t} \Delta\sigma_s\,1_{\{\|\Delta\sigma_s\|>1\}}$$

where M is a local martingale with $\|\Delta M_t\| \leq 1$, orthogonal to W, and $\langle M,M\rangle_t = \int_0^t a_s\,ds$ and the compensator of $\sum_{s\leq t} 1_{\{\|\Delta\sigma_s\|>1\}}$ is $\int_0^t \widetilde{a}_s\,ds$, with the following properties: the processes $\widetilde{b}$, $\widetilde{\sigma}$, $\widetilde{a}$ and a are progressively measurable, the processes $\widetilde{b}$, a and $\widetilde{a}$ are locally bounded, and the processes $\widetilde{\sigma}$ and b are càdlàg or càglàd.

Assumption (K-r) (for $r \in [0,1)$) We have (K) except for the càdlàg or càglàd property of b, and $\|\delta(\omega,t,z)\| \wedge 1 \leq \Gamma_n(z)$ for all (ω,t,z) with $t \leq \tau_n(\omega)$, where (τ_n) is a localizing sequence of stopping times and the Borel functions Γ_n on E satisfy $\int \Gamma_n(z)^r \lambda(dz) < \infty$. Moreover the process $b'_t = b_t - \int_{\{\|\delta(t,z)\|\leq 1\}} \delta(t,z)\,\lambda(dz)$ is càdlàg or càglàd.

Assumption (K') We have (K) and both processes c_t and c_{t-} take their values in $\mathcal{M}_{d\times d}^{++}$.

Assumption (K'-r) We have (K-r) and both processes c_t and c_{t-} take their values in $\mathcal{M}^{++}_{d\times d}$.

Concerning the discretization scheme, Assumption (D) is not enough. The limiting process θ_t of the intensities $r_n\theta^n_t$ is a kind of "volatility" of the inter-observation times, so we need it to satisfy the same properties as σ_t in (K), for example. Moreover, the rate in the CLT will be $1/\sqrt{r_n}$, because the number of observations up to time t is $N_n(t)$, whose order of magnitude is r_n. However, in the genuine irregular case we also have an intrinsic variability in the sampling times which might overcome the "statistical" variability, and we want to avoid this. Taking care of all of these problems leads us to impose the following assumption:

Assumption 14.3.1 (or **(E)**) The discretization scheme is a restricted scheme satisfying (D) and also

$$\sqrt{r_n}\,\big(r_n\,\theta^n - \theta\big) \stackrel{\text{u.c.p.}}{\Longrightarrow} 0, \qquad \sqrt{r_n}\,\big(m'_p(n) - m'_p\big) \to 0 \tag{14.3.1}$$

for all $p \geq 0$. Furthermore the process θ_t is an Itô semimartingale relative to the filtration $(\mathcal{F}_t)$, and it satisfies the same properties as σ_t does in Assumption (K).

Now we turn to the test function f, which is q-dimensional. As in Chap. 11, see (11.2.6), we have an integer $q' \in \{0,\dots,q\}$ and, when $q' < q$, a subset B of $(\mathbb{R}^d)^k$ which is a finite union of affine hyperplanes, and we suppose that

$$\begin{aligned} &j \le q' \;\Rightarrow\; x \mapsto f^j(x) \text{ is } C^1 \text{ on } \big(\mathbb{R}^d\big)^k \\ &j > q' \;\Rightarrow\; x \mapsto f^j(x) \text{ is continuous on } \big(\mathbb{R}^d\big)^k \text{ and } C^1 \text{ outside } B. \end{aligned} \tag{14.3.2}$$

We denote by $d(z,B)$ the distance between $z \in \mathbb{R}^d$ and B, and we reproduce the conditions (11.2.7)–(11.2.10) (with p'' instead of p, because here p is already used in the definition of $V'^n(f,p,X)$, and $k=1$):

$$\big\|f(x)\big\| \le K\big(1+\|x\|^{p''}\big) \tag{14.3.3}$$

$$\big|\nabla f^j(x)\big| \le \begin{cases} K\big(1+\|x\|^{p''}\big) & \text{if } j \le q' \\ K\big(1+\|x\|^{p''}\big)\big(1+\frac{1}{d(x,B)^{1-w}}\big) & \text{if } j > q' \text{ and } x \in B^c \end{cases} \tag{14.3.4}$$

$$\begin{aligned} &x \in B^c,\ \|y\| \le 1 \bigwedge \frac{d(x,B)}{2},\ j > q' \\ &\quad\Rightarrow \big|\nabla f^j(x+y) - \nabla f^j(x)\big| \le K\,\|y\|\left(1+\frac{1}{d(x,B)^{2-w}}\right)\big(1+\|x\|^{p''}\big) \end{aligned} \tag{14.3.5}$$

$$\big\|f(x+y)-f(x)\big\| \le K\,\big(\|y\|^s + \|y\|^{s'}\big)\big(1+\|x\|^{p'}\big). \tag{14.3.6}$$

Recall that $0 < w \le 1$ and $0 < s \le s'$ and $p'', p' \ge 0$.

Before stating the result, it is perhaps worthwhile to recall that we have a basic filtered space $(\Omega, \mathcal{F}, (\mathcal{F}_t)_{t\geq 0}, \mathbb{P})$ on which X is defined, and also a σ-field $\mathcal{G}$ which is bigger than $\mathcal{F}$ and on which the probability measure $\mathbb{P}$ is also well defined. However, the stable convergence in law proved below *is relative to the σ-field $\mathcal{F}$*. This is in accordance with the standing notation of this book, see for example (2.2.8).

Theorem 14.3.2 *Suppose that we have a restricted discretization scheme satisfying* (E) *(Assumption* 14.3.1*). Suppose also that the test function f is continuous, globally even, with the property* (14.3.2), *and satisfies at least* (14.3.3) *and* (14.3.4).

(i) *Assume further any one of the following five sets of hypotheses*:

(a) *We have $q' = q$ and* (K) *and X is continuous.*
(b) *We have* (K') *and* (14.3.5) *and X is continuous.*
(c) *We have $q' = q$ and* (K-1)*, and f and ∇f are bounded.*
(d) *We have $q' = q$ and* (K-*r*) *for some $r < 1$ and* (14.3.6) *with $r \leq s \leq s' < 1$ and $p' \leq 2$.*
(e) *We have* (K'-*r*) *for some $r \in (0, 1)$, and* (14.3.5) *and* (14.3.6) *with $r \leq s \leq s' < 1$ and $p' \leq 2$.*

Then, for any $p \geq 0$, the q-dimensional processes $\overline{V}'^n(f, p, X)$ converges stably in law (relative to $\mathcal{F}$) to a limiting process $\overline{V}'(f, p, X)$ which is a continuous process defined on a very good filtered extension $(\widetilde{\Omega}, \widetilde{\mathcal{F}}, (\widetilde{\mathcal{F}}_t)_{t\geq 0}, \widetilde{\mathbb{P}})$ of $(\Omega, \mathcal{F}, (\mathcal{F}_t)_{t\geq 0}, \mathbb{P})$, and conditionally on $\mathcal{F}$ is a continuous centered Gaussian martingale with independent increments satisfying

$$\widetilde{\mathbb{E}}\big(\overline{V}'(f^i, p, X)_t\, \overline{V}'(f^j, p, X)_t \mid \mathcal{F}\big) = \int_0^t \widehat{R}_{c_s}(p; f^i, f^j)\, (\theta_s)^{2p-1}\, ds,$$

where $\widehat{R}_a(p; g, h)$ is defined for any two real-valued functions g and h by

$$\widehat{R}_a(p; g, h) = m'_{2p}\, \rho_a(gh) - \big(2m'_{p+1}\, m'_p - (m'_p)^2\, m'_2\big)\rho_a(g)\rho_a(h). \tag{14.3.7}$$

(ii) *The same is true of the (truncated) processes $\overline{V}'^n(f, p, v_n-, X)$, provided we have either* (a) *or* (b) *above, or any one of the following two sets of hypotheses*:

(f) *We have $q' = q$ and* (K-*r*) *for some $r \in (0, 1]$, and* (14.3.6) *with $r \leq s \leq 1 \leq s'$, and*

$$\begin{aligned} &r = 1 \;\Rightarrow\; s = s' = 1 \\ &r < 1 \;\Rightarrow\; s \geq \frac{r}{2-r}, \quad \varpi \geq \frac{s'-1}{2(s'-r)}. \end{aligned} \tag{14.3.8}$$

(g) *We have* (K'-*r*) *with some $r \in (0, 1]$, and* (14.3.5) *and* (14.3.6) *with $r \leq s \leq 1 \leq s'$, as well as* (14.3.8).

14.3.2 Preliminaries

In this subsection we perform our usual localization procedure and we eliminate the jumps: this means that we reduce the problem to proving (i) of the theorem in cases (a) and (b).

1) First, by the localization lemma, we can replace (K), (K'), (K-r) or (K'-r), according to the case, by (SK), (SK'), (SK-r) or (SK-r) (for which the processes X_t, b_t, $\widetilde{b}_t$, σ_t, $\widetilde{\sigma}_t$, a_t, a'_t, and also c_t^{-1} in cases of (SK') and (SK'-r), are bounded, and all $\|\delta(\omega,t,z)\| \leq \Gamma(z)$ with Γ bounded and $\int \Gamma(z)^r \lambda(dz < \infty)$. By the same token, recalling that (E) implies that θ_t has a decomposition similar to the decomposition of σ_t in Assumption (K), we can assume that the coefficients of this decomposition satisfy the same boundedness hypotheses: we then say that we have (SE).

Next, exactly as in Lemma 14.1.5 we can assume that

$$\frac{1}{c} \leq \theta_t \leq c, \qquad \frac{1}{c} \leq r_n \theta_t^n \leq c \tag{14.3.9}$$

for some constant $c > 1$ and, in view of (14.3.1), that

$$\sqrt{r_n}\,\big|r_n \theta_t^n - \theta_t\big| \leq c. \tag{14.3.10}$$

Again as in the proof of Lemma 14.1.5, one may also assume that for some $\gamma \in (0, \frac{1}{8} \wedge \frac{1}{8p}]$ (which, apart from those bounds, is not specified now, but will only depend on the test function f), we have

$$\varepsilon(n,i) \leq r_n^\gamma, \quad \text{hence } r_n\,\Delta(n,i) \leq C r_n^\gamma \leq C r_n^{\frac{1}{8} \wedge \frac{1}{8p}}. \tag{14.3.11}$$

Finally, exactly as in the proof of Theorem 14.2.1 we may assume

$$N_n(t) \leq C t r_n, \qquad u \in U \Rightarrow H(u)_t^n \leq K_u t r_n^{1-u}, \tag{14.3.12}$$

for any given finite set U: this is the same as (14.2.11) for $p = 0$ and all $p \in U$, and below we need U to contain $3/2$, 2, 3 and $p+1$ (in fact, using Hölder's inequality, one might prove that if the above holds for $U = \{u_0\}$, then it holds for $U = [0, u_0]$, but we do not need this refinement here).

2) The next step consists in eliminating the jumps, for claim (i). Since we have at least (SK-1) we may write, as in (5.3.13):

$$X = X' + X'' \quad \text{where } X'_t = X_0 + \int_0^t b'_s\,ds + \int_0^t \sigma_s\,dW_s, \qquad X'' = \delta * \mathfrak{p},$$

where $b'_t = b_t - \int_{\{z:\|\delta(t,z)\| \leq 1\}} \delta(t,z)\,\lambda(dz)$ is bounded. The elimination of jumps amounts to the following, analogous to Lemma 11.2.4:

Lemma 14.3.3 *Under the assumptions of Theorem* 14.3.2-(i), *plus the reinforced assumptions of Step* 1 *above, we have*

$$\sqrt{r_n}\left(V'^n(f,p,X) - V'^n(f,p,X')\right) \stackrel{\text{u.c.p.}}{\Longrightarrow} 0.$$

Proof We follow the proof of Lemma 5.3.13, page 150. It suffices to show that

$$\eta_i^n = r_n^{p-1/2}\,\Delta(n,i)^p\left(f\left(\Delta_i^n X/\sqrt{\Delta(n,i)}\right) - f\left(\Delta_i^n X'/\sqrt{\Delta(n,i)}\right)\right)$$

satisfies

$$\mathbb{E}\left(\sum_{i=1}^{N_n(t)} \|\eta_i\|\right) \to 0. \tag{14.3.13}$$

Below we use the notation

$$\alpha_i^n = 1 + \frac{\|\Delta_i^n X\|^2}{\Delta(n,i)}, \qquad \alpha_i'^n = 1 + \frac{\|\Delta_i^n X'\|^2}{\Delta(n,i)}, \qquad \overline{\alpha}_i'^n = \frac{\|\Delta_i^n X''\|}{\sqrt{\Delta(n,i)}}.$$

In case (c), f is C^1 with bounded derivatives. In cases (d) and (e) we have (14.3.6) with $p' \le 2$ and $s = r \le s' < 1$, hence

$$\|\eta_i^n\| \le \begin{cases} K r_n^{p-1/2}\Delta(n,i)^p\,(1 \wedge \overline{\alpha}_i^n) & \text{in case (c)} \\ K r_n^{p-1/2}\Delta(n,i)^p\,((\overline{\alpha}_i^n)^r + (\overline{\alpha}_i^n)^{s'})\,\alpha_i^n & \text{in cases (d), (e).} \end{cases}$$

Next, we recall the σ-fields $\mathcal{G}_i^n$ of (14.1.6). Observing that conditionally on $\mathcal{G}_{i-1}^n$ the increment $\Delta_i^n X''$ is independent of $\Delta(n,i)$ (since the latter is $\mathcal{G}_{i-1}^n$-measurable), we can apply (2.1.47) with $p = r = 1$ and $q = 1/2$ to the increment $\Delta_i^n X''$, with $\mathcal{F}_{T(n,i-1)}$ replaced by $\mathcal{G}_{i-1}^n$. This gives us that under (SK-1), and for some sequence $\phi_n \to 0$,

$$\mathbb{E}\left(1 \wedge \overline{\alpha}_i^n \mid \mathcal{G}_{i-1}^n\right) \le K\sqrt{\Delta(n,i)}\,\phi_n. \tag{14.3.14}$$

In the same way using (2.1.40) and (2.1.44), we get that under (SK-r) and $r \le s' < 1$:

$$\mathbb{E}\left(\alpha_i^n + \alpha_i'^n \mid \mathcal{G}_{i-1}^n\right) \le K, \quad \mathbb{E}\left((\overline{\alpha}_i^n)^{s'} \mid \mathcal{G}_{i-1}^n\right) \le K\Delta(n,i)^{1-s'/2}.$$

Since $\mathbb{E}(\Delta(n,i)^v \mid \mathcal{F}_{T(n,i-1)}^n) \le K_v/r_n^v$ for all $v \ge 0$ by (14.3.9) and $m_v'(n) \le K_v$, we can use (14.1.7) to get that $\mathbb{E}(\|\eta_i^n\| \mid \mathcal{G}_{i-1}^n) \le K\phi_n/r_n$, where ϕ_n is as in (14.3.14) in case (c) and $\phi_n = 1/r_n^{(1-s')/2}$ in cases (d,e). Then the left side of (14.3.13) is smaller than $K\phi_n\,\mathbb{E}(N_n(t))$, which by (14.3.12) is smaller than $Kt\phi_n$. Thus (14.3.13) follows because in all cases $\phi_n \to 0$. □

3) In this step we show that, under our strengthened assumptions, (ii) follows from (i). To this end, we essentially reproduce the proof of Theorem 13.2.1, pages 384–386.

Suppose first that X is continuous, so $\mathbb{E}(\|\Delta_i^n X\|^m \mid \mathcal{G}_{i-1}^n) \leq K_m \Delta(n,i)^{m/2}$ for all $m > 0$. Since $\varpi < 1/2$, Markov's inequality yields

$$\mathbb{P}\big(\|\Delta_i^n X\| > \alpha\Delta(n,i)^{\varpi} \mid \mathcal{G}_{i-1}^n\big) \leq K\Delta(n,i)^2.$$

Hence the set $\Omega_t^n = \{\|\Delta_i^n X\| \leq \alpha\Delta(n,i)^{\varpi} : i = 1, 2, \ldots, N_n(t)\}$ satisfies

$$\begin{aligned}\mathbb{P}\big((\Omega_t^n)^c\big) &\leq \mathbb{E}\left(\sum_{i=1}^{N_n(t)} 1_{\{\|\Delta_i^n X\| > \alpha\Delta(n,i)^{\varpi}\}}\right) \leq K\mathbb{E}\left(\sum_{i=1}^{N_n(t)} \Delta(n,i)^2\right) \\ &= K\mathbb{E}\big(H(2)_t^n\big) \leq \frac{K}{r_n}\end{aligned}$$

by (14.3.12). Since on the set Ω_t^n we have $V'^n(f, p, v_n-, X)_s = V'^n(f, p, X)_s$ for all $s \leq t$, we deduce (ii) from (i).

Next we consider the discontinuous case, so (SK-1) holds. Assuming (i), what precedes shows that (ii) holds true for the continuous process $X' = X - \delta * \mathfrak{p}$ (which has the same volatility σ_t as X), hence it remains to prove that

$$\frac{1}{\sqrt{r_n}}\big(V'^n(f, p, v_n-, X) - V'^n\big(f, p, v_n-, X'\big)\big) \stackrel{\text{u.c.p.}}{\Longrightarrow} 0.$$

We modify (13.2.20) (with $k = 1$, so the index j is absent)) as follows:

$$\eta_i^n = f\left(\frac{\Delta_i^n X}{\sqrt{\Delta(n,i)}}\right) 1_{\{\|\Delta_i^n X\| \leq v(n,i)\}} - f\left(\frac{\Delta_i^n X}{\sqrt{\Delta(n,i)}}\right).$$

The proof of Lemma 13.2.6, page 384, works with obvious changes of notation. This gives us, by taking $m = 1$ in this lemma, and under (SH-r) for some $r \leq 1$ and (14.3.6) for some $p' \geq 0$ and $r \leq s \leq 1 \leq s'$:

$$\mathbb{E}\big(\|\eta_i^n\|^m \mid \mathcal{G}_{i-1}^n\big) \leq \big(\Delta(n,i)^{\frac{2-r}{2}} + \Delta(n,i)^{(1-r\varpi)-s'\frac{1-2\varpi}{2}}\big)\phi_n$$

where ϕ_n is a sequence of non random numbers going to 0; note that θ does not appear here because we assume $r \leq 1$, and there is no condition on ϖ (other than being inside $(0, 1/2)$, of course) because here $k = 1$. Then if (14.3.8) holds, we have

$$\mathbb{E}\big(\|\eta_i^n\| \mid \mathcal{G}_{i-1}^n\big) \leq K\sqrt{\Delta(n,i+j)}\,\phi_n. \tag{14.3.15}$$

Now, if $A_t^n = r_n^{p-1/2}\sum_{i=1}^{N_n(t)} \Delta(n,i)^p \|\eta_i^n\|$, we have

$$\sup_{s \leq t}\left|\frac{1}{\sqrt{r_n}}\big(V'^n(f, p, v_n-, X)_s - V'^n\big(f, p, v_n-, X'\big)_s\big)\right| \leq A_t^n.$$

Thus it remains to prove that $A_t^n \stackrel{\mathbb{P}}{\longrightarrow} 0$ for all $t > 0$. For this, we can use the fact that $T(n,i)$ is $\mathcal{G}_{i-1}^n$-measurable and (14.3.15) to get

$$\begin{aligned}\mathbb{E}\big(A_t^n\big) &\le r_n^{p-1/2} \sum_{i\ge 1} \mathbb{E}\big(\Delta(n,i)^p \, \|\eta_i^n\| \, 1_{\{T(n,i)\le t\}}\big)\\ &\le K\phi_n \, r_n^{p-1/2} \, \mathbb{E}\big(H(p+1/2)_t^n\big) \;\le\; Kt\,\phi_n,\end{aligned}$$

where the last inequality comes from (14.3.12). Since $\phi_n \to 0$, the result is proved.

14.3.3 The Scheme of the Proof when X is Continuous

In view of the preliminaries above, it remains to prove (i) in cases (a) and (b), and under the strengthened assumptions. Thus below X is *continuous*, and we are in the setting of Sect. 5.3.3, page 151, that is ($\mathcal{M}_A$ is the set of all $d \times d'$ matrices such that $\|\alpha\| \le A$, and $\mathcal{M}'_A$ the set of all $\alpha \in \mathcal{M}_A$ such that $\alpha\alpha^*$ is invertible and $\|(\alpha\alpha^*)^{-1}\| \le A$):

$$\begin{aligned}&\text{(a) } q'=q, \quad \text{(SK)}, \; f \text{ is } C^1 \text{ and } \nabla F \text{ has polynomial growth}, \;\; \sigma_t \in \mathcal{M}_A\\ &\text{(b) } q'<q, \quad \text{(SK')}, \; f \text{ satisfies (14.3.3), (14.3.4), (14.3.5)}, \;\; \sigma_t \in \mathcal{M}'_A.\end{aligned} \tag{14.3.16}$$

We then set

$$\begin{aligned}&\sigma_i^n = \sigma_{T(n,i)}, \qquad c_i^n = c_{T(n,i)}, \qquad \overline{\theta}_i^n = r_n \theta_{T(n,i)}^n\\ &w_i^n = \frac{\Delta_i^n W}{\sqrt{\Delta(n,i)}}, \qquad \beta_i^n = \sigma_{i-1}^n w_i^n.\end{aligned}$$

Let us also introduce the variables

$$\begin{aligned}&\chi_i^n = r_n^{p-1/2} \, \Delta(n,i)^p \, \big(f\big(X_i^n\big) - f\big(\beta_i^n\big)\big)\\ &\chi_i'^n = \mathbb{E}\big(\chi_i^n \mid \mathcal{G}_{i-1}^n\big), \qquad \chi_i''^n = \chi_i^n - \chi_i'^n\\ &\zeta_i^n = r_n^{p-1/2} \, \Delta(n,i)^p \, f\big(\beta_i^n\big) - \sqrt{r_n}\, m'_p(n)\, \Delta(n,i)\, \big(\overline{\theta}_{i-1}^n\big)^{p-1} \rho_{c_{i-1}^n}(f),\end{aligned}$$

and the processes

$$Y_t^n = \sum_{i=1}^{N_n(t)} \zeta_i^n, \quad A^n(0)_t = \sum_{i=1}^{N_n(t)} \chi_i''^n, \quad A^n(1)_t = \sum_{i=1}^{N_n(t)} \chi_i'^n,$$

$$A^n(2)_t = \sqrt{r_n}\left(m'_p(n) \sum_{i=1}^{N_n(t)} \Delta(n,i)\, \big(\overline{\theta}_{i-1}^n\big)^{p-1} \rho_{c_{i-1}}(f) - m'_p \int_0^t \rho_{c_s}(f)\, (\theta_s)^{p-1}\, ds\right).$$

We see that $\overline{V}'^n(f,p,X) = Y^n + A^n(0) + A^n(1) + A^n(2)$, and the claim (i) will follow from the next two lemmas:

Lemma 14.3.4 *Under the assumptions* (14.3.16) *and* (14.3.9)–(14.3.12) *the processes* Y^n *converge stably in law to* $\overline{V}'(f, p, X)$.

Lemma 14.3.5 *Under the assumptions* (14.3.16) *and* (14.3.9)–(14.3.12) *we have for* $j = 0, 1, 2$:

$$A^n(j) \stackrel{\text{u.c.p.}}{\Longrightarrow} 0. \tag{14.3.17}$$

14.3.4 Proof of Lemma 14.3.4

Step 1) The proof is based on the general criterion for stable convergence of triangular arrays, given by Theorem 2.2.15. The discrete-time filtrations which we have to consider are not the $(\mathcal{G}_i^n)$, as defined by (14.1.7), but $(\mathcal{F}^n_{T(n,i)})_{i\in\mathbb{N}}$, with respect to which the stopping rules $N_n(t)$ are not stopping times. To overcome this difficulty, instead of Y^n we consider the following processes

$$Y_t'^n = \sum_{i=1}^{N_n(t)+1} \zeta_i^n,$$

and use the fact that $N_n(t) + 1$ is an $(\mathcal{F}^n_{T(n,i)})$-stopping time. This is in the same spirit as the replacement of $V'^n(F, p, X)$ by $\mathcal{V}'^n(F, p, X)$ in the proof of Theorem 14.2.1 and, exactly as in that proof, the property $Y'^n \stackrel{\mathcal{L}\text{-s}}{\Longrightarrow} \overline{V}'(f, p, X)$ implies $Y^n \stackrel{\mathcal{L}\text{-s}}{\Longrightarrow} \overline{V}'(f, p, X)$.

Observe that ζ_i^n is $\mathcal{F}^n_{T(n,i)}$ measurable, and

$$\mathbb{E}\big(\zeta_i^n \mid \mathcal{G}_{i-1}^n\big) = r_n^{p-1/2}\big(\Delta(n,i)^p - m_p'(n)\,\Delta(n,i)\,\big(\theta^n_{T(n,i-1)}\big)^{p-1}\big)\rho_{c^n_{i-1}}(f),$$

and since $\mathbb{E}(\Delta(n,i)^q \mid \mathcal{F}_{T(n,i-1)}) = m_q'(n)(\theta^n_{T(n,i-1)})^q$ and $m_1'(n) = 1$, we deduce $\mathbb{E}(\zeta_i^n \mid \mathcal{F}^n_{T(n,i-1)}) = 0$. Moreover, by Lemma 14.1.5 we have (2.2.39) with $(\Omega_n, \mathcal{G}^n, (\overline{\mathcal{F}}^n_t), \mathbb{P}_n) = (\Omega, \mathcal{G}, (\overline{\mathcal{F}}^n_t), \mathbb{P})$. Therefore, by virtue of Theorem 2.2.15, it is enough to prove the following three properties:

$$\sum_{i=1}^{N_n(t)+1} \mathbb{E}\big(\|\zeta_i^n\|^4 \mid \mathcal{F}^n_{T(n,i-1)}\big) \stackrel{\mathbb{P}}{\longrightarrow} 0 \tag{14.3.18}$$

$$\sum_{i=1}^{N_n(t)+1} \mathbb{E}\big(\zeta_i^{n,l}\,\zeta_i^{n,r} \mid \mathcal{F}^n_{T(n,i-1)}\big) \stackrel{\mathbb{P}}{\longrightarrow} \int_0^t \widehat{R}\big(p; f^l, f^r\big)\,(\theta_s)^{2p-1}\,ds \tag{14.3.19}$$

$$\sum_{i=1}^{N_n(t)+1} \mathbb{E}\big(\zeta_i^n\,\Delta_i^n N \mid \mathcal{F}^n_{T(n,i-1)}\big) \stackrel{\mathbb{P}}{\longrightarrow} 0 \tag{14.3.20}$$

for all $t > 0$ and $N = W^j$ for some $j = 1, \dots, d'$ or when N belongs to the set $\mathcal{N}$ of all bounded $(\mathcal{F}_t)$-martingales which are orthogonal to W.

Step 2) Because f is of polynomial growth and c_t is bounded, and upon using (14.3.9) and (14.3.11), we see that

$$\|\zeta_i^n\| \leq K\big(r_n^{p-1/2}\Delta(n,i)\big(1+\|w_i^n\|^v\big)+\sqrt{r_n}\Delta(n,i)\big) \leq \frac{K}{r_n^{1/2-p\gamma}}\big(1+\|w_i^n\|^v\big)$$

for some $v \geq 0$, hence the left side of (14.3.18) is smaller than $K(N_n(t)+1)/r_n^{2-4p\gamma}$, which by (14.3.12) is smaller than $Kt/r_n^{1-4p\gamma}$. Since $1 > 4p\gamma$ we deduce the property (14.3.18).

Next, (14.3.20) holds if each summand vanishes, and this is true if we have $\mathbb{E}(\zeta_i^n\, \Delta_i^n N \mid \mathcal{G}_{i-1}^n) = 0$. Observe that, since N is an $(\mathcal{F}_t)$-martingale,

$$\mathbb{E}\big(\zeta_i^n\, \Delta_i^n N \mid \mathcal{G}_i^n\big) = r_n^{p-1/2}\, \Delta(n,i)^p\, \mathbb{E}\big(f(\beta_i^n)\, \Delta_i^n N \mid \mathcal{G}_{i-1}^n\big).$$

Now, the conditional expectation $\mathbb{E}(f(\beta_i^n)\, \Delta_i^n N \mid \mathcal{G}_{i-1}^n)$ vanishes when $N \in \mathcal{N}$ by exactly the same argument as for 4.2.10, page 108. When $N = W^j$, this conditional expectation is simply $\sqrt{\Delta(n,i)}\,\mathbb{E}(f(U)U^j)$, where U is a standard d'-dimensional Gaussian random vector. Since f is globally even, this expectation vanishes, and thus (14.3.20) holds in all cases.

Step 3) Now we turn to (14.3.19). First, we have

$$\begin{aligned}\mathbb{E}\big(\zeta_i^{n,l}\zeta_i^{n,r} \mid \mathcal{G}_{i-1}^n\big) = {} & r_n^{2p-1}\, \Delta(n,i)^{2p}\, \rho_{c_{i-1}^n}(f^l f^r)\\ & - 2r_n^p\, m'_p(n)\, \Delta(n,i)^{p+1}\, \big(\overline{\theta}_{i-1}^n\big)^{p-1}\, \rho_{c_{i-1}^n}(f^r)\\ & + r_n\, m'_p(n)^2\, \Delta(n,i)^2\, \big(\overline{\theta}_{i-1}^n\big)^{2p-2}\, \rho_{c_{i-1}^n}(f^l)\, \rho_{c_{i-1}^n}(f^r).\end{aligned}$$

This yields

$$\begin{aligned}\mathbb{E}\big(\zeta_i^{n,l}\zeta_i^{n,r} \mid \mathcal{F}_{T(n,i-1)}^n\big) = {} & \frac{1}{r_n}\,\big(\overline{\theta}_{i-1}^n\big)^{2p}\big(m'_{2p}(n)\, \rho_{c_{i-1}^n}(f^l f^r)\\ & - \big(2m'_p(n)\, m'_{p+1}(n) - m'_p(n)^2\, m'_2(n)\big)\rho_{c_{i-1}^n}(f^r)\big).\end{aligned} \tag{14.3.21}$$

Now, (14.1.10) yields, for any càdlàg process Z:

$$\frac{1}{r_n}\sum_{i=1}^{N_n(t)+1} Z_{T(n,i-1)} \stackrel{\mathbb{P}}{\longrightarrow} \int_0^t \frac{Z_s}{\theta_s}\, ds,$$

from which we deduce that if $Z^n \stackrel{\text{u.c.p.}}{\Longrightarrow} Z$, then

$$\frac{1}{r_n}\sum_{i=1}^{N_n(t)+1} Z_{T(n,i-1)}^n \stackrel{\mathbb{P}}{\longrightarrow} \int_0^t \frac{Z_s}{\theta_s}\, ds.$$

In view of (14.3.21), the above applied with $Z_t^n = (r_n\theta_t^n)^{2p}(m'_{2p}(n)\rho_{c_t}(f^l f^r) - (2m'_p(n)m'_{p+1}(n) - m'_p(n)^2 m'_2(n)\rho_{c_t}(f^l)\rho_{c_t}(f^r))$, which converges in the *u.c.p.* sense to $Z_t = (\theta_t)^{2p}\,\widehat{R}_{c_t}(p; f^l, f^r)$, due to (14.3.1) and (14.3.7).

14.3.5 Proof of Lemma 14.3.5

For Lemma 14.3.5 we can and will assume that f is one-dimensional.

The case $j = 2$. The proof of (14.3.17) for $j = 2$ exhibits some significant differences with Lemma 11.2.7. We set

$$
\begin{aligned}
Z_s &= \rho_{c_s}(f), \qquad Z'_s = (\theta_s)^{p-1} Z_s \\
A^n(3)_t &= m'_p\sqrt{r_n}\left(\sum_{i=1}^{N_n(t)} Z'_{T(n,i-1)}\,\Delta(n,i) - \int_0^t Z'_s\,ds\right) \\
A^n(4)_t &= \left(m'_p(n) - m'_p\right)\sqrt{r_n}\sum_{i=1}^{N_n(t)} Z'_{T(n,i-1)}\,\Delta(n,i) \\
A^n(5)_t &= m'_p(n)\sqrt{r_n}\sum_{i=1}^{N_n(t)} Z_{T(n,i-1)}\left(\left(\overline{\theta}_{i-1}^n\right)^{p-1} - (\theta_{T(n,i-1)})^{p-1}\right)\Delta(n,i).
\end{aligned}
$$

Then $A^n(2) = A^n(3) + A^n(4) + A^n(5)$, and it is enough to prove (14.3.17) for $j = 3, 4, 5$.

Since $\sum_{i=1}^{N_n(t)} \Delta(n,i) \le t$ and Z_t and θ_t are (uniformly) bounded, (14.3.17) for $j = 4$ and $j = 5$ follow from (14.3.1).

Exactly as in Part B of Sect. 5.3.3 (pages 153–154), the function $\psi(\alpha) = \rho_{\alpha\alpha^*}(f)$ is C_b^∞ on the set $\mathcal{M} = \mathcal{M}_A$ in case (a) and $\mathcal{M} = \mathcal{M}'_A$ in case (b). The function $\overline{\psi}$ defined on $[1/C, C] \times \mathcal{M}$ by $\overline{\psi}(x,\alpha) = x^{p-1}\psi(\alpha)$ is also C_b^∞, so with the notation $V_t = (\theta_t, \sigma_t)$ we have, instead of (5.3.27),

$$
\begin{aligned}
\left|\overline{\psi}(V_t)\right| + \left\|\nabla\overline{\psi}(V_t)\right\| &\le K \\
\left|\overline{\psi}(V_t) - \overline{\psi}(V_s)\right| &\le K\,\|V_t - V_s\| \\
\left|\overline{\psi}(V_t) - \overline{\psi}(V_s) - \nabla\overline{\psi}(V_s)(V_t - V_s)\right| &\le \Psi\left(\|V_t - V_s\|\right)\|V_t - V_s\|
\end{aligned}
$$

for some constant K and some increasing function Ψ on $\mathbb{R}_+$, continuous and null at 0. Then $Z'_t = \overline{\psi}(V_t)$ and $A^n(3)_t^n = m'_p(-\overline{\eta}_t^n - \sum_{i=1}^{N_n(t)}(\eta_i^n + \eta_i'^n))$, where (recalling $I(n,i) = (T(n,i-1), T(n,i)]$):

$$
\overline{\eta}_t^n = \sqrt{r_n}\int_{T(n,N_n(t))}^t Z_s\,ds
$$

$$\eta_i^n = \sqrt{r_n}\, \nabla\overline{\psi}(V_{T(n,(i-1))}) \int_{I(n,i)} (V_u - V_{T(n,i-1)})\, du$$

$$\eta_i'^n = \sqrt{r_n} \int_{I(n,i)} \big(\overline{\psi}(V_u) - \overline{\psi}(V_{T(n,i-1)}) - \nabla\overline{\psi}(V_{T(n,i-1)})(V_u - V_{T(n,i-1)})\big)\, du.$$

Our assumptions (SK) and (SE) imply that the process V_t is an $(\mathcal{F}_t)$-Itô semimartingale whose "integrands" satisfy the same boundedness as those of σ_t in (SK). So exactly as on page 154 we obtain

$$\big|\mathbb{E}(\eta_i^n \mid \mathcal{G}_{i-1}^n)\big| \le K\sqrt{r_n}\, \Delta(n,i)^2, \qquad \mathbb{E}\big(|\eta_i^n|^2 \mid \mathcal{G}_{i-1}^n\big) \le K r_n \Delta(n,i)^3.$$

Following the same proof of Sect. 5.3.3, and with the notation $H(q)^n$ of (14.1.5), we then get

$$\mathbb{E}\bigg(\sup_{s\le t}\bigg|\sum_{i=1}^{N_n(s)} \eta_i^n\bigg|^2\bigg) \le K r_n\, \mathbb{E}\big(H(3)_t^n + (H(2)_t^n)^2\big) \le \frac{K_t}{r_n}, \tag{14.3.22}$$

where the last inequality comes from (14.3.12). We also obtain, for some nonnegative function $\overline{\Psi}$ which is continuous and vanishes at 0, that for all $\varepsilon > 0$

$$\mathbb{E}\big(|\eta_i'^n| \mid \mathcal{G}_{i-1}^n\big) \le K\sqrt{r_n}\bigg(\overline{\Psi}(\varepsilon)\Delta(n,i)^{3/2} + \frac{\Delta(n,i)^2}{\varepsilon}\bigg).$$

This yields by (14.3.12) again:

$$\begin{aligned}
\mathbb{E}\bigg(\sum_{i=1}^{N_n(t)} |\eta_i'^n|\bigg) &\le K\sqrt{r_n}\bigg(\overline{\Psi}(\varepsilon)\mathbb{E}\big(H(3/2)_t^n\big) + \frac{1}{\varepsilon}\,\mathbb{E}\big((H(2)_t^n)\big)\bigg)\\
&\le K_t\bigg(\overline{\Psi}(\varepsilon) + \frac{1}{\varepsilon\sqrt{r_n}}\bigg).
\end{aligned}$$

Here, $\varepsilon > 0$ is arbitrary and $\overline{\Psi}(\varepsilon) \to 0$ as $\varepsilon \to 0$. So, letting first $n \to \infty$ and then $\varepsilon \to 0$, we deduce from this and (14.3.22) and $|\overline{\eta}_t^n| \le K/r_n^{1/4}$ (which follows from (14.3.11)), that (14.3.17) holds for $j = 3$, hence also for $j = 2$.

The case $j = 0$. We start with a notation similar to (5.3.30), except that we use Θ_i^n instead of θ_i^n which in this chapter is used in the description of the sampling scheme:

$$\Theta_i^n = \frac{\Delta_i^n X}{\sqrt{\Delta(n,i)}} - \beta_i^n.$$

By the properties of restricted schemes and (SK) we get as for (5.3.31):

$$l > 0 \Rightarrow \mathbb{E}\big(\|\beta_i^n\|^l \mid \mathcal{G}_{i-1}^n\big) \le K_l, \quad \mathbb{E}\big(\|\Theta_i^n\|^l \mid \mathcal{G}_{i-1}^n\big) \le K_l\, \Delta(n,i)^{(l/2)\wedge 1}. \tag{14.3.23}$$

We have (11.2.19) (page 315) with some $p'' \geq 1$ (instead of p), hence in view of the previous estimates we have

$$\begin{aligned}\mathbb{E}\Big(\sum_{i=1}^{N_n(t)} |\chi_i^n|^2\Big) &\leq K\, r_n^{2p-1}\, \mathbb{E}\Big(\Big(\phi'_C(\varepsilon)^2 + \frac{1}{C^2}\Big) H(2p)_t^n + \frac{C^{2p''}}{\varepsilon^2} H(p+1)_t^n\Big) \\ &\leq K t\Big(\phi'_C(\varepsilon)^2 + \frac{1}{C^2} + \frac{C^{2p''}}{r_n\,\varepsilon^2}\Big),\end{aligned}$$

where the last inequality follows from (14.3.12). Then, taking first C large and then ε small, then n large, we deduce that $\mathbb{E}(\sum_{i=1}^{N_n(t)} |\chi_i^n|^2) \to 0$. Since the $\chi_i''^n$ are martingale increments relative to the filtration $(\mathcal{G}_i^n)_{i\in\mathbb{N}}$, the convergence $A^n(0) \stackrel{\text{u.c.p.}}{\Longrightarrow} 0$ follows.

The case $j = 1$. Here we reproduce the proof of (5.3.25), pages 154–160, with a vanishing limit.

1) We have $\Theta_i^n = \frac{1}{\sqrt{\Delta(n,i)}} \sum_{j=1}^4 \zeta(j)_i^n$, where

$$\begin{aligned}\zeta(1)_i^n &= \Delta(n,i+j-1)\, b_{T(n,i-1)} \\ \zeta(2)_i^n &= \int_{I(n,i)} \big(\widetilde{\sigma}_{T(n,i-1)}(W_s - W_{T(n,i-1)})\big)\, dW_s \\ \zeta(3)_i^n &= \int_{I(n,i)} \big(M'_s - M'_{T(n,i-1)}\big)\, dW_s \\ \zeta(4)_i^n &= \int_{I(n,i)} (b_s - b_{T(n,i-1)})\, ds + \int_{I(n,i)} \Big(\int_{T(n,i-1)}^s \widetilde{b}'_u\, du\Big)\, dW_s \\ &\quad + \int_{I(n,i)} \Big(\int_{T(n,i-1)}^s (\widetilde{\sigma}_u - \widetilde{\sigma}_{T(n,i-1)})\, dW_u\Big)\, dW_s.\end{aligned}$$

We set $A_i^n = \{\|\Theta_i^n\| > d(\beta_i^n, B)/2\}$ (with the convention $B = \emptyset$ when f is C^1 everywhere, as in case (a)). Then, upon substituting θ_i^n with Θ_i^n, we have the decomposition (5.3.32), hence also (5.3.33): that is, $A^n(1) = \sum_{r=1}^7 D^n(r)$, where now

$$\begin{aligned}D^n(r)_t \;&=\; \sum_{i=1}^{N_n(t)} \delta(r)_i^n, \quad \text{where } \delta(r)_i^n \;=\; \mathbb{E}\big(\delta'(r)_i^n \mid \mathcal{G}_{i-1}\big) \quad \text{and} \\ \delta'(r)_i^n &= r_n^{p-1/2}\, \Delta(n,i)^{p-1/2}\, \nabla f\big(\beta_i^n\big)\, \zeta(r)_i^n \quad \text{for } r = 1,2,3,4 \\ \delta'(5)_i^n &= -r_n^{p-1/2}\, \Delta(n,i)^p\; \nabla f\big(\beta_i^n\big)\, \Theta_i^n\, 1_{A_i^n} \\ \delta'(6)_i^n &= r_n^{p-1/2}\, \Delta(n,i)^p\, \big(\nabla f\big(\beta_i^n + u_i^n \Theta_i^n\big) - \nabla f\big(\beta_i^n\big)\big)\, \Theta_i^n\, 1_{(A_i^n)^c} \\ \delta'(7)_i^n &= r_n^{p-1/2}\, \Delta(n,i)^p\, \big(f\big(\beta_i^n + \Theta_i^n\big) - f\big(\beta_i^n\big)\big)\, 1_{A_i^n}\end{aligned}$$

and u_i^n is some [0, 1]-valued random variable (we use vector notation above).

2) Next, we replace (5.3.35) by

$$\alpha_i^n = \Delta(n,i)^{3/2} + \mathbb{E}\left(\int_{I(n,i)} \left(\|b_s - b_{T(n,i-1)}\|^2 + \|\widetilde{\sigma}_s - \widetilde{\sigma}_{T(n,i-1)}\|^2\right) ds \mid \mathcal{G}_{i-1}^n\right), \tag{14.3.24}$$

and (5.3.36) and (5.3.37) become, for all $l \geq 2$:

$$\begin{aligned}
\mathbb{E}\left(\|\zeta(1)_i^n\|^l + \|\zeta(2)_i^n\|^l \mid \mathcal{G}_{i-1}^n\right) &\leq K_l\, \Delta(n,i)^l \\
\mathbb{E}\left(\|\zeta(4)_i^n\|^l \mid \mathcal{G}_{i-1}^n\right) &\leq K_l\, \Delta(n,i)^{l-1}\, \alpha_i^n \\
\mathbb{E}\left(\|\zeta(3)_i^n\|^l \mid \mathcal{G}_{i-1}^n\right) &\leq K_l\, \Delta(n,i)^{l/2+(1\wedge(l/2))}.
\end{aligned}$$

We also use the same notation

$$\gamma_i^n = \begin{cases} 1 & \text{if } w = 1 \\ \phi_B(\beta_{i,j}^n) & \text{if } w < 1, \end{cases} \qquad \phi_B(x) = 1 + \frac{1}{d(x,B)},$$

as in (5.3.38), where w is as in (14.3.4) and (14.3.5) in case (b), and $w = 1$ in case (a).

At this stage, we reproduce the proof of Lemma 5.3.15, with the conditional expectations relative to $\mathcal{G}_{i-}^n$ instead of ordinary expectations. This gives us, for any sequence of variables Φ_i^n satisfying

$$s > 0 \quad \Rightarrow \quad \sup_{n,i} \mathbb{E}\left(|\Phi_i^n|^r \mid \mathcal{G}_{i-1}^n\right) < \infty,$$

the following estimates:

$$\begin{aligned}
& r = 1,2,3,\ l < 2 \Rightarrow \mathbb{E}\left(|\Phi_i^n|^s\, \|\zeta(r)_i^n\|^l\, (\gamma_i^n)^m \mid \mathcal{G}_{i-1}^n\right) \leq K_{s,l,m}\, \Delta(n,i)^l \\
& u \in \left(0, (1-m) \wedge \frac{l}{2}\right) \Rightarrow \mathbb{E}\left(|\Phi_i^n|^s\, \|\zeta(4)_i^n\|^l\, (\gamma_i^n)^m \mid \mathcal{G}_{i-1}^n\right) \\
& \qquad \leq K_{s,l,m,u}\, \Delta(n,i+j)^{l-u}\, (\alpha_i^n)^u \leq K_{s,l,m,u} \Delta(n,i)^l.
\end{aligned} \tag{14.3.25}$$

3) Note that, when $r = 1,2,3$, by exactly the same argument as in Step 5 of the proof of Lemma 11.2.7, we see that indeed $\delta(r)_i^n = 0$ for all i, due to the fact that f is globally even. So it remains to prove $D^n(r) \stackrel{\text{u.c.p.}}{\Longrightarrow} 0$ for $r = 4,5,6,7$. This is Lemma 5.3.16, page 158, whose proof is followed below with the relevant modifications.

We first have to state the analogue of (5.3.46). Letting $v \in (0,1/2)$, we set

$$L(v)_t^n = r_n^{p-1/2} \sum_{i=1}^{N_n(t)} \Delta(n,i)^{p+1/2-v} \left(\alpha_i^n\right)^v.$$

By Hölder's inequality, and if $L_t'^n = \mathbb{E}(\sum_{i=1}^{N_n(t)} \alpha_i^n)$, we get

$$\mathbb{E}\left(L(u)_t^n\right) \leq r_n^{p-1/2} \left(L_t'^n\right)^v \left(\mathbb{E}\left(H\left(\frac{2p+1-2v}{2(1-v)}\right)_t^n\right)\right)^{1-v} \leq K t^{1-v} \left(L_t'^n\right)^v,$$

where the last inequality follows from (14.3.12). Since $N_n(t)$ is a $(\mathcal{G}_i^n)$-stopping time, we have

$$L_t'^n = \mathbb{E}\left(H(3/2)_t^n + \int_0^{T(n,N_n(t))} \left(\|b_s - b_{T(n,N_n(s))}\|^2 + \|\widehat{\sigma}_s - \widehat{\sigma}_{T(n,N_n(s))}\|^2\right) ds \right).$$

The variable whose expectation is taken in the right side above goes pointwise to 0 (use (14.3.12) and the càdlàg or càglàd properties of b' and $\widetilde{\sigma}$) and is bounded, hence $L_t'^n \to 0$, and we deduce

$$\mathbb{E}\big(L(v)_t^n\big) \to 0. \tag{14.3.26}$$

Next, the property (5.3.21) is valid here, hence (5.3.48) as well. Then (14.3.25) applied with $l = 1$ and $m = 1 - w$ and some $v \in (0, 1/2)$ yields

$$\mathbb{E}\left(\sum_{i=1}^{N_n(t)} |\delta(4)_i^n| \right) \le \mathbb{E}\left(\sum_{i=1}^{N_n(t)} |\delta'(4)_i^n| \right) \le K\,\mathbb{E}\big(L(v)_t^n\big)$$

with the notation of the previous step. Hence (14.3.26) yields $D^n(4) \stackrel{\text{u.c.p.}}{\Longrightarrow} 0$.

In case (a) we have $D^n(5) = D^n(7) = 0$ (as in Lemma 5.3.16, page 158), and now we prove $D^n(6) \stackrel{\text{u.c.p.}}{\Longrightarrow} 0$ in case (a). We can reproduce the corresponding proof in Lemma 5.3.16, with Θ_i^n instead of θ_i^n, and using (14.3.23). Instead of (5.3.49) we end up, for all $C > 1$ and $\varepsilon \in (0, 1)$ and for some $p'' \ge 1$, with

$$\mathbb{E}\big(|\delta'(6)_i^n| \mid \mathcal{G}_{i-1}^n\big) \le K r_n^{p-1/2}\,\mathbb{E}\left(\Delta(n,i)^{p+1/2} \left(\phi_C'(\varepsilon) + \frac{1}{C} + \frac{C^{p''}}{\varepsilon} \sqrt{\Delta(n,i)} \right) \right)$$

Then, since

$$\mathbb{E}\Big(\sup_{s\le t} |D^n(6)_s|\Big) \le \mathbb{E}\left(\sum_{i=1}^{N_n(t)} |\delta(6)_i^n| \right) \le \mathbb{E}\left(\sum_{i=1}^{N_n(t)} |\delta'(6)_i^n| \right)$$

(using again the property of $N_n(t)$ to be a $(\mathcal{G}_i^n)$-stopping time), we deduce

$$\begin{aligned}
\mathbb{E}\Big(\sup_{s\le t} |D^n(6)_s|\Big) &\le K r_n^{p-1/2} \left(\phi_C'(\varepsilon) + \frac{1}{C} \right) \mathbb{E}\big(H(p+1/2)_t^n\big) \\
&\quad + K r_n^{p-1/2} \frac{C^{p''}}{\varepsilon}\,\mathbb{E}\big(H(p+1)_t^n\big) \\
&\le Kt \left(\phi_C'(\varepsilon) + \frac{1}{C} + \frac{C^{p''}}{\varepsilon\sqrt{r_n}} \right),
\end{aligned}$$

where the last inequality follows from (14.3.12). Taking C large, and then ε small, then n large, gives $\mathbb{D}^n(6) \stackrel{\text{u.c.p.}}{\Longrightarrow} 0$.

Next we turn to case (b). Again we reproduce the corresponding proof in Lemma 5.3.16, and instead of the estimate $E(|\delta'(j)_i^n|) \le K\Delta_n^{1+w/4}$ which is obtained in that proof, the same argument (based on (14.3.25) here) gives

$$\mathbb{E}\big(\big|\delta'(j)_i^n\big| \mid \mathcal{G}_{i-1}^n\big) \le K\, r_n^{p-1/2}\, \Delta(n,i)^{p+1+w/4}.$$

Hence, just as above, we deduce

$$\mathbb{E}\Big(\sup_{s\le t}\big|D^n(j)_s\big|\Big) \le K r_n^{p-1/2}\,\mathbb{E}\big(H(p+1/2+w/4)_t^n\big) \le \frac{Kt}{r_n^{w/4}}.$$

Therefore $\mathbb{D}^n(j) \stackrel{\text{u.c.p.}}{\Longrightarrow} 0$ for $j = 5,6,7$ in case (b), completing the proof of Lemma 14.3.5, hence of Part (i) of our theorem.

14.4 Application to Volatility

We are still interested here in the estimation of the volatility, but now in the context of an irregular discretization scheme, and more specifically when we have a restricted discretization scheme which satisfies at least Assumption (D), that is Assumption 14.1.3. As said before, this setting is not very realistic for a multivariate process, hence we consider only the case where $d = d' = 1$ below: for a more realistic situation in the multivariate case, when the sampling times are different for different components, see Hayashi and Yoshida ([44] and [46]).

We again want to estimate

$$A(q)_t = \int_0^t |\sigma_s|^q\, ds$$

for some $q > 0$, typically $q = 2$. The laws of large numbers of Sect. 14.2 provide a whole family of estimators, based on power or multipower variations, truncated or not. For simplicity, and also because of the asymptotic comparison between power and multipower variations made in the previous chapter, we consider below estimators based only on a single increment in each summand.

However, we have another parameter coming in naturally, that is the power $p \ge 0$ which affects the length of the successive intervals $\Delta(n,i)$. That is, in accordance with (14.2.2) and (14.2.4), applied with $k = 1$ and $F(x) = |x|^q$, we can set

$$\overline{D}(X;p,q)_t^n = \sum_{i=1}^{N_n(t)} \Delta(n,i)^{p-q/2}\, \big|\Delta_i^n X\big|^q$$

and, with the notation $v(n,i) = \alpha\Delta(n,i)^{\varpi}$ of (14.2.3) with $\alpha > 0$ and $\varpi \in (0,\frac{1}{2})$, we have the truncated versions:

$$\overline{D}(X;p,q,v_n-)_t^n = \sum_{i=1}^{N_n(t)} \Delta(n,i)^{p-q/2}\, \big|\Delta_i^n X\big|^q\, 1_{\{\|\Delta_i^n X\| \le v(n,i)\}}. \tag{14.4.1}$$

First, an application of Theorem 14.2.1 gives us

$$r_n^{p-1}\overline{D}(X;p,q)^n \stackrel{\text{u.c.p.}}{\Longrightarrow} m_q\, m'_p\, \overline{A}(p,q), \quad \text{where } \overline{A}(p,q)_t = \int_0^t |\sigma_s|^q\, (\theta_s)^{p-1}\, ds. \tag{14.4.2}$$

Here, we assume (D) about the discretization scheme, and (H-2) for X, and when X has jumps this is true only when $q<2$. Similarly if we assume (H-r) for some $r\in[0,2]$ and either $q\le 2$ or $q>2$ and $1<r<2$ and $\varpi\ge\frac{q(-2}{2(q-r)}$, or $q>2$ and $r\le 1$ (without condition on ϖ, other than being in $(0,1/2)$), we also have

$$r_n^{p-1}\overline{D}(X;p,q,v_n-)^n \stackrel{\text{u.c.p.}}{\Longrightarrow} m_q\, m'_p\, \overline{A}(p,q). \tag{14.4.3}$$

Second, the associated central limit theorem 14.3.2 gives us

$$\sqrt{r_n}\left(r_n^{p-1}\overline{D}(X;p,q)^n - m_q m'_p \overline{A}(p,q)\right)$$
$$\stackrel{\mathcal{L}\text{-}s}{\Longrightarrow} \sqrt{m_{2q}m'_{2p} - m_q^2\left(2m'_{p+1}m'_p - \left(m'_p\right)^2 m'_2\right)}\int_0^t |\sigma_s|^q\,(\theta_s)^{p-1/2}\,dB_s \tag{14.4.4}$$

where the stable convergence in law is relative to the σ-field $\mathcal{F}$, and B is a Wiener process on a very good filtered extension, independent of $\mathcal{F}$ (use (14.3.7) to identify the limit). For this, we need Assumption (E) on the discretization scheme, and (K) if $q>1$ and (K') if $q\le 1$ when X is continuous; in the discontinuous case we need $q<1$ and (K'-q). The truncated version $\overline{D}(X;p,q,v_n-)^n$ also satisfies (14.4.4) under relaxed conditions in the discontinuous case:

$$\begin{aligned} q>1 \ &\Rightarrow\ \text{(K-}r\text{) for some } r<1 \text{ and with } \varpi\ge\tfrac{q-1}{2(q-r)}\\ q=1 \ &\Rightarrow\ \text{(K'-1)}\\ q<1 \ &\Rightarrow\ \text{(K'-}r\text{) for some } r\le\tfrac{2q}{1+q}. \end{aligned}$$

These results, as such, are still far from being applicable. First, we are usually not interested in estimating $\overline{A}(p,q)_t$, but rather $A(q)_t$. Recalling $m'_1=1$, this is easily done by taking $p=1$, that is (14.4.1) and (14.4.2) yield

$$\overline{D}(X;1,q)^n \stackrel{\text{u.c.p.}}{\Longrightarrow} m_q\, A(q), \qquad \overline{D}(X;1,q,v_n-)^n \stackrel{\text{u.c.p.}}{\Longrightarrow} m_q\, A(q)$$

under the conditions stated above.

Next, to be practically feasible, the associated CLT requires an estimation of the asymptotic variance. This asymptotic variance is $(m_{2q}-m_q^2)m'_2\overline{A}(2,2q)_t$, which can be estimated by $r_n\frac{m_{2q}-m_q^2}{m_{2q}}\overline{D}(X;2,2q)_t^n$ or by $r_n\frac{m_{2q}-m_q^2}{m_{2q}}\overline{D}(X;2,2q,v_n-)_t^n$. This leads to the following result:

Theorem 14.4.1 *Let $q>0$ and let X be a (possibly discontinuous) one-dimensional Itô semimartingale satisfying* (K'-r) *when $q\le 1$ and* (K-r) *otherwise, for some*

$r \in [0, 1]$. *Assume also* (E) *for the discretization scheme. Then for each* $t > 0$ *the random variables*

$$\frac{\sqrt{m_{2q}}\,(\overline{D}(X; 1, q; v_n-)_t^n - m_q\, A(q)_t)}{\sqrt{(m_{2q} - m_q^2)\,\overline{D}(X; 2, 2q, v_n-)_t^n}}$$

converge stably in law to a limit which is $\mathcal{N}(0, 1)$ *and independent of* $\mathcal{F}$, *in restriction to the set* $\Omega_t^W = \{A(q)_t > 0\}$ *(which is* Ω *when* (K'-1) *holds), provided either* X *is continuous, or we have*

$$r = 1 \Rightarrow q = 1, \qquad r < 1 \Rightarrow q \geq \frac{r}{2-r}, \quad \varpi \geq \frac{q-1}{2(q-r)}$$

(in addition to $0 < \varpi < \frac{1}{2}$). *If* X *is continuous, we have the same results with the non-truncated versions* $\overline{D}(X; 1, q)_t^n$ *and* $\overline{D}(X; 2, 2q)_t^n$.

In practice the time steps $\Delta(n, i)$ are observed, and thus the quantities $\overline{D}(X; p, q, v_n-)_t^n$ are computable from the data set. On the other hand, neither the "rate" r_n, nor the moments m'_p, are known to the statistician (in general). So for example not only (14.4.2) and (14.4.3) do not provide estimators for interesting quantities when $p \neq 1$, but they are not even "statistics" in the usual sense because r_n is unknown. In contrast, when $p = 1$, the previous theorem allows us to find confidence intervals for $A(q)_t$, for example, on the basis of completely observable quantities. In this sense, the previous theorem is quite remarkable (and even "too good" in a sense, because it gives an answer, regardless of whether the required assumption (E) is satisfied or not).

Bibliographical Notes

The problem of irregular discretization schemes has recently attracted much attention, because of its practical relevance. An early Central Limit Theorem can be found in Jacod [53], but in the somewhat unrealistic setting of "strongly predictable" observation times. The content of this chapter is mainly taken from Hayashi, Jacod and Yoshida [47], and is based on ideas of Barndorff-Nielsen and Shephard [9] and Hayashi and Yoshida [44–46], and the latter authors also consider the more interesting case of several components observed at non-synchronous times. One should also mention some recent work by Phillips and Yu [79], by Fukasawa [33, 34], by Robert and Rosenbaum [85] and Fukasawa and Rosenbaum [35].

Chapter 15
Higher Order Limit Theorems

The Central Limit Theorems expounded in the previous chapters are sometimes "degenerate", in the sense that the limit vanishes identically. This of course happens when the test function is for example a constant. But, apart from such trivial cases, this may occur in a variety of interesting situations, in connection with the properties of the underlying process X itself.

This chapter is concerned with examples of such situations, and indeed only very few of them. The typology of degenerate cases is vast, and very few have been considered so far in the literature.

In the whole chapter we assume that we have a regular discretization scheme and that the d-dimensional semimartingale X, defined on the filtered space $(\Omega, \mathcal{F}, (\mathcal{F}_t)_{t\geq 0}, \mathbb{P})$, has the Grigelionis decomposition

$$X_t = X_0 + \int_0^t b_s\, ds + \int_0^t \sigma_s\, dW_s + (\delta 1_{\{\|\delta\|\leq 1\}}) \star (\mathfrak{p} - \mathfrak{q})_t + (\delta 1_{\{\|\delta\|>1\}}) \star \mathfrak{p}_t, \quad (15.0.1)$$

where W is a d'-dimensional Wiener process and $\mathfrak{p}$ is a Poisson measure with compensator $\mathfrak{q}(dt, dz) = dt \otimes \lambda(dz)$, and $c = \sigma\sigma^\star$. Moreover μ is the jump measure of X, and ν its compensator.

We will analyze the behavior of the following functionals

$$\begin{aligned} V^n(F, X)_t &= \sum_{i=1}^{[t/\Delta_n]-k+1} F\big(\Delta_i^n X, \dots, \Delta_{i+k-1}^n X\big) \\ V'^n(F, X)_t &= \Delta_n \sum_{i=1}^{[t/\Delta_n]-k+1} F\big(\Delta_i^n X/\sqrt{\Delta_n}, \dots, \Delta_{i+k-1}^n X/\sqrt{\Delta_n}\big), \end{aligned}$$

where the test function F on $(\mathbb{R}^d)^k$ presents some degeneracy, in connection with the properties of X itself.

J. Jacod, P. Protter, *Discretization of Processes*,
Stochastic Modelling and Applied Probability 67,
DOI 10.1007/978-3-642-24127-7_15,

15.1 Examples of Degenerate Situations

1) In all the previous chapters, the limit of the normalized functionals $V'^n(F,X)$ is always a process which, apart from F itself, depends only on the process c_t. Moreover *all* the limits in the CLTs, for both normalized and non-normalized functionals, involve this process c_t and actually vanish identically when c_t does.

In other words, when in (15.0.1) the continuous martingale part $\int_0^t \sigma_s\,dW_s$ is absent, then none of the previous results has any significant content, apart form the first LLN given in Theorem 3.3.1.

In some cases, this is irremediable. Suppose for example that we have a "pure jump" process

$$X_t \;=\; X_0 + \sum_{p\geq 1} \Delta X_{S_p} 1_{\{S_p\leq t\}}$$

for a sequence S_p of stopping times increasing strictly to infinity, and consider $V^n(f,X)$ (take $k=1$ and f a function on $\mathbb{R}^d$). Then for all $t>0$, and all n large enough (depending on t) we have

$$s\leq t \;\Rightarrow\; V^n(f,X)_z = \sum_{p\geq} f(\Delta X_{S_p})\, 1_{\{S_p\leq \Delta_n[s/\Delta_n]\}} = f*\mu_{\Delta_n[s/\Delta_n]}.$$

So, assuming for example $\Delta_n = 1/n$, we see that $V^n(f,X)_t$ is actually equal to $V(f,X)_1 = f*\mu_1$ for all n large: the LLN is obvious, and there is no non-trivial associated CLT.

In other cases we do have some kind of CLT, and we give below an example.

Example 15.1.1 We suppose that X is a standard symmetrical (one-dimensional) stable process with index $\alpha\in(0,2)$. This is the Lévy process with characteristic function $\mathbb{E}(e^{iuX_t}) = \exp -t|u|^\alpha$. We consider the normalized functionals $V'^n(f,X)_1$, and with f of the form $f(x)=|x|^p$ for some $p>0$. We also consider below a generic variable U having the same law as X_1.

For each n, the variables $f(\Delta_i^n X/\sqrt{\Delta_n})$ are i.i.d., with the same law as $f(U\Delta_n^{1/\alpha-1/2}) = \Delta_n^{p(1/\alpha-1/2)}|U|^p$, because for all $s,t\geq 0$ the variable $X_{t+s}-X_t$ has the same law as $s^{1/\alpha}U$. If moreover $p<\alpha$ the variable $|U|^p$ has a finite mean, say m'_p. In this case, and since $V'^n(f,X)_t$ is the sum of $[t/\Delta_n]$ variables as above, the usual law of large numbers readily gives

$$\Delta_n^{p(\frac{1}{2}-\frac{1}{\alpha})}\, V'^n(f,X)_1 \;\stackrel{\text{u.c.p.}}{\Longrightarrow}\; m'_p\, t. \tag{15.1.1}$$

By the way, this suggests that the normalized functionals, with the inside normalizing factor $1/\sqrt{\Delta_n}$, is not the proper one; one should rather use here

$$V''^n(f,X)_t \;=\; \Delta_n \sum_{i=1}^{[t/\Delta_n]} f\big(\Delta_i^n X/\Delta_n^{1/\alpha}\big)$$

and then (15.1.1) would read as $V''^n(f,X)_1 \stackrel{\text{u.c.p.}}{\Longrightarrow} m'_p t$.

In other words, in this situation the LLN of Theorem 3.4.1 is degenerate, and should be replaced by (15.1.1), with the normalizing factor $\Delta_n^{p(\frac{1}{2}-\frac{1}{\alpha})}$ going to infinity. The usual Donsker's theorem also yields an associated CLT, under the additional assumption that $2p < \alpha$. It reads as follows:

$$\frac{1}{\sqrt{\Delta_n}}\left(\Delta_n^{p(\frac{1}{2}-\frac{1}{\alpha})}\, V'^n(f,X)_1 - m'_p t\right) \stackrel{\mathcal{L}}{\Longrightarrow} \left(m'_{2p} - m'^2_p\right)B,$$

where B is a standard Wiener process. And it turns out that this is even a stable convergence in law, with B being defined on an extension of the space and independent of $\mathcal{F}$.

When $p \geq \alpha$ the situation is quite different, because then $m'_p = \infty$. However, in this case the law of $|U|^p$ belongs to the domain of attraction of a stable distribution with index α/p. Hence a result like (15.1.1) simply cannot hold, even with another normalization. What happens is that, when $p > \alpha$,

$$\Delta_n^{p/2-1}\, V'^n(f,X)_1 \stackrel{\mathcal{L}}{\Longrightarrow} Z \tag{15.1.2}$$

where Z is a stable subordinator (that is, an increasing Lévy process which is also a stable process) with index α/p. When $p = \alpha$ the situation is more complicated, and the normalizing factor involves a logarithmic term $\log(1/\Delta_n)$. Moreover, the convergence in (15.1.2) also holds stably in law, with Z being defined on an extension of the space and again being independent of $\mathcal{F}$.

This shows in particular that the "first order" limit theorem with a non-trivial limit is a CLT-type result, since we have a (stable) convergence in law, but not in probability.

This simple example shows the need of a theory for semimartingales with no continuous martingale part. But nothing more will be said about this here.

2) We come back to the situation where c_t in (15.0.1) is not identically vanishing. Suppose that $d = 2$ with X continuous, but that the rank of the matrix c_t is always 0 or 1, and consider the normalized functionals $V'^n(F,X)$ with $k = 2$ and

$$\begin{aligned} F(x,y) &= \left((x^1)^2 + (y^1)^2\right)\left((x^2)^2 + (y^2)^2\right) - \left(x^1x^2 + y^1y^2\right)^2 \\ &= \left(x^1y^2\right)^2 + \left(x^2y^1\right)^2 - 2x^1x^2y^1y^2, \end{aligned}$$

which is the determinant of the 2×2 matrix $xx^* + yy^*$. Then if $a \in \mathcal{M}^+_{2\times 2}$ we have $\rho_a^{2\otimes}(F) = 3\det(a)$, with our standing notation $\rho_a = \mathcal{N}(0,a)$. Hence our assumption on c_t yields, by Theorem 8.4.1, that $V'^n(F,X) \stackrel{\text{u.c.p.}}{\Longrightarrow} 0$.

If further σ_t is itself an Itô semimartingale, and more precisely if (K), that is Assumption 4.4.3, holds, we can apply Theorem 11.2.1. However, a (tedious) computation shows that the number $R_a(F,F)$ defined by (11.2.4) vanishes as soon as $\det(a) = 0$. Therefore, not only the limit of $V'^n(F,X)$ vanishes, but also the limit in the associated CLT.

In this setting, one could show $\frac{1}{\Delta_n} V'^n(F,X)$ converges stably in law (but not in probability) to a non-trivial limit, under some additional assumptions, and in particular the assumption that the volatility $\widetilde{\sigma}$ of the process σ_t in its decomposition (5.3.2) is itself an Itô semimartingale, and σ_t is continuous.

We will not pursue here how to find a substitute of the CLT of Theorem 11.2.1 for the normalized functionals in the degenerate case, mainly because it is largely an open problem for the time being. We will however study in the next section a degenerate case for the functionals $V^n(F,X)$ when X jumps.

15.2 Functionals of Non-normalized Increments

1) We consider below a situation where Theorem 11.1.2 gives a degenerate result. For further use, we recall in some detail the notation of this theorem. First the assumption on the process X is (H), that is:

Assumption (H) In (15.0.1), b_t is locally bounded and σ_t is càdlàg, and $\|\delta(\omega,t,z)\| \wedge 1 \leq \Gamma_n(z)$ for all (ω,t,z) with $t \leq \tau_n(\omega)$, where (τ_n) is a localizing sequence of stopping times and each function Γ_n satisfies $\int \Gamma_n(z)^2 \lambda(dz) < \infty$.

Next, the first and second partial derivatives of the q-dimensional test function F are globally denoted as ∇F and $\nabla^2 F$, and we define the following $\mathbb{R}^q$-valued functions on $\mathbb{R}^d$:

$$\left.\begin{array}{l} f_j(x) = F(0,\dots,0,x,0,\dots,0) \\ \partial_i f_{(l);j}(x) = \frac{\partial F}{\partial x_l^i}(0,\dots,0,x,0,\dots,0) \\ \partial^2_{i,i'} f_{(l,l');j}(x) = \frac{\partial^2 F}{\partial x_l^i\, \partial x_{l'}^{i'}}(0,\dots,0,x,0,\dots,0) \end{array}\right\} \text{ with } x \text{ at the } j\text{th place.} \tag{15.2.1}$$

Next, let $\mathcal{K}_- = \{-k+1,-k+2,\dots,-1\}$ and $\mathcal{K}_+ = \{1,2,\dots,k-1\}$ and $\mathcal{K} = \mathcal{K}_- \cup \mathcal{K}_+$. We choose an arbitrary weakly exhausting sequence $(T_n)_{n\geq 1}$ for the jumps of X, and a very good filtered extension $(\widetilde{\Omega},\widetilde{\mathcal{F}},(\widetilde{\mathcal{F}}_t)_{t\geq 0},\widetilde{\mathbb{P}})$ of $(\Omega,\mathcal{F},(\mathcal{F}_t)_{t\geq 0},\mathbb{P})$ on which are defined some $\widetilde{\mathcal{F}}_{T_n}$-measurable variables $((\Psi_{n,j})_{j\in\mathcal{K}},\Psi_{n-},\Psi_{n+},\kappa_n)$ for $n \geq 1$. Those variables are, conditionally on $\mathcal{F}$, all independent and with the following laws:

- $\Psi_{n,j}, \Psi_{n-}, \Psi_{n-}$ are d'-dimensional, $\mathcal{N}(0, I_{d'})$,
- κ_n is uniform on $[0,1]$.

Finally, we define the d-dimensional random variables

$$R_{n,j} = \begin{cases} \sigma_{T_n-}\Psi_{n,j} & \text{if } j \in \mathcal{K}_- \\ \sqrt{\kappa_n}\,\sigma_{T_n-}\Psi_{n-} + \sqrt{1-\kappa_n}\,\sigma_{T_n}\Psi_{n+} & \text{if } j = 0 \\ \sigma_{T_n}\Psi_{n,j} & \text{if } j \in \mathcal{K}_+. \end{cases}$$

Theorem 11.1.2 says that, if F is C^2 with $F(0)=0$ and $\nabla F(0)=0$ and $\|\nabla^2 F(x)\| = \mathrm{o}(\|x\|)$ as $x \to 0$, and with the notation

$$\overline{V}(F,X)_t = \sum_{n=1}^{\infty} \left(\sum_{j=1}^{k} \sum_{l=1}^{k} \sum_{i=1}^{d} \partial_i f_{(l);j}(\Delta X_{T_n})\, R^i_{n,l-j} \right) 1_{\{T_n \le t\}}, \tag{15.2.2}$$

for each t we have

$$\frac{1}{\sqrt{\Delta_n}} \left(V^n(F,X)_t - \sum_{j=1}^{k} f_j \star \mu_t \right) \stackrel{\mathcal{L}\text{-s}}{\longrightarrow} \overline{V}(F,X)_t.$$

When $k=1$ (so $F=f$) we also have $\frac{1}{\sqrt{\Delta_n}}(V^n(f,X)_t - f \star \mu_{\Delta_n[t/\Delta_n]}) \stackrel{\mathcal{L}\text{-s}}{\Longrightarrow} \overline{V}(f,X)$.

2) It may happen that the limit $\overline{V}(F,X)$ is identically 0. This occurs in some "trivial" situations, when $\sigma_t = 0$ identically or $\Delta X_t = 0$ identically, and more generally if $\sigma_t = \sigma_{t-} = 0$ whenever $\Delta X_t \neq 0$. It also happen when the multipliers $\partial_i f_{(l);j}(\Delta X_{T_n})$ in (15.2.2) vanish identically, and this can occur in some non-trivial situations, as Theorem 15.2.4 below shows.

With the notation (15.2.1), an assumption implying degeneracy is

$$1 \le j,l \le k,\ 1 \le i \le d, \quad \Rightarrow \quad \sum_{s \ge 0} \left\| \partial_i f_{(l);j}(\Delta X_s) \right\| = 0. \tag{15.2.3}$$

Under (15.2.3) the limit $\overline{V}(F,X)$ in (15.2.2) vanishes, and one may look for another normalization, for which a non-trivial limit exists. It turns out that it is possible, under some more smoothness of the function F.

The limiting process can be constructed as follows, using the same ingredients R_n as above:

$$\widetilde{V}(F,X)_t = \frac{1}{2} \sum_{n=1}^{\infty} \left(\sum_{j,l,l'=1}^{k} \sum_{i,i'=1}^{d} \partial^2_{i,i'} f_{(l,l'),j}(\Delta X_{T_n})\, R^i_{n,l-j}\, R^{i'}_{n,l'-j} \right) 1_{\{T_n \le t\}}. \tag{15.2.4}$$

The same proof as for Proposition 11.1.1, page 299, except that we use Proposition 4.1.3, page 101, instead of Proposition 4.1.4 (because the products $R^i_{n,l-j}\, R^{i'}_{n,l'-j}$ are no longer $\mathcal{F}$-conditionally centered), shows that this process is well defined as soon as $\|\partial^2_{i,i'} f_{(l,l'),j}\| \le K\|x\|^2$ when $\|x\| \le 1$ for all $l, l', j \le k$ and $i, i' \le d$. Note that, conditionally on $\mathcal{F}$, the process $\widetilde{V}(F,X)$ has again independent increments and is of finite variation, but neither centered nor Gaussian. Its $\mathcal{F}$-conditional law, again, does not depend upon the chosen exhausting sequence (T_n). Computing the $\mathcal{F}$-conditional variance is possible but not extremely useful. The $\mathcal{F}$-conditional mean, however, is useful, and it takes the form

$$\widetilde{\mathbb{E}}(\widetilde{V}(F,X)_t \mid \mathcal{F}) = \frac{1}{2} \sum_{s \le t} \sum_{i,i'=1}^{d} \sum_{j,l=1}^{k} \partial^2_{i,i'} f_{(l,l),l}(\Delta X_s)$$
$$\times \left(c^{ii'}_{s-} \left(1_{\{j>l\}} + \frac{1}{2} 1_{\{j=l\}} \right) + c^{ii'}_{s} \left(1_{\{j<l\}} + \frac{1}{2} 1_{\{j=l\}} \right) \right). \tag{15.2.5}$$

Note that the rth component of this process is increasing (in t) as soon as the rth component F^r is a convex function of each of its k arguments.

It turns out that the degeneracy condition (15.2.3) is not enough to have a "higher order" CLT, and we need an additional condition on the behavior of F near 0. So we begin with a simple result, in which we simply assume that F vanishes on a neighborhood of 0:

Theorem 15.2.1 *Assume* (H), *and let F be a C^2 function from $(\mathbb{R}^d)^k$ into $\mathbb{R}^q$ which vanishes on a neighborhood of* 0 *and satisfies the "degeneracy condition"* (15.2.3). *Then for each t, we have the following stable convergence in law*:

$$\frac{1}{\Delta_n} \left(V^n(F,X)_t - \sum_{j=1}^{k} f_j \star \mu_t \right) \xrightarrow{\mathcal{L}\text{-s}} \widetilde{V}(F,X)_t \tag{15.2.6}$$

where $\widetilde{V}(F,X)_t$ is defined by (15.2.4).

When $k = 1$, one could also prove the "functional" stable convergence in (15.2.6), whereas this functional convergence does *not* hold when $k \ge 2$ in general. Note that in the above situation all summands in (15.2.4) vanish except finitely many of them.

Of course, this result is interesting only when, in addition to (15.2.3), we do not have

$$1 \le j,l \le k,\ 1 \le i,i' \le d, \ \Rightarrow \ \sum_{s \ge 0} \left\| \partial^2_{i,i'} f_{(l,l),j}(\Delta X_s) \right\| = 0,$$

otherwise $\widetilde{V}(F,X)_t = 0$ identically. In this case, assuming that F is C^3 and still vanishing in a neighborhood of 0, we can derive a "third order" CLT of the same type, with rate $1/\Delta_n^{3/2}$ and involving the third derivatives of F. And if this is also with a vanishing limit we can go further to get a CLT with rate $1/\Delta_n^2$. Once the above theorem established, this is a boring but easy task.

Remark 15.2.2 The condition (15.2.3) is rather extreme, an more typical is the situation where the set

$$\Omega_t \;=\; \cap_{1 \le j,l \le k,\ 1 \le i,i' \le d} \left\{ \sum_{s \le t} \left\| \partial_i f_{(l),j}(\Delta X_s) \right\| = 0 \right\}$$

satisfies $0 < \mathbb{P}(\Omega_t) < 1$. Then, by combining Theorems 11.1.2, page 301, and 15.2.1 and using the properties of the stable convergence in law, we easily obtain

$$\begin{aligned} &\bullet \ \frac{1}{\sqrt{\Delta_n}} \Big(V^n(F,X)_t - \sum_{j=1}^k f_j \star \mu_t \Big) \stackrel{\mathcal{L}\text{-s}}{\longrightarrow} \overline{V}(F,X)_t \quad \text{in restriction to } (\Omega_t)^c \\ &\bullet \ \frac{1}{\Delta_n} \Big(V^n(F,X)_t - \sum_{j=1}^k f_j \star \mu_t \Big) \stackrel{\mathcal{L}\text{-s}}{\longrightarrow} \widetilde{V}(F,X)_t \quad \text{in restriction to } \Omega_t. \end{aligned} \tag{15.2.7}$$

This is justified in the same way as in Corollary 3.3.4 for example: the first part of (15.2.7) always holds, and we easily construct another semimartingale X' satisfying (H), such that $X'_s = X_s$ for all $s \le t$ on the set Ω_t, and which satisfies (15.2.3) identically for the function F, thus obtaining the second part by applying Theorem 15.2.1 to X'.

Remark 15.2.3 Another problem arises with the conditions of the previous theorem. In applications, one may start with a C^2 function F satisfying (15.2.3) but not vanishing on a neighborhood of 0 in $(\mathbb{R}^d)^k$. The latter property is obtained by multiplying F by a function G which vanishes around 0 and equals 1 outside another neighborhood of 0: for example $G(z) = \psi_\varepsilon(z)$, where as usual ψ_ε is defined by (3.3.16).

If we take G to be C^2, like $G = \psi_\varepsilon$, the problem is that in general the product FG does *not* satisfies (15.2.3). Another possibility consists in simply "truncating" F, by taking for example F_ε to be

$$\begin{aligned} &\text{either} \quad F_\varepsilon(x_1,\dots,x_k) = F(x_1,\dots,x_k)\, 1_{\{\|(x_1,\dots,x_k)\| > \varepsilon\}} \\ &\text{or} \quad F_\varepsilon(x_1,\dots,x_k) = F(x_1,\dots,x_k)\Big(1 - \prod_{j=1}^k 1_{\{\|x_j\| \le \varepsilon\}}\Big) \end{aligned} \tag{15.2.8}$$

for some $\varepsilon > 0$ (and where $z = (x_1,\dots,x_k)$ in the last formula). Doing so, we loose the C^2 property. However, as the proof of the theorem will show, the only requirement is really that F be twice continuously differentiable at each point of a set $D \subset (\mathbb{R}^d)^k$, with the following property: for almost all ω, it contains the point $(0,\dots,0,\Delta X_s(\omega),0,\dots,0)$ for all $s > 0$ and any position for ΔX_s between 1 and k.

Thus if $\varepsilon > 0$ is such that almost surely we have $\|\Delta X_s\| \ne \varepsilon$ for all $s > 0$, we deduce that if F is C^2 and satisfies (15.2.3), then (15.2.6) holds for F_ε for any one of the two versions in (15.2.8). Moreover this property of ε is satisfied for Lebesgue-almost all values in $(0,\infty)$.

3) When F does not vanish in a neighborhood of 0, things are much more complicated. To understand why, and since the rate will still be $1/\Delta_n$, we can consider for a moment the case where X is continuous: in this case we want $\frac{1}{\Delta_n} V^n(F,X)$ to converge, because $\mu = 0$; when F is globally homogeneous of degree p, that is to say that $F(az) = a^p F(z)$ for all $a \in \mathbb{R}$ and $z \in (\mathbb{R}^d)^k$, we have $\frac{1}{\Delta_n} V^n(F,X) = \Delta_n^{p/2-2}\, V'^n(F,X)$; these processes converge to 0 when $p > 4$, to $\int_0^t \rho_{c_s}^{k\otimes F}(F)\, ds$ when $p = 4$, and do *not* converge in general if $p < 4$.

Coming back to the case with jumps, we see that, instead of $F(z)=\mathrm{o}(\|z\|^3)$, we need either $F(z)=\mathrm{o}(\|z\|^4)$ (as $z\to 0$), or $F(z)=\mathrm{O}(\|z\|^4)$ and in this second case we need further F to behave as an homogeneous function of degree 4 near the origin. Even under these stronger assumptions, it is not known in general whether a CLT holds with the rate $1/\Delta_n$. Thus, instead of a "general" result, we simply give a result in a very special case.

Namely, on the one hand we suppose that X is 2-dimensional ($d=2$), with the following special structure:

$$\Delta X_t^1\,\Delta X_t^2 = 0 \quad \text{identically.} \tag{15.2.9}$$

In other words, the two components X^1 and X^2 *never jump at the same times*. On the other hand, we take for F a 2-dimensional function on $(\mathbb{R}^2)^k$ with the following components (below, a vector x in $\mathbb{R}^2$ has components x^1 and x^2, and $x_1,\dots,x_k$ all belong to $\mathbb{R}^2$):

$$F^j(x_1,\dots,x_k)=\begin{cases} f(x_1) & \text{if } j=1\\ f(x_1+\cdots+x_k) & \text{if } j=2\end{cases} \quad \text{where } f(x)=\left(x^1x^2\right)^2. \tag{15.2.10}$$

This function is C^2 with $F(z)=\mathrm{O}(\|z\|^4)$ as $z\to 0$, and it is also globally homogeneous with degree 4. We obviously have

$$\begin{array}{lll} \partial_1 f(x)=2x^1\left(x^2\right)^2, & \partial_2 f(x)=2\left(x^1\right)^2x^2 & \\ \partial_{11}^2 f(x)=2\left(x^2\right)^2, & \partial_{22}^2 f(x)=2\left(x^1\right)^2, & \partial_{12}^2 f(x)=4x^1x^2, \end{array} \tag{15.2.11}$$

and the functions associated with F in (15.2.1) become

$$f_1^1=f_j^2=f, \qquad \partial_i f_{(1);1}^1=\partial_i f_{(l);j}^2=\partial_i f, \qquad \partial_{i,i'}^2 f_{(1,1);1}^1=\partial_{i,i'}^2 f_{(l,l');j}^2=\partial_{ii'}^2 f, \tag{15.2.12}$$

with all other functions f_j^1, $\partial_i f_{(l);j}^1$ and $\partial_{i,i'}^2 f_{(l,l');j}^1$ being 0.

Under (15.2.9), the function F satisfies the degeneracy condition (15.2.3), and also $f_j*\mu=0$ identically.

Theorem 15.2.4 *Assume* (H) *with $d=2$ and the condition* (15.2.9). *Let F be the function defined by* (15.2.10). *Then for each t, we have the following stable convergence in law*

$$\frac{1}{\Delta_n}V^n(F,X)_t \stackrel{\mathcal{L}\text{-s}}{\longrightarrow} \widetilde{V}(F,X)_t+\overline{C}(F)_t, \tag{15.2.13}$$

where $\widetilde{V}(F,X)_t$ is defined by (15.2.4) *and $\overline{C}(F)$ is the following 2-dimensional process*:

$$\overline{C}(F)_t^j = \begin{cases} H_t & \textit{if } j=1\\ k^2H_t & \textit{if } j=2,\end{cases} \quad \textit{where } H_t=\int_0^t \left(c_s^{11}c_s^{22}+2\left(c_s^{12}\right)^2\right)ds. \tag{15.2.14}$$

A simple calculation, based on (15.2.5), (15.2.11) and (15.2.12), plus the property (15.2.9), shows that the $\mathcal{F}$-conditional expectation of $\widetilde{V}(F^j, X)_t$ is

$$\widetilde{\mathbb{E}}\big(\widetilde{V}(F^j,X)_t \mid \mathcal{F}\big) = \begin{cases} H'_t & \text{if } j=1 \\ k^2 H'_t & \text{if } j=2, \end{cases} \qquad \text{where}$$
$$H'_t = \frac{1}{2}\sum_{s\le t}\big((\Delta X^1_s)^2(c^{11}_{s-}+c^{11}_s)+(\Delta X^2_s)^2(c^{22}_{s-}+c^{22}_s)\big). \tag{15.2.15}$$

Remark 15.2.5 The same result holds if we assume that F has the form (15.2.10) on a neighborhood of the origin only, provided it also satisfies (15.2.3). But then we do not necessarily have $f_j * \mu = 0$, and in (15.2.13) the left side should then be the same as in (15.2.6).

Remark 15.2.6 If in (15.2.10) we take $f(x) = |x^1x^2|^p$ with $p > 2$ instead of $p = 2$, we are in a case of degeneracy for the above result, that is $\frac{1}{\Delta_n} V^n(F,X)_t \stackrel{\mathbb{P}}{\longrightarrow} 0$. When p is an integer we do have a CLT with the rate $1/\Delta_n^{p/2}$, and otherwise the precise behavior of $V^n(F,X)$ is unknown.

Proof of Theorem 15.2.1 The whole proof is copied from the proof of Theorem 11.1.2, which we will use freely below: see pages 302–304.

To begin with, by localization we can replace (H) by the strengthened assumption (SH). Next, we use the notation A_m, $b(m)$, $X(m)$, $X'(m)$ given by (11.1.15), and $\Omega_n(T,m)$ of (11.1.16), so (11.1.17) holds. The exhausting sequence for the jumps of X is (S_p), defined before (11.1.15) (or equivalently by (4.3.1)), and $\mathcal{P}_m$ denotes the set of all indices p such that $S_p = S(m', j)$ for some $j \ge 1$ and some $m' \le m$, and $i(n,p)$ is the unique integer with $(i(n,p)-1)\Delta_n < S_p \le i(n,p)\Delta_n$, and $L(n,p)$ and $R(n,p,j)$ are given by (11.1.18). We also set

$$\widetilde{V}^n(F,X)_t = \frac{1}{\Delta_n}\left(V^n(F,X)_t - \sum_{j=1}^k f_j * \mu^X_t\right).$$

With $\zeta(j)^n_p$ being as before (11.1.22), we set

$$Z^n(m)_t = \sum_{p\in\mathcal{P}_m:\, S_p\le t} \zeta^n_p, \quad \text{where } \zeta^n_p = \frac{1}{\Delta_n}\sum_{j=1}^k \zeta(j)^n_p.$$

Then, similar to (11.1.22), we obtain

$$\widetilde{V}^n(F,X)_t = \widetilde{V}^n\big(F,X(m)\big)_t + Z^n(m)_t \quad \text{on the set } \Omega_n(t,m). \tag{15.2.16}$$

To evaluate $\zeta(j)^n_p$ we use a Taylor expansion. Here, in view of (15.2.3), we need a Taylor's expansion up to the second order, around the point $(0,\dots,0,\Delta X_{S_p},0,$

$\dots, 0)$, with ΔX_{S_p} at the jth place (this is the only occurrence of the C^2 property of F, and it explains why the extension of the theorem, as mentioned in Remark 15.2.3, is true). Using the tightness of the sequences $R(n,p,j)$, which follows from (11.1.19), we obtain

$$\frac{1}{\Delta_n}\zeta(j)^n_p - \frac{1}{2}\sum_{l,l'=1}^{k}\sum_{i,i'=1}^{d}\partial^2_{i,i'}f_{(l,l'),j}(\Delta X_{S_p})\,R(n,p,l-j)^i\,R\big(n,p,l'-j\big)^{i'} \stackrel{\mathbb{P}}{\longrightarrow} 0.$$

Then the convergence (11.1.19) again yields that $(\zeta^n_p)_{p\ge 1}$ converges stably in law to $(\zeta_p)_{p\ge 1}$, where the q-dimensional variables ζ_p are defined on the extended space by

$$\zeta_p = \frac{1}{2}\sum_{j,l,l'=1}^{k}\sum_{i,i'=1}^{d}\partial^2_{i,i'}f_{(l,l'),j}(\Delta X_{S_p})\,R^i_{p,l-j}\,R^{i'}_{p,l'-j}.$$

Since the set $\{S_p : p\in\mathcal{P}_m\}\cap[0,t]$ is finite, we deduce that, as $n\to\infty$:

$$Z^n(m) \stackrel{\mathcal{L}\text{-}s}{\Longrightarrow} \widetilde{V}\big(F, X'(m)\big), \tag{15.2.17}$$

where $\widetilde{V}(F,X'(m))$ is associated with the process $X'(m)$ by (15.2.4).

Now we use the hypothesis that $F(z)=0$ if $\|z\|\le\varepsilon$ for some $\varepsilon>0$. We then choose $m>1/\varepsilon$. First, since $\|\Delta X(m)_s\|\le\frac{1}{m}$ by construction, we have $\widetilde{V}(F,X(m))=\widetilde{V}(F,X)$ and $f_j*\mu^{X(m)}=0$. Second, recalling $\|\Delta^n_i X(m)\|\le\frac{2}{m}$ for all $i\le[t/\Delta_n]$ on the set $\Omega_n(t,m)$ by its very definition, we deduce that on this set we have $V^n(F,X(m))_t=0$, hence also $\widetilde{V}^n(F,X(m))_t=0$. Combining these facts with (15.2.16) and the property $\mathbb{P}(\Omega_n(t,m))\to 1$ as $n\to\infty$ allows us to deduce (15.2.6) from (15.2.17). □

Proof of Theorem 15.2.4 *Step 1)* As for the previous theorem we may assume (SH), and the previous proof works in exactly the same way down to (15.2.17), with further $\widetilde{V}^n(F,X)=\frac{1}{\Delta_n}V^n(F,X)$ and $\widetilde{V}^n(F,X(m))=\frac{1}{\Delta_n}V^n(F,X(m))$, because $F*\mu\equiv 0$ here.

(15.2.12) implies that the nth summand in the right side of (15.2.4) for $\widetilde{V}(F,X(m))_t$ is smaller than $K\|\Delta X(m)_s\|^2\sum_{l=-k}^{k}\|R_{n,l}\|^2$. Now, we have $\widetilde{V}(F,X)-\widetilde{V}(F,X'(m))=\widetilde{V}(F,X(m))$ and $\mathbb{E}(\|R_{n,l}\|^2\mid\mathcal{F})\le K$ by (SH), so we deduce

$$\widetilde{\mathbb{E}}\Big(\sup_{s\le t}\big\|\widetilde{V}(F,X)_s-\widetilde{V}\big(F,X'(m)\big)_s\big\|\Big)\le K\mathbb{E}\bigg(\sum_{s\le t}\big\|\Delta X(m)_s\big\|^2\bigg)$$
$$\le t\int_{A_m^c}\Gamma(z)^2\lambda(dz),$$

which goes to 0 as $m\to\infty$. Hence

$$\widetilde{V}\big(F,X'(m)\big) \stackrel{\text{u.c.p.}}{\Longrightarrow} \widetilde{V}(F,X). \tag{15.2.18}$$

At this stage, combining (15.2.16), (15.2.17), (15.2.18), plus the property $\mathbb{P}(\Omega_n(t,m)) \to 1$ as $n \to \infty$, we deduce from Proposition 2.2.4, page 52, that it remains to show that, for all $\eta > 0$, we have for $j = 1, 2$:

$$\lim_{m\to\infty} \limsup_{n\to\infty} \mathbb{P}\left(\left|\frac{1}{\Delta_n} V^n\big(F^j, X(m)\big)_t - \overline{C}(F)^j_t\right| > \eta\right) = 0. \tag{15.2.19}$$

Note that when $k = 1$ we have $F^2 = F^1$ and $\overline{C}(F)^2 = \overline{C}(F)^1$, so it is indeed enough to show this for $j = 2$. So below we forget about F^1 and simply write $F = F^2$ and also $\overline{C}(F) = \overline{C}(F)^2$.

Step 2) We can write $X(m) = \overline{X}'(m) + \overline{X}''(m)$, where

$$\overline{X}'(m)_t = X_0 + \int_0^t b(m)_s\, ds + \int_0^t \sigma_s\, dW_s, \quad \overline{X}''(m) = (\delta\, 1_{A_m^c}) * (\mathfrak{p} - \mathfrak{q}).$$

Since $F = F^2$ is homogeneous of degree 4, we have $V^n(F, \overline{X}'(m)) = \Delta_n\, V'^n(F, \overline{X}'(m))$. Moreover, with the usual notation $\rho_a = \mathcal{N}(0, a)$, we easily check that $\rho_{c_s}^{k\otimes}(F) = k^2(c_s^{11} c_s^{22} + 2(c_s^{12})^2)$. Hence, since $\overline{X}'(m)$ is continuous and satisfies (H), we deduce from Theorem 8.4.1 that, as $n \to \infty$,

$$\frac{1}{\Delta_n} V^n\big(F, \overline{X}'(m)\big) \overset{\text{u.c.p.}}{\Longrightarrow} \overline{C}(F).$$

Thus, instead of (15.2.19) (for $j = 2$), it suffices to prove

$$\lim_{m\to\infty} \limsup_{n\to\infty} \mathbb{E}\left(\frac{1}{\Delta_n} \left|V^n\big(F, X(m)\big)_t - V^n\big(F, \overline{X}'(m)\big)_t\right|\right) = 0.$$

For any process we write

$$\Delta_i'^n Y = Y_{(i+k-1)\Delta_n} - Y_{(i-1)\Delta_n}$$

(this is our usual increment $\Delta_i^n Y$, but with the discretization step $k\Delta_n$ and a starting time $(i-1)\Delta_n$ which is not necessarily a multiple of $k\Delta_n$). Then, recalling (15.2.10) for $j = 2$ and the function $f(x) = (x^1 x^2)^2$, we have $V^n(F, Y)_t = \sum_{i=1}^{[t/\Delta_n]-k+1} f(\Delta_i'^n Y)$ and thus

$$\frac{1}{\Delta_n} \left|V^n\big(F, X(m)\big)_t - V^n\big(F, \overline{X}'(m)\big)_t\right| \le \sum_{i=1}^{[t/\Delta_n]} \zeta(m)_i^n,$$

where

$$\zeta(m)_i^n = \frac{1}{\Delta_n} \left| f\big(\Delta_i'^n X(m)\big) - f\big(\Delta_i'^n \overline{X}'(m)\big)\right|.$$

Therefore we are left to prove

$$\lim_{m\to\infty} \limsup_{n\to\infty} \mathbb{E}\left(\sum_{i=1}^{[t/\Delta_n]} \zeta(m)_i^n\right) = 0. \tag{15.2.20}$$

Step 3) The form of f implies that for each $\varepsilon > 0$ there is a constant K_ε with

$$\left|f(x+y) - f(x)\right| \le \varepsilon\|x\|^4 + K_\varepsilon\|x\|^2\,\|y\|^2 + K_\varepsilon f(y).$$

Hence

$$\begin{aligned}
\zeta(m)_i^n &\le \varepsilon\, U(m)_i^n + K_\varepsilon\, U'(m)_i^n + K_\varepsilon\, U''(m)_i^n, \quad \text{where}\\
U(m)_i^n &= \frac{1}{\Delta_n}\left\|\Delta_i'^n\overline{X}'(m)\right\|^4\\
U'(m)_i^n &= \frac{1}{\Delta_n}\left\|\Delta_i'^n\overline{X}'(m)\right\|^2\left\|\Delta_i'^n\overline{X}''(m)\right\|^2\\
U''(m)_i^n &= \frac{1}{\Delta_n}\, f\left(\Delta_i'^n\overline{X}''(m)\right).
\end{aligned} \tag{15.2.21}$$

Below, ϕ_m denotes a sequence which may change from line to line, but always goes to 0 as $m \to \infty$. Recalling that σ_t is bounded and $\|b(m)_t\| \le Km$, we obtain from (2.1.44), and for all $q > 0$:

$$\begin{aligned}
&\mathbb{E}\Big(\sup_{s\le k\Delta_n}\left\|\overline{X}'(m)_{(i-1)\Delta_n+s} - \overline{X}'(m)_{(i-1)\Delta_n}\right\|^q\Big) \le K_q\Delta_n^{q/2} + K_q m^q\,\Delta_n^q\\
&\mathbb{E}\Big(\sup_{s\le k\Delta_n}\left\|\overline{X}''(m)_{(i-1)\Delta_n+s} - \overline{X}''(m)_{(i-1)\Delta_n}\right\|^q\Big) \le K_q\Delta_n^{(q/2)\wedge 1}\,\phi_m.
\end{aligned} \tag{15.2.22}$$

This yields

$$\mathbb{E}\left(U(m)_i^n\right) \le K\Delta_n + Km^4\Delta_n^3. \tag{15.2.23}$$

Step 4) In this step we study $U'(m)_i^n$. We apply Itô's formula to the pair of processes $(\overline{X}'(m), \overline{X}''(m))$, between the times $S = (i-1)\Delta_n$ and $S' = S + k\Delta_n$, to obtain

$$\Delta_n\, U'(m)_i^n = M'(n,m,i)_{S'} + \int_S^{S'} H_m'\left(\overline{X}'(m)_s - \overline{X}'(m)_S, \overline{X}''(m)_s - \overline{X}''(m)_S\right)_s ds,$$

where

$$H_m'(x,y)_s = 2\|y\|^2\sum_{j=1}^2 b(m)_s^j\, x^j + \|y\|^2\sum_{j=1}^2 c_s^{jj} + \|x\|^2\int_{A_m^c}\left\|\delta(s,z)\right\|^2\lambda(dz), \tag{15.2.24}$$

and $M'(n,m,i)$ is a locally square-integrable martingale vanishing on $[0,S]$, and whose predictable bracket is

$$\int_S^{S'} \overline{H}'_m\big(\overline{X}'(m)_s - \overline{X}'(m)_S, \overline{X}''(m)_s - \overline{X}''(m)_S\big)_s\, ds$$

for a (random) function $\overline{H}'_m(x,y)_s$ which is a polynomial in (x,y) with bounded (random) coefficients, exactly as $\overline{H}_m(x,y)_s$. By virtue of (15.2.22) we deduce that $\mathbb{E}(\overline{H}'_m(x,y)_s) \le K$, hence $M'(m,n,i)$ is indeed a martingale, and thus

$$\mathbb{E}\big(U'(m)_i^n\big) = \frac{1}{\delta_n} \int_S^{S'} \mathbb{E}\big(H'_m\big(\overline{X}'(m)_s - \overline{X}'(m)_S, \overline{X}''(m)_s - \overline{X}''(m)_S\big)_s\big)\, ds.$$

The definition (15.2.24) gives $|H'_m(x,y)_s| \le K(m\|y\|^2\,\|x\| + \|y\|^2 + \phi_m\|x\|^2)$ because again $\|b(m)_t\| \le Km$ and $\|c_t\| \le K$, and also $\|\delta(t,z)\| \le \Gamma(z)$. Then (15.2.22) and Hölder's inequality yield

$$\mathbb{E}\big(U'(m)_i^n\big) \le K\Delta_n\big(m\Delta_n^{1/4} + \phi_m\big) \tag{15.2.25}$$

(recall that ϕ_m changes from line to line).

Step 5) In the last step we study $U''(m)_i^n$. We use again Itô's formula to get

$$\Delta_n U''(m)_i^n = M''(n,m,i)_{S'} + \int_S^{S'} H''_m\big(\overline{X}''(m)_s - \overline{X}''(m)_S\big)_s\, ds,$$

where, for the same reason as in Step 4, $M''(n,m,i)$ is a martingale, and

$$H''_m(x)_s = \int_{A_m^c} \big(f\big(x+\delta(s,z)\big) - f(x) - \nabla f(x)\delta(s,z)\big)\, \lambda(dz).$$

Now, (15.2.9) implies that $\delta^1\delta^2$ vanishes $\mathfrak{p}$ almost everywhere, hence $\mathfrak{q}$ almost everywhere as well (because it is a predictable function). This implies that, for $\mathbb{P}(d\omega)\otimes ds$ almost all (ω,s), we have $\int |\delta(\omega,s,z)^1|^u |\delta(\omega,s,z)^2|^v \lambda(dz) = 0$ for any $u,v>0$. Coming back to the specific form of f and its derivatives given by (15.2.11), we see that in fact H''_m is also

$$H''_m(x)_s = \int_{A_m^c} \big(\big(x^1\big)^2\big(\delta(t,z)^2\big)^2 + \big(x^2\big)^2\big(\delta(t,z)^1\big)^2\big)\, \lambda(dz),$$

and thus $0 \le H''_m(x) \le \phi_m\,\|x\|^2$. At this stage, (15.2.22) allows us to deduce

$$\mathbb{E}\big(U'(m)_i^n\big) = \frac{1}{\delta_n} \int_S^{S'} \mathbb{E}\big(H''_m\big(\overline{X}''(m)_s - \overline{X}''(m)_S\big)_s\big)\, ds \le K\Delta_n\,\phi_m. \tag{15.2.26}$$

Now, we plug (15.2.23), (15.2.25) and (15.2.26) into (15.2.21) to get

$$\mathbb{E}\left(\sum_{i=1}^{[t/\Delta_n]} \zeta(m)_i^n\right) \leq t\big(K\varepsilon\big(1+m^4\Delta_n^2\big) + K_\varepsilon\, m\Delta_n^{1/4} + K_\varepsilon\, \phi_m\big).$$

This is true for all $\varepsilon > 0$. Hence, letting first $n \to \infty$, then $m \to \infty$, then $\varepsilon \to 0$, we obtain (15.2.20). This completes the proof. □

15.3 Applications

A natural application of the previous results, and especially of Theorem 15.2.4, concerns testing procedures for deciding whether two components of a given process have jumps at the same time, or not. This is clearly a 2-dimensional problem, since in the multidimensional situation one can perform the tests for any pair of components. So below we assume that $X = (X^1, X^2)$ is 2-dimensional. We also assume (H).

In contrast with the construction of test statistics for deciding whether one component has jumps or not, where the *same* test statistics may be used for testing both null hypotheses of "jumps" and "no jumps", here we need two different test statistics for the two possible null hypotheses. More specifically, these two hypotheses are

$$\begin{aligned}
\Omega_t^{cj} &= \big\{\omega:\ s \mapsto X_s^1(\omega) \text{ and } s \mapsto X_s^2(\omega) \text{ have common jumps on } [0,t]\big\} \\
\Omega_t^{dj} &= \big\{\omega:\ \text{both } s \mapsto X_s^1(\omega) \text{ and } s \mapsto X_s^2(\omega) \text{ have jumps, but they have} \\
&\qquad \text{no common jump, on } [0,t]\big\}.
\end{aligned}$$

Note that the time $t > 0$ is *fixed* here. The union of these two disjoint sets is not Ω, but their global complement is

$$\Omega_t^{cc} = \big\{\omega:\ \text{both } s \mapsto X_s^1(\omega) \text{ and } s \mapsto X_s^2(\omega) \text{ are continuous on } [0,t]\big\}.$$

All three sets above may have a positive probability. However, before testing for common jumps we should of course be (reasonably) sure in advance that both components have jumps on the interval of interest.

We consider the function $f(x) = (x^1x^2)^2$ on $\mathbb{R}^2$ and the function G on $(\mathbb{R}^d)^k$ for some integer $k \geq 2$, defined by $G(x_1,\dots,x_k) = f(x_1+\cdots+x_k)$ (so $G = F^2$, with the notation (15.2.10)). Then we introduce the following two statistics (below, $t > 0$ is fixed):

$$T(k)_n = \frac{V^n(G,X)_t}{V^n(f,X)_t}, \qquad T'_n = V^n(f,X)_t.$$

The motivation for these test statistics is the following result.

Theorem 15.3.1 *Assume* (H).

(a) *We have*

$$T(k)_n \xrightarrow{\mathbb{P}} k \quad \textit{in restriction to the set } \Omega_t^{cj}, \tag{15.3.1}$$

and

$$T(k)_n \xrightarrow{\mathcal{L}\text{-s}} T(k) := \frac{D_t + k^2 H_t}{D'_t + H_t} \quad \textit{in restriction to the set } \Omega_t^{dj}, \tag{15.3.2}$$

where H_t is given by (15.2.14) *and where D_t and D'_t are two variables defined on an extension of the space, and conditionally on $\mathcal{F}$ and on being in Ω_t^{dj} the pair (D_t, D'_t) admit a density, hence $T(k) \neq 1$ almost surely on Ω_t^{dj}.*

(b) *We have*

$$T'_n \xrightarrow{\mathbb{P}} \begin{cases} f \star \mu_t > 0 & \textit{in restriction to the set } \Omega_t^{cj} \\ 0 & \textit{in restriction to the set } \Omega_t^{dj}. \end{cases}$$

In fact (b) and the first claim of (a) need no assumption except that X is a semimartingale, plus the property $\Delta X_t = 0$ a.s. for (15.3.1). In contrast, (15.3.2) requires the full force of (H).

Proof (b) is easily deduced from our previous results: indeed Theorem 3.4.1, page 81, yields that $V^n(f, X)_t \xrightarrow{\mathbb{P}} f * \mu_t$ as soon as $\Delta X_t = 0$ a.s., and $f * \mu_t$ is positive on Ω_t^{cj} and vanishes on Ω_t^{dj}. By Theorem 8.2.1, page 230, we also know that $V^n(G, X)_t \xrightarrow{\mathbb{P}} kf * \mu_t$, so (15.3.1) follows.

Now we turn to (15.3.2). Suppose first that X^1 and X^2 have *no common jumps* at all, that is (15.2.9) holds. The function F of (15.2.11) has the components $F^2 = G$ and $F^1(x_1, \dots, x_k) = f(x_1)$, hence $V^n(F^2, X)_t = V^n(G, X)_t$, and $V^n(F^1, X)_t = V^n(f, X)_{t-(k-1)\Delta_n}$. A look at the proof of Theorem 15.2.4 shows that the convergence (15.2.13) also holds if we replace $V^n(F^1, X)_t$ by $V^n(F^1, X)_{t+(k-1)\Delta_n} = V^n(f, X)_t$. Therefore we have

$$\left(\frac{1}{\Delta_n} V^n(f, X)_t, \frac{1}{\Delta_n} V^n(F, X)_t\right) \xrightarrow{\mathcal{L}\text{-s}} \left(\widetilde{V}(F^1, X)_t + \overline{C}_t, \widetilde{V}(F^2, X)_t + k^2 \overline{C}_t\right). \tag{15.3.3}$$

At this point, (15.3.2) follows easily, upon setting $D_t = \widetilde{V}(F^2, X)_t$ and $D'_t = \widetilde{V}(F^1, X)_t$. The fact that the pair (D_t, D'_t) admits a density, $\mathcal{F}$-conditionally, is a consequence of the definition (15.2.4) and of the property that $\partial^2_{i,i'} f^r_{(l,l');j}$ vanishes identically when $r = 1$ if $l + l' + j \geq 4$, whereas it does not when $r = 2$. The last claim that $T(k) \neq 1$ a.s. then follows.

Finally, it may happen that $0 < \mathbb{P}(\Omega_t^{dj}) < 1$. In this case, we proceed exactly as in Theorem 3.5.1, page 94: we replace the definitions of u_s and v_s given before (3.5.11) by

$$u_s = \lambda\big(\{z : \delta(s,z)^1\delta(s,z)^2 \neq 0\}\big), \qquad v_s = \int_{\{\|\delta(s,z)\|\leq 1\}} \big\|\delta(s,z)\big\|^2 \lambda(dz),$$

and the rest of the proof is the same. □

This theorem suggests that $T(k)_n$ is a test statistic which allows us to test the null hypothesis Ω_t^{cj} that there are common jumps on the interval $[0,t]$, whereas T_n' is useful for the other null hypothesis Ω_t^{dj} that all jumps are "disjoint". Of course, exactly as in Chap. 11, in order to be able to construct a test with a prescribed (asymptotic) level, we need Central Limit Theorems for both $T(k)_n$ and T_n', under the relevant null hypotheses.

The CLT for $T(k)_n$ is in the "non-degenerate" case and uses Theorem 11.1.2. More precisely, exactly as for Proposition 11.4.2, page 332, we have:

Proposition 15.3.2 *Assume* (H). *With the above notation, we have*

$$\left(\frac{1}{\sqrt{\Delta_n}}\big(V^n(f,X)_t - f*\mu_t\big), \frac{1}{\sqrt{\Delta_n}}\big(V^n(G,X)_t - kf*\mu_t\big)\right) \stackrel{\mathcal{L}\text{-s}}{\longrightarrow} \big(Z_t, Z_t'\big)$$

where the pair (Z_t, Z_t') *(defined by* (11.1.7) *as* $(\overline{V}(F^1,X)_t, \overline{V}(F^2,X)_t)$ *with* F *given by* (15.2.11)) *is defined on an extension of the space and is* $\mathcal{F}$*-conditionally centered, and Gaussian if* X *and* σ *do not jump together, and with the* $\mathcal{F}$*-conditional variance-covariance given by*

$$\begin{aligned} \mathbb{E}\big((Z_t)^2 \mid \mathcal{F}\big) &= H_t'' \\ \mathbb{E}\big(Z_t\, Z_t' \mid \mathcal{F}\big) &= k\, H_t'' \\ \mathbb{E}\big((Z_t')^2 \mid \mathcal{F}\big) &= \frac{2k^3+k}{3}\, H_t'', \end{aligned}$$

where

$$\begin{aligned} H_t'' = 2\sum_{s\leq t} \big(\Delta X_s^1\big)^2 \big(\Delta X_s^2\big)^2 \big(\big(\Delta X_s^2\big)^2\big(c_{s-}^{11} + c_s^{11}\big) + 2\Delta X_s^1 \Delta X_s^2 \big(c_{s-}^{12} + c_s^{12}\big) \\ + \big(\Delta X_s^1\big)^2\big(c_{s-}^{22} + c_s^{11}\big)\big). \end{aligned}$$

At this point, it remains to use again the "delta method", to obtain the following theorem, which in turn can be used to derive a concrete test when the null hypothesis is "common jumps":

Theorem 15.3.3 *Assume* (H). *In restriction to the set* Ω_t^{cj} *the sequence* $\frac{1}{\sqrt{\Delta_n}}(T(k)_n - k)$ *converges stably in law to a variable* $T(k)$ *which, conditionally* *on $\mathcal{F}$, is centered with variance*

$$\widetilde{\mathbb{E}}\big((T(k))^2 \mid \mathcal{F}\big) = \frac{k(k-1)(2k-1)\,H''_t}{3(f*\mu_t)^2}$$

and is Gaussian conditionally on $\mathcal{F}$ if further the processes X and σ have no common jumps.

For the other null hypothesis of "disjoint jumps", we need a result saying at which rate T'_n goes to 0 on the set Ω_t^{dj}. This is essentially the same as for (15.3.2), and even simpler because we only have to worry about the single degenerate functional $V^n(f,X)$. In fact, it follows from (15.3.3) and also (15.2.5) that:

Theorem 15.3.4 *Assume* (H). *In restriction to the set Ω_t^{dj} the sequence $\frac{1}{\Delta_n} T'_n$ converges stably in law to a variable T' which, conditionally on $\mathcal{F}$, is positive with mean $\widetilde{\mathbb{E}}(T' \mid \mathcal{F}) = H_t$, as given by* (15.2.15).

Bibliographical Notes

The content of this chapter is new, but based on Jacod and Todorov [62], to which the application in Sect. 15.3 is borrowed, with some simplifications. We view the results presented in this chapter as only a start on this topic, and we hope that more thorough studies will be undertaken by interested researchers.

Chapter 16
Semimartingales Contaminated by Noise

Our last chapter is, even more than the others, motivated by applications. The setting can be sketchily described as follows: our process of interest is the semimartingale X, and as in most of what precedes it is sampled at regularly spaced observation times, with a "small" time step Δ_n. However, the observations are not totally accurate, but are contaminated with some kind of noise. That is, if an observation takes place at time t, instead of X_t we actually observe a variable of the form

$$Z_t = X_t + \chi_t \tag{16.0.1}$$

where χ_t is a "measurement error," or "noise." Then the question arises of what happens to functionals like $V^n(f, X)$ or $V'^n(f, X)$ if we substitute X with Z. Or, from a more applied viewpoint, and assuming that X is an Itô semimartingale, can we still estimate, say, the integrated volatility, and how could we do so?

The answer, of course, depends fundamentally on the properties of χ_t. At one end of the spectrum, the process χ_t may itself be a semimartingale. Then Z is a semimartingale and the previous theory applies to it: We can for example estimate the integrated volatility of the sum $Z = X + \chi$, but there is no way to estimate the integrated volatility of X itself. At the other end of the spectrum, we have a standard "additive white noise", meaning that the variables χ_t are i.i.d. centered and globally independent of X. Then the behavior of the functionals $V^n(f, Z)$ or $V'^n(f, Z)$ is typically dominated by the noise and, nevertheless, it is possible consistently to estimate the integrated volatility of X (and there is a vast literature on this central topic).

These two cases are by far not the only cases encountered in practice. We often have a significant rounding noise due to the fact that the observed values of X_t are recorded with a limited number of decimal points. For example, in financial applications, prices are given in cents, so for a stock at the average price of \$20 the rounding error is about 0.05% of the value: when the price is observed every second, one thus observes that it typically does not change for a few seconds in a row; if we use an Itô semimartingale model, this effect is due, in particular, to the rounding. In the case of a "pure rounding noise" when the observed value Z_t is the biggest multiple of a given rounding level $a > 0$ smaller than or equal to X_t (that is

J. Jacod, P. Protter, *Discretization of Processes*,
Stochastic Modelling and Applied Probability 67,
DOI 10.1007/978-3-642-24127-7_16, © Springer-Verlag Berlin Heidelberg 2012

$Z_t = a[X_t/a]$), there is again no way to estimate the integrated volatility of X. On the other hand when the level a is "small", that is $a = a_n$ goes to 0 as $\Delta_n \to 0$, this estimation becomes (theoretically) possible.

We devote the first section below to a precise description of the type of noise which will be considered, and this does *not* include "pure rounding," unfortunately: this case is so far very poorly understood, and is excluded here, although we allow for some kind of rounding below, because rounding is an everywhere present feature.

16.1 Structure of the Noise and the Pre-averaging Scheme

16.1.1 Structure of the Noise

The formulation (16.0.1) does not really fit reality, since the noise exists only when we observe something: that is, instead of observing $X_{i\Delta_n}$ we observe $X_{i\Delta_n} + \chi_i^n$ where χ_i^n is the noise at time $i\Delta_n$. However, it does no harm to suppose that χ_t is actually defined for all times, although it is immaterial outside the set of observation times.

Another preliminary remark should also be made: in (16.0.1) the "level" of noise does not depend on the sampling frequency $1/\Delta_n$. But in some applications it naturally does. For example a physicist may increase the sampling frequency together with the accuracy of measurement. Perhaps more to the point, the statistician does some asymptotic estimation by pretending that Δ_n goes to 0, but in practice Δ_n is maybe small but given; then if the (fixed) level of noise is also small, a rather natural asymptotic setting to consider is that $\Delta_n \to 0$ and *simultaneously* the noise level shrinks to 0. A shrinking noise is somewhat difficult to model in a realistic way, except in one situation: namely, the noise depends on n through a multiplicative factor going to 0. This is what we will do below.

Now we come to the description of the noise. Basically, χ_t should be, conditionally on the whole process X, a family of *independent, centered* random variables. This implies that the variables χ_t are (unconditionally) centered. However we emphasize the fact that, again unconditionally, they are *not* necessarily mutually independent, *nor independent of* X. This can be formalized in different ways, and here we use the following convenient method.

We have first a filtered probability space $(\Omega^{(0)}, \mathcal{F}^{(0)}, (\mathcal{F}_t^{(0)})_{t\geq 0}, \mathbb{P}^{(0)})$, on which our basic d-dimensional semimartingale X is defined. Second, for each time t we have a transition probability $Q_t(\omega^{(0)}, dz)$ from $(\Omega^{(0)}, \mathcal{F}_t^{(0)})$ into $\mathbb{R}^d$. We endow the space $\Omega^{(1)} = (\mathbb{R}^d)^{[0,\infty)}$ with the product Borel σ-field $\mathcal{F}^{(1)}$ and the "canonical process" $(\chi_t : t \geq 0)$ and the canonical filtration $\mathcal{F}_t^{(1)} = \bigcap_{s>t} \sigma(\chi_r : r \leq s)$, and the probability $\mathbb{Q}(\omega^{(0)}, d\omega^{(1)})$ which is the product $\otimes_{t\geq 0} Q_t(\omega^{(0)}, .)$. We then define a filtered probability space $(\Omega, \mathcal{F}, (\mathcal{F}_t)_{t\geq 0}, \mathbb{P})$ and a (bigger) filtration $(\mathcal{H}_t)$ as follows:

$$\left.\begin{array}{ll} \Omega = \Omega^{(0)} \times \Omega^{(1)}, & \mathcal{F} = \mathcal{F}^{(0)} \otimes \mathcal{F}^{(1)} \\ \mathcal{F}_t = \cap_{s>t} (\mathcal{F}_s^{(0)} \otimes \mathcal{F}_s^{(1)}) & \mathcal{H}_t = \mathcal{F}^{(0)} \otimes \mathcal{F}_t^{(1)} \\ \mathbb{P}(d\omega^{(0)}, d\omega^{(1)}) = \mathbb{P}^{(0)}(d\omega^{(0)})\, \mathbb{Q}(\omega^{(0)}, d\omega^{(1)}). & \end{array}\right\} \tag{16.1.1}$$

Any variable or process defined on either $\Omega^{(0)}$ or $\Omega^{(1)}$ is considered in the usual way as a variable or a process on Ω. Since $\omega^{(0)} \mapsto \mathbb{Q}(\omega^{(0)}, A)$ is $\mathcal{F}_t^{(0)}$ measurable if $A \in \sigma(\chi_s : s \in [0,t))$, by (2.1.28) and (2.1.29) the filtered extension $(\Omega, \mathcal{F}, (\mathcal{F}_t)_{t\geq 0}, \mathbb{P})$ of $(\Omega^{(0)}, \mathcal{F}^{(0)}, (\mathcal{F}_t^{(0)})_{t\geq 0}, \mathbb{P}^{(0)})$ is very good, and X is a semimartingale on the extension with the same characteristics as on the original space, and also the same Grigelionis form when it is an Itô semimartingale.

The (non-normalized) error process is the canonical process χ defined above, with components denoted $(\chi_t^j)_{1\leq j\leq d}$. By construction, the variables χ_t are mutually independent, *conditionally on* $\mathcal{F}^{(0)}$, but we need more:

Assumption 16.1.1 (or **(N)**): For each $q > 0$ the process $\int Q_t(\omega^{(0)}, dz)\, \|z\|^q$ is locally bounded on the space $(\Omega^{(0)}, \mathcal{F}^{(0)}, (\mathcal{F}_t^{(0)}), \mathbb{P}^{(0)})$, and we have

$$\int Q_t\big(\omega^{(0)}, dz\big)\, z = 0. \tag{16.1.2}$$

The second and third conditional moments will be explicitly used below, and are denoted as:

$$\begin{aligned} \Upsilon_t^{ij}\big(\omega^{(0)}\big) &= \int Q_t\big(\omega^{(0)}, dz\big)\, z^i z^j \\ \Upsilon_t^{\prime ijk}\big(\omega^{(0)}\big) &= \int Q_t\big(\omega^{(0)}, dz\big)\, z^i z^j z^k. \end{aligned} \tag{16.1.3}$$

Finally, the (potentially) observed process at stage n is the following process:

$$Z_t^n = X_t + (\Delta_n)^\eta\, \chi_t, \quad \text{where } \eta \geq 0. \tag{16.1.4}$$

Once more, the most useful case is when $\eta = 0$, so $Z^n = Z = X + \chi$. When $\eta > 0$ we have a shrinking noise with shrinking factor $u_n = (\Delta_n)^\eta$. For the shrinking factor one could use any sequence $u_n > 0$, but the relative behavior of Δ_n and u_n is fundamental here, so assuming $u_n = \Delta_n^\eta$ is a significant simplification.

Remark 16.1.2 The local boundedness of the all moments of the noise is not a serious practical restriction. It would be possible to require local boundedness for the moments up to order q only, with q depending on the specific results one wants to prove.

On the other hand, the conditional centering property (16.1.2) is quite restrictive, see Example 16.1.4, but cannot be dispensed with. It could be replaced by the fact that $Y_t = \int Q_t(\omega^{(0)}, dz)\, z$ is a semimartingale but, in this case, and when $\eta = 0$ in (16.1.4), it would be impossible to disentangle X from $X + Y$.

Example 16.1.3 Additive noise: This is when the χ_t are (unconditionally) mutually independent and independent of X. It means that $Q_t(\omega^{(0)}, dz) = Q_t(dz)$ does not depend on $\omega^{(0)}$. In most cases with additive noise we have that $Q_t = Q$ does not depend on t either, so $\Upsilon_t^{ij} = \Upsilon^{ij}$ and $\Upsilon_t'^{ijk} = \Upsilon'^{ijk}$ are just numbers. This case is by far the most frequent case considered in the literature, probably because it is the easiest to handle, and also because it corresponds to the common idea about "measurement errors". However, this type of noise does not properly account for the so-called "microstructure noise" in mathematical finance: in this application, the rounding noise explained below is probably dominant.

Example 16.1.4 Pure rounding noise: Letting $\alpha > 0$ be the rounding level, we observe $Z_t = \alpha[X_t/\alpha]$, say (rounding from below). Then conditionally on $\mathcal{F}^{(0)}$ the variables $\chi_t = \alpha[X_t/\alpha] - X_t$ are independent (because they are indeed $\mathcal{F}^{(0)}$-measurable) and with bounded moments. However (16.1.2) fails, since χ_t is $\mathcal{F}^{(0)}$ measurable and Q_t is the Dirac mass sitting at χ_t. Hence pure rounding *is excluded* in what follows.

This is of course not surprising. If for example $X = \sigma W$ with $\sigma > 0$ and W a Brownian motion, we will see later that the discrete observation of Z when $\Delta_n \to 0$ allows us to recover σ, under Assumption 16.1.1. However if Z is the rounded version of X, the σ-field generated by the variables Z_t is the σ-field generated by the upcrossings and downcrossings by X of all levels $k\alpha$ for $k \in \mathbb{Z}$. So even if the whole path of the noisy process Z were known over some finite interval $[0, t]$, one could not infer the value of σ.

Example 16.1.5 Additive noise plus rounding: Here we again fix a rounding level $\alpha > 0$. For each t we consider two variables L_t and U_t, where U_t is uniform over $[0, 1]$ and L_t is $\mathbb{Z}$-valued centered with moments of all order bounded in t, and L_t, U_t are mutually independent and independent of X. We observe the process

$$Z_t = \alpha\big(L_t + [U_t + X_t/\alpha]\big).$$

This fits our general model. Assumption (N) holds here, and more precisely $Q_t(\omega^{(0)}, dz)$ is the law of $\alpha(L_t + [X_t(\omega^{(0)})/\alpha + U_t])$ (checking (16.1.2), which amounts to $\mathbb{E}([u + U_t]) = u$, is straightforward). This model amounts to having an additive noise of the form $L_t + \alpha U_t$, followed by rounding.

This also accommodate shrinking noise: indeed the normalized noise $\Delta_n^\eta \chi_t$ follows a model of the same type, with $\Delta_n^\eta \alpha$ instead of α.

16.1.2 The Pre-averaging Scheme

Recall that our aim is to obtain estimators for quantities connected with X, and we are not *a priori* interested by the noise. For example, consider the case $X = \sigma W$ and a non-shrinking additive noise with law independent of time. We want to estimate $c = \sigma^2$.

The first (and wrong) idea is to use the approximate quadratic variation of the noisy process, that is $\sum_{i=1}^{[t/\Delta_n]}(\Delta_i^n Z)^2$. The ith summand here has the same law as $(\sigma\sqrt{\Delta_n}\,U + \chi - \chi')^2$, where U, χ, χ' are independent, U is $\mathcal{N}(0,1)$ and χ and χ' have the same law as all χ_t's. Hence $(\Delta_i^n Z)^2$ is approximately the same as $(\chi - \chi')^2$ in law when n is large, and the summands are independent as soon as they are separated by more than one time step. So it is easy to see that $\Delta_n \sum_{i=1}^{[t/\Delta_n]}(\Delta_i^n Z)^2 \xrightarrow{\mathbb{P}} 2t\Upsilon$ (here, according to (16.1.3), Υ is the second moment of the noise, so $\mathbb{E}((\chi - \chi')^2) = 2\Upsilon$). In other words, *the approximate quadratic variation explodes, and once properly normalized its limit depends on the noise but not on the quantity of interest σ.*

Another simple idea consists in taking the average of k_n successive values, say $Y_{i\Delta_n} = \frac{1}{k_n}\sum_{j=0}^{k_n-1} Z_{(i+j)\Delta_n}$, and then take $\sum_{i=0}^{[t/\Delta_n]-k_n}(Y_{(i+k_n)\Delta_n} - Y_{i\Delta_n})^2$. Since the χ_t are i.i.d. centered, the variable $Y_{i\Delta_n}$ is close to the average $\frac{1}{k_n}\sum_{j=0}^{k_n-1} X_{(i+j)\Delta_n}$ when k_n is large, and thus the above sum is close to the variable $\frac{1}{k_n^2} V(\Phi, 2k_n, X)_t$, where $V^n(\Phi, 2k_n, X)$ is given by (8.1.9) and with $\Phi(y) = 4(\int_{1/2}^1 y(t)\,dt - \int_0^{1/2} y(t)\,dt)^2$: so we may hope to obtain, after proper normalization, an approximation of c.

The second idea works in our more general setting, due to the fact that conditionally on X the variables χ_t are independent and centered, although with a law depending on t and on the process X. Taking the plain average as above may not be the best idea, so we introduce a "weighted average" below, which includes the previous one and is called "pre-averaging". It is also another name for the so-called "kernel methods", although there are some slight differences (mainly about the treatment of the first few and last few data in the time interval of interest). As one will see, the pre-averaging scheme is also a special case of the "average functions" introduced in Chap. 12, see (12.0.3).

We need two ingredients:

- a sequence of integers k_n satisfying

$$k_n = \frac{1}{\theta\,\Delta_n^{\eta'}}\left(1 + o\left(\Delta_n^{(1-\eta')/2}\right)\right) \quad \text{where } \theta > 0,\ \eta' \in (0,1), \tag{16.1.5}$$

  and we write $u_n = k_n \Delta_n$;
- a real-valued (weight) function g on $[0,1]$, satisfying

$$\begin{aligned} &g \text{ is continuous, piecewise } C^1 \text{ with a piecewise Lipschitz derivative } g', \\ &g(0) = g(1) = 0, \qquad \int_0^1 g(s)^2\,ds > 0. \end{aligned} \tag{16.1.6}$$

The sequence k_n is fixed throughout, the weight function g will sometimes vary, but always satisfies (16.1.6). Note that g can be written as in (12.0.4) for a signed measure G on $[0,1]$, which has no atom and mass $G([0,1]) = 0$, but here we require some additional smoothness. This leads us to use and generalize the notation

(12.2.1) for $p>0$ and any function h on [0, 1]:

$$\Lambda(h,p)=\int_0^1 |h(s)|^p\,ds, \qquad \Lambda(h)=\Lambda(h,2)=\int_0^1 h(s)^2\,ds. \tag{16.1.7}$$

It is convenient to extend g to the whole of $\mathbb{R}$ by setting $g(s)=0$ if $s\notin[0,1]$. We associate with g the following numbers, where $p\in(0,\infty)$ and $i\in\mathbb{Z}$:

$$\begin{array}{ll} g_i^n\ =\ g(i/k_n), & g_i'^n\ =\ g_i^n-g_{i-1}^n, \\ \Lambda_n(g,p)\ =\ \sum_{i=1}^{k_n}|g_i^n|^p, & \Lambda_n'(g,p)_n\ =\ \sum_{i=1}^{k_n}|g_i'^n|^p \end{array} \tag{16.1.8}$$

($g_i'^n$ is *not* $g'(i/k_n)$ for the derivative g' of g, which in general is not a weight function anyway. In fact, $g_i'^n$ is "close" to $g'(i/k_n)/k_n$, at least when g' is everywhere continuous). (16.1.6) yields, as $n\to\infty$,

$$\Lambda_n(g,p)=k_n\Lambda(g,p)+\mathrm{O}(1), \qquad \Lambda_n'(g,p)=k_n^{1-p}\Lambda\big(g',p\big)+\mathrm{O}\big(k_n^{-p}\big). \tag{16.1.9}$$

If $U=(U_t)_{t\geq 0}$ is a q-dimensional process, we rewrite the notation (12.1.10) and extend it as follows, for $i\geq 1$ (recall that $g_0^n=g_{k_n}^n=0$, hence $\sum_{j=1}^{k_n}g_j'^n=0$):

$$\begin{aligned} \overline{U}(g)_i^n &= \sum_{j=1}^{k_n-1} g_j^n\,\Delta_{i+j-1}^n U = -\sum_{j=1}^{k_n} g_j'^n\,U_{(i+j-2)\Delta_n} \\ &= -\sum_{j=1}^{k_n} g_j'^n\,(U_{(i+j-2)\Delta_n}-U_{(i-1)\Delta_n}) \\ \widehat{U}(g)_i^{n,lm} &= \sum_{j=1}^{k_n}\big(g_j'^n\big)^2\,\Delta_{i+j-1}^n U^l\,\Delta_{i+j-1}^n U^m. \end{aligned} \tag{16.1.10}$$

The variables $\overline{U}(g)_i^n$ are q-dimensional, and $\widehat{U}(g)_i^n=(\widehat{U}(g)_i^{n,lm})_{1\leq l,m\leq q}$ is $\mathcal{M}_{q\times q}^+$-valued. We also recall the notation (8.1.7) and (8.1.8), which are

$$\begin{aligned} t\in[0,1]\ \mapsto\ Y(n,i)_t &= Y_{(i-1)\Delta_n+tu_n}-Y_{(i-1)\Delta_n} \\ Y(n,i)_t^{(n)} &= Y(n,i)_{[k_nt]/k_n}. \end{aligned} \tag{16.1.11}$$

Now we can define some of the processes of interest for us. We consider a continuous function f on $\mathbb{R}^d$ and the function $\Phi=\Phi_{f,g}$ on $\mathbb{D}_1^d$ defined by

$$\Phi(x)\ =\ f\big(\Psi(x)\big), \quad \text{where } \Psi(x)\ =\ -\int_0^1 x(s)\,g'(s)\,ds, \tag{16.1.12}$$

which is the same as in (12.0.3). Taking into account the last equality in the first line of (16.1.10), plus (16.1.11), one readily checks that $\Psi(U(n,i)^{(n)})=\overline{U}(g)_i^n$.

Therefore the processes $V^n(\Phi, k_n, U)$ and $V'^n(\Phi, k_n, U)$ of (8.1.9) and (8.1.11) take the form:

$$\begin{aligned} V^n(\Phi, k_n, U)_t &= \sum_{i=1}^{[t/\Delta_n]-k_n+1} f\big(\overline{U}(g)_i^n\big) \\ V'^n(\Phi, k_n, U)_t &= \Delta_n \sum_{i=1}^{[t/\Delta_n]-k_n+1} f\big(\overline{U}(g)_i^n/\sqrt{u_n}\big), \end{aligned} \tag{16.1.13}$$

and when we plug in the noisy process, then $U = Z^n = X + \Delta_n^\eta \chi$.

Remark 16.1.6 It is possible to introduce different weight functions for each component of the process. Up to more cumbersome notation, the mathematical treatment and results would be the same.

Example 16.1.7 The simplest weight function is $g(s) = 2(s \wedge (1-s))$ for $s \in [0,1]$. Then $\Lambda(g, p) = \frac{1}{2(p+1)}$ and $\Lambda(g', p) = 2^p$ and also, when $k_n = 2k'_n$ is even,

$$\begin{aligned} \overline{U}(g)_i^n &= \frac{1}{k'_n}\big(U_{(i+k'_n-1)\Delta_n} + \cdots + U_{(i+2k'_n-2)\Delta_n}\big) \\ &\quad - \frac{1}{k'_n}\big(U_{(i-1)\Delta_n} + \cdots + U_{(i+k'_n-2)\Delta_n}\big) \\ \widehat{U}(g)_i^{n,lm} &= \frac{4}{k'^2_n} \sum_{j=1}^{2k'_n} \Delta_{i+j-1}^n U^l \, \Delta_{i+j-1}^n U^m. \end{aligned}$$

In this case, $\overline{U}(g)_i^n$ is simply the difference between two successive (non-overlapping) averages of k'_n values of U.

16.2 Law of Large Numbers for General (Noisy) Semimartingales

Throughout the rest of the chapter, we have a regular discretization and the integers k_n satisfy (16.1.5) and the weight function g satisfies (16.1.6), and Φ is associated by (16.1.12), and further the setting (16.1.1) is in force.

Here we consider an arbitrary d-dimensional semimartingale. Recalling the noisy process Z^n of (16.1.4), we give a law of large numbers for $V^n(\Phi, k_n, Z^n)$ of (16.1.13).

Theorem 16.2.1 *Let X be a d-dimensional semimartingale with jump measure μ, and assume* (N) *for the noise. Let f be a continuous function on $\mathbb{R}^d$ satisfying*

$f(x)=o(\|x\|^p)$ as $x\to 0$ for some $p\geq 2$. Let $Z^n=X+(\Delta_n)^\eta\chi$ with $\eta\geq 0$ and take k_n satisfying (16.1.5) *with $\eta'>0$ and also*

$$\eta' \geq 2\frac{1-p\eta}{2+p}. \tag{16.2.1}$$

Then, for each $t>0$, we have the following:

$$\frac{1}{k_n}V^n\big(\Phi,k_n,Z^n\big)_t \xrightarrow{\mathbb{P}} \overline{\Phi}*\mu_{t-}, \quad \text{where } \overline{\Phi}(z)=\int_0^1 f\big(zg(s)\big)\,ds. \tag{16.2.2}$$

When there is no noise (that is, $Z^n=X$), this result is Theorem 8.3.1, except that here we additionally assume that k_n behaves like a power of $1/\Delta_n$, plus the structure (16.1.12) for Φ: indeed in this case the definitions (8.3.1) and (16.2.2) for $\overline{\Phi}$ agree, and we can take η arbitrarily large and so (16.2.1) is satisfied. And of course, as in Theorem 8.3.1, the convergence (16.2.2) does *not* hold for the Skorokhod topology.

When (16.2.1) fails, one does not know the precise behavior of the processes $V^n(\Phi,k_n,Z^n)$, but presumably the noise "dominates".

Remark 16.2.2 The condition (16.2.1) means that k_n goes fast enough to ∞, the rate depending on the behavior of f at 0 and also on the rate with which the noise shrinks: the faster it shrinks, the slower k_n needs to increase. Two extreme cases are worth emphasizing:

- When $\eta\geq\frac{1}{2}$, that is the noise shrinks fast enough, (16.2.1) is fulfilled for all $p\geq 2$, as soon as $\eta'>0$. In other words, and as anticipated, a very small noise has no influence.
- When $\eta=0$, that is the noise is not shrinking, the condition becomes $\eta'\geq\frac{2}{2+p}$, which is satisfied for all $p\geq 2$ when $\eta'\geq\frac{1}{2}$.

Observe also that the right side of (16.2.1) decreases when p increases, and is never bigger than $\frac{1}{2}$ when $p\geq 2$.

Before giving the proof, we introduce a strengthened version of Assumption (N):

Assumption 16.2.3 (or **(SN)**): We have (N), and for all $q>0$ we also have $\sup_{\omega^{(0)},t}\int Q_t(\omega^{(0)},dz)\,\|z\|^q<\infty$.

We also give estimates on the noise, which hold under (SN) and will be of constant use. Recalling the filtration $(\mathcal{H}_t)$ defined in (16.1.1), and since $|g_j'^n|\leq K/k_n$, and conditionally on $\mathcal{F}^{(0)}$ the χ_t's are independent and centered, we deduce from the Burkholder-Davis-Gundy and Hölder's inequalities that, for all $p>0$,

$$\mathbb{E}\big(\|\overline{\chi}(g)_i^n\|^p\mid\mathcal{H}_{(i-1)\Delta_n}\big)\leq K_p\,k_n^{-p/2}, \qquad \mathbb{E}\big(\|\widehat{\chi}(g)_i^n\|^p\mid\mathcal{H}_{(i-1)\Delta_n}\big)\leq K_p\,k_n^{-p} \tag{16.2.3}$$

(here, $\overline{\chi}(g)_i^n$ and $\widehat{\chi}(g)_i^n$ are associated with the noise process χ by (16.1.10)).

Proof of Theorem 16.2.1 1) We essentially copy the proof of Theorem 8.3.1 (see pages 235–237), to which we make constant reference. By localization it is no restriction to assume (SN). Below, we heavily use the notation (16.1.11), (16.1.12), and also $\Phi(U(n,i)^{(n)}) = f(\overline{U}(g)_i^n)$ and $\overline{Z}^n(g)_i^n = \overline{X}(g)_i^n + (\Delta_n)^\eta\,\overline{\chi}(g)_i^n$.

Before starting, we observe that (16.2.3) yields for any $q \geq 0$:

$$\mathbb{E}\Big(\sup_{1\leq i\leq [t/\Delta_n]} \|\overline{\chi}(g)_i^n\|^q\Big) \leq \mathbb{E}\left(\sum_{i=1}^{[t/\Delta_n]} \|\overline{\chi}(g)_i^n\|^q\right) \leq K t \Delta_n^{q\eta'/2-1}. \qquad (16.2.4)$$

2) For any $\varepsilon > 0$ we have the functions ψ_ε and $\psi'_\varepsilon = 1 - \psi_\varepsilon$ of (3.3.16), in which $\psi_\varepsilon(x) = \psi(\|x\|/\varepsilon)$ and ψ is C^∞ on $\mathbb{R}$ with $1_{[1,\infty)} \leq \psi \leq 1_{[1/2,\infty)}$. We set $\Phi_\varepsilon(x) = (f\psi_\varepsilon)(\Psi(x))$ and $\Phi'_\varepsilon(x) = (f\psi'_\varepsilon)(\Psi(x))$ for $x \in \mathbb{D}_1^d$ (warning: they are not the same as in the proof of Theorem 8.3.1; here, Ψ is given by (16.1.12)). We also associate $\overline{\Phi}_\varepsilon$ and $\overline{\Phi}'_\varepsilon$ with Φ_ε and Φ'_ε by (16.2.2), and we still have (8.3.10), that is $\overline{\Phi}_\varepsilon * \mu_{t-} \xrightarrow{\mathbb{P}} \overline{\Phi} * \mu_{t-}$ as $\varepsilon \to 0$. Therefore we are left to prove the following two properties:

$$\varepsilon \in (0,1] \quad \Rightarrow \quad \frac{1}{k_n} V^n\big(\Phi_\varepsilon, k_n, Z^n\big)_t \xrightarrow{\mathbb{P}} \overline{\Phi}_\varepsilon * \mu_{t-}, \qquad (16.2.5)$$

$$\zeta > 0 \quad \Rightarrow \quad \lim_{\varepsilon\to 0}\ \limsup_n\ \mathbb{P}\left(\frac{1}{k_n}\big|V^n\big(\Phi'_\varepsilon, k_n, Z^n\big)_t\big| > \zeta\right) = 0. \qquad (16.2.6)$$

3) Here, following the proof of Lemma 8.3.4, we show (16.2.5). In this step, $\varepsilon \in (0,1]$ is fixed and $A = \sup(|g'(s)|)$. The successive jump times of X with size bigger than $\varepsilon/8A$ are $S_1, S_2, \ldots$, and we set $Q_t = \sum_{q\geq 1} 1_{\{S_q < t\}}$ and $X'_t = X_t - \sum_{q\geq 1: S_q\leq t} \Delta X_{S_q}$.

Since X' has no jump bigger than $\varepsilon/8A$, the set $\Omega_t'^n$, on which $\|X'_v - X'_w\| \leq \varepsilon/4A$ for all $v, w \in [0,t]$ with $|v-w| \leq 2u_n$, satisfies $\Omega_t'^n \to \Omega$ as $n \to \infty$. In view of (16.1.11) and $\overline{X'}(g)_i^n = \Psi(X'(n,i)^{(n)})$ and $|g'| \leq A$, we deduce

$$\big\|\overline{X'}(g)_i^n\big\| \leq \frac{\varepsilon}{4} \quad \text{on } \Omega_t'^n, \text{ if } i \leq [t/\Delta_n]. \qquad (16.2.7)$$

Now, we modify the definition of $\alpha(n,q,j)$ and A_t^n in the proof of Lemma 8.3.4 as follows:

$$\alpha(n,q,j) = (f\psi_\varepsilon)\big(\overline{X'}(g)^n_{i(n,q)+1-j} + g_j^n \Delta X_{S_q} + (\Delta_n)^\eta\,\overline{\chi}(g)^n_{i(n,q)+1-j}\big),$$

$$A_t^n = \begin{cases} \frac{1}{k_n}(f\psi_\varepsilon)(\overline{X}(g)^n_{t/\Delta_n} + (\Delta_n)^\eta\,\overline{\chi}(g)^n_{t/\Delta_n}) & \text{if } t/\Delta_n \text{ is an integer} \\ & \quad \text{and } S_{Q_t+1} = t \\ 0 & \text{otherwise,} \end{cases}$$

and we still set $\zeta_q^n = \frac{1}{k_n}\sum_{j=1}^{k_n\wedge i(n,q)} \alpha(n,q,j)$. By (16.2.4) and (16.2.7) and $\Omega_t'^n \to \Omega$ we see that the set $\Omega_t^n = \Omega_t'^n \cap \{\sup_{1\leq i\leq[t/\Delta_n]} \|(\Delta_n)^\eta\overline{\chi}(g)_i^n\| \leq \frac{\varepsilon}{4}\}$ satisfies

$\mathbb{P}(\Omega_t^n) \to 1$ and also that, since $(f\psi_\varepsilon)(x) = 0$ when $\|x\| \leq \varepsilon/2$,

$$\frac{1}{k_n} V^n(\Phi_\varepsilon, k_n, Z^n)_t = \sum_{q=1}^{Q_t} \zeta_q^n + A_t^n \quad \text{on the set } \Omega_t^n.$$

As in Lemma 8.3.4 we have $\zeta_q'^n := \frac{1}{k_n} \sum_{j=1}^{k_n} f(g_j^n \Delta X_{S_q}) \to \overline{\Phi}_\varepsilon(\Delta X_{S_q})$ by Riemann approximation. Hence, using $\mathbb{P}(\Omega_t^n) \to 1$, it remains to prove that

$$A_t^n 1_{\Omega_t^n} \to 0, \quad q \leq Q_t \Rightarrow (\zeta_q^n - \zeta_q'^n) 1_{\Omega_t^n} \xrightarrow{\mathbb{P}} 0. \tag{16.2.8}$$

As seen before, we have $\|\overline{X'}(g)_i^n + (\Delta_n)^\eta \overline{\chi}(g)_i^n\| \leq \varepsilon/2$ on Ω_t^n if $i \leq [t/\Delta_n]$, and also $\overline{X'}(g)_{i(n,q)+1-j}^n + (\Delta_n)^\eta \overline{\chi}(g)_{i(n,q)+1-j}^n \xrightarrow{\mathbb{P}} 0$ for all $q \leq Q_t$ and $j \geq 0$ (use (16.2.4) again and the continuity of X' at each time S_q). Then, since f is continuous and thus locally bounded, we immediately deduce the first part of (16.2.8), and also $\alpha(n,q,j) - f(g_j^n \Delta X_{S_q}) \xrightarrow{\mathbb{P}} 0$ and $|\alpha(n,q,j)| \leq K$ for all q, j, which in turn implies the second part of (16.2.8).

4) Finally we prove (16.2.6). Our assumption on f yields $|f(x)| \leq \phi(\|x\|)^2 \|x\|^p$ for some increasing continuous function ϕ satisfying $\phi(0) = 0$. By singling out the two cases $\|y\| \leq 2\|x\|$ and $\|y\| > 2\|x\|$ and since $\psi_\varepsilon'(x) = 0$ when $\|x\| \geq \varepsilon$, we see that $|(f\psi_\varepsilon')(x+y)| \leq K\phi(\varepsilon)(\phi(3\|x\|)\|x\|^p + \|y\|^p)$. Therefore, if $\widetilde{f}(x) = \phi(3\|x\|)\|x\|^p$ and $\widetilde{\Phi}(x) = \widetilde{f}(\Psi(x))$, we deduce

$$\left|V^n(\Phi_\varepsilon', k_n, Z^n)_t\right| \leq K\phi(\varepsilon)\left(V^n(\widetilde{\Phi}, k_n, X)_t + \Delta_n^{p\eta} \sum_{i=1}^{[t/\Delta_n]} \|\overline{\chi}(g)_i^n\|^p\right).$$

On the one hand, the function $\widetilde{f}$ is continuous and is $\mathrm{o}(\|x\|^2)$ as $x \to 0$, so Theorem 8.3.1 yields that $\frac{1}{k_n} V^n(\widetilde{\Phi}, k_n, X)_t$ converges in probability to a finite limit. On the other hand (16.2.4) yields $\frac{\Delta_n^{p\eta}}{k_n} \mathbb{E}(\sum_{i=1}^{[t/\Delta_n]} \|\overline{\chi}(g)_i^n\|^p) \leq Kt$ as soon as (16.2.1) holds. Since $\phi(\varepsilon) \to 0$ as $\varepsilon \to 0$, we deduce (16.2.6). □

16.3 Central Limit Theorem for Functionals of Non-normalized Increments

We now consider the central limit theorem associated with Theorem 16.2.1, when X is an Itô semimartingale with Grigelionis decomposition

$$X_t = X_0 + \int_0^t b_s\,ds + \int_0^t \sigma_s\,dW_s + (\delta 1_{\{\|\delta\| \leq 1\}}) \star (\mathfrak{p} - \mathfrak{q})_t + (\delta 1_{\{\|\delta\| > 1\}}) \star \mathfrak{p}_t. \tag{16.3.1}$$

(W is a d'-dimensional Wiener process and $\mathfrak{p}$ is a Poisson measure with compensator $\mathfrak{q}(dt,dz) = dt \otimes \lambda(dz)$, and $c = \sigma\sigma^\star$.) The noise is as before, and the noisy process is still $Z^n = X + \Delta_n^\eta \chi$, as in (16.1.4). Recall the following assumption (H-r):

Assumption (H-r) In (16.3.1), b_t is locally bounded and σ_t is càdlàg, and $\|\delta(\omega,t,z)\| \wedge 1 \le \Gamma_n(z)$ if $t \le \tau_n(\omega)$, where (τ_n) is a localizing sequence of stopping times and each function Γ_n satisfies $\int \Gamma_n(z)^r \lambda(dz) < \infty$.

Below, we always assume at least (H) = (H-2).

16.3.1 The Results

Just as Theorem 16.2.1 is an extension of Theorem 8.3.1 (page 234), with the same limit, we find here a result very similar to Theorem 12.1.2. As usual we give a multidimensional version, with q weight functions g^j satisfying (16.1.6), and q functions f^j on $\mathbb{R}^d$, and the functions Ψ^j and Φ^j associated by (16.1.12). We cannot do any better than in Theorem 12.1.2, and thus we suppose that all functions f^j are C^2 and are linear combinations of *positively homogeneous* functions on $\mathbb{R}^d$ with an homogeneity degree bigger than 3: recall that h is positively homogeneous with degree p if

$$x \in \mathbb{R}^d,\ \lambda > 0 \quad \Rightarrow \quad f^j(\lambda x) = \lambda^p f^j(x).$$

Recall the notation $u_n = k_n \Delta_n$. We are interested in the following q-dimensional variables with components:

$$Y^n(Z^n)^j_t = \frac{1}{\sqrt{u_n}} \left(\frac{1}{k_n} V^n(\Phi^j, k_n, Z^n)_t - \sum_{s<t} \int_0^1 f^j\big(g^j(u)\Delta X_s\big)\, du \right). \tag{16.3.2}$$

The results depends on how fast k_n goes to ∞; if it is fast enough we have the same limit as without noise, otherwise there is an additional term. To describe the limit we need to recall and extend the notation (12.1.3)–(12.1.6). We set for $x \in \mathbb{R}^d$ and $t \in [0,1]$ and $j, j' = 1, \dots, q$ and $i, i' = 1, \dots, d$, and with $(g^j)'$ denoting the derivative of g^j, which is defined almost everywhere:

$$\begin{aligned}
h_-(x,t)^j_i &= \int_0^t \partial_i f^j\big(g^j(s+1-t)x\big)\, g^j(s)\, ds \\
h'_-(x,t)^j_i &= \int_0^t \partial_i f^j\big(g^j(s+1-t)x\big)\, \big(g^j\big)'(s)\, ds \\
h_+(x,t)^j_i &= \int_t^1 \partial_i f^j\big(g^j(s-t)x\big)\, g^j(s)\, ds \\
h'_-(x,t)^j_i &= \int_t^1 \partial_i f^j\big(g^j(s-t)x\big)\, \big(g^j\big)(s)\, ds \\
H_-(x)^{jj'}_{ii'} &= \int_0^1 h_-(x,t)^j_i\, h_-(x,t)^{j'}_{i'}\, dt
\end{aligned} \tag{16.3.3}$$

$$H'_-(x)^{jj'}_{ii'} = \int_0^1 h'_-(x,t)^j_i \, h'_-(x,t)^{j'}_{i'} \, dt$$

$$H_+(x)^{jj'}_{ii'} = \int_0^1 h_+(x,t)^j_i \, h_+(x,t)^{j'}_{i'} \, dt$$

$$H'_+(x)^{jj'}_{ii'} = \int_0^1 h'_+(x,t)^j_i \, h'_+(x,t)^{j'}_{i'} \, dt.$$

For i, i' fixed, the functions $H_{\pm ii'}$ and $H'_{\pm ii'}$ on $\mathbb{R}^d$ are $\mathcal{M}^+_{d\times q}$-valued, and are $o(\|x\|^4)$ as $x \to 0$ because $\partial_i f^j(x) = o(\|x\|^2)$ by the fact that each f^j is a linear combination of positively homogeneous functions with degree bigger than 3. Recalling $c = \sigma\sigma^*$ and assuming that the conditional variance process Υ in (16.1.3) is càdlàg, we associate the $\mathcal{M}^+_{q\times q}$-valued processes ξ_s, ξ'_s, Ξ_t and Ξ'_t defined, component by component, as follows (this extends (12.1.4)):

$$\begin{aligned} \xi^{jj'}_s &= \sum_{i,i'=1}^d \big(c^{ii'}_{s-} H_-(\Delta X_s)^{jj'}_{ii'} + c^{ii'}_s H_+(\Delta X_s)^{jj'}_{ii'}\big), \qquad & \Xi_t &= \sum_{s\le t} \xi_s \\ \xi'^{jj'}_s &= \sum_{i,i'=1}^d \big(\Upsilon^{ii'}_{s-} H'_-(\Delta X_s)^{jj'}_{ii'} + \Upsilon^{ii'}_s H'_+(\Delta X_s)^{jj'}_{ii'}\big), \qquad & \Xi'_t &= \sum_{s\le t} \xi'_s. \end{aligned} \tag{16.3.4}$$

The following characterizes the two q-dimensional processes $\overline{Y}(X)$ and $\overline{Y}'(\chi)$:

$$\left.\begin{array}{l} \overline{Y}(X) \text{ and } \overline{Y}'(\chi) \text{ are defined on a very good filtered extension} \\ (\widetilde{\Omega}, \widetilde{\mathcal{F}}, (\widetilde{\mathcal{F}}_t)_{t\ge 0}, \widetilde{\mathbb{P}}) \text{ of } (\Omega, \mathcal{F}, (\mathcal{F}_t)_{t\ge 0}, \mathbb{P}) \text{ and, conditionally on } \mathcal{F}, \\ \text{they are two independent purely discontinuous centered Gaussian} \\ \text{martingales with} \\ \widetilde{\mathbb{E}}\big(\overline{Y}(X)^j_t \, \overline{Y}(X)^{j'}_t \mid \mathcal{F}\big) = \Xi^{jj'}_t, \qquad \widetilde{\mathbb{E}}\big(\overline{Y}'(\chi)^j_t \, \overline{Y}'(\chi)^{j'}_t \mid \mathcal{F}\big) = \Xi'^{jj'}_t. \end{array}\right\}$$

Note that $\overline{Y}(X)$ is exactly as in (12.1.5).

Furthermore, we can "realize" the pair $(\overline{Y}(X), \overline{Y}'(\chi))$ as

$$\overline{Y}(X)_t = \sum_{n\ge 1:\, T_n\le t} \alpha_{T_n} \Psi_n, \qquad \overline{Y}'(\chi)_t = \sum_{n\ge 1:\, T_n\le t} \alpha'_{T_n} \Psi'_n \tag{16.3.5}$$

where α_t and α'_t are optional $q \times q$-dimensional square-roots of the processes ξ_t and ξ'_t, and (T_n) is a weakly exhausting sequence for the jumps of X, and $(\Psi_n, \Psi'_n : n \ge 1)$ are independent $\mathcal{N}(0, I_q)$ variables on an auxiliary space $(\Omega', \mathcal{F}', \mathbb{P}')$, and $(\widetilde{\Omega}, \widetilde{\mathcal{F}}, (\widetilde{\mathcal{F}}_t)_{t\ge 0}, \widetilde{\mathbb{P}})$ is the product extension:

$$\left.\begin{array}{l} \widetilde{\Omega} = \Omega \times \Omega', \qquad \widetilde{\mathcal{F}} = \mathcal{F} \otimes \mathcal{F}', \qquad \widetilde{\mathbb{P}} = \mathbb{P} \otimes \mathbb{P}' \\ (\widetilde{\mathcal{F}}_t) \text{ is the smallest filtration containing } (\mathcal{F}_t) \text{ and such that} \\ \qquad \Psi_n \text{ and } \Psi'_n \text{ are } \widetilde{\mathcal{F}}_{T_n} \text{ measurable for all } n. \end{array}\right\}$$

Finally, we need an assumption which connects the exponents η and η' occurring in (16.1.4) and (16.1.5):

$$\eta+\eta' \geq \frac{1}{2}, \quad \text{and we set } \theta' = \begin{cases} \theta & \text{if } \eta+\eta'=1/2 \\ 0 & \text{if } \eta+\eta'>1/2. \end{cases} \tag{16.3.6}$$

The notation θ' is introduced for unifying the results below. Recall that $\eta'>0$, so necessarily $\theta'=0$ when $\eta \geq \frac{1}{2}$, that is when the noise shrinks fast enough.

Theorem 16.3.1 *Assume* (H) = (H-2) *for* X *and* (N) *for the noise. We let* k_n *satisfy* (16.1.5) *and* $Z^n = X + (\Delta_n)^\eta \chi$, *with* (16.3.6). *When* $\theta'=\theta>0$, *assume further that the process* Υ *admits a càdlàg* $d\times d$*-dimensional square-root* υ_t *(that is,* $\Upsilon_t = \upsilon_t\,\upsilon^*$*), and let* f *be a* q*-dimensional* C^2 *function on* $\mathbb{R}^d$ *whose components are linear combinations of positively homogeneous functions with degree (strictly) bigger than 3. Then for each* $t\geq 0$ *the variables* $Y^n(Z^n)_t$ *of* (16.3.2) *converge stably in law to the following limit:*

$$Y^n\big(Z^n\big)_t \xrightarrow{\mathcal{L}\text{-s}} \overline{Y}(X)_t + \theta'\,\overline{Y}'(\chi)_t.$$

When $\theta'=0$ the limit is $\overline{Y}(X)_t$, the same as in Theorem 12.1.2, without noise. So at this point the reader may wonder why we insist on having a result also when $\theta'=\theta>0$, since the case $\theta'=0$ is simpler to state (and much simpler to prove). The reason is that the rate of convergence of $V^n(\Phi,k_n,Z^n)_t$, which reflects the rate of convergence of the estimators for jumps which we might construct, is $\sqrt{k_n\Delta_n}$: this sequence goes to 0, but it should do so as fast as possible (as a function of Δ_n) for better estimators. This leads us to take η' as small as possible, that is $\eta'=\frac{1}{2}-\eta$ ($=\frac{1}{2}$ when the noise is not shrinking), and to investigate what happens in this limiting case.

Remark 16.3.2 When $\theta'=\theta>0$, the limit $\overline{Y}''=\overline{Y}(X)+\theta'\,\overline{Y}'(\chi)$ is, conditionally on $\mathcal{F}$, a purely discontinuous Gaussian martingale with

$$\mathbb{E}\big(\overline{Y}_t^j\,\overline{Y}_t^{j'} \mid \mathcal{F}\big) = \Xi_t^{jj'} + \theta'^2\,\Xi_t'^{jj'}. \tag{16.3.7}$$

It can be realized as $\overline{Y}_t = \sum_{n\geq 1,\,T_n\leq t} \alpha''_{T_n}\,\Phi''_n$, where (Φ''_n) is a sequence like (Φ'_n) above, and where α''_t is a $q\times q$-dimensional optional square root of the process $\xi_t + \theta'^2\,\xi'_t$. We prefer the formulation $\overline{Y}(X)+\theta'\,\overline{Y}'(\chi)$, which emphasizes the respective importance of the limit $\overline{Y}(X)$ due to X itself and the limit $\overline{Y}'(\chi)$ due to the noise, and the role of the constant θ introduced in (16.1.5).

Remark 16.3.3 The behavior of f near 0 is the same as in Theorem 12.1.2, page 343, and in accordance with all usual CLTs for functionals of non-normalized increments.

Remark 16.3.4 In the most interesting case of a non-shrinking noise $\eta = 0$, the cutoff level for η' is $1/2$, and if $\eta' < 1/2$ nothing is known, but probably (as in Theorem 16.2.1 when (16.2.1) fails) the noise is the leading term.

16.3.2 A Local Stable Convergence Result

This subsection is, even more than usual, very technical. It aims towards a stable convergence result, joint for the Brownian motion and the noise, and it extends Proposition 5.1.1 (page 127) in two directions, in addition to incorporating the noise: it is "local" in time, right before or after a given stopping time, and it is for the conditional laws knowing the past.

This needs some preparation. For some integer $J \geq 1$ and each j between 1 and J we have numbers $(h(j)_r^n, h'(j)_r^n : 1 \leq r \leq k_n)$ and piecewise Lipschitz (hence bounded) functions h^j and h'^j on $[0,1]$ such that for some finite integer N:

$$\left|h(j)_r^n - h^j\left(\frac{r}{k_n}\right)\right| + \left|h'(j)_r^n - h'^j\left(\frac{r}{k_n}\right)\right| \leq \begin{cases} K & \text{always} \\ K/k_n & \text{for all } r, \text{ except} \\ & \text{at most } N \text{ of them.} \end{cases} \tag{16.3.8}$$

For example if the g^j's are weight functions satisfying (16.1.6) the terms $h(j)_r^n = (g^j)_r^n$ and $h'(j)_r^n = k_n (g^j)_r'^n$, together with $h^j = g^j$ and $h'^j = (g^j)'$, satisfy this; but we will also encounter other situations, where h'^j is *not* the derivative of h^j. Associated with this, we consider the following variables:

$$\widetilde{W}(j)_i^n = \sum_{r=1}^{k_n} h(j)_r^n \Delta_{i+r-1}^n W, \qquad \widetilde{\chi}(j)_i^n = \sum_{r=1}^{k_n} h'(j)_r^n \chi_{(i+r-2)\Delta_n}.$$

Note that in the example above, and by virtue of (16.1.10), we have $\widetilde{W}(j)_i^n = \overline{W}(g^j)_i^n$ and $\widetilde{\chi}(j)_i^n = -k_n \overline{\chi}(g^j)_i^n$.

We will be interested in the limits of some functionals of the above variables, but before stating the problem we introduce some processes which are analogous to $L(g^j)$ in (12.2.8). We consider an auxiliary filtered space $(\Omega', \mathcal{F}', (\mathcal{F}'_t)_{t\geq 0}, \mathbb{P}')$ supporting two independent Brownian motions $\check{W}$ and $\check{W}'$, with respective dimensions d' and d, and we define two $J \times d'$ and $J \times d$-dimensional processes indexed by $\mathbb{R}_+$, and with components

$$L_t^{jl} = \int_t^{t+1} h^j(s-t)\, d\check{W}_s^l, \qquad L_t'^{jl} = \int_t^{t+1} h'^j(s-t)\, d\check{W}_s'^l. \tag{16.3.9}$$

We also use the following notation:

$$S(dx, dy) = \text{the law on } \mathbb{D}^{Jd'+Jd} \text{ of the pair } \big(L, L'\big). \tag{16.3.10}$$

Lemma 16.3.5 *The $(Jd' + Jd)$-dimensional process (L, L') is Gaussian centered and stationary, L and L' are independent, and the variables (L_t, L'_t) and (L_{t+s}, L'_{t+s}) are independent if $s \geq 1$. Moreover (L, L') has a continuous version, and for all $p > 0$ we have*

$$\mathbb{E}'\big(\|L_t\|^p\big) \leq K_p, \qquad \mathbb{E}'\big(\big\|L'_t\big\|^p\big) \leq K_p. \tag{16.3.11}$$

Proof All claims are simple, except (16.3.11) and the stationarity and the existence of a continuous version. For those properties it is enough to consider separately L and L' (by independence) and, both having the same structure, it is enough to consider, say, the process L. If $t, s \geq 0$ we have

$$\mathbb{E}'\big(L_t^{jl} L_{t+s}^{j'l'}\big) = \begin{cases} \int_{t+s}^{t+1} h^j(r-t) h^{j'}(r-t-s)\, dr & \text{if } s < 1, l = l' \\ 0 & \text{otherwise.} \end{cases}$$

The integral on the right is also $\int_s^1 h^j(r) h^{j'}(r-s)\, dr$, which does not depend on t, and this proves the stationarity. Moreover we also deduce when $s \in [0, 1]$:

$$a_s := \mathbb{E}'\big(\big(L_{t+s}^{jl} - L_t^{jl}\big)^2\big) = 2\int_0^s \big(h^j(r)\big)^2 dr + 2\int_s^1 h^j(r)\big(h^j(r) - h^j(r-s)\big)\, dr.$$

The piecewise Lipschitz property of h^j implies that $|h^j(r) - h^j(r-s)| \leq Ks$ for all r except on a finite union of intervals with length $2s$, hence, since we also have $|h^j| \leq K$, we get $a_s \leq Ks$. Since $L_{t+s}^{jl} - L_t^{jl}$ is Gaussian centered, we deduce $\mathbb{E}'((L_{t+s}^{jl} - L_t^{jl})^4) = 3a_s^2 \leq Ks^2$, and by Kolmogorov's criterion the process L^{jl} admits a continuous version. Finally (16.3.11) follows from the stationarity and from the fact that the process is Gaussian. □

Next, we fix a sequence of—possibly random—integers $i_n \geq 1$, and we consider the càdlàg processes L^n and L'^n indexed by $\mathbb{R}_+$, with components:

$$\left.\begin{array}{llll} 1 \leq j \leq J,\ 1 \leq l \leq d' & \mapsto & L_t^{n,jl} & = \frac{1}{\sqrt{u_n}}\, \widetilde{W}(j)_{i_n+[k_n t]}^{n,l} \\ 1 \leq j \leq J,\ 1 \leq l \leq d & \mapsto & L_t^{\prime n,jl} & = \frac{1}{\sqrt{k_n}}\, \widetilde{\chi}(j)_{i_n+[k_n t]}^{n,l}. \end{array}\right\} \tag{16.3.12}$$

The sequence i_n does not show in the notation, but those processes clearly depend on it. For the next statement, let us recall that if T is an $(\mathcal{F}_t)$-stopping time, then for each $\omega^{(0)} \in \Omega^{(0)}$ the variable $T(\omega^{(0)}, .)$ on $\Omega^{(1)}$ is an $(\mathcal{F}_t^{(1)})$-stopping time.

Lemma 16.3.6 *Assume* (SN) *and, in the above setting, that $T_n = (i_n - 1)\Delta_n$ is a stopping time relative to the filtration $(\mathcal{F}_t)$. Then for all $p \geq 1$ and $s, t \geq 0$ and $\omega^{(0)}$ we have, with the notation $\mathbb{Q} = \mathbb{Q}(\omega^{(0)}, .)$:*

$$\begin{aligned} &\mathbb{E}\big(\big\|L_t^n\big\|^p \mid \mathcal{F}_{T_n}\big) \leq K_p \\ &\mathbb{E}\big(\big\|L_{t+s}^n - L_t^n\big\|^p \mid \mathcal{F}_{T_n}\big) \leq K_p\Big(s \vee \frac{1}{k_n}\Big)^{p/2} \end{aligned} \tag{16.3.13}$$

$$\begin{aligned}&\mathbb{E}_{\mathbb{Q}}\big(\big\|L_t'^n\big\|^p \mid \mathcal{F}^{(1)}_{T_n(\omega^{(0)},.)}\big) \le K_p\\ &\mathbb{E}_{\mathbb{Q}}\big(\big\|L_{t+s}'^n - L_t'^n\big\|^p \mid \mathcal{F}^{(1)}_{T_n(\omega^{(0)},.)}\big) \le K_p\Big(s \vee \frac{1}{k_n}\Big)^{p/2}.\end{aligned} \tag{16.3.14}$$

Note that we cannot replace $(s \vee \frac{1}{k_n})^{p/2}$ by $s^{p/2}$: otherwise this would imply the continuity of the paths of L^n and L'^n, which are in fact piecewise constant and thus discontinuous.

Proof We begin with (16.3.14). By hypothesis, $\omega^{(1)} \mapsto T_n(\omega^{(0)}, \omega^{(1)})$ is an $(\mathcal{F}_t^{(1)})$-stopping time taking its values in $\{i\Delta_n : i \in \mathbb{N}\}$. Therefore, although $t \mapsto \chi_t$ is not measurable, the quantities $\chi_{T_n+i\Delta_n}$ for $i \ge 1$ are (measurable) random variables. With the convention that $h'(j)_l^n = 0$ when $l < 1$ or $l > k_n$, we see that

$$L_t'^{n,jl} = \frac{1}{\sqrt{k_n}} \sum_{i\ge 0} \delta_i^{n,j}(t)\chi^l_{T_n+i\Delta_n}, \quad \text{where } \delta_i^{n,j}(t) = h'(j)^n_{i+1-[k_nt]}.$$

The variables $\chi_{T_n+i\Delta_n}$ for $i = 0, 1, \dots$ are mutually independent, and independent of $\mathcal{F}^{(1)}_{T_n}$, centered, with bounded moments of any order. Then we deduce from the Burkholder-Gundy inequality for $p > 1$ that

$$\begin{aligned}&\mathbb{E}_{\mathbb{Q}}\big(\big\|L_t'^n\big\|^p \mid \mathcal{F}^{(1)}_{T_n}\big) \le K_p\,\mathbb{E}_{\mathbb{Q}}\bigg(\bigg(\frac{1}{k_n}\sum_{i\ge 0}\big\|\delta_i^n(t)\big\|^2\,\|\chi_{T_n+i\Delta_n}\|^2\bigg)^{p/2} \mid \mathcal{F}^{(1)}_{T_n}\bigg)\\ &\mathbb{E}_{\mathbb{Q}}\big(\big\|L_{t+s}'^n - L_t'^n\big\|^p \mid \mathcal{F}^{(1)}_{T_n}\big)\\ &\quad\le K_p\,\mathbb{E}_{\mathbb{Q}}\bigg(\bigg(\frac{1}{k_n}\sum_{i\ge 0}\big\|\delta_i^n(t+s) - \delta_i^n(t)\big\|^2\,\|\chi_{T_n+i\Delta_n}\|^2\bigg)^{p/2} \mid \mathcal{F}^{(1)}_{T_n}\bigg).\end{aligned} \tag{16.3.15}$$

The condition (16.3.8) implies $\|\delta_i^n(t)\| \le K$, and $\delta_i^n(t) = 0$ for all i except those in a (random and $\mathcal{F}^{(1)}_{T_n}$ measurable) set A_n of cardinality not bigger than k_n. Hence the first part of (16.3.14), Hölder's inequality and (SN) give

$$\mathbb{E}_{\mathbb{Q}}\big(\big\|L_t'^n\big\|^p \mid \mathcal{F}^{(1)}_{T_n}\big) \le K_p\,\mathbb{E}_{\mathbb{Q}}\bigg(\frac{1}{k_n}\sum_{i\in A_n}\big\|\delta_i^n(t)\big\|^p\,\|\chi_{T_n+i\Delta_n}\|^p \mid \mathcal{F}^{(1)}_{T_n}\bigg) \le K_p,$$

thus giving the first parts of (16.3.14).

From (16.3.8) again, $\alpha_i^n(t,s) = \|\delta_i^n(t+s) - \delta_i^n(t)\|$ vanishes for all i outside an $\mathcal{F}^{(1)}_{T_n}$ measurable) set A_n of cardinality not bigger than $2k_n$, and is always smaller than some constant K. Moreover there is a subset $A_n' \subset A_n$ with $\mathrm{card}(A_n \backslash A_n') \le 2N$, such that $\alpha_i^n(t,s) \le K/k_n + \|h'((i+1-[k_n(t+s)])/k_n) - h'((i+1-[k_nt])/k_n)\|$ (where h' is the function with components h'^j), for all $i \in A_n'$. Finally, the piecewise Lipschitz property of all h'^j yields that $\|h'((i+1-[k_n(t+s)])/k_n) - h'((i+1-[k_nt])/k_n)\| \le K(s \vee \frac{1}{k_n})$ for all i in a subset $A_n'' \subset A_n'$ such that $\mathrm{card}(A_n' \backslash A_n'') \le (C+1)(1+[k_ns])$, where $C \ge 1$ is the number of subintervals of [0, 1] on which h' is Lipschitz. Therefore $\alpha_i^n(t,s) \le K(s+\frac{1}{k_n})$ if $i \in A_n''$ and $\alpha_i^n(t,s) \le K$ if $i \in B_n = A_n \backslash A_n''$, hence (16.3.15) yields

$$\begin{aligned}
&\mathbb{E}_{\mathbb{Q}}\big(\|L'^n_{t+s} - L'^n_t\|^p \mid \mathcal{F}^{(1)}_{T_n(\omega^{(0)},.)}\big)\\
&\quad\le K_p\Big(s+\frac{1}{k_n}\Big)^{p/2}\,\mathbb{E}_{\mathbb{Q}}\bigg(\bigg(\frac{1}{k_n}\sum_{i\in A_n''}\|\chi_{T_n+i\Delta_n}\|^2\bigg)^{p/2} \mid \mathcal{F}^{(1)}_{T_n}\bigg)\\
&\qquad+K_p\,\mathbb{E}_{\mathbb{Q}}\bigg(\bigg(\frac{1}{k_n}\sum_{i\in B_n}\|\chi_{T_n+i\Delta_n}\|^2\bigg)^{p/2} \mid \mathcal{F}^{(1)}_{T_n}\bigg).
\end{aligned}$$

The first term on the right is smaller than $K_p(s+\frac{1}{k_n})^{p/2}$ because $\mathrm{card}(A_n'') \le 2k_n$ and by (SN) and Hölder's inequality. The second term satisfies the same inequality, because $\mathrm{card}(B_n) \le K(1+k_n s)$. Since $s+\frac{1}{k_n} \le 2(s \vee \frac{1}{k_n})$, we deduce the second part of (16.3.14).

The proof of (16.3.13) is similar: we argue under $\mathbb{P}$, we use the independence of the increments $\Delta_i^n W$ for $i \ge i_n$ from $\mathcal{F}_{T_n}$, and the normalization by $1/\sqrt{u_n}$ is easily seen to be the proper one because the moments of $\Delta_i^n W$ behave like $\Delta_n^{p/2}$. □

As with (L, L'), each process (L^n, L'^n) may be viewed as a $\mathbb{D}^{Jd'+Jd}$-valued random variable. We need a family of semi-norms on this space, depending on n and on some "time horizon" $m \in \mathbb{N}$, and defined as follows:

$$\|x\|_{(n,m)} = \|x(0)\| + \frac{1}{k_n}\sum_{j=1}^{mk_n-1}\|x(j/k_n)\| + \|x(m)\|.$$

Then, we are given a sequence (f_n) of functions on $\mathbb{D}^{Jd'+Jd}$ which satisfies for some $m \in \mathbb{N}$ and $w \ge 0$, and for all $n \ge 1$ and all $A > 0$:

$$\begin{aligned}
&|f_n(x)| \le K\big(1+\|x\|^w_{(n,m)}\big)\\
&\lim_{\varepsilon\to 0}\sup\big(|f_n(x)-f_n(y)| : \|x\|_{(n,m)} \le A,\ \|x-y\|_{(n,m)} \le \varepsilon, n \ge 1\big) = 0.
\end{aligned} \tag{16.3.16}$$

(Note that these conditions imply that each $f_n(x)$ depends on x only through its restriction to the interval $[0, m]$.)

In the next result, we write $f_n(x, y)$ where x, y are respectively in $\mathbb{D}^{Jd'}$ and $\mathbb{D}^{Jd}$. We also use the following convention: if $x \in \mathbb{D}^{Jd'}$ and $y \in \mathbb{D}^{Jd}$ and if α and β are $d \times d'$ and $d \times d$ matrices, then $\alpha x \in \mathbb{D}^{Jd}$ and $\beta y \in \mathbb{D}^{Jd}$ are defined, component by component, by

$$(\alpha x)(t)^{jl} = \sum_{r=1}^{d'}\alpha^{lr}x(t)^{jr}, \qquad (\beta y)(t)^{jl} = \sum_{r=1}^{d'}\beta^{lr}y(t)^{jr}. \tag{16.3.17}$$

For example $\upsilon_{S_n} L'^n$ below is the Jd-dimensional process with the (j,l)th component given by $t \mapsto \sum_{r=1}^{d}\upsilon_{S_n}(\omega)^{lr}L'^{n,jr}_t(\omega)$.

Lemma 16.3.7 *Assume* (SN) *and that the process* Υ *of* (16.1.3) *admits a càdlàg adapted square-root* υ_t, *and also that* $T_n = (i_n - 1)\Delta_n$ *is a stopping time. Finally, assume that the sequence* (f_n) *satisfies* (16.3.16) *with some* $m \in \mathbb{N}$ *and converges pointwise to a limit* f, *and that* $T_n \to T$ *with* T *finite-valued, and that we are in either one of the following two cases (recall* $u_n = k_n \Delta_n$*):*

$$\begin{array}{l} (1)\ T > 0 \quad \textit{and} \quad T_n \leq (T - (m+1)u_n)^+ \quad \textit{for all } n, \\ \qquad\qquad \textit{in which case we set } \upsilon_{(T)} = \upsilon_{T-} \textit{ and } \mathcal{F}_{(T)} = \mathcal{F}_{T-} \\ (2)\ T_n \geq T \ \ \forall n, \textit{ in which case we set } \upsilon_{(T)} = \upsilon_T \textit{ and } \mathcal{F}_{(T)} = \mathcal{F}_T. \end{array} \tag{16.3.18}$$

Then for every bounded random variable Z, *and if* $S(dx,dy)$ *is given by* (16.3.10), *we have*

$$\mathbb{E}\big(Z f_n\big(L^n, L'^n\big) \mid \mathcal{F}_{T_n}\big) \xrightarrow{\mathbb{P}} \mathbb{E}\bigg(Z \int f\big(x, \upsilon_{(T)} y\big)\, S(dx,dy) \mid \mathcal{F}_{(T)}\bigg). \tag{16.3.19}$$

This can be interpreted as a kind of "stable convergence in law" for the pair (L^n, L'^n), under the conditional expectations.

Proof *Step 1)* We start with an auxiliary result. As in the previous lemma, we fix $\omega^{(0)} \in \Omega^{(0)}$ and consider the probability space $(\Omega^{(1)}, \mathcal{F}^{(1)}, \mathbb{Q})$, where $\mathbb{Q} = \mathbb{Q}(\omega^{(0)}, .)$. Our aim in this step is to show that under $\mathbb{Q}$,

$$L'^n \stackrel{\mathcal{L}}{\Longrightarrow} \upsilon_{(T)}\big(\omega^{(0)}\big) L', \tag{16.3.20}$$

(functional convergence in law), where we use the notation (16.3.17).

We begin with the finite-dimensional convergence. Let $0 \leq t_1 < \cdots < t_r$. By (16.3.12) and (16.1.10) the rJd-dimensional variable $Y_n = (L'^n_{t_i} : 1 \leq i \leq r)$ is

$$Y_n = \sum_{w \geq 1} y_w^n, \quad \text{where } y_w^{n,ijl} = a_w^{n,ij} \chi^l_{(i_n+w-2)\Delta_n}$$

$$\text{and } a_w^{n,ij} = \begin{cases} \frac{1}{\sqrt{k_n}} h'(j)^n_{w-[k_n t_i]} & \text{if } 1 + [k_n t_i] \leq w \leq k_n + [k_n t_i] \\ 0 & \text{otherwise.} \end{cases}$$

By (SN), and under $\mathbb{Q}$, the variables y_w^n are independent when w varies and satisfy (use the consequence $|h'(j)^n_i| \leq K$ of (16.3.8)):

$$\mathbb{E}_Q\big(y_w^n\big) = 0, \quad \mathbb{E}_Q\big(\|y_w^n\|^4\big) \leq K k_n^{-2}, \quad \sum_{w \geq 1} \mathbb{E}_Q\big(\|y_w^n\|^4\big) \to 0,$$

$$\sum_{w \geq 1} \mathbb{E}_Q\big(y_w^{n,ijl} y_w^{n,i'j'l'}\big) = \frac{1}{k_n} \sum_{m \geq 1} \Upsilon^{ll'}_{(i_n+m-2)\Delta_n}\big(\omega^{(0)}\big) a_m^{n,ij} a_m^{n,i'j'}.$$

On the one hand $\upsilon_{(i_n+w-1)\Delta_n}(\omega^{(0)})$ converges uniformly in $w \leq k_n + [t_r k_n]$ to $\upsilon_{(T)}(\omega^{(0)})$ by (16.3.18). On the other hand, (16.3.8) and a Riemann approximation yield for $i \le i'$:

$$\sum_{w\ge1} a_w^{n,ij}\, a_w^{n,i'j'} \;\to\; c^{i,j,i',j'} \;:=\; \int_{t_{i'}\wedge(t_i+1)}^{t_i+1} h'^j(s-t_i)\, h'^{j'}(s-t_{i'})\, ds.$$

Hence

$$\sum_{w\ge1} \mathbb{E}_Q\big(y_w^{n,ijl}\, y_w^{n,i'j'l'}\big) \to \overline{c}^{ijl,i'j'l'} := c^{i,j,i',j'}\,\big(\upsilon_{(T)}\upsilon_{(T)}^*\big)^{ll'}\big(\omega^{(0)}\big). \tag{16.3.21}$$

Then Theorem 2.2.14 yields that Y_n converges in law under $\mathbb{Q}$ to a centered Gaussian variable with covariance matrix $(\overline{c}^{ijl,i'j'l'})$. In view of (16.3.9), this matrix is the covariance of the centered Gaussian vector $(\upsilon_{(T)}(\omega^{(0)})L'_{t_i} : 1 \le i \le r)$, and the finite-dimensional convergence in (16.3.20) is proved.

To obtain the functional convergence in (16.3.20) it remains to prove that the processes L'^n are C-tight, and for this we use a special case of Kolmogorov's criterion, see Ibragimov and Has'minski [49], as already explained in (12.1.32) for example: namely, a sequence of processes V^n indexed by $\mathbb{R}_+$ is C-tight as soon as

$$t \ge 0,\ v \in (0,1] \quad\Rightarrow\quad \mathbb{E}\big(|V_{t+v}^n - V_t^n|^p\big) \le K v^{p/2} \tag{16.3.22}$$

for some $p > 2$. Of course, as seen in Step 1, this fails for L'^n because otherwise those processes would be continuous. However the C-tightness of the sequence (L'^n)'s is implied by the C-tightness of the sequence $(\overline{L}'^n)$ of the "linearized versions", defined as follows: $\overline{L}'^n$ is continuous, coincides with L'^n on the grid $G_n = \{j/k_n : j \in \mathbb{N}\}$ and each of its components is affine between successive points of the grid. Now, (16.3.14) yields (16.3.22) for $\overline{L}'^n$. So the sequence $\overline{L}'^n$ is C-tight, and so also is the sequence L'^n, hence completing the proof of (16.3.20).

Step 2) Here we essentially do the same job for the processes L^n, proving that

$$L^n \stackrel{\mathcal{L}}{\Longrightarrow} L. \tag{16.3.23}$$

For this, we see that if $0 \le t_1 < \cdots < t_r$ the variables $Y_n = (L_{t_i}^n : 1 \le i \le r)$ still satisfy $Y_n = \sum_{w\ge1} y_w^n$, but now

$$y_w^{n,ijl} = a_w^{n,ij}\,\Delta_{i_n+w-1}^n W^l, \quad \text{where}$$

$$a_w^{n,ij} = \begin{cases} \frac{1}{\sqrt{u_n}}\, h(j)_{w-[k_nt_i]}^n & \text{if } 1+[k_nt_i] \le w \le k_n+[k_nt_i] \\ 0 & \text{otherwise.} \end{cases}$$

Then the proof of (16.3.23), both for the finite-dimensional convergence and the C-tightness, is exactly the same as for (16.3.20) (when $i \le i'$, the right side of (16.3.21) is replaced by $\delta_{ll'} \int_{t_{i'}\wedge(t_i+1)}^{t_i+1} h^j(s-t_i)h^{j'}(s-t_{i'})\,dv$, which is the covariance matrix of $(L_{t_i} : 1 \le i \le r)$).

Step 3) We prove again an auxiliary result, which is more or less classical and holds in a much more general context and under weaker conditions, but we stick to our setting of the convergence of processes. For each n let Y^n be a q-dimensional processes satisfying $Y^n \stackrel{\mathcal{L}}{\Longrightarrow} Y$, where Y is a continuous process. Let also g_n be a sequence of functions on $\mathbb{D}^q$ which satisfies

$$\begin{aligned}&\big|g_n(x)\big| \le K\\&\lim_{\varepsilon\to 0}\sup\big(\big|g_n(x)-g_n(y)\big|: \|x\|_{(n,m)}\le A,\ \|x-y\|_{(n,m)}\le\varepsilon, n\ge 1\big)=0\end{aligned} \tag{16.3.24}$$

(the same as (16.3.16) with $w=0$, except that the space is $\mathbb{D}^q$). Then we have

$$\left.\begin{array}{l}\mathbb{E}(g_n(Y^n))-\mathbb{E}(g_n(Y))\to 0\\ g_n\to g \text{ pointwise}\end{array}\right\} \Rightarrow \mathbb{E}\big(g_n(Y^n)\big)\to\mathbb{E}\big(g(Y)\big). \tag{16.3.25}$$

For proving this we use the Skorokhod representation theorem, see for example (2.2.19); namely, there are processes Y'^n and Y', all defined on the same probability space, with the same laws as Y^n and Y (so $\mathbb{E}(g_n(Y'^n))=\mathbb{E}(g_n(Y^n))$, and the same for Y, Y'), and such that $Y'^n\to Y'$ a.s. Now, if $y_n\to y$ in the Skorokhod space, we have $\|y_n-y\|_{(n,m)}\to 0$ if y is continuous, and $g_n(y_n)-g_n(y)\to 0$ follows from (16.3.24). Therefore $g_n(Y'^n)-g_n(Y')\to 0$ a.s., and

$$\mathbb{E}\big(g_n(Y^n)\big)-\mathbb{E}\big(g_n(Y)\big) = \mathbb{E}\big(g_n(Y'^n)\big)-\mathbb{E}\big(g_n(Y')\big)\to 0$$

by the dominated convergence theorem, because $|g_n|\le K$.

When $g_n\to 0$ pointwise, we have $\mathbb{E}(g_n(Y))\to\mathbb{E}(g(Y))$ by the dominated convergence theorem, hence (16.3.25) follows from the above.

Step 4) In this step we start the proof of (16.3.19) by showing that it is enough to prove it under the additional assumption that $|f_n|\le K$ identically. We denote by $\Psi_n(f_n,Z)$ and $\Psi(f,Z)$ the left and right sides of (16.3.19). For each $A>1$ we consider the functions $\xi_A(z)=A\wedge((-A)\vee z)$ and $\xi'_A(z)=z-\xi_A(z)$ on $\mathbb{R}$, and we observe that the functions $\xi_A\circ f_n$ satisfy (16.3.16) and converge pointwise to $\xi_A\circ f$. So in this step we suppose that

$$\Psi_n(\xi_A\circ f_n,Z)\stackrel{\mathbb{P}}{\longrightarrow}\Psi(\xi_A\circ f,Z)\qquad\forall A>0,$$

and to prove the result, that is $\Psi_n(f_n,Z)\stackrel{\mathbb{P}}{\longrightarrow}\Psi(f,Z)$, it is then enough to show that

$$\lim_{A\to\infty}\sup_n\mathbb{E}\big(\big|\Psi_n(\xi'_A\circ f_n,Z)\big|\big)=0,\qquad \lim_{A\to\infty}\mathbb{E}\big(\big|\Psi(\xi'_A\circ f,Z)\big|\big)=0. \tag{16.3.26}$$

The second property of (16.3.16) and Hölder's inequality yield

$$\big|f_n(x)\big|\le K\left(1+\big\|x(0)\big\|^w+\frac{1}{k_n}\sum_{j=1}^{mk_n-1}\big\|x(j/k_n)\big\|^w+\big\|x(m)\big\|^w\right). \tag{16.3.27}$$

Since $f_n(x) \to f(x)$, by passing to the limit we deduce

$$|f(x)| \leq K\left(1 + \|x(0)\|^w + \int_0^m \|x(s)\|^w\,ds + \|x(m)\|^w\right). \tag{16.3.28}$$

Recalling that Z is bounded and the definition of $\Psi_n(f_n, Z)$, we have

$$\mathbb{E}\big(\big|\Psi_n(\xi'_A \circ f_n, Z)\big|\big) \leq K\mathbb{E}\big(\big|\xi'_A \circ f_n(L^n, L'^n)\big|\big) \leq \frac{K}{A}\,\mathbb{E}\big(\big|f_n(L^n, L'^n)\big|\big).$$

Combining this with (16.3.27) and Lemma 16.3.6, we readily obtain the first part of (16.3.26). In the same way, we have with $\widehat{L} = (L, L')$:

$$\begin{aligned}\mathbb{E}\big(\big|\Psi(\xi'_A \circ f, Z)\big|\big) &\leq \frac{K}{A}\mathbb{E}\left(\int \big|f(x, \upsilon_{(T)}y)\big|^2\,S(dx, dy)\right)\\ &\leq \frac{K}{A}\left(1 + \mathbb{E}'\big(\|\widehat{L}_0\|^w\big) + \int_0^m \mathbb{E}'\big(\|\widehat{L}_s\|^w\big)\,ds + \mathbb{E}'\big(\|\widehat{L}_m\|^w\big)\right),\end{aligned}$$

where the second inequality comes from (16.3.28) and the boundedness of the variable $\upsilon_{(T)}$ and the definition of the measure $S(dx, dy)$. Then (16.3.11) gives $\mathbb{E}(|\Psi(\xi'_A \circ f, Z)|) \leq K/A$, hence the second part of (16.3.26) holds as well.

Step 5) In view of what precedes we may and will assume further on that $|f_n| \leq K$. In this step we show that it is enough to prove

$$\Psi_n(f_n, 1) \xrightarrow{\mathbb{P}} \Psi(f, 1). \tag{16.3.29}$$

Indeed, assume this, and take an arbitrary bounded variable Z. We consider the càdlàg version of the bounded martingale $Z_t = \mathbb{E}(Z \mid \mathcal{F}_t)$.

Suppose first that we are in Case (1). Then T is a predictable time and the variable $\Psi(f, 1)$ is $\mathcal{F}_{T-}$ measurable, and so is $f_n(L^n, L'^n)$ in restriction to the set $\Omega_n = \{T > (m+1)u_n\}$. We deduce $\Psi(f, Z) = Z_{T-}\Psi(f, 1)$, and also $\Psi_n(f_n, Z) = \Psi(f_n, Z_{T-})$ on Ω_n because $Z_{T-} = \mathbb{E}(Z \mid \mathcal{F}_{T-})$. We also obviously have $\Psi_n(f_n, Z_{T_n}) = \Psi_n(f_n, 1)\,Z_{T_n} \xrightarrow{\mathbb{P}} \Psi(f, 1)\,Z_{T-}$ by (16.3.29) and $Z_{T_n} \to Z_{T-}$. So it remains to observe that $\Omega_n \to \Omega$, and that $\Psi_n(f_n, Z_{T_n}) - \Psi_n(f_n, Z_{T-}) \xrightarrow{\mathbb{P}} 0$, which follows from the fact that $\mathbb{E}(|\Psi_n(f_n, Z_{T_n}) - \Psi_n(f_n, Z_{T-})|) \leq K\mathbb{E}(|Z_{T_n} - Z_{T-}|)$ (because $|f_n| \leq K$), whereas $\mathbb{E}(|Z_{T_n} - Z_{T-}|) \to 0$ by $Z_{T_n} \to Z_{T-}$ again, and also by the boundedness of the sequence Z_{T_n}.

Suppose now that we are in Case (2). Then $\Psi(f, Z) = Z_T\,\Psi(f, 1)$ and $\Psi_n(f_n, Z_T) = Z_T\,\Psi_n(f_n, 1)$. Moreover if $T''_n = T_n + (m+1)u_n$ (those are again stopping times), $f_n(L^n, L'^n)$ is $\mathcal{F}_{T''_n}$ measurable, so $\Psi_n(f_n, Z) = \Psi_f(f_n, Z_{T''_n})$. Since $T''_n \to T$ and $T''_n > T$, we have $Z_{T''_n} \to Z_T$, and the same argument as above shows that $\Psi_n(f_n, Z_{T''_n}) - \Psi_n(f_n, Z_T) \xrightarrow{\mathbb{P}} 0$. The result follows.

Step 6) We are now ready to prove the convergence (16.3.29). We set for $y \in \mathbb{D}^{Jd}$:

$$h^n_{\omega^{(0)}}(y) = f\big(L^n(\omega^{(0)}), y\big)$$

$$A^n_j(\omega^{(0)}) = \begin{cases} \int h^n_{\omega^{(0)}}(L'^n(\omega^{(1)}))\,\mathbb{Q}(\omega^{(0)}, d\omega^{(1)}), & \text{if } j=1 \\ \int h^n_{\omega^{(0)}}(\upsilon_{T_n}(\omega^{(0)})y)\; S(dx,dy), & \text{if } j=2 \\ \int h^n_{\omega^{(0)}}(\upsilon_{(T)}(\omega^{(0)})y)\; S(dx,dy), & \text{if } j=3. \end{cases}$$

We also define the following function of the $d \times d$ matrix β:

$$F(\beta) = \int f(x, \beta y)\, S(dx,dy), \qquad F_n(\beta) = \int \mathbb{E}\big(f_n(L^n, \beta y)\big)\, S(dx,dy).$$

These are related to our quantities of interest as follows: using the $\mathcal{F}^{(0)}$-conditional independence of the χ_t's for the first relation below, we have

$$\Psi_n(f_n, 1) = \mathbb{E}\big(A^n_1 \mid \mathcal{F}_{T_n}\big), \qquad \Psi(f,1) = F(\upsilon_{(T)}).$$

We will apply (16.3.25) several times:
1) First, with $Y^n = L^n$ and the functions $g_n(x) = f_n(x, \beta_n y)$, which converge pointwise to $g(x,y) = f(x, \beta y)$ if $\beta_n \to \beta$ and satisfy (16.3.24); taking (16.3.23) into consideration, this gives $\mathbb{E}(f_n(L^n, \beta_n y)) \to \mathbb{E}'(f(x, \beta y))$. Then, due to the definition of S (which is indeed a product $S(dx,dy) = S_1(dx)S_2(dy)$), we deduce

$$\beta_n \to \beta \quad \Rightarrow \quad F_n(\beta_n) \to F(\beta).$$

2) Second, with $Y^n = L'^n$ and $g_n = h^n_{(\omega^{(0)})}$; taking (16.3.20) into consideration, this gives $A^n_1(\omega^{(0)}) - A^n_3(\omega^{(0)}) \to 0$.
3) Third, with $Y^n = \upsilon_{T_n}(\omega^{(0)})L'$, which converges to $Y = \upsilon_{(T)}(\omega^{(0)})L'$, and $g_n = h^n_{(\omega^{(0)})}$; we then get $\mathbb{E}'(g_n(\upsilon_{T_n}(\omega^{(0)})L')) - \mathbb{E}'(g_n(\upsilon_{(T)}(\omega^{(0)})L'))$, which is also $A^n_2(\omega^{(0)}) - A^n_3(\omega^{(0)}) \to 0$.

Putting all these together, and recalling that all variables under consideration here are uniformly bounded, we deduce on the one hand that $F_n(\upsilon_{T_n}) \to \Psi(f,1)$, and on the other hand that $\Psi_n(f_n,1) - \mathbb{E}(A^n_2 \mid \mathcal{F}_{T_n}) \xrightarrow{\mathbb{P}} 0$. It remains to observe that indeed $\mathbb{E}(A^n_2 \mid \mathcal{F}_{T_n}) = F_n(\upsilon_{T_n})$, which follows from the fact that $(W_{T_n+t} - W_{T_n})_{t\geq 0}$ is independent of $\mathcal{F}_{T_n}$: the proof is thus complete. □

The previous result gives the behavior of $\overline{W}(g)$ and $\overline{\chi}(g)$ for a weight function g, but not the behavior of $\widehat{\chi}(g)$. For this, we have the following result:

Lemma 16.3.8 *Let g be a weight function, and assume* (SN) *and that Υ_t is càdlàg. If i_n is a random integer satisfying the condition of Lemma* 16.3.7 *for some $m \in \mathbb{N}$, and with $\Upsilon_{(T)}$ equal to Υ_{T-} or Υ_T, according to the case of* (16.3.18), *and with the notation $\Lambda(g')$ of* (16.1.7), *for any $\omega^{(0)}$ we have, with $\mathbb{Q} = \mathbb{Q}(\omega^{(0)}, .)$:*

$$\sup_{t\in[0,m]} \Big| k_n\, \widehat{\chi}(g)^{n,lm}_{i_n+[k_n t]} - 2\,\Lambda(g')\,\Upsilon^{lm}_{(T)}\Big| \xrightarrow{\mathbb{Q}} 0. \tag{16.3.30}$$

Proof 1) We argue under the probability $\mathbb{Q} = \mathbb{Q}(\omega^{(0)}, .)$. We also fix the indices l, m. We make the convention $g_r'^n = 0$ if $r < 1$ or $r > k_n$, and set $T_n = (i_n - 1)\Delta_n$. We also set

$$\begin{aligned}
\zeta_i^n &= \chi^l_{T_n + i\Delta_n} \chi^m_{T_n + i\Delta_n} - \Upsilon^{lm}_{T_n + i\Delta_n} \\
\zeta_i'^n &= \chi^l_{T_n + i\Delta_n} \chi^m_{T_n + (i-1)\Delta_n} + \chi^l_{T_n + (i-1)\Delta_n} \chi^m_{T_n + i\Delta_n} \\
\delta_i^n(t) &= k_n \big(g'^n_{i - [k_n t]}\big)^2.
\end{aligned}$$

Observe that

$$k_n \widehat{\chi}(g)^{n,lm}_{i_n + [k_n t]} - 2\Lambda(g')\, \Upsilon^{lm}_{(T)} = \sum_{j=1}^{5} A^n(j)_t,$$

where

$$\begin{aligned}
A^n(1)_t &= \sum_{i \ge 0} \delta_i^n(t)\, \zeta_i^n, & A^n(2)_t &= \sum_{i \ge 0} \delta_{i+1}^n(t)\, \zeta_i^n \\
A^n(3)_t &= -\sum_{i \ge 0} \delta_{2i}^n(t)\, \zeta_{2i}'^n, & A^n(4)_t &= -\sum_{i \ge 0} \delta_{2i+1}^n(t)\, \zeta_{2i+1}'^n
\end{aligned}$$

$$A^n(5)_t = k_n \sum_{r=1}^{k_n} \big(g_r'^n\big)^2 \big(\Upsilon^{lm}_{T_n + ([k_n t] + r)\Delta_n} + \Upsilon^{lm}_{T_n + ([k_n t] + r - 1)\Delta_n}\big) - 2\Lambda(g')\Upsilon^{lm}_{(T)}.$$

2) Recalling that $T_n \le (T - (m+1)u_n)^+$ in case (1) and $T_n \ge T$ in case (2), and $T_n \to T$ always, we see that $\Upsilon^{lm}_{T_n + ([k_n t] + r - j)\Delta_n)}$ converges to $\Upsilon^{lm}_{(T)}$ for $j = 0, 1$, uniformly in $t \in [0, m]$. Therefore it readily follows from $k_n \Lambda_n(g', 2) \to \Lambda(g')$, recall (16.1.9), that

$$\sup_{t \in [0,m]} \big|A^n(j)_t\big| \stackrel{\mathbb{Q}}{\longrightarrow} 0 \tag{16.3.31}$$

when $j = 5$ (in this case, it is even a "sure" convergence, since $A^n(j)_t$ only depends on ω through $\omega^{(0)}$, which here is fixed). It thus remains to prove this for $j = 1, 2, 3, 4$. Under $\mathbb{Q}$ the variables ζ_i^n are independent centered with bounded moments of all order, and the same holds for the two sequences $\zeta_{2i}'^n$ and $\zeta_{2i+1}'^n$ (but not for the "full" sequence $\zeta_i'^n$, of course). The argument for proving (16.3.31) is based on the above properties and on the form of $\delta_i^n(t)$, so it works in the same way for $j = 1, 2, 3, 4$ and we only consider the case $j = 1$ below.

Using these properties, we can apply Burkholder-Gundy and Hölder's inequalities, plus the fact that $\delta_i^n(t) = 0$ is always smaller than K/k_n and vanishes when $i \le [k_n t]$ or $i > [k_n t] + k_n$, to get for $p \ge 2$:

$$\mathbb{E}_{\mathbb{Q}}\big(\big(A^n(1)_t\big)^p\big) \le \mathbb{E}_{\mathbb{Q}}\bigg(\bigg(\sum_{i \ge 1} \big|\delta_i^n(t)\big|^2 \big|\zeta_i^n\big|^2\bigg)^{p/2}\bigg) \le \frac{K_p}{k_n^{p/2}}. \tag{16.3.32}$$

Hence $A^n(1)_t \stackrel{\mathbb{Q}}{\longrightarrow} 0$, and it remains to prove that the sequence $A^n(1)$ is C-tight. For this, we observe that (16.3.32) *a fortiori* implies $\mathbb{E}_{\mathbb{Q}}(|A^n(1)_{t+s} - A^n(1)_t|^p) \leq K_p/k_n^{p/2}$, hence by the same argument as in Step 3 of the previous proof we deduce the C-tightness. This completes the proof. □

We can now combine the previous two results, at least when we consider weight functions $g^1, \dots, g^q$. That is, we look for the joint behavior of the three processes

$$\left.\begin{array}{lll} 1 \leq j \leq q,\ 1 \leq l \leq d' & \mapsto & L_t^{n,jl} = \frac{1}{\sqrt{u_n}}\, \overline{W}(g^j)^{n,l}_{i_n+[k_n t]} \\ 1 \leq j \leq q,\ 1 \leq l \leq d & \mapsto & L_t^{\prime n,jl} = \sqrt{k_n}\, \overline{\chi}(g^j)^{n,l}_{i_n+[k_n t]} \\ 1 \leq j \leq q,\ 1 \leq l, m \leq d & \mapsto & \widehat{L}_t^{n,jlm} = k_n\, \widehat{\chi}(g^j)^{n,lm}_{i_n+[k_n t]}. \end{array}\right\} \tag{16.3.33}$$

The first two processes above are the same as in (16.3.12), upon choosing $J = q$ and $h^j = g^j$, $h(j)^n_i = (g^j)^n_i$, $h'^j = (g^j)'$, $h'(j)^n_i = k_n (g^j)'^n_i$ (which all satisfy (16.3.8)).

We will also multiply the Brownian motion by the volatility, and use the notation (16.3.17). We assume that Υ_t has a càdlàg square-root υ_t, and also (16.3.18) for the sequences i_n and $T_n = (i_n - 1)\Delta_n$. Then in case 1 we set $\Upsilon_{(T)} = \Upsilon_{T-}$ as above and also $\sigma_{(T)} = \sigma_{T-}$, and in case 2 we set $\Upsilon_{(T)} = \Upsilon_T$ and $\sigma_{(T)} = \sigma_T$. Finally, in view of (16.3.30), we consider the following "constant" element $z_0 \in \mathbb{D}^{qd^2}$ defined by

$$1 \leq j \leq q,\ 1 \leq l, m \leq d,\ t \geq 0 \ \mapsto\ \left(\Lambda(g')\Upsilon_{(T)}\right)(t)^{jlm} = \Lambda\left((g^j)'\right) \Upsilon^{lm}_{(T)}. \tag{16.3.34}$$

Lemma 16.3.9 *In the previous setting, and under the assumptions of Lemma* 16.3.7 *plus* (SH), *and if the sequence of functions f_n on $\mathbb{D}^{qd+qd+qd^2}$ satisfies* (16.3.16) *and converges pointwise to some limit f, and with $S(dx,dy)$ as defined in* (16.3.10), *we have for any bounded random variable Z:*

$$\begin{aligned} &\mathbb{E}\left(Z\, f_n\left(\sigma_{T_n} L^n, L'^n, \widehat{L}^n\right) \mid \mathcal{F}_{T_n}\right) \\ &\quad \stackrel{\mathbb{P}}{\longrightarrow} \mathbb{E}\left(Z \int f\left(\sigma_{(T)} x, \upsilon_{(T)} y, 2\Lambda(g')\Upsilon_{(T)}\right) S(dx,dy) \mid \mathcal{F}_{(T)}\right). \end{aligned} \tag{16.3.35}$$

Proof First, exactly as in Step 4 of the proof of Lemma 16.3.7, one can show that it is enough to consider the case where $|f_n| \leq K$ identically, for some constant K (we additionally use the facts that Υ is bounded under (SN) and that $\mathbb{E}(\|\widehat{L}^n_t\|^p) \leq K_p$ for all t, which in turn readily follows from (16.2.3)).

If α is a $d \times d'$ matrix we define the following functions on $\mathbb{D}^{qd^2} \times \mathbb{D}^{qd'} \times \mathbb{D}^{qd}$:

$$G_n(\alpha; z, x, y) = f_n(\alpha x, y, z), \qquad G(\alpha; z, x, y) = f(\alpha x, y, z).$$

Observe that by (16.3.16) we have for all $A > 0$:

$$\lim_{\varepsilon \to 0} \sup\left(\left|G_n(\alpha'; z', x, y) - G_n(\alpha; z, x, y)\right| :\right.$$

$$n \geq 1,\ \|\alpha - \alpha'\| + \|z - z'\|_{(n,m)} \leq \varepsilon,$$
$$\|x\|_{(n,m)} + \|y\|_{(n,m)} + \|z\|_{(n,m)} + \|\alpha\| \leq A\big) = 0.$$

First, Lemma 16.3.7 yields that the two sequences L^n and L'^n are tight, so the above property of G_n yields $G_n(\alpha_n; z_n, L^n, L'^n) - G_n(\alpha; z, L^n, L'^n) \stackrel{\mathbb{P}}{\longrightarrow} 0$ if $\alpha_n \to \alpha$ and $\|z - z'\|_{(n,m)} \to 0$. Second, the same lemma yields

$$F_n(\alpha, z) := \mathbb{E}\big(ZG_n\big(\alpha; z, L^n, L'^n\big) \mid \mathcal{F}_{T_n}\big)$$
$$\stackrel{\mathbb{P}}{\longrightarrow} F(\alpha, z) := \mathbb{E}\left(Z \int G(\alpha; z, x, \upsilon_{(T)} y) S(dx, dy) \mid \mathcal{F}_{(T)}\right)$$

and thus

$$\alpha_n \to \alpha,\ \|z - z'\|_{(n,m)} \to 0 \quad \Rightarrow \quad F_n(\alpha_n, z_n) \stackrel{\mathbb{P}}{\longrightarrow} F(\alpha, z). \qquad (16.3.36)$$

Third, let $\Phi'_n = \mathbb{E}(Zf_n(\sigma_{T_n} L^n, L'^n, 2\Lambda(g')\Upsilon_{T_n}) \mid \mathcal{F}_{T_n})$ and Φ_n be the left side of (16.3.35). In view of the assumptions (16.3.16) and (16.3.18), of $|f_n| \leq K$ and of the convergences (16.3.30) and $\Upsilon_{T_n} \to \Gamma_{(T)}$, we see that $\mathbb{E}(|\Phi_n - \Phi'_n|) \to 0$, so

$$\Phi_n - \Phi'_n \stackrel{\mathbb{P}}{\longrightarrow} 0. \qquad (16.3.37)$$

Now since σ_{T_n} and Υ_{T_n} are $\mathcal{F}_{T_n}$ measurable, $\Phi'_n = F_n(\sigma_{T_n}, 2\Lambda(g')\Upsilon_{T_n})$. By virtue of $\sigma_{T_n} \to \sigma_{(T)}$ and $\Upsilon_{T_n} \to \Upsilon_{(T)}$, (16.3.36) yields that Φ'_n converges to $F(\sigma_{(T)}, 2\Lambda(g')\Upsilon_{(T)})$, which is the right side of (16.3.35). Then we conclude by (16.3.37). □

16.3.3 A Global Stable Convergence Result

This subsection, with almost the same title as Sect. 12.1.2, is devoted to proving a result analogous to Proposition 12.1.6, but in a noisy setting. We constantly refer to that proposition and its proof, and thus will use freely the notation of Sect. 12.1.2, for example (12.1.11)–(12.1.17), pages 344–345, to begin with.

We will assume below the strengthened assumption (SH) = (SH-2), where (SH-r) is (H-r) plus the facts that b_t and σ_t and X_t are bounded and $\|\delta(\omega, t, z)\| \leq \Gamma(z)$ with Γ bounded and $\int \Gamma(z)^r \lambda(dz) < \infty$. As for notation, we recall only that $i(n, p)$ is the unique integer with $i(n, p)\Delta_n < S_p \leq (i(n, p) + 1)\Delta_n$, where the S_p's are a reordering of the successive jump times $S(m, q)$ of $1_{A_m} * \mathfrak{p}$ when $m, q \geq 1$ vary, and where $A_m = \{z : \Gamma(z) \geq 1/m\}$. Then instead of the variables $R(n, p)$ of (12.1.16) we use the following ones:

$$R''(n, p)^j = \frac{1}{k_n \sqrt{u_n}} \sum_{l=(i(n,p)-k_n+1)\vee 1}^{i(n,p)} \left(f^j\big(\overline{X}(g^j)^n_l + (\Delta_n)^\eta\, \overline{\chi}(g^j)^n_i\big)\right.$$
$$\left. - f^j\left(g^j\left(\frac{i(n, p) - l + 1}{k_n}\right)\Delta X_{S_p}\right)\right).$$

Then, with Ψ_p, Ψ'_p, α_{T_p} and α'_{T_p} as in (16.3.5) (so R_p is as in (12.1.17)), we set

$$R_p = \alpha_{T_p} \Psi_p, \quad R'_p = \alpha'_{T_p} \Psi'_p, \quad R''_p = R_p + \theta' R'_p. \tag{16.3.38}$$

Our aim in this subsection is the following result:

Proposition 16.3.10 *Assume* (SH), (SN) *and the hypotheses of Theorem* 16.3.1 *about f and η', and that Υ has a càdlàg square-root if $\theta' = \theta > 0$. Then the sequence $(R''(n,p))_{p\geq 1}$ converges stably in law to $(R''_p)_{p\geq 1}$.*

As usual, it is enough to prove that for any finite integer P we have

$$\big(R''(n,p)\big)_{1\leq p\leq P} \xrightarrow{\mathcal{L}\text{-s}} \big(R''_p\big)_{1\leq p\leq P}.$$

Below we fix P and consider the smallest integer m such that for any $p = 1, \dots, P$ we have $S_p = S(m', j)$ for some $m' \leq m$ and some $j \geq 1$. Recall that $g^j(l/k_n)$ is denoted by $(g^j)^n_l$, and $(g^j)'^n_l = g^j(l/k_n) - g^j((l-1)/k_n)$, and that $\chi^n_i = \chi_{i\Delta_n}$.

We use the notation (12.1.19)–(12.1.21), complemented with

$$\begin{aligned}
h'_-(x,j,l)^n_r &= \sum_{u=1}^{r} \partial_l f^j\Big(g^j\Big(\frac{u+k_n-r}{k_n}\Big)x\Big)\,(g^j)'^n_u \\
h'_+(x,j,l)^n_r &= \sum_{u=1+r}^{k_n} \partial_l f^j\Big(g^j\Big(\frac{u-r}{k_n}\Big)x\Big)\,(g^j)'^n_u \\
z'^{m,jlv}_{p-}(x) &= \frac{1}{\sqrt{k_n}} \sum_{r=1}^{k_n} h'_-(x,j,l)^n_r\, \chi^v_{(i(n,p)-k_n+r-1)\Delta_n} \\
z'^{m,jlv}_{p+}(x) &= \frac{1}{\sqrt{k_n}} \sum_{r=1}^{k_n-1} h'_+(x,j,l)^n_r\, \chi^v_{(i(n,p)+r-1)\Delta_n}.
\end{aligned} \tag{16.3.39}$$

Finally $(\mathcal{G}_t)$ is the smallest filtration which contains $(\mathcal{F}_t)$ and such that S_p is $\mathcal{G}_0$ measurable for all $p \geq 1$, so the variables $i(n,p)$ are $\mathcal{G}_0$ measurable, and W is a $(\mathcal{G}_t)$-Brownian motion, and the representation (12.1.13) for $X(m)$ is the same relative to three filtrations $(\mathcal{F}^{(0)}_t)$ and $(\mathcal{F}_t)$ and $(\mathcal{G}_t)$.

Observe that $|h'_\pm(x,j,i)^n_r| \leq K_x$ for any x, with $x \mapsto K_x$ being locally bounded. Thus we have estimates similar to (16.2.3), for all $q, A > 0$ (we argue under each probability $\mathbb{Q}(\omega^{(0)}, .)$):

$$\|x\| \leq A,\ q \geq 1 \quad \Rightarrow \quad \mathbb{E}\big(\big\|z'^n_{p\pm}(x)\big\|^q\big) \leq K_{q,A}. \tag{16.3.40}$$

Lemma 16.3.11 *Under the assumptions of Proposition* 16.3.10, *for each $p \leq P$ we have*

$$R''(n,p)^j - \left(\overline{z}_{p-}^{n,j}(\Delta X_{S_p}) + \overline{z}_{p+}^{n,j}(\Delta X_{S_p})\right) + \frac{\Delta_n^{\eta-1/2}}{k_n}\left(z_{p-}^{\prime n,j}(\Delta X_{S_p}) + z_{p+}^{\prime n,j}(\Delta X_{S_p})\right) \xrightarrow{\mathbb{P}} 0. \tag{16.3.41}$$

Proof The proof is as in Lemma 12.1.7, with the following changes. We substitute $R(n,p)$ with $R''(n,p)$, and (12.1.23) becomes

$$B_l^n = \sum_{v=1}^d \partial_v f^j\left(g^j\left(\frac{i(n,p)-l+1}{k_n}\right)\Delta X_{S_p}\right)\left(\overline{X(m)}(g^j)_l^{n,v} + (\Delta_n)^\eta\,\overline{\chi}(g^j)_l^{n,v}\right),$$

and (12.1.24) is replaced by the following:

$$\left|B_l^{\prime n}\right| \le K\left(\left\|\overline{X(m)}(g^j)_l^n\right\|^2 + \left\|\overline{X(m)}(g^j)_l^n\right\|^{2+w} + \left((\Delta_n)^\eta\left\|\overline{\chi}(g^j)_l^n\right\|\right)^2 + \left((\Delta_n)^\eta\left\|\overline{\chi}(g^j)_l^n\right\|\right)^{2+w}\right).$$

Then instead of (12.1.25), the variable $\frac{1}{k_n\sqrt{u_n}}\sum_{l=i(n,p)-k_n+1}^{i(n,p)} B_l^n$ equals

$$\overline{z}_{p-}^{n,j}(\Delta X_{S_p}) + \overline{z}_{p+}^{n,j}(\Delta X_{S_p}) - \frac{\Delta_n^{\eta-1/2}}{k_n}\left(z_{p-}^{\prime n,j}(\Delta X_{S_p}) + z_{p+}^{\prime n,j}(\Delta X_{S_p})\right) + B^{\prime n}$$

with the same $B^{\prime n}$ as in (12.1.25). Therefore, since we have (12.1.26), it remains to prove that

$$\frac{1}{k_n\sqrt{u_n}}\sum_{l=i(n,p)-k_n+1}^{i(n,p)}\left(\left((\Delta_n)^\eta\left\|\overline{\chi}(g^j)_l^n\right\|\right)^2 + \left((\Delta_n)^\eta\left\|\overline{\chi}(g^j)_l^n\right\|\right)^{2+w}\right) \xrightarrow{\mathbb{P}} 0. \tag{16.3.42}$$

We can use (16.2.3) to obtain that the expectation of the left side of (16.3.42) is smaller than $K\Delta_n^{2\eta}/k_n\sqrt{u_n}$, so (16.3.42) follows from $\eta+\eta'\ge 1/2$. □

Now we extend Lemma 12.1.9 to the quadruple $(z_{p-}^n, z_{p+}^n, z_{p-}^{\prime n}, z_{p+}^{\prime n})$. Recall that $z_{p\pm}^n$ are processes indexed by $\mathbb{R}^d$, and the same is true of $z_{p\pm}^{\prime n}$. We describe the joint limit on an auxiliary space $(\Omega', \mathcal{F}', \mathbb{P}')$ which supports the d'-dimensional Brownian motions $W(p-)$ and $W(p+)$, and also another independent family $(W'(p-), W'(p+))$ of standard d-dimensional Brownian motions, all independent. Then the limits $z_\pm$ of $z_{p\pm}^n$ are described by (12.1.30), and we set

$$z_{p-}^{\prime jlv}(x) = \int_0^1 h'_-(x,t)_l^j\,dW'(p-)_t^v, \quad z_{p+}^{jlv}(x) = \int_0^1 h'_+(x,t)_l^j\,dW'(p+)_t^v.$$

All those are still independent centered Gaussian, with paths in $\mathbf{C}' = \mathbf{C}(\mathbb{R}^d, \mathbb{R}^{qdd})$, defined on $(\Omega', \mathcal{F}', \mathbb{P}')$, and we take the product extension (see e.g. (4.1.16)). We begin with a "finite-dimensional" version of the result, namely:

Lemma 16.3.12 *Assume* (SH), (SN), *and that Υ_t admits a càdlàg adapted square-root υ_t. Then for any family $x_1,\dots,x_N$ in $\mathbb{R}^d$ we have*

$$\begin{aligned}&\big(z^n_{p-}(x_i), z^n_{p+}(x_i), z'^n_{p-}(x_i), z'^n_{p+}(x_i)\big)_{1\le p\le P,\,1\le i\le N}\\&\quad\xrightarrow{\mathcal{L}\text{-s}} \big(z_{p-}(x_i), z_{p+}(x_i), \upsilon_{S_p-}z'_{p-}(x_i), \upsilon_{S_p}z_{p+}(x_i)\big)_{1\le p\le P,\,1\le i\le N}.\end{aligned}$$

Proof 1) We write $\overline{L}^n_{p-}$ for the $Nqdd'$-dimensional variable with components $\overline{L}^{n,ijlv}_{p-}=z^{n,jlv}_{p-}(x_i)$, and likewise for $\overline{L}^n_{p+}$, $\overline{L}'^n_{p-}$ and $\overline{L}'^n_{p+}$, the last two variables being $Nqdd$-dimensional. Accordingly, we set $\overline{L}^{ijlv}_{p\pm}=z^{jlv}_{p\pm}(x_i)$ and $\overline{L}'^{ijlv}_{p\pm}=z'^{jlv}_{p\pm}(x_i)$; then the limits of $\overline{L}^n_{\pm}$ will be $\overline{L}_{\pm}$, and with the notation (16.3.17) the limits of $\overline{L}'^n_{p-}$ and $\overline{L}'^n_{p+}$ will be $\upsilon_{S_p-}\overline{L}'_{p-}$ and $\upsilon_{S_p}\overline{L}'_{p+}$.

We need to show that, for all bounded $\mathcal{F}$ measurable variables Z and bounded Lipschitz functions F_{p-} and F_{p+} on $\mathbb{R}^{Nqdd'+Nqdd}$, and with

$$Y^n_{p\pm} = F_{p\pm}\big(\overline{L}^n_{p\pm}, \overline{L}^n_{p\pm}\big),$$

then

$$\mathbb{E}\Bigg(Z\prod_{p=1}^P Y^n_{p-}Y^n_{p+}\Bigg)\to\widetilde{\mathbb{E}}\Bigg(Z\prod_{p=1}^P F_{p-}(\overline{L}_{p-},\upsilon_{S_p-}\overline{L}_{p-})\,F_{p+}(\overline{L}_{p+},\upsilon_{S_p}\overline{L}_{p+})\Bigg).\tag{16.3.43}$$

2) In this step we explain how the variables $\overline{L}^n_{p\pm}$ and $\overline{L}'^n_{p\pm}$ are connected with the processes studied in Lemma 16.3.7. The index j in that lemma, denoted here by $\mathcal{J}$, will be the triple $\mathcal{J}=(ijl)$, so $J=Nqd$. When considering $\overline{L}^n_{p-}$ and $\overline{L}'^n_{p-}$, the functions and sequences showing in (16.3.8) are, with the notation (12.1.3), (12.1.19), (16.3.3) and (16.3.39):

$$\begin{aligned}&h^{\mathcal{J}}(t)=h_-(x_i,t)^j_l\\&h'^{\mathcal{J}}(t)=h'_-(x_i,t)^j_l\\&h(\mathcal{J})^n_r=h_-(x_i,j,l)^n_r\,1_{\{1,\dots,k_n-1\}}(r)\\&h'(\mathcal{J})^n_r=k_n\,h'_-(x_i,j,l)^n_r\,1_{\{1,\dots,k_n\}}(r).\end{aligned}$$

In view of the properties (16.1.6) of the weight functions g^j and the fact that f is C^2, checking that (16.3.8) holds is a simple matter. Now, recall that the processes L^n and L'^n of (16.3.12) depend also on a sequence i_n of integers. Then we have:

$$i_n=i(n,p)-k_n+1\quad\Rightarrow\quad L^n_0=\overline{L}^n_{p-},\quad L'^n_0=\overline{L}'^n_{p-}.$$

In the same way, when we consider $\overline{L}^n_{p+}$ and $\overline{L}'^n_{p+}$, we take

$$h^{\mathcal{J}}(t)=h_+(x_i,t)^j_l$$

$$h'^{\mathcal{J}}(t) = h'_+(x_i, t)^j_l$$
$$h(\mathcal{J})^n_r = h_+(x_i, j, l)^n_r \, 1_{\{1,\dots,k_n-1\}}(r)$$
$$h'(\mathcal{J})^n_r = k_n \, h'_+(x_i, j, l)^n_r \, 1_{\{0,\dots,k_n-1\}}(r),$$

and we get

$$i_n = i(n,p)+1 \quad \Rightarrow \quad L^n_0 = \overline{L}^n_{p+}, \quad L'^n_0 = \overline{L}'^n_{p+}.$$

Now, with $\overline{S}_\pm(dx,dy)$ denoting the law on $\mathbb{R}^{Nqd'} \times \mathbb{R}^{Nqd}$ of the pair $(\overline{L}_{p\pm}, \overline{L}'_{p\pm})$, we set

$$Y_{p-} = \int F_{p-}(x, \upsilon_{S_p -} y) \, \overline{S}_-(dx,dy), \qquad Y_{p+} = \int F_{p+}(x, \upsilon_{S_p} y) \, \overline{S}_+(dx,dy).$$

Then with the notation of Lemma 16.3.7, and if we take $i_n = i(n,p) - k_n + 1$, we see that $Y_{p-} = \int f(x, \upsilon_{S_p -} y) \, S(dx,dy)$ with the function $f(x,y) = F_{p-}(x(0), y(0))$, and likewise for Y_{p+} when $i_n = i(n,p)+1$.

At this stage, we use Lemma 16.3.7 with the filtration $(\mathcal{G}_t)$, with respect to which $T_n = (i(n,p)-k_n)^+ \Delta_n$ (resp. $T_n = i(n,p)\Delta_n$) are stopping times and case 1 (resp. 2) of (16.3.18) holds. Since F_{p-} and F_{p+} are bounded Lipschitz, this lemma applied with $f_n(x,y) = F_{p\pm}(x(0), y(0))$ yields that for all bounded variables Z,

$$\begin{aligned} &\mathbb{E}\big(Z\,Y^n_{p-} \mid \mathcal{G}_{(i(n,p)-k_n)\Delta_n}\big) \stackrel{\mathbb{P}}{\longrightarrow} \mathbb{E}(Z\,Y_{p-} \mid \mathcal{G}_{S_p -}) \\ &\mathbb{E}\big(Z\,Y^n_{p+} \mid \mathcal{G}_{i(n,p)\Delta_n}\big) \stackrel{\mathbb{P}}{\longrightarrow} \mathbb{E}(Z\,Y_{p+} \mid \mathcal{G}_{S_p}). \end{aligned} \tag{16.3.44}$$

Using the $\mathcal{G}_{i(n,p)\Delta_n}$ measurability of $(\overline{L}^n_{p-}, \overline{L}'^n_{p-})$ and the $\mathcal{G}_{S_p}$ measurability of Y_{p-}, we have

$$\mathbb{E}\big(Z\,Y^n_{p-}\,Y^n_{p+} \mid \mathcal{G}_{(i(n,p)-k_n)\Delta_n}\big) = \mathbb{E}\big(Y^n_{p-}\,\mathbb{E}\big(Z\,Y^n_{p+} \mid \mathcal{G}_{i(n,p)\Delta_n}\big) \mid \mathcal{G}_{(i(n,p)-k_n)\Delta_n}\big)$$
$$\mathbb{E}(Z\,Y_{p-}\,Y_{p+} \mid \mathcal{G}_{S_p -}) = \mathbb{E}\big(Y_{p-}\,\mathbb{E}(ZY_{p+} \mid \mathcal{G}_{S_p}) \mid \mathcal{G}_{S_p -}\big).$$

Since all variables are bounded, the convergence in (16.3.44) also takes place in $\mathbb{L}^1$. Then if we use the second convergence with Z, and then the first convergence with $\mathbb{E}(ZY_{p+} \mid \mathcal{G}_{S_p})$ instead of Z, we readily deduce from the above that

$$\mathbb{E}\big(Z\,Y^n_{p-}\,Y^n_{p+} \mid \mathcal{G}_{(i(n,p)-k_n)\Delta_n}\big) \stackrel{\mathbb{L}^1}{\longrightarrow} \mathbb{E}(Z\,Y_{p-}\,Y_{p+} \mid \mathcal{G}_{S_p -}). \tag{16.3.45}$$

3) We are now ready to prove (16.3.43). On the set $\{S_P - S_{P-1} > 3u_n\}$, which converges to Ω, the variables $Y^n_{p\pm}$ for $p \le P-1$ are $\mathcal{G}_{(i(n,P)-k_n)\Delta_n}$ measurable. Hence the left side of (16.3.43) is equal to

$$\mathbb{E}\Bigg(\prod_{p=1}^{P-1} \big(Y^n_{p-}\,Y^n_{p+}\big)\,\mathbb{E}\big(Z\,Y^n_{P-}\,Y^n_{P+} \mid \mathcal{G}_{(i(n,P)-k_n)\Delta_n}\big)\Bigg) + R_n,$$

where the remainder term R_n satisfies $|R_n| \le K\mathbb{P}(S_P - S_{P-1} \le 3u_n)$. Then if $Z' = \mathbb{E}(Z\,Y_{p-}\,Y_{p+} \mid \mathcal{G}_{S_p-})$, we deduce from (16.3.45) and $R_n \to 0$ that

$$\mathbb{E}\left(Z\prod_{p=1}^{P}\left(Y^n_{p-}\,Y^n_{p+}\right)\right) - \mathbb{E}\left(Z'\prod_{p=1}^{P-1}\left(Y^n_{p-}\,Y^n_{p+}\right)\right) \to 0.$$

Now we observe that the right side of (16.3.43) is equal to

$$\mathbb{E}\left(Z\prod_{p=1}^{P}(Y_{p-}\,Y_{p+})\right) = \mathbb{E}\left(Z'\prod_{p=1}^{P-1}(Y_{p-}\,Y_{p+})\right).$$

Therefore if we have (16.3.43) for $P-1$ and all bounded Z, we deduce that it holds for P as well. Since it trivially holds for $P=0$, it is proved by induction for all P. □

The functional version is as follows:

Lemma 16.3.13 *Assume* (SH), (SN), *and that Υ_t admits a càdlàg $(\mathcal{F}^{(0)}_t)$-adapted square-root υ_t. Then*

$$\left(z^n_{p-}, z^n_{p+}, z'^n_{p-}, z'^n_{p+}\right)_{1\le p\le P} \xrightarrow{\mathcal{L}\text{-s}} \left(z_{p-}, z_{p+}, \upsilon_{S_p-}z'_{p-}, \upsilon_{S_p}z'_{p+}\right)_{1\le p\le P} \qquad (16.3.46)$$

for the product topology on $(\mathbf{C}(\mathbb{R}^d,\mathbb{R}^{qdd'}) \times \mathbf{C}(\mathbb{R}^d,\mathbb{R}^{qdd'}) \times \mathbf{C}(\mathbb{R}^d,\mathbb{R}^{qdd}) \times \mathbf{C}(\mathbb{R}^d,\mathbb{R}^{qdd}))^P$.

Proof Let $\mathcal{Z}^n$ and $\mathcal{Z}$ be the left and right sides of (16.3.46). They can be considered as taking values in $\mathbf{C} = \mathbf{C}(\mathbb{R}^d,\mathbb{R}^{2Pqdd'+2Pqdd})$, and we need to prove that

$$\mathbb{E}\left(Z\,F(\mathcal{Z}^n)\right) \to \widetilde{\mathbb{E}}\left(Z\,F(\mathcal{Z})\right), \qquad (16.3.47)$$

where Z is bounded $\mathcal{F}$ measurable and F is bounded continuous on $\mathbf{C}$ for the local uniform topology. It is of course enough to show this when $Z \ge 0$ and $\mathbb{E}(Z) = 1$, in which case (16.3.47) amounts to

$$\mathbb{E}_{\mathbb{P}'}\left(F(\mathcal{Z}^n)\right) \to \mathbb{E}_{\widetilde{\mathbb{P}}'}\left(F(\mathcal{Z})\right), \qquad (16.3.48)$$

for all continuous bounded F, where $\mathbb{P}'$ and $\widetilde{\mathbb{P}}'$ are the probability measures admitting the Radon-Nikodym density Z with respect to $\mathbb{P}$ and $\widetilde{\mathbb{P}}$.

Now, (16.3.48) is the convergence in law $\mathcal{Z}^n \stackrel{\mathcal{L}}{\Longrightarrow} \mathcal{Z}$ under $\mathbb{P}'$, and the previous lemma shows that we have the finite-dimensional convergence in law under $\mathbb{P}'$. Therefore, in order to get (16.3.48) it only remains to prove the C-tightness of the sequence $\mathcal{Z}^n$, which is a property of each of their components. In other words it remains to show that any given component of $\mathcal{Z}^n$ forms a C-tight sequence under $\mathbb{P}'$, for any measure $\mathbb{P}'$ absolutely continuous with respect to $\mathbb{P}$, and this property is clearly implied by the C-tightness under $\mathbb{P}$.

The C-tightness of each component $z_{p\pm}^{n,jlr}$ has been proved in Lemma 12.1.9. For the components $z_{p-}^{\prime n,jlr}$ and $z_{p+}^{\prime n,jlr}$ it is proved in exactly the same way: it is enough to show that they satisfy (12.1.32) for some $v>0$ and $v'>d$.

This is simple: we have $\|h'_-(x,j,l)^n - h'_-(x',j,l)^n\| \le K_A\|x-x'\|$ if $\|x\|$, $\|x'\| \le A$ by our assumptions on f and g^j, and the Burkholder-Davis-Gundy inequality yields for $v\ge 1$:

$$\begin{aligned}
&\mathbb{E}\big(\big|z_{p-}^{\prime n,jlm}(x) - z_{p-}^{\prime n,jlm}(x')\big|^v\big)\\
&\quad\le \frac{K}{k_n^{v/2}}\,\mathbb{E}\Bigg(\bigg(\sum_{r=1}^{k_n}\big|h'_-(x,j,l)^n_r - h'_-(x',j,l)^n_r\big|^2\,\big|\chi^m_{(i(n,p)-k_n+r-1)\Delta_n}\big|^2\bigg)^{v/2}\Bigg)\\
&\quad\le \frac{K_{A,v}\|x-x'\|^v}{k_n}\,\mathbb{E}\bigg(\sum_{r=1}^{k_n}\big|\chi^m_{(i(n,p)-k_n+r-1)\Delta_n}\big|^v\bigg)\\
&\quad\le K_{A,v}\|x-x'\|^v,
\end{aligned}\tag{16.3.49}$$

where the second inequality uses Hölder's inequality, and the last one comes from (SN) and the fact that $(i(n,p)-k_n-1)\Delta_n$ is a stopping time for the filtration $(\mathcal{H}_t)$ of (16.1.1), hence the variables $\chi^m_{(i(n,p)-k_n+r-1)\Delta_n}$ for $r\ge 1$ are $\mathcal{H}_{(i(n,p)-k_n-1)\Delta_n}$-conditionally independent with bounded moments of all orders. The same estimate (16.3.49) holds for $z_{p+}^{\prime n}$, hence we get (12.1.32), upon choosing $v=v'>d$. □

Proof of Proposition 16.3.10 First, observe that the estimates (16.3.40) and (16.3.49) require no assumption on Υ_t other than being bounded, and they imply that the processes $x\mapsto z_{p-}^{\prime n}(x)$ are C-tight, and the same for $z_{p+}^{\prime n}$. Therefore the variables $z_{p\pm}^{\prime n}(\Delta X_{S_p})$ are bounded in probability. Under (16.3.6) with $\theta'=0$ we have $\Delta_n^{\eta-1/2}/k_n\to 0$, and thus the last term on the left of (16.3.41) goes to 0 in probability. In other words we have (12.1.22), and our proposition reduces to Proposition 12.1.6.

It remains to consider the case where $\theta'=\theta>0$. In this case $\Delta_n^{\eta+1/2}\to\theta$, so (16.3.41) implies

$$R''(n,p)^j - \big(\overline{z}_{p-}^{n,j}(\Delta X_{S_p}) + \overline{z}_{p+}^{n,j}(\Delta X_{S_p})\big) + \theta\,\big(z_{p-}^{\prime n,j}(\Delta X_{S_p}) + z_{p+}^{\prime n,j}(\Delta X_{S_p})\big) \stackrel{\mathbb{P}}{\longrightarrow} 0.$$

Now we can use Lemma 12.1.8 (this is where, again, one uses the fact that the f^j's are linear combination of positively homogeneous functions) to deduce

$$\begin{aligned}
&R''(n,p)^j - \bigg(\sum_{l=1}^{d}\sum_{v=1}^{d'}\big(\sigma^{lv}_{S_p-}\,z_{p-}^{n,jlv}(\Delta X_{S_p}) + \sigma_{S_p}\,z_{p+}^{n,jlv}(\Delta X_{S_p})\big)\bigg)\\
&\qquad + \theta\,\big(z_{p-}^{\prime n,j}(\Delta X_{S_p}) + z_{p+}^{\prime n,j}(\Delta X_{S_p})\big) \stackrel{\mathbb{P}}{\longrightarrow} 0.
\end{aligned}$$

The same argument as in the proof of Proposition 12.1.6 allows us to deduce from (16.3.46), instead of (12.1.33), that

$$\big(R''(n,p)^j\big)_{j,p} \xrightarrow{\mathcal{L}\text{-s}} \Bigg(\sum_{l=1}^{d}\sum_{v=1}^{d'}\big(\sigma^{lv}_{S_p-}\, z^{jlv}_{p-}(\Delta X_{S_p}) + \sigma_{S_p}\, z^{jlv}_{p+}(\Delta X_{S_p})\big)$$
$$-\,\theta\sum_{l=1}^{d}\sum_{v=1}^{d}\big(\upsilon^{lv}_{S_p-}\, z'^{jlv}_{p-}(\Delta X_{S_p}) + \upsilon^{lv}_{S_p}\, z'^{jlv}_{p+}(\Delta X_{S_p})\big)\Bigg)_{j,p},$$

where p ranges from 1 to P. It is straightforward to check that the right side above has the same $\mathcal{F}$-conditional law as the family $(R''^j_p)_{p,j}$, as defined by (16.3.38), and this completes the proof. □

16.3.4 Proof of Theorem 16.3.1

The proof is the same as for Theorem 12.1.2, whose notation is used, and we mainly emphasize the few changes to be made. The time t is fixed throughout.

By localization we can assume (SH) and (SN). Recalling (16.3.2), we replace (12.1.35) by

$$Y^n(Z^n)_t = Y^n\big(X(m) + (\Delta_n)^\eta \chi\big)_t + Z^n(m)_t \quad \text{on the set } \Omega_n(t,m), \tag{16.3.50}$$

where $Z^n(m)_t = \sum_{p\in\mathcal{P}_m:\, S_p\le t} \zeta^n_p$, and $\zeta^n_p = (\zeta^{n,j}_p)_{1\le j\le q}$ is given by

$$\zeta^{n,j}_p = \frac{1}{\sqrt{u_n}}\Bigg(\frac{1}{k_n}\sum_{l=(i(n,p)-k_n+1)\vee 1}^{i(n,p)} \Big(f^j\big(\overline{X}(g^j)^n_l + (\Delta_n)^\eta \overline{\chi}(g^j)^n_l\big)$$
$$-\, f^j\big(\overline{X(m)}(g^j)^n_l + (\Delta_n)^\eta \overline{\chi}(g^j)^n_l\big)\Big) - \int_0^1 f^j\big(g^j(u)\Delta X_{S_p}\big)\,du\Bigg).$$

Hence, again on the set $\Omega_n(t,m)$, we have $\zeta^{n,j}_p - R''(n,p)^j = \gamma(j,p)_n + \gamma'(j,p)_n$, where

$$\gamma(j,p)_n = -\frac{1}{k_n\sqrt{u_n}}\sum_{l=(i(n,p)-k_n+1)\vee 1}^{i(n,p)} f^j\big(\overline{X(m)}(g^j)^n_l + (\Delta_n)^\eta \overline{\chi}(g^j)^n_l\big)$$

$$\gamma'(j,p)_n = -\frac{1}{\sqrt{u_n}}\int_0^1 \Big(f^j\big(g^j(u)\Delta X_{S_p}\big) - f^j\big(g^j\big(k_n(1+[u/k_n])\Delta X_{S_p}\big)\big)\Big)\,du.$$

That $\gamma'(j,p)_n \xrightarrow{\mathbb{P}} 0$ is explicitly proved in Theorem 12.1.2. On the other hand,

$$\big|f^j\big(\overline{X(m)}(g^j)^n_l + (\Delta_n)^\eta \overline{\chi}(g^j)^n_l\big)\big| \le K\big(\|\overline{X(m)}(g^j)^n_l\|^2 + (\Delta_n)^\eta\|\overline{\chi}(g^j)^n_l\|^2\big)$$

on the set $\Omega_n(t,m)\cap\{S_p\le t\}$. In view of (12.1.15) and (12.1.26) and (16.2.3) we deduce that $\gamma(j,p)_n\stackrel{\mathbb{P}}{\longrightarrow}0$, hence $\zeta_p^{n,j}-R''(n,p)^j\stackrel{\mathbb{P}}{\longrightarrow}0$. Then for each m, t fixed, it follows from Proposition 16.3.10 that

$$Z^n(m)_t\stackrel{\mathcal{L}\text{-s}}{\longrightarrow}\overline{Y}\big(X'(m)\big)_t+\theta'\overline{Y}'(m,\chi)_t,\tag{16.3.51}$$

where $\overline{Y}(X'(m))$ is associated with the process $X'(m)$ by (12.1.6), and $\overline{Y}'(m,\chi)$ is as $\overline{Y}'(\chi)$ in (16.3.5), except that the sum is extended over the stopping times S_p (instead of T_n) for $p\in\mathcal{P}_m$ only. Since obviously $\overline{Y}(X'(m))_t\to\overline{Y}(X)_t$ and $\overline{Y}'(m,\chi)_t\to\overline{Y}'(\chi)_t$ in probability as $m\to\infty$, we are left to prove that for all $\eta>0$ we have

$$\lim_{m\to\infty}\limsup_{n\to\infty}\mathbb{P}\big(\big|Y^n\big(X(m)+(\Delta_n)^\eta\chi\big)_t\big|>\eta\big)=0.$$

Since we have (12.1.36), it is thus enough to prove the following, for each j:

$$\lim_{m\to\infty}\limsup_{n\to\infty}\mathbb{E}\bigg(\frac{1}{k_n\sqrt{u_n}}\sum_{i=1}^{[t/\Delta_n]}\big|f^j\big(\overline{X(m)}(g^j)_i^n+(\Delta_n)^\eta\overline{\chi}(g^j)_i^n\big)-f^j\big(\overline{X(m)}(g^j)_i^n\big)\big|\bigg)=0.\tag{16.3.52}$$

For proving (16.3.52) we can assume $d=1$, so we omit the index j. Recall that f is C^2 with $\nabla f(x)=\mathrm{o}(\|x\|^2)$ and $f(x)=\mathrm{o}(\|x\|^3)$ as $x\to0$, and f and ∇f are of polynomial growth with powers $p'+1$ and p', for some $p'>2$. Then, by singling out the cases $\|y\|\le\|x\|$ and $\|y\|>\|x\|$, we deduce that for some function ϕ such that $\phi(s)\to0$ as $s\to0$, we have for all $\varepsilon\in(0,1)$:

$$\begin{aligned}&\big|f(x+y)-f(x)\big|\\&\quad\le K\bigg(\|y\|\bigg(\phi(\varepsilon)\|x\|^2+\frac{\|x\|^3}{\varepsilon}+\|x\|^{p'}\bigg)+\phi(\varepsilon)\|y\|^3+\frac{\|y\|^4}{\varepsilon}+\|y\|^{p'+1}\bigg).\end{aligned}$$

By the representation (12.1.12) for $\overline{X(m)}(g)_i^n$ and (2.1.44) and (SH), we have

$$\begin{aligned}&\mathbb{E}\big(\big\|\overline{X(m)}(g)_i^n\big\|^2\big)\le Ku_n\\&q>2\quad\Rightarrow\quad\mathbb{E}\big(\big\|\overline{X(m)}(g)_i^n\big\|^q\big)\le K_qu_n\rho_m,\quad\text{with }\rho_m\to0.\end{aligned}$$

Since $p'>0$, by combining these estimates and (16.2.3) and doing successive conditioning, we deduce that

$$\begin{aligned}&\mathbb{E}\big(\big|f\big(\overline{X(m)}(g)_i^n+(\Delta_n)^\eta\overline{\chi}(g)_i^n\big)-f\big(\overline{X(m)}(g)_i^n\big)\big|\big)\\&\quad\le K\phi(\varepsilon)\bigg(\frac{u_n\Delta_n^\eta}{\sqrt{k_n}}+\frac{\Delta_n^{3\eta}}{k_n^{3/2}}\bigg)+\frac{K}{\varepsilon}\bigg(\frac{u_n\Delta_n^\eta\rho_m}{\sqrt{k_n}}+\frac{\Delta_n^{4\eta}}{k_n^2}\bigg).\end{aligned}$$

The expectation in (16.3.52) is then smaller than

$$K\,t\biggl(\biggl(\phi(\varepsilon)+\frac{\rho_m}{\varepsilon}\biggr)\Delta_n^{\eta+\eta'-1/2}+\phi(\varepsilon)\Delta_n^{3\eta+5\eta'/2-3/2}+\frac{1}{\varepsilon}\,\Delta_n^{4\eta+4\eta'-3/2}\biggr).$$

The powers of Δ_n appearing above are all nonnegative under (16.3.6). Then, letting first $n\to\infty$, then $m\to\infty$, then $\varepsilon\to 0$, we deduce (16.3.52), and the proof is complete.

16.4 Laws of Large Numbers for Normalized Functionals and Truncated Functionals

16.4.1 Statement of Results

Here we turn to the normalized functionals $V'^n(\Phi,k_n,Z^n)$, as given by (16.1.13) for the processes $Z^n=X+(\Delta_n)^\eta\,\chi$. Here, Φ is associated with some function f on $\mathbb{R}^d$ and some weight function g by (16.1.12). As a matter of fact, functionals depending only on the averages $\overline{Z^n}(g)_i^n$ are not enough, and we need functionals that depend on $\widehat{Z^n}(g)_i^n$ as well, see (16.1.10). Therefore we will consider the following functionals, where g is a weight function and f a function on $\mathbb{R}^d\times\mathbb{R}^{d^2}$:

$$V'^n(f,g,k_n,Y)_t\;=\;\Delta_n\sum_{i=1}^{[t/\Delta_n]-k_n+1}f\biggl(\frac{\overline{Y}(g)_i^n}{\sqrt{u_n}},\,\frac{\widehat{Y}(g)_i^n}{u_n}\biggr).$$

Our aim is to extend Theorem 8.4.2 to the noisy case, and we suppose that X is an Itô semimartingale, with the Grigelionis decomposition (16.3.1).

It is convenient to extend the notation ρ_a where $a\in\mathcal{M}_{d\times d}^+$, which comes in under the form ρ_{c_s} in our LLNs. Namely, if a' is another matrix in $\mathcal{M}_{d\times d}^+$, and recalling (16.1.7), we use the notation:

$$\overline{\rho}_{g;a,a'}(dx,dy)=\text{the law of two independent variables with respective laws}$$
$$\mathcal{N}\bigl(0,\Lambda(g)a\bigr)\quad\text{and}\quad\mathcal{N}\bigl(0,\Lambda(g')a'\bigr).$$

Theorem 16.4.1 *Assume that X satisfies* (H) *and that the noise satisfies* (N) *with a square-root υ_t of Υ_t which is càdlàg, and $Z^n=X+(\Delta_n)^\eta\,\chi$, and let g be a weight function. Let f be a continuous function on $\mathbb{R}^d\times\mathbb{R}^{d^2}$ satisfying*:

$$\bigl|f(x,y)\bigr|\;\leq\;K\bigl(1+\|x\|^p+\|y\|^{p/2}\bigr)\tag{16.4.1}$$

for some $p\geq 0$ with further $p<2$ when X has jumps. Then, if k_n satisfies (16.1.5) *and* (16.3.6) *holds, we have*

$$V'^n\bigl(f,g,k_n,Z^n\bigr)_t\;\stackrel{\text{u.c.p.}}{\Longrightarrow}\;\int_0^t ds\int f\bigl(x+\theta'\,y,2\theta'^2\,\Lambda(g')\,\Upsilon_s\bigr)\,\overline{\rho}_{g,c_s,\Upsilon_s}(dx,dy).\tag{16.4.2}$$

As in Theorem 16.3.1, we have two different limits according to whether $\theta'=0$ or $\theta'=\theta>0$ in (16.3.6). When $\theta'=0$ the limit is exactly the same as in Theorem 8.4.2, as a simple calculation shows. However, as for Theorem 16.3.1 the rate in the associated CLT is $\sqrt{u_n}$, so for practical purposes we need to have k_n, hence η', as small as possible: this is why again the case $\eta'=\frac{1-2\eta}{2}$ is of importance.

Unless $\theta'=0$, the limit above is a complicated function of c_s and Υ_s, whereas in practice one wants some information on c_s, and more rarely on Υ_s to get some idea about the size of the noise. This seems a very arduous task in general, but it is tractable in the one-dimensional case, as we will see in the next section.

When X jumps, the growth condition on f in Theorem 16.4.1 is quite stringent, although it is (almost) the same as in Theorem 8.4.2. To remove these restrictions, we may use a truncation procedure, as in Chap. 9. We need in fact two truncation levels, one for the variables $\overline{Z}(g)^n_i$ and another one for the variables $\widehat{Z}(g)^n_i$, which do not have the same order of magnitude. These truncation levels have a form similar to (9.0.3):

$$v_n=\alpha\,u_n^{\varpi},\quad v_n'=\alpha'\,\Delta_n^{1-\eta'-\varpi'}\quad \text{for some } \alpha,\alpha'>0,\ \varpi\in\left(0,\frac{1}{2}\right),\ \varpi'>0. \tag{16.4.3}$$

The choice of v_n is as in (9.0.3) with Δ_n substituted with u_n, and this is motivated by the fact that the typical order of magnitude of $\overline{Z^n}(g)^n_i$ under (16.3.6) is $\sqrt{u_n}$ when X is continuous. As for v_n', the number ϖ' is of course not arbitrary: its range will be specified in the next theorem. Then we introduce the downwards truncated functionals by

$$\begin{aligned}&V'^n(f,g,v_n-,k_n,Y)_t\\&\quad=\Delta_n\sum_{i=1}^{[t/\Delta_n]-k_n+1} f\left(\frac{\overline{Y}(g)^n_i}{\sqrt{u_n}},\frac{\widehat{Y}(g)^n_i}{u_n}\right)1_{\{\|\overline{Y}(g)^n_i\|\le v_n\}}\,1_{\{\|\widehat{Y}(g)^n_i\|\le v_n'\}}.\end{aligned} \tag{16.4.4}$$

Theorem 16.4.2 *Assume that X satisfies* (H-r) (*that is, Assumption* 6.1.1) *for some $r\in[0,2]$ and that the noise satisfies* (N) *with a square-root υ_t of Υ_t which is càdlàg, and $Z^n=X+(\Delta_n)^\eta\,\chi$, and let g be a weight function. Let f be a continuous function on $\mathbb{R}^d\times\mathbb{R}^{d2}$ satisfying* (16.4.1). *Then, if k_n satisfies* (16.1.5) *and* (16.3.6) *holds, we have*

$$V'^n\left(f,g,v_n-,k_n,Z^n\right)_t\overset{\text{u.c.p.}}{\Longrightarrow}\int_0^t ds\int f\left(x+\theta' y,2\theta'^2\Lambda\left(g'\right)\Upsilon_s\right)\overline{\rho}_{g,c_s,\Upsilon_s}(dx,dy) \tag{16.4.5}$$

in the following two cases:

(a) *X is continuous*;
(b) *X may be discontinuous and either $p\le 2$ or*

$$p>2,\ r<2,\ \varpi\ge\frac{p-2}{2(p-r)},\ \varpi'<\frac{2}{p-2}. \tag{16.4.6}$$

When the function $f(x, y)$ does not depend on y, there is of course no reason to truncate the second argument, and the second indicator function in (16.4.4) can be dropped. Note also that the conditions on (p, r, ϖ) above are exactly those of Theorem 9.2.1.

16.4.2 The Proofs

As usual, by localization we may assume (SN) and (SH), or (SH-r) for Theorem 16.4.2. We then have the decomposition $X = X' + X''$ as in (8.4.6), with $X'' = \delta * (\mathfrak{p} - \mathfrak{q})$. We unify the different cases by putting:

$$w = \begin{cases} 0 & \text{if } X \text{ is continuous} \\ 1 & \text{otherwise} \end{cases} \tag{16.4.7}$$

and introduce some new notation:

$$\begin{aligned} &\overline{\beta}_i^n = \frac{1}{\sqrt{u_n}}\, \sigma_{(i-1)\Delta_n}\, \overline{W}(g)_i^n, && \widehat{\beta}_i^n = \frac{\Delta_n^{2\eta}}{u_n}\, \widehat{\chi}(g)_i^n \\ &\overline{\beta}_i'^n = \frac{1}{\sqrt{u_n}}\, \overline{X}(g)_i^n - \overline{\beta}_i^n, && \widehat{\beta}_i'^n = \frac{1}{u_n}\, \widehat{Z^n}(g)_i^n - \widehat{\beta}_i^n \\ &\xi_t^n = \sup_{s:|t-s|\le u_n} \|\sigma_t - \sigma_s\|^2, && \xi_t'^n(\rho) = \int_{\{\|\delta(t,z)\|\le u_n^\rho\}} \big\|\delta(t,z)\big\|^2 \lambda(dz). \end{aligned}$$

Some of the estimates below were already proved before and are recalled for convenience, and some will be used in the next sections only. Below, Γ is the function occurring in (SH).

Lemma 16.4.3 *Under* (SH) *and* (SN) *and* (16.3.6), *and with the notation* (16.4.7), *we have for some* $\rho \in (0, 1)$ *and all* $q > 0$:

$$\mathbb{E}\big(\|\overline{\beta}_i^n\|^q \mid \mathcal{F}_{(i-1)\Delta_n}\big) \le K_q \tag{16.4.8}$$

$$\begin{aligned} &\mathbb{E}\bigg(\bigg(\frac{\|\overline{Z^n}(g)_i^n\|}{\sqrt{u_n}}\bigg)^q \mid \mathcal{F}_{(i-1)\Delta_n}\bigg) \\ &\quad \le K_q\bigg(1 + \frac{w}{u_n^{(q/2-1)^+}}\bigg(\int \Gamma(z)^{q\vee 2}\lambda(dz)\bigg)^{(q/2)\wedge 1}\bigg) \le K_q\bigg(1 + \frac{w}{u_n^{(q/2-1)^+}}\bigg) \end{aligned} \tag{16.4.9}$$

$$\begin{aligned} &\mathbb{E}\big(\|\overline{\beta}_i'^n\|^2 \wedge 1 \mid \mathcal{F}_{(i-1)\Delta_n}\big) \\ &\quad \le Ku_n + Kw\,u_n^\rho + \frac{K}{u_n}\,\mathbb{E}\bigg(\int_{(i-1)\Delta_n}^{(i-1)\Delta_n+u_n} \big(\xi_s^n + w\,\xi_s'^n(\rho)\big)\,ds \mid \mathcal{F}_{(i-1)\Delta_n}\bigg) \end{aligned} \tag{16.4.10}$$

$$\mathbb{E}\bigg(\bigg(\frac{\|\widehat{Z^n}(g)^n_i\|}{u_n}\bigg)^q \mid \mathcal{F}_{(i-1)\Delta_n}\bigg) \leq K_q\big(\Delta_n^{2q\eta'} + \Delta_n^{(2\eta+2\eta'-1)q} + w\Delta_n^{q\wedge 1+2q\eta'-q}\big) \tag{16.4.11}$$

$$\mathbb{E}\big(\|\widehat{\beta}'^n_i\|^q \mid \mathcal{F}_{(i-1)\Delta_n}\big) \leq K_q\big(\Delta_n^{2q\eta'} + w\Delta_n^{q\wedge 1+2q\eta'-q}\big). \tag{16.4.12}$$

Proof Recalling (12.1.12) (page 344), which we apply with $U = X'$ and $U = X''$ and g_n given by (12.1.11) and thus satisfying $|g_n(s)| \leq K$, and by (SH), we deduce exactly as for (8.4.9), (8.4.10) and (8.4.11) (page 240) that for all $q > 0$ and for some $\rho \in (0,1)$ we have (16.4.8), and also

$$\mathbb{E}\big(\|\overline{X'}(g)^n_i\|^q \mid \mathcal{F}_{(i-1)\Delta_n}\big) \leq K_q u_n^{q/2}$$

$$\mathbb{E}\big(\|\overline{X''}(g)^n_i\|^q \mid \mathcal{F}_{(i-1)\Delta_n}\big) \leq K_q\bigg(u_n^{q/2} + \bigg(u_n\int \Gamma(z)^{q\vee 2}\lambda(dz)\bigg)^{(q/2)\wedge 1}\bigg)$$

$$\mathbb{E}\bigg(\bigg\|\frac{\overline{X'}(g)^n_i}{\sqrt{u_n}} - \overline{\beta}^n_i\bigg\|^q \mid \mathcal{F}_{(i-1)\Delta_n}\bigg)$$

$$\leq K_q u_n^{q/2} + \frac{K_q}{u_n}\mathbb{E}\bigg(\int_{(i-1)\Delta_n}^{(i-1)\Delta_n+u_n} \|\sigma_s - \sigma_{(i-1)\Delta_n}\|^2\, ds \mid \mathcal{F}_{(i-1)\Delta_n}\bigg)$$

$$\mathbb{E}\bigg(\frac{\|\overline{X''}(g)^n_i\|^2}{u_n} \wedge 1 \mid \mathcal{F}_{(i-1)\Delta_n}\bigg)$$

$$\leq K u_n^{\rho} + \frac{K}{u_n}\mathbb{E}\bigg(\int_{(i-1)\Delta_n}^{(i-1)\Delta_n+u_n} \xi'^n_s(\rho)\, ds \mid \mathcal{F}_{(i-1)\Delta_n}\bigg).$$

These estimates, plus $\xi^n_s + \xi'^n_s(\rho) \leq K$ and (16.2.3) and the property $(\Delta_n)^\eta/\sqrt{u_n} \leq K/\sqrt{k_n}$ which is implied by (16.3.6), yield (16.4.9) and (16.4.10).

Next,

$$\mathbb{E}\big(\|\Delta^n_i X\|^q \mid \mathcal{F}_{(i-1)\Delta_n}\big) \;\leq\; K_q\big(\Delta_n^{q/2} + w\Delta_n^{(q/2)\wedge 1}\big) \tag{16.4.13}$$

for all $q > 0$. Then by Hölder's inequality and $|g'^n_i| \leq K/k_n$ we deduce

$$\mathbb{E}\bigg(\bigg(\frac{\|\widehat{X}(g)^n_i\|}{u_n}\bigg)^q \mid \mathcal{F}_{(i-1)\Delta_n}\bigg)$$

$$\leq \frac{K_q}{(k_n u_n)^q}\big(\Delta_n^q + w\Delta_n^{q\wedge 1}\big) \leq K_q\big(\Delta_n^{2q\eta'} + w\Delta_n^{q\wedge 1+2q\eta'-q}\big). \tag{16.4.14}$$

This, together with 16.2.3) and $\widehat{Z^n}(g)^n_i \leq K(\widehat{X}(g)^n_i + \Delta_n^{2\eta}\widehat{\chi}(g)^n_i)$, and $\Delta_n^{2\eta} \leq K k_n u_n$, yield (16.4.11).

Finally, we have $\widehat{\beta}_i^{\prime n} = \frac{1}{u_n}\widehat{X}(g)_i^n + a_i^n + a_i^{\prime n}$, where

$$a_i^{n,lm} = \frac{\Delta_n^\eta}{u_n}\sum_{j=1}^{[k_n/2]}(g_{2j}^{\prime n})^2\big(\Delta_{i+2j-1}^n X^l\,\Delta_{i+2j-1}^n\chi^m + \Delta_{i+2j-1}^n\chi^l\,\Delta_{i+2j-1}^n X^m\big)$$

$$a_i^{\prime n,lm} = \frac{\Delta_n^\eta}{u_n}\sum_{j=0}^{[(k_n-1)/2]}(g_{2j+1}^{\prime n})^2\big(\Delta_{i+2j}^n X^l\,\Delta_{i+2j}^n\chi^m + \Delta_{i+2j}^n\chi^l\,\Delta_{i+2j}^n X^m\big).$$

The jth summand in a_i^n is a martingale increment, relative to the filtration $(\mathcal{H}_{(i+2j-1)\Delta_n})_{j\ge0}$. Then by the Burkholder-Davis-Gundy and Hölder inequalities, plus (SN) and (16.4.13), we get

$$\mathbb{E}\big(\|a_i^n\|^q \mid \mathcal{F}_{(i-1)\Delta_n}\big) \le K_q\big(\Delta_n^{(\eta+5\eta'/2-1/2)q} + w\Delta_n^{(\eta+5\eta'/2-1)q+(q/2)\wedge1}\big)$$

for $q\ge1$, hence also for $q\in(0,1)$ by Hölder's inequality again. The same holds for $a_i^{\prime n}$. Then, upon using (16.4.14) and $\eta+5\eta'/2-1/2\ge2\eta'$, we obtain (16.4.12). □

For further reference, we need slightly more general results than what is strictly needed for Theorems 16.4.1 and 16.4.2. We have integers $m_n\ge0$ and $l_n\ge k_n$, with possibly m_n random, such that the variables $R_n=m_n\Delta_n$ are stopping times, hence $R'_n=R_n+l_n\Delta_n$ as well, and we consider three cases, where $t>0$ is a real number and R a stopping time:

$$\begin{array}{lll} \text{case (1):} & m_n=0, & l_n=[t/\Delta_n]\vee k_n \\ \text{case (2):} & R_n\to R \ \text{ and } \ R'_n\le R & \text{for all } n \\ \text{case (3):} & R'_n\to R \ \text{ and } \ R_n\ge R & \text{for all } n. \end{array} \tag{16.4.15}$$

We need the following process, which is càdlàg under the assumptions of the two theorems:

$$\gamma_t = \int f\big(x+\theta'y,2\theta'^2\Lambda(g')\Upsilon_t\big)\overline{\rho}_{g,c_t,\Upsilon_t}(dx,dy), \tag{16.4.16}$$

and we associate the variable

$$\widehat{\gamma} = \begin{cases} \frac{1}{t}\int_0^t\gamma_s\,ds & \text{in case (1)} \\ \gamma_{R-} & \text{in case (2)} \\ \gamma_R & \text{in case (3)} \end{cases} \tag{16.4.17}$$

Lemma 16.4.4 *Under the assumptions of Theorem* 16.4.1, *and if* $l_n/k_n\to\infty$, *we have*

$$\frac{1}{l_n\Delta_n}\big(V'^n(f,g,k_n,Z^n)_{R'_n} - V'^n(f,g,k_n,Z^n)_{R_n}\big) \xrightarrow{\mathbb{P}} \widehat{\gamma}.$$

Proof 1) We can assume (SH) and (SN). In a first step we prove that, under (16.3.6), and with the additional notation $\widetilde{\beta}_i^n = \frac{(\Delta_n)^\eta}{\sqrt{u_n}}\,\overline{\chi}(g)_i^n$, we have

$$H_t^n := \mathbb{E}\bigg(\frac{1}{l_n}\sum_{i=m_n+1}^{m_n+l_n}\bigg|f\bigg(\frac{\overline{Z^n}(g)_i^n}{\sqrt{u_n}},\frac{\widehat{Z^n}(g)_i^n}{u_n}\bigg)-f\big(\overline{\beta}_i^n+\widetilde{\beta}_i^n,\widehat{\beta}_i^n\big)\bigg|\bigg)\to 0. \quad (16.4.18)$$

We let $p'=2$ if $w=1$ and $p'=2p$ if $w=0$, where p is the exponent appearing in the assumption (16.4.1). This assumption and the continuity of f yield

$$\begin{aligned}&\big|f\big(x+x',y+y'\big)-f(x,y)\big|\\&\quad\le\psi_A(\varepsilon)+K\bigg(\frac{A^p\,\|x'\|^2\wedge 1}{\varepsilon^2}+\frac{A^p\,\|y'\|}{\varepsilon}+\frac{\|x+x'\|^{p'}+\|y+y'\|^{p'/2}}{A^{p'-p}}\bigg).\end{aligned}$$

for all $\varepsilon\in(0,1]$, $A\ge 1$, $x,x'\in\mathbb{R}^d$ and $y,y'\in\mathbb{R}^{d^2}$, and where each ψ_A is a continuous increasing function ψ_A null at 0. We apply this with $x=\overline{\beta}_i^n+\widetilde{\beta}_i^n$, $x'=\overline{\beta}_i'^n$, $y=\widehat{\beta}_i^n$ and $y'=\widehat{\beta}_i'^n$ and use (16.2.3), (16.4.9), (16.4.10), (16.4.11) and (16.4.12) plus the fact that $p'=2$ when $w=1$ and the property $(\Delta_n)^\eta/\sqrt{u_n}\to 0$, to get

$$\begin{aligned}H_t^n&\le\psi_A(\varepsilon)+K\bigg(\frac{A^p u_n^\rho}{\varepsilon^2}+\frac{A^p}{\varepsilon^2}\mathbb{E}\bigg(\frac{1}{R_n'-R_n}\int_{R_n}^{R_n'}\big(\xi_s^n+\xi_s'^n(\rho)\big)\,ds\bigg)\\&\quad+\frac{A^p\,\Delta_n^{2\eta'}}{\varepsilon}+\frac{1}{A^{p'-p}}\bigg).\end{aligned}$$

We have $\eta'>0$ and $\rho>0$, and also $\xi_s^n+\xi_s'^n(\rho)\le K$, whereas $\xi_t^n\to 0$ and $\xi_t'^n(\rho)\to 0$ for almost all t. Therefore $\limsup_n H_t^n\le\psi_A(\varepsilon)+K/A^{p'-p}$. By choosing first A large, and then ε small, we obtain (16.4.18).

2) If $\zeta_i^n=f(\overline{\beta}_i^n+\widetilde{\beta}_i^n,\widehat{\beta}_i^n)$, we deduce from (16.4.18) that it remains to prove

$$\frac{1}{l_n}\sum_{i=1}^{l_n}\zeta_{m_n+i}^n\ \stackrel{\mathbb{P}}{\longrightarrow}\ \widehat{\gamma}. \quad (16.4.19)$$

Set $\zeta_i'^n=\mathbb{E}(\zeta_i^n\mid\mathcal{F}_{(i-1)\Delta_n})$ and $\zeta_i''^n=\zeta_i^n-\zeta_i'^n$. Since $(m_n+i-1)\Delta_n$ is a stopping time for each $i\ge 1$ and $\zeta_{m_n+i}^n$ is $\mathcal{F}_{(m_n+i+k_n-1)\Delta_n}$ measurable, we have

$$\mathbb{E}\bigg(\bigg(\frac{1}{l_n}\sum_{i=1}^{l_n}\zeta_{m_n+i}''^n\bigg)^2\bigg)\le\frac{2}{l_n^2}\sum_{i=1}^{l_n}\sum_{j=0}^{k_n-1}\big|\mathbb{E}\big(\zeta_{m_n+i}^n\,\zeta_{m_n+i+j}^n\big)\big|\le\frac{2k_n}{l_n^2}\sum_{i=1}^{l_n+k_n}\mathbb{E}\big(\big|\zeta_{m_n+i}^n\big|^2\big).$$

(16.2.3), (16.4.1) and (16.4.8), plus $(\Delta_n)^\eta\le K\sqrt{k_n u_n}$, yield $\mathbb{E}(|\zeta_i^n|^2)\le K$ and $|\zeta_i'^n|\le K$. In particular the right side above goes to 0 because $k_n/l_n\to 0$, and instead of (16.4.19) it is then enough to prove

$$\frac{1}{l_n}\sum_{i=1}^{l_n}\zeta'^n_{m_n+i} \xrightarrow{\mathbb{P}} \widehat{\gamma}. \tag{16.4.20}$$

With $\gamma^n_t = \zeta'^n_i$ for $(i-1)\Delta_n \le t < i\Delta_n$, the left side of (16.4.20) is

$$\frac{1}{R'_n - R_n}\int_{R_n}^{R'_n} \gamma^n_s\,ds = \int_0^1 \gamma^n_{R_n+s(R'_n-R_n)}\,ds.$$

In case (1) we have $\widehat{\gamma} = \int_0^1 \gamma_{st}\,ds$ and $R_n = 0$ and $R'_n \to t$. Since γ_s is càdlàg and $|\gamma^n_s| \le K$ and $|\gamma_s| \le K$, and by using the dominated convergence theorem, we see that, for obtaining (16.4.20), it is enough to prove the following:

$$s \in (0,1) \Rightarrow \gamma^n_{R_n+s(R'_n-R_n)} \xrightarrow{\mathbb{P}} \begin{cases} \gamma_{(st)-} & \text{in case (1)} \\ \gamma_{R-} & \text{in case (2)} \\ \gamma_R & \text{in case (3)} \end{cases} \tag{16.4.21}$$

3) Below we fix $s \in [0,1)$, and we consider the (random) sequence i_n of integers, defined by $i_n = i$ if $(i-1)\Delta_n \le R_n + s(R'_n - R_n) < i\Delta_n$, so $T_n = (i_n - 1)\Delta_n$ satisfies (16.3.18)-(1) with $T = st$ in case (1) and $T = R$ in case (2), and (16.3.18)-(2) with $T = R$ in case (3), and $m = 0$ in all three cases. So we apply Lemma 16.3.9 with this sequence i_n and with $Z = 1$, $J = 1$, $h(1)^n_i = g^n_i$ and $h'(1)^n_i = k_n g'^n_i$, so (16.3.8) holds with $h(1) = g$ and $h'(1) = g'$. With the notation (16.3.33) we have $\sigma_{T_n} L^n_0 = \beta^n_{i_n}$ and $L'^n_0 = \sqrt{k_n}\,\overline{\chi}(g)^n_{i_n}$ and $\widehat{L}^n_0 = k_n \widehat{\chi}(g)^n_{i_n}$. We also take the functions f_n on $\mathbb{D}^{d'+d+d^2}$ defined by

$$f_n(x,y,z) = f\left(x(0) - \frac{\Delta_n^{\eta}}{\sqrt{k_n u_n}}\,y(0),\ \frac{\Delta_n^{2\eta}}{k_n u_n} z(0)\right),$$

They satisfy (16.3.16) with some $w \ge 0$ and $m = 0$ by (16.4.1), and converge pointwise to a limit denoted by F here, and which is $F(x,y,z) = f(x(0) - \theta' y(0), \theta'^2 z(0))$. Moreover, with the notation (16.3.34) for $\Lambda(g')\Upsilon_{(T)}$ and (16.3.10) for $S(dx,dy)$, we have

$$\gamma^n_{R_n+s(R'_n-R_n)} = \zeta'^n_{i_n} = \mathbb{E}\big(f_n\big(\sigma_{T_n}L^n, L'^n, \widehat{L}^n\big) \mid \mathcal{F}_{T_n}\big)$$

$$\gamma_{(T)} = \int F\big(\sigma_{(T)}x, y, 2\Lambda(g')\,\Upsilon_{(T)}t\big)\,S(dx,dy),$$

(for the second part above, we use the fact that the image of the measure S by the mapping $(x,y) \mapsto (\sigma_{(T)}x(0), \upsilon_{(T)}y(0))$ is $\overline{\rho}_{g;c_{(T)},\Upsilon_{(T)}}$. Then (16.4.21) follows from Lemma 16.3.9. □

Proof of Theorem 16.4.1 In the setting of the previous lemma, case (1), we have $R_n = 0$ and $R'_n = [t/\Delta_n]\Delta_n$. Therefore we have $V'^n(f,g,k_n,Z^n)_{R_n} = 0$ and

$V'^n(f,g,k_n,Z^n)_{R'_n} = V'^n(f,g,k_n,Z^n)_t$. Hence Lemma 16.4.4 yields that the convergences (16.4.2), under the appropriate conditions, in probability for any fixed t. The limit being continuous, they also hold uniformly in $t \in [0,N]$ for any N, when $f \geq 0$. Then by difference the local uniform convergence holds for f taking positive and negative values. □

Lemma 16.4.5 *In the setting of Lemma* 16.4.4, *and under the assumptions of Theorem* 16.4.2 *and* $l_n/k_n \to \infty$, *we have*

$$\frac{1}{l_n\,\Delta_n}\Big(V'^n(f,g,v_n-,k_n,Z^n)_{R'_n} - V'^n(f,g,v_n-,k_n,Z^n)_{R_n}\Big) \stackrel{\mathbb{P}}{\longrightarrow} \widehat{\gamma}. \qquad (16.4.22)$$

Proof 1) We may assume (SH-r) and (SN), and also $p \geq 2$. We set

$$\begin{aligned} U_n(h) &= \frac{1}{l_n\,\Delta_n}\Big(V'^n(h,g,k_n,Z^n)_{R'_n} - V'^n(h,g,k_n,Z^n)_{R_n}\Big) \\ U'_n(h) &= \frac{1}{l_n\,\Delta_n}\Big(V'^n(h,g,v_n-,k_n,Z^n)_{R'_n} - V'^n(h,g,v_n-,k_n,Z^n)_{R_n}\Big) \end{aligned}$$

for any function h. We basically reproduce the proof of Theorem 9.2.1. We use the function ψ'_ε of (3.3.16) (the same notation is used when the argument is in $\mathbb{R}^d$ or in $\mathbb{R}^{d^2}$), with $\varepsilon = M$ an integer. Instead of (9.2.8) we set

$$f_M(x,y) = f(x,y)\,\psi'_M(x)\,\psi'_{M^2}(y),$$

which is a continuous bounded function, so for each M Theorem 16.4.1 yields $U_n(f_M) \stackrel{\mathbb{P}}{\longrightarrow} \widehat{\gamma}(M)$ as $n \to \infty$, where $\widehat{\gamma}(M)$ is associated by (16.4.16) and (16.4.17) with the function f_M. Moreover $f_M \to f$ pointwise and $|f_M(x,y)| \leq K(1 + \|x\|^p + \|y\|^{p/2})$ uniformly in M, so $\widehat{\gamma}(M) \to \widehat{\gamma}$ as $M \to \infty$. Therefore we are left to prove that

$$\lim_{M\to\infty}\ \limsup_{n\to\infty}\ \mathbb{E}\big(\big|U'_n(f) - U_n(f_M)\big|\big) = 0.$$

2) We set $\overline{v}_n = v_n/\sqrt{u_n}$ and $\overline{v}'_n = v'_n/u_n$, which both go to ∞, so $\overline{v}_n \geq M$ and $\overline{v}'_n \geq M^2$ for all n bigger than some n_M. Hence if $n < n_M$, (16.4.1) implies

$$\begin{aligned} &\big|f(x,y)\,1_{\{\|x\|\leq\overline{v}_n,\,\|y\|\leq\overline{v}'_n\}} - f_M(x,y)\big| \\ &\quad \leq K\Big(\|x\|^p\,1_{\{\frac{M}{2}<\|x\|\leq\overline{v}_n\}} + \|y\|^{p/2}\,1_{\{\frac{M^2}{2}<\|y\|\leq\overline{v}'_n\}}\Big), \end{aligned}$$

and thus

$$\big|U'_n(f) - U_n(f_M)\big| \leq \frac{K}{l_n}\sum_{i=1}^{l_n}\zeta(M)^n_{m_n+i}, \qquad (16.4.23)$$

where, with the notation $\overline{A}_i^n = \|\overline{Z^n}(g)_i^n\|/\sqrt{u_n}$ and $\widehat{A}_i^n = \|\widehat{Z^n}(g)_i^n\|/u_n$,

$$\zeta(M)_i^n = \big(\overline{A}_i^n\big)^p\, 1_{\{\frac{M}{2}<\overline{A}_i^n\le\overline{v}_n\}} + \big(\widehat{A}_i^n\big)^{p/2}\, 1_{\{\frac{M^2}{2}<\widehat{A}_i^n\le\overline{v}'_n\}}. \tag{16.4.24}$$

3) Exactly as for (9.2.13) (page 254), since $\overline{X''}(g)_i^n = \int_{(i-1)\Delta_n}^{(i-1)\Delta_n+u_n} g_n(s-(i-1)\Delta_n)\,dX''_s$ with $|g_n(s)|\le K$, we deduce from (SH-r) and (2.1.45) that for some sequence $\phi_n\to 0$:

$$\mathbb{E}\big(\big\|\overline{X''}(g)_i^n/u_n^{\varpi}\big\|^2\wedge 1\mid\mathcal{F}_{(i-1)\Delta_n}\big) \le K u_n^{1-r\varpi}\,\phi_n.$$

Moreover $\mathbb{E}(\|\overline{X'}(g)_i^n\|^q)\le K_q u_n^{q/2}$ and $\mathbb{E}(\|\overline{\chi}(g)_i^n\|^q)\le K/k_n^{q/2}$ for all $q>0$. Then, recalling $\overline{Z^n}(g)_i^n = \overline{X'}(g)_i^n + \overline{X''}(g)_i^n + (\Delta_n)^\eta\overline{\chi}(g)_i^n$, and as in the proof of Theorem 9.2.1,

$$\begin{aligned}
&E\big((\overline{A}_i^n)^p\, 1_{\{\frac{M}{2}<\overline{A}_i^n\le\overline{v}_n\}}\mid\mathcal{F}_{(i-1)\Delta_n}\big)\\
&\quad\le \frac{K}{M}\,\mathbb{E}\bigg(\bigg(\frac{\|\overline{X'}(g)_i^n\|}{\sqrt{u_n}}\bigg)^{p+1} + \bigg(\frac{\Delta_n^\eta\,\|\overline{\chi}(g)_i^n\|}{\sqrt{u_n}}\bigg)^{p+1}\bigg)\\
&\qquad + wK\overline{v}_n^p\,\mathbb{E}\bigg(\bigg(\frac{\|\overline{X''}(g)_i^n\|}{u_n^{\varpi}}\bigg)^2\wedge 1\mid\mathcal{F}_{(i-1)\Delta_n}\bigg)\\
&\quad\le \frac{K}{M} + \frac{K}{M}\bigg(\frac{\Delta_n^\eta}{k_n\sqrt{\Delta_n}}\bigg)^{p+1} + w\,K\,\overline{v}_n^p\,u_n^{1-r\varpi}\,\phi_n\\
&\quad\le K\bigg(\frac{1}{M} + w\,u_n^{p(\varpi-1/2)+1-r\varpi}\,\phi_n\bigg),
\end{aligned}$$

where we have used (16.3.6) for the last inequality.

On the other hand,

$$0\le z\le z_1+z_2+z_3 \Rightarrow z^{p/2}1_{\{\frac{M^2}{2}<z\le\overline{v}'_n\}} \le K\bigg(\frac{z_1^{p/2+1}}{M^2} + \frac{z_2^{p/2+1}}{M^2} + \overline{v}_n'^{\,p/2-1}\,z_3\bigg).$$

Since $\|\widehat{Z^n}(g)_i^n\|\le K(\|\widehat{X'}(g)_i^n\| + w\|\widehat{X''}(g)_i^n\| + \Delta_n^{2\eta}\|\widehat{\chi}(g)_i^n\|)$, by using (16.2.3) and $\Delta_n^{2\eta}\le Kk_nu_n$, plus (16.4.14) (once with $q=p/2+1$ and $w=0$, once with $q=1$), we get

$$\begin{aligned}
&E\big((\widehat{A}_i^n)^p\, 1_{\{\frac{M^2}{2}<\widehat{A}_i^n\le\overline{v}'_n\}}\mid\mathcal{F}_{(i-1)\Delta_n}\big)\\
&\quad\le \frac{K}{M^2}\,\mathbb{E}\bigg(\bigg(\frac{\|\widehat{X'}(g)_i^n\|}{u_n}\bigg)^{p/2+1} + \bigg(\frac{\Delta_n^{2\eta}\,\|\widehat{\chi}(g)_i^n\|}{u_n}\bigg)^{p/2+1}\bigg)\\
&\qquad + wK\overline{v}_n'^{\,p/2-1}\,\mathbb{E}\bigg(\frac{\|\widehat{X''}(g)_i^n\|}{u_n}\mid\mathcal{F}_{(i-1)\Delta_n}\bigg)
\end{aligned}$$

$$\leq \frac{K}{M^2}\big(\Delta_n^{(p+2)\eta'}+1\big)+w\,K\,\overline{v}_n^{p/2-1}\,\Delta_n^{2\eta'}\;\leq\;K\bigg(\frac{1}{M^2}+w\,\Delta_n^{1-(p/2-1)\varpi'}\bigg).$$

If we combine these estimates with (16.4.24), we obtain

$$\mathbb{E}\big(\zeta(M)_i^n\mid\mathcal{F}_{(i-1)\Delta_n}\big)\;\leq\;K\bigg(\frac{1}{M}+w\,u_n^{p(\varpi-1/2)+1-r\varpi}\,\phi_n+w\,\Delta_n^{1-(p/2-1)\varpi'}\bigg).$$

Then in view of (16.4.23) plus the fact that $(m_n+i-1)\Delta_n$ is a stopping time for each $i\geq 1$, we see that

$$\limsup_n\,\mathbb{E}\big(\big|U_n'(f)-U_n(f_M)\big|\big)\;\leq\;K/M$$

for all p when $w=0$, whereas when $w=1$ this holds provided $p(\varpi-1/2)+1-r\varpi\geq 0$ and $1-(p/2-1)\varpi'>0$: these conditions are satisfied when $p=2$, and under (16.4.6) otherwise. This ends the proof. □

Proof of Theorem 16.4.2 In the setting of the previous lemma, case (1), we now have $V'^n(f,g,v_n-,k_n,Z^n)_{R_n}=0$ and $V'^n(f,g,v_n-,k_n,Z^n)_{R'_n}=V'^n(f,g,v_n-,k_n,Z^n)_t$, so this lemma yields the convergences (16.4.5) for any fixed t. We conclude the local uniform convergence as for Theorem 16.4.1. □

16.5 Laws of Large Numbers and Central Limit Theorems for Integral Power Functionals

16.5.1 The Laws of Large Numbers

In this section we suppose that X is an Itô semimartingale with dimension $d=1$, and in (16.3.1) it is no restriction to suppose that the dimension of W is $d'=1$ as well. Recall that $c_t=\sigma_t^2$ in this case, and the noise is of course one-dimensional as well.

Our aim is to infer the value $\int_0^t c_s^p\,ds$ for some p, or perhaps of $\int_0^t c_s^p\,\Upsilon_s^q\,ds$. In contrast with the no-noise case, the theorems of the previous section do not *a priori* provide estimators for these quantities. However, if we take for f a polynomial, the limit in the right side of (16.4.2) is a linear combination of the variables $\int_0^t c_s^p\,\Upsilon_s^q\,ds$ for various (integer) values of p and q. Therefore an appropriate choice of the polynomial f may result in a limit which is exactly $\int_0^t c_s^p\,ds$ or $\int_0^t c_s^p\Upsilon_s^q\,ds$, but of course *only when p and q are integers.*

This is what we do in this section, through a rather complicated procedure which can probably be extended to the multi-dimensional case, although the extension does not seem totally straightforward! As of this writing, such an extension has not been done.

To this end, for each integer $p \geq 1$ we introduce the numbers $\zeta_{p,l}$ for $l = 0, \dots, p$ which are the solutions of the following triangular system of linear equations ($C_p^q = \frac{p!}{q!(p-q)!}$ are the binomial coefficients):

$$\begin{aligned} &\zeta_{p,0} = 1, \\ &\sum_{l=0}^{j} 2^l \, m_{2j-2l} \, C_{2p-2l}^{2p-2j} \, \zeta_{p,l} = 0, \quad j = 1, 2, \dots, p. \end{aligned} \tag{16.5.1}$$

These could be explicitly computed, and for example

$$\zeta_{p,1} = -\frac{1}{2} \, C_{2p}^2, \qquad \zeta_{p,2} = \frac{3}{4} \, C_{2p}^4, \qquad \zeta_{p,3} = -\frac{15}{8} \, C_{2p}^6.$$

When q is also an integer, we define the functions $f_{p,q}$ and f_p on $\mathbb{R}^2$ by

$$f_{p,q}(x,z) = \sum_{l=0}^{p} \zeta_{p,l} \, |x|^{2p-2l} \, |z|^{q+l}, \qquad f_p = f_{p,0}. \tag{16.5.2}$$

Theorem 16.5.1 *Assume $d = 1$ and (N) with Υ_t càdlàg, and $Z^n = X + (\Delta_n)^\eta \chi$, and let k_n satisfy (16.1.5) and (16.3.6). Let $p, q \in \mathbb{N}$.*

(a) *If X is continuous and satisfies (H), we have*

$$V'^n\big(f_{p,q}, g, k_n, Z^n\big)_t \stackrel{\text{u.c.p.}}{\Longrightarrow} m_{2p} \, 2^q \, \theta'^{2q} \, \Lambda(g)^p \, \Lambda\big(g'\big)^q \int_0^t c_s^p \, \Upsilon_s^q \, ds.$$

(b) *Assume (H-r) for some $r \in [0,2]$, and (16.4.3). Then if X is continuous, or if X jumps and either $p + q \leq 1$ or*

$$p + q > 1, \quad r < 2, \quad \varpi \geq \frac{p+q-1}{2p+2q-r}, \quad \varpi' < \frac{1}{p+q-1}, \tag{16.5.3}$$

we have

$$V'^n\big(f_{p,q}, g, v_n-, k_n, Z^n\big)_t \stackrel{\text{u.c.p.}}{\Longrightarrow} m_{2p} \, 2^q \, \theta'^{2q} \, \Lambda(g)^p \, \Lambda\big(g'\big)^q \int_0^t c_s^p \, \Upsilon_s^q \, ds.$$

In particular, when $q = 0$, and under the above conditions, we have

$$V'^n\big(f_p, g, k_n, Z^n\big)_t \stackrel{\text{u.c.p.}}{\Longrightarrow} m_{2p} \, \Lambda(g)^p \int_0^t c_s^p \, ds, \tag{16.5.4}$$

$$V'^n\big(f_p, g, v_n-, k_n, Z^n\big)_t \stackrel{\text{u.c.p.}}{\Longrightarrow} m_{2p} \, \Lambda(g)^p \int_0^t c_s^p \, ds. \tag{16.5.5}$$

When $q > 0$ this is useful when $\theta' > 0$, otherwise the limits vanish, whereas when $q = 0$ and $\theta' = 0$ one would rather use the fact that

$$V'^n\big(f, g, k_n, Z^n\big)_t \stackrel{\text{u.c.p.}}{\Longrightarrow} m_{2p} \, \Lambda(g)^p \int_0^t c_s^p \, ds$$

if we simply take $f(x)=|x|^{2p}$ (this is then true even if p is not an integer). Note that $\theta'>0$ implies $\eta<\frac{1}{2}$, because $\eta'>0$. In fact, when $\eta\geq\frac{1}{2}$ the noise is too small to be disentangled from the process X itself, and is in fact "negligible".

Proof The function $f_{p,q}$ satisfies (16.4.1) with $2p+2q$ instead of p. Hence by virtue of Theorem 16.4.1 and 16.4.2 it is enough to prove that

$$\int f_{p,q}\big(x+\theta' y, 2\theta'^2 \Lambda(g')\Upsilon_s\big)\overline{\rho}_{g,c_s,\Upsilon_s}(dx,dy) \\ = m_{2p}\,2^q\,\theta'^{2q}\,\Lambda(g)^p\,\Lambda(g')^q\,c_s^p\,\Upsilon_s^q. \tag{16.5.6}$$

In view of (16.5.2), the left side above is equal to

$$\begin{aligned}
&\sum_{l=0}^{p}\zeta_{p,l}\sum_{j=0}^{p-l}C_{2p-2l}^{2j}\,m_{2j}\,m_{2p-2l-2j}\,\Lambda(g)^j\,\Lambda(g')^{q+p-j}2^{q+l}\,\theta'^{2q+2p-2j}\,c_s^j\,\Upsilon_s^{q+p-j}\\
&\quad=\sum_{j=0}^{p}m_{2j}\,2^q\,\Lambda(g)^j\,\Lambda(g')^{q+p-j}\,\theta'^{2q+2p-2j}\,c_s^j\,\Upsilon_s^{q+p-j}\\
&\qquad\times\sum_{l=0}^{p-j}2^l\,\zeta_{p,l}\,C_{2p-2l}^{2j}\,m_{2p-2l-2j}.
\end{aligned}$$

The last sum above vanishes for $j=0,\dots,p-1$ and equals 1 for $j=p$, by (16.5.1), and (16.5.6) follows. □

The particular case $p+q=1$ of the previous theorem is worth a special mention, given in the following corollary, for which we use the fact that $\zeta_{1,1}=-1/2$:

Corollary 16.5.2 *Assume* $d=1$ *and* (H) *and* (N) *with* Υ_t *càdlàg. Let* $Z^n=X+(\Delta_n)^\eta\,\chi$, *and let* k_n *satisfy* (16.1.5) *and* (16.3.6). *Recalling the notation* (16.1.10) *for* $\overline{Z^n}(g)_i^n$ *and* $\widehat{Z^n}(g)_i^n$, *we have*

$$\begin{aligned}
&\frac{1}{k_n}\sum_{i=1}^{[t/\Delta_n]-k_n+1}\Big(\big(\overline{Z^n}(g)_i^n\big)^2-\frac{1}{2}\widehat{Z^n}(g)_i^n\Big)1_{\{|\overline{Z^n}(g)_i^n|\leq v_n,\,\widehat{Z^n}(g)_i^n\leq v_n'\}}\\
&\quad\stackrel{\text{u.c.p.}}{\Longrightarrow}\Lambda(g)\int_0^t c_s\,ds\\
&\frac{1}{k_n}\sum_{i=1}^{[t/\Delta_n]-k_n+1}\widehat{Z^n}(g)_i^n\,1_{\{|\overline{Z^n}(g)_i^n)|\leq v_n,\,\widehat{Z^n}(g)_i^n\leq v_n'\}}\stackrel{\text{u.c.p.}}{\Longrightarrow}2\theta'^2\Lambda(g')\int_0^t\Upsilon_s\,ds
\end{aligned}$$

and also, when further X is continuous,

$$\frac{1}{k_n}\sum_{i=1}^{[t/\Delta_n]-k_n+1}\Big(\big(\overline{Z^n}(g)_i^n\big)^2-\frac{1}{2}\widehat{Z^n}(g)_i^n\Big)\overset{\text{u.c.p.}}{\Longrightarrow}\Lambda(g)\int_0^t c_s\,ds \tag{16.5.7}$$

$$\frac{1}{k_n}\sum_{i=1}^{[t/\Delta_n]-k_n+1}\widehat{Z^n}(g)_i^n\overset{\text{u.c.p.}}{\Longrightarrow}2\theta'^2\Lambda\big(g'\big)\int_0^t \Upsilon_s\,ds.$$

Next, when X is discontinuous and $p\geq 2$, it is possible to specify the behavior of the non-truncated functionals $V'^n(f_p,g,k_n,Z^n)$, even when X is not an Itô semimartingale:

Theorem 16.5.3 *Let X be an arbitrary one-dimensional semimartingale, and assume* (N) *and $Z^n=X+(\Delta_n)^\eta\chi$. Let k_n satisfy* (16.1.5) *and $p\geq 2$ be an integer, such that*

$$\eta' > 2\frac{1-2p\eta}{1+2p}. \tag{16.5.8}$$

Then for all $t\geq 0$ we have, with the notation $\Lambda(g,2p)=\int_0^1 g(s)^{2p}\,ds$, and recalling (16.5.2) *for f_p:*

$$u_n^{p-1}\,V'^n\big(f_p,g,k_n,Z^n\big)_t\overset{\mathbb{P}}{\longrightarrow}\Lambda(g,2p)\sum_{s<t}|\Delta X_s|^{2p}. \tag{16.5.9}$$

Proof 1) By localization we can assume (SN). We write $\overline{h}_q(x)=|x|^q$ for all real $q>0$, and $h_{p,l}(x,z)=|x|^{2p-2l}|z|^l$. We have $f_p=\sum_{l=0}^p\zeta_{p,l}h_{p,l}$ and $h_{p,0}=\overline{H}_{2p}$, hence

$$V'^n\big(f_p,g,k_n,Z^n\big)=V'^n\big(\overline{h}_{2p},g,k_n,Z^n\big)+\sum_{l=1}^p\zeta_{p,l}V'^n\big(h_{p,l},g,k_n,Z^n\big). \tag{16.5.10}$$

First, $V'^n(\overline{h}_{2p},g,k_n,Z^n)=\frac{\Delta_n}{u_n^p}V^n(\Phi,k_n,Z^n)$, where Φ is associated with $x\mapsto|x|^{2p}$ by (16.1.12), so in (16.2.2) we have $\overline{\Phi}(z)=|z|^{2p}\Lambda(g,2p)$. Thus, since (16.5.8) for p implies (16.2.1) for some $q\in[2,2p)$, so $\overline{h}_{2p}(x)=\mathrm{o}(\|x\|^q)$ as $x\to 0$, we deduce from Theorem 16.2.1 that

$$u_n^{p-1}\,V'^n\big(\overline{h}_{2p},g,k_n,Z^n\big)_t\overset{\mathbb{P}}{\longrightarrow}\Lambda(g,2p)\sum_{s<t}|\Delta X_s|^{2p}. \tag{16.5.11}$$

We are thus left to prove that $u_n^{p-1}\,V'^n(h_{p,l},g,k_n,Z^n)_t\overset{\mathbb{P}}{\longrightarrow}0$ for $l=1,\dots,p$.

2) For further reference, we prove a more general result. Namely, we look for conditions implying, for some given $\alpha\geq 0$:

$$l=1,\dots,p\ \Rightarrow\ u_n^{p-1-\alpha}\,V'^n\big(h_{p,l},g,k_n,Z^n\big)_t\overset{\mathbb{P}}{\longrightarrow}0. \tag{16.5.12}$$

For any $q \geq 1$, we set $U(q)_t^n = \frac{\Delta_n}{u_n^q} \sum_{i=1}^{[t/\Delta_n]-k_n+1} |\widehat{Z^n}(g)_i^n|^q$. If $r \in (0,1)$ and $1 \leq l < p$, Hölder's inequality yields

$$u_n^{p-1-\alpha}\, V'^n\big(h_{p,l}, g, k_n, Z^n\big)_t \leq \Big(u_n^{\frac{p-l}{1-r}-1}\, V'^n\big(\overline{h}_{2\frac{p-l}{1-r}}, k_n, Z^n\big)_t\Big)^{1-r} \Big(u_n^{\frac{l-\alpha}{r}-1}\, U(l/r)_t^n\Big)^r.$$

Exactly as for (16.5.11), we deduce from Theorem 16.2.1 that the variables $u_n^{q-1}\, V'^n(\overline{h}_{2q}, k_n, Z^n)_t$ converge in probability to a finite limit, as soon as $q \geq 1$ and $\eta' > (1-2q\eta)/(1+q)$. When $q = \frac{p-l}{1-r}$, the latter condition amounts to $\eta'(p - l + r - 1) > 1 - r - 2(p-l)\eta$. On the other hand, we have $V'^n(h_{p,p}, g, k_n, Z^n) = U(p)^n$. Therefore the property (16.5.12) will hold if we have

$$u_n^{\frac{l-\alpha}{r}-1}\, U(l/r)_t^n \stackrel{\mathbb{P}}{\longrightarrow} 0 \tag{16.5.13}$$

for some r satisfying

$$\begin{array}{lll} \text{if } l = p: & r = 1 \\ \text{if } 1 \leq l \leq p-1: & 0 < r < 1, \ \eta' > \frac{1-r-2(p-l)\eta}{1-r+p-l}. \end{array} \tag{16.5.14}$$

On the one hand we have $\widehat{Z^n}(g)_i^n \leq 2\widehat{X}(g)_i^n + 2\Delta_n^{2\eta}\widehat{\chi}(g)_i^n$. On the other hand, $\widehat{X}(g)_i^n \leq \frac{K}{k_n^2} \sum_{j=1}^{k_n} (\Delta_{i+j-1}^n X)^2$ because $|g_i'^n| \leq K/k_n$, hence $|\widehat{X}(g)_i^n|^q \leq \frac{K}{k_n^{q+1}} \sum_{j=1}^{k_n} (\Delta_{i+j-1}^n X)^{2q}$ when $q \geq 1$. It follows that $U(q)^n \leq K(U'(q)^n + U''(q)^n)$, where

$$U'(q)_t^n = \frac{\Delta_n}{k_n^q\, u_n^q} \sum_{i=1}^{[t/\Delta_n]} |\Delta_i^n X|^{2q}, \qquad U''(q)_t^n = \frac{\Delta_n^{1+2q\eta}}{u_n^q} \sum_{i=1}^{[t/\Delta_n]-k_n+1} |\widehat{\chi}(g)_i^n|^q.$$

As soon as $q \geq 1$ the sequence $\sum_{i=1}^{[t/\Delta_n]} |\Delta_i^n X|^{2q}$ converges in probability to a finite limit by Theorem 3.3.1, whereas (SN) and (16.2.3) imply $\mathbb{E}(U''(q)_t^n) \leq Kt\Delta_n^{2q\eta}/k_n^q u_n^q$. Therefore (16.5.13) holds provided $k_n^{1+l/r} u_n^{\alpha/r}$ and $k_n^{1+l/r} u_n^{\alpha/r} \times \Delta_n^{1-2\eta l/r}$ go to ∞, which amounts to

$$\eta'(l + \alpha + r) > \alpha \vee (\alpha - 2\eta l + r). \tag{16.5.15}$$

Hence we will have (16.5.12) as soon as for each $l = 1, \dots, p$ we can find r, such that both (16.5.14) and (16.5.15) are true. A tedious but straightforward calculation shows that this is the case if

$$\eta' > \frac{1+\alpha-2p\eta}{1+\alpha+p} \bigvee \frac{\alpha}{1+\alpha}. \tag{16.5.16}$$

In particular, when $\alpha = 0$ this reduces to (16.5.8), hence the theorem is proved. □

For statistical reasons, we also need approximations for the processes Ξ_t and Ξ'_t showing in (16.3.4), at least when the test functions f^j which are used to construct those processes, through (12.1.3) and (16.3.3), are even polynomials. The associated functions $H_\pm$ and $H'_\pm$ are then even polynomials as well. In other words, we want to approximate the following processes, for any integer $p \geq 1$:

$$\begin{aligned} \Xi^-(p)_t &= \sum_{s\leq t} c_{s-}|\Delta X_s|^{2p}, & \Xi^+(p)_t &= \sum_{s\leq t} c_s|\Delta X_s|^{2p} \\ \Xi'^-(p)_t &= \sum_{s\leq t} \Upsilon_{s-}|\Delta X_s|^{2p}, & \Xi'^+(p)_t &= \sum_{s\leq t} \Upsilon_s|\Delta X_s|^{2p}, \end{aligned} \tag{16.5.17}$$

when Υ_t is càdlàg for the last two ones, of course. This is accomplished by a method analogous to what is done in Theorems 9.5.1 and 16.5.1: we start with a choice of truncation levels v_n and v'_n, as in (16.4.3), plus another sequence k'_n of integers satisfying

$$k'_n/k_n \to \infty, \qquad k'_n\Delta_n \to 0. \tag{16.5.18}$$

Then we set, for a given weight function g:

$$\begin{aligned} \Xi^{n+}(p,g)_t &= \frac{1}{k_n k'_n u_n} \sum_{i=1}^{[t/\Delta_n]-k'_n-k_n+1} \big|\overline{Z^n}(g)^n_i\big|^{2p}\, 1_{\{|\overline{Z^n}(g)^n_i|>v_n\}} \\ &\quad\times \sum_{j=1}^{k'_n}\Big(\big(\overline{Z^n}(g)^n_{i+j}\big)^2 - \frac{1}{2}\widehat{Z^n}(g)^n_{i+j}\Big) 1_{\{|\overline{Z^n}(g)^n_{i+j}|\leq v_n,\, \widehat{Z^n}(g)^n_{i+j}\leq v'_n\}} \\ \Xi^{n-}(p,g)_t &= \frac{1}{k_n k'_n u_n} \sum_{i=k_n+k'_n}^{[t/\Delta_n]-k_n+1} \big|\overline{Z^n}(g)^n_i\big|^{2p}\, 1_{\{|\overline{Z^n}(g)^n_i|>v_n\}} \\ &\quad\times \sum_{j=k_n}^{k_n+k'_n-1}\Big(\big(\overline{Z^n}(g)^n_{i-j}\big)^2 - \frac{1}{2}\widehat{Z^n}(g)^n_{i-j}\Big) 1_{\{|\overline{Z^n}(g)^n_{i-j}|\leq v_n,\, \widehat{Z^n}(g)^n_{i-j}\leq v'_n\}} \\ \Xi'^{n+}(p,g)_t &= \frac{1}{k_n k'_n u_n} \sum_{i=1}^{[t/\Delta_n]-k'_n-k_n+1} \big|\overline{Z^n}(g)^n_i\big|^{2p}\, 1_{\{|\overline{Z^n}(g)^n_i|>v_n\}} \\ &\quad\times \sum_{j=1}^{k'_n} \widehat{Z^n}(g)^n_{i+j}\, 1_{\{|\overline{Z^n}(g)^n_{i+j}|\leq v_n,\, \widehat{Z^n}(g)^n_{i+j}\leq v'_n\}} \\ \Xi'^{n-}(p,g)_t &= \frac{1}{k_n k'_n u_n} \sum_{i=k_n+k'_n}^{[t/\Delta_n]-k_n+1} \big|\overline{Z^n}(g)^n_i\big|^{2p}\, 1_{\{|\overline{Z^n}(g)^n_i|>v_n\}} \\ &\quad\times \sum_{j=k_n}^{k_n+k'_n-1} \widehat{Z^n}(g)^n_{i-j}\, 1_{\{|\overline{Z^n}(g)^n_{i-j}|\leq v_n,\, \widehat{Z^n}(g)^n_{i-j}\leq v'_n\}}. \end{aligned} \tag{16.5.19}$$

Theorem 16.5.4 *Assume $d=1$ and* (H) *and* (N) *with Υ_t càdlàg, and $Z^n = X + (\Delta_n)^\eta \chi$, and* (16.3.6) *and* (16.5.18). *Then for any $p>1$ we have*

$$\begin{aligned}
\Xi^{n+}(p,g)_t &\stackrel{\mathbb{P}}{\longrightarrow} \Lambda(g)\,\Lambda(g,2p)\,\Xi^{+}(p)_t\\
\Xi'^{n+}(p,g)_t &\stackrel{\mathbb{P}}{\longrightarrow} 2\theta'^2\Lambda\big(g'\big)\,\Lambda(g,2p)\,\Xi'^{+}(p)_t\\
\Xi^{n-}(p,g)_t &\stackrel{\mathbb{P}}{\longrightarrow} \Lambda(g)\,\Lambda(g,2p)\,\Xi^{-}(p)_t\\
\Xi'^{n-}(p,g)_t &\stackrel{\mathbb{P}}{\longrightarrow} 2\theta'^2\Lambda\big(g'\big)\,\Lambda(g,2p)\,\Xi'^{-}(p)_t.
\end{aligned}$$

Remark 16.5.5 We could dispense with the truncation from below in this result. That is, the theorem remains true if in (16.5.19) we replace everywhere $(\overline{Z^n}(g)^n_i)^{2p}\,1_{\{|\overline{Z^n}(g)^n_i|>v_n\}}$ by $(\overline{Z^n}(g)^n_i)^{2p}$. The proof is exactly the same.

We do not need here p to be an integer, the result holds for any real $p>1$.

Proof *Step 1)* By localization, we may assume (SH) and (SN). We only prove the statements about $\Xi^{n-}(p,g)$ and $\Xi'^{n-}(p,g)$, the others being similar. To unify the proof we write $\Xi^{n-}(p,g)=\Xi(1)^n$ and $\Xi'^{n-}(p,g)=\Xi(2)^n$, and also $\Lambda(g)\,\Lambda(g,2p)\,\Xi^{-}(p)=\Xi(1)$ and $2\theta'^2\Lambda(g')\,\Lambda(g,2p)\,\Xi'^{-}(p)=\Xi(2)$. We introduce some notation, where $m=1$ or $m=2$:

$$\begin{aligned}
\rho(1)^n_i &= \Big(\big(\overline{Z^n}(g)^n_i\big)^2 - \frac{1}{2}\,\widehat{Z^n}(g)^n_i\Big)1_{\{|\overline{Z^n}(g)^n_i|\le v_n,\ \widehat{Z^n}(g)^n_i\le v'_n\}}\\
\rho(2)^n_i &= \widehat{Z^n}(g)^n_i\,1_{\{|\overline{Z^n}(g)^n_i|\le v_n,\ \widehat{Z^n}(g)^n_i\le v'_n\}}\\
\eta^n_i &= \big|\overline{Z^n}(g)^n_i\big|^{2p}\,1_{\{|\overline{Z^n}(g)^n_i|>v_n\}}\\
\zeta(m)^n_i &= \frac{1}{k'_n u_n}\sum_{j=k_n}^{(k_n+k'_n-1)\wedge(i-1)} \rho(m)^n_{i-j}.
\end{aligned}$$

Then we have

$$\Xi(m)^n_t = \frac{1}{k_n}\sum_{i=k_n+k'_n}^{[t/\Delta_n]-k_n+1} \zeta(m)^n_i\,\eta^n_i.$$

Step 2) In this step we use the notation of Step 2 of the proof of Theorem 9.5.1, with $r=2$: We pick $\varepsilon>0$ and denote by $(S_q)_{q\ge 1}$ the successive jump times of the Poisson process $1_{\{\Gamma>\varepsilon/2\}}*\mathfrak{p}$, where Γ is the dominating function occurring in (SH), and $S_0=0$. Then $i(n,q)$ is the random integer such that $(i(n,q)-1)\Delta_n < S_q\le i(n,q)\Delta_n$, and $(\mathcal{G}_t)$ is the smallest filtration containing $(\mathcal{F}_t)$ and such that all S_q are $\mathcal{G}_0$ measurable. We also denote by Ω^n_t the set on which all $S_q\le t$ satisfy

$S_{q-1} + 3(k_n + k'_n) < S_q < t - 3(k_n + k'_n)$, so that $\mathbb{P}(\Omega^n_t) \to 1$. We set

$$X(\varepsilon)_t = X_t - \sum_{q:S_q \le t} \Delta X_{S_q},$$

which is a $(\mathcal{G}_t)$-semimartingale satisfying (SH).

The aim of this step is to prove the analogue of (9.5.5), namely that, for all $q \ge 1$,

$$\begin{aligned} &\zeta(m)^n_{i(n,q)} \xrightarrow{\mathbb{P}} \zeta^*(m)_q, \quad \text{where} \\ &\zeta^*(1)_q = \Lambda(g)c_{S_q-}, \quad \zeta^*(2)_q = 2\theta'^2 \Lambda(g')\Upsilon_{S_q-}. \end{aligned} \tag{16.5.20}$$

Fix $a > 0$, and take n large enough for having $(k_n + k'_n)\Delta_n < a$. The integers $m_n = i(n,q) \vee [a/\Delta_n] - k'_n$ and $l_n = k'_n$ satisfy the conditions before and in (16.4.15), case (2) with $R = S_q \vee a$, and relative to the filtration $(\mathcal{G}_t)$. The variable $\widehat{\gamma}$ of (16.4.17) is $\Lambda(g)c_{T-}$ if $f = f_2$ (notation (16.5.2)) and $2\theta'^2\Lambda(g')\Upsilon_{T-}$ if $f = f_{0,1}$. In restriction to $\{a < S_q \le t\} \cap \Omega^n_t$, and since on this set we have $m_n = i(n,q) - k'_n$, we see that the variable $\zeta(m)^n_{i(n,q)}$ equals the left side of (16.4.22) for $f = f_2$ if $m = 1$ and for $f = f_{0,1}$ if $m = 2$, computed with $X(\varepsilon)$ instead of X. Hence Lemma 16.4.5 applied with $(\mathcal{G}_t)$ and $X(\varepsilon)$, plus $\mathbb{P}(\Omega^n_t) \to 1$, yield (16.5.20) in restriction to $\{a < S_q \le t\}$. Since $a, t > 0$ are arbitrary, (16.5.20) holds.

Step 3) Now we give some estimates. If we combine (16.4.9) and (16.4.11) for X and the filtration $(\mathcal{F}_t)$, and for $X(\varepsilon)$ and the filtration $(\mathcal{G}_t)$ plus the fact that $i(n,q)$ is $\mathcal{G}_0$ measurable, we obtain for $m = 1, 2$ and all $i \ge 1$ and $j = 1, \dots, k_n + k'_n$:

$$\mathbb{E}\big(\big|\rho(m)^n_i\big|\big) \le K u_n, \quad \mathbb{E}\big(\big|\rho(m)^n_{(i(n,q)-j)^+}\big| 1_{\{S_q \le t\} \cap \Omega^n_t}\big) \le K u_n, \tag{16.5.21}$$

where for the latter we also use the fact that on $\{S_q \le t\} \cap \Omega^n_t$ the variables $\rho(m)^n_{(i(n,q)-j)^+}$ are the same for the processes X and $X(\varepsilon)$ if $1 \le j \le k_n + k'_n$. Therefore

$$\mathbb{E}\big(\big|\zeta(m)^n_i\big|\big) \le K, \quad \mathbb{E}\big(\big|\zeta(m)^n_{(i(n,q)-j)^+}\big| 1_{\{S_q \le t\} \cap \Omega^n_t}\big) \le K. \tag{16.5.22}$$

Next, on the set $\{S_q \le t\} \cap \Omega^n_t$ we have for $i(n,q) - k_n + 1 \le i \le i(n,q)$:

$$\begin{aligned} &\zeta(m)^n_i - \zeta(m)^n_{i(n,q)} \\ &\quad = \frac{1}{k'_n u_n} \left(\sum_{j=k_n+k'_n}^{i(n,q)-i+k_n+k'_n-1} \rho(m)^n_{i(n,q)-j} - \sum_{j=k_n}^{i(n,q)-i+k_n-1} \rho(m)^n_{i(n,q)-j} \right), \end{aligned}$$

and each sum above has at most k_n summands. Therefore (16.5.21) yields

$$\mathbb{E}\big(\big|\zeta(m)^n_i - \zeta(m)^n_{i(n,q)}\big| 1_{\{S_q \le t\} \cap \Omega^n_t}\big) \le K k_n / k'_n.$$

Since $k_n/k'_n \to 0$ and $\mathbb{P}(\Omega^n_t) \to 1$, we then deduce from (16.1.9) and (16.5.20) that

$$\frac{1}{k_n} \sum_{i=i(n,q)-k_n+1}^{i(n,q)} \zeta(m)^n_i \, \big|g^n_{i(n,q)-i+1}\big|^{2p} \stackrel{\mathbb{P}}{\longrightarrow} \Lambda(g,2p)\, \zeta^*(m)_q. \tag{16.5.23}$$

Our last estimates are about η^n_i and $\eta(\varepsilon)^n_i = (\overline{Z^n(\varepsilon)(g)^n_i})^{2p}\, 1_{\{|\overline{Z^n(\varepsilon)(g)^n_i}|>v_n\}}$, where $Z^n(\varepsilon) = X(\varepsilon) + (\Delta_n)^\eta \chi$, and they readily follow from (16.4.9), upon using $(\mathcal{G}_t)$ and $X(\varepsilon)$ instead of $(\mathcal{F}_t)$ and X (so Γ is replaced by $\Gamma\, 1_{\{\Gamma \le \varepsilon/2\}}$) for the second estimate:

$$\mathbb{E}\big(\eta^n_i \mid \mathcal{F}_{(i-1)\Delta_n}\big) \le K u_n, \qquad \mathbb{E}\big(\eta(\varepsilon)^n_i \mid \mathcal{G}_{(i-1)\Delta_n}\big) \le K\big(u^p_n + u_n \varepsilon^{2p-2}\big). \tag{16.5.24}$$

Step 4) For $m = 1, 2$ and with the notation (16.5.20), we set

$$\begin{aligned}
\Xi(m,\varepsilon)_t &= \sum_{q\ge 1} \zeta^*(m)_q \, |\Delta X_{S_p}|^{2p}\, 1_{\{S_q \le t\}} \\
B(m,\varepsilon)^n_t &= \frac{1}{k_n} \sum_{i=k_n+k'_n}^{[t/\Delta_n]-k_n+1} \zeta(m)^n_i \, \eta(\varepsilon)^n_i \\
\Xi(m,\varepsilon)^n_t &= \Xi(m)^n_t - B(m,\varepsilon)^n_t.
\end{aligned}$$

Observing that $\zeta(m)^n_i$ is $\mathcal{F}_{(i-1)\Delta_n}$ measurable, by successive conditioning we deduce from (16.5.22) and (16.5.24):

$$\mathbb{E}\big(\big|B(m,\varepsilon)^n_t\big|\big) \le K t \,\big(u^{p-1}_n + \varepsilon^{2p-2}\big).$$

Hence, since $p > 1$,

$$\lim_{\varepsilon\to 0} \limsup_{n\to\infty} \mathbb{E}\big(\big|B(m,\varepsilon)^n_t\big|\big) = 0.$$

Since c_s and Υ_s are bounded, by the Lebesgue theorem and the definition of $\Xi(m)_t$ we get $\Xi(m,\varepsilon)_t \to \Xi(m)_t$, as $\varepsilon \to 0$. Therefore it remains to prove that, for $m = 1, 2$ and ε fixed and $n \to \infty$,

$$\Xi(m,\varepsilon)^n_t \stackrel{\mathbb{P}}{\longrightarrow} \Xi(m,\varepsilon)_t. \tag{16.5.25}$$

Step 5) In restriction to the set Ω^n_t, we have

$$\Xi(m,\varepsilon)^n_t = \sum_{q\ge 1} \xi(m)^n_q \, 1_{\{S_q\le t\}}, \quad \text{where}$$

$$\xi(m)^n_q = \frac{1}{k_n} \sum_{i=i(n,q)-k_n+1}^{i(n,q)} \zeta(m)^n_i \big(\eta^n_i - \eta(\varepsilon)^n_i\big)$$

because $\eta_i^n = \eta(\varepsilon)_i^n$ when i is not between $i(n,q) - k_n + 1$ and $i(n,q)$ for some q. Hence (16.5.25) will follow if we prove $\xi(m)_q^n \xrightarrow{\mathbb{P}} \Lambda(g,2p)\widehat{\zeta}(m)_q\,|\Delta X_{S_q}|^{2p}$. In view of (16.5.23), it thus remains to show

$$\xi'(m)_q^n := \frac{1}{k_n} \sum_{i=i(n,q)-k_n+1}^{i(n,q)} \zeta(m)_i^n \left(\eta_i^n - \eta(\varepsilon)_i^n - \left|g_{i(n,q)-i+1}^n\, \Delta X_{S_q}\right|^{2p}\right) \xrightarrow{\mathbb{P}} 0. \tag{16.5.26}$$

When $i(n,q) - k_n + 1 \le i \le i(n,q)$, and on the set $\{S_q \le t\} \cap \Omega_t^n$, we have $\overline{Z^n}(g)_i^n = \overline{Z(\varepsilon)^n}(g)_i^n + w_i^n$, where $w_i^n = g_{i(n,q)+1-i}^n \Delta X_{S_q}$. For any $x, y \in \mathbb{R}$ we have the estimate

$$\begin{aligned}
&\left||x+y|^{2p}\, 1_{\{|x+y|>v_n\}} - |x|^{2p}\, 1_{\{|x|>v_n\}} - |y|^{2p}\right| \\
&\quad \le K\left(v_n|y|^{2p-1} + |x|^{2p} + |y|^{2p-1}\,|x|\right).
\end{aligned}$$

We apply this with $x = \overline{Z(\varepsilon)^n}(g)_i^n$ and $y = w_i^n$, which is uniformly bounded, and we use (16.4.9) for $X(\varepsilon)$ and the filtration $(\mathcal{G}_t)$, to get for i as above:

$$\mathbb{E}\left(\left|\eta_i^n - \eta(\varepsilon)_i^n - \left|w_i^n\right|^{2p}\right| \mid \mathcal{G}_{(i-1)\Delta_n}\right) 1_{\{S_q \le t\} \cap \Omega_t^n} \le K\left(v_n + \sqrt{u_n}\right).$$

Since $\zeta(m)_i^n$ is $\mathcal{G}_{(i-1)\Delta_n}$ measurable for i as above, by successive conditioning we deduce from the above and from (16.5.22) that

$$\mathbb{E}\left(\left|\xi'(m)_i^n\right| 1_{\{S_q \le t\} \cap \Omega_t^n}\right) \le K\left(v_n + \sqrt{u_n}\right),$$

and (16.5.26) follows because $\mathbb{P}(\Omega_t^n) \to 1$ and $u_n \to 0$ and $v_n \to 0$. □

16.5.2 Central Limit Theorems: The Results

Our next task is to exhibit the Central Limit Theorems associated with the Laws of Large Numbers proved in the previous subsection. One could of course look for a CLT associated with Theorem 16.4.1, but this is probably quite complicated and so far not proved, and also of dubious practical relevance. So we concentrate on the one-dimensional case for X and test functions like f_p in (16.5.2) for p an integer. Even in this case, we only consider the CLTs associated with the convergence (16.5.4) when X is continuous, and with the convergence (16.5.9).

In the whole section we suppose that X is an Itô semimartingale, written as in (16.3.1), and with $d = d' = 1$. We are given a family $g = (g^j)_{1\le j\le q}$ of weight functions, and a family $f = (f^j)_{1\le j\le q}$ of functions on $\mathbb{R} \times \mathbb{R}$, each component being $f^j = f_{p_j}$, as given by (16.5.2) for some integer $p_j \ge 1$.

The first, simplest, result is the CLT associated with Theorem 16.5.3: in this case, our processes of interest are the q-dimensional processes defined component by component by

$$\overline{Y}_t^{n,j} = \frac{1}{\sqrt{u_n}}\left(u_n^{p_j-1}\, V'^n\big(f_{p_j}, g^j, k_n, Z^n\big)_t - \Lambda\big(g^j, 2p_j\big)\sum_{s\le t}|\Delta X_s|^{2p_j}\right). \tag{16.5.27}$$

The result is exactly the same as the result of Theorem 16.3.1 for the following test functions:

$$h = \big(h^j\big)_{1\le j\le q}, \qquad h^j(x) = \overline{h}_{2p_j}(x) = x^{2p_j}. \tag{16.5.28}$$

This means that the functionals $V'^n(f, g, k_n, Z^n)$ behave exactly alike when the test function is $f = f_p$ and when it is $f = \overline{h}_{2p}$, as far as the LLN and CLT are concerned: among all summands like $(\overline{Z^n}(g)_i^n)^{2p-2l}\,(\widehat{Z^n}(g)_i^n)^l$, only those with $l = 0$ really matter when the process X has jumps.

Theorem 16.5.6 *Let X be a one-dimensional Itô semimartingale satisfying* (H), *and assume that the noise satisfies* (N) *with Υ_t càdlàg, and that $Z^n = X + (\Delta_n)^\eta\chi$. Assume that* (16.1.5) *and* (16.3.6) *hold and, for all $j = 1, \dots, q$,*

$$p_j \ge 2, \quad \eta' > \frac{3 - 4p_j\eta}{3 + 2p_j} \bigvee \frac{1}{3}. \tag{16.5.29}$$

Then for each $t > 0$ the q-dimensional variables $\overline{Y}_t^n$ defined by (16.5.27) *converge stably in law to the variable $\overline{Y}(X)_t + \theta'\overline{Y}'(\chi)_t$, associated as in Theorem* 16.3.1 *with the test function $h = (h^j)$ of* (16.5.28).

When the noise is not shrinking ($\eta = 0$) the second condition in (16.5.29) amounts to $\eta' > \frac{3}{7}$ when $p_j = 2$ for at least one j, and to $\eta' > \frac{1}{3}$ when $p_j \ge 3$ for all j.

Conditionally on $\mathcal{F}$, the two q-dimensional processes $\overline{Y}(X)$ and $\overline{Y}'(\chi)$ are independent and centered Gaussian, with conditional covariances Ξ_t and Ξ'_t given by (12.1.4) and (16.3.4). Here the functions $H_\pm^{jj'}$ and $H_\pm'^{jj'}$ involved in the definition of Ξ_t and Ξ'_t are

$$H_\pm(x)^{jj'} = c(p_j, p_{j'}, x)\,\alpha(p_j, p_{j'})^{jj'}, \quad H'_\pm(x)^{jj'} = c(p_j, p_{j'}, x)\,\alpha'(p_j, p_{j'})^{jj'},$$

where $c(p, p', x) = 4pp'\,x^{2p+2p'-2}$ and where and $\alpha(p_j, p_{j'})^{jj'}$ and $\alpha'(p_j, p_{j'})^{jj'}$ are constants, easily computable in terms of the functions g^j and $g^{j'}$ and their first derivatives. So with the notation (16.5.17) and $\overline{Y}'' = \overline{Y}(X) + \theta'\overline{Y}'(\chi)$ we have

$$\begin{aligned}\widetilde{\mathbb{E}}\big(\overline{Y}_t''^j\,\overline{Y}_t''^{j'} \mid \mathcal{F}\big) = 4p_jp_{j'}\big(&\alpha(p_j, p_{j'})^{jj'}\,\big(\Xi^-(p_j + p_{j'} - 2) + \Xi^+(p_j + p_{j'} - 2)\big)\\ &+ \alpha'(p_j, p_{j'})^{jj'}\,\big(\Xi'^-(p_j + p_{j'} - 2) + \Xi'^+(p_j + p_{j'} - 2)\big)\big).\end{aligned}$$

Therefore Theorem 16.5.4 provides an estimator for this conditional covariance.

The proof of this theorem is rather simple, and we give it right away:

Proof We recall the decomposition (16.5.10), and observe that $h_{p_j,0}$ in that decomposition is the function $h^j = \overline{h}_{2p_j}$ of (16.5.28). Therefore we have $u_n^{p_j-1} V'^n(h^j, g^j, k_n, Z^n) = \frac{1}{k_n} V^n(\Phi^j, g^j, k_n, Z^n)$, where Φ^j is associated with h^j by (16.1.12). Since each h^j is positively homogeneous with degree $2p_j > 3$, we see that Theorem 16.5.6 reduces to Theorem 16.3.1, provided we have (16.5.12) with $\alpha = 1/2$, for each p_j and weight function g^j. This property has been proved under the condition (16.5.16), which amounts to (16.5.29). □

Now we turn to the CLT associated with (16.5.4), when X is continuous. The setting is as before, but we are now interested in the process $\overline{Y}'^n$ with components

$$\overline{Y}_t'^{n,j} = \frac{1}{\sqrt{u_n}} \left(V'^n\big(f_{p_j}, g^j, k_n, Z^n\big)_t - m_{2p_j} \Lambda\big(g^j\big)^{p_j} \int_0^t c_s^{p_j} \, ds \right). \tag{16.5.30}$$

Of course, this is a CLT for the same processes $V'^n(f_{p_j}, g^j, k_n, Z^n)$ which occur in (16.5.27), but with different normalization and centering: so this is possible only when the limit in Theorem 16.5.6 is trivial, that is when X is *continuous*.

For a description of the limit, we use the processes

$$L\big(g^j\big)_t = \int_t^{t+1} g^j(s-t) \, d\overline{W}_s, \qquad L'\big(g^j\big)_t = \int_t^{t+1} \big(g^j\big)'(s-t) \, d\overline{W}'_s$$

associated with two independent one-dimensional Brownian motions $\overline{W}$ and $\overline{W}'$ defined on an auxiliary space $(\Omega', \mathcal{F}', (\mathcal{F}'_t)_{t \ge 0}, \mathbb{P}')$ (those are the same as in (16.3.9) with $h^j = g^j$ and $h'^j = (g^j)'$ and $d = d' = 1$). The processes $L = (L(g^j))_{1 \le j \le q}$ and $L' = (L'(g^j))_{1 \le j \le q}$ are independent, stationary, centered, Gaussian, with the covariance structure

$$\begin{aligned} \mathbb{E}'\big(L\big(g^j\big)_t L\big(g^l\big)_s\big) &= \int_{t \vee s}^{(t+1)\wedge(s+1)} g^j(u-t) \, g^l(u-s) \, du \\ \mathbb{E}'\big(L'\big(g^j\big)_t L'\big(g^l\big)_s\big) &= \int_{t \vee s}^{(t+1)\wedge(s+1)} \big(g^j\big)'(u-t) \big(g^l\big)'(u-s) \, du. \end{aligned} \tag{16.5.31}$$

Next, we set (the stationarity of (L, L') implies that the first expression below does not depend on s):

$$\begin{aligned} \mu\big(v, v'\big)^j &= \mathbb{E}'\big(f^j\big(vL\big(g^j\big)_s + v'L'\big(g^j\big)_s, 2v'^2 \Lambda\big(\big(g^j\big)'\big)\big)\big) \\ \mu'\big(v, v'; s, s'\big)^{jl} &= \mathbb{E}'\big(f^j\big(vL\big(g^j\big)_s + v'L'\big(g^j\big)_s, 2v'^2 \Lambda\big(\big(g^j\big)'\big)\big) \\ &\quad \times f^l\big(vL\big(g^l\big)_{s'} + v'L'\big(g^l\big)_{s'}, 2v'^2 \Lambda\big(\big(g^l\big)'\big)\big)\big) \\ R\big(v, v'\big)^{jl} &= \int_0^2 \big(\mu'\big(v, v'; 1, s\big)^{jl} - \mu\big(v, v'\big)^j \mu\big(v, v'\big)^l\big) \, ds. \end{aligned} \tag{16.5.32}$$

The matrix $(R(v,v')^{jl}: 1 \le j,l \le q)$ is symmetric nonnegative. We will see more explicit forms for those quantities at the end of this subsection. When $v'=0$, we recover the covariance R_a^{jl} as given by (12.2.9) (with $d=d'=1$), namely

$$R(v,0)^{jl} = R_{v^2}^{jl}.$$

Finally, before stating the result, we recall Assumption (K), or 4.4.3:

Assumption (K) We have (H) and

$$\sigma_t = \sigma_0 + \int_0^t \widetilde{b}_s\,ds + \int_0^t \widetilde{\sigma}_s\,dW_s + M_t + \sum_{s\le t} \Delta\sigma_s\,1_{\{\|\Delta\sigma_s\|>1\}}, \tag{16.5.33}$$

where

- M is a local martingale with $\|\Delta M_t\| \le 1$, orthogonal to W, and an angle bracket of the form $\langle M,M\rangle_t = \int_0^t a_s\,ds$
- the compensator of $\sum_{s\le t} 1_{\{\|\Delta\sigma_s\|>1\}}$ has the form $\int_0^t \widetilde{a}_s\,ds$.

Moreover, the processes $\widetilde{b}$, a and $\widetilde{a}$ are locally bounded, and the processes $\widetilde{\sigma}$ and b are càdlàg or càglàd.

Theorem 16.5.7 *Let X be a one-dimensional continuous Itô semimartingale satisfying* (K), *and assume that the noise satisfies* (N) *with Υ_t càdlàg, and we set $\upsilon_t = \sqrt{\Upsilon_t}$. We also assume that $Z^n = X + (\Delta_n)^\eta \chi$, and we let k_n satisfy* (16.1.5) *and with* (16.3.6) *and*

$$\eta' > \frac{1}{3}. \tag{16.5.34}$$

Then the q-dimensional processes $\overline{Y}'^n$ of (16.5.30) *converge stably in law to a process $\overline{Y}'$ defined on a very good extension $(\widetilde{\Omega}, \widetilde{\mathcal{F}}, (\widetilde{\mathcal{F}}_t)_{t\ge 0}, \widetilde{\mathbb{P}})$ of the space $(\Omega, \mathcal{F}, (\mathcal{F}_t)_{t\ge 0}, \mathbb{P})$, and which conditionally on $\mathcal{F}$ is a continuous centered Gaussian process with independent increments satisfying*

$$\widetilde{\mathbb{E}}\big(\overline{Y}_t'^j\,\overline{Y}_t'^l \mid \mathcal{F}\big) = \int_0^t R\big(\sigma_s, \theta'\upsilon_s\big)^{jl}\,ds. \tag{16.5.35}$$

Remark 16.5.8 As previously, when $\theta'=0$, that is $\eta+\eta' > \frac{1}{2}$, the limit above is exactly the same as in Theorem 12.2.1. However, even when there is no noise, we need the extra condition (16.5.34); this is not really restrictive for applications (it is implied by (16.3.6) when the noise is not shrinking, for example). This extra condition is needed because we have the correcting terms $\widehat{Z^n}(g^j)_i^n$, which are of course superfluous when there is no noise, and indeed potentially harmful in this case: one could show that this correcting term introduces a bias when $\eta' = \frac{1}{3}$, and changes the rate of convergence when $\eta' < \frac{1}{3}$.

For concrete applications, one also needs an estimator for the (conditional) variance-covariance in (16.5.35). For this we may use the explicit form of f^j and expand the polynomials in (16.5.32). We thus obtain that $R(v,v')^{jl}$ is a polynomial in (v,v'). First, using (16.5.2) with $q_j=0$ allows us to deduce from (16.5.1) (as in the proof of Theorem 16.5.1) and from (16.5.31) that $\mu(v,v')^j$ does not depend on v' and is

$$\mu\big(v,v'\big)^j = m_{2p_j}\,\Lambda\big(g^j\big)^{p_j}\,v^{2p_j}.$$

Next, we have

$$\mu'\big(v,v';s,s'\big)^{jj'} = \sum_{w=0}^{p_j+p_{j'}} a\big(w;s,s'\big)^{jj'}\,v^{2w}\,v'^{2p_j+2p_{j'}-2w}, \quad \text{where}$$

$$\mu'\big(v,v';s,s'\big)^{jj'} = \sum_{w=0}^{p_j+p_{j'}} a\big(w;s,s'\big)^{jj'}\,v^{2w}\,v'^{2p_j+2p_{j'}-2w}, \quad \text{with}$$

$$\begin{aligned} a\big(w;s,s'\big)^{jj'} = & \sum_{l=0}^{p_j}\sum_{l'=0}^{p_{j'}}\sum_{w'=(2w-2p_j-2l)^+}^{(2p_{j'}-2l')\wedge(2w)} \\ & \zeta_{p_j,l}\zeta_{p_{j'},l'}\,C_{2p_j-2l}^{2w-w'}\,C_{2p_{j'}-2l'}^{w'}\,2^{l+l'}\,\Lambda\big(\big(g^j\big)'\big)^l\,\Lambda\big(\big(g^{j'}\big)'\big)^{l'} \\ & \mathbb{E}'\big(L\big(g^j\big)_s^{2w-w'}\,L\big(g^{j'}\big)_{s'}^{w'}\big)\,\mathbb{E}'\big(L'\big(g^j\big)_s^{2p_j-2w+w'}\,L'\big(g^{j'}\big)_{s'}^{2p_{j'}-w'}\big). \end{aligned}$$

Finally,

$$R\big(v,v'\big)^{jj'} = \sum_{w=0}^{p_j+p_{j'}} A_w^{jj'}\,v^{2w}\,v'^{2p_j+2p_{j'}-2w}, \quad \text{where}$$

$$A_w^{jj'} = \int_0^2 a(w;1,s)^{jj'}\,ds + 2m_{2p_j}m_{2p_{j'}}\,\Lambda\big(g^j\big)^{p_j}\,\Lambda\big(g^{j'}\big)^{p_{j'}}\,1_{\{w=p_j+p_{j'}\}}.$$

Therefore we deduce from Theorem 16.5.1 that:

Proposition 16.5.9 *Assume* (H) *with* $d=1$ *and* X *continuous, and* (N), *and* $Z^n = X+(\Delta_n)^\eta\chi$, *and* k_n *satisfying* (16.1.5) *and* (16.3.6). *Let* g *be a weight function and*

$$f(x,z) = \sum_{w=0}^{p_j+p_{j'}} A_w'^{jj'} \sum_{l=0}^{p_j+p_{j'}} \zeta_{p_j+p_{j'},l}\,|x|^{2w-2l}\,|z|^{p_j+p_{j'}-w+l}$$

$$\textit{where}\ A_w'^{jj'} = A_w^{jj'}\,/\,\big(m_{2w}\,2^{p_j+p_{j'}-w}\,\Lambda(g)^w\,\Lambda\big(g'\big)^{p_j+p_{j'}-w}\big).$$

Then we have, with the notation (16.4.3) *and* (16.4.4) *for the second claim*:

$$\overline{V}'^n\left(f,g,k_n,Z^n\right)_t \stackrel{\text{u.c.p.}}{\Longrightarrow} \int_0^t R\left(\sigma_s,\theta'\upsilon_s\right)^{jj'}\,ds,$$

$$\overline{V}'^n\left(f,g,v_n-,k_n,Z^n\right)_t \stackrel{\text{u.c.p.}}{\Longrightarrow} \int_0^t R\left(\sigma_s,\theta'\upsilon_s\right)^{jj'}\,ds.$$

Moreover the latter also holds when X is discontinuous, provided we have (16.5.3) *with $p_j+p_{j'}$ instead of $p+q$.*

Remark 16.5.10 The previous proposition gives the behavior of the estimators of the conditional covariance (16.5.35), even when X has jumps, under appropriate conditions. This is useful in some testing questions, for example for testing whether there are jumps or not. Under the statistical hypothesis that there is no jump, one uses the CLT of Theorem 16.5.7, and it is also necessary to determine the behavior of the test statistics under the alternative hypothesis that there are jumps.

However, we do not give a CLT for the LLN (16.5.5) which would allow one to estimate $\int_0^t c_s^p\,ds$ when there are jumps by the means of the truncated functionals. Such a CLT is so far unknown.

The subsequent subsections are devoted to the proof of Theorem 16.5.7.

16.5.3 Some Estimates

By localization we can assume (SN), and also the strengthened assumption (SK), according to which one can rewrite the equation for σ as

$$\sigma_t = \sigma_0 + \int_0^t \widetilde{b}_s\,ds + \int_0^t \widetilde{\sigma}_s\,dW_s + M_t,$$

where now M (which is not necessarily the same as in (K) is a square-integrable martingale with bounded jumps and $\langle M,M\rangle_t = \int_0^t a_s\,ds$, and $\widetilde{b}_t, \widetilde{\sigma}_t, a_t$ are bounded. We then have for all $s,t\geq 0$ and $p\geq 2$:

$$\mathbb{E}\Big(\sup_{r\in[0,s]} |\sigma_{t+r}-\sigma_t|^p \mid \mathcal{F}_t\Big) \leq K_p\,s. \tag{16.5.36}$$

Next, we introduce some notation. Recalling the number $\zeta_{p,l}$ of (16.5.1), with any process Y and weight function g and integer $p\geq 1$ we associate the variables

$$\phi(g,p,Y)_i^n = \sum_{l=0}^p \zeta_{p,l}\left(\overline{Y}(g)_i^n/\sqrt{u_n}\right)^{2p-2l}\left(\widehat{Y}(g)_i^n/u_n\right)^l. \tag{16.5.37}$$

These are such that

$$V'^n\big(f^j, g^j, k_n, Y\big)_t = \Delta_n \sum_{i=1}^{[t/\Delta_n]-k_n+1} \phi\big(g^j, p_j, Y\big)_i^n.$$

We also set for all integers $i \geq 1$ and $0 \leq j \leq i-1$:

$$\begin{aligned} \overline{\kappa}(g)_{i,j}^n &= \sigma_{(i-j-1)\Delta_n} \overline{W}(g)_i^n + \Delta_n^\eta \overline{\chi}(g)_i^n \\ \phi(g,p)_{i,j}^n &= \sum_{l=0}^{p} \zeta_{p,l} \big(\overline{\kappa}(g)_{i,j}^n / \sqrt{u_n}\big)^{2p-2l} \big(\Delta_n^{2\eta} \widehat{\chi}(g)_i^n / u_n\big)^l. \end{aligned} \tag{16.5.38}$$

With any process Y, we associate the variables

$$\begin{aligned} \Gamma(Y)_i^n &= \sup_{t \in [(i-1)\Delta_n, i\Delta_n + u_n]} \|Y_t - Y_{(i-1)\Delta_n}\| \\ \Gamma'(Y)_i^n &= \big(\mathbb{E}\big(\big(\Gamma(Y)_i^n\big)^4 \mid \mathcal{F}_{(i-1)\Delta_n}\big)\big)^{1/4}. \end{aligned} \tag{16.5.39}$$

The rest of the subsection is devoted to various technical results. The first one is about the (conditional) moments of the noise, and more specifically of the associated variables $\overline{\chi}(g)_i^n$ and $\widehat{\chi}(g)_i^n$. These moments will be expressed in terms of the following variables, whose notation is similar to $\widehat{Y}(g)_i^n$ in (16.1.10) (recall $d = 1$):

$$\widehat{\Upsilon}(g)_i^n = \sum_{j=1}^{k_n} \big(g_j'^n\big)^2 \Upsilon_{i+j-2}. \tag{16.5.40}$$

For random variables U_γ and V_γ indexed by a parameter γ (for example $\gamma = (n,i)$ just below), with $V_\gamma > 0$, we write $U_\gamma = \mathrm{O}_u(V_\gamma)$ if the family U_γ / V_γ is bounded in probability.

Lemma 16.5.11 *Assume* (SN) *and let* v *and* r *be integers with* $v + r \geq 1$. *Recall that* m_p *is the* p *absolute moment of* $\mathcal{N}(0,1)$, *and* $(\mathcal{H}_t)$ *is the filtration defined in* (16.1.1). *Then we have the following estimates, uniform in* $i \geq 1$, *as* $n \to \infty$:

a) *If* v *is even,*

$$\begin{aligned} &\mathbb{E}\big(\big(\overline{\chi}(g)_i^n\big)^v \big(\widehat{\chi}(g)_i^n\big)^r \mid \mathcal{H}_{(i-1)\Delta_n}\big) \\ &\quad = m_v 2^r \big(\widehat{\Upsilon}(g)_i^n\big)^{r+v/2} + \mathrm{O}_u\bigg(\frac{1}{k_n^{r+1+v/2}}\bigg) \end{aligned} \tag{16.5.41}$$

$$\begin{aligned} &\quad = \frac{m_v 2^r}{k_n^{r+v/2}} \Lambda\big(g'\big)^{r+v/2} \big(\Upsilon_{(i-1)\Delta_n}\big)^{r+v/2} \\ &\qquad + \mathrm{O}_u\bigg(\frac{1}{k_n^{r+v/2}}\bigg(\frac{1}{k_n} + \Gamma(\Upsilon)_i^n\bigg)\bigg), \end{aligned} \tag{16.5.42}$$

b) *If v is odd, and recalling Υ'_t in* (16.1.3),

$$\mathbb{E}\big(\big(\overline{\chi}(g)_i^n\big)^v\,\big(\widehat{\chi}(g)_i^n\big)^r\mid\mathcal{H}_{(i-1)\Delta_n}\big)$$

$$= \frac{6r+v-1}{6k_n^{r+v/2+1/2}}\,\Lambda(g')^{r+v/2-3/2}\int_0^1 g'(s)^3\,ds\,(\Upsilon_{(i-1)\Delta_n})^{r+v/2-3/2}\,\Upsilon'_{(i-1)\Delta_n}$$

$$+\,\mathrm{O}_u\bigg(\frac{1}{k_n^{r+v/2+1/2}}\bigg(\frac{1}{k_n}+\Gamma(\Upsilon)_i^n+\Gamma(\Upsilon')_i^n\bigg)\bigg) \tag{16.5.43}$$

$$= \mathrm{O}_u\bigg(\frac{1}{k_n^{r+v/2+1/2}}\bigg). \tag{16.5.44}$$

Proof *Step 1)* Recalling the notation (16.1.8), we have

$$\big|\widehat{\Upsilon}(g)_i^n-\Lambda'_n(g,2)\Upsilon_{(i-1)\Delta_n}\big|\le\Lambda'_n(g,2)\Gamma(\Upsilon)_i^n,$$

so (16.5.42) follows from (16.5.41) and (16.3.7) and the fact that Υ_t is bounded. Since Υ'_t is also bounded, (16.5.44) follows from (16.5.43). Hence it is enough to prove (16.5.41) and (16.5.43).

Step 2) We write $\chi_j^n=\chi_{i\Delta_n}$. Using the definition (16.1.10), we see that the product $(\overline{\chi}(g)_i^n)^v(\widehat{\chi}(g)_i^n)^r$ is the sum of all the terms of the form

$$\Phi(J,s)^n=(-1)^{v+s}\prod_{l=1}^{v} g_{j_l}'^n\,\chi_{i+j_l-2}^n\prod_{l=1}^{s}\big(g_{j'_l}'^n\,\chi_{i+j'_l+j''_l-2}^n\big)^2$$

$$\times\prod_{l=s+1}^{r} 2\big(g_{j'_l}'^n\big)^2\chi_{i+j'_l-1}^n\,\chi_{i+j''_l-2}^n, \tag{16.5.45}$$

extended over all $s\in\{0,\dots,r\}$ and $J=(j_1,\dots,j_v,j'_1,\dots,j'_r,j''_1,\dots,j''_s)$ in the set $\mathcal{J}_s^n=\{1,\dots,k_n\}^{v+r}\times\{0,1\}^s$. We denote by $I(J)$ the set of all distinct indices j of the form j_l for $1\le l\le v$, or $j'_l+j''_l$ for $1\le l\le s$, or j'_l+1 or j'_l for $s+1\le l\le r$. For any given s, we also denote by $D(u,s)^n$ the class of all $J\in\mathcal{J}_s^n$ such that $\#I(J)=u$ and that each index appears at least twice. Note that $D(u,s)^n=\emptyset$ if $u>r+v/2$, because in (16.5.45) χ_j^n appears for $v+2r$ times, for various values of j.

By the $\mathcal{F}^{(0)}$-conditional independence of the χ_t's and (16.1.2), plus $|g_j'^n|\le K/k_n$ and (SN), we obtain that $|\mathbb{E}(\Phi(J,s)^n\mid\mathcal{H}_{(i-1)\Delta_n})|$ is smaller than K/k_n^{v+2r}, and vanishes if $J\in\mathcal{J}_s^n\setminus\cup_{u=1}^{r+[v/2]}D(u,s)^n$. Hence

$$\mathbb{E}\big(\big(\overline{\chi}(g)_i^n\big)^v\big(\widehat{\chi}(g)_i^n\big)^r\mid\mathcal{H}_{(i-1)\Delta_n}\big)=\sum_{u=1}^{r+[v/2]}\overline{\Phi}_u^n,\qquad\text{where}$$

$$\overline{\Phi}_u^n=\sum_{s=0}^{r}\overline{\Phi}(u,s)^n,\qquad\overline{\Phi}(u,s)^n=\sum_{J\in D(u,s)^n}\mathbb{E}\big(\Phi(J,s)^n\mid\mathcal{G}_i^n\big).$$

Moreover $\#D(u,s)^n \le K k_n^u$, so $|\overline{\Phi}_u^n| \le K k_n^{u-v-2r} \le K K k_n^{-r-1-v/2}$ if $u \le r-1+v/2$. We deduce that it is enough to prove that $\overline{\Phi}_u^n$ equals the right side of (16.5.41) for $u=r+v/2$ when v is even, and the right side of (16.5.43) for $u=r+v/2-1/2$ when v is odd.

Step 3) In this step we prove that $\overline{\Phi}_u^n$ equals the right side of (16.5.41) when v is even and $u=r+v/2$. By definition of $D(u,s)^n$, if $J \in D(u,s)^n$ there is an integer w such that the variable $\Phi(J,s)^n$ is the product of $\frac{v+s+r-w}{2}$ terms, of three types, all for different indices j for χ_j^n:

(1) $s-w+v/2$ terms as $(g_j'^n \chi_{i+j-2}^n)^2$ or $(g_j'^n \chi_{i+j-1}^n)^2$,
(2) w terms as $-2(g_j'^n)^3 g_{j+1}'^n (\chi_{i+j-2}^n \chi_{i+j-1}^n)^2$,
(3) $\frac{r-s-w}{2}$ terms as $4(g_j'^n)^4 (\chi_{i+j-2}^n \chi_{i+j-1}^n)^2$.

When $s<r$ the number of those terms is not bigger than $r+\frac{v}{2}-\frac{1}{2}$, hence than $r+\frac{v}{2}-1$ because it is an integer and v is even. Since the indices range from 1 to k_n we thus have $\#D(u,s)^n \le K k_n^{r+v/2-1}$ (this improves upon the bound given in Step 2), and we deduce

$$s<r \;\Rightarrow\; \left|\overline{\Phi}(u,s)^n\right| \le K/k_n^{r+v/2+1}. \tag{16.5.46}$$

It thus remains to prove that $\overline{\Phi}(u,r)^n$ is equal to the right side of (16.5.41). If $J \in D(u,r)^n$ then $\Phi(J,r)^n$ contains only terms of type (1). In fact $D(u,r)^n$ contains exactly the families $J \in \mathcal{J}_r^n$ for which $I(J)$ is the pairwise disjoint union $J_1 \cup J_2 \cup J_3$, where J_1 is the set of all distinct $j_1,\dots,j_v$ (there are $v/2$ of them, each one appearing twice), and $J_2=\{j_l'+j_l'': 1\le l\le r,\ j_l''=0\}$ and $J_3=\{j_l'+j_l'': 1\le l\le r,\ j_l''=1\}$ have distinct indices. With this notation, we have (with u terms all together in the products):

$$\mathbb{E}\big(\Phi(J,r)^n \mid \mathcal{H}_{(i-1)\Delta_n}\big) = \prod_{j\in J_1\cup J_2} \big(g_j'^n\big)^2 \Upsilon_{(i+j-2)\Delta_n}^n \prod_{j\in J_3} \big(g_{j-1}'^n\big)^2 \Upsilon_{(i+j-2)\Delta_n}^n. \tag{16.5.47}$$

Observe that when J ranges through $D(u,r)^r$, so $k_n \ge u$, then $I(J)$ ranges trough the set $\mathcal{L}_u^n$ of all subsets of $\{1,\dots,k_n\}$ having u points. If $L \in \mathcal{L}_u^n$ we use the notation

$$w(L) \;=\; \text{the number of all } J \in D(u,r)^n \text{ such that } I(J)=L.$$

This number does not depend on n, and can be evaluated as follows: we fix L. There are C_u^r many ways of choosing the two complementary subsets J_1 and $J_2\cup J_3$ of L. Next, with J_1 given, there are $(v/2)!\,(v-1)(v-3)\cdots 3\cdot 1$ ways of choosing the indices j_l so that $j_1,\dots,j_v$ contains $v/2$ paired distinct indices which are the indices in J_1, and we recall that $(v-1)(v-3)\cdots 3\cdot 1=m_v$ (if $v=0$ then J_1 is empty and there is $m_0=1$ ways again of choosing J_1). Finally with $J_2\cup J_3$ fixed, there are $2^r\,r!$ ways of choosing the indices $j_l'+j_l''$, all different, when the smallest index in

$J_2 \cup J_3$ is bigger than 1, and $2^{r-1} r!$ ways if this smallest index is 1. Summarizing, and with $\mathcal{L}_u'^n$ denoting the subset of all $L \in \mathcal{L}_u^n$ such that $1 \notin L$, we get

$$L \in \mathcal{L}_u^n \Rightarrow w(L) \leq m_v 2^r u!, \qquad L \in \mathcal{L}_u'^n \Rightarrow w(L) = m_v 2^r u!. \tag{16.5.48}$$

Now we come back to $\Phi(J,r)^n$. The property (16.1.6) yields $|g_j'^n - g_{j-1}'^n| \leq K/k_n^2$, except for j belonging to the set Q_n of indices for which g' fails to exist or to be Lipschitz on the interval $[(j-1)/k_n, jk_n]$, so $\#Q_n \leq K$. Since Υ is bounded, we thus have

$$\mathbb{E}\big(\Phi(J,r)^n \mid \mathcal{H}_{(i-1)\Delta_n}\big)$$
$$= \begin{cases} \prod_{j\in I(J)} (g_j'^n)^2 \Upsilon_{(i+j-2)\Delta_n)} + \mathrm{O}_u(k_n^{-2u-1}) & \text{if } Q_n \cap I(J) = \emptyset \\ \mathrm{O}_u(k_n^{-2u}) & \text{always.} \end{cases}$$

Since $\#D(u,r)^n \leq Kk_n^u$ and $\sup_n \#Q_n < \infty$, the number of $J \in D(u,r)^n$ such that $Q_n \cap I(J) \neq \emptyset$ is smaller than Kk_n^{u-1}, and we also have $\#(\mathcal{L}_u^n \backslash \mathcal{L}_u'^n) \leq k_n^{u-1}$. Hence we deduce from the above that

$$\overline{\Phi}(u,r)^n = \sum_{L\in\mathcal{L}_u'^n} w(L) \prod_{j\in L} (g_j'^n)^2 \Upsilon_{(i+j-2)\Delta_n} + \mathrm{O}_u\Big(\frac{1}{k_n^{u+1}}\Big). \tag{16.5.49}$$

On the other hand (16.1.6) and (16.5.40) yield

$$\big(\widehat{\Upsilon}(g)_i^n\big)^u = u! \sum_{L\in\mathcal{L}_u^n} \prod_{j\in L} (g_j'^n)^2 \Upsilon_{(i+j-2)\Delta_n} + \mathrm{O}_u\Big(\frac{1}{k_n^{u+1}}\Big)$$
$$= u! \sum_{L\in\mathcal{L}_u'^n} \prod_{j\in L} (g_j'^n)^2 \Upsilon_{(i+j-2)\Delta_n} + \mathrm{O}_u\Big(\frac{1}{k_n^{u+1}}\Big). \tag{16.5.50}$$

Therefore by (16.5.48), and comparing (16.5.49) and (16.5.50), we deduce that $\Phi(u,r)^n$ is equal to $m_r 2^r (\widehat{\Upsilon}(g)_i^n)^u + \mathrm{O}_u(1/k_n^{u+1})$. So the proof of (16.5.41) is complete.

Step 4) Finally we prove that $\overline{\Phi}_u^n$ equals the right side of (16.5.43) when v is odd and $u = r + v/2 - 1/2$. If $J \in D(u,s)^n$, there is a number w' in $\{0,1\}$ and a nonnegative integer w as in Step 4, such that $\Phi(J,s)^n$ is the product of $\frac{v+s+r-w-1}{2}$ terms, all for different indices j for χ_j^n, with $s - w + w' + \frac{v-3}{2}$ terms of type 1, w terms of type 2, $\frac{r-s-w-2w'}{2}$ terms of type 3, and $1-w'$ and w' terms respectively of the types (4) and (5) described below:

(4) terms as $(g_j'^n \chi_{i+j-2}^n)^3$ or $(g_j'^n)^2 g_{j+1}'^n (\chi_{i+j-1}^n)^3$,
(5) terms as $-2(g_j'^n)^4 g_{j+1}'^n (\chi_{i+j-2}^n)^3 (\chi_{i+j-1}^n)^2$,
or as $-2(g_j'^n)^3 (g_{j+1}'^n)^2 (\chi_{i+j-2}^n)^2 (\chi_{i+j-1}^n)^3$,

the whole product being multiplied by -1. It follows that $\#D(u,s)^n \leq K/k_n^{(v+r+s-1)/2}$ and $\mathbb{E}(|\Phi(J,s)^n| \mid \mathcal{H}_{(i-1)\Delta_n}) \leq K/k_n^{v+2r}$ still holds. Hence, instead of (16.5.46) we get $|\overline{\Phi}(u,s)^n| \leq K/k_n^{r+v/2+1+(r-s-1)/2}$, and it thus enough to prove that $\overline{\Phi}(u,r)^n$ is equal to the right side of (16.5.43).

If $J \in D(u,r)^n$ then $\Phi(J,n)$ has $u-1$ terms of type (1) and one of type (4) (one has $w=w'=0$). Therefore, $D(u,r)^n$ contains exactly the families $J \in \mathcal{J}_u^n$ with the following properties: there is some $p \in \{1,\dots,v\}$ such that either $j_p = j_{p'} = j_{p''}$ for p, p', p'' distinct (case (a), with $v \geq 3$, and we set $(l,l') = (j_p, 0)$), or $j_p = j'_q + j''_q$ for some $q \in \{1,\dots,r\}$ (case (b), with $r \geq 1$, and we set $(l,l') = (j'_q, j''_q)$); moreover $I(J)$ is the pairwise disjoint union $J'_0 \cup J'_1 \cup J'_2 \cup J'_3$, where $J'_0 = \{l+l'\}$ and J'_1 is like in J_1 in Step 3, except that $v/2$ is substituted with $v/2-1/2$ and j_p is omitted in case (b), and with $v/2-3/2$ and $j_p, j_{p'}, j_{p''}$ omitted in case (a), and with J'_2 and J'_3 are like J_1 and J_2 except that r is substituted with $r-1$ and $j'_q + j''_q$ is omitted in case (b). Therefore, instead of (16.5.47) we have

$$\begin{aligned}\mathbb{E}\big(\Phi(J,r)^n \mid \mathcal{H}_{(i-1)\Delta_n}\big) = &-\big(g_l'^n\big)^2 g_{l+l'}'^n\, \Upsilon'_{(i+l+l'-2)\Delta_n} \\ &\times \prod_{j\in J'_1\cup J'_2} \big(g_j'^n\big)^2 \Upsilon_{(i+j-2)\Delta_n} \prod_{j\in J'_3} \big(g_{j-1}'^n\big)^2 \Upsilon_{(i+j-2)\Delta_n}.\end{aligned} \tag{16.5.51}$$

Now if $j \in \{1,\dots,k_n\}$ we denote by $\mathcal{L}^n_{m,u-1}$ the set of all $L \in \mathcal{L}^n_{u-1}$ which do not contain m, and for such an L we set

$$w'(L,j) = \text{the number of } J \in D(u,r)^n \text{ for which } l=j \text{ and } J'_1 \cup J'_2 \cup J'_3 = L.$$

Then by the same argument as for (16.5.48) we obtain

$$\begin{aligned}&L \in \mathcal{L}^n_{j,u-1} \quad \Rightarrow \quad w'(L,j) \leq m_{v+1}\, 2^r\, (u-1)!\, \frac{6r+v-1}{6}\\ &L \in \mathcal{L}'^n_{j,u-1},\ j \neq 1,\ 1 \notin L \ \Rightarrow \ w'(L,j) = m_{v+1}\, 2^r\, (u-1)!\, \frac{6r+v-1}{6}.\end{aligned}$$

Below, we denote by R_n a quantity changing from line to line, but which is similar to the last term in (16.5.43) with $A=1$, that is $O_u(\frac{1}{k_n^{u+1}}(\frac{1}{k_n} + \Gamma(\Upsilon,1)_i^n + \Gamma(\Upsilon',1)_i^n))$. We also set $\overline{\Upsilon}_i^n = \Upsilon'_{(i-1)\Delta_n}(\Upsilon_{(i-1)\Delta_n})^{u-1}$. Coming back to $\Phi(u,r)^n$, we deduce from (16.5.51) and from the properties of g that

$$\begin{aligned}&\mathbb{E}\big(\Phi(J,r)^n \mid \mathcal{H}_{(i-1)\Delta_n}\big)\\ &\quad = \begin{cases} -\overline{\Upsilon}_i^n \prod_{j\in J'_1\cup J'_2\cup J'_3} (g_j'^n)^2 + R_n/k_n^u & \text{if } Q_n \cap I(J) = \emptyset\\ R_n/k_n^{u-1} & \text{otherwise.}\end{cases}\end{aligned}$$

Then, exactly as for (16.5.49), we deduce

$$\begin{aligned}\overline{\Phi}(u,r)^n &= -\overline{\Upsilon}_i^n \sum_{j=1}^{k_n} (g_j'^n)^3 \sum_{L\in\mathcal{L}^n_{m,u-1}} w'(L,m) \prod_{l\in L} (g_l'^n)^2 + R_n \\ &= -m_{v+1}\, 2^r\, (u-1)!\, \frac{6r+v-1}{6}\, \overline{\Upsilon}_i^n \sum_{j=1}^{k_n} (g_j'^n)^3 \sum_{L\in\mathcal{L}^n_{u-1}} \prod_{l\in L} (g_l'^n)^2 + R_n \\ &= -m_{v+1}\, 2^r\, \frac{6r+v-1}{6}\, \overline{\Upsilon}_i^n \sum_{j=1}^{k_n} (g_j'^n)^3 \Big(\sum_{l=1}^{k_n} (g_l'^n)^2\Big)^{u-1} + R_n.\end{aligned}$$

Finally we have (16.1.9) and also $\sum_{j=1}^{k_n} (g_j'^n)^3 = k_n^{-2} \int_0^1 g'(s)^3\, ds + \mathrm{O}(k_n^{-3})$. Then $\overline{\Phi}(u,r)^n$ is equal to the right side of (16.5.43). □

Now we apply the previous lemma to the variables $\phi(g,p)^n_{i,r}$ of (16.5.38). We use below the notation (16.5.39), and also

$$\begin{aligned}\overline{\beta}(g)^n_{i,r} &= \frac{1}{\sqrt{u_n}}\, \sigma_{(i-r-1)\Delta_n} \overline{W}(g)^n_i \\ \Psi(g,p)^n_{i,r} &= \mathbb{E}\big(\phi(g,p)^n_{i,r} \mid \mathcal{H}_{(i-1)\Delta_n}\big) - \big(\overline{\beta}(g)^n_{i,r}\big)^{2p}.\end{aligned} \tag{16.5.52}$$

Lemma 16.5.12 *Assume* (SN) *and* σ_t *bounded. Then*

$$\big|\mathbb{E}\big(\Psi(g,p)^n_{i,r} \mid \mathcal{F}_{(i-1)\Delta_n}\big)\big| \le \frac{K}{k_n} + \frac{K\Delta_n^{\eta+\eta'-1/2}}{\sqrt{k_n}} \big(\Gamma'(\Upsilon)^n_i + \Gamma'(\Upsilon')^n_i\big), \tag{16.5.53}$$

$$\mathbb{E}\big(\big|\Psi(g,p)^n_{i,r}\big|^2 \mid \mathcal{F}_{(i-1)\Delta_n}\big) \le K/k_n. \tag{16.5.54}$$

Proof The index i is fixed, and for simplicity we write $\rho_n = \Delta_n^{2\eta}/u_n$, and also

$$A_n(u,v) = \mathbb{E}\big(\big(\overline{\chi}(g)^n_i\big)^u \big(\widehat{\chi}(g)^n_i\big)^v \mid \mathcal{H}_{(i-1)\Delta_n}\big).$$

In view of (16.5.38) and since $\overline{\beta}(g)^n_{i,r}$ is $\mathcal{H}_{(i-1)\Delta_n}$ measurable, we have

$$\Psi(g,p)^n_{i,r} = \sum_{l=0}^{p} \zeta_{p,l} \sum_{w=0}^{2p-2l} C^w_{2p-2l} \big(\overline{\beta}(g)^n_{i,r}\big)^w \rho_n^{p-w/2} A_n(2p-2l-w,l) - \big(\overline{\beta}(g)^n_i\big)^{2p}.$$

By (16.5.1) and a change of the order of summation, we easily get

$$\sum_{l=0}^{p} \sum_{u=0}^{p-l} C^{2u}_{2p-2l}\, \zeta_{p,l}\, 2^l\, m_{2p-2l-2u} \big(\overline{\beta}(g)^n_{i,r}\big)^{2u} \big(\rho_n \widehat{\Upsilon}(g)^n_{i,r}\big)^{p-v} = \big(\overline{\beta}(g)^n_{i,r}\big)^{2p},$$

hence $\Psi(g,p)_i^n = B_n + B'_n$, where

$$B_n = \sum_{l=0}^{p}\sum_{u=0}^{p-l} C_{2p-2l}^{2u}\,\zeta_{p,l}\big(\overline{\beta}(g)_{i,r}^n\big)^{2u}\,\rho_n^{p-u}$$
$$\times\big(A_n(2p-2l-2u,l) - 2^l\,m_{2p_j-2l-2u}\big(\widehat{\Upsilon}(g)_{i,r}^n\big)^{p-u}\big)$$
$$B'_n = \sum_{l=0}^{p}\sum_{u=0}^{p-l-1} C_{2p-2l}^{2u+1}\,\zeta_{p,l}\big(\overline{\beta}(g)_{i,r}^n\big)^{2u+1}\,\rho_n^{p-u-1/2}\,A_n(2p-2l-2u-1,l).$$

It is then enough to show that both B_n and B'_n satisfy (16.5.53) and (16.5.54). For this we apply Lemma 16.5.13. First, (16.5.41) yields

$$B_n = \sum_{u=0}^{p}\big(\overline{\beta}(g)_{i,r}^n\big)^{2u}\,\mathrm{O}_u\big(\rho_n^{p-u}/k_n^{p+1-u}\big) = \sum_{u=0}^{p}\big(\overline{\beta}(g)_{i,r}^n\big)^{2u}\,\mathrm{O}_u(1/k_n),$$

where the last estimate comes from the fact that $\rho_n/k_n \le K$. Since we have $\mathbb{E}((\overline{\beta}(g)_{i,r}^n)^q \mid \mathcal{F}_{(i-1)\Delta_n}) \le K_q$ by (16.4.8), we obtain $\mathbb{E}(|B_n|^q \mid \mathcal{F}_{(i-1)\Delta_n}) \le K_q/k_n^q$ for all $q>0$ and the two results are proved for B_n.

Now we turn to B'_n. The same argument, now based on (16.5.44), shows that (16.5.54) is satisfied. For (16.5.53) we use (16.5.43), which yields $B'_n = \overline{B}_n + \overline{B}'_n$, where $\overline{B}_n = \sum_{u=0}^{p}\gamma_n^{2u+1}\,\alpha_i^n$ for some $\mathcal{F}_{(i-1)\Delta_n}$ measurable variables α_i^n, and

$$\overline{B}'_n = \sum_{u=0}^{p-1}\big(\overline{\beta}(g)_{i,r}^n\big)^{2u+1}\,\mathrm{O}_u\bigg(\frac{\rho_n^{p-u-1/2}}{k_n^{p-u}}\Big(\frac{1}{k_n} + \Gamma(\Upsilon)_i^n + \Gamma(\Upsilon')_i^n\Big)\bigg).$$

Again, the same argument, plus the Cauchy-Schwarz inequality and the property that $\rho_n/k_n \le K\Delta_n^{2\eta+2\eta'-1} \le K$, imply that $\mathbb{E}(|\overline{B}'_n| \mid \mathcal{F}_{(i-1)\Delta_n})$ is smaller than the right side of (16.5.53). Finally, $\mathbb{E}((\overline{\beta}(g)_{i,r}^n)^q \mid \mathcal{F}_{(i-1)\Delta_n}) = 0$ for all odd integers q because, conditionally on $\mathcal{F}_{(i-1)\Delta_n}$ the variable $\overline{W}(g)_{i,r}^n$ has a law which is symmetrical about 0. Hence $\mathbb{E}(\overline{B}_n \mid \mathcal{F}_{(i-1)\Delta_n}) = 0$, and this finishes the proof. □

In the next lemma, we compare $\phi(g,p,Z^n)_i^{n,j}$ and $\phi(g,p)_i^n$.

Lemma 16.5.13 *Assume* (SH) *with X continuous, and* (SN) *and $\eta+\eta' \ge \frac{1}{2}$. Then for all $q>0$ we have*

$$\mathbb{E}\big(|\phi(g,p,Z^n)_i^n|^q + |\phi(g,p)_{i,0}^n|^q \mid \mathcal{F}_{(i-1)\Delta_n}\big) \le K_q \tag{16.5.55}$$
$$\mathbb{E}\big(|\phi(g,p,Z^n)_i^{n,j} - \phi(g,p)_{i,0}^n|^2 \mid \mathcal{F}_{(i-1)\Delta_n}\big) \le K\big(u_n + \big(\Gamma'(\sigma)_i^n\big)^2 + \Delta_n^{4\eta'}\big). \tag{16.5.56}$$

Proof First, (16.4.9), (16.4.11), (16.2.3) and $\Delta_n^{2\eta} \le K k_n u_n$ yield for all $q > 0$:

$$\begin{aligned}
&\mathbb{E}\big(\big|\overline{Z^n}(g)^n_i/\sqrt{u_n}\,\big|^q \mid \mathcal{F}_{(i-1)\Delta_n}\big) + \mathbb{E}\big(\big|\widehat{Z^n}(g)^n_i/u_n\big|^q \mid \mathcal{F}_{(i-1)\Delta_n}\big) \le K_q \\
&\mathbb{E}\big(\big|\overline{\kappa}(g)^n_i/\sqrt{u_n}\,\big|^q \mid \mathcal{F}_{(i-1)\Delta_n}\big) + \mathbb{E}\big(\big|\Delta_n^{2\eta}\widehat{\chi}(g)^n_i/u_n\big|^q \mid \mathcal{F}_{(i-1)\Delta_n}\big) \le K_q
\end{aligned} \tag{16.5.57}$$

because $w = 0$ here. Next, (16.4.12) gives

$$\mathbb{E}\big(\big|\widehat{Z^n}(g)^n_i/u_n - \Delta_n^{2\eta}\,\widehat{\chi}(g)^n_i/u_n\big|^q \mid \mathcal{F}_{(i-1)\Delta_n}\big) \le K\Delta_n^{2q\eta'}.$$

Moreover we know that, with $g_n(s) = \sum_{j=1}^{k_n} g_j^n 1_{((j-1)\Delta_n, j\Delta_n]}(s)$ (hence $|g_n(s)| \le K$),

$$\overline{Z^n}(g)^n_i - \overline{\kappa}(g)^n_{i,0} = \int_{(i-1)\Delta_n}^{(i-1)\Delta_n + u_n} g_n\big(s - (i-1)\Delta_n\big)\big(b_s\,ds + (\sigma_s - \sigma_{(i-1)\Delta_n})\,dW_s\big).$$

Then we deduce from (2.1.33), (2.1.34) and the boundedness of b_t and σ_t that

$$\mathbb{E}\big(\big|\big(\overline{Z^n}(g)^n_{i,0} - \overline{\kappa}(g)^n_i\big)/\sqrt{u_n}\,\big|^4 \mid \mathcal{F}_{(i-1)\Delta_n}\big) \le K_q\, u_n^2\big(u_n^2 + \big(\Gamma'(\sigma)^n_i\big)^4\big).$$

These estimates and Hölder's inequality give for $l = 0, \ldots, p$:

$$\begin{aligned}
&\mathbb{E}\left(\left|\left(\frac{\overline{Z^n}(g)^n_i}{\sqrt{u_n}}\right)^{2p-2l}\left(\frac{\widehat{Z^n}(g)^n_i}{u_n}\right)^l\right|^q \mid \mathcal{F}_{(i-1)\Delta_n}\right) \le K_p \\
&\mathbb{E}\left(\left|\left(\frac{\overline{Z^n}(g^j)^n_i}{\sqrt{u_n}}\right)^{2p-2l}\left(\frac{\widehat{Z^n}(g)^n_i}{u_n}\right)^l - \left(\frac{\overline{\kappa}(g)^n_i}{\sqrt{u_n}}\right)^{2p-2l}\left(\frac{\Delta_n^{2\eta}\overline{\chi}(g)^n_i}{u_n}\right)^l\right|^2 \mid \mathcal{F}_{(i-1)\Delta_n}\right) \\
&\quad \le K\big(u_n + \big(\Gamma'(\sigma)^n_i\big)^2 + \Delta_n^{4\eta'}\big)
\end{aligned}$$

in view of (16.5.37) and (16.5.38), we deduce (16.5.55) and (16.5.56). □

Unfortunately, (16.5.56) is still not enough for us, and we need a more sophisticated estimate, which uses the full force of (SK).

Lemma 16.5.14 *Assume* (SK), (SN) *and* $\eta + \eta' \ge \frac{1}{2}$. *Then*

$$\begin{aligned}
&\big|\mathbb{E}\big(\phi(g,p,Z^n)^n_i - \phi(g,p)^n_{i,0} \mid \mathcal{F}_{(i-1)\Delta_n}\big)\big| \le K\sqrt{u_n}\,\delta^n_i, \quad \textit{where} \\
&\delta^n_i = \sqrt{u_n} + \Delta_n^{\eta'/2} + \Delta_n^{2\eta'-1/2} + \Gamma'(b)^n_i + \Gamma'(\widetilde{\sigma})^n_i + \Gamma'(\Upsilon)^n_i.
\end{aligned} \tag{16.5.58}$$

Proof 1) In view of (16.5.37) and (16.5.38), it is enough to prove that for $l = 0, \ldots, p$,

$$\begin{aligned}
&\big|\mathbb{E}\big(\big(\overline{Z^n}(g)^n_i\big)^{2p-2l}\big(\widehat{Z^n}(g)^n_i\big)^l - \big(\overline{\kappa}(g)^n_{i,0}\big)^{2p-2l}\big(\Delta_n^{2\eta}\widehat{\chi}(g)^n_i\big)^l \mid \mathcal{F}_{(i-1)\Delta_n}\big)\big| \\
&\quad \le K u_n^{p+1/2}\,\delta^n_i.
\end{aligned}$$

We fix i and l and use the simplifying notation $S=(i-1)\Delta_n$. It is enough to show that, for $r=1,2,3$,

$$\left|\mathbb{E}\left(F_r^n \mid \mathcal{F}_S\right)\right| \leq K u_n^{p+1/2}\,\delta_i^n, \tag{16.5.59}$$

where

$$F_r^n = \begin{cases} (\overline{\kappa}_{i,0}^n)^{2p-2l}((\widehat{Z^n}(g)_i^n)^l - (\Delta_n^{2\eta}\widehat{\chi}(g)_i^n)^l) & \text{if } r=1 \\ (\Delta_n^{2\eta}\widehat{\chi}(g)_i^n)^l((\overline{Z^n}(g)_i^n)^{2p-2l} - (\overline{\kappa}(g)_i^n)^{2p-2l}) & \text{if } r=2 \\ ((\overline{Z^n}(g)_i^n)^{2p-2l} - (\overline{\kappa}(g)_{i,0}^n)^{2p-2l})((\widehat{Z^n}(g)_i^n)^l - (\Delta_n^{2\eta}\widehat{\chi}(g)_i^n)^l) & \text{if } r=3. \end{cases}$$

When $r=1,3$ we have $F_r^n=0$ if $l=0$, and $|\mathbb{E}(F_r^n \mid \mathcal{F}_S)| \leq K u_n^p \Delta_n^{2\eta'}$ if $l\geq 1$ (apply (16.4.12) with $w=0$ and (16.5.57) and the Cauchy-Schwarz inequality), hence (16.5.59) holds.

2) It remains to consider the case $r=2$, with $l<p$ because $F_2^n=0$ when $l=p$. Recalling (16.5.33), we have $\overline{Z^n}(g)_i^n = \overline{\kappa}(g)_{i,0}^n + \overline{\lambda}_n$, where $\overline{\lambda}_n = \overline{X}(g)_i^n - \sigma_{(i-1)\Delta_n}\overline{W}(g)_i^n$, hence

$$F_2^n = \sum_{u=1}^{2p-2l} C_{2p-2l}^u\, G_u^n, \qquad G_u^n = \left(\Delta_n^{2\eta}\widehat{\chi}(g)_i^n\right)^l \left(\overline{\kappa}(g)_{i,0}^n\right)^{2p-2l-u} (\overline{\lambda}_n)^u,$$

and we prove (16.5.59) separately for each G_u^n. First, we have $\overline{\lambda}_n = \xi_n + \xi_n'$, where

$$\xi_n = \int_S^{S+u_n} g_n(s-S)\left((b_s - b_S)\,ds + \left(\int_S^s \left(\widetilde{b}_r\,dr + (\widetilde{\sigma}_r - \widetilde{\sigma}_S)\,dW_r\right)\right) dW_s\right),$$

$$\xi_n' = \int_S^{S+u_n} g_n(s-S)\left(b_S\,ds + \widetilde{\sigma}_S(W_s - W_S)\,dW_s + (M_s - M_S)\,dW_s\right),$$

with g_n as in the previous proof. Then for $q\geq 1$ and since $|g_n(s)|\leq K$ and $b_t, \widetilde{b}_t, \widetilde{\sigma}_t, a_t$ are bounded, where $\langle M,M\rangle_t = \int_0^t a_s\,ds$, we have for $q\geq 2$:

$$\begin{aligned} \mathbb{E}\left(|\xi_n|^q \mid \mathcal{F}_S\right) &\leq K_q u_n^q \left(u_n^{q/2} + \left(\Gamma'(b)_i^n\right)^2 + \left(\Gamma'(\widetilde{\sigma})_i^n\right)^2\right) \\ \mathbb{E}\left(|\xi_n'|^q \mid \mathcal{F}_S\right) &\leq K_q u_n^{q/2+1}, \qquad \mathbb{E}\left(|\overline{\lambda}_n|^q \mid \mathcal{F}_S\right) \leq K_q u_n^{q/2+1}. \end{aligned} \tag{16.5.60}$$

3) Next we prove that, for u an odd integer,

$$\mathbb{E}\left(\left(\overline{W}(g)_i^n\right)^u \xi_n' \mid \mathcal{F}_S\right) = 0. \tag{16.5.61}$$

We prove this separately for each of the three terms constituting ξ_n'. Since $x\mapsto x^u$ is an odd function, this is obvious for the term involving b_S, and also for the term involving $\widetilde{\sigma}_i^n$ (in both cases the corresponding variable whose conditional expectation is taken is an odd function of the path of $s\mapsto W_{S+s}-W_S$ and is integrable).

For the term involving M, we have $(\overline{W}(g)_i^n)^u = Y + \int_S^{S+u_n} \gamma_s \, dW_s$ for some $\mathcal{F}_S$ measurable variable Y and a process γ adapted to the filtration $(\mathcal{F}_t^W)$ generated by the Brownian motion. Since this term is a martingale increment we are left to prove $\mathbb{E}(U_{S+u_n} \mid \mathcal{F}_S) = 0$, where

$$U_t = \left(\int_S^t \gamma_s \, dW_s \right) \left(\int_S^{S+u_n} g_n(s-S)(M_s - M_S) \, dW_s \right).$$

Itô's formula yields $U_t = M'_t + \int_S^t g_n(s-S)\gamma_s(M_s - M_S)\, ds$ for $t \geq S$, where M' is a martingale with $M'_S = 0$, so it is enough to prove that

$$\mathbb{E}\big(\gamma_t(M_t - M_S) \mid \mathcal{F}_S\big) = 0.$$

For any fixed $t \geq T$ we again have $\gamma_t = Y'_t + \int_S^t \gamma'_s \, dW_s$ where Y'_t is $\mathcal{F}_S$ measurable. Hence the above follows from the orthogonality of W and M, and we have (16.5.61).

4) At this stage, we use the form of G_u^n as a product of three terms at the respective powers l, $r = 2p - 2l - u$ and u. Hölder's inequality with the respective exponents l', r', u' with $\frac{1}{l'} + \frac{1}{r'} + \frac{1}{u'} = 1$ and (16.5.57) and (16.5.60) yield $\mathbb{E}(|G_u^n| \mid \mathcal{F}_S) \leq K u_n^{p+u''}$, where $u'' = (u/2) \wedge (1/u')$. When $u \geq 2$ we can choose $u'' = 1/u' = 1 - \varepsilon$, hence (16.5.59) holds for G_u^n when $u \geq 2$.

It remains to study G_1^n, which is the sum $G'_n + G''_n$, where

$$G'_n = \big(\Delta_n^{2\eta} \widehat{\chi}(g)_i^n\big)^l \, \big(\overline{\kappa}(g)_{i,0}^n\big)^{2p-2l-1} \, \xi_n$$

$$G''_n = \big(\Delta_n^{2\eta} \widehat{\chi}(g)_i^n\big)^l \, \big(\overline{\kappa}(g)_{i,0}^n\big)^{2p-2l-1} \, \xi'_n.$$

By (16.5.57) and (16.5.60) and Hölder's inequality as above, we get $\mathbb{E}(|G'_n| \mid \mathcal{F}_S) \leq K u_n^{p+1/2} \delta_i^n$. Finally $G''_n = \sum_{w=0}^{2p-2l-1} C_{2p-2l-1}^w a_w^n$, where

$$a_w^n = \big(\Delta_n^{2\eta} \widehat{\chi}(g)_i^n\big)^l \, \big(\sigma_S \overline{W}_i^n\big)^{2p-2l-1-w} \, \big(\Delta_n^{\eta} \overline{\chi}(g)_i^n\big)^w \, \xi'_n.$$

By successive conditioning, (16.5.44), (16.5.57) and (16.5.60) yield that $\mathbb{E}(|a_w^n| \mid \mathcal{F}_S)) \leq K u_n^{p+1/2} \Delta_n^{\eta'/2}$ when w is odd. When w is even, the same argument with (16.5.42), plus (16.5.61) and the fact that $p - 2l - 1 - w$ is then odd yields

$$\big|\mathbb{E}\big(a_w^n \mid \mathcal{F}_S\big)\big| \leq K u_n^{p+1/2} \big(\Delta_n^{\eta} + \Gamma'(\Upsilon)_i^n\big).$$

If we put together all these estimates, we deduce that G_1^n also satisfies (16.5.59), and this finishes the proof. □

Our last result is about asymptotically negligible arrays, and more specifically those which satisfy

$$0 \leq \delta_i^n \leq K, \qquad \Delta_n \, \mathbb{E}\left(\sum_{i=1}^{[t/\Delta_n]} \delta_i^n \right) \to 0 \;\; \forall t > 0. \tag{16.5.62}$$

Lemma 16.5.15 a) *If an array* (δ_i^n) *satisfies* (16.5.62) *then for any* $q>0$ *the array* $(|\delta_i^n|^q)$ *also satisfies* (16.5.62).

b) *If* Y *is a càdlàg bounded process, then the two arrays* $(\Gamma(Y)_i^n)$ *and* $(\Gamma'(Y)_i^n)$ *satisfy* (16.5.62).

Proof a) If $q>1$, we have $\sum_{i=1}^{[t/\Delta_n]} |\delta_i^n|^q \leq K \sum_{i=1}^{[t/\Delta_n]} \delta_i^n$, so $(|\delta_i^n|^q)$ satisfies (16.5.62). When $q<1$ we have by a repeated use of Hölder's inequality:

$$\left(\Delta_n \mathbb{E}\left(\sum_{i=1}^{[t/\Delta_n]} \delta_i^n\right)\right)^{1/q} \leq \Delta_n^{1/q} \mathbb{E}\left(\left(\sum_{i=1}^{[t/\Delta_n]} \delta_i^n\right)^{1/q}\right)$$
$$\leq \Delta_n^{1/q} \mathbb{E}\left(\Delta_n^{1-1/q} \sum_{i=1}^{[t/\Delta_n]} \delta_i^n\right).$$

Then again $(|\delta_i^n|^q)$ satisfies (16.5.62).

b) Let $\delta_i^n = \Gamma(Y)_i^n$. If $\varepsilon>0$, let $N(\varepsilon)_t$ be the number of jumps of Y with size bigger than ε on the interval $[0,t]$, and let $\upsilon(\varepsilon,t,\eta)$ be the supremum of $|Y_s - Y_r|$ over all pairs (r,s) with $s \leq r \leq s+\eta$ and $s \leq t$ and such that $N(\varepsilon)_s - N(\varepsilon)_r = 0$. Since Y is bounded,

$$\Delta_n \mathbb{E}\left(\sum_{i=1}^{[t/\Delta_n]} \delta_i^n\right) \leq \mathbb{E}\big(t\, \upsilon(\varepsilon, t+1, 2u_n) + (Kt) \wedge \big(K\, u_n\, N(\varepsilon)_{t+1}\big)\big)$$

as soon as $2u_n \leq 1$. Since $\limsup_{n\to\infty} \upsilon(\varepsilon, t+1, 2u_n) \leq \varepsilon$, Fatou's lemma implies that the limsup of the left side above is smaller than $Kt\varepsilon$, so (16.5.62) holds because ε is arbitrarily small.

By (a), the array $(\Gamma(Y)_i^n)^4$ satisfies (16.5.62). Since $\mathbb{E}((\Gamma'(Y)_i^n)^4) = \mathbb{E}((\Gamma(Y)_i^n)^4)$, we deduce that the array $(\Gamma'(Y)_i^n)^4$ also satisfies (16.5.62), and another application of (a) implies that $\Gamma'(Y)_i^n$ satisfies (16.5.62) as well. □

16.5.4 Proof of Theorem 16.5.7

At this stage, the proof is analogous to the proof of Theorem 12.2.1, under the simplest assumption (a) of this theorem because we have (K) and the test function f is a polynomial. As before, we may assume (SK) and (SN).

Step 1) To begin with, and in view of the specific form of f and of (16.5.37), (16.5.38) and (16.5.52), we have the decomposition

$$\overline{Y}'^n(Z^n) = \sum_{l=1}^{4} \overline{H}(l)^n, \qquad \text{where}$$

$$\overline{H}(1)_t^{n,j} = \frac{\Delta_n}{\sqrt{u_n}} \sum_{i=1}^{[t/\Delta_n]-k_n+1} \left(\phi(g^j, p_j)_{i,0}^n - \mathbb{E}\left(\phi(g^j, p_j)_{i,0}^n \mid \mathcal{F}_{(i-1)\Delta_n}\right)\right)$$

$$\overline{H}(2)_t^{n,j} = \frac{\Delta_n}{\sqrt{u_n}} \sum_{i=1}^{[t/\Delta_n]-k_n+1} \left(\phi(g^j, p_j, Z^n)_i^n - \phi(g^j, p_j)_{i,0}^n\right)$$

$$\overline{H}(3)_t^{n,j} = \frac{\Delta_n}{\sqrt{u_n}} \sum_{i=1}^{[t/\Delta_n]-k_n+1} \mathbb{E}\left(\Psi(g^j, p_j)_{i,0}^n \mid \mathcal{F}_{(i-1)\Delta_n}\right)$$

$$\overline{H}(4)_t^n = \frac{1}{\sqrt{u_n}} \left(\Delta_n \sum_{i=1}^{[t/\Delta_n]-k_n+1} \mathbb{E}\left(|\overline{\beta}_i^{n,j}|^{2p_j} \mid \mathcal{F}_{(i-1)\Delta_n}\right) - m_{2p_j} \Lambda(g^j)^{p_j} \int_0^t c_s^{p_j}\, ds\right).$$

Then the theorem will follow from the next two lemmas:

Lemma 16.5.16 *Under* (SK) *and* (SN) *and* $\eta' > \frac{1}{3}$ *the processes* $\overline{H}(1)^n$ *converge stably in law to the process* $\overline{Y}'$ *described in Theorem* 16.5.7, *page* 533.

Lemma 16.5.17 *Under* (SK) *and* (SN) *and if* $\eta' > \frac{1}{3}$ *and* $\eta + \eta' \geq \frac{1}{2}$ *we have* $\overline{H}(l)^n \stackrel{\text{u.c.p.}}{\Longrightarrow} 0$ *for* $l = 2, 3, 4$.

Step 2)

Proof of Lemma 16.5.17 We begin with $\overline{H}(2)^n \stackrel{\text{u.c.p.}}{\Longrightarrow} 0$, and we set

$$\zeta_i^n = \frac{\Delta_n}{\sqrt{u_n}} \left(\phi(g^j, p_j, Z^n)_i^{n,j} - \phi(g^j, p_j)_{i,0}^n\right)$$

$$\zeta_i'^n = \mathbb{E}\left(\zeta_i^n \mid \mathcal{F}_{(i-1)\Delta_n}\right), \qquad \zeta_i''^n = \zeta_i^n - \zeta_i'^n.$$

Thus the result follows if the two arrays $(\zeta_i'^n)$ and $(\zeta_i''^n)$ are asymptotically negligible. For $(\zeta_i'^n)$ this is easy: indeed (16.5.58) yields $\mathbb{E}(\|\zeta_i'^n\|) \leq K\Delta_n \mathbb{E}(\delta_i^n)$, so Lemma 16.5.15 and our assumptions on $b, \widetilde{\sigma}, \Upsilon$ yield that $\sum_{i=1}^{[t/\Delta_n]} \mathbb{E}(\|\zeta_i'^n\|) \to 0$ as soon as $\eta' > \frac{1}{5}$.

For $(\zeta_i''^n)$ we operate as in Step 3 of Sect. 12.2.3. We observe that $\sum_{i=1}^{[t/\Delta_n]-k_n+1} \times \zeta_i''^n = \sum_{r=0}^{k_n-1} H(2,r)_t^n$ where, with the notation $l_n(r,t) = [([t/\Delta_n]-r)/k_n] - 1$, we take $H(2,r)_t^n = \sum_{i=0}^{l_n(r,t)} \zeta_{ik_n+r+1}''^n$. The variables $\zeta_{ik_n+r+1}''^n$ are martingale increments for the discrete time filtration $(\mathcal{F}_{((i+1)k_n+r)\Delta_n})_{i\geq 0}$. Therefore by Doob's inequality

$$\mathbb{E}\left(\sup_{s\leq t} \left\| \sum_{i=0}^{[s/\Delta_n]-k_n+1} \zeta_i''^n \right\|^2\right) \leq 4\mathbb{E}\left(\sum_{i=0}^{l_n(r,t)} \|\zeta_{ik_n+r+1}^n\|^2\right)$$

$$\leq K \frac{\Delta_n^2}{u_n} \sum_{i=0}^{l_n(r,t)} \big(u_n + \mathbb{E}\big(\big(\Gamma'(\sigma)_{ik_n+r+1}^n\big)^2\big) + \Delta_n^{4\eta'}\big)$$

$$\leq K t \frac{\Delta_n^2}{u_n}\left(1 + \frac{\Delta_n^{4\eta'}}{u_n}\right) \tag{16.5.63}$$

where the second inequality comes from (16.5.56) and the last one from (16.5.36) and $l_n(r,t) \leq Kt/u_n$. Taking the square-root and summing up over r yields

$$\mathbb{E}\left(\sup_{s\leq t}\left\| \sum_{i=0}^{[s/\Delta_n]-k_n+1} \zeta_i''^n \right\|\right) \leq K\sqrt{t}\,\big(\sqrt{u_n} + \Delta_n^{2\eta'}\big). \tag{16.5.64}$$

This finishes the proof of Lemma 16.5.17 for $l = 2$. □

Second, we observe that $\overline{H}(3)^n \stackrel{\text{u.c.p.}}{\Longrightarrow} 0$ is an immediate consequence of (16.5.53) and Lemma 16.5.15, and the assumptions $\eta' > \frac{1}{3}$ and $2\eta + 2\eta' \geq 1$.

Finally, we observe that the variable $\overline{\beta}(g^j)_i^n$ is nothing else than the variable $\overline{\beta}_i^{n,j}$ defined in (12.2.12). Therefore with the notation (12.2.15) and the function $f^j(x) = |x|^{2p_j}$ (instead of the function f^j associated with p_j by (16.5.2)), we see that $\overline{H}(4)_t^{n,j} = H(3)_t^{n,j} + H(6)_t^{n,j}$. Therefore $\overline{H}(4)^n \stackrel{\text{u.c.p.}}{\Longrightarrow} 0$ follows from Lemma 12.2.4.

Step 3) (block splitting) We rewrite $\overline{H}(1)^n$ as

$$\overline{H}(1)_t^n = \sum_{i=1}^{[t/\Delta_n]-k_n+1} \zeta_i^n, \quad \text{where}$$

$$\zeta_i'^{n,j} = \frac{\Delta_n}{\sqrt{u_n}}\,\psi(g^j)_{i,0}^n, \quad \zeta_i''^{n,j} = \mathbb{E}\big(\zeta_i'^{n,j} \mid \mathcal{F}_{(i-1)\Delta_n}\big), \quad \zeta_i^{n,j} = \zeta_i'^{n,j} - \zeta_i''^{n,j}.$$

The variables ζ_i^n are *not* martingale differences. So to prove Lemma 16.5.16, and for the very same reason of forcing the martingale property, we do as in the proof of Theorem 12.2.1: we fix an integer $m \geq 1$, and we divide the summands in the definition of $\overline{H}(1)^n$ into blocks of sizes mk_n and k_n.

The lth big block, of size mk_n, contains the indices between $I(m,n,l) = (l-1)(m+1)k_n + 1$ and $I(m,n,l) + mk_n - 1$. The number of such blocks before time t is $l_n(m,t) = [\frac{[t/\Delta_n]-1}{(m+1)k_n}]$. These big blocks are separated by small blocks of size k_n, and the "real" time corresponding to the beginning of the lth big block is $t(m,n,l) = (I(m,n,l)-1)\Delta_n$. Then we introduce the following q-dimensional variables and processes:

$$\zeta(m)_i^n = \sum_{r=0}^{mk_n-1} \zeta_{I(m,n,i)+r}^n, \qquad Z(m)_t^n = \sum_{i=1}^{l_n(m,t)} \delta(m)_i^n.$$

We end this step by proving that $\overline{H}^n(1) - Z(m)^n$ is asymptotically negligible:

Lemma 16.5.18 *Under* (SK) *and* (SN) *and* $\eta + \eta' \geq \frac{1}{2}$ *we have for all* $t > 0$:

$$\lim_{m\to\infty} \limsup_{n\to\infty} \mathbb{E}\Big(\sup_{s\le t} \big\|\overline{H}^n(1)_s - Z^n(m)_s\big\|\Big) = 0.$$

Proof The proof is exactly the same as for Lemma 12.2.5, the only difference being that the inequality $\mathbb{E}(\|\zeta_i^n\|^2) \leq K\Delta_n^2/u_n$ is now implied by (16.5.57). □

Step 4) As for Theorem 12.2.1, we modify the process $Z(m)$ in such a way that each summand involves the volatility at the beginning of the corresponding large block. Set

$$\begin{aligned}
\eta_{i,r}^{n,j} &= \frac{\Delta_n}{\sqrt{u_n}} \Big(\big(\phi(g^j,p_j)\big)_{i,r}^n - \mathbb{E}\big(\big(\phi(g^j,p_j)\big)_{i,r}^n \mid \mathcal{F}_{(i-r-1)\Delta_n}\big)\Big)\\
\eta_{i,r}^{\prime n,j} &= \frac{\Delta_n}{\sqrt{u_n}} \Big(\mathbb{E}\big(\big(\phi(g^j,p_j)\big)_{i,r}^n \mid \mathcal{F}_{(i-r-1)\Delta_n}\big) - \mathbb{E}\big(\big(\phi(g^j,p_j)\big)_{i,r}^n \mid \mathcal{F}_{(i-1)\Delta_n}\big)\Big)\\
\eta(m)_i^n &= \sum_{r=0}^{mk_n-1} \eta_{i+r,r}^n, \qquad \eta(m)_i^{\prime n} = \sum_{r=0}^{mk_n-1} \eta_{i+r,r}^{\prime n}\\
M^n(m)_t &= \sum_{i=1}^{l_n(m,t)} \eta(m)_{I(m,n,i)}^n, \qquad M^{\prime n}(m)_t = \sum_{i=1}^{l_n(m,t)} \eta(m)_{I(m,n,i)}^{\prime n}.
\end{aligned} \tag{16.5.65}$$

Lemma 16.5.19 *Under* (SK) *and* (SN) *and* $\eta + \eta' \geq \frac{1}{2}$ *we have*

$$\lim_{n\to\infty} \mathbb{E}\Big(\sup_{s\le t} \big\|Z^n(m)_s - M^n(m)_s - M^{\prime n}(m)_s\big\|\Big) = 0.$$

Proof Here again, the proof is the same as for Lemma 12.2.6, with the following changes: we substitute $M^n(m)$ with $M^n(m) + M^{\prime n}(m)$ and, instead of (12.2.29), we use (16.5.36) and (16.5.57). Then we obtain the better estimate $\mathbb{E}(\|\theta_i^n\|^2) \leq Km\Delta_n^2 = Kmu_n^2/k_n^2$ (this is because the test function is polynomial here). □

Lemma 16.5.20 *Under* (SK) *and* (SN) *and* $\eta + \eta' \geq \frac{1}{2}$ *we have*

$$\lim_{n\to\infty} \mathbb{E}\Big(\sup_{s\le t} \big\|M^{\prime n}(m)_s\big\|\Big) = 0.$$

Proof The variables $\eta(m)_i^{\prime n}$ are martingale increments relative to the discrete time filtration $\mathcal{G}_i^n = \mathcal{F}_{t(m,n,i+1)}$. Then it is enough to prove that $\sum_{i=1}^{l_n(m,t)} \mathbb{E} \times (\|\eta(m)_{I(m,n,i)}^{\prime n}\|^2) \to 0$ as $n \to \infty$ by Doob's inequality, and for this it suffices to show that

$$\mathbb{E}\big(\big\|\eta(m)_i^{\prime n}\big\|^2\big) \leq Km^2\Delta_n \tag{16.5.66}$$

because $l_n(m,t) \le t/u_n$ and $\Delta_n/u_n \to 0$. Now we observe that

$$\eta'^{m,j}_{i,r} = \frac{\Delta_n}{\sqrt{u_n}} \big(\mathbb{E}\big(\Psi\big(g^j, p_j\big)^n_{i,r} \mid \mathcal{F}_{(i-r-1)\Delta_n}\big) - \mathbb{E}\big(\Psi\big(g^j, p_j\big)^n_{i,r} \mid \mathcal{F}_{(i-1)\Delta_n}\big)\big),$$

because $\overline{W}(g^j)^n_i$ is independent of $\mathcal{F}_{(i-1)\Delta_n}$. Then (16.5.54) yields that $\mathbb{E}(|\Psi(g^j, p^j)^n_{i,r}|^2) \le K/k_n$. Since there are mk_n summands in $\eta(m)'^n_i$, (16.5.66) follows. □

Step 5) At this stage we prove a CLT for the processes $M^n(m)$, for each fixed m. We use the notation (16.5.32) and we set

$$\gamma(m)^j_t = m\,\mu\big(\sigma_t, \theta'\upsilon_t\big)^j, \qquad \gamma'(m)^{jj'}_t = \int_0^m ds \int_0^m ds'\, \mu'\big(\sigma_t, \theta'\upsilon_t; s, s'\big)^{jj'}.$$

Lemma 16.5.21 *Under* (SK) *and* (SN) *and* $\eta + \eta' \ge \frac{1}{2}$, *and for each* $m \ge 1$, *the processes* $M^n(m)$ *converge stably in law to a limit* $\overline{Y}'(m)$ *defined on a very good extension* $(\widetilde{\Omega}, \widetilde{\mathcal{F}}, (\widetilde{\mathcal{F}}_t)_{t\ge0}, \widetilde{\mathbb{P}})$ *of* $(\Omega, \mathcal{F}, (\mathcal{F}_t)_{t\ge0}, \mathbb{P})$, *and which conditionally on* $\mathcal{F}$ *is a continuous centered Gaussian process with independent increments with*

$$\widetilde{\mathbb{E}}\big(M(m)^j_t\, M(m)^l_t \mid \mathcal{F}\big) = \frac{1}{m+1} \int_0^t \big(\gamma'(m)^{jj'}_s - \gamma(m)^j_s\, \gamma(m)^{j'}_s\big)\, ds.$$

Proof Once more, the proof is analogous to the proof of Lemma 12.2.7, whose notation is used. We will apply Theorem 2.2.15 to the array $(\eta(m)^n_i)$, with $N_n(t) = [t/u_n(m+1)]$ and $T(n,i) = t(m,n,i+1)$ and the discrete-time filtration $\mathcal{G}^n_i = \mathcal{F}_{t(m,n,i+1)}$. We have $\mathbb{E}(\eta(m)^n_i \mid \mathcal{G}^n_{i-1}) = 0$, and also (12.2.31) by applying (16.5.57), hence it remains to prove

$$\sum_{i=1}^{l_n(m,t)} \mathbb{E}\big(\eta(m)^{n,j}_{I(m,n,i)}\, \eta(m)^{n,j'}_{I(m,n,i)} \mid \mathcal{G}^n_{i-1}\big) \stackrel{\mathbb{P}}{\longrightarrow} \frac{1}{m+1} \int_0^t \big(\gamma'(m)^{jj'}_s - \gamma(m)^j_s\, \gamma(m)^{j'}_s\big)\, ds \tag{16.5.67}$$

and also, for any bounded martingale N,

$$\sum_{i=1}^{l_n(m,t)} \mathbb{E}\big(\eta(m)^n_{I(m,n,i)} (N_{t(m,n,i+1)} - N_{t(m,n,i)}) \mid \mathcal{G}^n_{i-1}\big) \stackrel{\mathbb{P}}{\longrightarrow} 0. \tag{16.5.68}$$

1) We start by proving (16.5.67), in a similar fashion as in the proof of Theorem 16.4.1. The integer $m \ge 1$ is fixed, and we set

$$\zeta''^{m,j}_i = \frac{1}{k_n} \sum_{r=0}^{mk_n-1} \phi\big(g^j, p_j\big)^n_{i+r,r},$$

$$\zeta_i^{n,j} = \mathbb{E}\big(\zeta_i^{\prime\prime n,j} \mid \mathcal{F}_{(i-1)\Delta_n}\big), \qquad \zeta_i^{\prime n,jj'} = \mathbb{E}\big(\zeta_i^{\prime\prime n,j}\,\zeta_i^{\prime\prime n,j'} \mid \mathcal{F}_{(i-1)\Delta_n}\big),$$

$$t(m,n,i) \le t < t(n,m,i+1) \;\Rightarrow\; \gamma_t^{\prime n,jj'} = \zeta_i^{\prime n,jj'}, \quad \gamma_t^{n,j} = \zeta_i^{n,j}\,\zeta_i^{n,j'}.$$

The convergence (16.5.67) amounts to

$$u_n \sum_{i=1}^{l_n(m,t)} \big(\zeta_i^{\prime n,jj'} - \zeta_i^{n,j}\,\zeta_i^{n,j'}\big) \stackrel{\mathbb{P}}{\longrightarrow} \frac{1}{m+1}\int_0^t \big(\gamma'(m)_s^{jj'} - \gamma(m)_s^j\,\gamma(m)_s^{j'}\big)\,ds.$$

Exactly as for (16.4.20), and because $u_n\,l_n(m,t) \to \frac{t}{m+1}$, this will be satisfied as soon as, for all t, we have

$$\gamma_t^{n,j} \stackrel{\mathbb{P}}{\longrightarrow} \gamma(m)_t^j, \qquad \gamma_t^{\prime n,jj'} \stackrel{\mathbb{P}}{\longrightarrow} \gamma'(m)_t^{jj'}. \tag{16.5.69}$$

We fix t and we apply Lemma 16.3.9 with the sequence $i_n = i+1$ if $(i-1)\Delta_n \le t < i\Delta_n$ (so $T_n = (i_n-1)\Delta_n$ satisfies (2) of (16.3.18) with $T = t$), and with $Z = 1$. With the notation (16.3.33) we have $\zeta_{i_n}^{\prime\prime n,j} = F_n^j(\sigma_{T_n} L^{n,j}, L^{\prime n,j}, \widehat{L}^{n,j})$, where F_n^j is the function on $\mathbb{D}^q \times \mathbb{D}^q \times \mathbb{D}^q$ defined by

$$F_n^j(x,y,z) = \frac{1}{k_n}\sum_{r=0}^{mk_n-1} f^j\bigg(x\Big(\frac{r}{k_n}\Big)^j - \frac{\Delta_n^\eta}{\sqrt{k_n u_n}}\,y\Big(\frac{r}{k_n}\Big), \frac{\Delta_n^{2\eta}}{k_n u_n}\,z\Big(\frac{r}{k_n}\Big)\bigg).$$

The functions F_n^j and $F_n^j F_n^{j'}$ satisfy (16.3.16) with $m+1$ instead of m, and they converge pointwise to F^j and $F^j F^{j'}$, where

$$F^j(x,y,z) = \int_0^m f^j\big(x(s) - \theta' y(s), \theta'^2 z(s)\big)\,ds.$$

At this stage, we deduce (16.5.69) from (16.3.35) with $Z = 1$ (so the conditional expectation in the right side disappears), plus the fact that under S_g the laws of $(y(s): s \ge 0)$ and $(-y(s): s \ge 0)$ are the same, and the result is proved.

2) The proof of (16.5.68) is more complicated than in Lemma 12.2.7, because of the noise. For any process Y we write $D_i^n(Y) = Y_{t(m,n,i+1)} - Y_{t(m,n,i)}$. In view of the definition (16.5.65) of $\eta(m)_i^n$, and since we only consider bounded martingales N which thus satisfy $\mathbb{E}(D(N)_i^n \mid \mathcal{G}_{i-1}^n) = 0$, it is enough to prove that for any weight function g and any integer $p \ge 1$ we have

$$\sum_{i=1}^{l_n(m,t)} \mathbb{E}\big(\zeta_{I(m,n,i)}^n\,D(N)_i^n \mid \mathcal{G}_{i-1}^n\big) \stackrel{\mathbb{P}}{\longrightarrow} 0, \quad \text{where } \zeta_i^n = \frac{\Delta_n}{\sqrt{u_n}}\sum_{r=0}^{mk_n-1} \phi(g,p)_{i+r,r}^n. \tag{16.5.70}$$

By (16.5.57) we have $\mathbb{E}(|\phi(g,p)_i^n|^2) \leq K$, hence $\mathbb{E}(|\zeta_i'^n|^2) \leq Ku_n$ and, if N is a square-integrable martingale, the Cauchy-Schwarz and Doob inequalities yield:

$$\sum_{i=1}^{l_n(m,t)} \mathbb{E}\big(\big|\zeta^n_{I(m,n,i)} D_i^n(N)\big|\big) \leq \Bigg(\sum_{i=1}^{l_n(m,t)} \mathbb{E}\big(\big|\zeta^n_{I(m,n,i)}\big|^2\big) \sum_{i=1}^{l_n(m,t)} \mathbb{E}\big(\big|D_i^n(N)\big|^2\big)\Bigg)^{1/2}$$
$$\leq K\sqrt{\mathbb{E}(N_t^2)}. \tag{16.5.71}$$

With the notation (16.5.52) and $\zeta_i'^n = \frac{\Delta_n}{\sqrt{u_n}} \sum_{r=0}^{mk_n-1} \Psi(g,p)^n_{i+r,r}$, the same argument and (16.5.54) also yield

$$\sum_{i=1}^{l_n(m,t)} \mathbb{E}\big(\big|\zeta'^n_{I(m,n,i)} D_i^n(N)\big|\big) \leq \frac{K}{k_n}\sqrt{\mathbb{E}(N_t^2)}. \tag{16.5.72}$$

By virtue of (16.5.71), the set of all square-integrable martingales N satisfying (16.5.70) is closed under $\mathbb{L}^2$-convergence. Suppose now that (16.5.68) holds when N belongs to the set of $\mathcal{N}^{(0)}$ of all bounded $(\mathcal{F}_t^{(0)})$-martingales, and when N belongs to the set $\mathcal{N}^{(1)}$ of all martingales having $N_\infty = h(\chi_{t_1},\dots,\chi_{t_w})$, where h is a Borel bounded function on $\mathbb{R}^w$ and $t_1 < \cdots < t_w$ and $w \geq 1$. Since $\mathcal{N}^{(0)} \cup \mathcal{N}^{(1)}$ is total in the set $\mathcal{N}$ of all bounded $(\mathcal{F}_t)$-martingales for the $\mathbb{L}^2$-convergence, we deduce that (16.5.70), hence (16.5.68) as well, holds for all $N \in \mathcal{N}$.

It thus remains to prove (16.5.70) for $N \in \mathcal{N}^{(i)}$, $i = 0,1$. We start with $N \in \mathcal{N}^{(0)}$, in which case $D(N)_i^n$ is $\mathcal{H}_\infty$ measurable. Therefore $\mathbb{E}(\zeta^n_{I(m,n,i)} D(N)_i^n \mid \mathcal{G}^n_{i-1})$ is equal to

$$\mathbb{E}\big(\zeta'^n_{I(m,n,i)} D(N)_i^n \mid \mathcal{G}^n_{i-1}\big) + \mathbb{E}\Bigg(\sum_{r=0}^{mk_n-1} \big(\overline{\beta}(g)^n_{I(m,n,i)+r}\big)^{2p} D(N)_i^n \mid \mathcal{G}^n_{i-1}\Bigg),$$

with again the notation (16.5.52). We have seen in the proof of Lemma 12.2.5 that the last conditional expectation above vanishes: so in view of (16.5.72) we have the following inequality, which implies the result:

$$\mathbb{E}\Bigg(\sum_{i=1}^{l_n(m,t)} \big|\mathbb{E}\big(\zeta^n_{I(m,n,i)} D(N)_i^n \mid \mathcal{G}^n_{i-1}\big)\big|\Bigg) \leq \frac{K}{k_n}\sqrt{\mathbb{E}(N_t^2)}. \tag{16.5.73}$$

Finally, let $N \in \mathcal{N}^{(1)}$ be associated with h and w and the t_i's. In view of (16.1.1) it is easy to check that N takes the following form (by convention $t_0 = 0$ and $t_{w+1} = \infty$):

$$t_l \leq t < t_{l+1} \quad \Rightarrow \quad N_t = M(l; Z_{t_1},\dots,Z_{t_l})_t$$

for $l = 0,\dots,w$, and where $M(l; z_1,\dots,z_l)$ is a version of the martingale

$$M(l; z_1,\dots,z_l)_t = \mathbb{E}^{(0)}\Bigg(\int \prod_{r=l+1}^{w} Q_{t_r}(dz_r) h(z_1,\dots,z_l,z_{l+1},\dots,z_w) \mid \mathcal{F}_t^{(0)}\Bigg)$$

(with obvious conventions when $l=0$ and $l=w$). Each $M(l;z_1,\dots,z_l)$ is in $\mathcal{N}^{(0)}$ and those martingales are bounded uniformly in $(l;z_1,\dots,z_l,\omega^{(0)},t)$, so if we apply (16.5.73) for these martingales we obtain for each l:

$$\mathbb{E}\Bigg(\sum_{i\geq 1:\,t_l\leq t(m,n,i)<t(m,n,i+1)<t_{l+1}} \big|\mathbb{E}\big(\zeta^n_{I(mn,i)}\,D(N)^n_i\mid\mathcal{G}^n_{i-1}\big)\big|\Bigg)\leq\frac{K}{k_n}.$$

Furthermore, each summand on the left side of (16.5.71) is smaller than $K\sqrt{u_n}$ because $\mathbb{E}(|\zeta^n_i|^2)\leq Ku_n$. Therefore

$$\mathbb{E}\Bigg(\sum_{i=1}^{l_n(m,t)}\big|\mathbb{E}\big(\zeta^n_{I(m,n,i)}\,D(N)^n_i\mid\mathcal{G}^n_{i-1}\big)\big|\Bigg)\leq Kw\bigg(\frac{1}{k_n}+\sqrt{u_n}\bigg),$$

and we deduce (16.5.70). This finishes the proof of our lemma. □

Step 6) We are now ready to prove Lemma 16.5.16, and this will end the proof of Theorem 16.5.7.

Proof of Lemma 16.5.16 If we combine Lemmas 16.5.18, 16.5.19 and 16.5.20 with the general criterion of Proposition 2.2.4, we see that the only thing left to prove is the stable convergence in law $\overline{Y}'(m)\stackrel{\mathcal{L}\text{-}s}{\Longrightarrow}\overline{Y}'$, as $m\to\infty$. For this, and as in the proof of Lemma 12.2.3, we only need to show that

$$\frac{1}{m+1}\int_0^t\big(\gamma'(m)^{jj'}_s-\gamma(m)^j_s\,\gamma(m)^{j'}_s\big)\,ds\to\int_0^t R\big(\sigma_t,\theta' \upsilon_t\big)^{jj'}\,ds \tag{16.5.74}$$

for all t. Recall once more that the process (L,L') is stationary, and the variables (L_t,L'_t) and (L_s,L'_s) are independent if $|s-t|\geq 1$, so $\mu'(v,v';s,s')^{jj'}=\mu(v,v')^j\,\mu(v,v')^{j'}$ when $|s-s'|\geq 1$, and $\mu'(v,v';s,s')^{jj'}_t=\mu'(v,v';1,s'+1-s)^{jj'}$ for all $s,s'\geq 0$ with $s'+1-s\geq 0$. Then if $m\geq 2$ and $\mu^j=\mu(\sigma_t,\theta'\upsilon_t)^j$ and $\mu'(s,s')^{jj'}=\mu'(\sigma_t,\theta'\upsilon_t;s,s')$ we have

$$\begin{aligned}
\gamma'(m)^{jj'}_t-\gamma(m)^j_t\,\gamma(m)^{j'}_t &= \int_0^m ds\int_0^m \mu'(s,s')^{jj'}\,ds'-m^2\,\mu^j\,\mu^{j'}\\
&= \int_0^m ds\int_{(s-1)^+}^{m\wedge(s+1)}\big(\mu'(1,s'+1-s)^{jj'}-\mu^j\mu^{j'}\big)\,ds'\\
&= (m-1)\int_0^2\big(\mu'(1,s')^{jj'}-\mu^j\mu^{j'}\big)\,ds'\\
&\quad+\int_0^1 ds\int_{1-s}^2\big(\mu'(1,s'+1-s)^{jj'}-\mu^j\mu^{j'}\big)\,ds'.
\end{aligned}$$

Since $\mu(\sigma_t,\theta'\upsilon_t)^j$ and $\mu'(\sigma_t,\theta'\upsilon_t;s,s')$ are bounded, (16.5.74) follows. □

16.6 The Quadratic Variation

Finally we give a central limit theorem for the quadratic variation, again in the one-dimensional case only. When X is continuous the approximation of the quadratic variation is given by (16.5.7), in which we see the necessity of a "de-biasing" term. We will do the same when, possibly, there are jumps.

We assume that X is an Itô semimartingale X written as (16.3.1), and we take $d'=1$ as well, and we want to estimate the brackets $[X,X]$. We fix a weight function g. The observed process at stage n is again $Z^n = X + (\Delta_n)^\eta\,\chi$, and the approximate quadratic variation is $\overline{V}'^n(f_1,g,k_n,Z^n)$, which below is written as

$$U_t^n = \frac{1}{k_n}\sum_{i=1}^{[t/\Delta_n]-k_n+1}\Big(\big(\overline{Z^n}(g)_i^n\big)^2 - \frac{1}{2}\,\widehat{Z^n}(g)_i^n\Big). \tag{16.6.1}$$

Theorem 16.6.1 *Under* (H) *and* (N), *and if* $\eta+\eta'\geq \frac{1}{2}$, *we have* $U_t^n \stackrel{\mathbb{P}}{\longrightarrow} \Lambda(g)\,[X,X]_t$ *for any fixed* t.

In the continuous case this is (16.5.7), and in the general case it is easy to prove. In fact for us it will be a consequence of the following associated Central Limit Theorem (albeit under slightly stronger conditions on η' than necessary). This CLT is about the normalized difference

$$\overline{U}_t^n = \frac{1}{\sqrt{u_n}}\big(U_t^n - \Lambda(g)\,[X,X]_t\big). \tag{16.6.2}$$

The limit will of course be a mixture of the limit for the "jump part" as given in Theorem 16.5.6 (although the condition $p\geq 2$ is not satisfied, since here $p=1$) and the limit for the "continuous part", as given in Theorem 16.5.7. So on the one hand we define the processes Ξ and Ξ' by (16.3.4) (they are one-dimensional here), where

$$\begin{aligned}
H_-(x) &= 4x^2\int_0^1 dt\bigg(\int_0^t g(s+1-t)\,g(s)\,ds\bigg)^2\\
H_+(x) &= 4x^2\int_0^1 dt\bigg(\int_0^t g(s-t)\,g(s)\,ds\bigg)^2\\
H'_-(x) &= 4x^2\int_0^1 dt\bigg(\int_0^t g(s+1-t)\,g'(s)\,ds\bigg)^2\\
H'_+(x) &= 4x^2\int_0^1 dt\bigg(\int_0^t g(s-t)\,g'(s)\,ds\bigg)^2.
\end{aligned} \tag{16.6.3}$$

On the other hand we have $R(v,v')$ as given by (16.5.32), with $j=l=1$ here. Then we can define three processes $\overline{Y}(X)$, $\overline{Y}'(\chi)$ and $\overline{Y}''$ on a very good extension $(\widetilde{\Omega},\widetilde{\mathcal{F}},(\widetilde{\mathcal{F}}_t)_{t\geq 0},\widetilde{\mathbb{P}})$ of $(\Omega,\mathcal{F},(\mathcal{F}_t)_{t\geq 0},\mathbb{P})$, which conditionally on $\mathcal{F}$ are independent, centered Gaussian with independent increments, the first two ones being purely discontinuous and the last one being continuous, and with (conditional) variances (where as usual $\upsilon_t = \sqrt{\Upsilon_t}$)

$$\begin{aligned}
&\widetilde{\mathbb{E}}\big((\overline{Y}(X)_t)^2 \mid \mathcal{F}\big) = \Xi_t, \qquad \widetilde{\mathbb{E}}\big((\overline{Y}'(\chi)_t)^2 \mid \mathcal{F}\big) = \Xi'_t \\
&\widetilde{\mathbb{E}}\big((\overline{Y}''_t)^2 \mid \mathcal{F}\big) = \int_0^t R(\sigma_s, \theta'\upsilon_s)\, ds.
\end{aligned} \tag{16.6.4}$$

Theorem 16.6.2 *Assume* (H) *and* (N) *with* Υ_t *càdlàg and let* k_n *satisfy* (16.1.5), *with further* (16.3.6) *and* (16.5.34). *Then for each* t *the variables* $\overline{U}^n_t$ *of* (16.6.2) *converges stably in law to the variable* $\overline{Y}(X)_t + \theta'\overline{Y}'(\chi)_t + \overline{Y}''_t$, *as described above.*

When further X *is continuous, the processes* $\overline{U}^n$ *converge stably (in the functional sense) to the process* $\overline{Y}''$.

The second claim is a particular case of Theorem 16.5.7, except that (K) is not needed here, exactly as in Chap. 5 (and $\overline{Y}''$ here is the same as $\overline{Y}'$ in Theorem 16.5.7). As said before, we will prove only the second theorem, which implies the first one under the extra assumption (16.5.35).

Remark 16.6.3 Although we will not do it here, we could prove a similar result for the cross quadratic variation $[X^j, X^l]$ between any two components X^j and X^l, when X is multi-dimensional. Then (16.6.1) would be substituted with

$$U^{n,jl}_t = \frac{1}{k_n} \sum_{i=1}^{[t/\Delta_n]-k_n+1} \Big(\overline{Z^n}(g)^{n,j}_i\, \overline{Z^n}(g)^{n,l}_i - \frac{1}{2}\, \widehat{Z^n}(g)^{n,jl}_i \Big),$$

and (16.6.2) with

$$\overline{U}^{n,jl}_t = \frac{1}{\sqrt{u_n}} \big(U^{n,jl}_t - \Lambda(g)\big[X^j, X^l\big]_t \big).$$

One even has the multi-dimensional stable convergence in law for the variables $(\overline{U}^{n,jl}_t, 1 \le j, l \le d)$. The proof is basically the same.

Proof of Theorem 16.6.2 The proof is done through several steps, and as usual we may assume (SH) and (SN). As a rule, we omit to mention the weight function g. In contrast, we will vary the process X in the course of the proof: so we mention it explicitly by writing $U^n(X)$ and $\overline{U}^n(X)$ (warning: those processes also depend on the noise, of course). Analogously, for the limit we write $\overline{Y}(X)$ and $\overline{Y}'(X, \chi)$ (despite the notation, the process $\overline{Y}'(\chi)$, related to the noise, is in fact dependent on Υ_t but not on the noise *per se*, and it depends on X as well through its jumps), and $\overline{Y}''(X)$ (which depends on X through σ_t). Finally the three conditional variances in (16.6.4) will be denoted by $\Xi(X)$ and $\Xi'(X)$, and also $\Xi''(X)$ for the third one.

Step 1) In this step we prove the result when the process X has locally finitely many jumps, and when σ is piecewise constant. More specifically, we fix some integer

$m \geq 1$ and use our customary notation $A_m = \{z : \Gamma(z) > 1/m\}$, where Γ is the function appearing in (SH), and $S(m, j)$ for the successive jump times of the process $1_{A_m} * \mathfrak{p}$. According to (2.1.48), $(\mathcal{G}_t^{A_m})$ is the smallest filtration containing $(\mathcal{F}_t)$ and such that the restriction of the measure $\mathfrak{p}$ to the set $\mathbb{R}_+ \times A_m$ is $\mathcal{G}_0^{A_m}$-measurable.

Then, in addition to (SH), we suppose the following:

$$\left.\begin{array}{l} \Gamma(z) \leq 1/m \quad \Rightarrow \quad \delta(\omega, t, z) = 0 \\ \sigma_s = \sum_{q \geq 1} \sigma_{T_q} 1_{[T_q, T_{q+1})}(t) \end{array}\right\} \tag{16.6.5}$$

for some $m \geq 1$ and where $0 = T_0 < T_1 < \cdots$ is a sequence of random times, increasing to infinity, and such that each T_q is $\mathcal{G}_0^{A_m}$-measurable.

1) We set $X_t'' = \sum_{s \leq t} \Delta X_s$ and $X_t' = X_t - X_t'' = X_0 + \int_0^t b_s' \, ds + \int_0^t \sigma_s \, dW_s$, where $b_s' = b_s - \int \delta(t, x) 1_{\{|\delta(y,z)| \leq 1\}} \lambda(dz)$. With the notation of the proof of Theorem 16.3.1 (or Theorem 12.1.2), the process $X(m)$ of (12.1.13) is equal to X', and it is a continuous $(\mathcal{G}_t)$-Itô semimartingale. Similar to (16.3.50), we have

$$\overline{U}^n(X)_t = \overline{U}^n(X')_t + Y_t^n - \frac{1}{2} Y_t'^n$$

on the set $\Omega_n(t, m)$, where $Y_t^n = \sum_{p \in \mathcal{P}_m : S_p \leq t} \widehat{\zeta}_p^n$ and $Y_t'^n = \sum_{p \in \mathcal{P}_m : S_p \leq t} \widehat{\zeta}_p'^n$, and where $\widehat{\zeta}_p^n$ is as ζ_p^n after (16.3.50) with $f^j(x) = x^2$ (so the last term in the definition of $\widehat{\zeta}_p^n$ is $-\Lambda(g)|\Delta X_{S_p}|^2$) and

$$\widehat{\zeta}_p'^n = \frac{\Lambda_n'(g, 2)}{k_n \sqrt{u_n}} \big((\Delta X_{S_p})^2 + 2\Delta X_{S_p} \, \Delta_{i(n,p)}^n (X' + \Delta_n^\eta \chi)\big).$$

We have $|\Lambda_n'(g, 2)| \leq K/k_n$, and as soon as $\eta' > \frac{1}{5}$ we obtain $\mathbb{E}(|\widehat{\zeta}_p'^n|) \stackrel{\mathbb{P}}{\longrightarrow} 0$ by the estimates already used (recall that the jumps of X are bounded). Therefore $Y_t'^n \stackrel{\mathbb{P}}{\longrightarrow} 0$, whereas exactly as in (16.3.51) we have

$$Y_t^n \stackrel{\mathcal{L}\text{-s}}{\longrightarrow} \overline{Y}(X)_t + \theta' \overline{Y}'(X, \chi)_t. \tag{16.6.6}$$

2) Next we prove that $\overline{U}^n(X') \stackrel{\mathcal{L}\text{-s}}{\Longrightarrow} \overline{Y}''(X)$. Since X' is driven by W only and since W is a $(\mathcal{G}_t^{A_m}$-Brownian motion, we can argue relatively to the filtration $(g_t^{A_m})$ instead of $(\mathcal{F}_t)$, and also conditionally on $\mathcal{G}_9^{A_m}$: in other words, the times T_r can be considered as being non-random.

The property $\overline{U}^n(X') \stackrel{\mathcal{L}\text{-s}}{\Longrightarrow} \overline{Y}''(X)$ is Theorem 16.5.7 for $p = 1$, except that here we do not assume (SK), but only (SH). However a look at the proof of this theorem shows that (SK) instead of (SH) is used in the proof of Lemma 16.5.17 only, in three places: to prove the asymptotic negligibility of the array $(\zeta_i'^n)$ of this lemma (for which (16.5.58) is used), for the asymptotic negligibility of the arrays $(\zeta_i''^n)$ of the same lemma (because one deduces (16.5.64) from (16.5.63)), and for the proof in the case $l = 4$. For the latter occurrence, this is as in Lemma 12.2.4 (case

$l = 6$) and the Itô semimartingale property of σ can be replaced by the fact that σ_t is piecewise constant (the result is then much simpler to prove).

In the present situation, (16.5.58) may fail. However because of (16.6.5), and with again the notation of the proof of Lemma 16.5.17, we indeed have $\zeta_i^n = 0$ for all i except those for which $i\Delta_n \in [T_r - u_n, T_r]$ for some r, whereas $\mathbb{E}(|\zeta_i^n|t) \leq K$ by (16.5.55). It follows that

$$\mathbb{E}\left(\sum_{i=1}^{[t/\Delta_n]-k_n+1} |\zeta_i^n| \mid \mathcal{G}_0^{A_m}\right) \leq K \frac{\Delta_n}{\sqrt{u_n}} r_t k_n \leq K r_t \sqrt{u_n},$$

where $r_t = \sup(r : T_r \leq t)$. Then we deduce the asymptotic negligibility of the array (ζ_i^n) without using the decomposition $\zeta_i^n = \zeta_i'^n + \zeta_i''^n$. This ends the proof of $\overline{U}^n(X') \stackrel{\mathcal{L}\text{-s}}{\Longrightarrow} \overline{Y}''(X)$.

3) So far we have $\overline{U}^n(X') \stackrel{\mathcal{L}\text{-s}}{\Longrightarrow} \overline{Y}''(X)$ and (16.6.6). However, by exactly the same kind of argument as in Theorem 4.2.1 one may show the joint stable convergence in law (the formal argument is rather tedious to develop, but the idea is exactly the same as in the afore-mentioned proposition). This implies that for all t we have

$$\overline{U}^n(X)_t \stackrel{\mathcal{L}\text{-s}}{\longrightarrow} \overline{Y}(X)_t + \theta' \overline{Y}'(X, \chi)_t + \overline{Y}''(X)_t.$$

Step 2) From now on, we suppose that X satisfies (SH), but not (16.6.5) any longer. In this step we construct an approximation of X satisfying (16.6.5).

Since Γ is bounded, we recall the decomposition

$$X_t = X_0 + \int_0^t b'_s\, ds + \int_0^t \sigma_s\, dW_s + (\delta\, 1_{(A_m)^c}) * (\mathfrak{p} - \mathfrak{q}) + (\delta\, 1_{A_m}) * (\mathfrak{p} - \mathfrak{q})_t,$$

where $b'_t = b_t + \int_{\{|\delta(t,z)|>1\}} \delta(t,z)\lambda(dz)$. We let $0 = T(m)_0 < T(m)_1 < \cdots$ be the successive times in the set $F(m) = \{i/2^m : i \in \mathbb{N}\} \cup \{S(m,i) : i \geq 1\}$, and we set

$$\sigma(m)_t = \sigma_{T(m)_i} \quad \text{if } T(m)_i \leq t < T(m)_{i+1}.$$

Since σ is càdlàg bounded and $F(m)$ increases as m increases, we have

$$\varepsilon(m)_t := \mathbb{E}\left(\int_0^t \left|\sigma(m)_s - \sigma_s\right|^2 ds\right) \to 0 \text{ as } m \to \infty \tag{16.6.7}$$

for all t. Then, for any $m \geq 1$ we set

$$X(m)_t = X_0 + \int_0^t b'_s\, ds + \int_0^t \sigma(m)_s\, dW_s + (\delta 1_{A_m}) * (\mathfrak{p} - \mathfrak{q})_t. \tag{16.6.8}$$

By construction $X(m)$ satisfies (16.6.5), so Step 1 gives us

$$\overline{U}^n\big(X(m)\big)_t \stackrel{\mathcal{L}\text{-s}}{\longrightarrow} \overline{Y}\big(X(m)\big)_t + \theta' \overline{Y}'\big(X(m), \chi\big)_t + \overline{Y}''\big(X(m)\big)_t \tag{16.6.9}$$

for any t and $m \geq 1$, and the convergence even holds in the functional sense when X (hence $X(m)$) is continuous.

Step 3) Now we consider the $\mathcal{F}$-conditional variances (16.6.4). In view of their definitions in (16.3.4) and of the fact that the functions in (16.6.3) are of the form $a_{\pm}x^2$ and $a'_{\pm}x^2$, and with the notation $D_m = \{(\omega,t) : \mathfrak{p}(\omega,\{t\}\times\{z : 0 < \Gamma(z) \leq 1/m\}) = 0\}$, we have

$$\Xi\big(X(m)\big)_t = \sum_{s\leq t}\big(a_-\sigma(m)^2_{s-} + a_+\sigma(m)^2_s\big)\,(\Delta X_s)^2 1_{D_m}(s)$$

$$\Xi'\big(X(m)\big)_t = \sum_{s\leq t}\big(a'_-\Upsilon_{s-} + a'_+\Upsilon_s\big)\,(\Delta X_s)^2 1_{D_m}(s)$$

(because $\Delta X(m)_s$ equals ΔX_s when $s \in D_m$ and vanishes otherwise), and similar formulas for $\Xi(X)$ and $\Xi'(X)$, with $\sigma(m)$ and D_m substituted with σ and $\Omega\times\mathbb{R}_+$. Now, D_m increases to $\Omega\times\mathbb{R}_+$ and $\sum_{s\leq t}|\Delta X_s|^2 < \infty$ and $|\sigma(m)| \leq K$ and $\Upsilon \leq K$; moreover, for all $s \in D_m$ such that $\Delta X_s \neq 0$ we have $\sigma(m)_s = \sigma_s$ and $\sigma(m)_{s-} \to \sigma_{s-}$ by construction. Then by the dominated convergence theorem we obtain, as $m \to \infty$:

$$\Xi\big(X(m)\big) \stackrel{\text{u.c.p.}}{\Longrightarrow} \Xi(X), \qquad \Xi'\big(X(m)\big) \stackrel{\text{u.c.p.}}{\Longrightarrow} \Xi'(X).$$

Next, we recall that $\Xi''(X)_t = \int_0^t R(\sigma_s, \theta' \upsilon_s)\,ds$, whereas the function R of (16.5.32) is continuous. Hence (16.6.7) yields $\Xi''(X(m)) \stackrel{\text{u.c.p.}}{\Longrightarrow} \Xi''(X)$ as $m \to \infty$. Putting this together with the previous convergence, we deduce

$$\big(\Xi\big(X(m)\big), \Xi'\big(X(m)\big), \Xi''\big(X(m)\big)\big) \stackrel{\text{u.c.p.}}{\Longrightarrow} \big(\Xi(X), \Xi'(X), \Xi''(X)\big). \qquad (16.6.10)$$

Conditionally on $\mathcal{F}$, the three processes $\overline{Y}(X(m))$, $\overline{Y}'(X(m),\chi)$ and $\overline{Y}''(X(m))$ are centered Gaussian with independent increments and respective covariances $\Xi(X(m))_t$, $\Xi'(X(m))_t$ and $\Xi''(X(m))_t$ at time t, and they are $\mathcal{F}$-conditionally independent. Then it follows from (16.6.10) that

$$\begin{aligned}&\overline{Y}\big(X(m)\big) + \theta'\overline{Y}'\big(X(m),\chi\big) + \overline{Y}''\big(X(m)\big)\\ &\quad \stackrel{\mathcal{L}\text{-s}}{\Longrightarrow} \overline{Y}(X) + \theta'\overline{Y}'(X,\chi) + \overline{Y}''(X)\end{aligned}$$

as $m \to \infty$. Hence, in view of (16.6.9), the theorem will be proved if we show the following: for all $t, \eta > 0$ we have

$$\lim_{m\to\infty}\limsup_{n\to\infty}\mathbb{P}\Big(\sup_{s\leq t}\big|\overline{U}^n\big(X(m)\big)_s - \overline{U}^n(X)_s\big| > \eta\Big) = 0 \qquad (16.6.11)$$

when X is continuous, a case referred to as case C, whereas in the general case where X may be discontinuous, referred as case D, it is enough to show

$$\lim_{m\to\infty}\limsup_{n\to\infty}\mathbb{P}\big(\big|\overline{U}^n\big(X(m)\big)_t - \overline{U}^n(X)_t\big| > \eta\big) = 0. \qquad (16.6.12)$$

Step 4) Recall that we usually omit g, writing for example $\overline{X}_i^n = \overline{X}(g)_i^n$. We have

$$u_n\big(\phi(g,2,Z^n)_i^n - \phi\big(g,2,X(m)+(\Delta_n)^\eta\,\chi\big)_i^n\big)$$
$$= \big(\overline{X}_i^n\big)^2 - \big(\overline{X(m)}_i^n\big)^2 + 2(\Delta_n)^\eta\,\overline{\chi}_i^n\big(\overline{X}_i^n - \overline{X(m)}_i^n\big) - \frac{1}{2}\,v_i^n$$

where

$$v_i^n = \sum_{j=1}^{k_n}\big(g_j'^n\big)^2\big(\big(\Delta_{i+j-1}^n X\big)^2 - \big(\Delta_{i+j-1}^n X(m)\big)^2$$
$$+ 2(\Delta_n)^\eta\,\Delta_{i+j-1}^n\chi\big(\Delta_{i+j-1}^n X - \Delta_{i+j-1}^n X(m)\big)\big).$$

Therefore

$$\overline{U}^n(X)_t - \overline{U}^n\big(X(m)\big)_t = G^1(m)_t^n + G^2(m)_t^n - \frac{1}{2}\,V_t^n,$$

where

$$V_t^n = \frac{1}{k_n\sqrt{u_n}}\sum_{i=1}^{[t/\Delta_n]-k_n+1} v_i^n$$

$$G^1(m)_t^n = \frac{1}{\sqrt{u_n}}\bigg(\frac{1}{k_n}\sum_{i=1}^{[t/\Delta_n]-k_n+1}\big(\big(\overline{X}_i^n\big)^2 - \big(\overline{X(m)}_i^n\big)^2\big)$$
$$- \Lambda(g)\big([X,X]_t - \big[X(m),X(m)\big]_t\big)\bigg)$$

$$G^2(m)_t^n = \frac{2\Delta_n^\eta}{k_n\sqrt{u_n}}\sum_{i=1}^{[t/\Delta_n]-k_n+1}\overline{\chi}_i^n\big(\overline{X}_i^n - \overline{X(m)}_i^n\big).$$

Since $\mathbb{E}(|\Delta_i^n X|^2) + \mathbb{E}'(|\Delta_i^n X(m)|^2) \le K\Delta_n$ and $|g_j'^n| \le K/k_n$ and $\mathbb{E}(|\Delta_i^n\chi|^2) \le K$, we have $\mathbb{E}(|v_i^n|) \le K(\Delta_n + \Delta_n^{\eta+1/2})/k_n$. We deduce that $\mathbb{E}(\sup_{s\le t}|V_s^n|) \le Kt(\Delta_n^{(5\eta'-1)/2} + \Delta_n^{(5\eta'+2\eta-2)/2})$, which goes to 0 because $\eta+\eta' \ge 1/2$ and $\eta' > 1/3$. Therefore, instead of (16.6.11), we are left to prove for $l=1,2$, and according the case C (continuous) or D (possibly discontinuous):

$$\begin{aligned} &\text{C:} \quad \lim_{m\to\infty}\limsup_{n\to\infty}\mathbb{P}\Big(\sup_{s\le t}\big|G^l(m)_s^n\big| > \eta\Big) = 0 \\ &\text{D:} \quad \lim_{m\to\infty}\limsup_{n\to\infty}\mathbb{P}\big(\big|G^l(m)_t^n\big| > \eta\big) = 0. \end{aligned} \tag{16.6.13}$$

Step 5) We begin proving (16.6.13) for $l=2$. We split the sum in the definition of $G^2(m)_t^n$ into two parts: $G^3(m)_t^n$ is the sum over those i's for which the fractional part of $i/2k_n$ is in $[0,1/2)$, and $G^4(m)_t^n$ which is the sum when the fractional part is in $[1/2, 1)$. It enough to show (16.6.13) for $l=3$ and $l=4$, and we do it for $l=3$ only. We have

$$G^3(m)^n_t = \sum_{j=0}^{J_n(t)+1} \zeta(m,r)^n_i,$$

$$\zeta(m)^n_j = \frac{2\Delta_n^\eta}{k_n\sqrt{u_n}} \sum_{i=(2jk_n)\vee 1}^{(2jk_n+k_n-1)\wedge([t/\Delta_n]-k_n+1)} \overline{\chi}^n_i \big(\overline{X}^n_i - \overline{X(m)}^n_i\big),$$

where $J_n(t)$ is the integer part of $([t/\Delta_n]-k_n)/2k_n$ (all $\zeta(m)^n_j$ have at most k_n summands). Note that $\zeta(m)^n_j$ is $\mathcal{F}^n_{2(j+1)k_n}$ measurable, and by successive conditioning $\mathbb{E}(\zeta(m)^n_j \mid \mathcal{F}^n_{2jk_n}) = 0$. Therefore by a martingale argument (16.6.13) will follow, if we prove

$$\lim_{m\to\infty} \limsup_{n\to\infty} \mathbb{E}\bigg(\sum_{j=0}^{J_n(t)} |\zeta(m)^n_j|^2\bigg) = 0. \tag{16.6.14}$$

Recalling (16.6.7), we set

$$\varepsilon(m)^n_i = \varepsilon(m)_{i\Delta_n+u_n} - \varepsilon(m)_{i\Delta_n}, \quad \varepsilon_m = \int_{\{z:\Gamma(z)\le 1/m\}} \Gamma(z)^2\,\lambda(dz).$$

Then (16.6.7) and (16.6.11), together with (16.2.3) and our usual estimates for Itô semimartingales applied to $X - X(m)$ (as for (16.4.9)) and successive conditioning, yield

$$\mathbb{E}\big((\overline{\chi}^n_i)^2\,(\overline{X}^n_i - \overline{X(m)}^n_i)^2\big) \le \frac{K}{k_n}\big(\varepsilon(m)^n_i + u_n\,\varepsilon_m\big),$$

and so the expectation in (16.6.14) is smaller than $K\Delta_n^{2\eta+2\eta'-1}(\varepsilon(m)_t + t\varepsilon_m)$. Hence (16.6.14) holds by (16.6.7) and $2\eta+2\eta'\ge 1$ and the property $\varepsilon_m \to 0$ as $m\to\infty$.

Step 6) Now we turn to the case $l=1$ in (16.6.13). We write $G^1(m)^n_t = G^5(m)^n_t + G^6(m)^n_t$ where, with g_n as in the proof of Lemma 16.5.13 and $A(m) = [X,X] - [X(m),X(m)]$,

$$G^5(m)^n_t = \sum_{i=1}^{[t/\Delta_n]-k_n+1} \vartheta(m)^n_i$$

$$\vartheta(m)^n_i = \frac{1}{k_n\sqrt{u_n}}\bigg(\big(\overline{X}^n_i\big)^2 - \big(\overline{X(m)}^n_i\big)^2 - \int_{i\Delta_n}^{i\Delta_n+u_n} g_n(s-i\Delta_n)^2\,dA(m)_s\bigg)$$

$$G^6(m)^n_t = \frac{1}{\sqrt{u_n}}\bigg(\frac{1}{k_n}\sum_{i=1}^{[t/\Delta_n]-k_n+1}\int_{i\Delta_n}^{i\Delta_n+u_n} g_n(s-i\Delta_n)^2\,dA(m)_s - \Lambda(g)A(m)_t\bigg).$$

In this step we prove that $G^6(m)^n$ satisfies (16.6.13). A simple calculation shows

$$G^6(m)_t^n = \frac{1}{\sqrt{u_n}} \int_0^t \left(\frac{\Lambda_n(g,2)}{k_n} - \Lambda(g) \right) dA(m)_s + v(m,r)_t^n,$$

where the remainder term $v(m,r)_t^n$ satisfies, with $A'(m)$ being the variation process of $A(m)$:

$$\left|v(m)_t^n\right| \leq \frac{K}{\sqrt{u_n}} \left(A'(m)_{u_n} + \left(A'(m)_t - A'(m)_{t-2u_n}\right)\right).$$

In the continuous case C we have $A'(m)_{s+u_n} - A'(m)_s \leq K u_n$, hence $\sup_{s \leq t} |v(m)_s^n| \leq K\sqrt{u_n}$. In the discontinuous case D this fails, but we have $\mathbb{E}(A'(m)_{s+u_n} - A'(m)_s) \leq K u_n$, so $v(m)_t^n \xrightarrow{\mathbb{P}} 0$ as $n \to \infty$. Then if we apply (16.1.9) we obtain (16.6.13) for $l = 6$.

Step 7) It remains to prove (16.6.13) for $l = 5$. For this we use Itô's formula, to get, with the notation $Y_t^{n,i} = \int_{i\Delta_n}^t g_n(s - i\Delta_n)\, dY_s$ for any semimartingale Y, and for $t \geq i\Delta_n$:

$$\left(\overline{X}_i^n\right)^2 - \int_{i\Delta_n}^{i\Delta_n+u_n} g_n(s - i\Delta_n)^2\, d[X,X]_s = 2 \int_{i\Delta_n}^{i\Delta_n+u_n} X_s^{n,i}\, g_n(s - i\Delta_n)\, dX_s,$$

and a similar expression for $(\overline{X(m)}_i^n)^2$. Therefore, recalling that $X^\#(m)_t := X_t - X(m)_t = \int_0^t (\sigma_s - \sigma(m)_s)\, dW_s + (\delta 1_{(A_m)^c}) * (\mathfrak{p} - \mathfrak{q})_t$, we see that

$$\vartheta(m)_i^n = \frac{2}{k_n \sqrt{u_n}} \sum_{j=1}^5 \eta(m,j)_i^n,$$

where, with the notation $I(n,i) = (i\Delta_n, i\Delta_n + u_n]$,

$$\eta(m,1)_i^n = \int_{I(n,i)} X^\#(m)_s^{n,i}\, g_n(s - i\Delta_n)\, b'_s\, ds$$

$$\eta(m,2)_i^n = \int_{I(n,i)} X^\#(m)_s^{n,i}\, g_n(s - i\Delta_n)\, \sigma_s\, dW_s$$

$$\eta(m,3)_i^n = \int_{I(n,i)} X(m)_s^{n,i}\, g_n(s - i\Delta_n) \left(\sigma_s - \sigma(m)_s\right) dW_s$$

$$\eta(m,4)_i^n = \int_{I(n,i)} \int_{(A_m)^c} X_{s-}^{n,i}\, g_n(s - i\Delta_n)\, \delta(s,z)\, (\mathfrak{p} - \mathfrak{q})(ds,dz)$$

$$\eta(m,5)_i^n = \int_{I(n,i)} \int_{A_m} X^\#(m)_{s-}^{n,i}\, g_n(s - i\Delta_n)\, \delta(s,z)\, (\mathfrak{p} - \mathfrak{q})(ds,dz).$$

We also set

$$a(m,j,q)^n_t \;=\; \mathbb{E}\Bigg(\sum_{i=1}^{[t/\Delta_n]-k_n+1} \big|\eta(m,j)^n_i\big|^q\Bigg).$$

Since for $j=2,3,4,5$ the variables $\eta(m,j)^n_i$ are $\mathcal{F}_{(i+k_n)\Delta_n}$ measurable and satisfy $\mathbb{E}(\eta(m,j)^n_i \mid \mathcal{F}_{(i-1)\Delta_n})=0$, we see that (16.6.13) for $l=5$ will follow if we prove that for all $t>0$:

$$\begin{aligned} j=1 \;&\Rightarrow\; \lim_{m\to\infty}\limsup_{n\to\infty}\frac{1}{k_n\sqrt{u_n}}\,a(m,j,1)^n_t=0\\ j=2,3,4,5 \;&\Rightarrow\; \lim_{m\to\infty}\limsup_{n\to\infty}\frac{1}{k_n^2u_n}\,a(m,j,2)^n_t=0. \end{aligned} \tag{16.6.15}$$

Below, we use the notation $\varepsilon(m)^n_i$ and ε_m of Step 4. Our usual estimates yield for $s\in I(n,i)$ and $p\ge 2$:

$$\begin{aligned} \mathbb{E}\Big(\sup_{t\le s}\big|X^{\#}(m)^{n,i}_t\big|^2\Big) \;&\le\; K\big(\varepsilon(m)^n_i+u_n\varepsilon_m\big)\\ \mathbb{E}\Big(\sup_{t\le s}\big|X(m)^{n,i}_t\big|^p\Big)+\mathbb{E}\Big(\sup_{t\le s}\big|X^{n,i}_t\big|^p\Big) \;&\le\; K_pu_n \end{aligned}$$

and since $\int_{A_m}|\delta(s,z)|^2\lambda(dz)\le\int\Gamma(z)^2\lambda(dz)<\infty$ and $|g_n|\le K$ and $\varepsilon(m)^n_i\le K$ and $\varepsilon_m\le K$, it follows that

$$\begin{aligned} j=1 \;&\Rightarrow\; \mathbb{E}\big(\big|\eta(m,j)^n_i\big|\big)\le Ku_n\Big(\sqrt{\varepsilon(m)^n_i}+\sqrt{u_n\,\varepsilon_m}\Big)\\ j=2,3,4,5 \;&\Rightarrow\; \mathbb{E}\big(\big|\eta(m,j)^n_i\big|^2\big)\le Ku_n\Big(\sqrt{\varepsilon(m)^n_i}+u_n\Big). \end{aligned}$$

By Hölder's inequality

$$\Bigg(\sum_{i=1}^{[t/\Delta_n]-k_n+1}\sqrt{\varepsilon(m)^n_i}\Bigg)^2 \;\le\; \frac{t}{\Delta_n}\sum_{i=1}^{[t/\Delta_n]-k_n+1}\varepsilon(m)^n_i \;\le\; \frac{k_nt}{\Delta_n}\,\varepsilon(m)_t.$$

Therefore

$$a(m,1,1)^n_t\le Kk_n\sqrt{u_n}\,\Big(\sqrt{t\,\varepsilon(m)_t}+t\sqrt{\varepsilon_m}\Big)$$

and, for $j=2,3,4,5$,

$$a(m,j,2)^n_t\le Kk_n\sqrt{u_n}\,\Big(\sqrt{t\,\varepsilon(m)_t}+t\,\sqrt{u_n}\Big).$$

Since $\varepsilon(m)_t\to 0$ and $\varepsilon_m\to 0$ as $m\to\infty$ and $\eta'\ge\frac{1}{3}$, we deduce (16.6.15), and the proof is complete. □

Bibliographical Notes

The question of noisy observations has long been considered in statistics. Historically, for financial data (with which this chapter is really concerned), researchers have considered first the case where the noise is additive i.i.d. centered, and quite often Gaussian, see for example Bandi and Russell [7], Aït-Sahalia, Mykland and Zhang [3]. Various methods have been proposed in this case, like the "two-scales" method of Zhang, Mykland and Aït-Sahalia [96], the "multi-scale" approach of Zhang [97], the "kernel" approach of Barndorff-Nielsen, Hansen, Lunde and Shephard [13], or Podolskij and Vetter [80].

Now, as explained in the chapter itself, additive noise is often not adequate to model the reality of financial data. Round-off errors are prevalent, and in the case of a shrinking noise this has been studied by Delattre and Jacod [26] in the diffusion setting and by Rosenbaum [86] in a more realistic framework. A fundamental paper about the possible structure of the noise is [69] by Li and Mykland.

The pre-averaging method has been introduced in the continuous case by Jacod, Li, Mykland, Podolskij and Vetter [61] and by Podolskij and Vetter [81] when there are jumps, and the account given here closely follows Jacod, Podolskij and Vetter [63], except that in this book we also allow for a shrinking noise. This approach is in a sense another name for the so-called realized kernels approach of [13]. Finally, the paper [14] of Barndorff-Nielsen, Hansen, Lunde and Shephard mixes the multivariate non-synchronous observation times with a noise which may be more general than an additive noise.

Appendix

In the Appendix we prove the results left unproved in Chap. 2, plus a few others. For clarity, we re-state the results before their proofs.

A.1 Estimates for Itô Semimartingales

Here we prove the lemmas of Sect. 2.1.5. The first one is essentially contained in Protter and Talay [82], although the setting here is more general, and the proof somewhat simpler. Recall that δ is a d-dimensional predictable function, and we associate with it and with $q \geq 0$ the variables

$$\begin{aligned}
\widehat{\delta}(q,a)_{t,s} &= \frac{1}{s}\int_t^{t+s} du \int_{\{\|\delta(u,z)\|\leq a\}} \|\delta(u,z)\|^q \,\lambda(dz), \qquad \widehat{\delta}(q) = \widehat{\delta}(q,\infty) \\
\widehat{\delta}'(q)_{t,s} &= \widehat{\delta}(q,1) + \frac{1}{s}\int_t^{t+s} du \int_{\{\|\delta(u,z)\|>1\}} \|\delta(u,z)\| \,\lambda(dz) \\
\widehat{\delta}''(q)_{t,s} &= \widehat{\delta}(q,1) + \frac{1}{s}\int_t^{t+s} \lambda\big(\{z: \|\delta(u,z)\|>1\}\big)\,du.
\end{aligned} \tag{A.1}$$

Lemma 2.1.5 *Suppose that* $\int_0^t ds \int \|\delta(s,z)\|^2\lambda(dz) < \infty$ *for all* t. *Then the process* $Y = \delta \star (\mathfrak{p} - \mathfrak{q})$ *is a locally square integrable martingale, and for all finite stopping times* T *and* $s > 0$ *and* $p \in [1,2]$ *we have*

$$\mathbb{E}\Big(\sup_{0\leq u\leq s} \|Y_{T+u} - Y_T\|^p \mid \mathcal{F}_T\Big) \leq K_p\, s\, \mathbb{E}\big(\widehat{\delta}(p)_{T,s} \mid \mathcal{F}_T\big) \tag{A.2}$$

and also for $p \geq 2$:

$$\mathbb{E}\Big(\sup_{0\leq u\leq s} \|Y_{T+u} - Y_T\|^p \mid \mathcal{F}_T\Big) \leq K_p\big(s\,\mathbb{E}\big(\widehat{\delta}(p)_{T,s} \mid \mathcal{F}_T\big) + s^{p/2}\mathbb{E}\big(\widehat{\delta}(2)_{T,s}^{p/2} \mid \mathcal{F}_T\big)\big). \tag{A.3}$$

J. Jacod, P. Protter, *Discretization of Processes*,
Stochastic Modelling and Applied Probability 67,
DOI 10.1007/978-3-642-24127-7, © Springer-Verlag Berlin Heidelberg 2012

Proof 1) Upon arguing component by component, we can assume that the dimension is $d=1$. The fact that Y is a locally square-integrable martingale has already been mentioned in (2.1.31) for example. The finite stopping time T is fixed, and for all $w\geq 0$ we introduce the increasing processes

$$Z(w)=\big(|\delta|^w\,1_{(T,\infty)}\big)*\mathfrak{p},\qquad \widetilde{Z}(w)=\big(|\delta|^w\,1_{(T,\infty)}\big)*\mathfrak{q},$$

which are $[0,\infty]$-valued, and $\widetilde{Z}(w)_{T+s}=s\widehat{\delta}(w)_{T,s}$.

By the Burkholder-Davis-Gundy inequality (2.1.32) we have when $p\geq 1$:

$$\mathbb{E}\Big(\sup_{0\leq u\leq s}|Y_{T+u}-Y_T|^p\mid\mathcal{F}_T\Big)\;\leq\;K_q\,\mathbb{E}\big(Z(2)^{p/2}_{T+s}\mid\mathcal{F}_T\big).\tag{A.4}$$

2) Suppose $p\in[1,2]$. Then $(x^2+y^2)^{p/2}\leq x^p+y^p$ for all $x,y\geq 0$, and thus $Z(2)^{p/2}\leq Z(p)$. Moreover, $\mathfrak{q}$ being the predictable compensator of $\mathfrak{p}$, we have $\mathbb{E}(Z(w)_{T+s}\mid\mathcal{F}_T)=\mathbb{E}(\widetilde{Z}(w)_{T+s}\mid\mathcal{F}_T)$, those variables being finite or infinite, and (A.2) follows.

3) From now on we assume $p>2$. In view of (A.4), we need to prove that

$$\mathbb{E}\big(Z(2)^{p/2}_{T+s}\mid\mathcal{F}_T\big)\;\leq\;K_p\,\mathbb{E}\big(\widetilde{Z}(p)_{T+s}+\widetilde{Z}(2)^{p/2}_{T+s}\mid\mathcal{F}_T\big).\tag{A.5}$$

Set $S_n=\inf(t:\widetilde{Z}(2)_t+\widetilde{Z}(p)_t\geq n)$, which satisfies $S_n\geq T$. Suppose that we have shown

$$\mathbb{E}\big(Z(2)^{p/2}_{(T+s)\wedge S_n}\mid\mathcal{F}_T\big)\leq K_p\,\mathbb{E}\big(\widetilde{Z}(p)_{(T+s)\wedge S_n}+\widetilde{Z}(2)^{p/q}_{(T+s)\wedge S_n}\mid\mathcal{F}_T\big)\tag{A.6}$$

for all n. Let $S=\lim_n S_n$ and $B=\{\mathbb{P}(S\leq T+s\mid\mathcal{F}_T)=0\}$. As $n\to\infty$, the right side of (A.6) increases to the right side of (A.5), which is infinite on B^c. On the set B, we have $S>T+s$ almost surely, so the left side of (A.6) increases a.s. to the left side of (A.5) on this set: thus (A.3) holds on B, and also (trivially) on B^c.

We are thus left to proving (A.6) for any given n. Upon stopping all processes at time S_n, this amounts to proving (A.5) under the additional assumption that

$$\widetilde{Z}(p)_\infty\;\leq\;K,\qquad \widetilde{Z}(2)_\infty\;\leq\;K,$$

and by Hölder's inequality we also have $\widetilde{Z}(w)_\infty\leq K_w$ for all $w\in[2,p]$. For those w's, the process $\widetilde{Z}(w)$ is the predictable compensator of $Z(w)$, and $M(w)=Z(w)-\widetilde{Z}(w)$ is a martingale with quadratic variation $Z(2w)$, and the Burkholder-Davis-Gundy inequality yields for $r\geq 1$:

$$\mathbb{E}\big(\big|M(w)_{T+s}\big|^r\mid\mathcal{F}_T\big)\;\leq\;K_r\,\mathbb{E}\big(Z(2w)^{r/2}_{T+s}\mid\mathcal{F}_T\big).$$

Then, writing $Z(w)=M(w)+\widetilde{Z}(w)$, we get for $q\leq w\leq p$ and $r\geq 1$:

$$\mathbb{E}\big(Z(w)^r_{T+s}\mid\mathcal{F}_T\big)\;\leq\;K_r\,\mathbb{E}\big(\widetilde{Z}(w)^r_{T+s}+Z(2w)^{r/2}_{T+s}\mid\mathcal{F}_T\big).\tag{A.7}$$

4) We have $2^n < p \le 2^{n+1}$ for some integer $n \ge 1$. Applying (A.7) repeatedly, we deduce that

$$\begin{aligned}
\mathbb{E}\big(Z(2)_{T+s}\big)^{p/2} \mid \mathcal{F}_T) &\le K_p\,\mathbb{E}\big(\widetilde{Z}(2)^{p/2}_{T+s} + Z(4)^{p/4}_{T+s} \mid \mathcal{F}_T\big) \\
&\le K_p\,\mathbb{E}\big(\widetilde{Z}(2)^{p/2}_{T+s} + \widetilde{Z}(4)^{p/4}_{T+s} + Z(8)^{p/8}_{T+s} \mid \mathcal{F}_T\big) \\
&\le \cdots \text{(repeat the argument)} \\
&\le K_p\,\mathbb{E}\Bigg(\sum_{j=1}^n \widetilde{Z}\big(2^j\big)^{p/2^j}_{T+s} + Z\big(2^{n+1}\big)^{p/2^{n+1}}_{T+s} \mid \mathcal{F}_T\Bigg). \qquad \text{(A.8)}
\end{aligned}$$

Now, by the definition of $\widetilde{Z}(w)$ and Hölder's inequality, we have

$$\widetilde{Z}\big(2^j\big)_{T+s} \le \widetilde{Z}(2)^{(p-2^j)/(p-2)}_{T+s}\, \widetilde{Z}(p)^{(2^j-2)/(p-2)}_{T+s},$$

for $j = 1, \dots, n$, and another application of Hölder's inequality yields

$$\begin{aligned}
&\mathbb{E}\big(\widetilde{Z}\big(2^j\big)^{p/2^j}_{T+s} \mid \mathcal{F}_T\big) \\
&\quad\le \mathbb{E}\big(\widetilde{Z}(2)^{p/2}_{T+s} \mid \mathcal{F}_T\big)^{2(p-2^j)/2^j(p-2)}\, \mathbb{E}\big(\widetilde{Z}(p)_{T+s} \mid \mathcal{F}_T\big)^{p(2^j-2)/2^j(p-2)} \\
&\quad\le \mathbb{E}\big(\widetilde{Z}(2)^{p/2}_{T+s} \mid \mathcal{F}_T\big) + \mathbb{E}\big(\widetilde{Z}(p)_{T+s} \mid \mathcal{F}_T\big),
\end{aligned}$$

where the last inequality comes from $x^u y^v \le x + y$ when $x, y, u, v \ge 0$ and $u + v = 1$. On the other hand since $p/2^{n+1} \le 1$ we have $Z(2^{n+1})^{p/2^{n+1}} \le Z(p)$, hence $\mathbb{E}(Z(2^n)^{p/2^n} \mid \mathcal{F}_T) \le \mathbb{E}(\widetilde{Z}(p) \mid \mathcal{F}_T)$. Plugging all these in (A.8) gives us (A.5), and the proof is finished. □

Lemma 2.1.7 a) *If $\int_0^t \lambda(\{z : \delta(r, z) \ne 0\})\,dr < \infty$ for all t, the process $Y = \delta \star \mathfrak{p}$ has finitely many jumps on any finite interval.*

b) *Suppose that $\int_0^t ds \int \|\delta(s, z)\|\lambda(dz) < \infty$ for all t. Then the process $Y = \delta \star \mathfrak{p}$ is of locally integrable variation, and for all finite stopping times T and $s > 0$ and $p \in (0, 1]$ we have*

$$\mathbb{E}\Big(\sup_{0\le u\le s} \|Y_{T+u} - Y_T\|^p \mid \mathcal{F}_T\Big) \le K_p\, s\, \mathbb{E}\big(\widehat{\delta}(p)_{T,s} \mid \mathcal{F}_T\big), \qquad \text{(A.9)}$$

and also for $p \ge 1$

$$\mathbb{E}\Big(\sup_{0\le u\le s} \|Y_{T+u} - Y_T\|^p \mid \mathcal{F}_T\Big) \le K_p\big(s\,\mathbb{E}\big(\widehat{\delta}(p)_{T,s} \mid \mathcal{F}_T\big) + s^p\,\mathbb{E}\big(\widehat{\delta}(1)^p \mid \mathcal{F}_T\big)\big). \qquad \text{(A.10)}$$

Proof We still can assume that δ is one-dimensional.

a) The assumption amounts to saying that the continuous increasing process $1_{\{\delta\ne 0\}} \star \mathfrak{q}$ is finite-valued, hence locally integrable, thus $1_{\{\delta\ne 0\}} \star \mathfrak{p}$ is also finite-valued and the result follows.

b) By hypothesis $|\delta| \star \mathfrak{q}$ is finite-valued, and being continuous it is also locally integrable. Then $|\delta| \star \mathfrak{p}$ is also locally integrable, and thus Y is of locally integrable variation.

We use the notation of the previous proof, and the left sides of (A.9) and (A.10) are smaller than $\mathbb{E}(Z(1)^p_{T+s} \mid \mathcal{F}_T)$. When $p \in (0,1]$ we have $Z(1)^p \le Z(p)$, hence (A.9) follows. When $p > 1$, the proof of (A.10) is exactly the same as in the previous lemma for (A.3), except that $Z(2)$ and $p/2$ are substituted with $Z(1)$ and p. We have $2^n < p \le 2^{n+1}$ for some integer $n \ge 0$, and (A.8) is replaced by

$$\mathbb{E}\big((Z(1)_{T+s})^p \mid \mathcal{F}_T\big) \le K_p\, \mathbb{E}\bigg(\sum_{j=0}^{n} \widetilde{Z}(2^j)^{p/2^j}_{T+s} + Z(2^{n+1})^{p/2^{n+1}}_{T+s} \mid \mathcal{F}_T\bigg).$$

The rest of the proof goes in the same way. □

Next, we prove Lemmas 2.1.6 and 2.1.8 and Corollary 2.1.9.

Lemma 2.1.6 *Let $r \in [1,2]$. There exists a constant $K > 0$ depending on r, d, such that for all $q \in [0, 1/r]$ and $s \in [0,1]$, all finite stopping times T, and all d-dimensional processes $Y = \delta \star (\mathfrak{p} - \mathfrak{q})$, we have*

$$\mathbb{E}\bigg(\sup_{u\le s}\bigg(\frac{\|Y_{T+u} - Y_T\|}{s^q} \bigwedge 1\bigg)^r \mid \mathcal{F}_T\bigg)$$
$$\le K s^{1-qr}\, \mathbb{E}\big(\widehat{\delta}(r, s^{\frac{q}{2}})_{T,s} + s^{\frac{q(r-1)}{2}} \widehat{\delta}'(r)_{T,s} \mid \mathcal{F}_T\big), \qquad \text{(A.11)}$$

where $\widehat{\delta}(r,a)$ and $\widehat{\delta}'(r)$ are associated with δ by (A.1).

Proof 1) Again we argue in the one-dimensional case. The stopping time T, the numbers r, q and the time $s \in (0,1]$ are fixed throughout, and we use the simplifying notation

$$\gamma(a) = \mathbb{E}\big(\widehat{\delta}(r,a)_{T,s} \mid \mathcal{F}_T\big), \qquad \gamma' = \mathbb{E}\big(\widehat{\delta}'(r) \mid \mathcal{F}_T\big).$$

The set $\Omega_s = \{\gamma' < \infty\}$ is $\mathcal{F}_T$ measurable, and outside this set there is nothing to prove.

We put $\delta'(\omega,u,z) = \delta(\omega,u,z) 1_{\Omega_s}(\omega) 1_{(T(\omega),T(\omega)+s)}(u)$. The function δ' is predictable and by definition of Ω_s and $r \le 2$ it satisfies $(|\delta'|^2 \wedge |\delta'|) * \mathfrak{q}_t < \infty$ for all t. We deduce that for any $\varepsilon \in (0,1]$ the following processes are well defined:

$$N(\varepsilon) = 1_{\{|\delta'|>\varepsilon\}} \star \mathfrak{p}, \qquad M(\varepsilon) = \big(\delta' 1_{\{|\delta'|\le\varepsilon\}}\big) \star (\mathfrak{p} - \mathfrak{q}), \qquad B(\varepsilon) = -\big(\delta' 1_{\{|\delta'|>\varepsilon\}}\big) \star \mathfrak{q},$$

and they vanish at all times $t \le T$, and on the set Ω_s we have

$$N(\varepsilon)_{T+s} = 0,\ u \in [0,s] \quad \Rightarrow \quad Y_{T+u} - Y_T = M(\varepsilon)_{T+u} + B(\varepsilon)_{T+u}.$$

Therefore, since $(|x+y|\wedge 1)^r \le K(|x|^r+|y|)$ because $r>1$, we have on Ω_s:

$$\sup_{u\le s}\Big(\frac{|Y_{T+u}-Y_T|}{s^q}\bigwedge 1\Big)^r$$
$$\le K\Big(1_{\{N(\varepsilon)_{T+s}>0\}}+s^{-qr}\sup_{u\le s}\big|M(\varepsilon)_{T+u}\big|^r+s^{-q}\sup_{u\le s}\big|B(\varepsilon)_{T+u}\big|\Big). \quad \text{(A.12)}$$

2) In the second step we give a few simple estimates. First, $N(\varepsilon)$ is non-decreasing and integer-valued, with compensator $1_{\{|\delta'|>\varepsilon\}}\star\mathfrak{q}$, hence

$$\mathbb{P}\big(N(\varepsilon)_{T+s}>0\mid\mathcal{F}_T\big)$$
$$\le\mathbb{E}\big(N(\varepsilon)_{T+s}\mid\mathcal{F}_T\big)=\mathbb{E}\big((1_{\{|\delta'|>\varepsilon\}})\star\mathfrak{q}_{T+s}\mid\mathcal{F}_T\big)\le s\,\varepsilon^{-r}\gamma', \quad \text{(A.13)}$$

the last inequality coming from $1_{\{|\delta'|>\varepsilon\}}\le\varepsilon^{-r}(|\delta'|^r\,1_{\{|\delta'|\le 1\}}+|\delta'|\,1_{\{|\delta'|>1\}})$. Second, (A.2) yields

$$\mathbb{E}\Big(\sup_{u\le s}\big|M(\varepsilon)_{T+u}\big|^r\mid\mathcal{F}_T\Big)\ \le\ Ks\,\gamma(\varepsilon). \quad \text{(A.14)}$$

Third, $|\delta'|\le\varepsilon^{1-r}(|\delta'|^r\,1_{\{|\delta'|\le 1\}}+|\delta'|\,1_{\{|\delta'|>1\}})$, because $r\ge 1$, hence

$$\mathbb{E}\Big(\sup_{u\le s}\big|B(\varepsilon)_{T+u}\big|\mid\mathcal{F}_T\Big)\ \le\ s\,\varepsilon^{1-r}\gamma'. \quad \text{(A.15)}$$

3) We can now turn to the proof itself. By (A.12) and the estimates (A.13), (A.14) and (A.15), and with $\varepsilon=s^{q/2}$, we see that on the set Ω_s the left side of (A.11) is smaller than

$$Ks^{1-qr}\big(\gamma\big(s^{q/2}\big)+\big(s^{q(r-1)/2}+s^{qr/2}\gamma'\big)\big),$$

which completes the proof. ☐

Lemma 2.1.8 *Let $r\in(0,1]$. There exists a constant $K>0$ depending on r,d, such that for all $q\in[0,1/r]$ and $s\in[0,1]$, all finite stopping times T, and all d-dimensional processes $Y=\delta\star\mathfrak{p}$, we have*

$$\mathbb{E}\Big(\sup_{u\le s}\Big(\frac{\|Y_{T+u}-Y_T\|}{s^q}\bigwedge 1\Big)^r\mid\mathcal{F}_T\Big)$$
$$\le Ks^{1-qr}\,\mathbb{E}\big(\widehat{\delta}\big(r,s^{\frac{q}{2}}\big)_{T,s}+s^{\frac{rq}{2}}\widehat{\delta}''(r)_{T,s}\mid\mathcal{F}_T\big), \quad \text{(A.16)}$$

where $\widehat{\delta}(r,a)$ and $\widehat{\delta}''(r)$ are associated with δ by (A.1).

Proof The proof is similar—and slightly simpler—than for the previous lemma, to which we borrow the notation, and in particular $\gamma(a)$, whereas we set

$$\gamma''=\mathbb{E}\big(\widehat{\delta}''(r)\mid\mathcal{F}_T\big),\qquad \Omega_s=\big\{\gamma''<\infty\big\},$$

and again it suffices to prove the result on the set Ω_s.

We define δ' and $N(\varepsilon)$ as in the previous proof, and we set $A(\varepsilon) = (\delta' 1_{\{|\delta'|\leq\varepsilon\}})\star\mathfrak{p}$, which is well defined on the set Ω_s, and instead of (A.12) we have

$$N(\varepsilon)_{T+s} = 0,\ u \in [0,s] \quad \Rightarrow \quad Y_{T+u} - Y_T \ = \ A(\varepsilon)_{T+u},$$

which yields on Ω_s:

$$\sup_{u\leq s}\left(\frac{|Y_{T+u} - Y_T|}{s^q} \bigwedge 1\right)^r \leq 1_{\{N(\varepsilon)_{T+s}>0\}} + s^{-qr} \sup_{u\leq s} |A(\varepsilon)_{T+u}|^r. \tag{A.17}$$

Observing that $1_{\{|\delta'|>\varepsilon\}} \leq \varepsilon^{-r}|\delta'|^r\, 1_{\{|\delta'|\leq 1\}} + 1_{\{|\delta'|>1\}}$, we replace (A.13) by

$$\mathbb{P}\big(N(\varepsilon)_{T+s} > 0 \mid \mathcal{F}_T\big) \ \leq \ s\,\varepsilon^{-r}\,\gamma''.$$

Moreover, (A.9) implies

$$\mathbb{E}\Big(\sup_{u\leq s} \big|A(\varepsilon)_{T+u}\big|^r \mid \mathcal{F}_T\Big) \ \leq \ Ks\gamma(\varepsilon)_s.$$

These estimates and (A.17) yield, upon taking $\varepsilon = s^{q/2}$, that the left side of (A.16) is smaller on Ω_s than $Ks^{1-qr}(s^{qr/2}\widehat{\delta'} + \widehat{\delta}(s^{q/2}))$, hence (A.16) holds. □

Corollary 2.1.9 *Assume that the d-dimensional predictable function δ satisfy $\|\delta(\omega,t,z)\| \leq \Gamma(z)$ for some measurable function Γ on E, and let $p > 0$, $r \in (0,2]$ and $q \in [0, 1/r)$.*

a) *If $r \in (1,2]$ and $\int(\Gamma(z)^r \wedge \Gamma(z))\,\lambda(dz) < \infty$, the process $Y = \delta * (\mathfrak{p} - \mathfrak{q})$ satisfies*

$$\mathbb{E}\left(\sup_{u\leq s}\left(\frac{\|Y_{T+u} - Y_T\|}{s^q} \bigwedge 1\right)^p \mid \mathcal{F}_T\right) \leq \begin{cases} Ks^{p(1-qr)/r}\,\phi(s) & \text{if } p \leq r \\ Ks^{1-qr}\,\phi(s) & \text{if } p \geq r \end{cases} \tag{A.18}$$

for all $s \in (0,1]$ and all finite stopping times T, where K and ϕ depend on r, p, q, Γ and λ, and $\phi(s) \to 0$ as $s \to 0$ when $q > 0$, and $\sup\phi < \infty$ when $q = 0$.

b) *If $r \in (0,1]$ and $\int(\Gamma(z)^r \vee \Gamma(z))\,\lambda(dz) < \infty$, the process $Y = \delta * (\mathfrak{p} - \mathfrak{q})$ satisfies*

$$\mathbb{E}\left(\sup_{u\leq s}\left(\frac{\|Y_{T+u} - Y_T\|}{s^q} \bigwedge 1\right)^p \mid \mathcal{F}_T\right) \leq Ks^{1-qr}\,\phi(s) \quad \text{if } p > 1,\ q < \frac{p-1}{p-r}. \tag{A.19}$$

for all $s \in (0,1]$ and all finite stopping times T, with K and ϕ as in (a).

c) *If $r \in (0,1]$ and $\int(\Gamma(z)^r \wedge 1)\,\lambda(dz) < \infty$, the process $Y = \delta * \mathfrak{p}$ satisfies*

$$\mathbb{E}\left(\sup_{u\leq s}\left(\frac{\|Y_{T+u} - Y_T\|}{s^q} \bigwedge 1\right)^p \mid \mathcal{F}_T\right) \leq \begin{cases} Ks^{p(1-qr)/r}\,\phi(s) & \text{if } p \leq r \\ Ks^{1-qr}\,\phi(s) & \text{if } p \geq r. \end{cases} \tag{A.20}$$

for all $s \in (0,1]$ and all finite stopping times T, with K and ϕ as in (a).

Proof Despite the name "corollary", this result does not follow *stricto sensu* from the two previous lemmas, for the following reason: assuming $\|\delta(\omega,t,z)\| \le \Gamma(z)$, we have $\widehat{\delta}(q)_{t,s} \le \int \Gamma(z)^q\,\lambda(dz)$, but with the notation

$$\widetilde{\delta}(q,a) = \int_{\{z:\Gamma(z)\le a\}} \Gamma(z)^q\,\lambda(dz), \qquad \widetilde{\delta}'(q) = \widetilde{\delta}(q,1) + \int \Gamma(z) 1_{\{\Gamma(z)>1\}}\,\lambda(dz)$$
$$\widetilde{\delta}''(q) = \widetilde{\delta}(q,1) + \lambda\big(\{z:\Gamma(z)>1\}\big),$$

the inequalities $\widehat{\delta}(q,a)_{t,s} \le \widetilde{\delta}(q,a)$ or $\widehat{\delta'}(q)_{t,s} \le \widetilde{\delta}'(q)$ or $\widehat{\delta''}(q)_{t,s} \le \widetilde{\delta}''(q)$ may fail.

However, we can replace $N(\varepsilon)$, $M(\varepsilon)$ and $B(\varepsilon)$ in the proof of Lemma 2.1.6 by

$$N(\varepsilon) = 1_{\{\Gamma>\varepsilon\}} * \mathfrak{p}, \qquad M(\varepsilon) = (\delta 1_{\{\Gamma\le\varepsilon\}}) * (\mathfrak{p}-\mathfrak{q}), \qquad B(\varepsilon) = (\delta 1_{\{\Gamma>\varepsilon\}}) * \mathfrak{q}.$$

These processes are well defined if $\widetilde{\delta}'(r) < \infty$ for some $r \in [1,2]$, and in this case we have (A.12), because $\|\delta(\omega,t,z)\| \le \Gamma(z)$. Moreover, the estimates (A.13), (A.14) and (A.15) hold if we substitute γ' and $\gamma(\varepsilon)$ with $\widetilde{\delta}'(r)$ and $\widehat{\delta}(r,\varepsilon)$. Therefore, we have (A.11) with $\widetilde{\delta}(r,s^{q/2})$ and $\widetilde{\delta}'(r)$ instead of $\widehat{\delta}(r,s^{q/2})$ and $\widehat{\delta}'(r)$. In exactly the same way, (A.16) holds when $r \in [0,1]$ with $\widetilde{\delta}''(r)$ instead of $\widehat{\delta}''(r)$.

These new versions of (A.11) and (A.16) give (A.18) and (A.20) when $p = r$, upon taking $\phi(s) = \widetilde{\delta}(r,s^{q/2}) + s^{q(r-1)/2}$ in the first case, and $\phi(s) = \widetilde{\delta}(r,s^{q/2}) + s^{qr/2}$ in the second case. Now, (A.18) and (A.20) for $p < r$ follow by Hölder's inequality, and for $p > r$ from the fact that the left sides when $p > r$ is smaller than the left side for $p = r$.

It remains to prove (b), and this can be done separately for $Y = \delta * \mathfrak{p}$ and for $Y = \delta * \mathfrak{q}$. The case $Y = \delta * \mathfrak{p}$ reduces to (A.20) applied when $p > r$, because here $r \le 1 < p$. For $Y = \delta * \mathfrak{q}$ one observes that $\|Y_{t+s} - Y_t\| \le Ks$: the result follows, upon taking $\phi(s) = s^{p-pq-1+qr}$, which goes to 0 as $s \to 0$ (when $p \ge r$) if and only if $p > 1$ and $q < \frac{p-1}{p-r}$. □

Finally we prove Proposition 2.1.10. Recall that A is a measurable subset of E. We denote by $\mathcal{H}^A$ the σ-field generated by the restriction of the measure $\mathfrak{p}$ to $\mathbb{R}_+ \times A$, and by $\mathcal{H}^W$ the σ-field generated by the process W, and we set

$$\begin{aligned} &(\mathcal{G}_t^A) = \text{the smallest filtration containing } (\mathcal{F}_t) \text{ and with } \mathcal{H}^A \subset \mathcal{G}_0^A \\ &(\mathcal{G}_t^{A,W}) = \text{the smallest filtration containing } (\mathcal{F}_t) \text{ and with } \mathcal{H}^A \cup \mathcal{H}^W \subset \mathcal{G}_0^{A,W} \end{aligned} \tag{A.21}$$

Proposition 2.1.10 *In the above setting, we have:*

a) *The process W is a Brownian motion relative to the filtration $(\mathcal{G}_t^A)$, and* (2.1.34) *holds if σ is $(\mathcal{F}_t)$-optional and T is a stopping time relative to the filtration $(\mathcal{G}_t^A)$ and the conditional expectations are taken relative to $\mathcal{G}_T^A$.*

b) *The restriction $\mathfrak{p}'$ of $\mathfrak{p}$ to the set $\mathbb{R}_+ \times A^c$ is a Poisson random measure with respect to the filtration $(\mathcal{G}_t^{A,W})$, and its Lévy measure λ' is the restriction of λ to A^c. Moreover if δ is $(\mathcal{F}_t)$-predictable and satisfies $\delta(\omega,t,z) = 0$ for all (ω,t,z) with*

$z \in A$, *Lemmas* 2.1.5, 2.1.6, 2.1.7 *and* 2.1.8 *hold if* T *is a stopping time relative to the filtration* $(\mathcal{G}_t^{A,W})$ *and the conditional expectations are taken relative to* $\mathcal{G}_T^{A,W}$.

Proof a) Since W and $\mathfrak{p}$ are independent, the process W is still a Brownian motion with respect to the filtration $(\mathcal{G}_t^A)$. Since σ_t is $(\mathcal{F}_t)$-optional, it is *a fortiori* $(\mathcal{G}_t^A)$-optional, and the stochastic integral $\int_0^t \sigma_s \, dW_s$ is the same, relative to both filtrations. The result follows.

b) The first claim follows from the independence of $\mathfrak{p}'$ and the σ-field $\mathcal{H}^A \cup \mathcal{H}^W$, and both the $(\mathcal{F}_t)$ and $(\mathcal{G}_t^{A,W})$ compensators of $\mathfrak{p}'$ are equal to the restriction $\mathfrak{q}'$ of $\mathfrak{q}$ to $\mathbb{R}_+ \times A^c$. Now, if δ is $(\mathcal{F}_t)$-predictable and satisfies $\delta(\omega, t, z) = 0$ when $z \in A$, we have $\delta * (\mathfrak{p} - \mathfrak{q}) = \delta * (\mathfrak{p}' - \mathfrak{q}')$ as soon as any one of these two stochastic integrals exists, because they have the same jumps and are purely discontinuous local martingales vanishing at 0. This is all relative to $(\mathcal{F}_t)$, however by virtue of Exercice 9.4, p. 294, of [52], the stochastic integral $\delta * (\mathfrak{p}' - \mathfrak{q}')$ is the same, relative to $(\mathcal{F}_t)$ and to $(\mathcal{G}_t^{A,W})$. And, obviously, $\delta * \mu = \delta * \mu'$ as soon as any one of these two (ordinary) integrals exists, and in this case $\delta * \mu'$ does not depend on the filtration. Applying Lemmas 2.1.5, 2.1.6, 2.1.7 or 2.1.8 for the filtration $(\mathcal{G}_t^{A,W})$ gives the last claim. □

A.2 Convergence of Processes

Here we prove the convergence results of Sect. 2.2.2. First we study the processes having the decomposition (2.2.20), that is, we prove Propositions 2.2.1, 2.2.2 and 2.2.4. Recall that we have d-dimensional processes X^n and X with, for each $m \geq 1$, the decompositions

$$X^n = X(m)^n + X'(m)^n. \tag{A.22}$$

Proposition 2.2.1 *Let* X^n *and* X *be defined on the same probability space. For* $X^n \stackrel{\mathbb{P}}{\Longrightarrow} X$ *it is enough that there are decompositions* (A.22) *and also* $X = X(m) + X'(m)$, *with the following properties*:

$$\begin{gathered}
\forall m \geq 1, \quad X(m)^n \stackrel{\mathbb{P}}{\Longrightarrow} X(m), \quad \textit{as } n \to \infty \\
X(m) \stackrel{\text{u.c.p.}}{\Longrightarrow} X, \quad \textit{as } m \to \infty, \\
\forall \eta, t > 0, \quad \lim_{m \to \infty} \limsup_{n \to \infty} \mathbb{P}\Big(\sup_{s \leq t} \|X'(m)^n_s\| > \eta\Big) = 0.
\end{gathered} \tag{A.23}$$

Proof We will denote δ_U and δ_S two distances on the space $\mathbb{D}^d$ which are compatible with the local uniform topology and the Skorokhod topology, respectively. Among all possible choices, we can choose them in such a way that $\delta_S(x, y) \leq \delta_S(x, z) + \delta_U(z, y)$ for all $x, y, z \in \mathbb{D}^d$ (this is in line with the property (2.2.9)), and also such that $\delta_U(x, y) \leq 1/n$ when $x(t) = y(t)$ for all $t \leq n$. For example $\delta_U = \delta_{lu}$

and $\delta_S = \delta'$, as defined in (VI.1.2) and Remark VI.1.27 of [57], provides such a choice.

With this notation, and using (A.22), we can rewrite the three hypotheses above as

$$\forall m \geq 1, \quad \delta_S\big(X(m)^n, X(m)\big) \stackrel{\mathbb{P}}{\longrightarrow} 0, \quad \text{as } n \to \infty$$

$$\delta_u\big(X(m), X\big) \stackrel{\mathbb{P}}{\longrightarrow} 0, \quad \text{as } m \to \infty, \tag{A.24}$$

$$\forall \eta > 0, \quad \lim_{m\to\infty} \limsup_{n\to\infty} \mathbb{P}\big(\delta_U\big(X^n, X(m)^n\big) > \eta\big) = 0. \tag{A.25}$$

The properties of δ_U and δ_S give

$$\delta_S\big(X^n, X\big) \leq \delta_S\big(X(m)^n, X(m)\big) + \delta_U\big(X(m), X\big) + \delta_U\big(X^n, X(m)^n\big).$$

Therefore for all $\eta > 0$, the number $\mathbb{P}(\delta_S(X^n, X) > 3\eta)$ is not bigger than

$$\mathbb{P}\big(\delta_S\big(X(m)^n, X(m)\big) > \eta\big) + \mathbb{P}\big(\delta_U\big(X(m), X\big) > \eta\big) + \mathbb{P}\big(\delta_U\big(X^n, X(m)^n\big) > \eta\big).$$

By choosing first m large, and then N large, this quantity can be made arbitrarily small, uniformly in $n \geq N$. Hence $\mathbb{P}(\delta_S(X^n, X) > 3\eta) \to 0$ as $n \to \infty$. □

Proposition 2.2.2 *For* $X^n \stackrel{\mathcal{L}}{\Longrightarrow} X$ *it is enough that there are decompositions* (A.22) *satisfying* (A.23) *and*

$$\forall m \geq 1, \quad X(m)^n \stackrel{\mathcal{L}}{\Longrightarrow} X(m), \quad \text{as } n \to \infty$$

for some limiting processes $X(m)$, *which in turn satisfy*

$$X(m) \stackrel{\mathcal{L}}{\Longrightarrow} X, \quad \text{as } m \to \infty.$$

Proof The notation is as in the previous proof. As mentioned after (2.2.2), it suffices to check the convergence of the integrals of any bounded Lipschitz function. Hence here we only need to prove that $\mathbb{E}(f(X^n)) \to \mathbb{E}(f(X)$ for any bounded function f on $\mathbb{D}^d$ which is Lipschitz for the distance δ_S. The assumptions imply (A.24), (A.25), and

$$\left.\begin{array}{l} \forall m \geq 1, \quad \mathbb{E}\big(f\big(X(m)^n\big)\big) \to \mathbb{E}\big(f\big(X(m)\big)\big) \quad \text{as } n \to \infty, \\ \mathbb{E}\big(f\big(X(m)\big)\big) \to \mathbb{E}\big(f(X)\big) \quad \text{as } m \to \infty. \end{array}\right\} \tag{A.26}$$

The Lipschitz property of f, plus $\delta_S \leq \delta_U$, yield

$$\big|f\big(X^n\big) - f\big(X(m)^n\big)\big| \leq \delta_U\big(X^n, X(m)^n\big) \wedge K$$

where K is twice the bound of f. Therefore $|\mathbb{E}(f(X^n)) - \mathbb{E}(f(X))|$ is not bigger than

$$\big|\mathbb{E}\big(f\big(X(m)^n\big)\big) - \mathbb{E}\big(f\big(X(m)\big)\big)\big| + \mathbb{E}\big(\delta_U\big(X^n, X(m)^n\big) \wedge K\big)$$

$$+\big|\mathbb{E}\big(f\big(X(m)\big)\big)-\mathbb{E}\big(f(X)\big)\big|.$$

Exactly as in the previous proof, this quantity may be made as small as one wishes for all n large enough, by (A.25) and (A.26), and the result follows. □

Proposition 2.2.4 *For $X^n \stackrel{\mathcal{L}\text{-s}}{\Longrightarrow} X$ it is enough that there are decompositions* (A.22) *satisfying* (A.23), *and*

$$\forall m\geq 1,\quad X(m)^n \stackrel{\mathcal{L}\text{-s}}{\Longrightarrow} X(m),\quad \textit{as } n\to\infty$$

for some limiting processes $X(m)$, which in turn satisfy

$$X(m) \stackrel{\mathcal{L}\text{-s}}{\Longrightarrow} X,\quad \textit{as } m\to\infty.$$

Proof Recall that here X and $X(m)$ are defined on their own extensions $(\widetilde{\Omega},\widetilde{\mathcal{F}},\widetilde{\mathbb{P}})$ and $(\widetilde{\Omega}_m,\mathcal{F}_m,\widetilde{\mathbb{P}}_m)$ of the space $(\Omega,\mathcal{F},\mathbb{P})$. As to the processes X^n and $X(m)^n$, they are defined on an extension $(\overline{\Omega}_n,\overline{\mathcal{F}}_n,\mathbb{P}_n)$ of the same space $(\Omega,\mathcal{F},\mathbb{P})$. We need to prove that $\overline{\mathbb{E}}_n(Yf(X^n))\to\widetilde{\mathbb{E}}(Yf(X))$ for any function f as in the previous proof, and any $\mathcal{F}$ measurable variable Y which is bounded by 1. The assumptions imply (A.24), (A.25), and

$$\forall m\geq 1,\quad \overline{\mathbb{E}}_n\big(Yf\big(X(m)^n\big)\big)\to\widetilde{\mathbb{E}}_m\big(Yf\big(X(m)\big)\big)\quad \text{as } n\to\infty$$
$$\widetilde{\mathbb{E}}_m\big(Yf\big(X(m)\big)\big)\to\widetilde{\mathbb{E}}\big(Yf(X)\big)\quad \text{as } m\to\infty.$$

Then the proof goes exactly as in the previous proposition. □

Now we turn to the "martingale" results.

Proposition 2.2.5 *Let (M^n) be a sequence of local martingales on the filtered space $(\Omega,\mathcal{F},(\mathcal{F}_t)_{t\geq 0},\mathbb{P})$, with $M^n_0=0$. We have $M^n \stackrel{\text{u.c.p.}}{\Longrightarrow} 0$ as soon as one of the following two conditions holds*:

(i) *each M^n admits an angle bracket and $\langle M^n,M^n\rangle_t \stackrel{\mathbb{P}}{\longrightarrow} 0$ for all $t>0$,*
(ii) *we have $|\Delta M^n_s|\leq K$ for a constant K, and $[M^n,M^n]_t \stackrel{\mathbb{P}}{\longrightarrow} 0$ for all $t>0$.*

Proof We first prove the result under (i). Since $M^n_0=0$, our assumption implies the existence of a localizing sequence (T_n) of stopping times such that $\mathbb{E}((M^n_{T\wedge T_n})^2)=\mathbb{E}(\langle M^n,M^n\rangle_{T\wedge T_n})$ for any finite stopping time T, and by Fatou's lemma and the monotone convergence theorem we have $\mathbb{E}((M^n_T)^2)\leq\mathbb{E}(\langle M^n,M^n\rangle_T)$. That is, $(M^n)^2$ is Lenglart-dominated by $\langle M^n,M^n\rangle$, and the first part of (2.1.49) yields

$$\mathbb{P}\Big(\sup_{s\leq t}\big(M^n_s\big)^2\geq\varepsilon\Big)\leq\frac{\eta}{\varepsilon}+\mathbb{P}\big(\langle M^n,M^n\rangle_t\geq\eta\big)$$

for all $t,\varepsilon,\eta>0$. Choosing first η small and then n large, we see that for all $\varepsilon>0$ the left side above goes to 0 as $n\to\infty$, and the result follows.

Next, we suppose that we have (ii), which implies that $\langle M^n, M^n\rangle$ exists, so by what precedes it is enough to prove that $\langle M^n, M^n\rangle_t \stackrel{\mathbb{P}}{\longrightarrow} 0$. The process $\langle M^n, M^n\rangle$ is Lenglart-dominated by the process $[M^n, M^n]$, which is adapted increasing with jumps smaller than K^2. Hence the second part of (2.1.49) yields

$$\mathbb{P}\big(\langle M^n, M^n\rangle_t \geq \varepsilon\big) \leq \frac{\eta}{\varepsilon} + \frac{1}{\varepsilon}\,\mathbb{E}\big(K^2 \wedge [M^n, M^n]_t\big) + \mathbb{P}\big([M^n, M^n]_t \geq \eta\big)$$

for all $t, \varepsilon, \eta > 0$. The second term on the right side above goes to 0 as $n \to \infty$ by the dominated convergence theorem and (ii), as well as the third term. Then obviously $\mathbb{P}(\langle M^n, M^n\rangle_t \geq \varepsilon) \to 0$, and we have the result. □

Remark A.1 This proposition is stated when all local martingales are on the same filtered space $(\Omega, \mathcal{F}, (\mathcal{F}_t)_{t\geq 0}, \mathbb{P})$, but this is not necessary. Each process M^n can be a local martingale on a filtered space $(\Omega_n, \mathcal{F}^n, (\mathcal{F}^n_t)_{t\geq 0}, \mathbb{P}_n)$ depending on n: the results are not modified (we apply Lenglart's inequality on each filtered space). This slight extension will be used later.

In the next result, μ is the jump measure of a càdlàg d-dimensional process, hence $E = \mathbb{R}^d$, or it is a Poisson random measure on $\mathbb{R}_+ \times E$ for E a Polish space, and ν is its compensator.

Proposition 2.2.6 *Let (δ_n) be a sequence of predictable functions on $\Omega \times \mathbb{R}_+ \times E$, each δ_n satisfying* (2.1.16). *Then*

$$\big((\delta_n)^2 \wedge |\delta_n|\big) \star \nu_t \stackrel{\mathbb{P}}{\longrightarrow} 0 \quad \forall t > 0 \quad \Rightarrow \quad \delta_n \star (\mu - \nu) \stackrel{\text{u.c.p.}}{\Longrightarrow} 0. \tag{A.27}$$

Proof $M^n = \delta_n * (\mu - \nu)$ can be written as $M^n = M'^n + M''^n$, where

$$M'^n = (\delta_n 1_{\{|\delta_n|\leq 1\}}) \star (\mu - \nu), \qquad M''^n = (\delta_n 1_{\{|\delta|>1\}}) \star (\mu - \nu).$$

First, M'^n is a local martingale with bounded jumps, so its angle bracket exists and it turns out to be (see e.g. [57]):

$$\langle M'^n, M'^n\rangle_t = \big(\delta_n^2 1_{\{|\delta_n|\leq 1\}}\big) \star \nu_t - \sum_{s\leq t} \bigg(\int \delta_n(s,x)\, 1_{\{|\delta_n(s,x)|\leq 1\}} \nu(\{s\}, dx)\bigg)^2,$$

which is smaller than $(\delta_n^2 1_{\{|\delta_n|\leq 1\}}) \star \nu_t$, so our assumption yields $\langle M'^n, M'^n\rangle_t \stackrel{\mathbb{P}}{\longrightarrow} 0$, hence $M'^n \stackrel{\text{u.c.p.}}{\Longrightarrow} 0$ follows from the previous proposition.

Second, the process M''^n is of finite variation, and by the properties of the predictable compensators it is Lenglart-dominated by the increasing predictable process $B^n = 2(|x|\, 1_{\{|x|>1\}}) \star \nu^{M^n}$. Then the first part of (2.1.49) yields for all $t, \eta, \varepsilon > 0$:

$$\mathbb{P}\Big(\sup_{s\leq t} |M''^n_s| \geq \varepsilon\Big) \leq \frac{\eta}{\varepsilon} + \mathbb{P}\big(B^n_t \geq \eta\big).$$

The last term above goes to 0 as $n \to \infty$ by the assumption in (A.27), hence $M''^n \stackrel{\text{u.c.p.}}{\Longrightarrow} 0$. This finishes the proof. □

Proposition 2.2.7 *Let X be a semimartingale and (H^n) a sequence of predictable processes satisfying $|H^n| \le H'$ for some predictable and locally bounded process H'. If outside a null set we have $H^n_t \to H_t$ for all t, where H is another predictable process, then we have*

$$\int_0^t H^n_s\,dX_s \stackrel{\text{u.c.p.}}{\Longrightarrow} \int_0^t H_s\,dX_s.$$

Proof Using the decomposition (2.1.10), we can write $X = X_0 + A + M$, where A is a process of locally finite variation and M is a local martingale with bounded jumps. Obviously, $\int_0^t H^n_s\,dA_s \stackrel{\text{u.c.p.}}{\Longrightarrow} \int_0^t H_s\,dA_s$ by the "ordinary" Lebesgue dominated convergence theorem, so it remains to prove that

$$N^n \stackrel{\text{u.c.p.}}{\Longrightarrow} 0, \quad \text{where } N^n_t = \int_0^t H^n_s\,dM_s - \int_0^t H_s\,dM_s = \int_0^t \bigl(H^n_s - H_s\bigr)\,dM_s.$$

Since M has bounded jumps, $\langle M, M\rangle$ exists, as well as the angle bracket of N^n, which indeed is

$$\langle N^n, N^n\rangle_t = \int_0^t \bigl(H^n_s - H_s\bigr)^2\,d\langle M, M\rangle_s.$$

This goes to 0 a.s. by the ordinary Lebesgue theorem again, and we conclude by Proposition 2.2.5. □

Proposition 2.2.8 *Let X be a semimartingale and H be a càglàd adapted process. For each n let $(T(n,i) : i \ge 0)$ be a sequence of stopping times, which strictly increases to $+\infty$, and with $T(n,0) = 0$, and such that $\sup(T(n,i+1) \wedge t - T(n,i) \wedge t : i \ge 0)$ goes to 0 in probability for all t as $n \to \infty$. Then we have*

$$\sum_{i \ge 1,\ T(n,i) \le t} H_{T(n,i-1)}\bigl(X_{T(n,i)} - X_{T(n,i-1)}\bigr) \stackrel{\mathbb{P}}{\Longrightarrow} \int_0^t H_s\,dX_s \tag{A.28}$$

(convergence for the Skorokhod topology). If further X is continuous the same holds also when H is adapted càdlàg.

Proof (a) We first suppose H càglàg adapted, so $H'_t = \sup_{s \le t} \|H_s\|$ is a locally bounded predictable process. Moreover the processes H^n defined by $H^n_t = H_{T(n,i-1)}$ if $T(n,i-1) < t \le T(n,i)$ and $H^n_0 = H_0$ are predictable, with $\|H^n\| \le H'$, and converge pointwise to H as $n \to \infty$. Then Proposition 2.2.7 yields that $Y^n_t := \int_0^t H^n_s\,dX_s \stackrel{\text{u.c.p.}}{\Longrightarrow} Y_t := \int_0^t H_s\,dX_s$.

Below we use the notation (2.2.12) with $t(n,i) = T(n,i)$ and the process Y^n, thus giving rise to the "discretized process" $Y^{n,(n)}_t = Y^n_{T(n,i)}$ if $T(n,i) \le t <$

$T(n, i+1)$. The left side of (A.28) is $Y_t^{n,(n)}$. Then (2.2.13) and the subsequences principle (2.2.17) give us the convergence (A.28).

(b) Now suppose that X is continuous and H is adapted càdlàg. The process H' is no longer predictable and locally bounded, but its left limit process H'_- is. Moreover we have $\|H^n\| \le H'_-$, and $H^n \to H_-$ pointwise, so what precedes yields that the left side of (A.28) converges in the u.c.p. sense to $\int_0^t H_{s-} dX_s$. We conclude by observing that, since X is continuous, the integral $\int_0^t H_s\, dX_s$ is actually well defined and coincides with $\int_0^t H_{s-} dX_s$. □

A.3 Triangular Arrays

Here we prove Lemmas 2.2.10 and 2.2.11. We have a one-dimensional triangular array (ζ_i^n) and the stopping rules $N_n(t)$, and a discrete-time filtration $(\mathcal{G}_i^n)_{i\in\mathbb{N}}$, with the following basic assumption:

- $n \ge 1,\ i \ge 1 \ \Rightarrow\ \zeta_i^n$ is $\mathcal{G}_i^n$ measurable
- $n \ge 1,\ t \ge 0 \ \Rightarrow\ N_n(t)$ is a $(\mathcal{G}_i^n)$-stopping time.

We also recall the following property, stated in (2.2.31), and which plays an important role: assuming that the ζ_i^n's are either all integrable or all nonnegative, then, with $\overline{\mathcal{F}}_t^n = \mathcal{G}_{N_n(t)}^n$,

$$\text{the } \big(\overline{\mathcal{F}}_t^n\big)\text{-compensator of } S_t^n = \sum_{i=1}^{N_n(t)} \zeta_i^n \text{ is } S_t'^n = \sum_{i=1}^{N_n(t)} \mathbb{E}\big(\zeta_i^n \mid \mathcal{G}_{i-1}^n\big). \tag{A.29}$$

Lemma 2.2.10 *The array (ζ_i^n) is AN as soon as the array $(\mathbb{E}(|\zeta_i^n| \mid \mathcal{G}_{i-1}^n))$ is AN.*

Proof Without restriction we can assume $\zeta_i^n \ge 0$. We use the notation (A.29), which implies that S^n is Lenglart-dominated by the predictable increasing process S'^n. The first part of (2.1.49) yields

$$\mathbb{P}\big(S_t^n \ge \varepsilon\big) \le \frac{\eta}{\varepsilon} + \mathbb{P}\big(S_t'^n \ge \eta\big)$$

for all $t, \varepsilon, \eta > 0$. By (a) we have $S_t'^n \xrightarrow{\mathbb{P}} 0$, so we deduce from the above that $S_t^n \xrightarrow{\mathbb{P}} 0$, which is the result. □

Lemma 2.2.11 *Let (ζ_i^n) be a triangular array such that each ζ_i^n is square-integrable. Then the array $(\zeta_i^n - \mathbb{E}(\zeta_i^n \mid \mathcal{G}_{i-1}^n))$ is AN under each of the following three conditions*:

(a) *The array $(\mathbb{E}(|\zeta_i^n|^2 \mid \mathcal{G}_{i-1}^n))$ is AN.*

(b) *The sequence of variables $(\sum_{i=1}^{N_n(t)} \mathbb{E}(|\zeta_i^n|^2 1_{\{|\zeta_i^n|>1\}} \mid \mathcal{G}_{i-1}^n))_{n\ge 1}$ is bounded in probability for each $t > 0$, and the array $(|\zeta_i^n|^2)$ is AN.*

(c) *We have* $|\zeta_i^n| \le K$ *for a constant* K, *and the array* $(|\zeta_i^n|^2)$ *is AN.*
In particular if (ζ_i^n) *is a "martingale difference" array, that is* $\mathbb{E}(\zeta_i^n \mid \mathcal{G}_{i-1}^n) = 0$ *for all* $i, n \ge 1$, *then either one of the above conditions imply that it is AN.*

Proof We set $\zeta_i'^n = \zeta_i^n - \mathbb{E}(\zeta_i^n \mid \mathcal{G}_{i-1}^n)$. The process $M_t^n = \sum_{i=1}^{N_n(t)} \zeta_i'^n$ is a $(\overline{\mathcal{F}}_t^n)$-locally square-integrable martingale, with brackets given by

$$[M^n, M^n]_t = \sum_{i=1}^{N_n(t)} |\zeta_i'^n|^2,$$

$$\langle M^n, M^n\rangle_t = \sum_{i=1}^{N_n(t)} \mathbb{E}\big(|\zeta_i'^n|^2 \mid \mathcal{G}_{i-1}^n\big) \le A_t^n = \sum_{i=1}^{N_n(t)} \mathbb{E}\big(|\zeta_i^n|^2 \mid \mathcal{G}_{i-1}^n\big)$$

(the form of the angle bracket comes from (A.29)). Then we have $\langle M^n, M^n\rangle_t \stackrel{\mathbb{P}}{\longrightarrow} 0$ under (a), and also under (c) by the same proof as for Proposition 2.2.5 under (ii), so this proposition and Remark A.1 yield the result.

It remains to prove the result under (b), which we assume below. For any $a \ge 1$, set

$$\zeta_i^n(a-) = \zeta_i^n 1_{\{|\zeta_i^n| \le a\}}, \qquad \zeta_i'^n(a-) = \zeta_i^n(a-) - \mathbb{E}\big(\zeta_i^n(a-) \mid \mathcal{G}_{i-1}^n\big)$$

$$\zeta_i^n(a+) = \zeta_i^n 1_{\{|\zeta_i^n| > a\}}, \qquad \zeta_i^n(a+) = \zeta_i^n(a+) - \mathbb{E}\big(\zeta_i^n(a+) \mid \mathcal{G}_{i-1}^n\big)$$

$$A_t^n(a) = \sum_{i=1}^{N_n(t)} \mathbb{E}\big(|\zeta_i^n(a+)| \mid \mathcal{G}_{i-1}^n\big), \qquad B_t^n = \sum_{i=1}^{N_n(t)} \mathbb{E}\big(|\zeta_i^n|^2 1_{\{|\zeta_i^n| > 1\}} \mid \mathcal{G}_{i-1}^n\big).$$

For each fixed a the array $(\zeta_i^n(a-))$ satisfies (c) (with $K = a$), hence what precedes yields that the array $(\zeta_i'^n(a-))$ is AN. Since $\zeta_i'^n = \zeta_i'^n(a-) + \zeta_i'^n(a+)$, it is thus enough to prove

$$\lim_{a\to\infty} \sup_n \mathbb{P}\left(\sup_{s\le t} \left|\sum_{i=1}^{N_n(s)} \zeta_i'^n(a+)\right| > \varepsilon\right) = 0 \tag{A.30}$$

for all $t, \varepsilon > 0$. Observe that if $s \le t$,

$$\left|\sum_{i=1}^{N_n(s)} \zeta_i'^n(a+)\right| \le \sum_{i=1}^{N_n(t)} \Big(|\zeta_i^n(a+)| + \mathbb{E}\big(|\zeta_i^n(a+)| \mid \mathcal{G}_{i-1}^n\big)\Big),$$

which is Lenglart-dominated by the predictable increasing process $2A_t^n(a)$. Hence by the first part of (2.1.49),

$$\mathbb{P}\left(\sup_{s\le t} \left|\sum_{i=1}^{N_n(s)} \zeta_i'^n(a+)\right| > \varepsilon\right) \le \frac{\varepsilon}{\eta} + \mathbb{P}\big(2A_t^n(a) \ge \eta\big). \tag{A.31}$$

Now, we have $A_t^n(a) \leq B_t^n/a$ and by hypothesis the sequence $(B_t^n)_{n\geq 1}$ is bounded in probability, from which we deduce that $\mathbb{P}(2A_t^n(a) \geq \eta) \leq \mathbb{P}(B_t^n \geq a\eta/2)$, which for each $\eta > 0$ goes to 0 as $a \to \infty$, uniformly in n. Then (A.30) readily follows from (A.31). □

A.4 Processes of Finite Variation

This subsection is devoted to proving Proposition 3.1.2 of Chap. 3, to which we refer for the notation.

Proposition 3.1.2 *Suppose that the one-dimensional process X is of finite variation, and let $f(x) = x$. Then for any random discretization scheme we have*

$$\left.\begin{array}{ll} V^n(f,X) \stackrel{\mathbb{P}}{\Longrightarrow} \mathrm{Var}(X), & V^n(f,X)_t - \mathrm{Var}(X)_{T_n(t)} \stackrel{\text{u.c.p.}}{\Longrightarrow} 0 \\ V^n_{int}(f,X) \stackrel{\mathbb{P}}{\Longrightarrow} \mathrm{Var}(X), & V^n_{int}(f,X) \stackrel{\text{u.c.p.}}{\Longrightarrow} \mathrm{Var}(X). \end{array}\right\} \tag{A.32}$$

Proof 1) This result is a *pathwise* result, which holds when X and the $T(n,i)$'s are non-random, and then the extension as convergence in probability to the case when X is random and the subdivision scheme satisfies (3.2.1) becomes straightforward. So henceforth we suppose that X and $T(n,i)$ are non-random. There are two increasing right-continuous functions A and A', null at 0, such that $X - X_0 = A - A'$ and $Y = \mathrm{Var}(X) = A + A'$.

Since $V^n(f,X)_t = V^n_{int}(f,X)_{T_n(t)}$ is the discretized version of $V^n_{int}(f,X)$, the last claim in (A.32) implies the first three ones (use (2.2.14)). So below we fix $t > 0$, and we are left to prove that, if for short we write $V^n = V^n_{int}(f,X)$ in this proof,

$$\sup_{s\leq t} \left|V^n_s - Y_s\right| \to 0. \tag{A.33}$$

When $Y_t = 0$ the left side above vanishes and there is nothing to prove. Hence below we assume $Y_t > 0$.

2) Since t is fixed we clearly can assume $t = T(n, N_n(t))$ for each n. We consider the probability measure m on $(0,t]$ whose repartition function is $s \mapsto Y_s/Y_t$, and $\mathbb{E}_m$ is the expectation with respect to m. The set $(0,t]$ is endowed with the Borel σ-field $\mathcal{G}$, and also for each n with the σ-field $\mathcal{G}_n$ generated by the intervals $I(n,i)$ for $i = 1,\dots,N_n(t)$.

Let Z be a Borel bounded function on $(0,t]$. We set $U_s = \int_0^s Z(r)\,dY_r$ and, for each $n \geq 1$, we define the function Z_n on $(0,t]$ by

$$s \in I(n,i), \quad i \leq N_n(t) \;\Rightarrow\; Z_n(s) \;=\; \frac{\Delta_i^n U}{\Delta_i^n Y} \tag{A.34}$$

(with the convention $0/0 = 0$). If $\Delta_i^n Y > 0$ we have

$$\mathbb{E}_m(Z_n 1_{I(n,i)}) = \frac{\Delta_i^n U}{\Delta_i^n Y}\, m\big(I(n,i)\big) = \frac{1}{Y_t}\,\Delta_i^n U = \frac{1}{Y_t}\int_{I(n,i)} Z_r dY_r = \mathbb{E}_m(Z 1_{I(n,i)}),$$

whereas $\mathbb{E}_m(Z_n 1_{I(n,i)}) = \mathbb{E}_m(Z 1_{I(n,i)}) = 0$ if $\Delta_i^n Y = 0$. Hence by the definition of $\mathcal{G}_n$,

$$Z_n = \mathbb{E}_m(Z \mid \mathcal{G}_n). \tag{A.35}$$

3) In this step we prove that, for any Z as above, we have

$$Z_n \stackrel{\mathbb{L}^1(m)}{\longrightarrow} Z. \tag{A.36}$$

We have $\Delta_i^n U = Z(T(n,i-1))\Delta_i^n Y + O(w(Z,\varepsilon_n))$, where

$$w(Z,\theta) = \sup\bigl(\bigl|Z(s+r) - Z(s)\bigr| : 0 \le s \le s+r \le t, r \le \theta\bigr)$$

and $\varepsilon_n = \sup(\Delta(n,i) : 1 \le i \le N_n(t))$. If Z is continuous we have $w(Z,\varepsilon_n) \to 0$, hence (A.34) implies that $Z_n(s) \to Z(s)$ for all s, and (A.36) follows.

Suppose now that Z is Borel, bounded. One can find a sequence $(Z^p : p \ge 1)$ of continuous functions converging to Z in $\mathbb{L}^1(m)$. For each p, the sequence Z_n^p associated with Z^p by (A.34) converges to Z^p in $\mathbb{L}^1(m)$, from what precedes. On the other hand, (A.35) yields

$$\mathbb{E}_n\bigl(|Z_n - Z|\bigr) \le \mathbb{E}_m\bigl(|Z_n^p - Z^p|\bigr) + 2\mathbb{E}_m\bigl(|Z^p - Z|\bigr)$$

because the conditional expectation is a contraction on $\mathbb{L}^1(m)$. Therefore (A.36) follows.

4) We turn back to the situation at hand. There is a Borel function Z taking only the values ± 1, such that $A_s = \int_0^s Z(r)^+ dY_r$ and $A'_s = \int_0^s Z(r)^- dY_r$, and we associate the function Z_n by (A.34), so that $\alpha_n = \mathbb{E}_m(||Z_n| - 1|) \to 0$ by (A.36). Observing that $V_{T_n(s)}^n = \mathbb{E}_m(|Z_n| 1_{(0,T_n(s)]})$ and $Y_{T_n(s)} = m((0,T_n(s)])$, we see that

$$\sup_{s \le t} \bigl|V_{T_n(s)}^n - Y_{T_n(s)}\bigr| \le \alpha_n \to 0. \tag{A.37}$$

Let now $\eta > 0$ and $S_1, \dots, S_q$ the q successive times in $(0,t]$ at which X has a jump of size bigger than η (we may have $q = 0$), and $X'_s = X_s - \sum_{r=1} q \Delta X_{S_r} 1_{\{S_r \le s\}}$ and $Y' = \mathrm{Var}(X')$, and finally $w'(\theta) = \sup(|X'_{s+r} - X'_s| + Y'_{s+r} - Y'_s : 0 \le s \le s+r \le t, r \le \theta)$. Recalling the mesh ε_n of the subdivision at stage n over $[0,t]$, as defined in Step 3 above, and since $|\Delta X'_s| + \Delta Y'_s \le 2\eta$, we have $\limsup_n w'(\varepsilon_n) \le 3\eta$. In view of the definitions of $V^n = V_{int}^n(f,X)$ and of Y, we have for $s \in (0,t]$, and with $S_0 = 0$:

$$\inf_{1 \le r \le q} (S_r - S_{r-1}) > \varepsilon_n \Rightarrow \bigl||V_s^n - V_{T_n(s)}^n| - (Y_s - Y_{T_n(s)})\bigr| \le w'(\varepsilon_n)$$

(the condition on the left above implies that each interval $I(n,i)$ included into $(0,t]$ contains at most one time S_r). Therefore we deduce from (A.37) that

$$\limsup_n \sup_{s \le t} \bigl|V_s^n - Y_s\bigr| \le 3\eta.$$

Since $\eta > 0$ is arbitrarily small, we have (A.33). □

A.5 Some Results on Lévy Processes

In this last part, we prove some (classical) results about Lévy processes. We recall that $\Sigma(p,X)_t = \sum_{s\le t} \|\Delta X_s\|^p$.

Lemma A.2 *Let X be a d-dimensional Lévy process, with Lévy measure F, and $p \in [0,2)$. Then*

a) *If $\int(\|x\|^p \wedge 1)\, F(dx) < \infty$ we have $\Sigma(p,X)_t < \infty$ a.s. for all $t \ge 0$.*
b) *If $\int(\|x\|^p \wedge 1)\, F(dx) = \infty$ we have $\Sigma(p,X)_t = \infty$ a.s. for all $t > 0$.*

Proof (a) follows from general results about semimartingales, proved in Chap. 2. So we suppose now that $\int(\|x\|^p \wedge 1)\, F(dx) = \infty$. Recalling that the jump measure μ of X is a Poisson random measure with (deterministic) intensity measure $\nu(ds,dx) = ds \otimes F(dx)$, for any nonnegative function f on $\mathbb{R}_+ \times \mathbb{R}^d$ we have

$$\mathbb{E}\big(e^{-\int f d\mu}\big) = \exp - \int \big(1 - e^{-\lambda f}\big)\, d\nu$$

(by the form of the "Laplace functional" of a Poisson random measure). Applying this to the function $f(s,x) = \lambda(\|x\|^p \wedge 1) 1_{(0,t]}(s)$ and observing that $\int f d\mu = \sum_{s\le t}(\|\Delta X_s\|^p \wedge 1)$, we obtain

$$\mathbb{E}\big(e^{-\sum_{s\le t}(\|\Delta X_s\|^p \wedge 1)}\big) = \exp\Big(-t \int \big(1 - e^{-\|x\|^p \wedge 1}\big) F(dx)\Big). \tag{A.38}$$

Our assumption implies that $\int(1 - e^{-(\|x\|^p \wedge 1)}) F(dx) = \infty$. Therefore the left side of (A.38) vanishes, hence the result. □

Assumptions on the Process X

We list below the various assumptions which are made on the process X, at one time or another. They are listed in alphabetical order, except that a "strengthened" assumption is stated right after its non-strengthened version. They all suppose that X is an Itô semimartingale with Grigelionis form

$$X_t = X_0 + \int_0^t b_s\,ds + \int_0^t \sigma_s\,dW_s + (\delta 1_{\{\|\delta\|\le 1\}}) \star (\mathfrak{p} - \mathfrak{q})_t + (\delta 1_{\{\|\delta\|>1\}}) \star \mathfrak{p}_t.$$

Some involve a number r, always in $[0,2]$. We also set $c_t = \sigma_t \sigma_t^*$.

Assumption (H) The same as (H-2) below.

Assumption (SH) The same as (SH-2) below.

Assumption (H-r) (for $r \in [0,2]$) The process b is locally bounded, the process σ is càdlàg. and there is a localizing sequence (τ_n) of stopping times and, for each n, a *deterministic* nonnegative function Γ_n on E satisfying $\int \Gamma_n(z)^r \lambda(dz) < \infty$ (with the convention $0^0 = 0$) and such that $\|\delta(\omega,t,z)\| \wedge 1 \le \Gamma_n(z)$ for all (ω,t,z) with $t \le \tau_n(\omega)$.

Assumption (SH-r) (for $r \in [0,2]$) We have (H-r), and there are a constant A and a nonnegative function Γ on E, such that

$$\begin{array}{lll} \|b_t(\omega)\| \le A, & \|\sigma_t(\omega)\| \le A, & \|X_t(\omega)\| \le A \\ \|\delta(\omega,t,z)\| \le \Gamma(z), & \Gamma(z) \le A, & \int \Gamma(z)^r \lambda(dz) \le A. \end{array}$$

Assumption (K) We have (H) and

$$\sigma_t = \sigma_0 + \int_0^t \widetilde{b}_s\,ds + \int_0^t \widetilde{\sigma}_s\,dW_s + M_t + \sum_{s\le t} \Delta\sigma_s\, 1_{\{\|\Delta\sigma_s\|>1\}}$$

J. Jacod, P. Protter, *Discretization of Processes*,
Stochastic Modelling and Applied Probability 67,
DOI 10.1007/978-3-642-24127-7,

where M is a local martingale with $\|\Delta M_t\| \leq 1$, orthogonal to W, and $\langle M, M\rangle_t = \int_0^t a_s\,ds$ and the compensator of $\sum_{s\leq t} 1_{\{\|\Delta\sigma_s\|>1\}}$ is $\int_0^t \widetilde{a}_s\,ds$, with the following properties: the processes $\widetilde{b}$, $\widetilde{\sigma}$, $\widetilde{a}$ and a are progressively measurable, the processes $\widetilde{b}$, a and $\widetilde{a}$ are locally bounded, and the processes $\widetilde{\sigma}$ and b are càdlàg or càglàd.

Assumption (SK) We have (K), (SH-2), and there is a constant A such that

$$\|\widetilde{b}_t(\omega)\| + \|\widetilde{\sigma}_t(\omega)\| + \widetilde{a}_t(\omega) + \|a_t(\omega)\| \leq A.$$

Assumption (K-r) (for $r \in [0,1]$) We have (K) except for the càdlàg or càglàd property of b, and $\|\delta(\omega,t,z)\| \wedge 1 \leq \Gamma_n(z)$ for all (ω,t,z) with $t \leq \tau_n(\omega)$, where (τ_n) is a localizing sequence of stopping times and the Borel functions Γ_n on E satisfy $\int \Gamma_n(z)^r \lambda(dz) < \infty$. Moreover the process $b'_t = b_t - \int_{\{\|\delta(t,z)\|\leq 1\}} \delta(t,z)\,\lambda(dz)$ is càdlàg or càglàd.

Assumption (SK-r) (for $r \in [0,1]$) We have (K-r), (SH-r) and there is a constant A such that

$$\|\widetilde{b}_t(\omega)\| + \|\widetilde{\sigma}_t(\omega)\| + \widetilde{a}_t(\omega) + \|a_t(\omega)\| \leq A.$$

Assumption (K') We have (K) and both processes c_t and c_{t-} take their values in the set $\mathcal{M}^{++}_{d\times d}$ of all symmetric positive definite $d \times d$ matrices.

Assumption (SK') We have (SK) and (K') and the process c_t^{-1} is bounded.

Assumption (K'-r) (for $r \in [0,1]$) We have (K-r) and both processes c_t and c_{t-} take their values in the set $\mathcal{M}^{++}_{d\times d}$ of all symmetric positive definite $d \times d$ matrices.

Assumption (SK'-r) (for $r \in [0,1]$) We have (SK-r) and (K') and the process c_t^{-1} is bounded.

Assumption ($\overline{\text{K}}$-r) (for $r \in [0,1]$) The process X satisfies (K-r), and the process σ satisfies (H).

Assumption (Q) Either X is continuous and $\int_0^t (\|b_s\|^2 + \|c_s\|^2)\,ds < \infty$ for all $t > 0$, or X has jumps and then it satisfies (H).

Assumption (SQ) Either X is continuous and $\int_0^t (\|b_s\|^2 + \|c_s\|^2)\,ds \leq A$ for some constant A, or X has jumps and then it satisfies (SH).

References

1. Aït-Sahalia, Y. (2004). Disentangling diffusions from jumps. *J. of Financial Economics* **74**, 487–528.
2. Aït-Sahalia Y. and J. Jacod (2009). Testing for jumps in a discretely observed process. *Annals of Statistics* **37**, 184–222.
3. Aït-Sahalia Y., P. Mykland and L. Zhang (2011). Ultra high frequency volatility estimation with dependent microstructure noise. *Journal of Econometrics* **160**, 190–203.
4. Aldous D. and G.K. Eagleson (1978). On mixing and stability of limit theorems. *Annals Probab.* **6**, 325–331.
5. Alvarez A., F. Panloup, M. Pontier and N. Savy (2010). Estimation of the instantaneous volatility. Preprint.
6. Andersen, T.G., T. Bollerslev, F.X. Diebold and P. Labys (2001). The distribution of realized stock returns. *J. of Financial Economics* **1** 43–67.
7. Bandi, F.M. and J.R. Russell (2006). Separating microstructure noise from volatility. *J. of Financial Economics* **79**, 655–692.
8. Barndorff-Nielsen O.E. and N. Shephard (2004). Power and bipower variation with stochastic volatility and jumps. *J. of Financial Econometrics* **2**, 1–48.
9. Barndorff-Nielsen O.E. and N. Shephard (2005). Power variation and time change. *Theory Probab. and its Appl.* **50**, 1–15.
10. Barndorff-Nielsen O.E. and N. Shephard (2006). Econometrics of testing for jumps in financial economics using bipower variation. *J. of Financial Econometrics* **4**, 1–30.
11. Barndorff-Nielsen O.E., S.E. Graversen, J. Jacod, M. Podolskij and N. Shephard (2006). A central limit theorem for realised power and bipower variations of continuous semimartingales. In *From Stochastic Calculus to Mathematical Finance. Festschrift in Honour of A.N. Shiryaev*, Yu. Kabanov, R. Liptser and J. Stoyanov, eds., 33–68. Springer Verlag: Heidelberg.
12. Barndorff-Nielsen, O.E., N. Shephard and M. Winkel (2006). Limit theorems for multipower variation in the presence of jumps. *Stoch. Proc. Appl.* **116**, 796–806.
13. Barndorff-Nielsen O.E., P.R. Hansen, A. Lunde and N. Shephard (2008). Designing realised kernels to measure the ex-post variation of equity prices in the presence of noise. *Econometrica* **76**, 1481–1536.
14. Barndorff-Nielsen O.E., P.R. Hansen, A. Lunde and N. Shephard. Multivariate realised kernels: Consistent positive semi-definite estimators of the covariation of equity prices with noise and non-synchronous trading. Preprint.
15. Bertoin J. (1998). *Lévy Processes*. Cambridge University Press: Cambridge.
16. Beskos A. and G.O. Roberts (2005). Exact simulation of diffusions. *Annals Applied Probab.* **15**, 2422–2444.
17. Beskos A., O. Papaspiliopoulos and G.O. Roberts (2006). Retrospective exact simulation of diffusion sample paths with applications. *Bernoulli* **12**, 1077–1098.

J. Jacod, P. Protter, *Discretization of Processes*,
Stochastic Modelling and Applied Probability 67,
DOI 10.1007/978-3-642-24127-7, © Springer-Verlag Berlin Heidelberg 2012

18. Billingsley P. (1999). *Convergence of Probability Measures*, second edition. John Wiley and Sons: New York.
19. Blumenthal R.M. and R.K. Getoor (1961). Sample functions of stochastic processes with stationary independent increments. *Journal of Mathematics and Mechanics* **10**, 493–516.
20. Burkholder D. and R. Gundy (1970). Extrapolation and interpolation of quasi-linear operators on martingales. *Acta Mathematica* **124**, 249–304.
21. Burkholder D., B. Davis and R. Gundy (1972). Integral inequalities for convex functions of operators on martingales. In *Proceedings 6th Berkeley Symp. Math. Stat. Probab.* Vol. 2, 223–240.
22. Christensen K. and M. Podolskij (2007). Realized range-based estimation of integrated variance. *J. of Econometrics* **141**, 323–349.
23. Cont R. and C. Mancini (2009). Nonparametric tests for pathwise properties of semimartingales. Preprint.
24. Dellacherie C. (1972). *Capacités et Processus Stochastiques*. Springer Verlag: Heidelberg.
25. Dellacherie C. and P.-A. Meyer (1975–1992). *Probabilités et Potentiel*, Vol. I–V. Hermann: Paris.
26. Delattre S. and J. Jacod (1997). A central limit theorem for normalized functions of the increments of a diffusion process, in the presence of round-off errors. *Bernoulli* **3** 1–28.
27. Diop A. (2009). *Convergence fonctionnelle de semimartingales discrétisées*, Phd Thesis, Université Pierre et Marie Curie, Paris.
28. Diop A. (2010). Convergence of some random functionals of discretized semimartingales. *Bernoulli*, in press.
29. Dohnal G. (1987). On estimating the diffusion coefficient. *J. Applied Probab.* **24**, 105–114.
30. El Karoui N. and J.P. Lepeltier (1977). Représentation des processus ponctuels multivariés à l'aide d'un processus de Poisson. *Z. für Wahrscheinlichkeitstheorie Verw. Geb.* **59**, 111–134.
31. Ethier S.N. and T.G. Kurtz (1986). *Markov Processes: Characterization and Convergence*. John Wiley and Sons: New York.
32. Fan Y. and J. Fan (2008). Testing and detecting jumps based on a discretely observed process. Preprint.
33. Fukasawa M. (2010). Central limit theorem for the realized volatility based on tick time sampling. *Finance and Stochastics* **34**, 209–233.
34. Fukasawa M. (2010). Realized volatility with stochastic sampling. Preprint.
35. Fukasawa M. and M. Rosenbaum (2010). Central limit theorems for realized volatility under hitting times of an irregular grid. Preprint.
36. Garman M.B. and M.J. Klass (1980). On the estimation of security prices volatilities from historical data. *J. of Business* **53**, 67–78.
37. Genon-Catalot V. and J. Jacod (1993). On the estimation of the diffusion coefficient for multi-dimensional diffusion processes. *Annales Inst. H. Poincaré (B)* **29**, 119–151.
38. Greenwood P. and B. Fristedt (1972). Variation of processes with stationary independent increments. *Z. Wahrscheinlichkeitstheorie verw. Gebiete* **23**, 171–186.
39. Grigelionis B. (1971). On the representation of integer-valued random measures by means of stochastic integrals with respect to the Poisson measure. *Litovsk. Math. Sbornik* **11**, 93–108.
40. Grigelionis B. (1971). On the absolute continuity of measures corresponding to stochastic processes. *Litovsk. Math. Sbornik* **11**, 783–794.
41. Grigelionis B. (1975). The characterization of stochastic processes with conditionally independent increments. *Litovsk. Math. Sbornik* **15**, 53–60.
42. Grosswald E. (1950). Functions of bounded variation. *Duke Math. J.* **17**, 313–315.
43. Hall P. and C.C. Heyde (1980). *Martingale Limit Theory and its Applications*. Academic Press: New York.
44. Hayashi T. and N. Yoshida (2005). On covariance estimation of nonsynchronously observed diffusion processes. *Bernoulli* **11**, 359–379.
45. Hayashi T. and N. Yoshida (2006). Nonsynchronous covariance estimator and limit theorems. Preprint.

46. Hayashi T. and N. Yoshida (2008). Asymptotic normality of a covariance estimator for non-synchronously observed diffusion processes. *Annals of the Institute of Statistical Mathematics* **60**, 357–396.
47. Hayashi T., J. Jacod and N. Yoshida (2011). Irregular sampling and central limit theorems for power variations: The continuous case. *Annales Inst. H. Poincaré, Prob. Stat.* **47**, 1197–1218.
48. Huang X. and G. Tauchen (2005). The relative contribution of jumps to total price variance. *J. Financial Econometrics* **3**, 456–499.
49. Ibragimov I.A. and R.Z. Has'minski (1981). *Statistical Estimation: Asymptotic Theory*. Springer Verlag: Heidelberg.
50. Ikeda N. and S. Watanabe (1981). *Stochastic Differential Equations and Diffusion Processes*. North-Holland: Amsterdam.
51. Jacod J. and J. Mémin (1976). Caractéristiques locales et conditions de continuité absolue pour les semimartingales. *Z. für Wahrscheinlichkeitstheorie Verw. Geb.* **35**, 1–37.
52. Jacod J. (1979). *Calcul Stochastique et Problèmes de Martingales*. Lectures Notes in Mathematics 714. Springer Verlag: Heidelberg.
53. Jacod J. (1994). Limit of random measures associated with the increments of a Brownian semimartingale. Preprint.
54. Jacod J. (1997). On continuous Gaussian martingales and stable convergence in law. In *Séminaire de Probabilités XXXI*, Lectures Notes in Mathematics 851, 169–212. Springer Verlag: Heidelberg.
55. Jacod J. and P. Protter (1998). Asymptotic error distributions for the Euler method for stochastic differential equations. *Annals of Probab.* **26**, 297–307.
56. Jacod J., A. Jakubowski and J. Mémin (2003). On asymptotic errors in discretization of processes. *Annals of Probab.* **31**, 592–608.
57. Jacod J. and A.N. Shiryaev (2003). *Limit Theorems for Stochastic Processes*, second edition. Springer Verlag: Heidelberg.
58. Jacod J. (2004). The Euler scheme for Lévy driven stochastic differential equations: Limit theorems. *Ann. I.H. Poincaré (PS)* **32**, 1830–1872.
59. Jacod J., T.G. Kurtz, S. Méléard and P. Protter (2005). The approximate Euler method for Lévy driven stochastic differential equations. *Ann. I.H. Poincaré (PS)* **41**, 523–558.
60. Jacod J. (2008). Asymptotic properties of realized power variations and associated functionals of semimartingales. *Stoch. Proc. Appl.* **118**, 517–559.
61. Jacod J., Y. Li, P. Mykland, M. Podolskij and M. Vetter (2009). Microstructure noise in the continuous case: The pre-averaging approach. *Stoch. Proc. Appl.* **119**, 2249–2276.
62. Jacod J. and V. Todorov (2009). Testing for common arrivals of jumps for discretely observed multidimensional processes. *Annals of Statistics* **37**, 1792–1838.
63. Jacod J., M. Podolskij and M. Vetter (2010). Limit theorems for moving averages of discretized processes plus noise. *Annals of Statistics* **38**, 1478–1545.
64. Jacod J. and V. Todorov (2010). Do price and volatility jump together? *Annals of Applied Probab.* **20**, 1425–1469.
65. Jarrow R. and P. Protter (2004). A short history of stochastic integration and mathematical finance: The early years, 1880–1970. In *The Herman Rubin Festschrift*, Anirban Das Gupta, ed., IMS Lecture Notes 45, 75–91.
66. Kinnebrock S. and M. Podolskij (2008). A note on the central limit theorem for bipower variation of general functions. *Stoch. Proc. Appl.* **118**, 1056–1070.
67. Kosulajeff P. (1937). Sur le répartition de la partie fractionnaire d'une variable aléatoire. *Math. Sbornik* **2**, 1017–1019.
68. Kurtz T.G. and P. Protter (1991). Wong-Zakai corrections, random evolutions and numerical schemes for SDEs. In *Stochastic Analysis*, 331–346. Academic Press: New York.
69. Lee Z. and P.A. Mykland (1991). Jumps in financial markets: a new nonparametric test and jump dynamics. *Review of Financial Studies* **21**, 2535–2563.
70. Lenglart E. (1977). Relation de domination entre deux processus. *Annales Inst. H. Poincaré (B)* **13**, 171–179.
71. Lépingle D. (1976). La variation d'ordre p des semi-martingales. *Z. Wahrscheinlichkeitstheorie verw. Gebiete* **36**, 295–316.

72. Li, Y. and P. Mykland (2007). Are volatility estimators robust with respect to modeling assumptions? *Bernoulli* **13** (3), 601–622.
73. Mancini C. (2001). Disentangling the jumps of the diffusion in a geometric jumping Brownian motion. *Giornale dell'Istituto Italiano degli Attuari* **LXIV**, 19–47.
74. Mancini, C. (2009). Non-parametric threshold estimation for models with diffusion coefficient and jumps. *Scandinavian J. Statistics* **36**, 270–296.
75. Mancini C. (2010). Speed of convergence of the threshold estimator of integrated variance. Preprint.
76. Meyer P.A. (1967). Intégrales stochastiques – 1. In *Séminaire de Probabilités I*, Lectures Notes in Mathematics 39, 72–94. Springer Verlag: Heidelberg.
77. Parkinson M. (1980). The extreme value method for estimating the variance of the rate of return. *J. of Business* **53**, 61–65.
78. Parthasarathy K.S. (1967). *Probability Measures on Metric Spaces*. Academic Press: New York.
79. Phillips, P.C.B. and J. Yu (2008). Information loss in volatility measurement with flat price trading. Preprint.
80. Podolskij M. and M. Vetter (2009). Bipower-type estimation in a noisy diffusion setting. *Stoch. Proc. Appl.* **119**, 2803–2831.
81. Podolskij M. and M. Vetter (2009). Estimation of volatility functionals in the simultaneous presence of microstructure noise and jumps. *Bernoulli* **15**, 634–658.
82. Protter P. and D. Talay (1997). The Euler scheme for Lévy driven stochastic differential equations. *Annals Probab.* **25**, 393–423.
83. Protter P. (2005). *Stochastic Integration and Differential Equations*, second edition, version 2.1. Springer Verlag: Heidelberg.
84. Rényi A. (1963). On stable sequences of events. *Sankya, Series A* **25**, 293–302.
85. Robert C.Y. and M. Rosenbaum (2009). Volatility and covariation estimation when microstructure noise and trading times are endogeneous. *Math. Finance*, in press.
86. Rosembaum M. (2009). Integrated volatility and round-off errors. *Bernoulli* **15** 687–720.
87. Sato K.I. (1999). *Lévy Processes and Infinitely Divisible Distributions*. Cambridge University Press: Cambridge.
88. Shimizu Y. and N. Yoshida (2006). Estimation of parameters for diffusion processes with jumps from discrete observations. *Stat. Inf. for Stochastic Processes* **9**, 227–277.
89. Skorokhod A.V. (1956). Limit theorems for stochastic processes. *Theory Probab. Appl.* **1**, 261–290.
90. Skorokhod A.V. (1965). *Studies in the Theory of Random Processes*. Addison-Wesley: Reading.
91. Talay D. and L. Tubaro (1990). Expansion of the global error for numerical schemes solving stochastic differential equations. *Stochastic Analysis Appl.* **8**, 94–120.
92. Tuckey J.W. (1937). On the distribution of the fractional part of a statistical variable. *Mat. Sbornik* **4**, 561–563.
93. Vetter M. (2010). Limit theorems for bipower variation of semimartingales. *Stoch. Proc. Appl.* **120**, 22–38.
94. Woerner, J. (2005). Estimation of integrated volatility in stochastic volatility models. *Appl. Stochastic Models Bus. Ind.* **21**, 27–44.
95. Woerner J. (2006). Power and multipower variations: Inference for high frequency data. In *Stochastic Finance*, A.N. Shiryaev, M.R. Grossinho, P.E. Oliveira and M.L. Esquivel, eds., 343–354. Springer-Verlag: New-York.
96. Zhang, L., P.A. Mykland, and Y. Aït-Sahalia (2005). A tale of two time scales: Determining integrated volatility with noisy high-frequency data. *Journal of the American Statistical Association* **100**, 1394–1411.
97. Zhang, L. (2006). Efficient estimation of volatility using noisy observations. *Bernoulli* **12** (6), 1019–1043.

Index of Functionals

Increments and discretized processes

$$\Delta_i^n X = X_{T(n,i)} - X_{T(n,i-1)}, \qquad I(n,i) = \big(T(n,i-1), T(n,i)\big]$$

$$N_n(t) = \sum_{i\geq 1} 1_{\{T(n,i)\leq t\}}$$

$$X_t^{(n)} = \sum_{i\geq 0} X_{T(n,i)}\, 1_{\{T(n,i)\leq t<T(n,i+1)\}}$$

Non-normalized functionals

f a function on $\mathbb{R}^d$:

$$V^n(f,X)_t = \sum_{i=1}^{[t/\Delta_n]} f\big(\Delta_i^n X\big) \qquad \text{p. 65}$$

$$V_{int}^n(f,X)_t = \sum_{i\geq 1} f\big(X_{t\wedge(i\Delta_n)} - X_{t\wedge(i-1)\Delta_n}\big) \qquad \text{p. 65}$$

$$V^n(f,v_n+,X)_t = \sum_{i=1}^{[t/\Delta_n]} f\big(\Delta_i^n X\big)\, 1_{\{\|\Delta_i^n X\|>v_n\}} \qquad \text{p. 248}$$

$$V^n(f,v_n-,X)_t = \sum_{i=1}^{[t/\Delta_n]} f\big(\Delta_i^n X\big)\, 1_{\{\|\Delta_i^n X\|\leq v_n\}} \qquad \text{p. 248}$$

F a function on $(\mathbb{R}^d)^k$:

$$V^n(F,X)_t = \sum_{i=1}^{[t/\Delta_n]-k+1} F\big(\Delta_i^n X, \Delta_{i+1}^n X, \ldots, \Delta_{i+k-1}^n X\big) \qquad \text{p. 227}$$

$$\mathcal{V}^n(F,X)_t = \sum_{i=1}^{[N_n(t)/k]} F\big(\Delta_{ik-k+1}^n X, \ldots, \Delta_{ik}^n X\big) \qquad \text{p. 227}$$

J. Jacod, P. Protter, *Discretization of Processes*,
Stochastic Modelling and Applied Probability 67,
DOI 10.1007/978-3-642-24127-7, © Springer-Verlag Berlin Heidelberg 2012

Φ a function on the Skorokhod space $\mathbb{D}_1^d$:

$$V^n(\Phi, k_n, X)_t = \sum_{i=1}^{[t/\Delta_n]-k_n+1} \Phi\big(X(n,i)^{(n)}\big) \quad \text{p. 229}$$

F a function on $\Omega \times \mathbb{R}_+ \times \mathbb{R}^d$:

$$V^{n,l}(F, X)_t = \sum_{i=1}^{N_n(t)} F\big(\omega, T(n, i-1), \Delta_i^n X\big) \quad \text{p. 219}$$

$$V^{n,r}(F, X)_t = \sum_{i=1}^{N_n(t)} F\big(\omega, T(n, i), \Delta_i^n X\big) \quad \text{p. 219}$$

$$V^n(F, X)_t = \sum_{i=1}^{[t/\Delta_n]} F\big(\omega, (i-1)\Delta_n, \Delta_i^n X\big) \quad \text{p. 273}$$

$\overline{F}$ a function on $\mathbb{R}^d \times \mathbb{R}^d$:

$$V^n\big(\overline{F}(X), X\big)_t = \sum_{i=1}^{[t/\Delta_n]} \overline{F}\big(X_{(i-1)\Delta_n}, \Delta_i^n X\big) \quad \text{p. 279}$$

Normalized functionals (for regular schemes)
f a function on $\mathbb{R}^d$:

$$V'^n(f, X)_t = \Delta_n \sum_{i=1}^{[t/\Delta_n]} f\left(\frac{\Delta_i^n X}{\sqrt{\Delta_n}}\right) \quad \text{p. 80}$$

$$V'^n(f, v_n-, X)_t = \Delta_n \sum_{i=1}^{[t/\Delta_n]} f\left(\frac{\Delta_i^n X}{\sqrt{\Delta_n}}\right) 1_{\{\|\Delta_i^n X\| \le v_n\}} \quad \text{p. 251}$$

F a function on $(\mathbb{R}^d)^k$:

$$V'^n(F, X)_t = \Delta_n \sum_{i=1}^{[t/\Delta_n]-k+1} F\left(\frac{\Delta_i^n X}{\sqrt{\Delta_n}}, \frac{\Delta_{i+1}^n X}{\sqrt{\Delta_n}}, \dots, \frac{\Delta_{i+k-1}^n X}{\sqrt{\Delta_n}}\right) \quad \text{p. 227}$$

$$\mathcal{V}'^n(F, X)_t = \Delta_n \sum_{i=1}^{[t/k\Delta_n]} F\big(\Delta_{ik-k+1}^n X/\sqrt{\Delta_n}, \dots, \Delta_{ik}^n X/\sqrt{\Delta_n}\big) \quad \text{p. 227}$$

$$V'^n(F, v_n-, X)_t = \Delta_n \sum_{i=1}^{[t/\Delta_n]-k+1} F\left(\frac{\Delta_i^n X}{\sqrt{\Delta_n}}, \dots, \frac{\Delta_{i+k-1}^n X}{\sqrt{\Delta_n}}\right) \times 1_{\cap_{l=0}^{k-1}\{\|\Delta_{i+l}^n X\| \le v_n\}} \quad \text{p. 251}$$

Φ a function on the Skorokhod space $\mathbb{D}_1^d$:

$$V'^n(\Phi, k_n, X)_t = \Delta_n \sum_{i=1}^{[t/\Delta_n]-k_n+1} \Phi\left(\frac{1}{\sqrt{u_n}}\, X(n,i)^{(n)}\right) \quad \text{p. 230}$$

F a function on $\Omega \times \mathbb{R}_+ \times \mathbb{R}^d$:

$$V'^n(F, X)_t = \Delta_n \sum_{i=1}^{[t/\Delta_n]} F\left(\omega, \tau(n,i), \frac{\Delta_i^n X}{\sqrt{\Delta_n}}\right) \quad \text{p. 216}$$

$$V'^n(F, X)_t = \Delta_n \sum_{i=1}^{[t/\Delta_n]} F\left(\omega, (i-1)\Delta_n, \frac{\Delta_i^n X}{\sqrt{\Delta_n}}\right) \quad \text{p. 216}$$

$\overline{F}$ a function on $\mathbb{R}^d \times \mathbb{R}^d$:

$$V'^n(\overline{F}(X), X)_t = \Delta_n \sum_{i=1}^{[t/\Delta_n]} \overline{F}\left(X_{(i-1)\Delta_n}, \frac{\Delta_i^n X}{\sqrt{\Delta_n}}\right) \quad \text{p. 283}$$

Integrated error processes (for regular schemes, f on $\mathbb{R}^d$)

$$\widetilde{V}^n(f, X)_t = \int_0^{\Delta_n[t/\Delta_n]} \left(f\left(X_s^{(n)}\right) - f(X_s)\right) ds \quad \text{p. 187}$$

$$\widetilde{V}^n(f, p, X)_t = \int_0^{\Delta_n[t/\Delta_n]} \left|f\left(X_s^{(n)}\right) - f(X_s)\right|^p ds \quad \text{p. 187}$$

Local estimators for the volatility

$$\widehat{c}_i^n(k_n)^{jl} = \frac{1}{k_n \Delta_n} \sum_{m=0}^{k_n-1} \Delta_{i+m}^n X^j\, \Delta_{i+m}^n X^l \quad \text{p. 255}$$

$$\widehat{c}_i^n(k_n, v_n)^{jl} = \frac{1}{k_n \Delta_n} \sum_{m=0}^{k_n-1} \Delta_{i+m}^n X^j\, \Delta_{i+m}^n X^l\, 1_{\{\|\Delta_{i+m}^n X\| \le v_n\}} \quad \text{p. 255}$$

$$\left.\begin{aligned} \widehat{c}^n(k_n, t-) &= \widehat{c}_{i-k_n}^n(k_n) \\ \widehat{c}^n(k_n, t) &= \widehat{c}_{i+1}^n(k_n) \\ \widehat{c}^n(k_n, v_n, t-) &= \widehat{c}_{i-k_n}^n(k_n, v_n) \\ \widehat{c}^n(k_n, v_n, t) &= \widehat{c}_{i+1}^n(k_n, v_n) \end{aligned}\right\} \quad \text{if } t \in I(n,i) \quad \text{p. 256}$$

$$\begin{aligned} V^n(G; k_n, v_n, X)_t = & \sum_{i=k_n+1}^{[t/\Delta_n]-k_n} 1_{\{\|\Delta_i^n X\| > v_n\}} \\ & \times G\left(\Delta_i^n X, \widehat{c}_{i-k_n}^n(k_n, v_n), \widehat{c}_{i+1}^n(k_n, v_n)\right) 1_{\{\|\Delta_i^n X\| > v_n\}} \quad \text{p. 265} \end{aligned}$$

Index

Symbols

J. Jacod, P. Protter, *Discretization of Processes*,
Stochastic Modelling and Applied Probability 67,
DOI 10.1007/978-3-642-24127-7, © Springer-Verlag Berlin Heidelberg 2012